11th AUN/SEED-Net Regional Conference on Mechanical and Manufacturing Engineering (RCMEManuE 2020)

IOP Conference Series: Materials Science and Engineering
Volume 1109

Manila, Philippines
14 - 15 January 2021

Editors:

Alvin Chua
Andi IsraMahyuddin
Neil Stephen Lopez
Gerardo Augusto

ISBN: 978-1-7138-2568-5
ISSN: 1757-8981

Printed from e-media with permission by:

Curran Associates, Inc.
57 Morehouse Lane
Red Hook, NY 12571

proceedings.com

Some format issues inherent in the e-media version may also appear in this print version.

Institute of Physics
Dirac House, Temple Back
Bristol BS1 6BE UK

Phone: 44 1 17 929 7481
Fax: 44 1 17 920 0979

techtracking@iop.org

Additional copies of this publication are available from:

Curran Associates, Inc.
57 Morehouse Lane
Red Hook, NY 12571 USA
Phone: 845-758-0400
Fax: 845-758-2633
Email: curran@proceedings.com
Web: www.proceedings.com

TABLE OF CONTENTS

CONVENTIONAL TOPICS IN MECHANICAL ENGINEERING

MANUFACTURING AND MECHATRONICS ENGINEERING

ENERGY STUDIES

EMERGING TOPICS IN ENGINEERING ECONOMICS AND SUSTAINABILITY POLICY

IOP Conf. Series: Materials Science and Engineering 1109 (2021) 011001 doi:10.1088/1757-899X/1109/1/011001

Preface

The 11th AUN/SEED-Net Regional Conference on Mechanical and Manufacturing Engineering (RCMEManuE 2020) was held virtually on January 14-15, 2021. The conference was co-organized by the Mechanical Engineering Department of De La Salle University, Manila, Philippines and the Faculty of Mechanical and Aerospace Engineering of Institut Teknologi Bandung, Bandung, Indonesia, through funding and technical support from the ASEAN University Network / Southeast Asia Engineering Education Development Network (AUN/SEED-Net) and the Japan International Cooperation Agency (JICA). The theme of the conference was "Smart and cyber technologies for mechanical and manufacturing systems: Shifting to the new normal".

The AUN/SEED-Net Regional Conferences are annual academic events involving ten thematic engineering fields. Each event is organized by Member Institutions with technical assistance from Japanese Supporting Universities. For almost two decades, the program has successfully served as a platform for exchanging current engineering knowledge among AUN/SEED-Net members. The conference also promotes academic collaboration among ASEAN and Japanese universities, strengthens relationships among academicians in the region, and opens up future collaboration in the engineering field with the government and private sectors.

The opening and plenary sessions were held via Zoom Webinar, while the oral research presentations were conducted through the SDEWES Virtual Conference System. The conference showcased 73 research presentations from 166 authors across 10 different countries.

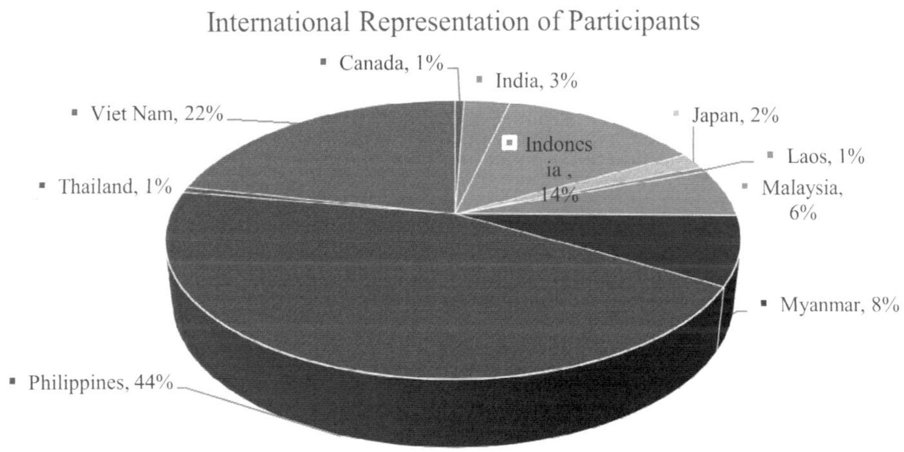

The plenary sessions were graced by 7 invited speakers from the government, industry and academe:

- Hon. Sec. Fortunato dela Peña – Secretary, Department of Science and Technology, Philippines
- Prof. Dr. Yoshio Yamamoto – Tokai University, Japan
- Prof. Dr. Alvin Chua – De La Salle University, Philippines
- Prof. Dr. Tatacipta Dirgantara – Dean, Faculty of Mechanical and Aerospace Engineering, Institut Teknologi Bandung, Indonesia
- Engr. Rico Mercado – President, ASHRAE-Philippines
- Mr. Quintin Pastrana – President, WEnergy Power Pilipinas

- Prof. Dr. Manuel Belino – Commission on Higher Education—Philippines Technical Panel Head for Mechanical Engineering

On behalf of the whole organizing committee, we would like to thank the AUN/SEED-Net and JICA for making this conference possible, our invited speakers, international scientific committee and participants for the highly successful regional conference.

Prof. Dr. Alvin Chua – De La Salle University, Philippines
Prof. Dr. Andi Isra Mahyuddin – Institut Teknologi Bandung, Indonesia
Dr. Neil Stephen Lopez – De La Salle University, Philippines
Dr. Gerardo Augusto – De La Salle University, Philippines
Editors

IOP Conf. Series: Materials Science and Engineering 1109 (2021) 011001 doi:10.1088/1757-899X/1109/1/011001

Conference Photographs

Photo 1. Speakers of Day 1 (January 14, 2021). **Top row, from left to right:** Br. Raymundo Suplido, PhD (President, De La Salle University), Prof. Emeritus Dr. Tamon Ueda (Chief Advisor and Acting Executive Director, AUN/SEED-Net), Ms. Wanichar Sukprasertchai (Project Manager, AUN/SEED-Net); **Bottom row, from left to right:** Prof. Dr. Andi Isra Mahyuddin (Section Editor-in-Chief, ASEAN Engineering Journal), Prof. Dr. Yoshio Yamamoto (Tokai University), Prof. Dr. Alvin Chua (De La Salle University).

Photo 2. Day 1 Participants (1 of 2).

IOP Conf. Series: Materials Science and Engineering 1109 (2021) 011001 doi:10.1088/1757-899X/1109/1/011001

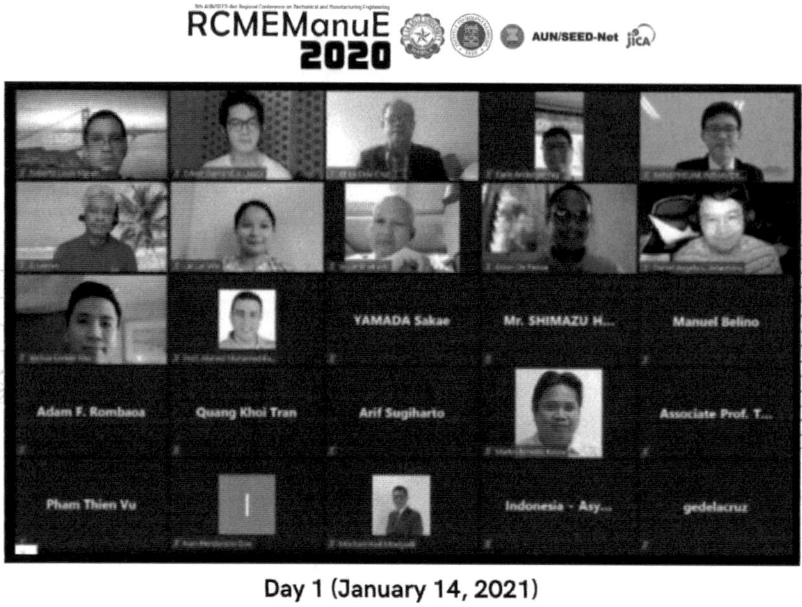

Photo 3. Day 1 Participants (2 of 2).

Photo 4. Speakers of Day 2 (January 15, 2021). **Top row, from left to right:** Prof. Dr. Ir. Tatacipta Dirgantara (Dean, Faculty of Mechanical and Aerospace Engineering, Institut Teknologi Bandung), Engr. Rico Mercado (President, ASHRAE-Philippines), Mr. Quintin Pastrana (President, WEnergy Power Pilipinas); **Bottom row, from left to right:** Prof. Dr. Manuel Belino (Commission on Higher Education—Philippines Technical Panel Head for Mechanical Engineering), Hon. Sec. Fortunato dela Peña (Secretary, Department of Science and Technology-Philippines), Prof. Dr. Jonathan Dungca (Dean, Gokongwei College of Engineering, De La Salle University).

IOP Conf. Series: Materials Science and Engineering 1109 (2021) 011001 doi:10.1088/1757-899X/1109/1/011001

Photo 5. Day 2 Participants (1 of 2).

Photo 6. Day 2 Participants (2 of 2).

IOP Conf. Series: Materials Science and Engineering 1109 (2021) 011001 doi:10.1088/1757-899X/1109/1/011001

Photo 7. Conference Organizers from De La Salle University, Philippines.

List of Committees

International Scientific Committee (Reviewers)

1. Prof. Dr. Alvin Chua – De La Salle University, Philippines
2. Prof. Dr. Alvin Culaba – De La Salle University, Philippines
3. Prof. Dr. Archie Maglaya – De La Salle University, Philippines
4. Prof. Dr. Aristotle Ubando – De La Salle University, Philippines
5. Prof. Dr. Laurence Gan Lim – De La Salle University, Philippines
6. Dr. Annisa Jusuf – Institut Teknologi Bandung, Indonesia
7. Dr. Bentang Arief Budiman – Institut Teknologi Bandung, Indonesia
8. Dr. Gerardo Augusto – De La Salle University, Philippines
9. Dr. Leonardo Gunawan – Institut Teknologi Bandung
10. Dr. Martin Ernesto Kalaw – De La Salle University, Philippines
11. Dr. Neil Stephen Lopez – De La Salle University, Philippines
12. Dr. Rith Monorom – Thammasat University, Thailand
13. Dr. Robby Manrique – De La Salle University, Philippines
14. Dr. Sri Raharno – Institut Teknologi Bandung, Indonesia
15. Dr. Yuli Indartono – Institut Teknologi Bandung, Indonesia
16. Dr. Yunendar Aryo Handoko – Institut Teknologi Bandung, Indonesia
17. Engr. Arvin Fernando, MSc – De La Salle University, Philippines
18. Engr. Charles Felix, MSc – De La Salle University, Philippines
19. Engr. Conrad Allan Jay Pantua, MSc – De La Salle University, Philippines
20. Engr. Isidro Antonio Marfori III, MSc – De La Salle University, Philippines
21. Engr. Ivan Henderson Gue, MSc – De La Salle University, Philippines
22. Engr. Jeremias Gonzaga, MSc – De La Salle University, Philippines
23. Engr. Nechoh Arbon, MSc – De La Salle University, Philippines

IOP Conf. Series: Materials Science and Engineering 1109 (2021) 011001 doi:10.1088/1757-899X/1109/1/011001

24. Engr. Niño Rigo Emil Lim, MSc – De La Salle University, Philippines

Session Chairs

1. Conventional Topics in Mechanical Engineering
    a. Dr. Martin Ernesto Kalaw – De La Salle University, Philippines
    b. Dr. Satrio Wicaksono – Institut Teknologi Bandung, Indonesia
2. Manufacturing and Mechatronics Engineering
    a. Dr. Sri Raharno – Institut Teknologi Bandung, Indonesia
    b. Prof. Dr. Laurence Gan Lim – De La Salle University, Philippines
3. Energy Studies
    a. Prof. Dr. Alvin Culaba – De La Salle University, Philippines
    b. Dr. Yuli Setyo Indartono – Institut Teknologi Bandung, Indonesia
4. Emerging Topics in Engineering Economics and Sustainability Policy
    a. Dr. Robby Manrique – De La Salle University, Philippines
    b. Dr. Poetro Sambegoro – Institut Teknologi Bandung, Indonesia

Organizing Committee

Mechanical Engineering Department, De La Salle University, Philippines

Conference General Chair	Prof. Dr. Alvin Chua
Conference Program Co-Chairs	Dr. Neil Stephen Lopez
	Engr. Efren Dela Cruz, MSc
Secretariat	Prof. Dr. Aristotle Ubando
	Engr. Jeremias Gonzaga, MSc
	Engr. Arvin Fernando, MSc
	Ms. Lexus May Cajucom
	Mr. Miguel Luis Lava
	Mr. Luis Antonio Gutierrez

Peer review declaration

All conference organisers/editors are required to declare details about their peer review. Therefore, please provide the following information:

- **Type of peer review:** Single-blind
- **Conference submission management system:** EasyChair Conference Management System
- **Number of submissions received:** 87 papers
- **Number of submissions sent for review:** 87 papers
- **Number of submissions accepted:** 70 papers
- **Acceptance Rate (Number of Submissions Accepted / Number of Submissions Received X 100):** 80.46%
- **Average number of reviews per paper:** 2 reviewers per paper
- **Total number of reviewers involved:** 24
- **Any additional info on review process:**
  1. The reviewers were asked to comment on the following aspects of the paper:
     - Theoretical soundness of the manuscript.
     - Literature review and quality of the references.
     - Writing, readability and overall quality of the manuscript.
  2. In addition to the reviews, each submission was also subject to a similarity check via Turnitin. The authors were required to revise their manuscript until a similarity rating of less than 20% was achieved.
- **Contact person for queries:** Dr. Neil Stephen Lopez (neil.lopez@dlsu.edu.ph)

Please submit this form along with the rest of your files on the submission date written in your publishing agreement.

The information you provide will be published as part of your proceedings.

Performance of a solar chimney configuration to achieve equal flow rate for ventilation of three-story building

T V Nguyen and Y Quoc Nguyen

Faculty of Engineering, Van Lang University, 45 Nguyen Khac Nhu St., Co Giang Ward, Dist. 1, Ho Chi Minh City, Viet Nam

Email: y.nq@vlu.edu.vn

Abstract. In energy-efficient buildings, natural ventilation using solar chimneys greatly reduces energy consumption. In this study, a solar chimney was installed in a three-story building to examine the flow rate of the floors while balancing the flow rate of each floor. To predict the flow rate, a fluid dynamics model was developed based on a computational fluid dynamics (CFD) method with a calculated model of RNG k-e, heat flux 600 W/m^2 . By changing design parameters, the air flow was investigated.

Keywords: Solar chimney, energy efficiency, natural ventilation, CFD

1. Introduction

Among the methods for reducing energy consumption in buildings, solar chimneys have been extensively studied by many researchers. Previous studies showed that solar chimneys could provide sufficient ventilation rate for buildings. Shi et al. [1] reported that a solar chimney can supply air at 7.42 ACH (Air changes per hour – the number of times in an hour that the air inside the room is replaced) for a two – story building in Australia. Tan and Wong [2] claimed that the solar chimneys on the roof of a three – story building in Singapore induced the maximum air speed of 0.49 m/s inside the building.

When a solar chimney is applied to a multi – story building, different forms have been proposed, such as on the roof [3] or along the wall [4, 13-14]. Its effectiveness has been reported in those studies. Using solar chimneys helped to reduce the room temperature by 0.5–3 °C in [14] or up to 5 °C in [3], to increase the air speed inside the buildings up to 0.5 m/s [14], and to maintain a daily ventilation rate of 4 times of air exchange per hour for a seven – floor office building [13].

In those previous studies, solar chimneys with constant gaps were employed. Although their effectiveness has been shown, one of their drawbacks was also reported. Both studies [3,4] showed that the induced flow rate decreased at higher floors and the maximum difference was up to 2 times. Zhou et al. [2019] reported that for a two – story building, the temperature of the upper room was higher than that of the lower one of up to 1 °C. A similar trend was also reported by Punyasompun et al. [3] for a three – story building. Therefore, the ventilation rate and thermal comfort conditions were not equal for all floors in those studies.

To achieve similar flow rates for similar thermal comfort conditions on all floors, in the previous study [5], we tested a solar chimney connected to all three floors of a three – story building with different inlet locations and sizes but also with a constant gap of the air channel. The results showed that the location of the air inlet (near the floor or the ceiling), the number of the inlet in each room, and the size of the inlet strongly influenced the induced flow rate through each room. However, an equal ventilation rate for all rooms has not been achieved.

In this study, the solar chimney in the previous study [5] was modified. Instead of using a constant air gap, different gaps were designed at different floors. As previous studies [3-5] reported that the higher floors had lower flow rates, it is deduced that the air gap should increase with the height of the chimney to achieve similar flow rates. Therefore, the proposed chimney configuration had larger air gaps at the higher floors. Floors are considered as a new room with empty space and no activity inside the room.

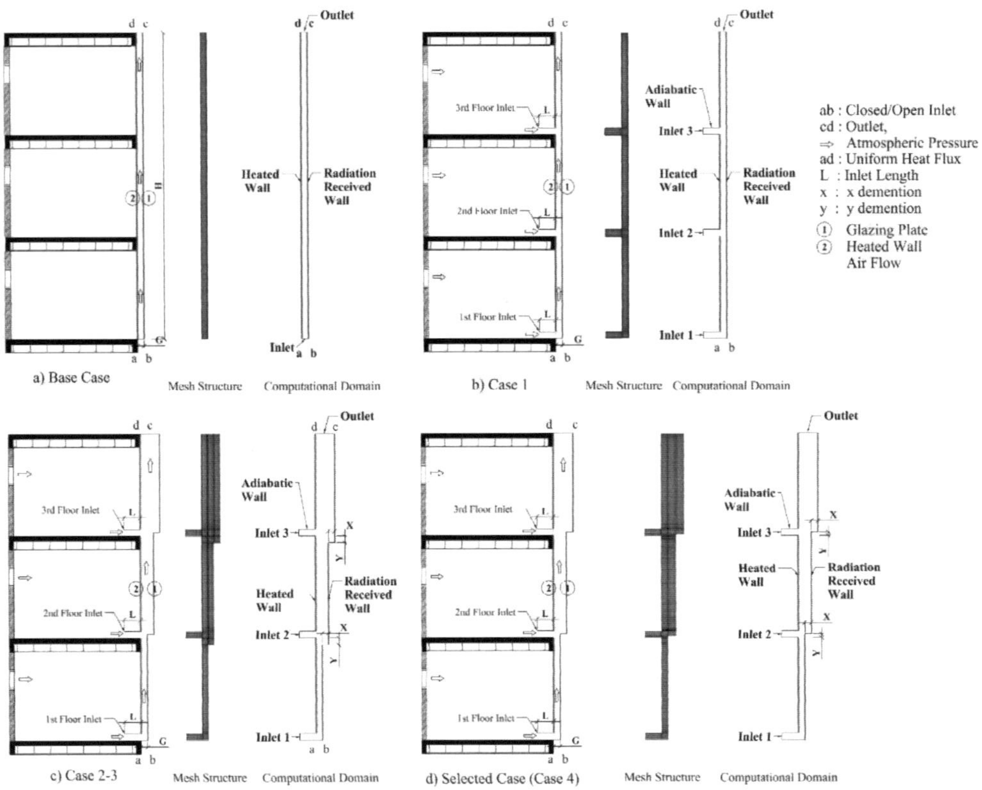

Figure 1. Schematic of a solar chimney for ventilation of a three – story building (H=9.0 m, G=0.2 m) with computational domain and mesh.

2. Problem formulation and Numerical Method

2.1. The solar chimney

The solar chimney consists of an air channel enclosed by an outer wall and a wall of the building. In this study, a solar chimney is assumed to be integrated on the right side of a three – story building, as sketched in figure 1 and tested under 4 configurations. In figure 1a, the solar chimney is vertically and measures $9mx0.2\ m$ in the height and gap, respectively. Solar chimney's right side is a wall and the heated wall is on the left side of the chimney. The two openings are located at the bottom (inlet) and the top (outlet) of the solar chimney, respectively. Then, to examine the internal flow of the solar chimney, inlets near the floor (lower inlet) on each floor of the building were installed as depicted in figures 1b and 1c. In addition, in figure 1c, in order to investigate the variation in the inlet flow rates of the floors, the dimensions of solar chimney are changed horizontally (case 2) and longitudinally (case 3). The width of the solar chimney on each floor is extended to size x (Case 2) and height to size y (Case 3). In all cases except the base case, the lower end of the solar chimney is closed so that the inlet flow at the floors is higher and the upper outlet is always opened to let the air flow out of the solar chimney.

2.2. *Assumptions and governing equations.*

For obtaining the induced air flow rate through the chimney, together with the assumption of steady flow, time – averaged flow variables can be described with the Reynolds – Averaged Navier Stokes equations [6,7]. The equations represent the conservation principles of the mass, momentum, and energy of the flow. Using the Einstein summation notation, the equations are as follows.

$$\frac{\partial u_j}{\partial x_j} = 0 \tag{1}$$

$$\frac{\partial (u_i u_j)}{\partial x_j} = -\frac{1}{\rho}\frac{\partial p}{\partial x_i} - g_i\beta(T - T_0) + \frac{\partial}{\partial x_j}\left(\nu\frac{\partial u_i}{\partial x_j} - \underline{u_i' u_j'}\right) \tag{2}$$

$$\frac{\partial (T u_j)}{\partial x_j} = \frac{\partial}{\partial x_j}\left(\frac{\nu}{Pr}\frac{\partial T}{\partial x_j} - \underline{T' u_j'}\right) \tag{3}$$

The turbulence tress $\underline{u_i' u_j'}$ in Equation (2) and the turbulent heat flux $\underline{T' u_j'}$ in equation (3) were solved with the standard k–ω model with modification for low – Reynolds number effects by a damping coefficient for the eddy viscosity. This turbulent model was also employed in previous studies of solar chimneys [8]. For the solar chimney in figure 1, the gap – to – height ratio, G/H, is 4.25%. Gan [12] showed that for solar chimneys with very low G/H, a computational domain fitted with the air channel was sufficient. For discretizing the governing equations, equations (1), (2) and (3), Finite Volume Method was used on a structured rectangular mesh with the ANSYS Fluent CFD Code (Academic version 2020R2).

The mesh density was higher near the walls of the chimney where the gradients of the flow variables are strong as displayed in figure 1b. To check effects of the mesh resolution, we ran the preliminary numerical model for the solar chimney in the experiment by Yilmaz and Fraser [9]. The number of elements in each direction was increased gradually and the according induced flow rate was plotted in figure 2. Denser meshes resulted in a change of the flow rate below 1.0%. The equivalent required non-dimensional distance of the first grid point from the solid surfaces, or y^+ was below 1.0, which agrees with the findings by Zamora and Kaiser [8].

For the boundary conditions, atmospheric pressure was applied at the inlets of the rooms and the outlet of the solar chimney. The size $L = 0.5m$ is the distance from the inlet to the solar chimney is an adiabatic wall. At all inlets, turbulent kinetic energy and a length scale were used. As the turbulence should be low [9], it was assumed a turbulent intensity of 2.0% and a length scale of half of the inlet height, i.e. $0.1\ m$.

2.3. *Validation of the numerical method*

The CFD model was applied to predict the induce flow rate through the solar chimney in the experiment by Yilmaz and Fraser [9]. The solar chimney is similar to the base case (figure 2). The height and the gap of the chimney were 3.0 m and 0.1 m, respectively. In the experiment, uniform heat fluxes were applied on a surface of the air channel and the corresponding wall temperatures were in the range of 60 – 130°C. Uniform wall temperatures were also applied in the CFD model. The predicted and measured flow rates are presented in figure 3. Difference in flow rate is negligible, less than 10%.

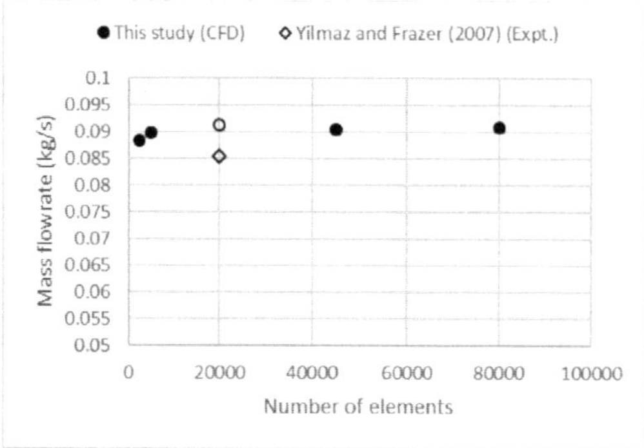

Figure 2. Variation of the flow rate versus the number of elements for simulation of the solar chimney in the experiment by Yilmaz and Fraser [9] (case of 100°C). The measured flow rate by Yilmaz and Fraser [9] was also plotted. The selected mesh is indicated with the open marker.

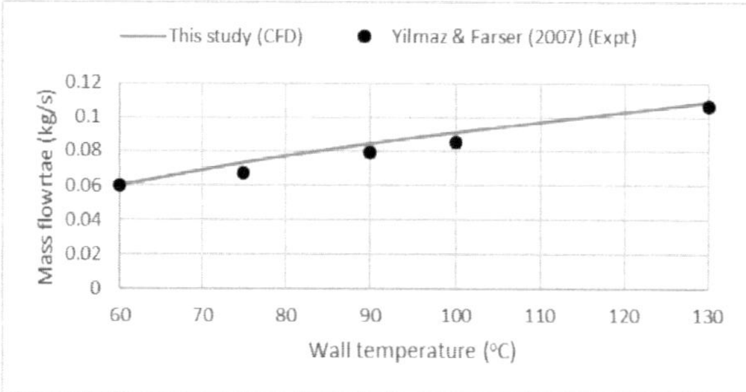

Figure 3. Comparison of the computed flowrate versus the measured data for the experiments by Yilmaz and Fraser [9].

3. Results and discussions

In this section, wind velocity and flow rate of solar chimney cases different from figure 1 are surveyed and presented. The heat flux applied to the walls is 600 W/m^2.

3.1. Base configuration (Base case)

In figure 1a, a solar chimney with 2 openings is integrated into the building, the opening below is the inlet and the top is the outlet in the air channel (figure 1a). Heated wall is placed on the left and right to the radiation received wall of a solar chimney. This configuration has been studied widely in the literature [9-11]. The temperature values in figure 7a show the temperature rising from the bottom up on both walls in the air channel while only the left wall is the heated wall. At the output of the base case, no flow reversal occurs. Air velocity increases and becomes higher when adjacent to a heated wall, as shown in figure 8a. Since there is no outlet flow reversal as shown in figure 8a, it is sufficient to use a computational domain that includes only air channels in this study.

Investigation of the mass flow rate of the base case was also studied under many effects on the variability of the width of the solar chimney as depicted in figure 4. The width of the solar chimney is changed from 0.1 to 0.4m and the result is that the flow rate increases with increasing the width of the air gap. This trend is similar to those of the experimental data of Chen et al. [10] and Burek and Habeb [11].

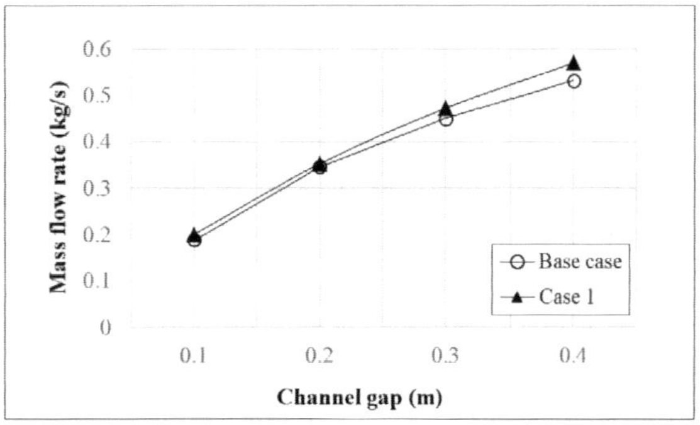

Figure 4. The induced flow rate of the Base case & case 1.

3.2. Testing configurations (case 1-3)

Survey with changes in the width of solar chimney in the base configuration shows the trend of increasing flow rate from 0.188 kg/s in the case of $Gap = 0.1m$ to 0.53 kg/s in case of $Gap = 0.4m$. Based on the base case, a vertical solar chimney case whose inlet values lie close to the floor of each floor are studied, as depicted in figure 1b (case 1). The total flow rate at the output of case 1 shows that the lower inlet in each floor can be improved compared to the base case, about $0.01 - 0.03\ kg/s$ as shown in figure 4e. When the gap of the solar chimney is changed, there is no big difference in the inlet traffic of the floors. The flow conservation trend is shown by the increase in the flow of inlet 1 and decrease in inlet 2 and 3 at the same time (figure 5).

Figure 5 shows that, with a constant gap, the higher floors had lower flow rate. This effect was also reported in previous studies [3,4]. It might be because the inlets at the higher floors had lower elevation difference between the inlet and the outlet of the air channel; hence weaker thermal effects and, consequently, lower induced flow rate.

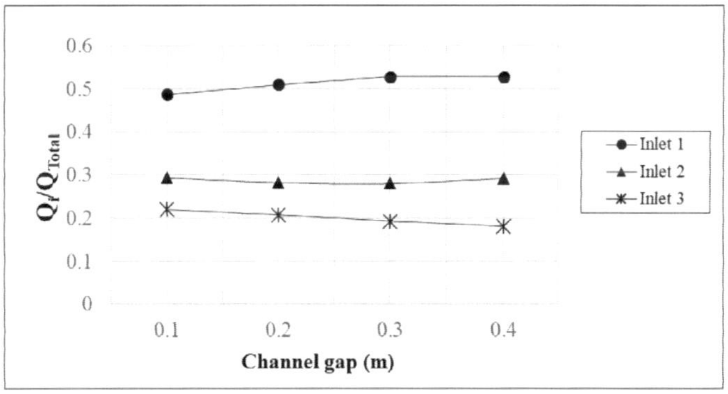

Figure 5. The induced flow rate of the case 1.

When investigating the change of channel gap in case 1, the variation in flow rate between floors is negligible. Based on this, two solar chimney models with gradual widening on each floor were computed. In case 2 (figure 1b), the model gap size remains $G = 0.2m$ on the first floor, on the second floor the dimensions are added xG and $2xG$ on the third floor where x is the variable in the computational domain, $x = 0 - 0.3m$.

From the research results of case 2, case $x = 0.0m$ is considered the case of vertical solar chimney of case 1. As the x size increases (figure 6a), outlet flow also increases gradually from 0.35 $kg/\ s$ in the case x = 0m to 0.69 kg / s in the case $x = 0.3m$. The flow rates in inlet 1 tended to decrease from

0.17 kg / s $(x = 0)$ to 0.24 kg/s $(x = 0.3m)$ and inlet 2 and inlet 3 increased the flow rate simultaneously. There is a change in the case $x = 0.3m$, the discharge in inlet 2 is higher than inlet 1, and inlet 1 continues to decrease.

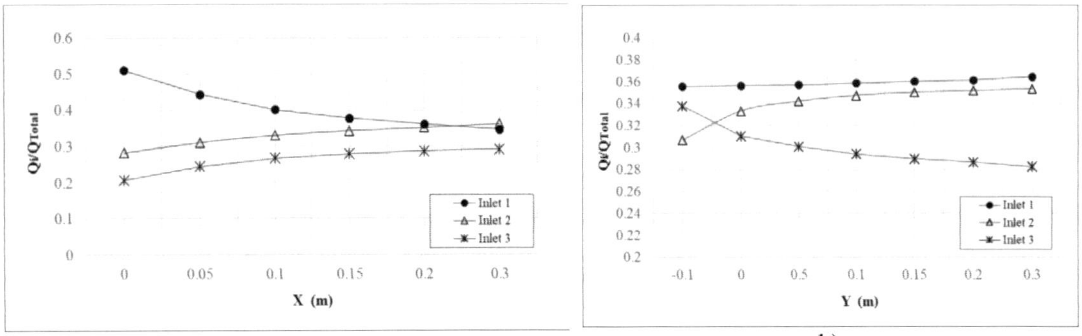

a) b)

Figure 6. The induced flow rate of: a) Case 2; b) Case 3.

To evaluate the behavior of the air inside the solar chimney, three different solar chimney-sized models were compared including base case, case 1 and case 2 with dimensions $x = y = G = 0.2m$.

Survey from 3 cases of temperature and velocity (figure 7) shows a clear difference within the solar chimney. The temperature field tends to decrease gradually from inlet to outlet most clearly in the base case. The solar chimney in case 1 has a more even temperature on all 3 floors. The velocity field has a clear difference in case 1 as the velocity increases dramatically as it passes through the inlet from each layer. The velocity in case 2 tends to decrease gradually and the difference in velocity when passing through the inlet of each layer is significantly lower than that of case 1. From the study in figure 7, it can be seen that case 2 is more suitable. with the need to balance flow and velocity on each floor.

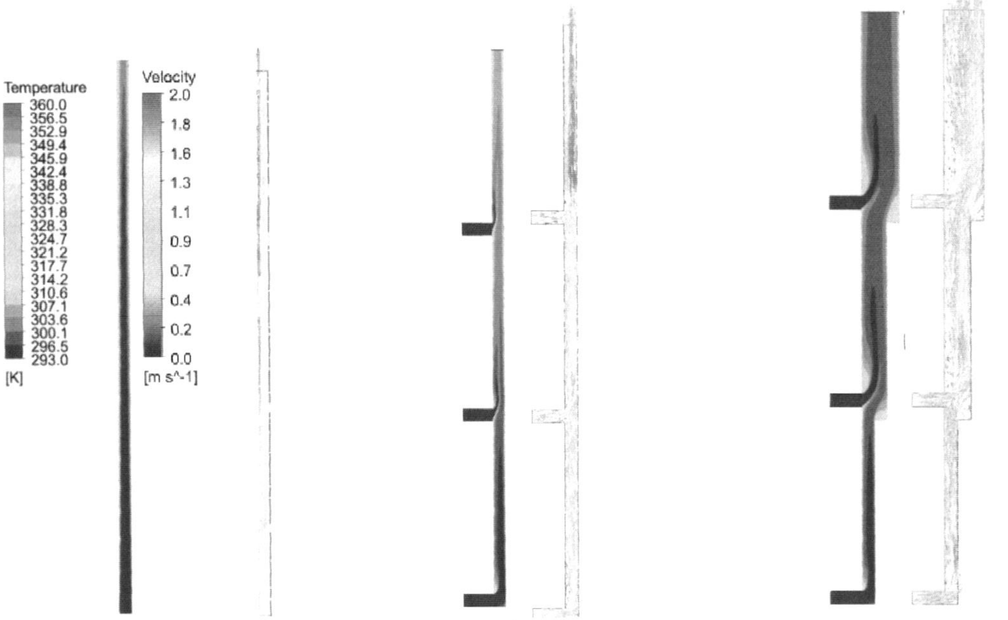

a b c

Figure 7. Temperature and flow field of: a) Base case; b) Case 1; c) Case 2.

3.3. *Selected configurations*

With the aim of achieving a flow balance between the floors of the building. The selected configuration (case 4) is a combination of case 2 & 3 with dimensions x and y of $x = 1.25G\ (0.25m)$, $y = -0.25G\ (0.05m)$. The computational domain and mesh features of the selected configuration are shown in figure 1d. Investigation of the flow rate of each floor during heat flux changes is also examined in two heated wall cases located on the left (LW) and the right (RW), as shown in figure 7.

The total output of the case with a heated wall on the left is always higher than on the right. In all 3 cases of change of heat flux, the output flow difference is less than 0.1 kg/s. The heat flux case located on the right shows that the discharge on the second floor is significantly lower than that of floors 1 and 3 and is also the lowest of the two LW and RW cases examined. In the case of LW, the flow difference between the floors is more stable than RW and tends to balance the flow more than the other cases.

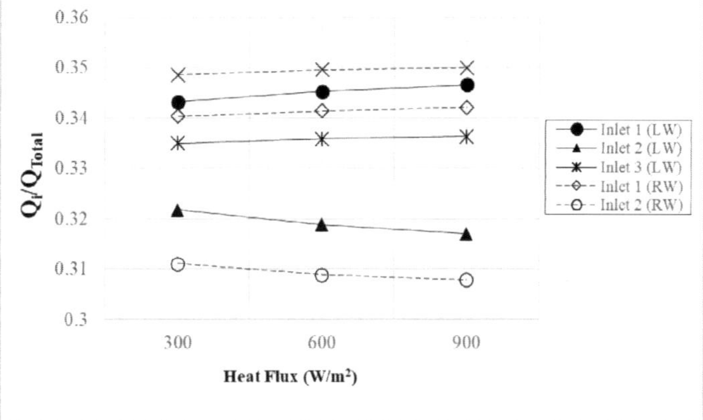

Figure 8. Heated wall induction current rate on left (LW) and right (RW) side of solar chimney

4. Conclusion

The survey results show that the optimum case (case 4) is the case with the best flow rate balance between the inner floors of the building. Survey of the discharge in each stratum using solar chimney in case 1 shows that the discharge in each stratum gradually decreases from bottom to top. Resizing x and y in case 2 and case 3 is to find the solar chimney configuration that best balances the inlet flow rate of the floors. The best configuration for natural ventilation balance surveyed is the one with dimensions x=1.25G (0.25m), y=-0.25G (0.05m). A survey of heated wall's heat flux increase and left-to-right position change is also evaluated in figure 8. Surveys show that there is still a gap in inlet flow rate for each floor but with a heated wall on the left it is more efficient when it is on the right.

This study shows the applicability of natural ventilation in sustainable architectural design. The balance of natural ventilation in the floors of the building through solar chimney positively impacts the comfort of use and wide application in multi-story projects. In the future, optimization techniques will be applied to optimize the flow balance through the resizing of solar chimneys.

References

[1] Shi L, Cheng X, Zhang L, Li Z, Zhang G, Huang D and Tu J 2019 Interaction effect of room opening and air inlet on solar chimney performance *Applied Thermal Engineering* **159** 113877

[2] Tan A Y K and Wong N H 2012 Natural ventilation performance of classroom with solar chimney system *Energy and Buildings* **53** 19–27

[3] Punyasompun S, Hirunlabh J, Khedari J and Zeghmati B 2009 Investigation on the application of solar chimney for multi-storey buildings *Renewable Energy* **34** (12)

2545–2561

[4] Mohammed H J, Jubear A J and Obaid H 2020 Natural ventilation in passive system of vertical two-stores solar chimney *Journal of Advanced Research in Fluid Mechanics and Thermal Sciences* **69** (2) 130–146

[5] Nguyen T V, Nguyen Y Q and Huynh T N 2020 A solar chimney for natural ventilation of a three – story building *The International Conference on Modern Mechanics and Applications*

[6] Shi L, Cheng X, Zhang L, Li Z, Zhang G, Huang D and Tu J 2019 Interaction effect of room opening and air inlet on solar chimney performance *Applied Thermal Engineering* **159** 113877

[7] Khanal R and Lei C 2014 An experimental investigation of an inclined passive wall solar chimney for natural ventilation *Solar Energy* **107** 461–474

[8] Zamora B and Kaiser A S 2009 Optimum wall-to-wall spacing in solar chimney shaped channels in natural convection by numerical investigation *Applied Thermal Engineering* **29** (4) 762–769

[9] Yilmaz T and Fraser S M 2007 Turbulent natural convection in a vertical parallel-plate channel with asymmetric heating *International Journal of Heat and Mass Transfer* **50** (13–14) 2612–2623

[10] Chen Z D, Bandopadhayay P, Halldorsson J, Byrjalsen C, Heiselberg P and Li Y 2003 An experimental investigation of a solar chimney model with uniform wall heat flux *Building and Environment* **38** (7) 893–906

[11] Burek S A M and Habeb A 2007 Air flow and thermal efficiency characteristics in solar chimneys and Trombe Walls *Energy and Buildings* **39** (2) 128–135

[12] Gan G 2010 Impact of computational domain on the prediction of buoyancy-driven ventilation cooling *Building and Environment* **45** (5) 1173–1183

[13] Asadi, S., Fakhari, M., Fayaz, R., & Mahdaviparsa, A 2016 The effect of solar chimney layout on ventilation rate in buildings. Energy and Buildings, **123** 71–78.

[14] Zhou, Y., Wang, Z., Yang, C., Xu, L., & Chen, W 2019 Influence of Trombe wall on indoor thermal environment of a two-story building in rural Northern China during summer. Science and Technology for the Built Environment, **25**(4) 438–449.

Optimization of 25-bar truss structure with the consideration of thermal expansion

K H Lin[1], N L Htet[1] and T M Htike[1,*]
[1]Department of Mechanical Engineering, Yangon Technological University, Insein 11101, Yangon

me.tmh@ytu.edu.mm

Abstract. This paper presents the change in the optimal cross-sectional areas of 25-bar truss structure due to thermal expansion. The 25-bar spatial truss is commonly used in structural engineering as a benchmark for structural optimization. Previous studies used different optimization methods to compute the optimal cross-sectional areas of the truss structures related to the minimal weight of each bar element. However, most of the previous studies conducted the optimization under mechanical load only and reported almost the same optimal areas and minimum weight with slight variations depending on the method used. However, very little has been done on the structural optimization with the consideration of thermal expansion. This study aimed to investigate the effect of thermal expansion in structural optimization and conducted optimization of this structure under the thermal load and the same mechanical load case with the previous studies. The difference in the optimal areas between with and without thermal expansion is analysed through optimization of both cases using MATLAB Optimization Toolbox. The optimization method used is fmincon with Sequential Quadratic Programming (SQP) . The result shows both an increase and decrease in optimal cross-sectional areas. The optimal weight of the structure increases by 12.26% at the surrounding temperature of *30°C*. Therefore, this study highlights the needs of consideration of thermal load in optimizing the cross-sectional areas of the structure for minimum weight and this is particularly of important consideration in designing the truss structures in tropical climate. One of the limitations of this study that the SQP may not be appropriate for large structure and hence the robustness of the method should be tested with more variable.

1. Introduction

The structural optimization has been conducted in structural engineering for more than 30 years. The optimization can reduce the quantity of material needed for the structure and hence cut down the expenses. Therefore, many researchers have used different algorithms to solve the optimization problems [1–7].

The majority of the research has focused on the actual applied forces such as loads on the structure including the ambient loads. In addition, the impact of earthquake has been considered [3]. For example, Pierezan et al. conducted a more efficient optimization method based on Coyote Optimization algorithm (COA) [1]. This method controls population system and as a result it is capable of finding the better minimum weight of the structures than original COA. The study showed that the adaptation leads to the improvement of the robustness of the existed algorithm. Li compared convergence rate of heuristic particle swarm optimization (HPSO) with two other algorithms based on pin-connected structures [7].

They have reported that HPSO converges more quickly than the particle swarm optimizer with passive congregation (PSOPC) and harmony search scheme (HS). This comparison is effective since it reduces the iteration number and saves as much time as the structure is large.

Degertekin et al. conducted structural optimization problem with 204 design variables covering size, layout and topology of different benchmark structures using Jaya algorithm [8]. They have reported that Jaya algorithm is better than heuristic and gradient-based optimization methods. The study demonstrated the robustness of JA algorithm to handle large scale structures with small standard deviation.

The 25-bar truss structure was originally designed by Fox and Schmit [9] as the pin-join transmission tower. It is a standard benchmark used for structural optimization specially to check if the optimization algorithm is efficient or not before it is applied to the optimization of larger structures. Many studies have used this structure as a benchmark in structural optimization under some standard load cases. Other studies have tested the performance of different optimization algorithms for structural optimization using this structure [9]. Most of them reported, almost the same results in minimum weight and optimal cross-section areas although there are some slight variations depending on the method used. Although different optimization schemes for structure problems have been reported widely, little has considered the thermal load in the optimization.

This study aims to evaluate the effect of surrounding temperature on the optimal areas and weight of the 25-bar truss structure. First, finite element model of the structure was built in MATLAB and optimizations were conducted using MATLAB optimization toolbox by Minimizer for constrained nonlinear function and Multiple-objective genetic algorithm. Results were confirmed by the result from the literature. For thermal consideration, ADINA model was used to validate results from the MATLAB-based FEM model under the same load and constraint functions. Optimization with consideration of thermal load would help determine if effect of thermal considerations in structural optimization is significant or not.

2. Methods

2.1. Optimization Procedure

The process of optimization can be approached from different programming platforms and efficient algorithms are chosen and adopted based on the nature of the problem. A quick tool for optimization with the wide variety of solvers has been developed by MathWorks and added to MATLAB software.

The required objective function and constraint function were created in MATLAB. Two different optimization methods were used in this study; Find minimum of constrained nonlinear multivariable function (fmincon) and Multiobjective genetic algorithm (gamultiobj) from MATLAB optimizer. fmincon with sequential quadratic programming (SQP) was used because these routines are more efficient in both memory usage and by using SQP, high net present value can be obtained which means the minimum is found within fewer iterations. Moreover, when the objective function is unbounded, the routines can take smaller step and the solution can converge. The algorithm also sets the penalty parameter for the merit function automatically. gamultiobj is useful in solving multi-objective optimization problems and is time efficient. Since the optimization in this study contains single-objective, fmincon performs better than gamultiobj.

Before the optimization process, the outputs of objective function, stress and displacement, are ensured by comparing with the results given out by ADINA software. The final optimization result is compared with the results from other research papers.

2.2. Model of truss optimization

If the truss structure of interest consists of *nb* bar elements and corresponding cross-sectional area of each element is A_i, the total weight of the structure, which is the objective to the optimization, can be formulated as:

$$\boldsymbol{W} = \sum_{i=1}^{nb} \boldsymbol{\rho l_i A_i} \tag{1}$$

where ρ is the density and l_i is the length of each corresponding element. The elements are subjected to both stress limit and nodal displacement limit. Loading applied to the elements includes ordinary carrying load and predicted external force, which are called real forces, f.

The main contribution of this study is incorporation of thermal load in the finite element model for the optimization problem. For handling temperature effect, thermal expansion occurred in each element is converted into virtual thermal force as:

$$|f_t| = A_i E \alpha \Delta T \tag{2}$$

where E is the modulus of elasticity and α is the coefficient of thermal expansion of the material. This virtual force is equivalent to the reaction to the stress caused by thermal expansion. Thus, the direction of this virtual force is opposite to that of thermal expansion. The system can be summarized as:

$$\{\boldsymbol{u}\} = [\boldsymbol{K}]^{-1}(\{\boldsymbol{f}\} - \{\boldsymbol{f_t}\}) \tag{3}$$

where $[K]$ is the global stiffness matrix of the structure. In consequence of the nodal displacement, the resultant stress in each bar element can now be calculated as:

$$\sigma_i = \frac{E}{l_i} \Delta u \tag{4}$$

As a summary, this optimization model contains objective function $W(A_i)$ and two constraints, nodal displacement (u) and stress (σ).

2.3. Structure, Loads and Constraints

A 25-bar spatial truss structure as shown in figure 1 is made up of aluminum metallic truss bars. The modulus of elasticity of the given material is 69 GPa and the density is given to be 2770 kgm^{-3}. The structure is subjected to two load conditions given in table 1. The normal temperature of the bar elements is set to 25°C. Both tensile and compressive stress limits are given in table 2. The nodal displacement of all nodes in all of the three spatial directions are limited to 8.9 mm.

Six optimizations were conducted under six different surrounding temperatures. The surrounding temperature for first optimization is 25°C, therefore the whole structure is in the same surrounding at 25°C. Then, the optimizations were repeated at different surrounding temperatures from 26°C to 30°C. Since the main aim of this study is to investigate if the thermal expansion affects the optimal areas, this study did not consider any special temperature range of practical interest. When considering the thermal expansion, its effect on the density was also taken into account by updating the density as temperature-dependent parameter.

Although there are 25 variables, areas of 25 bars, to consider in optimization, they can be divided into 8 groups, each consists of areas of symmetric elements having the same physical properties: (1) A_1, (2) A_2-A_5, (3) A_6-A_9, (4) A_{10}-A_{11}, (5) A_{12}-A_{13}, (6) A_{14}-A_{17}, (7) A_{18}-A_{21}, (8) A_{22}-A_{25}.

Table 1. Two load conditions for 25-bar spatial truss structure.

Node	Case 1 (Pa)			Case 2 (Pa)		
	f_x	f_y	f_z	f_x	f_y	f_z
1	0.0	8.8964e+04	-2.2241e+04	4.4482e+03	4.4482e+04	-2.2241e+04
2	0.0	-8.8964e+04	-2.2241e+04	0.0	4.4482e+04	-2.2241e+04
3	0.0	0.0	0.0	2.2241e+03	0.0	0.0
4	0.0	0.0	0.0	2.2241e+03	0.0	0.0

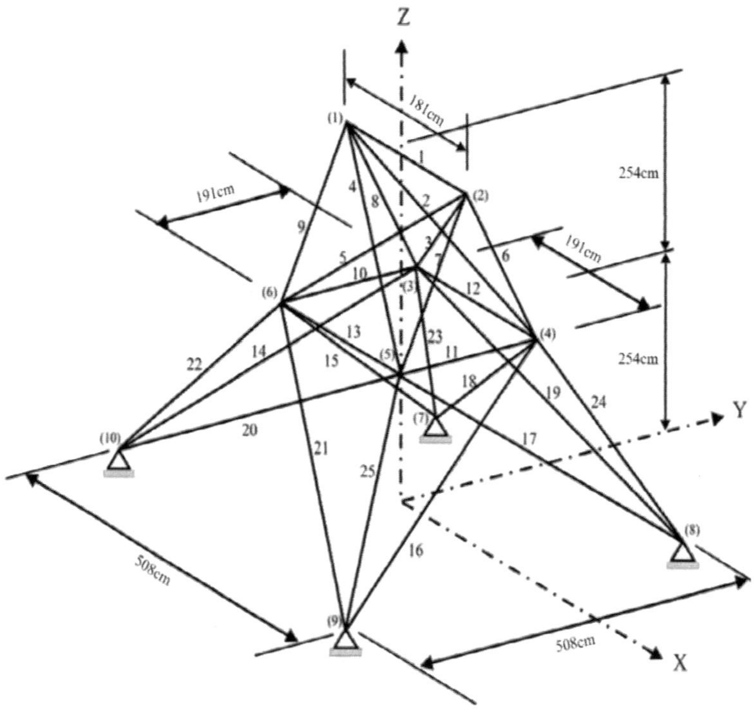

Figure 1. 25-bar spatial truss structure [8].

Table 2. Stress limits for bar elements of 25-bar spatial truss problem.

Element groups	Compressive stress limit	Tensile stress limit
1	241950822.82	275790291.60
2 – 5	79910236.99	275790291.60
6 – 9	119313774.90	275790291.60
10 – 11	241950822.82	275790291.60
12 - 13	247535576.22	275790291.60
14 – 17	46601664.52	275790291.60
18 – 21	47980615.98	275790291.60
22 - 25	81371925.53	275790291.60

3. Results

3.1. Verification of FEM model

The above mathematical model is confirmed by the 25 bars truss structure. It consists of two parts: (1) deformation with only real load case; and (2) deformation with only thermal load case. For the first part, load case 1, given in table 1, is taken. For the second part, the temperature difference is taken as 5°C for calculating thermal load. In this confirmation, area of every bar element is taken to be 0.001 m^3.

Figure 2 shows the comparison of deformed shape by MATLAB-based FEM model and ADINA model with scale factors of '30'. Table 3 shows the numerical values of displacements of both computations. The only the discrepancy found in 'x' direction of node 6 is very small. Therefore, the MATLAB model is verified for the load case.

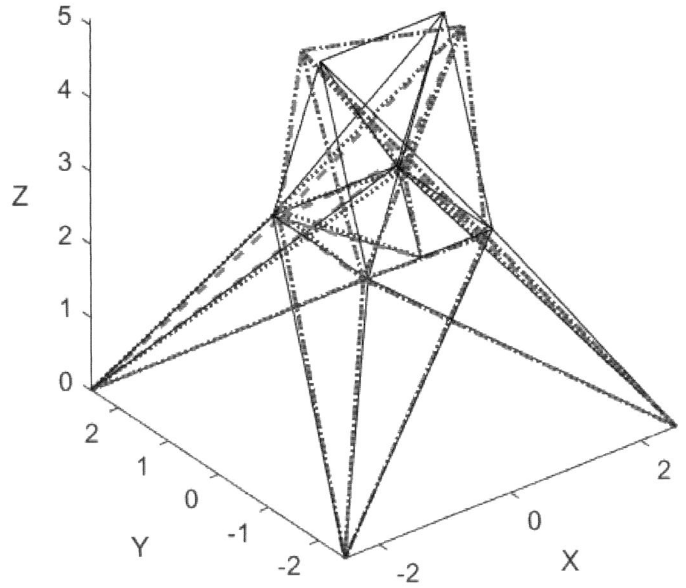

Figure 2. Deformed shape due to load case 1 with scale factor '30' (red dashed line represents the described model and blue dashed line represents ADINA)

Table 3. Comparison of displacement results due to load case 1

Node	MATLAB model			ADINA		
	u_x	u_y	u_z	u_x	u_y	u_z
1	-7.1800e-05	1.2460e-02	-8.8813e-04	-7.1800e-05	1.2460e-02	-8.8813e-04
2	7.1800e-05	-1.2460e-02	-8.8813e-04	7.1800e-05	-1.2460e-02	-8.8813e-04
3	2.9755e-03	-5.2321e-04	-2.2533e-03	2.9755e-03	-5.2321e-04	-2.2533e-03
4	2.9916e-03	5.7390e-04	1.1831e-03	2.9916e-03	5.7390e-04	1.1831e-03
5	-2.9755e-03	5.2321e-04	-2.2533e-03	-2.9755e-03	5.2321e-04	-2.2533e-03
6	-2.9916e-03	-5.7390e-04	1.1831e-03	-2.9915e-03	-5.7390e-04	1.1831e-03
7-10	0.0	0.0	0.0	0.0	0.0	0.0

Figure 3 shows the deformation of 25-bars structure due to thermal load calculated in MATLAB and ADINA. Table 4 shows the corresponding numerical values of nodal displacement. All the numerical results from both methods are same except y coordinate of node 1 and 2. This is because the nodal displacements of these nodes are very close to zero and their values are negligible. So, the model is verified with the results from the ADINA model for dealing with thermal load for further optimization.

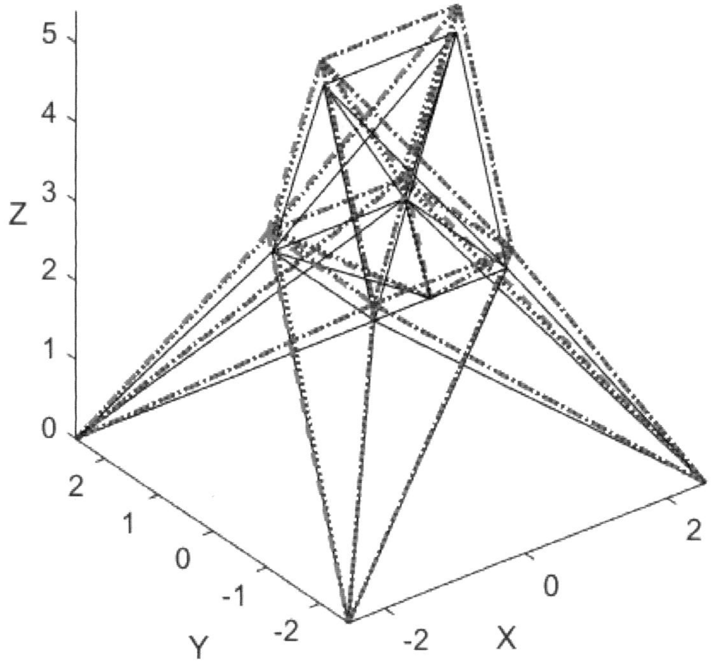

Figure 3. Deformed shape due to thermal load with scale factor '300' (red dashed line represents the described model and blue dashed line represents ADINA)

Table 4. Comparison of displacement results due to thermal load

Node	MATLAB model			ADINA		
	u_x	u_y	u_z	u_x	u_y	u_z
1	-1.1283e-04	3.3531e-20	1.0887e-03	-1.1283e-04	1.2460e-19	1.0887e-03
2	1.1283e-04	2.7384e-19	1.0887e-03	1.1283e-04	-1.2460e-19	1.0887e-03
3	-1.2774e-04	1.2833e-04	7.9236e-04	-1.2774e-04	1.2833e-04	7.9236e-04
4	1.2774e-04	1.2833e-04	7.9236e-04	1.2774e-04	1.2833e-04	7.9236e-04
5	1.2774e-04	-1.2833e-04	7.9236e-04	1.2774e-04	-1.2833e-04	7.9236e-04
6	-1.2774e-04	-1.2833e-04	7.9236e-04	-1.2774e-04	-1.2833e-04	7.9236e-04
7-10	0.0	0.0	0.0	0.0	0.0	0.0

3.2. Optimization results

3.2.1. Results under mechanical load only

Table 5 shows optimal areas for minimum weight by the fmincon and gamultiobj optimization algorithms in comparison with heuristic particle swarm optimization by [7]. As seen in table 5, the weight by both fmincon and gamultiobj methods agree very well with the value reported in the literature. The fmincon method results in better agreement with the HPSO results. Overall, these results indicate that both fmincon and gamultiobj algorithms can be verified with HPSO algorithm published in the literature. Moreover, for the next step optimization considering the same load case but subjected to different thermal loads, fmincon will be used because this method resulted in almost the same minimum weight by HPSO. Apart from that, the fmincon method (about 20 iterations) converges faster than gamultiobj method (about 800 iterations).

Table 5. Optimal areas subjected to minimizing total weight.

Element groups	Optimal areas (m^2)		
	HPSO [7]	fmincon	gamultiobj
1	6.4500e-04	1.0035e-06	4.894e-05
2 – 5	1.2707e-03	1.2755e-03	1.2081e-03
6 – 9	1.9453e-03	1.9278e-03	1.8838e-03
10 – 11	6.4500e-06	1.0025e-06	7.9984e-06
12 - 13	6.4500e-03	1.0046e-06	7.7493e-05
14 – 17	4.4763e-04	4.4102e-04	5.4176e-04
18 – 21	1.0842e-03	1.0856e-03	1.1723e-03
22 - 25	1.7047e-03	1.7146e-03	1.6078e-03
Weight	2.4261e+03	2.4217e+03	2.4500e+03

3.2.2. Results under mechanical and thermal loads

Apart from the static loads, thermal loads were considered for different temperatures from 26 to 30°C and optimizations were done at each temperature to evaluate the effect of temperature on the minimum weight of the structures under the same constraint conditions. Table 6 shows the optimal areas and minimum weights under thermal loads at different temperatures. As can be seen clearly, for element group 2-5, 14-17, 18-21 and 22-25, the higher the temperature, the larger the cross-sectional areas. However, for element 1, 6-9, 10-11 and 12-13 the cross-sectional areas are not affected by the temperature.

Percentage increases in the minimum weight are also reported in table 6. It can be clearly seen that consideration of thermal loads increase the minimum weight of the structure. At surrounding temperature of 30°C, the minimum weight of the structure increases by 12.26. Therefore, the results indicate that the effect of thermal loads on the optimal cross-sections and minimum weight is significant and hence it cannot be ignored especially optimizing the structure in the tropical regions. Moreover, the results also indicate that the higher the temperature, the higher the percentage increase in the minimum weight. This is expected since the larger thermal expansion at the higher temperature causes the higher stress and displacement resulting in the larger areas of some members and the increase in the minimum weight. Day time temperature in some tropical countries can raise up to around 40 C in summer. For optimizing and designing the truss structure in such locations, the results by this study indicate that effect of thermal expansion must be taken into account.

Table 6. Optimal areas and weight at various temperature.

Element groups	Optimal areas (m^2)					
	25°C (normal)	26°C	27°C	28°C	29°C	30°C
1	1.0035e-06	1.0000e-10	1.0000e-10	1.0000e-10	1.0000e-10	1.0000e-10
2 – 5	1.2755e-03	1.4236e-03	1.5343e-03	1.6444e-03	1.7552e-03	1.8649e-03
6 – 9	1.9278e-03	1.8659e-03	1.8020e-03	1.7429e-03	1.6886e-03	1.6392e-03
10 – 11	1.0025e-06	1.0000e-10	1.0000e-10	1.0000e-10	1.0000e-10	1.0000e-10
12 - 13	1.0046e-06	1.0000e-10	1.0000e-10	1.0000e-10	1.0000e-10	1.0000e-10
14 – 17	4.4102e-04	4.5160e-04	4.6154e-04	4.7010e-04	4.7761e-04	4.8354e-04
18 – 21	1.0856e-03	1.0673e-03	1.0867e-03	1.1070e-03	1.1270e-03	1.1476e-03
22 - 25	1.7146e-03	1.7935e-03	1.8601e-03	1.9222e-03	1.9802e-03	2.0352e-03
Weight	2.4217e+03	2.4818e+03	2.5419e+03	2.6013e+03	2.6602e+03	2.7186e+03
% increase	0	2.48	4.96	7.42	9.85	12.26

4. Conclusion

In this study, the effect of thermal load on the optimal cross-sectional areas for weight minimization was evaluated by using benchmark structure of 25-bar truss. Finite element model and optimization algorithms were first verified by comparing respective results on nodal displacements, element stresses with results from ADINA model and also optimal areas and weights by comparing with published values. Main conclusions of this study are:

(1) The results show that thermal expansion can lead the deformation of the structure to exceed the limited tolerances hence affecting the optimal cross-sectional areas. Thus, temperature is one of the necessary factors to consider in precision works.

(2) Because of thermal stresses and thermal expansion, the optimal weight increases. Surrounding temperature of $30°C$ increases the optimal weight by 12.26%.

The results from this study indicate that thermal considerations are substantially important in optimizing truss structures. This is particularly important in optimizing structures in tropical regions. Further studies should verify the results from this study by comparing them with empirical observations. The study can also be extended to optimize the structure to incorporate other important considerations such as the natural frequencies of vibration, global stability and critical buckling load.

References

[1] Pierezan J, dos Santos Coelho L, Cocco Mariani V, Hochsteiner de Vasconcelos Segundo E and Prayogo D 2021 Chaotic coyote algorithm applied to truss optimization problems *Comput. Struct.* **242** 106353

[2] Shih C J and Lee H W 2006 Modified double-cuts approach in 25-bar and 72-bar fuzzy truss optimization *Comput. Struct.* **84** 2100–4

[3] Wang H and Ohmori H 2013 Elasto-plastic analysis based truss optimization using Genetic Algorithm *Eng. Struct.* **50** 1–12

[4] Makris P A and Provatidis C G 2002 Weight minimisation of displacement-constrained truss structures using a strain energy criterion *Comput. Methods Appl. Mech. Eng.* **191** 2187–205

[5] Dede T, Bekiroğlu S and Ayvaz Y 2011 Weight minimization of trusses with genetic algorithm *Appl. Soft Comput.* **11** 2565–75
[6] Lemonge A C C, Carvalho J P G, Hallak P H and Vargas D E C 2021 Multi-objective truss structural optimization considering natural frequencies of vibration and global stability *Expert Syst. Appl.* **165** 113777
[7] Li L J, Huang Z B, Liu F and Wu Q H 2007 A heuristic particle swarm optimizer for optimization of pin connected structures *Comput. Struct.* **85** 340–9
[8] Degertekin S O, Lamberti L and Ugur I B 2018 Sizing, layout and topology design optimization of truss structures using the Jaya algorithm *Appl. Soft Comput.* **70** 903–28
[9] FOX R L and SCHMIT L A 1966 Advances in the integrated approach to structural synthesis. *J Spacecr. Rockets* **3** 858–66

IOP Conf. Series: Materials Science and Engineering 1109 (2021) 012003 doi:10.1088/1757-899X/1109/1/012003

Assessment of site-specific features to the unit cost of micro hydro power plant

I A V Marfori III , A B Culaba and A Ubando

Mechanical Engineering Department, De La Salle University, Philippines

isidro.marfori@dlsu.edu.ph

Abstract. Micro hydro power is a renewable source of energy and is commonly utilized for remote communities for electricity generation. The sustainability of a micro hydro system can be evaluated based on initial capital cost and the overall power output. The ratio of the capital cost and the overall power output is the unit-cost. In this paper, site specific characteristics specifically canal length and penstock slope are analysed to determine its effect in the unit-cost for micro hydro power plants. Calculations were conducted to quantify the total head loss that results from varying canal length and penstock slope. Capital cost is then evaluated using a disaggregated method based on literature. The results show that the canal length and penstock slope will give a significant effect on the micro hydro unit cost. Aside from the common micro hydro parameters of head and flowrate, the site-specific characteristic can also influence the sustainability of the system. It is therefore necessary to include the site-specific characteristics in the assessment of micro hydro systems.

1. Introduction

Micro hydro power or MHP are hydroelectric power plants with capacity between 1KW to 100KW and have been utilized for electricity generation in remote areas in many countries as practical solutions for both off-grid and grid connected systems [1]. Aggregated approach assumes that the total cost of the MHP is a function of its power and gross head while disaggregated approach estimates the total cost using the sum of individual component costs. [2]. Zema et al have developed methodology to evaluate the feasibility of low head MHP plants installed in irrigation systems [3]. Thorough investigation for aggregated cost estimates is presented by Aggidis et al [4]. Similarly, an aggregated approach to evaluate financial sustainability was used in the study of Marfori et al. [5]. In contrast to the mentioned study, research from Singal et al shows the disaggregated cost for small hydro power plants with capacity greater than 10,000 KW [6]. In the study of Singal et al, it was shown that the unit cost based on disaggregated approach considering, channel length, penstock dimensions, power house capacity and others was in the range of about Php10,000 per kilowatt to Php65,000 per kilowatt [7]. The unit cost presented by Singal et al can be considered as low compared to unit cost for hydro power plant [8]. Malhan and Mittal study shows unit cost for micro hydro power in the range of P80,000 per kilowatt to P150,000 per kilowatt [8]. Unit cost has wide range over the same capacity as shown in figure 1. The variation in unit cost can be attributed to site-specific characteristics. This paper aims to quantify the effect of site-specific characteristics of a micro hydro power plant and analyse its effect on the unit-cost.

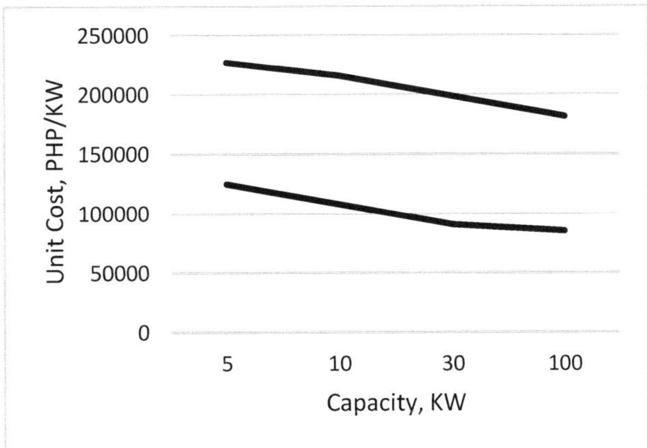

Figure 1. MHP Unit Cost adjusted with inflation of 1USD to 2.27USD for 1985 to 2017. [9]

2. Methodology

Figure 2 shows the basic components of a MHP system. Each component of the MHP has site-specific characteristics that affects the design, cost, and function. For this paper only the canal length and penstock slope are being evaluated as the two provide the biggest cost contribution and have significant effect on the power output.

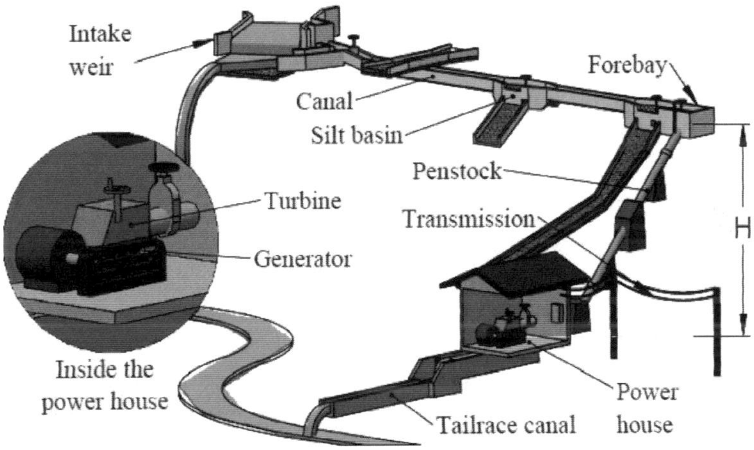

Figure 2. Components of a MHP system (Culaba & Marfori, 2020)

2.1 Power Output

The power generated by an MHP system is calculated based on the available head and flow rate given in equation 1.

$$P = \delta Q H \eta \quad (1)$$

δ is the specific weight of water at 9.8 kN/m^3, Q is net the flow rate of water, H is the net elevation and η is the overall efficiency of the MHP. The net head is the gross head H_g less the total head losses brought about the canal length and penstock slope. H_g is the vertical height from the intake and powerhouse. The elevation of the intake and powerhouse are often set to produce the highest H_g while minimizing flooding hazard to both components. For a particular MHP system, H_g may be the same but the canal length and penstock slope may vary. For this study, the gross head H_g will be held constant at 60m.

The flow rate throughout the year on an MHP site varies due to seasonal changes, gross flow rate Q_g is selected as a design input for the MHP components. The lowest flow rate is often selected to assure continued operation throughout the year. For this study, the gross flowrate is held constant at 0.2m^3/s. The overall efficiency is based on energy conversion from the turbine, transmission, and generator that are not affected by canal length nor penstock length. Therefore η is held constant at 60%. The canal length affects the head loss h_{LC} from the intake towards the forebay. The canal head loss h_{LC} is a function of the canal length L_C and canal slope S_C in equation 2. The canal slope S_C is calculated based on the canal width w_c and canal velocity v_c with a constant friction coefficient n in equation 3. For a smooth concrete canal, n = 0.015 and for a canal velocity to avoid siltation v_c should be greater than 2.0m/s [9]. The canal width w_c shown in figure 3 is calculated as a function of the flow rate Q, intake velocity v_i and intake height d in equation 4. Referring to figure 3, the intake velocity v_i should be greater than the canal velocity v_c and less than 4m/s to avoid siltation and erosion. For this study, intake velocity v_i is set at 2.5m/s. Intake height d is a design parameter and should be set less than the canal water height h_c. For this study, d is set as $d = 0.9h_c$.

$$h_{LC} = S_C L_C \quad (2)$$

$$S_C = \left(\frac{n v_C}{\left(\frac{w_C h_C}{w_C + 2h_C} \right)^{0.667}} \right)^2 \quad (3)$$

$$w_C = \frac{Q_g}{d\, v_i} \quad (4)$$

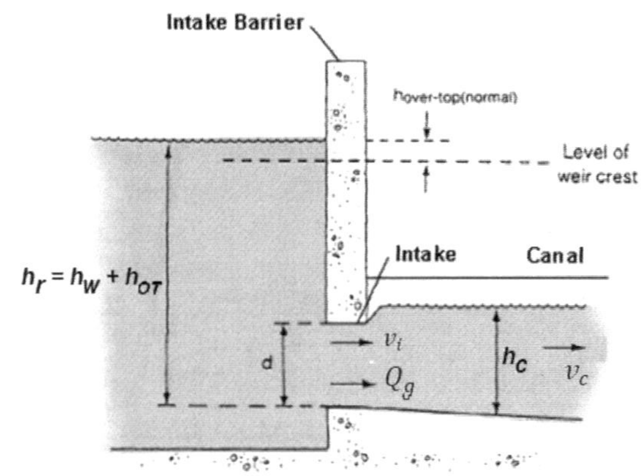

Figure 3. Intake and canal transition details

The penstock slope S_P dictates the length of the penstock in which affects the penstock head loss h_{LP}. The assumption is that the friction inside the penstock is assumed constant given a fixed penstock diameter. Although penstock diameter is selected to yield an acceptable penstock head loss, it is held constant for simplicity in calculation. The penstock head loss h_{LP} is calculated using equation 5 while the penstock length L_P is calculated using Pythagorean principle of the penstock vertical height and horizontal length in equation 6.

$$h_{LP} = \frac{f L_P v_p^2}{D 2g} \quad (5)$$

$$L_P = \sqrt{\left(H_g + h_{LC}\right)^2 + \left(S_P\left(H_g + h_{LC}\right)\right)^2} \tag{6}$$

$$D = \sqrt{\frac{4Q}{\pi v_P}} \tag{7}$$

Friction factor f is determined from Moody chart and can be held constant at 0.019 for steel since water velocity inside the pipe is always set at $v_p = v_C$. The penstock diameter D is calculated as a function of net flow rate Q and v_p in equation 7. While site-specific factors L_C affects the net head H, it also has effect on the net flow rate Q due to seepage loss happening in the entire length of the canal. The seepage loss Q_{LS} is a function of L_C, wetted perimeter P and seepage factor K shown in equation 8. Wetted perimeter is calculated based on the canal width and w_C and canal height h_C in equation 9. Canal height h_C is then calculated using equation 10. The height of the weir h_W in equation 10 is a design parameter and is set as 1.2m for this study while the intake coefficient of discharge C_d is held constant at 0.6 for a sharp rough finish intake. The seepage factor K is based on seepage rate equivalent to 1.4 x 10^{-5} m^3/s-m^2 [10]. The net flowrate is then $Q = Q_g - Q_{LS}$.

$$Q_{LS} = L_C P K \tag{8}$$

$$P = w_C + 2h_C \tag{9}$$

$$h_C = h_W - \frac{\left(\frac{v_i}{C_d}\right)^2}{2g} \tag{10}$$

2.2. Cost of MHP System

The cost of the MHP system is based on the summation of the cost of civil works, electro-mechanical cost, and engineering cost [11]. The total capital cost can be calculated based on the power output P and net head H. The electro-mechanical equipment composed of the turbine, generator and controller which are all designed based on the net flow rate Q and net head H. Given that Q and H is calculated as a function of the site-specific characteristic it can be concluded that the cost of electro-mechanical equipment incorporates the site-specific characteristics. The site-specific characteristic canal length and penstock slope was being considered in the power output and therefore its effects in the cost must also be included. This is accomplished by removing the cost of the canal and penstock from the capital cost. The partial capital cost in which excludes canal and penstock is shown in equation 11[6].

$$CP_P = 1100000\left(\frac{P}{H^{0.35}}\right)^{0.65} - 128075P^{-0.6189}H^{0.0307} - 7875P^{0.6194}H^{.3804} \tag{11}$$

The cost for the canal as well as the penstock is based on the disaggregated cost taken from the study of Singal et al. Both cost estimates are calculated as a function of canal length and penstock length shown in equation 12 and 13, respectively. Finally, the total capital cost CP is the summation of CP_P, C_c, and C_P. The unit cost is then calculated using equation 14.

$$C_c = 5118Q^{0.664}L_C \tag{12}$$

$$C_P = 6.017(1000D)^{1.239}L_P \tag{13}$$

$$UC = \frac{CP}{P} \tag{14}$$

3. Results and Discussion

The unit cost analysis is evaluated with varying canal length from 100m to 1000m with 100m increments. For the canal length variation, two sets of data were generated for slope of 1 and 6. A slope of 1 indicates that the penstock is at an angle of 45 degree yielding a short penstock length while a slope of 6 gives an angle approximately equal to 9.5 degree from a horizontal plane yielding a long penstock. Alternatively, the unit cost is also analysed for penstock slope from 1 to 6 with 0.5 increments. Two sets of data are also generated for the smallest canal length of 100m and longest canal length of 1000m.

3.1. Variation in Canal Length

In figure 4, the effect of the canal length to the capital cost is minimal as described by the curve for the short penstock resembling an almost horizontal line. It is not the same for longer penstock where the capital cost decreases with increasing canal length. This contradicts the fact that a longer canal should result in a higher capital cost due to the added material and labor. This trend can be attributed to two reasons: (1) the cost contribution for penstock is significantly larger as compared to the cost contribution of the canal and (2) the penstock length tends to become shorter as the canal becomes longer due to the canal head loss. Therefore, a longer canal will result in a shorter penstock. And shorter penstock will result in a lower capital cost. The power output with increasing canal length is shown in figure 5 and in contrast to capital cost, the canal length has greater impact on the power output. Further observation also reveals that the power output difference for short and long penstock is minimal. The length of the penstock is affected by its slope given the same head from the forebay to the powerhouse. This means that the penstock friction head loss is the only factor that affects the power output for penstocks with varying lengths. The design parameter for the penstock is the penstock diameter D which is set to have low penstock water velocity making the head loss less 5% of the net head. This can be seen in the small gap between the curves in figure 5. The unit cost in PHP/KW for varying canal length is shown in figure 6. The unit cost has an increasing trend with increasing canal length. Although the capital cost has minimal change with varying canal length, the unit cost is mostly affected by the power output.

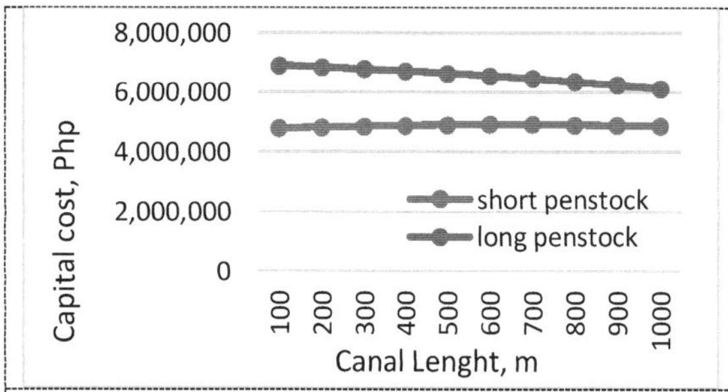

Figure 4. Capital cost with varying canal length

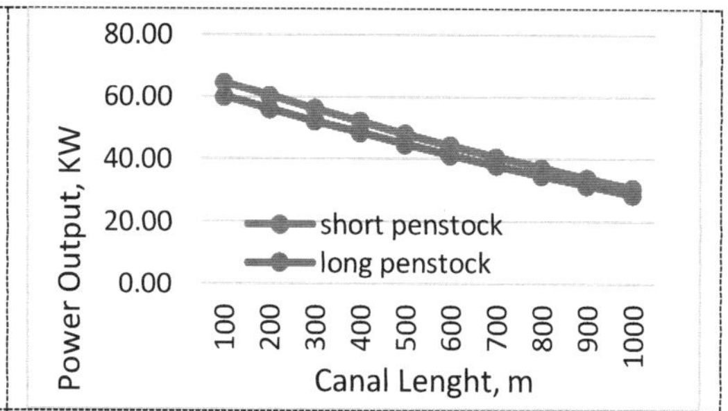

Figure 5. Power output at varying canal length

3.2. Variation in Penstock Slope

Figure 7 and 8 shows the power output and capital cost at varying penstock slope. In contrast to canal length, the penstock slope has more impact on the cost. As the penstock design parameter are set to have minimal penstock head loss, its power output has little difference at varying penstock slope. However, with a longer canal the power output has significant difference due to the increased canal head loss. The capital cost for increasing penstock slope or length, gives an increasing trend which coincides with typical MHP installations since longer penstock will be more expensive. Similarly, the unit cost at varying penstock slope also shows an increasing trend shown in figure 9.

IOP Conf. Series: Materials Science and Engineering 1109 (2021) 012003 doi:10.1088/1757-899X/1109/1/012003

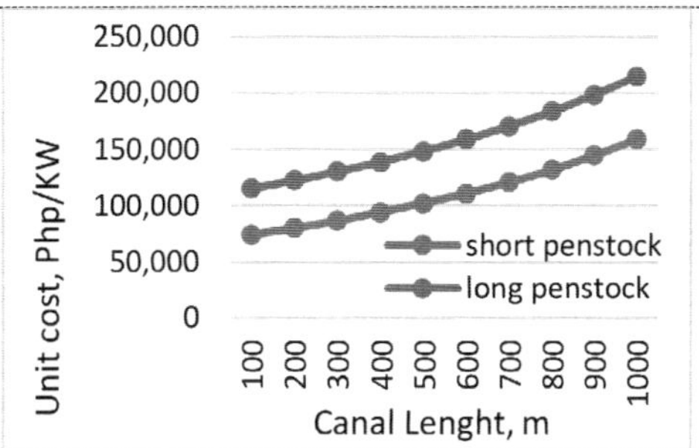

Figure 6. Unit cost at varying canal length

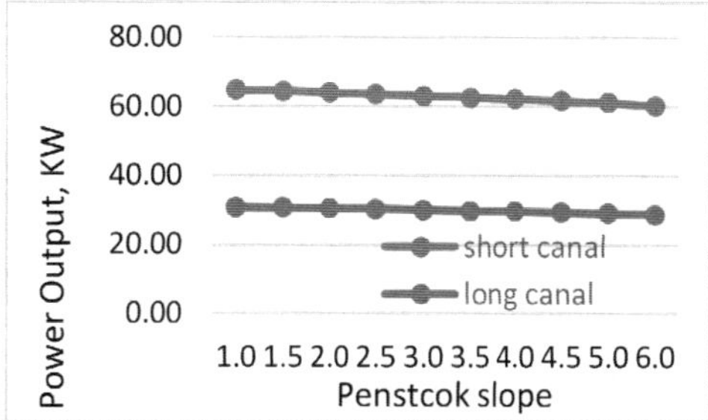

Figure 7. Power output at varying penstock slope

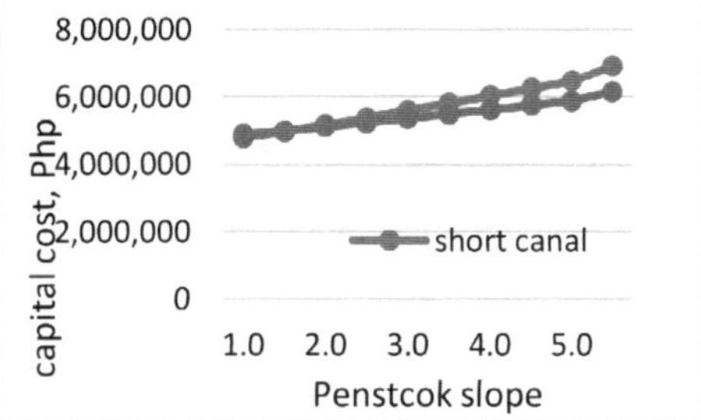

Figure 8. Capital cost with varying penstock slope

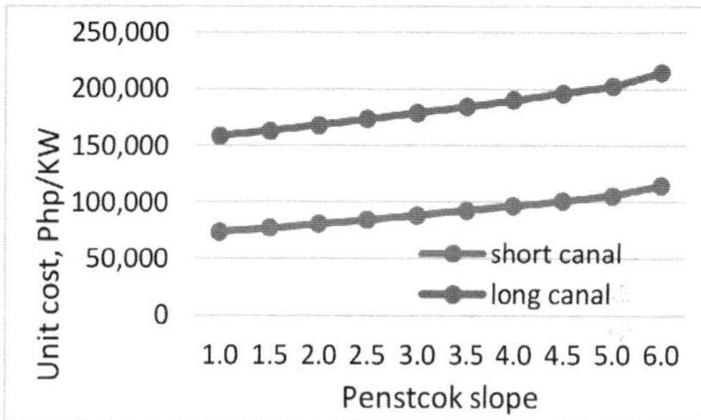

Figure 9. Unit cost with varying penstock slope

Comparing the unit cost of varying penstock to the unit cost with varying canal length, it can be seen that varying canal length has more impact to the unit cost. Figure 10 shows a graph of the unit cost at varying canal length and penstock slope. Penstock is often seen as one of the largest contributors to the cost of MHP because of the cost related to its materials and the difficulty in its construction. However, the data shows that canal length also gives significant effect on the unit cost as longer canal tends to shorten the net head resulting in lower power output.

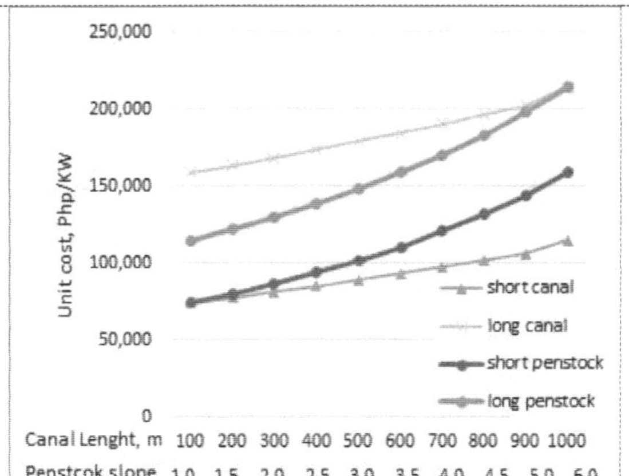

Figure 10. Unit cost at varying canal length and penstock slope

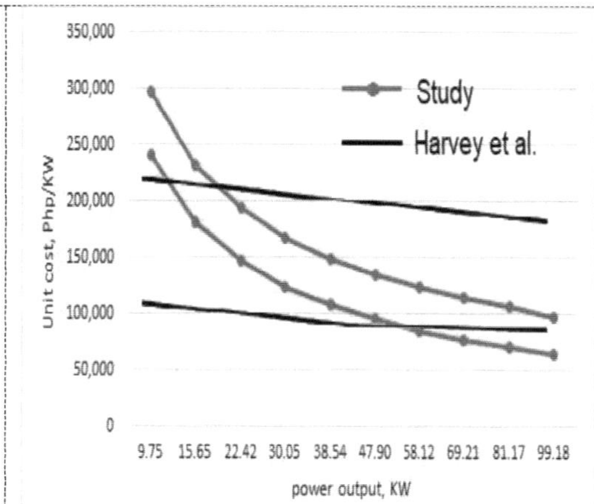

Figure 11. Unit cost comparison of study and Harvey et al.

4. Conclusions

The evaluation of the unit cost for MHP in the context of varying site-specific characteristics specifically canal length and penstock slope has been shown to be significant. Depending on site-specific characteristics, the unit cost for an MHP can vary from 70,000 Php/KW to as high as 225,000 Php/KW. By incorporating certain site-specific characteristics to the power and cost estimate, a more accurate result can be obtained. Comparing the unit-cost model in this study to that of Harvey et al. in figure 11, it can be seen that there is some variation. The lower limit difference at the 10KW power output is about 120,000 Php/KW while the upper limit has a difference of 70,000 Php/KW. This difference may be attributed to the cost estimate used in this study in which were derived from a small hydro system with power more than 1000KW while the study is intended for MHP with power less than 100KW. [6]. It can also be observed that the gap between the upper limit and lower limit in the unit cost shown in figure 10 has a lower gap compared to Harvey et al. This can be attributed to the other site-specific characteristics that are not included in this study such as weir length, earth retaining walls, and power house location. It is therefore recommended that more disaggregated approach is conducted for the cost estimate of the entire MHP specifically (1) assessment of other site-specific characteristic to the power output, (2) cost estimate based on composition of all materials of the MHP components, and (3) use of local material cost and labor rates. Although the capital cost calculation is derived from literature, the power output is based on a model developed solely for MHP applications. The model considers design approaches that are more favourable for MHP intended for community use and not necessarily for commercial use. Contrast to commercial hydro power, community MHP lean towards low-maintenance and low-cost components and not necessarily towards high efficiency components. This is evident in the selection of gross flow rate as an example where constant continuous flow is selected rather than seasonal maximum flow rate. The study is able to present an explanation in the high variation in unit-cost that is affected by canal length and penstock slope. This is important in the assessment of future MHP installations as the said parameters will have significant impact upon the power output and capital cost.

References

[1] Bracken L J, Bulkeley H A and Maynard C M 2014 Micro-hydro power in the UK: The role of communities in an emerging energy resource **68**, pp 92–101

[2] Lucio G, Filho T, Felipe I and Barros R M 2017 Cost estimate of small hydroelectric power plants based on the aspect factor **77** (April) pp 229–238

[3] Zema D A, Nicotra A, Tamburino V and Zimbone S M 2016 A simple method to evaluate the technical and economic feasibility of micro hydro power plants in existing irrigation systems *Renew. Energy* **85** pp 498–506

[4] Aggidis G A, Luchinskaya E, Rothschild R and Howard D C 2010 The costs of small-scale hydro power production: Impact on the development of existing potential *Renew. Energy* **35** (12) pp 2632–2638

[5] Marfori I A V, Culaba A B, Ubando A T, Anthony R and Chen W 2019 Determining the sustainability of a community micro hydro power system using real options analysis pp. 0–6

[6] Singal S K, Saini R P and Raghuvanshi C S 2010 Energy for sustainable development analysis for cost estimation of low head run-of-river small hydropower schemes *ESD* **14** (2) pp 117–126

[7] Kumar R and Singal S K 2015 Penstock material selection in small hydropower plants using MADM methods *Renew. Sustain. Energy Rev.* **52** pp 240–255, 2015.

[8] Malhan P and Mittal M 2021 Evaluation of different statistical techniques for developing cost correlations of micro hydro power plants *Sustain. Energy Technol. Assessments* **43** (May 2020) p 100904

[9] Harvey A, Brown A and Hettiarachi P 1993 *Micro Hydro Design Manual: a guide to small-scale water power scheme* (Southampton Row, London: Intermediate Technology Publications)
[10] Mohammadi A, Parvaresh A and Abbasi N2019 Field measurement and analysis of water losses at the main and tertiary levels of irrigation canals: Varamin irrigation *Glob. Ecol. Conserv.* **18** p e00646
[11] Santos L, Soares I, Mendes C and Ferreira P 2014 Real options versus traditional methods to assess renewable energy projects *Renew. Energy* **68** pp 588–594

Contact stress analysis and optimization of spur gears

N H Loc[1,2] **and L T Anh**[1,2]

[1] Faculty of Mechanical Engineering, Ho Chi Minh city University of Technology (HCMUT),
[2] Vietnam National University of Ho Chi Minh City (VNUHCM), Ho Chi Minh City, Vietnam.

nhloc@hcmut.edu.vn

Abstract. Gear drive is one of the most important components in mechanical power transmission systems. Gears in gearbox generally fail when the working contact stress exceeds the contact fatigue limit. In the scope of this paper, the researcher studies gear design methods based on ISO 6336 and draws a comparison with AGMA 9001. In addition, this paper also presents the optimal gear design results with 3D gear models designed in CAD software, then imported into CAE software for topology and shape optimization. More specifically, the amount of material that can be removed from the gear body without affecting the properties of the gear mesh is considered and calculated. Consequently, the aim of this research is to reduce the weight of the gears and choose the optimal parameters of gear pair by contact stress.

Keywords: *CAD/CAE Software, Contact stress, Shape Optimization, Spur Gears, Topology*

1. Introduction

The design and analysis of optimal structures play an increasingly important role in the mechanical field with the goals of saving costs, time, and resources while still ensuring the quality and costs. The development of CAD / CAE software is a great and urgent support tool to facilitate the optimal design calculation to be further completed in theory and applications. The optimization problems of gear drive include [3-8]:

- Optimization of geometry parameters for teeth according to contact stress.
- Topology, shape, and size optimization of gears.
- Optimization of gearboxes and reducer with the goals of achieving minimum total shaft distance, minimum gearbox mass, equality strength of contact and bending stress, maximum efficiency, and minimum angle error.
- Optimization of the gearbox and reducer housing.

This paper concentrates on topology, shape and size optimization of spur gears, and performs an analysis of those factors affecting the calculation of teeth design according to contact stress.

1.1. Topology Optimization Using CAD / CAE Software

The design process has been continuously improved over time to create the products that satisfy the increasing demands of people. Previously, through the utilization process, defects will appear on the products, so the manufacturer has to correct them to meet the needs of users. These improvements gradually create the optimal product design, shape, quality as well as price. Therefore, it takes a very long time, maybe a few years or even decades, to get the best results for the products. The overall design development time above can be significantly reduced when CAD and CAE tools are employed in the design process.

Currently, the product design process is a cycle utilizing CAD and CAE tools (Figure 1): from modeling on CAD software, to topology, shape and size optimizing designs on CAE. The design input involves designing task, loads, boundary condition, manufacturing method, cost criteria, design space, design style, designs with the same function, and much more. At the beginning, a new idea is given based on the designer's knowledge and experience. Next comes the calculation of optimal design (topology, shape and size) of details based on criteria or by analysis. The design is then re-modeled using CAD software and re-calculated by CAE software.

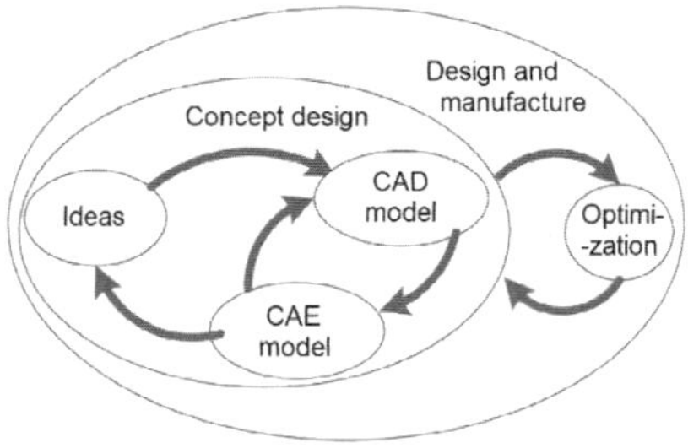

Figure 1. The design sequence of topology optimization using CAD/CAE software.

CAD systems build 3D parametric models, and then export them through CAE software which will mesh, set boundary conditions, loads, material properties and related information.

The method of structural design optimization plays an increasingly important role in product design and development. The basis of the design topology optimization process is finding the distribution of the material in a given area called the design space, so that the structure can respond to load conditions and vibration frequency, and at the same time satisfy the constraints on weight, volume, fabrication ability. Many topology optimization methods have been developed, some of which are homogenous method, density method, and evolution method.

In general, design optimization includes optimization of continuous structure design and optimal discrete structure design. Optimizing continuous structure design refers to the optimization of cubic structures made up of consecutive elements, which can be either solid or hollow. Optimizing discrete structure designs involves solving optimal problems for the structures formed by discrete elements such as frames or trusses.

The design topology optimization process follows the six main steps as shown in Figure 2:

1. Create initial shape space for the part and set boundary condition for load parameters.
2. Define the objective function, the constraint function, and control parameters.
3. Conduct calculation based on an algorithm to gradually remove unnecessary materials (according to the algorithm's shell type criteria).
4. Determine whether the structure satisfies the binding conditions and specifications.
5. If satisfactory, go to step 6; otherwise, go back to step 3.
6. View the outcomes.

Currently, commercial CAE software such as MSC Nastran, Ansys, Altair, and Hyberworks [9] are calculated based on density method, using isotropic material model and only one design variable which is material density. The structure is separated into elements by the finite element method. The optimized density method is based on the standard of structural hardness, whereby the found structural hardness will reach the maximum value corresponding to the mass of material used. 3D-printed objects are created from a digital file of designed products using 3D printing technology.

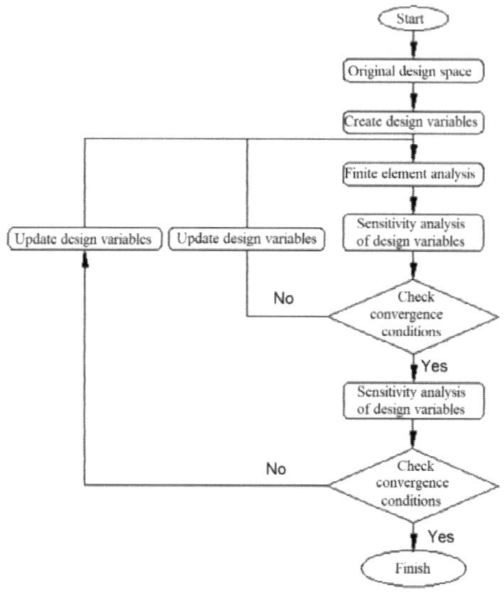

Figure 2. Topology optimization calculation process.

2. Gear design calculation based on ISO 6336, AGMA 9001

Currently in the world, there are many standards for calculating and designing gears, such as ISO 6336 (1996, 2006, 2019), AGMA 9001, Merrit, Legacy ANSI, Bach ... Depending on the purpose of utilization, the standard and suitable ones for design will be particularly chosen. In the scope of this paper, the researcher will only study gear design methods based on ISO 6336 [1, 9, 10] and comparison will be made with AGMA 9001 [2] for the specific case.

According to ISO 6336 the contact stresses are calculated from the following formula [1]:

$$\sigma_H = Z_B \sigma_{H0} \sqrt{K_A K_V K_{H\beta} K_{H\alpha}} \leq \sigma_{HP} \tag{1}$$

Where : $\sigma_{H0} = Z_H Z_E Z_\varepsilon Z_\beta \sqrt{\dfrac{F_t (u+1)}{d_1 b u}}$

Calculated contact stress σ_{HP} is determined as follows:

$$\sigma_{HP} = \frac{\sigma_{HG}}{S_{H\lim}} = \frac{\sigma_{H\lim} Z_{NT}}{S_{H\lim}} Z_L Z_V Z_R Z_W Z_X \tag{2}$$

The bending stress is calculated from the formula:

$$\sigma_F = \sigma_{F0} K_A K_V K_{F\beta} K_{F\alpha} \leq \sigma_{FP} \tag{3}$$

Where: $\sigma_{F0-C} = \dfrac{F_t}{b.m_n} Y_{Fa} Y_{Sa} Y_\varepsilon Y_\beta = \dfrac{F_t}{b.m_n} Y_{FS} Y_\varepsilon Y_\beta$

Calculated bending stress σ_{FP} is determined as follows:

$$\sigma_{FP} = \frac{\sigma_{F\lim} Y_{ST} Y_{NT}}{S_{F\min}} Y_{\delta relT} Y_{RrelT} Y_X = \frac{\sigma_{FG}}{S_{F\min}} \tag{4}$$

2.1. Analysis and design of spur gear in CAD software

To draw a comparison, the researcher conducts calculations for specific values, calculating with input parameters for compressor equipment with the following specific data: Power P (5.66kW), number of revolutions n (298 rpm), precision level 7, and working 8000 hours. Calculation is made with the standard transmission ratio u = 2.0; 2.5; 3.15; 4.0; 5.0; 6.3, using materials with medium and high contact fatigue strength: normalized 36Mn5 steel (with σ_{Hlim} = 520 MPa, σ_{Flim} = 372 MPa) and surface hardened 36Mn5 steel (with σ_{Hlim}= 1140 MPa, σ_{Flim} = 352 MPa).

With heat treatment the contact fatigue limit increases significantly, but the bending fatigue limit decreases. The remaining calculated parameters are selected according to the standard series. The results calculated according to ISO and AGMA standards are shown in the following tables and figures.

Table 1. Comparison of Calculation Results of Normalized 36Mn5 Steel Spur Gear.

Gear Radio	ISO						AGMA					
	a	Z_1	S_H	S_F	d_1	V	a	Z_1	S_H	S_F	d_1	V
2	250	37	1.237	8.74	166.5	1742	200	30	1.273	7.872	135	1145
2.5	280	35	1.228	8.205	157.5	1559	224	28	1.253	7.461	126	998
3.15	315	34	1.207	7.893	153	1471	250	27	1.213	7.066	121.5	928
4	400	35	1.283	8.12	157.5	1559	315	28	1.288	7.512	126	998
5	450	33	1.223	7.663	148.5	1386	355	26	1.229	6.953	117	860
6.3	560	34	1.268	7.766	153	1471	450	27	1.28	7.279	121.5	928

Table 2. Comparison of Calculation Results of Surface Hardened 36Mn5 Steel Spur Gear.

Gear Radio	ISO						AGMA					
	a	Z_1	S_H	S_F	d_1	V	a	Z_1	S_H	S_F	d_1	V
2	125	18	1.369	4.117	81	412	100	15	1.255	2.473	67.5	286
2.5	140	18	1.339	3.877	81	412	125	16	1.390	2.805	72	326
3.15	180	19	1.398	3.829	85.5	459	140	15	1.282	2.486	67.5	286
4	224	20	1.501	3.781	90	509	180	16	1.399	2.768	72	326
5	250	18	1.332	3.382	81	412	200	15	1.298	2.944	67.5	286
6.3	315	19	1.422	3.446	85.5	459	224	14	1.224	2.303	63	249

2.2. Compare Calculation Results

The charts below compare the results according to ISO 6336 and AGMA 2001, the chart on the left uses 36Mn5 normalized steel, and the right one uses 36Mn5 hardened steel (Figure 3-5).

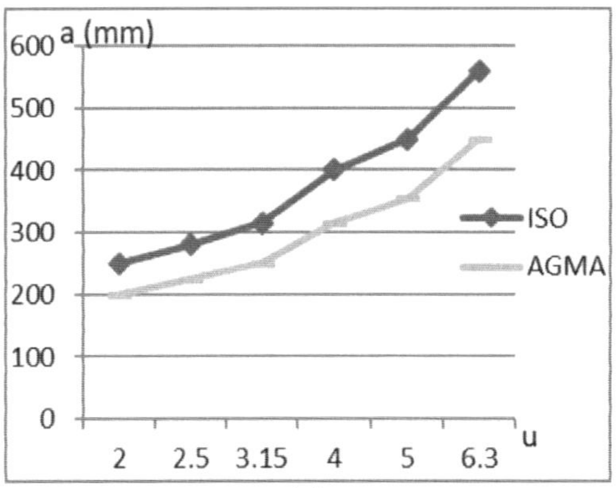

(a)

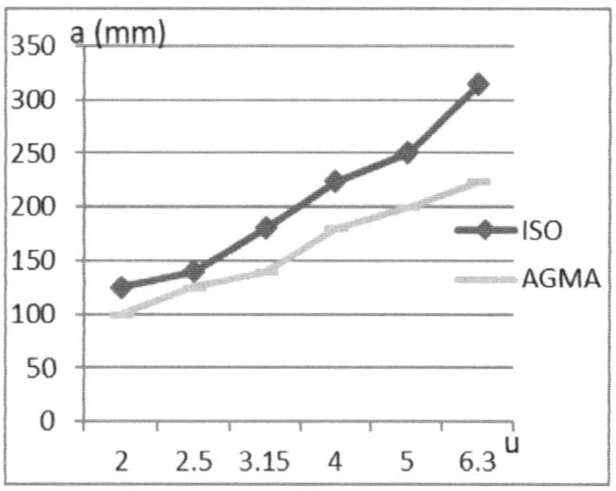

(b)

Figure 3. Comparison of shaft distance.

Dimensions calculated according to ISO 6336 (when using both types of steel) are higher than those calculated according to AGMA 2001 - D04: 2005. However, when the 36Mn5 surface hardened steel is used, the safety factor of contact strength when calculated according to ISO 6336 is about 10% higher and the safety factor for bending strength when calculated according to ISO 6336 is higher up to about 40%. Similarly, as researchers calculated according to AGMA 2001 - D04: 2005, the calculating contact stress will be higher than ISO 6336 with the corresponding rate. The safety factor for the bending strength is significantly reduced to nearly reach the minimum limit when surface hardened steel is used.

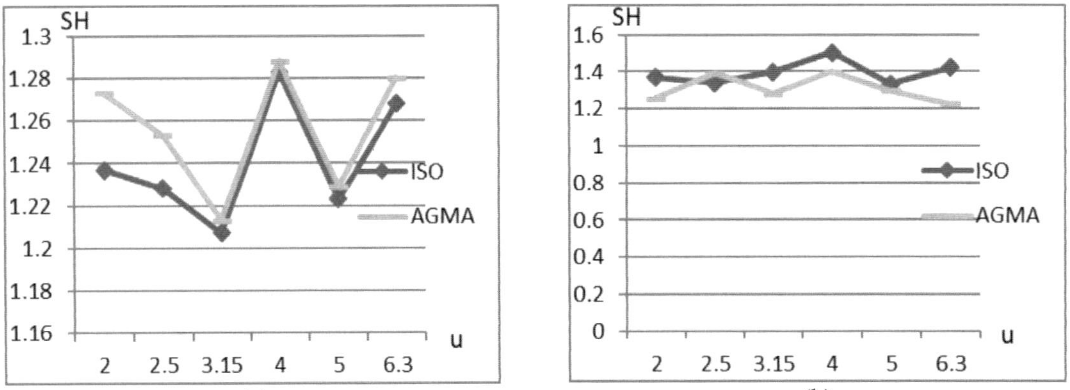

(a) (b)

Figure 4. Comparison of factor of safety S_H.

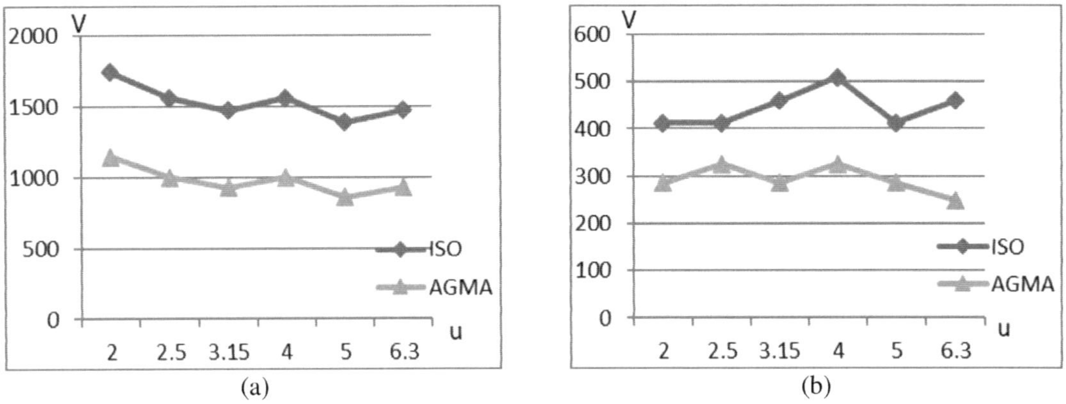

(a) (b)

Figure 5. Volume comparison V.

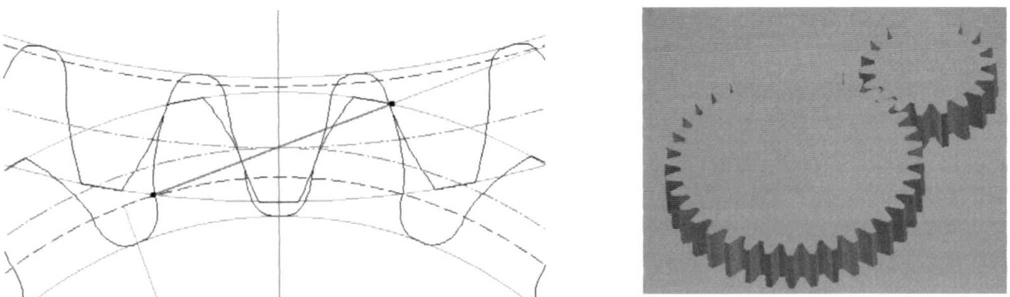

(a) (b)

Figure 6. Teeth mesh and the main generation of spur gears.

This assembly 3D model (Figure 6) can be transferred or edited back and forth in any CAD/CAE software such as: CATIA, Inventor, and Solidworks, with similar dimensions. Gear topology optimization is performed with the finite element method using ANSYS software.

3. Analysis and optimization of spur gear with Finite Element Method – ANSYS Workbench

3.1. Topology Optimization of Gears
After the driving gear has been designed in CAD software, the researcher will export to ANSYS software to analyse and optimize the design (Figure 7-10). We will proceed with keyhole cutting and disk creation in the Space Claim environment.

Figure 7. Meshed 2D Model of Spur gear.

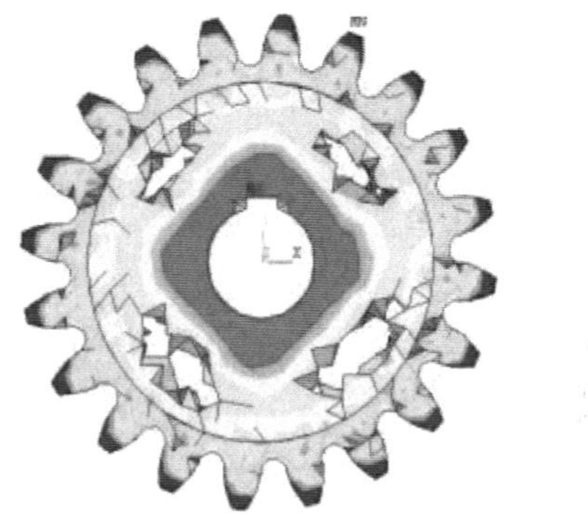

Figure 8. Model with 60% material removal.

A design calculation is performed with the model with 60% material removal (Figure 8). This is obtained after the material particles are removed and holes of the corresponding hole or groove shape are cut, as long as the fabrication conditions are fulfilled and aesthetic factors are guaranteed. It is then possible to replace the holes with corresponding large and small round holes, which is very convenient in machining, and optimal shape (Figure 9) is ensured.

Based on the results of the standard material density distribution in the above gear models, we will opt for the appropriate gear model option which is close to the optimal result (Figure 19). However, the gears of the given shape are intended to guarantee optimal geometry only, and may have non-technologic shapes despite satisfying the geometrical optimization. Hence, in the given shape, the form which can be fabricated will be chosen for utilization.

Contact stress test: The permissible contact stress is determined (2) with a value of 391.3 N/mm^2 (MPa). To ensure durable conditions, the force is put at in the most dangerous position on the teeth (right at the top of the teeth). When two gears in mesh, only 1 or 2 pair teeth are touching at a time. The value of the force acting on the tooth is equal to the value of the force during optimal design process.

3.2. Shape Optimization of Gears
In fact, for large gears (diameter > 400mm), 4-6 round holes are usually cut. Besides, in addition to the common round whole profile, other profiles are also cut, such as alternating grooves, large and small round holes (Figure 9). However, the number of holes must be an even number so as to avoid the eccentricity when the gears are operating.

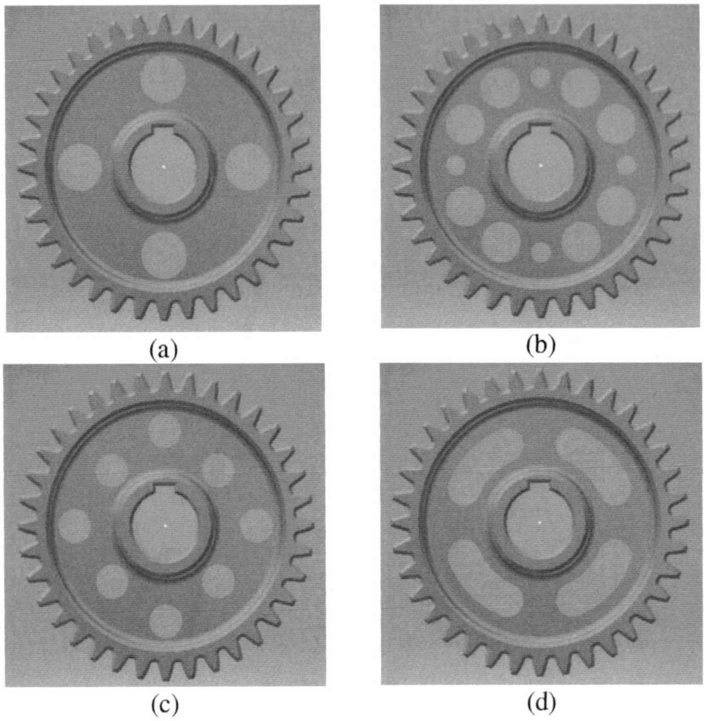

Figure 9. Common shape of hole in gear structure.

Based on the element removal results in Section 2.1 and practical experience, the researcher cuts the 3D gear model at the disk part with different shapes and rechecks the maximum stress (Figure 10).

Figure 10. Maximum stress results for different shapes.

To test the teeth strength, we use ISO 6336 [2]. Allowable contact stress is determined by formula (2) to obtain a value of 608.8 N/mm^2 (MPa). The volume percent reduction, compared to the original model according to the different shapes, is shown in Table 3.

Table 3. Shape of Holes and Material Reduction Percentage.

Shape	Holes	Material Reduction Percentage
1 (Fig. 14.a)	4 holes ∅70mm	$\%V_{decrease} = \frac{V_0 - V_1}{V_0} = \frac{\Delta V}{V_0} = \frac{4839427}{11097613} = 43.6\%$
2 (Fig. 14.b)	8 holes ∅60mm 4 holes ∅30mm	46,1%
3 (Fig. 14.c)	8 holes ∅50mm	43,7%
4 (Fig. 14.d)	4 grooves	43,8%

As mentioned above, during the operation of the gear, there is only one bearing tooth at a time when the gear is engaged. The force acting on the tooth is equal to the value of the force used in optimal process. Compared with the allowable stress value, the stress result after optimal calculation is satisfied.

The four types of holes and grooves are optimized and re-tested according to the contact strength. The volume of the gear is significantly reduced according to the shape. For those gears with extremely large or very small diameters (used in watches) a material reduction is essential. The whole profiles and grooves are remarkably diverse and widely used, depending on the increasingly modern machining methods which contribute to help the optimization process become more convenient and efficient.

From the aforementioned theory and applications, once again the correctness and the necessary role of optimization in engineering are confirmed. The optimal design not only finds the most technically optimal model, but also the most economically optimal. The application of the CAD / CAE systems in structural optimization brings high efficiency in the design processes. Considerably, the possibilities of structural optimization can go even further in the future as more and more people are engaged in the research and employment of design support software.

4. Study on the Effects of Factors on Contact Stress

4.1. Distribution of Maximum Stress at Point of Contact

The workability of gear teeth is determined according to the contact stress; therefore, the study into the parameters affecting the contact strength to choose the reasonable range of values is remarkably crucial. With gear transmission, there are many studies about the effects of geometric parameters on contact strength [4, 6, 10, and 11]. Currently, with such modern tools as CAD/CAE software with modules design of machine elements, we can quickly find out these effects of gear module m and pressure angle α on contact stress.

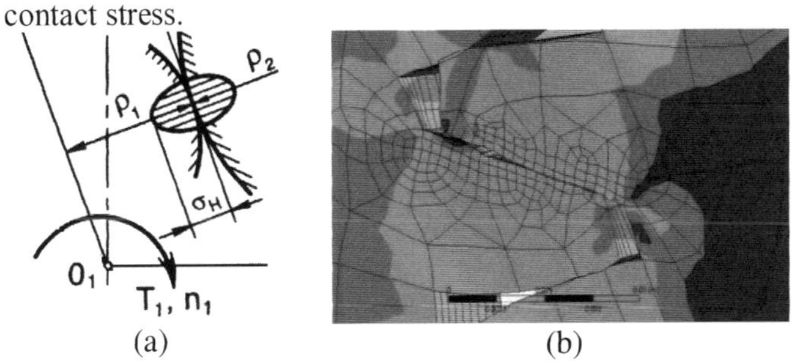

(a) (b)

Figure 11. Distribution of maximum stress at point of contact

First, we respectively compare the results calculated with the analytical formula according to ISO and with FEA (ANSYS software). When simulated on ANSYS, the maximum value of contact stress is equal to 371.84 N/m^2 (Figure 11). Then, another calculation is made with the analytical formula according to ISO 6336 using Equation (1) is 381,142 N/mm^2 (MPa). Compare the % difference between the two results:

$$\% \text{ERROR} = \frac{383{,}142 - 371{,}84}{383{,}142} \approx 2{,}9\% \quad (5)$$

There is no significant difference between the values obtained when calculations are done respectively by ISO 6336 standard and by FEA on ANSYS software. The stress determined when the researcher conducts calculation using ISO 6363: 1996 is slightly greater. However, according to [6], an error of less than 4% is acceptable; this shows that the simulation performed in ANSYS is compliant with ISO 6336.

4.2. Influence of Teeth Modulus and Angle of Pressure on Contact Strength

In this section, to study the influence of the gear module and the pressure angle on the contact stress of gear teeth, the gear transmission uses the parameters in gear transmission in item 2, with the pressure angle of 17.5°; 20°; 22.5° and the modules are respectively changed to 4, 6, 8 and 10. The results are illustrated in Figure 12.

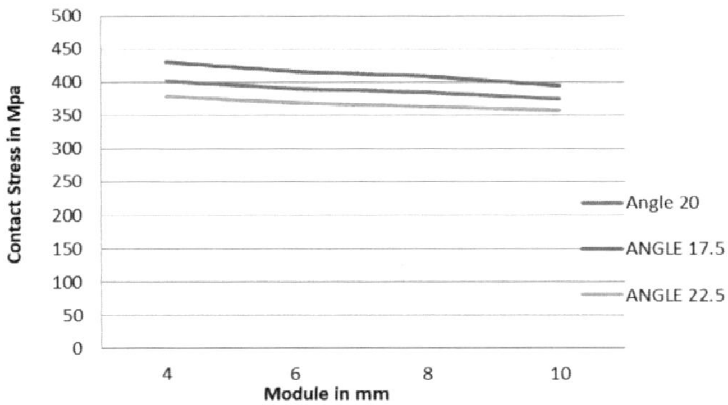

Figure 12. Diagram of contact stress with modules and pressure angle.

When the module increases, the number of teeth decreases correspondingly, the calculated contact stress also decreases. Then the amount of cutting volume of the teeth will rise; when the pressure angle increases, the contact stress also decreases. When the pressure angle is large, the undercut of gears when cutting a gear with a small number of teeth can be overcome, but it is easy to happen the tip relief of the gear tooth surface. When the pressure angle is small, the undercut of gears becomes apparent.

5. Conclusions

Thus, this article has studied the design calculation and the optimization of gears with some following conclusions and directions for development:

- Some studies have been conducted into the world standards for gear calculations, especially those calculations made according to the contact strength when Autodesk Inventor Professional software is utilized to calculate the data and compare calculation results. When calculated according to ISO, the size is about 20 to 30% larger than when AGMA is used; this difference is due to the different calculation coefficients between the two standards. In addition, other sets of standards can be studied and in fact all of these standards design gears based on bending and contact strength.
- ANSYS application is utilized for topology optimal design. With its optimized shape, the gear saves about 43.6 to 46.1% of the material.

- The gears are optimized for shape by cutting holes on large diameter conventional discs. When the diameter is large (usually more than 500mm), it is possible to create grooves or holes by stamping, then, the work surfaces can be additionally improved while the teeth are machined by envelope or copying methods. In some cases, it is possible to use castings.
- Design software is used to create and test a design 3D model after the optimization of a shape. In addition, it is necessary to conduct experimental research to verify the theoretical and simulation results.
- The researcher studies on calculating and simulating the effects of geometric parameters on gear contact strength. In fact, when the coupling angle increases, the contact stress decreases. When the researcher calculates according to ISO standards and compares the results with mathematical simulation software, the difference is about 2.9%.
- If more time is allowed, the researcher can learn more about the factors that influence the contact strength and bending strength, such as helical angle, number of teeth, rim width, pitch diameter, etc. to form a basis for gear design research.
- More different shapes of the disc profile can be expanded. Besides, it is possible to optimize the coronary region. In addition to the trunk profile, some more other profiles can be studied.

Acknowledgement

We acknowledge the support of time and facilities from Ho Chi Minh City University of Technology (HCMUT), VNU-HCM for this study.

References

[1] International Organization for Standardization 2016 *Calculation of load capacity of spur and helical gears – Part 2: Calculation of surface durability (pitting)* ISO 6336-2.
[2] American National Standards 2005 *Fundamental Rating Factors and Calculation Methods for Involute Spur and Helical Gear Teeth* AGMA 2001-D04.
[3] Panda S, Biswal B B, Jena S D and Mishra D 2017 An approach to weight optimization of a spur gear *Proceedings of the Institution of Mechanical Engineers Part J: Journal of Engineering Tribology*.
[4] Balaji D S, Prabhakaran S and Kumar J H 2017 Analysis of surface contact stress for a spur gear of material steel 15NI2CR1MO28 *ARPN Journal of Engineering and Applied Sciences* **12**.
[5] Zhen Q, Yu T W and Sung K L 2018 A Review of Recent Advances in Design Optimization of Gearbox *International Journal of Precision Engineering and manufacturing* **19** p 1753-62.
[6] M V Murali and S L Ajit Prasad 2016 Influence of Module and Pressure Angle on Contact Stresses in Spur Gears *International Journal of Mechanical Engineering and Robotics Research* **5**
[7] Sarfraz A N Q and Dhanajay R D 2015 Mass Reduction of Involute Spur Gear under Static Loading *American Journal of Mechanical Engineering and Automation* **2** pp 40-43.
[8] Paridhi R and Aman A M 2018 Volume optimization of helical gear with profile shift using real coded genetic algorithm *Procedia Computer Science* **133** p 718-724.
[9] Marco A M, Fábio K, Urbano R J and Carlos S 2012 The Influence of Contact Stress Distribution and Specific Film Thickness on the Wear of Spur Gears During Pitting Tests *Journal of the Brazilian Society of Mechanical Sciences and Engineering* p 135-144.
[10] Putti S R, Nadipalli S and Mohammad F 2015 Contact Stress Analysis of Spur Gear for Different Materials using ANSYS and Hertz Equation *International Journal of Modern Studies in Mechanical Engineering* (IJMSME) **1** p 45-52.
[11] Adis J M, Adil M, Elmedin M, Isad S and Nedim P 2019 Spur Gear Tooth Topology Optimization: Finding Optimal Shell Thickness for Spur Gear Tooth produced using Additive Manufacturing *TEM Journal* **8** p 788-794.

Design and fabrication of a transmission mechanism arm for rotational molding

H N Thi [1*], N V Dang [2] and Ha Thai Thi Thu[3]

[1,2] Faculty of Engineering, Van Lang University, Ho Chi Minh City, Vietnam

[3] Faculty of Mechanical Engineering, Ho Chi Minh City University of Technology (HCMUT)Ho Chi Minh City, Vietnam

*hoa.ngo@vlu.edu.vn

Abstract: Compared with blow molding and thermoforming technology, rotary die technology has many advantages for manufacturing hollow plastic products. The purpose of this paper is to compute and design the transmission crank mechanism of the rotary die machine. The authors chose a reasonable solution to design the experimental model to produce flower pots with dimensions of 1000 x 400 x 400 mm. The principle diagram and the dynamic diagram of the machine are presented. Computation and design of main clusters and hand fabrication of the rotary die machine are investigated. The model of the transmission crank mechanism is fabricated to verify the validity of the proposed model.

Key words: *Transmission crank, thermal forming, Additive interactions; Antioxidants; Rotational Molding.*

1. Introduction

Rotational molding, also known as rotocasting or rotomolding, is a method of fabricating hollow plastic products with low pressure, high temperature. Many centuries ago, it was known the method of coating a layer of material inside a rotating mold to create products, but it was not until the 1940s of the twentieth century that it was recognized as french rotary molds for plastic products. Since then, scientists have embarked on the research and design process of the first rotary die machine models. In particular, some countries have developed quite strongly in this field such as: Italy (Rotomachinery Group), Brazil (Rotoline company), India (Naroto firm), etc.

Rotational molding is a highly competitive method compared to methods such as blow molding and injection molding in the production of hollow plastic products. Rotational molding network offers more economic value when plastic products require no residual stress, wide size range, uniform thickness, complex shape and high detail.

Currently in Vietnam, there are not many studies on the topic "Research and design of rotary molding machine". This is a relatively new research direction in Vietnam, opening up an overview of Rotational molding. Since then, proposing the manufacturing process to meet the needs of the market, bringing new development directions for businesses, replacing imported products, this is also the urgency of the topic.

IOP Conf. Series: Materials Science and Engineering 1109 (2021) 012005 doi:10.1088/1757-899X/1109/1/012005

2. Analysis and selection of plan

2.1 The Process

During the molding process, the plastic powder is heated and melted in the mold cavity, and the shape, size, and surface of the product are copied from the inner surface of the mold. During heating and cooling, the mold rotates in two axes perpendicular to each other. The most common heating is done in a heating furnace, convection with a stream of hot air, and the plastic melts, covering the entire inner surface of the mold, followed by cooling (with air, mist or water), lower the temperature to allow the resin to solidify. Finally, remove the mold and take the product (Figure 1).

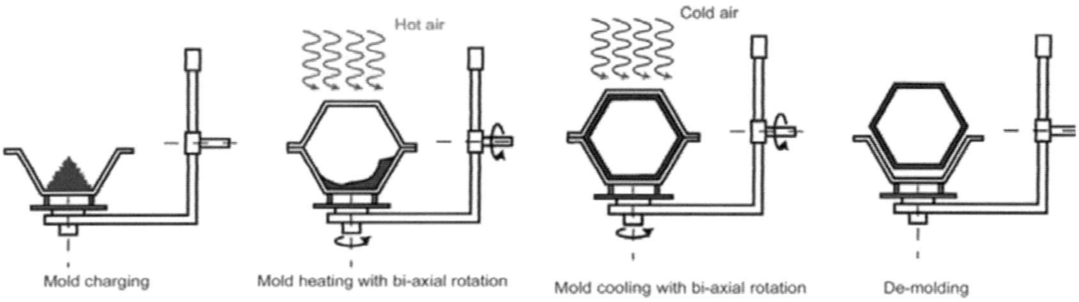

Figure 1. Principle of rotational molding.

2.2 Analysis and selection of plans

Rotating Mold Technology products have a very wide range of volumes from 0.05 liters to over 10,000 liters, so the rotary die systems are very diverse. The basic common feature of this process is that molds need to be rotated, heated, and cooled. Depending on product size, productivity, allowable area, heating and cooling characteristics, there are different rotary die systems, which are commonly used today as shown in Figure 2.

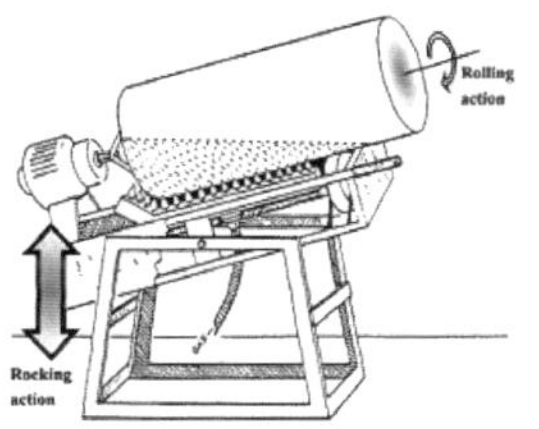

a. Typical rock-and-roll machine

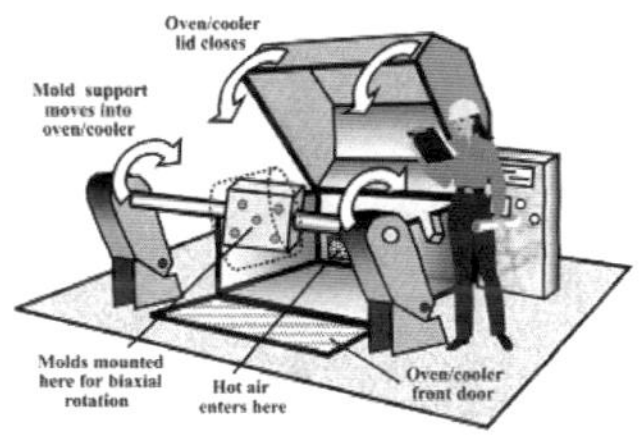

b. Clamshell type rotational molding

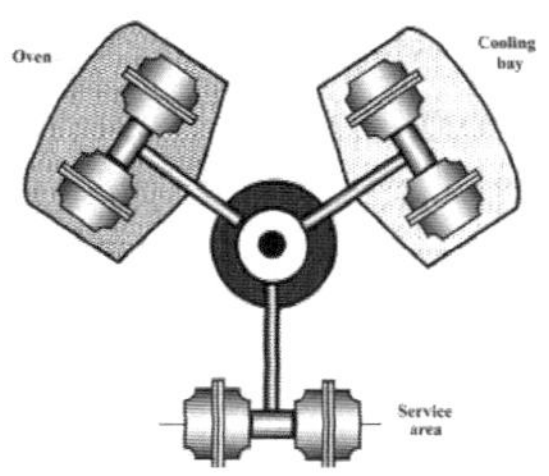

c. Fixed-arm carousel machine

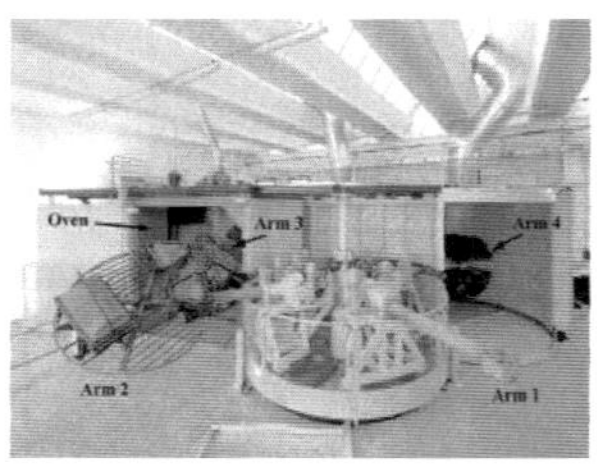

d. Independent-arm rotational molding

Figure 2. Types of Rotational Molding Machines, courtesy of Polivinil, Italy- The Queen's University, Belfast

After learning about the product to be manufactured, the product's size parameters, materials, Rotational molding as well as the principle, operation method, scope of use of the machines used in mold technology turned. We choose the independent rotary arm molding machine to design the machine for making flower pots products.

3. Design calculation of mold driving system

After analyzing the product, the principle of the rotational mold technology, the design team evaluates and chooses a suitable plan for the flower pot product. The following is the drive diagram of the designed rotary die system (Figure 3).

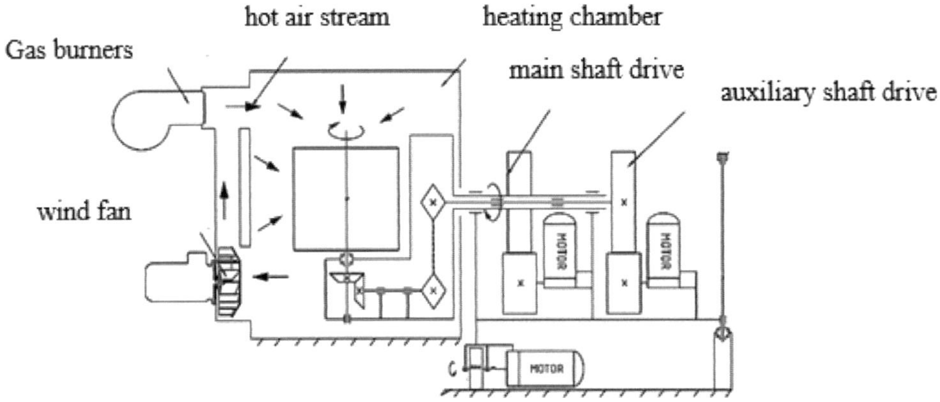

Figure 3. Drive diagram of rotary die system

3.1. Calculation of the drive on the auxiliary shaft

The drive on the auxiliary shaft is shown on Figure 4 below.

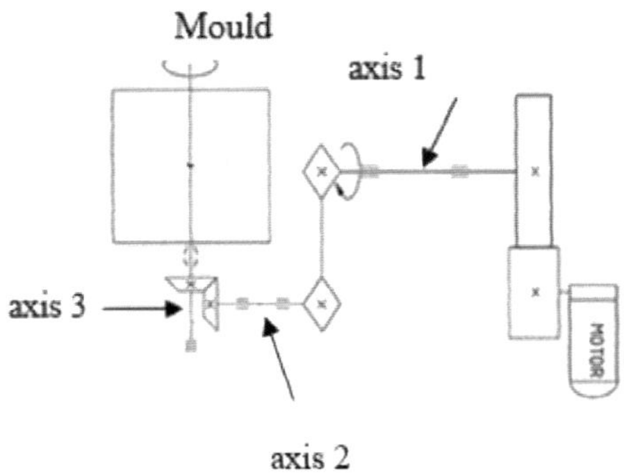

Figure 4. Drive diagram of auxiliary shaft

3.1.1. Select the engine and distribute the gear ratio

Based on the DRS catalog, the design team chose the K67 modular single-phase motor with the following parameters: P_{dc} = 1.1 kW; n_{dc} = 20 rpm.

Table 1. Engine size K67 DRS 80M4.

CODE	AC	AD	ADS	L	LS	LB	LBS
DR80M	156	128	139	530	611	261	342

Based on the table 1, we calculate the number of revolutions, power and torque on the axes (Table 2):

Table 2. Auxiliary shaft characteristics.

Axes / Parameter	geared motor	1	2	3
Power P (kW)	0.64	0.49	0.44	0.41
Transmission ratio u	2		1	1
Revolutions n (rpm)	20	10	10	10
Torque T (Nmm)	305600	467950	420200	391550

3.1.2. Calculate the drive on the auxiliary shaft

Axis 1

Material selection: Improved 45 steel with $\sigma_b = 600MPa$; $\tau = 20$ MPa

Diameter of shaft determined by torque:

$$d \geq \left(\frac{T}{0.2[\tau]}\right)^{1/3} = \left(\frac{467950}{0.2x20}\right)^{1/3} = 48.9mm\ (1) \quad (1)$$

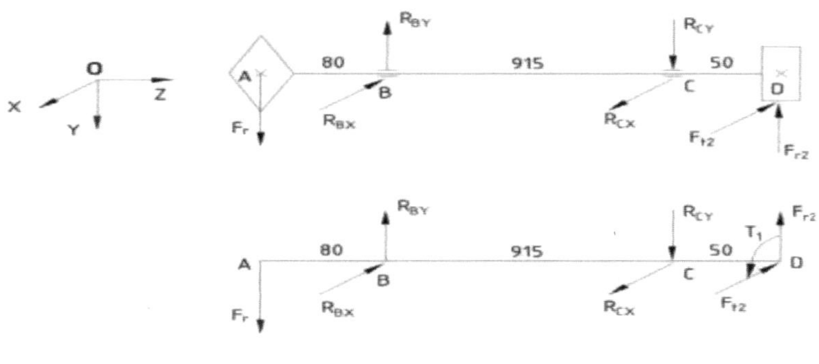

Figure 5. Force exerted on axis 1.

Analysis of force and torque on the shaft determines the dangerous section at B. Shaft diameter at B:

$$d \geq \left(\frac{M_{td}}{0.2[\sigma]}\right)^{1/3} = \left(\frac{700619}{0.1x85}\right)^{1/3} = 43.52mm\ (2) \quad (2)$$

According to the diameter standard at the sections: $d_A = d_D = 42$ mm; $d_B = d_C = 45$ mm.

Axis 2

Material Selection: Improved 45 steel with $\sigma_b = 600MPa$; $\tau = 20$ MPa.

The shaft diameter is determined by the torque:

$$d \geq \left(\frac{T}{0.2[\tau]}\right)^{1/3} = \left(\frac{420200}{0.2x20}\right)^{1/3} = 47.2mm\ (3) \quad (3)$$

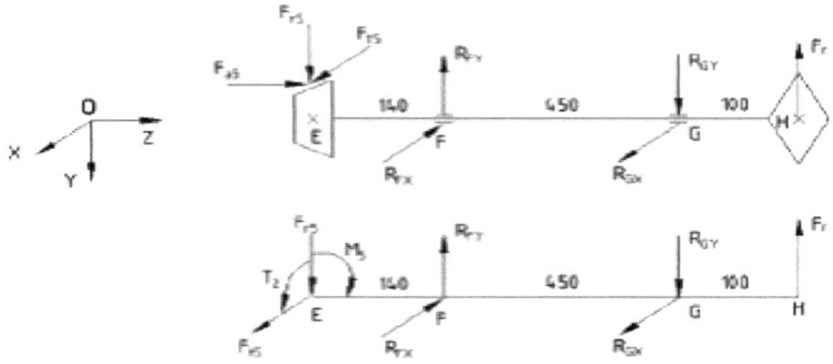

Figure 6. Force exerted on axis 2

Analysis of force and torque on the shaft determines the dangerous section at F. The shaft diameter at F

$$d \geq \left(\frac{M_{td}}{0.1[\sigma]}\right)^{1/3} = \left(\frac{1137330}{0.1x85}\right)^{1/3} = 49.6\, mm\ (4) \quad (4)$$

According to the diameter standard at the sections: $d_F = d_G = 50$ mm; $d_E = d_H = 45$ mm.

Axis 3

Material Selection: improved 45 steel with b $\sigma_b = 600$MPa; $\tau = 20$ MPa.

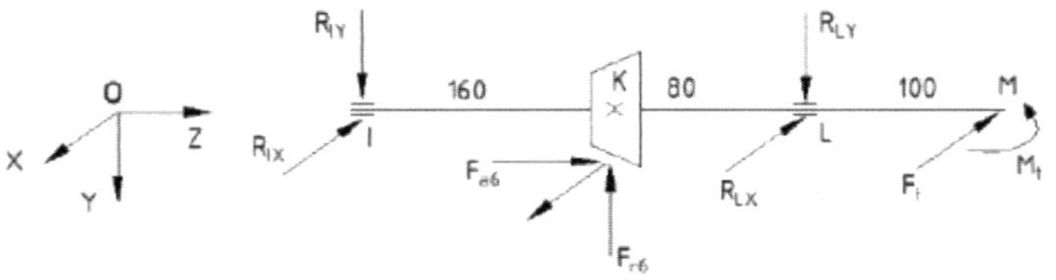

Figure 7. Force exerted on axis 3

Calculating similar to the above two axes, the design team determined the diameter of axis 3 at the sections: $d_I = d_K = d_L = 50$ mm; $d_M = 45$ mm

3.2. Calculate the drive on the main shaft (Figure 8)

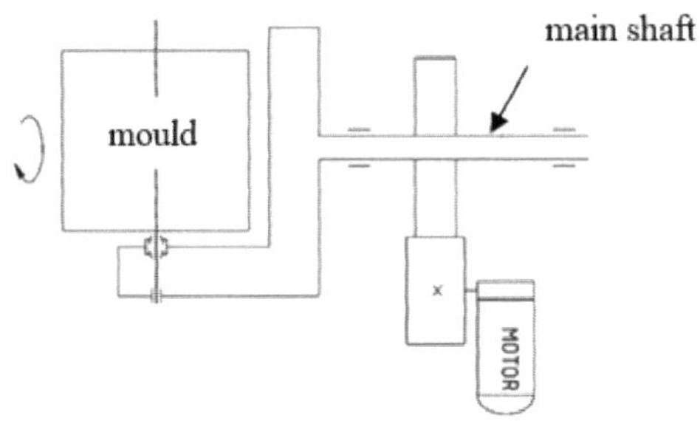

Figure 8. Drive diagram on the main shaft

3.2.1. Select the engine and distribute the gear ratio
Based on the DRS catalog, we chose the K67 modular single-phase motor with the following parameters: P_{dc} = 1.5 kW; n_{dc} = 16 rpm.

Table 3. Engine size K67 DR90M

CODE	AC	AD	ADS	L	LS	LB	LBS
DR90M	179	140	150	532	611	261	342

Based on the table 3, we calculate the number of revolutions, power and torque on the axes (table 4):

Table 4. Main shaft characteristic

Parameter \ Axes	geared motor	4
Power P (kW)	1	0.96
Transmission ratio u	2	
Revolutions n (rpm)	16	8
Torque T (Nmm)	596875	1146000

3.2.2. Calculation of cylindrical gear transmission
Based on the calculation results and look up the table, we chose the gear transmission, chain with the following basic parameters (Table 5)

Table 5. Parameters of gear transmission and chain

Parameters	**value**
Spindle distance	aw = 150 mm
Module	m = 2 mm
The width of the rim	b_{w7} = 60mm; b_{w8} = 50 mm
Ratio	u = 2
Number of teeth	Z_7 = 50; Z_8 = 100
Shift coefficient	X_7 = 0; X_8 = 0
Diameter of divider	d_7 = 100 mm; d_8 = 200 m

3.2.3. Design calculation of main shaft
Material Selection: Improved 45 steel with σ_b = 600MPa; τ = 20 MPa.
The shaft diameter is determined by the torque:

$$d \geq \left(\frac{T}{0.2[\tau]}\right)^{1/3} = \left(\frac{1146000}{0.2\,x20}\right)^{1/3} = 66\ mm \tag{5}$$

Analysis of force and torque on the shaft determines the dangerous section at P. Shaft diameter at P:

$$d \geq \left(\frac{M_{td}}{0.1[\sigma]}\right)^{1/3} = \left(\frac{3028730}{0.1\,x85}\right)^{1/3} = 77.5\ mm \tag{6}$$

According to the diameter standard at the sections: d_P =d_Q =100 mm, d_O =110 mm, d_N = 115 mm

3.3. Durability and displacement test of some important details
Critical parts, susceptible to destructive deformation, should be simulated to check for the following properties: Ensure the deformation is within the allowable range, thereby ensuring the working requirements of the parts. Ensure that the stress generated by the force exerted on the part does not exceed the strength limit of steel σ_b <598 MPa (TCVN 8301).

IOP Conf. Series: Materials Science and Engineering 1109 (2021) 012005 doi:10.1088/1757-899X/1109/1/012005

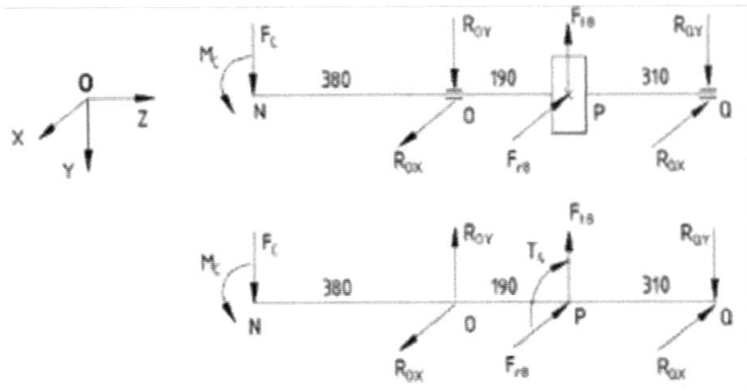

Figure 9. Force exerted on axis 4.

3.3.1. Large shaft details (Figure 10)

Input parameters: m = 40.68kg; steel C45, V = 0.0051772 m^3, F = 2452N

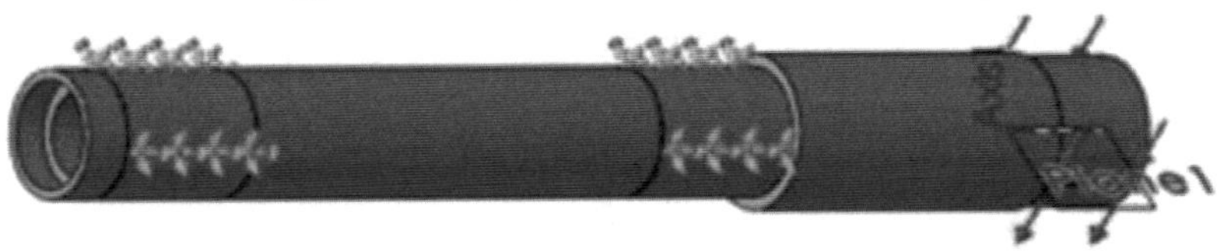

Figure 10. Placing of mounting and force on part.

4. Results

The meshing and simulation results are shown on Figure 11.

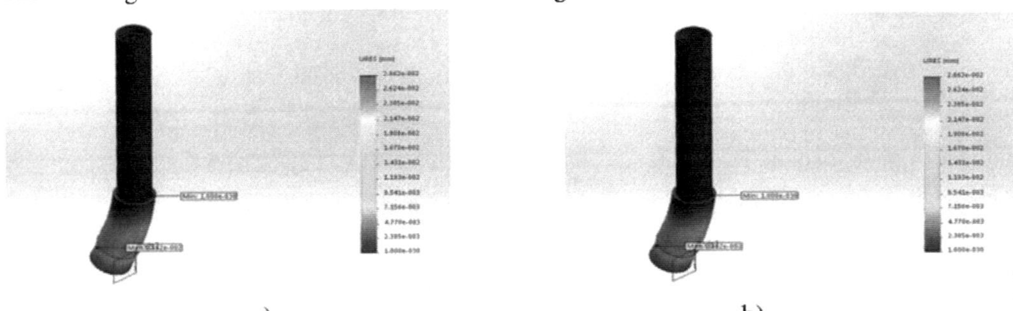

a) b)

Figure 11. Results of stress analysis (a) and strain (b) product on large shaft

Table 6. stress and deformation of large shaft

Stress	Deformation	The analytical results show
Min : $4.2618.10^{-9}$ N/mm^2	Min: 0 mm	Part displacement is relatively small 0.02 mm
Max : 12.86 N/mm^2	Max: 0.0286 mm	Ensure the surface of the part 12.85 MPa <598 Mpa

The horizontal part is shown on Figure 12.
Input parameters: m = 46.37kg; steel C45; V = 0.00590125 m3; F = 1776N

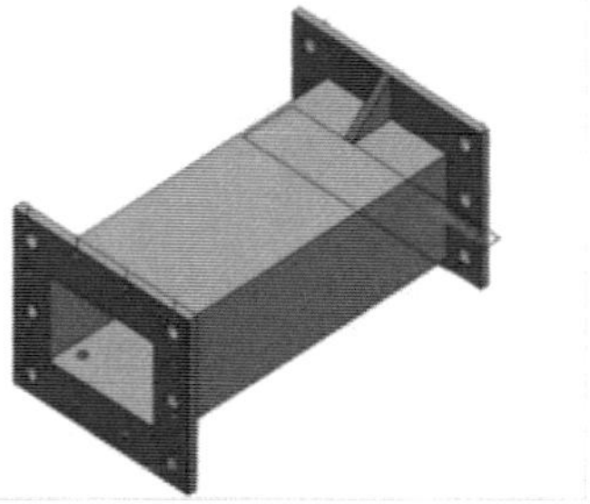

Figure 12. Placing mount and force on part

The meshing and simulation results of the part is shown on Figure 13.

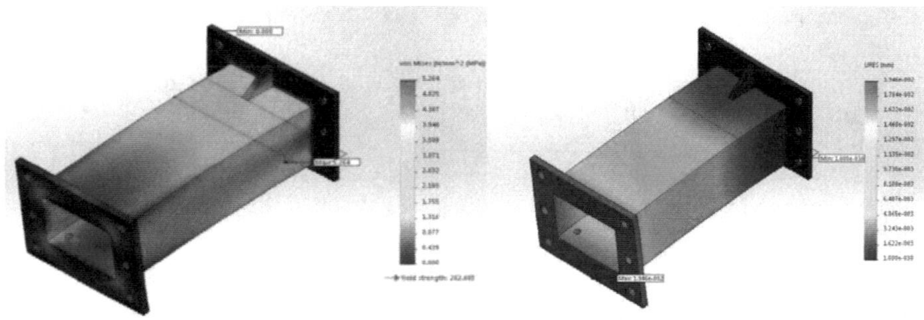

a b

Figure 13. Results of stress (a) and deformation (b) analysis on the details

Table 7. Stress and deformation of more horizontal

Stress	Deformation	The analytical results show
Min : $9.10462.10^{-5}$ N/mm^2	Min: 0 mm	Part displacement is relatively small 0.019 mm
Max : 5.264 N/mm^2	Max: 0,0194604 mm	Ensure the surface of the part 12.85 MPa <598 MPa

The design team calculated, designed and manufactured the rotary die machine with the following main technical parameters (Table 8).

Table 8. Main technical parameters of rotary die machine

Parameters:	**Result**
Spindle rotation speed	8 rpm
Auxiliary shaft rotation speed	10 rpm
Productivity	26 products / 8 hours
Product size	1000 x 400 x 400 mm

Figure 14. Photos of the actual fabricated machine.

5. Conclusion

Rotational molding is a highly competitive method compared to methods such as blow molding and injection molding in the production of hollow plastic products. The authors studied the set-up of the drive diagram of the rotary die system. In addition, the analysis and calculation of the force acting on the drive axes are also presented in this study. We designed and manufactured the rotary die machine a rotating mold for flower pots products with dimensions 1000x400x400 mm. The machine can change its two-axis speed, with a productivity of 26 products / 8 hours. However, it is necessary to continue with in-depth studies such as the effects of technological parameters on product quality in terms of size and sample strength on different materials, and researches to improve accuracy of product.

References

[1] Jachowicz T, Janusz W, Sikora J 2013 Investigation of the influence of mold rotational speed on the cast wall thickness in the rotational molding process *Advances in Science and Technology Research Journal* **7**.

[2] Crawford R J, Throne J L 2002 Rotational molding technology *Plastics Design Library / William Andrew Publishing*.

[3] Crawford R J and Kearns M P 2003 Practical Guide to Rotational Moulding. *Rapra Technology Limited.*

[4] Sharifi P, Henwood N, Liauw C, Lees G and Quaratino A 2012 Studies of degradation effects during rotational molding *Journal of Thermoplastic Composite Materials*.

[5] Costa L, Cramez M C, Pontes A J 2013 A Study on Shrinkage and Warpage of Rotational Moulded Polyethylene *Materials Science Forum* **730-732** p 957-962.

[6] Monzón M D, Ortega Z, Benítez A Z, Ortega F, Díaz N, Marrero M D 2012 Development Towards a More Sustainable Rotational Moulding Process. *In Proceedings of ECCM15 – 15th European Conference on Composite Materials, Venice, Italy*, (24-28 June 2012).

[7] Jachowicz T, Krasinskyi V 2017 Influence of Rotational Moulding Process Parameters on Accuracy of Polymer Casts *Acta Mechanica Slovaca* **21**(1) p 40-46.

[8] Ogila K O, Shao M, Yang W, Tan J 2017 Rotational molding: A review of the models and materials. *https://doi.org/10.3144/expresspolymlett.*

[9] Joshi M G, Bhatt P B, Chaniyara P N 2014 Review on Automation of Independent Four Arm Bi Axial Machine in Rotational Moulding Industry *International Journal of Engineering Development and Research.*

[10] Mongaraj S, Mahajan E, Gupta S, Sanap K, Patil S K 2017 Review on Fabrication of Roto-Moulding Machine *International Journal for Research Trends and Innovation.*

[11] Bergamo L, Spa P 2013 Measurement and Control of Prosseure Inside Rotacional Moulds. *www.rotoworldmag.com*.
[12] Perot E, Lamnawar K, Maazouz A 2008 Optimization and modelling of rotational molding process *International Journal of Material Forming*.
[13] Barhoumi N, Lamnawar K, Maazouz A 2008 A reactive rotational molding process of PP/PA6 bilayer systems: experimental investigations *International Journal of Material Forming*.
[14] Ogila K O, Shao M, Yang W, Tan J 2017 Rotational molding: A review of the models and materials. https://doi.org/10.3144/expresspolymlett.2017.75.

IOP Conf. Series: Materials Science and Engineering 1109 (2021) 012006 doi:10.1088/1757-899X/1109/1/012006

Design investigations on a 2-speed gear reducer for agricultural application

J L Luces[1] and M C E Manuel[2]

[1]Department of Science and Technology - Metals Industry Research and Development Center, Gen. Santos Ave., Bicutan, Taguig City, 1631, Philippines
[2]School of Mechanical and Manufacturing Engineering, Mapua University, Muralla St., Intramuros, Manila 1002, Philippines

joein.mirdc@yahoo.com, mcemanuel.mapua@gmail.com

Abstract. A dual-speed gear reducer was designed to offer a broader application to the common single-speed reducers. However, in order to provide additional value to the design, an investigation was made to determine the effects of varying the normal module, center distance, and facewidth on the gear components of the dual-speed gear reducer. KISSsoft gear simulation software was used to perform the computation for each variation in the mentioned parameters. The optimized values were based on the calculated parameters of gears with the root and flank safeties of not less than one or greater than 3000 hours service life as minimum standard for designing gears used for agricultural application. Based on the investigations made the increase in normal module, center distance, and facewidth also increases the flank and root safeties of both drive and driven gears. However, increasing further the normal module decreases the flank safeties of both gears. Nevertheless, increase in normal module with corresponding increase in center distance considerably increases both the root and flank safeties of gears. Finally, a dual-speed gear reducer was fabricated that proved the data generated using KISSsoft was useful for efficient manufacturing and assembly of the gears and gearbox, respectively.

1. Introduction

Torque and speed requirements in the industrial facilities, fishery, and agricultural machineries paved the way for the development of speed reducers. In agricultural processes, ranging from land preparation until post-harvest, machines performing the operations require a gear reducer to decrease the speed of normally high rpm prime movers. Zimmerman [1] established that for tractor plowing, the lower speeds are more desirable than the higher speeds. Primary tillage should be done at a slower speed of 4-6 km/hr and the secondary working such as offset disking is best done at higher speed of 7- 8km/hr (IRRI). This is supported by the Philippine Agricultural Engineering Standards (PAES) 118: 2001 which requires tractors to pull field implements at a maximum speed of 8 km/h. In post-harvest operation, such as mechanical drying using rotary type paddy dryer, the machine requires angular speeds of approximately 15-20 rpm which is far from the usual speed of electric motors which ranges from 900 - 3600 rpm and that of single cylinder engines which have more or less 3600 rpm [2]. Another agricultural machine that requires slower speed is the juice extractor for sugar cane. In a patent filed by Velumani Aravind [3], the machine used a gearbox to reduce the speed of the two different motors of about 900 and 1440 rpm to an output of 10 to 50 rpm of a juice extractor.

From these, it is apparent that reduction of speed for high-speed prime movers is required which will consequently increase torque of low power engines or motors. Nevertheless, in a more practical sense, it would be of paramount importance to optimize the use of a speed reducer by providing a dual speed output to the apparatus. This will reduce the cost of acquiring two gear reducers to serve two different purposes. This would be beneficial particularly among low income earner farmers.

One feature in gear reducer which deserved to be taken into consideration is the importance of helix angle among helical gears. Helical gears have several teeth engaged constantly that transmits considerably high torque and reduce stress among gear teeth due to the load distribution through the gear teeth. It also produces less noise to about 10-12 dB more quiet than spur gearing [4]. It is therefore essential to conduct investigation on the optimum helix angle by which a helical gear will be on its optimum performance. Facewidth on the other hand is another factor that influences the life cycle of a gear. In the study conducted by Lateno [5], he observed that the maximum bending stress and contact stress decreases with increasing face width, number of teeth and root fillet radius relative to spur gear set weight. Module also deserves to be considered in designing gears. It is known that the larger the module, the stronger the gear hence, the higher the load it could carry. Lastly, center distance between gear pairs should also be given equally utmost attention. It is important to note that center distance plays a vital role in gear designs since larger center distance also means larger gears.

In order to have an efficient and speedy calculation using a combination of certain gear parameters of specified range, simulation or computational softwares are employed in the process. Kiekbusch et al. [6] presented a paper on the development of detailed two- and three-dimensional finite element models used to determine the torsional mesh stiffness of spur gears. He made use of the finite element software ANSYS to generate different pairs of spur gears which included the adaptive meshing algorithm for the contact zones. He used 2D models to simulate varied gear pairs in a short period of time. He concluded that using a 3D model it is possible to simulate helical gears or apply modifications in tooth face such as face angle corrections or crowning. He added that the effect of meshing interferences can be studied. Likewise, he concluded that the resulting values from the finite element models and the mesh stiffness formula can be used in the dynamic simulations such as multi body simulation of gearboxes. The advantage of the developed formula is that only the basic gearing conditions are needed to obtain the torsional mesh stiffness. The finite element models presented the option to evaluate stresses in crucial areas of gears. The results showed that both the finite element models and the mesh stiffness formula are credible.

Jia, Howard and Wang [7] presented a paper regarding the dynamic modeling of multiple pairs of spur gears in mesh, including friction and geometrical errors. A dynamic model of two pair of gears in mesh with three shafts, 26 degrees of freedom, friction, effects of variable tooth stiffness, pitch and profile errors, and a localized tooth crack on one of the gears were presented in the article. It also explained how to include in the model the geometrical errors in teeth which also incorporates the effects of variations in torsional mesh stiffness in gear teeth by using a common formula to elucidate stiffness that occurs during the meshing of gears. The determination of the occurrence of geometrical errors in teeth was made using MatLab and Simulink models, developed out of the equations of motion. It also discussed the effects of pitch and profile errors on the resultant input pinion angular velocity. The simulation results revealed that pitch and profile errors of 10 microns have a substantial impact on the vibration of gearbox. The geometrical errors are noticed to provide extra energy into the harmonics of the tooth mesh and associated sidebands, shadowing the changes in modulation caused by localised damage in tooth like tooth crack. It was observed that tooth crack of 5mm in length was difficult to detect using 10 microns geometrical pitch and profile errors.

Lahtivirta, J. and A. Lehtovaara [8] emphasized the important role of finite element method (FEM) in the modelling and simulating of the gear pair in order to approximate deformations, stresses, and damage risks in different working conditions as part of the design process. To analyze stresses in the gear root and the gear contact, they developed a parameterized calculation model. The local adaptive FE mesh they used sped up the computation. They created in Matlab environment the rotation of the gear pair, the adaptive FE mesh, and the accurate surface profile based on gear hobbing process to obtain a good control of the flank profile and the meshing parameters. They then integrated the method they used to a commercial FEM software to calculate stresses and deformations. They successfully validated the FE mesh approach they developed against analytical Hertzian theory. They also compared the developed spur gear model with ISO 6336 gear standard and a commercial gear simulation software (KISSsoft) and turned out a relatively good conformity results with the maximum contact pressure and the maximum tooth root stresses.

Yu-Ren Wu and Van-The Tran [9] mentioned that the tooth surfaces of gears are usually crowned with twist-free tooth flanks to reduce the vibration and noise of the gear systems in the high-speed rotational gearbox. With this they proposed an innovative method for longitudinal crowning tooth flank of helical gear. They constructed a linear crossed-angle function for honing external gear. They used a variable pressure angle honing cutter for free-twist of the crowned tooth flank. They investigated transmission and load characteristics of meshing gear pair with twist-free tooth flank. They developed 3-D models of twisted and twist-free meshing gear pairs and analyzed them using KISSsoft. Their study revealed that the working conditions of the meshing gear pair have substantial improvements using the proposed gear honing method.

The study of Jani and Shah [10] focused on the theoretical and finite element method to calculate bending and contact stresses on the tooth geometry of major effected parameters on helical gear pair. They used various approaches to calculate the wear and bending effects of gear failure cause in static condition using finite element analysis, AGMA standard and ANSYS. They checked the root and flank safety of a helical gear by calculating their strength manually using Lewis's equation and KISSsoft software for the gear system.

Cosmin Petra [11] used KISSsoft to study noise reduction in spur gears transmission. His motivation focused on several characteristics of transmissions such as high efficiency, high load carrying capacity, and low sound excitation behavior. His aim was to optimize the helical gear pair to achieve the best possible noise and contact ratio. He stressed that the well-known source of vibration was the disparity of the tooth pair stiffness during meshing and appropriate measures should be applied to reduce the excitation.

In this research, a dual-speed gear reducer for a 12 HP diesel engine developed by MIRDC [12] for agricultural application was designed. Investigations on the effects of varying the normal module, center distance, and facewidth for each of the gear components was done numerically using KISSsoft gear software [13]. This was done to achieve the optimal sizes of gear components that would result to the minimal size of gear box which consequently reduce material and manufacturing costs.

2. Design Constraints and Limitations

The gear reducer was developed to increase the torque while reducing the speed of a generally higher rpm prime movers. A dual speed function was introduced to the gear reducer for broader scope of applications. The gear reducer was made of helical gears for smooth and quiet operation, higher efficiency, lesser manufacturing cost, and improved gear life. The design parameters were basically centered on the number of gear pairs used along with its corresponding number of shafts and bearings. The parameters tested for the gear components was normal module, center distance, and facewidth. Said gear components were designed using KISSsoft gear software while the modelling and drafting made use of NX 3D modelling software. The optimized values were based on the calculated parameters of gears with the root and flank safeties of not less than 1 or greater than 3000 hours service life. The power input, pressure angle, and helix angle were limited to 12 HP, 20 degrees, and 30 degrees, respectively. The maximum diameter of gears was limited only to 250 mm. The dual speed ratios were 1:6 and 1:12. The material used in this study was 1045 medium carbon steel. The gear reducer was developed for 12 HP-rated agricultural machines using helical gears.

3. Design Development and Gear Pairs Optimization

The number of gears, shafts, and bearings, as conceptualized, were added to the module in KISSsys. These are among the basic requirements of the software in order for it to perform the necessary calculation. Preliminary analysis on three (3), five (5), and seven (7) gear pairs was conducted to determine which design was appropriate, that also served as the basis of the selection and further investigation. The computation revealed that the five gear pairs was the most applicable given the limitations of the study and considering its weight. One gear of the three gear pairs have exceeded the tip diameter of 250 mm and the set also have heavy gears. On the other hand, the tip diameters of the seven gear pairs are all within the range but the weight is greater than that of the five gear pairs. Table 1 depicts data on the preliminary calculation made using KISSsoft. Additional investigations were performed to choose the optimal set of parameters. The lightest gear pairs with root and flank service life greater than 3,000 hours was desired for this study. The normal module was set from 1 to 5 with 0.25 mm increment after completion of each set of calculation. The center distance, on the other hand, was set from 100 to 200mm with 1mm increment after each set of calculation. Furthermore, facewidth was set from 30 to 80mm with 1mm increment after calculating each set. The summary of the optimal set of parameters is presented in Table 2 which include the center distance, face width, normal module, root safeties of drive and driven gears, flank safeties of drive and driven gears, root service life and flank service life. The mock-up of an experimental apparatus for further testing is shown in Figure 1.

Table 1. Preliminary calculation using KISSsoft.

Design	Gear Pair	Ratio			Center Distance (mm)	Facewidth (mm)	Normal Module (mm)	Tip Diameter1 (mm)	Tip Diameter2 (mm)	Root Safety Gear1	Root Safety Gear2	Flank Safety Gear1	Flank Safety Gear2	Mass (kg)	Root Service Life (hr)	Flank Service Life (hr)
1	1		6		263	100	1.75	79.68	453.31	2.84	2.66	1.00	1.16	127.72	1000000	1000000
	2a	1			200	25	1.25	200.97	203.89	1.41	1.41	1.09	1.09	12.26	1000000	1000000
	2b			2	200	40	1.25	134.43	270.51	1.50	1.48	1.04	1.04	21.90	1000000	1000000
	Final	6		12										**161.87**		
2	1		1.5		200	38	2	146.88	216.01	3.07	3.13	1.01	1.02	18.86	1000000	3940.22
	2		2		186	69	1	128.53	248.36	2.70	2.65	1.01	1.05	31.52	1000000	3795.25
	3		2		186	65	1	128.53	248.36	1.67	1.68	1.00	1.05	31.05	1000000	3303.78
	4a	1			200	36	1.75	147.55	146.40	2.89	2.88	1.08	1.08	16.17	1000000	3198.41
	4b			2	186	60	1	106.59	210.34	1.67	1.68	1.00	1.05	31.05	1000000	3303.78
	Final	6		12										**128.64**		
3	1		1.5		178	30	1.75	146.88	216.01	2.31	2.34	1.00	1.00	12.07	1000000	1000000
	2		1.5		197	35	1.5	160.75	239.20	1.75	1.75	1.00	1.04	17.31	1000000	1000000
	3		1.5		207	55	1.5	170.10	249.88	1.50	1.49	1.01	1.06	29.96	1000000	3474.64
	4		2		185	100	2	128.35	249.65	1.49	1.44	0.84	0.90	46.39	1000000	602.194
	5		2		106	100	5	81.73	150.05	1.76	1.76	1.00	1.06	14.93	1000000	3012.38
	6a	1			219	100	2.5	223.50	224.50	1.72	1.72	0.82	0.82	58.65	1000000	3002.09
	6b			2	106	100	5	81.73	150.05	1.76	1.76	1.00	1.06	14.93	1000000	3012.38
	Final	6		12										**194.24**		

Table 2. The optimized gear parameters based on simulation results.

Pair	Center Distance (mm)	Facewidth (mm)	Normal Module (mm)	Root Safety Gear1	Root Safety Gear2	Flank Safety Gear1	Flank Safety Gear2	Root Service Life (hr)	Flank Service Life (hr)
1	198	37	2	3.02	3.05	1.002	1.015	1000000	3162.40
2	184	67	1.75	3.19	3.27	1.003	1.047	1000000	3351.03
3	185	66	1.25	2.14	2.15	1.001	1.044	1000000	3054.88
4a	185	33	2	3.16	3.18	1.003	1.003	1000000	3321.40
4b	185	66	1.25	2.14	2.15	1.001	1.044	1000000	3054.88

Figure 1. The 3D view of the dual speed gear reducer.

4. Design Investigations

Investigations on how the parameters interact with the responses were further performed by varying the center distance, normal module, and facewidth. The effect to the root and flank safeties of gears where then observed. Sample plots from the investigations are shown in Figures 2, 3, and 4. Results show the forward-leaning curve which means that the increase in center distance improves the root safety of gears. In addition, it could be noticed that as the module increases, it further enhances the root safeties of both gears. Center distance had similar effects to flank safety of gears at increasing module. However, varying the module has no significant effect to the flank safeties of both gears.

Moreover, as the normal module increased further, given a certain limit in center distance, the flank safety of gears decreases. This means that an optimum normal module should be selected in order to have a better design of gears. Furthermore, increase in facewidth improves both the flank and root safeties of both gears. However, an optimum facewidth should also be investigated to optimize the size of gears for reduced weight and consequently, cost. Preliminary investigation also revealed that increase in helix angle also improved the root and flank safeties of gears however, due to machine restriction, the maximum helix angle tested was limited until 30°.

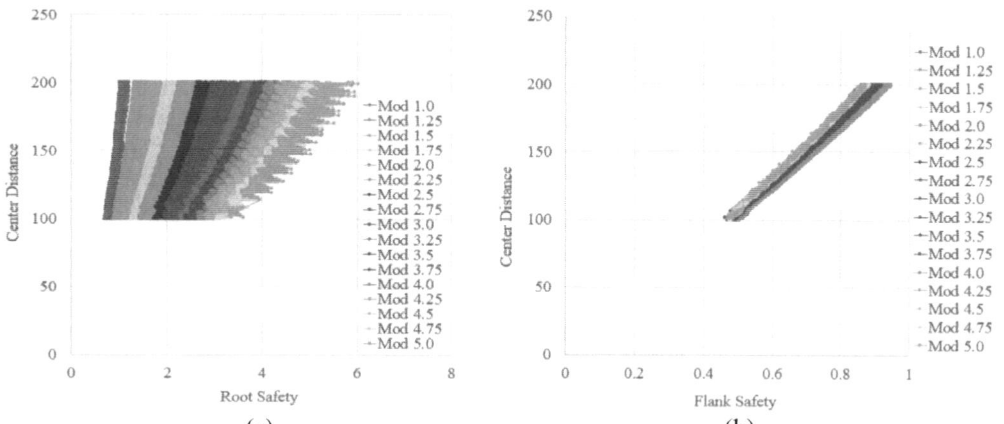

Figure 2. Effect of center distance to (a) root safety and (b) flank safety of drive gear at increasing module (facewidth=30).

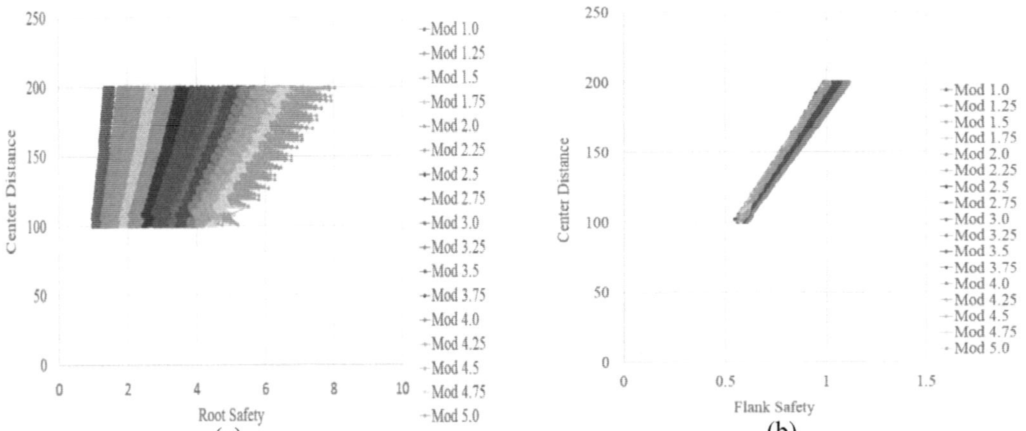

(a) (b)

Figure 3. Effect of center distance to (a) root safety and (b) flank safety of drive gear at increasing module (facewidth=55).

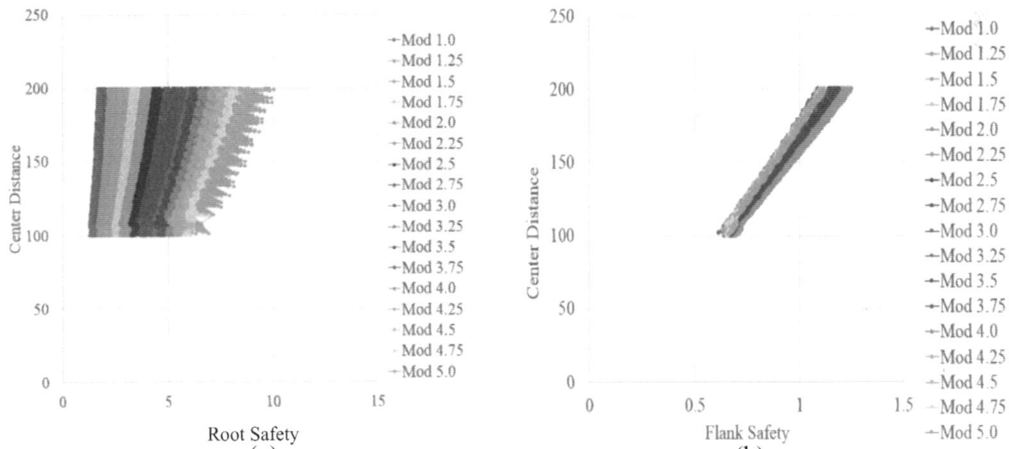

(a) (b)

Figure 4. Effect of center distance to (a) root safety and (b) flank safety of drive gear at increasing module (facewidth=80).

Figure 5. The fabricated 2-speed gear reducer.

5. Conclusions

A dual-speed gear reducer for 12 HP single cylinder diesel engine was successfully designed using KISSsoft/KISSsys gear software and NX 3D modelling software. Based on the investigations made the following conclusions are drawn:

1. Increase in facewidth, center distance and normal module also increases the root safeties of both drive and driven gears.
2. Increase in facewidth and center distance increases the flank safeties of both drive and driven gears however, increase in normal module has no significant effect on the flank safety of gears.
3. Further increase in normal module decreases the flank safeties of both gears therefore, an optimal normal module should be determined.
4. Increase in normal module with corresponding increase in center distance considerably increase both the root and flank safeties of gears.
5. Increase in helix angle until 30° also increases root and flank safeties.
6. The computation process in KISSsoft was useful in determining the optimized normal module, center distance and facewidth which further helped in determining gear pairs with lesser weight.
7. A dual-speed gear reducer was developed that proved the data generated was useful for efficient manufacturing and assembly of the gears and gearbox, respectively.

References

[1] Zimmerman O B 1920 *Tractor Plowing Speeds* (SAE Trans.) 471–85
[2] Johnson K G E, Mollenhauer K, Tschöke H 2010 *Handbook of Diesel Engines* (Springer Science & Business Media)
[3] Aravind V 2008 Sugar cane crushing and juice vending machine (WIPO PCT WO2009128086A1)
[4] Mallenkamp A 2017 Machine design: Getting the most out of gearboxes penton Last updated February 9, 2012 Accessed April 5, 2017 http://www.machinedesign.com/mechanical-drives/getting-most-out-gearboxes
[5] Lateno E 2014 Effect of change of spur gear tooth parameters on bending and contact stresses MS thesis (Ethiopia: Addis Ababa University, Addis Ababa Institute of Technology School of Mechanical and Industrial Engineering) http://etd.aau.edu.et/bitstream/123456789/5397/1/Esayas%20Lateno.pdf
[6] Kiekbusch et al. 2011 Calculation of the combined torsional mesh stiffness of spur gears with two- and three-dimensional parametrical FE models *Strojniski Vestnik/Journal of Mechanical Engineering* **57** 11 810–818 doi:10.5545/sv-jme.2010.248
[7] Jia S, Howard I, and Wang J 2003 The dynamic modeling of multiple pairs of spur gears in mesh, including friction and geometrical errors *The International Journal of Rotating Machinery* **9** 6 437–442 doi:10.1155/S1023621X03000423
[8] Lahtivirta1 J and Lehtovaara1 A 2016 Modelling of spur gear contact using a local adaptive finite element mesh *TRIBOLOGIA Finnish Journal of Tribology* **34** no 1-2
[9] Wu Y and Tran V 2016 Transmission and load analysis for a crowned helical gear pair with twist-free tooth flanks generated by an external gear honing machine *Mechanism and Machine Theory* **98** 36-47 doi:10.1016/j.mechmachtheory.2015.11.014
[10] Jani S and Shah J 2017 Design, modelling and analysis of helical gear pair using ANSYS and AGMA standards for calculating a bending and contact stress on gear profiles *International Journal of Advance Research and Innovative Ideas in Education* **3** 3813-3821
[11] Cosmin P 2011 Numerical research in kisssoft for noise reduction in spur gears transmissions *The 5th Edition of the Interdisciplinarity in Engineering International Conference*
[12] Puerto J Q, Luces J L, Limson A J S, Dime F C, Liza F P An approach for the development of a locally made single cylinder diesel Engine *Philippine Metals* 2016;3:46–50
[13] KISSsoft AG 2016 KiSSsoft Release 03/2016 - User Manual no. 03/2016: 1209

Design of the double pass solar collector for drying

Duy Tue Nguyen and Tri Nhut Do*

Van Lang University, 45 Nguyen Khac Nhu Street, Co Giang Ward, District No.#1, Ho Chi Minh City 700000, Vietnam

*trinhutdo@gmail.com or nhut.dt@vlu.edu.vn

Abstract. Solar energy is the free and nearly endless source which can be applied in various fields. Viet Nam locates near the equation and has enormous solar radiation, so this is the good chance for us to harness it. Moreover, Viet Nam is also the agricultural country with the large number of export product. With the large advantage in having enormous solar radiation, we can save up the large amount of energy for drying and other applications. In this paper, the 2 double pass solar collectors which have fins and without fins for drying were designed and simulated by CFD Ansys in some different conditions. The result displays that with the same size, the outlet temperature of the solar collectors with fins is 2 times higher than the without fins one. This temperature can reach 94.6°C with solar radiation 1026.32W/m2, mass flow rate 0.12kg/s and can reach 68.63°C with the same radiation and mass flow rate 0.18kg/s, so it is eligible for agricultural drying.

1. Introduction

The sun is a giant thermal ball that always reacts in the core, providing enormous heat. The sun's surface temperature of 6000°C corresponds to a radiation intensity of 70,000 to 80,000 kW per m^2, but the earth receives only a small fraction of this radiation intensity. If the total energy received from the sun to the earth of 174 petawatts (PW) per year could be completely exploited, that would meet more 10,000 times than the total energy used on earth in 1 year. The solar radiation intensity outside the atmosphere reaches to 1360 W/m^2. However, it is lost when penetrating the atmosphere, reduced to about 700 to 1000 W/m^2 depending on the region and season. Solar energy can be harvested for heating [1], lighting, cooking and a wide range of applications in life.

The direct drying process applied for drying agricultural products has the advantage of simplicity, low-cost. However, it is influenced by factors such as dust, insects and rain. Therefore, the chamber is used for drying in order to help minimize these effects. Moreover, the solar energy can be used during day time and resistors can be used during times of lack of sunlight, rain storms or at night in order to save energy. When using a solar-based drying system in a closed chamber, the internal temperature will be higher due to the greenhouse effect [2]. In the solar-based drying system, air is passed through the absorber plate, heated and fed into the drying chamber. However, a backup heat source is required in order to avoid product deterioration due to the effects of the weather [3]. This system achieves high operating efficiency due to the important part named the heat absorbing plate and increasing the heat transfer area by using metal and wavy fin.

A solar-based drying system with V-profile collector for drying tea and chili was experimented by Fudholit in [4]. The system is tested with solar radiation conditions of 700 W/m^2, total heat transfer area of

15m^2, mass flow of 0.3 kg/s and average outlet air temperature of 50°C. In this system, an auxiliary resistor of 10 kW is used in condition of solar energy absence. Fresh tea leaves with an initial moisture content of 87%, must be reduced to moisture content of 54% in order to maintain the tea color. The auxiliary resistance is used when the temperature is lower than 50°C and the air flow through the system of 15m^3/min. The initial weight of the fresh tea leaves is 10.03 kg before drying and is 2.86 kg after drying. Another solar-based drying system for corn grain was designed by BM Santos, MR Queiroz and TPF Borges in [5]; with a flat collector of cobblestones with a thickness of 20 cm, heat transfer area of 1.8m^2, glass thickness of 5mm and air flow when operating of 2.1m^3/min. It is operated under condition of outlet temperature of 50°C in the city of Campinas, Brazil.

Azharul Karim [6] pointed out that when the double pass collector is used for the flat-profile, it has significantly higher efficiency than the single pass collector and the double pass collector should not be used for V-profile collectors. K. J. Grecia designed and evaluated the mango solar dryer with thermal energy storage and recirculated air [7]. After the comparison with other results, they claimed that their hybrid solar dryer can reduce the drying time from 7.17 hours to 5.32 hours. L. Diana, A. G. Safitra, D. Ichsani and S. Nugroho simulated the air flow through the prism obstacles inside the solar heater channel, they concluded that the obstacle can create the back flow, so air temperature can be increased inside the channel [8]. Ekadewi A. Handoyo, Djatmiko Ichsani, Prabowo and Sutardi conducted the experiment the solar air heater having V-corrugated absorber plate with obstacles bent vertically at any angle [9]. They found that the obstacles raised the solar air heating performance, but pressure drop also went up. The optimum angle in their experiment was found 30°. Based on the basic dimensions of 0.9m by 1.9m, two models of flat and finned collector are designed by Inventor software in this paper. Then, the CFD Ansys software is used in order to calculate simulation parameters of two models at several different times in order to compare two performances.

The rest of this paper is organized as follows. Designs of simulation model for solar-based drying system are described in section 2. In section 3, an evaluation of experimental results is presented. Section 4 concludes the paper.

2. Simulation Model for Solar-Based Drying System

The model was designed using the Inventor software, version 2018 and was simulated using Ansys Fluent software, version 2020. The first model is a flat collector (as shown in figure 1) with the dimension of 1000mm by 2000mm by 5mm (2m^2 of the upper surface and 2m^2 of the under surface); the material of steel; an absorption coefficient of ε=0.9; coated glass sheet thickness of 5mm; a penetration coefficient of τ=0.9; the insulation wooden wall thickness of 40mm and the dimension of inlet and outlet of 200mm by 1000mm.

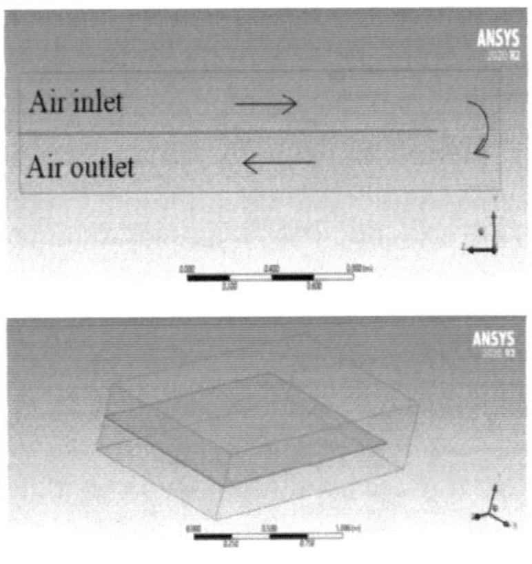

Figure 1. The flat collector model

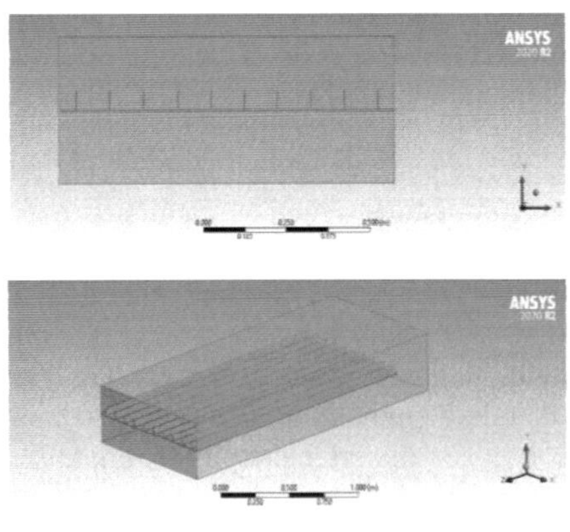

Figure 2. The finned collector model

The second model (as shown in figure 2) has the same dimension, but the flat collector is replaced with the one with steel fins with a total of 10 blades. Each fin has size of 3mm thickness, 50mm height and 100mm fin spacing.

The two models were simulated by the diagram as shown in figure 3 and meshed calculation with 174,930 nodes and 162,950 elements as shown in figure 4.

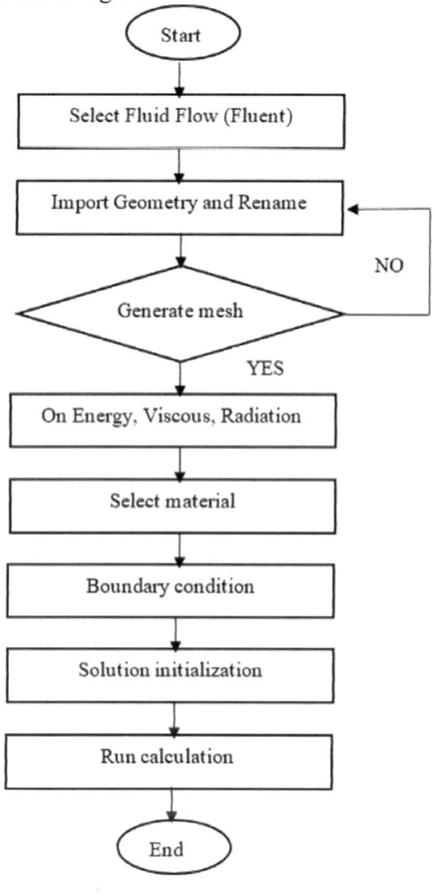

Figure 3. The diagram for simulation.

Figure 4. Meshing calculation on the models.

The simulation was performed with Roseland radiation model and Solar Ray Tracing. The coordinates are simulated with latitude of 10.769° and longitude of 106.662° in Ho Chi Minh City. The air flow in the system is calculated with a k-epsilon model. All residual values of parameters consist of continuity, velocity, energy and k-epsilon are set to $10e^{-6}$. Solar radiation simulated at 9am, 13pm, and 16pm on January 15 in order to evaluate system parameters. The materials used in the system have parameters listed as in Table 1.

Table 1. Material parameters.

Material type	Density (kg/m3)	Specific heat (J/kg.K)	Coefficient of thermal conductivity (W/m.K)
Glass	2321	750	1.15
Steel	8030	502.48	16.27
Wood	700	2310	0.173

According to [10], the collector plate radiation is obtained as in Eq. (1).

$$q_l=\varepsilon.I_{solar} \quad (1)$$

where ε is the absorption coefficient of collector and I_{solar} is the intensity of solar radiation through the glass.

Therefore, we have the following collector heat equilibrium as in Eq. (2).

$$\eta = q_l - q_2 \quad (2)$$

where η is heat increase in air temperature, q_l is absorbed collector heat and q_2 is heat loss.

There are 3 types of heat losses could be listed as follows: loss through coated glass, side loss and heat loss through bottom. The boundary conditions for determining the value of heat loss through the surrounding environment in this model are: convection heat coefficient from the wall, from the cover glass to the surrounding environment α = 20W/m².K; ambient air temperature of 29°C. The surrounding walls are considered to receive no solar radiation.

3. Evaluation of Experimental Results

Radiation values of 948.917 (W/m²), 1026.32 (W/m²), 848.967 (W/m²) are calculated by Ansys program at 9am, 13pm, 16pm on Jan 15, respectively. The outlet air temperatures at every moment are also have different values. The inlet air velocity is also changed. According to [9,10], the mass flow rate can be calculated based on air velocity v (m/s), duct area (m²) and density δ (kg/m³) by using Eq. (3).

$$G=v \times F \times \delta \quad (3)$$

The mass flow rates have the results of 0.12 (kg/s) and 0.18 (kg/s) according to the inlet dimension of 900mm by 1000mm and simulated inlet velocity of 0.5m/s and 0.75m/s, respectively.

3.1 The outlet temperature

The outlet temperature values at 9:00, 13:00 and 16:00 with mass flow rate of 0.12kg/s and 0.18kg/s, respectively are displayed in figure 5.

With the flow rate 0.12 kg/s (figure 5a), the model 1 which has the highest outlet temperature at 13:00 is 44.8°C due to having the highest number of radiation. However, according to the model 2 at the same period, the higher outlet temperature acquired is 94.6°C as a result of being up 1.5 times of area (4m² of the upper surface and 2m² of the under surface) as the first one. At 9:00, the outlet temperature is lower than the one is at 13:00, reach 41.5°C, 89.6°C correspond to the model 1 and model 2, respectively, because the radiation is 948.97 W/m². At 16:00, radiation is lowest with its value 848.967 (W/m²), so the outlet temperatures which are also low are 35.36°C, 73.66°C correspond to the model 1 and model 2.

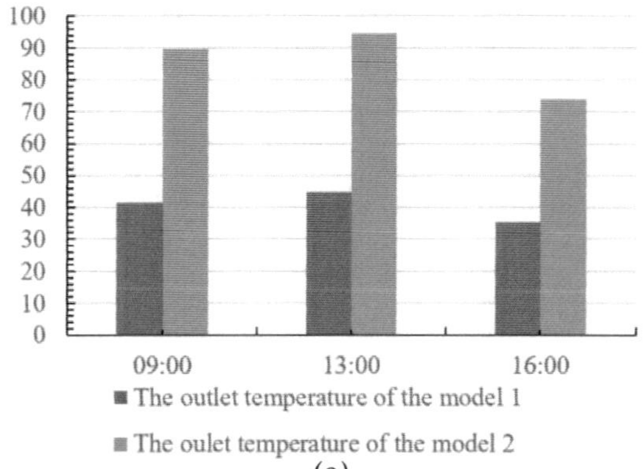

(a)

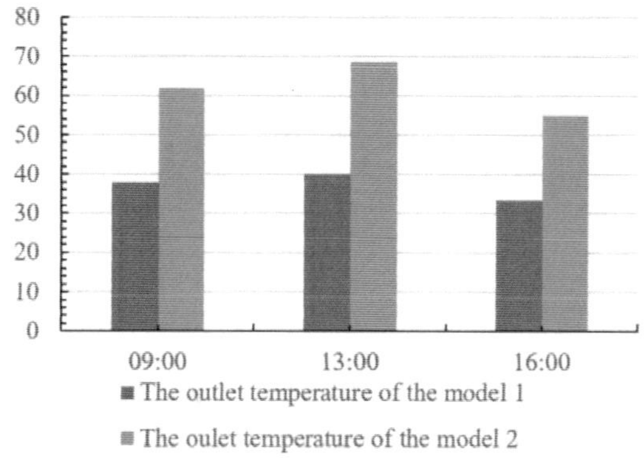

(b)

Figure 5. The outlet air temperature with the mass flow rate 0.12 kg/s and 0.18 kg/s.

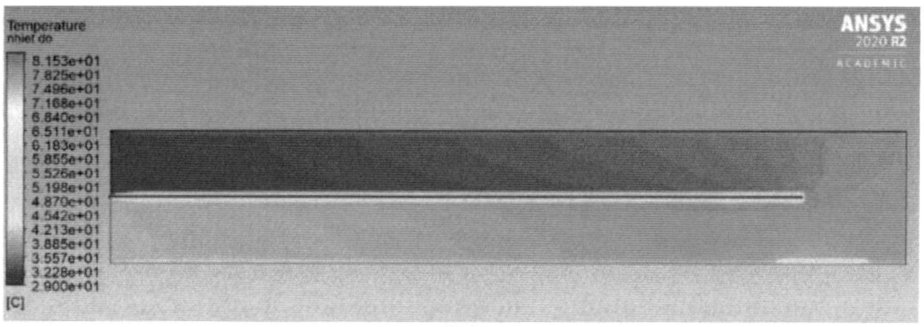

(a)

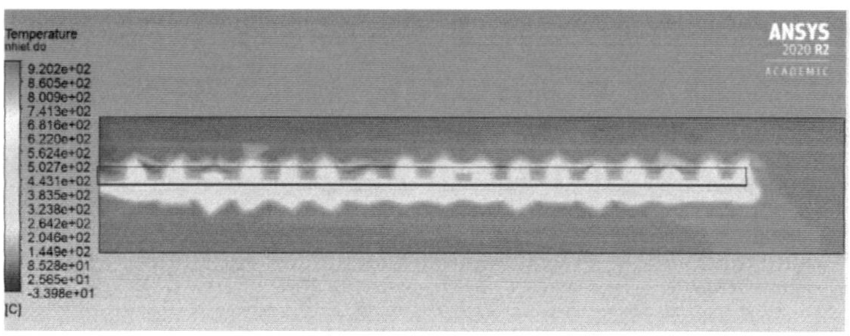

(b)

Figure 6. The temperature variation in plane of model 1 (a) and model 2 (b).

When the mass flow is up to 0.18 kg/s (figure 5b), the outlet temperature is down. Obviously, at 13:00, the outlet temperature of the model 1 is 40°C and the model 2 is 68.6°C. At 9:00, the outlet temperature is 37.79°C, 61.77°C correspond to the model 1 and model 2, respectively. At 16:00, the outlet temperature of model 1 and model 2 are 33.43°C and 55°C, respectively. The variation of temperature in the model 1 and model 2's plane at 13:00 with the highest radiation and mass flow rate 0.18 kg/s is predicted in figure 6.

Obviously, air temperature is highest near the collector surface. In term of the finned collector, because the heat transfer area is two times higher than that of the flat plate, the radiation is significantly concentrated at there, result in high temperature on collector's surface. Therefore, the outlet temperature is higher than that of the plate collector.

3.2 Glass surface's temperature

When solar sunlight strikes on the glass surface, a large portion of it transmit through glass to heat the collector, other portions will heat the surface and reflect from it. The higher glass transmission coefficient, the more radiation will penetrate the glass and heat collector. The glass surface's temperature is displayed in figure 7.

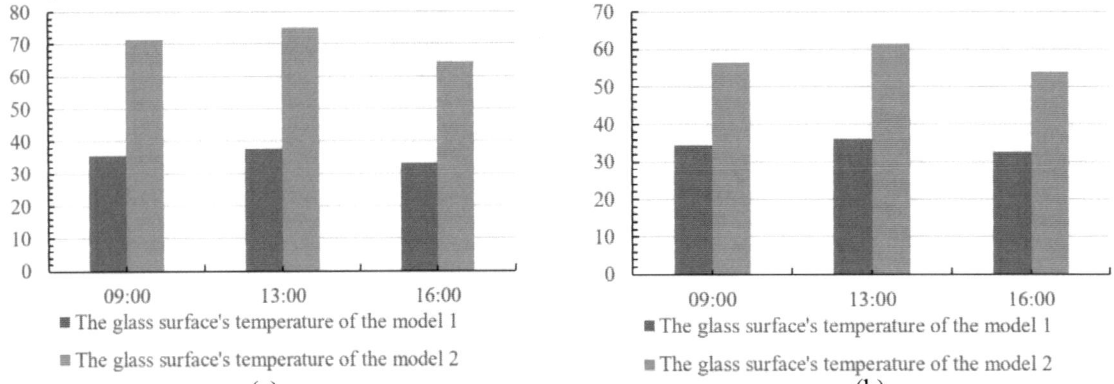

(a) (b)

Figure 7. The glass surface's temperature with the mass flow rate of (a) 0.12 kg/s and (b) 0.18 kg/s.

The glass surface's temperature variation depends on the outlet temperature and radiation at the period time. In term of finned collector, glass surface's temperature which reaches the highest value at 13:00 is 75°C with the mass flow rate 0.12kg/s and the surface temperature of model 1 also reaches 37.6°C with the same mass flow rate. When the mass flow rate increases to 0.18kg/s, the outlet temperature is low and the glass surface's temperature is also low, namely 61.4°C of model 1 and 36.15°C of model 2 at 13:00.

3.3 Air velocity variation

The air velocity varying along equipment correspond to 0.5 m/s and 0.75 m/s is depicted in figure (8a, 8b) and figure (8c, 8d), respectively.

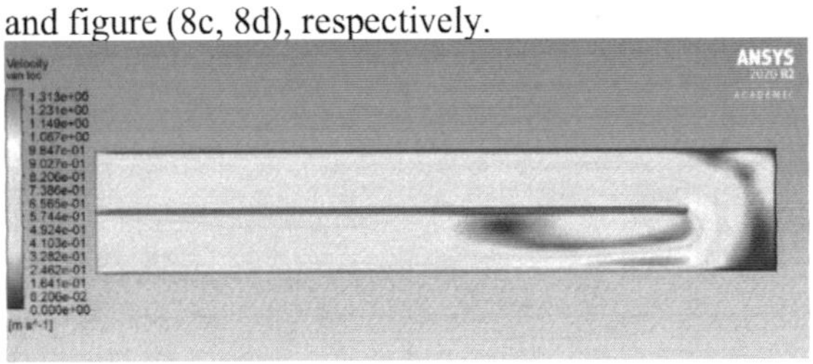

(a) (b)

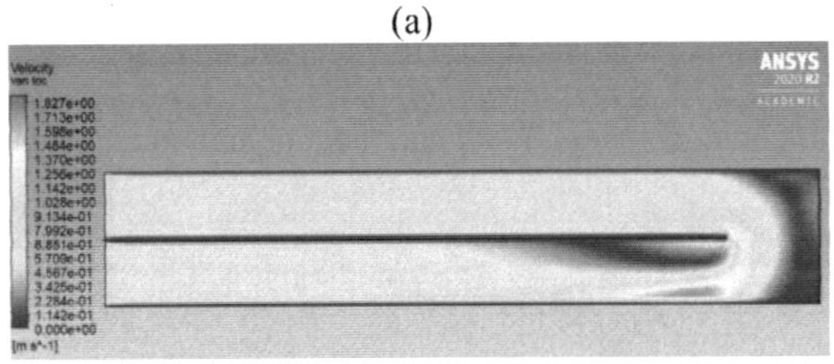

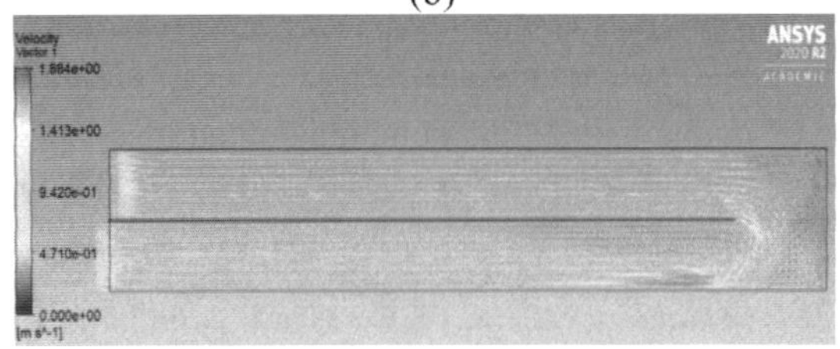

(c) (d)

Figure 8. Air velocity along the equipment of model 1 correspond to 0.5 m/s (a,b) and 0.75 m/s (c,d).

Air velocity variation inside the finned collector corresponds to 0.5 m/s and 0.75m/s is displayed in figure (9a, 9b) and figure (9c, 9d), respectively.

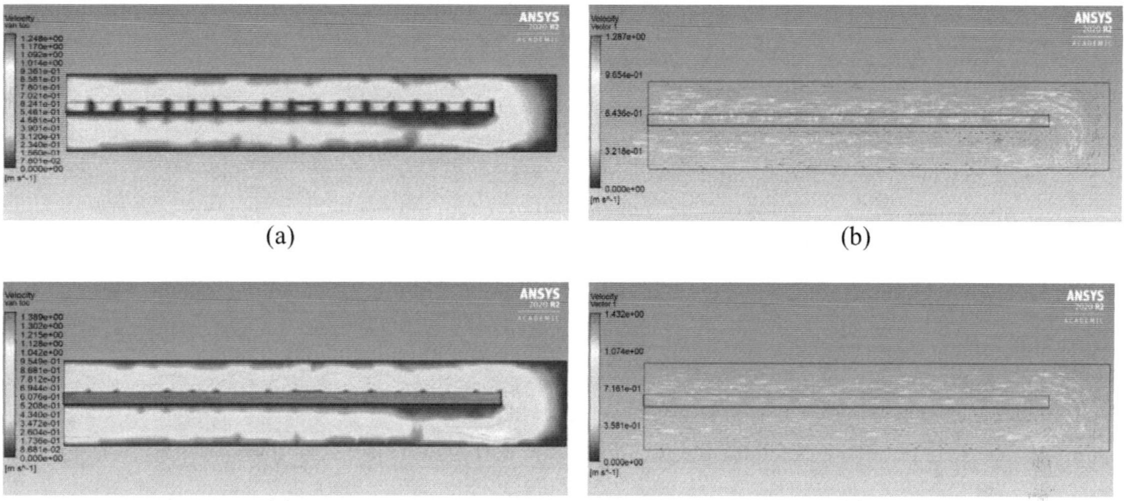

(a) (b) (c) (d)

Figure 9. Air velocity along the equipment of model 2 correspond to 0.5 m/s (a,b) and 0.75 m/s (c,d).

As a result of using finned collector, air flow will be turbulent. Therefore, heat transfer coefficient will not only be surge, but the efficiency is also increased. For that reason, the aim of fin is to raise heat transfer area and heat transfer coefficient.

4. Conclusion

In this paper, the authors design 2 models of flat plate collector and finned collector by using Inventor software and then simulate heat and aerodynamic transmission. The use of a finned collector improves heat transfer efficiency up to twice as much as a flat panel. The highest air temperature coming out of the finned collector can reach up to 94.6°C by 13:00, with a flow of 0.12 kg/s. When using the finned type, the original size of the device does not change, thus saving the usable area.

In addition, the paper shows that the application of simulation software helps designers save time in designing products, as well as a preliminary assessment of equipment efficiency in operation.

Acknowledgments

This project is supported by Van Lang University (VLU), Electronics and Electrical Engineering Division, Ho Chi Minh City 700000, Vietnam.

References

[1] Nguyen D T and Do T N 2020 The efficiency of solar water heating system with heat pump software application designed for resorts in Vietnam *Vietnam Journal of Science, Technology and Engineering* vol **62** issue 2 pp 56-64

[2] Rajkumar P 2007 Drying kinetics of tomato slices in vacuum assisted solar and open sun drying methods *Dry Technol.* vol **25** pp 1349-1357

[3] Fadhel M I, Sopian K, Daud W R W, and Alghoul M A 2010 Performance analysis of solar-assisted chemical heat-pump dryer *Solar Energ.* vol **84** pp 1920-1928

[4] Fudholi A, Sopian K, Othman M Y, Ruslan M H, AlGhoul M A, et al. 2008 Heat transfer correlation for the V-Groove solar collector *8th WSEAS International Conference on SIMULATION, MODELLING and OPTIMIZATION (SMO '08)* Santander, Cantabria, Spain, September 23-25, 2008.

[5] Santos B M, Queiroz M R and Borges T P F 2005 A solar collector design procedure for crop drying *Brazilian Journal of Chemical Engineering* vol **22** issue 02 pp 277 – 284

[6] Karim A Performance investigation of flat plate, v-corrugated and finned air collectors *Solar Energ.* vol **31** issue 4 pp 452-470

[7] Grecia K J, Albert Luce A, Buenaventura M A, Ubando A, and Gue I H V 2019 Design and evaluation of a mango solar dryer with thermal energy storage and recirculated air *IEEE 11th International Conference on Humanoid, Nanotechnology, Information Technology, Communication and Control, Environment, and Management (HNICEM)* pp 1–5 (Laoag: IEEE)

[8] Diana L, Safitra A G, Ichsani D, and Nugroho S 2019 CFD analysis of airflow through prism obstacles inside solar air heater channel *Journal of Physics: Conference Series 1577 012038*

[9] Handoyo E A, Ichsani D, Prabowo and Sutardi 2014 Experimental studies on a solar air heater having V-corrugated absorber plate with obstacles bent vertically *Applied Mechanics and Materials Vol. 493* pp 86-92

[10] El-Sebaii A A, Aboul-Enein S, Ramadan M R I, Shalaby S M and Moharram B M Investigation of thermal performance of-double pass-flat and v-corrugated plate solar air heaters *Solar Energ.* **36** pp 1076-1086

Design, fabrication, and testing of an automated machine for the processing of dried water hyacinth stalks for handicrafts.

K Amante[1], L Ho[1], A Lay[1], J Tungol[1], A Maglaya[1], A Fernando[1]

[1] Mechanical Engineering Department, De La Salle University, Manila Philippines

arvin.fernando@dlsu.edu.ph

Abstract. In this paper a portable machine that is capable of pressing and cutting dried water hyacinth stalks was designed, fabricated and tested. This prototype will help automate the laborious manual cutting of the water hyacinth dried stalks. The machine will flatten the dried stalks and cut stalks into three different widths of 0.5in, 0.25in, and 0.75in. This machine might improve the consistency and quality of the pre-cut stalks. This will also help workers focus on other handicraft processes such as weaving thus improving productivity and helping the local handicraft industry by providing a low-cost alternative equipment.

1. Introduction

The common water hyacinth, also known by its scientific name Eichhornia crassipes, is a perennial aquatic plant that will be referred to simply as water hyacinth. Being one of the fastest growing plants in the world, the water hyacinth spreads using specialized reproductive parts called stolons. Through the stolons, it can produce seeds which may still grow even after 28 years of stagnancy. Water hyacinth is known to be one of the world's sturdiest and most resilient plants [1]. Though the plant is not much of a problem in its native habitat in the Amazon basin, it is an invasive species in all other parts of the world, especially in developing countries and areas such as those in Southeast Asia, where some specimen water hyacinths grow between two to five meters a day and can double their water surface area, or mats, in two weeks [2]. Once introduced to local waters, water hyacinth can acclimate easily to new waters, especially in waterways found in tropical countries such as the Philippines. When growth is introduced and not controlled in areas, water hyacinths will be able to cover vast areas of water in lakes, ponds, swamps, and rivers.

In the Philippines, Water hyacinth affects the Philippines' lake tributaries, canals, rivers and waterways by deoxygenating the water, promoting water retention which may cause flooding, and may shelter various bacteria, viruses, and insects [3]. The ecological and economic impact of water hyacinth growth may be mitigated by harvesting the water hyacinths for use as raw materials for handicrafts, furniture, bags, and other products that are similar to products made from dried coconut leaves. Improving the existing water hyacinth processing through improving production processes and techniques may increase the output of available fiber strips for use, improve consistency and quality, decrease the cost of production, increase the speed and efficiency of the dried water hyacinth processing as well as improving and recognizing the potential growth of the water hyacinth handicrafts and furniture industry which can impact the economy and culture of the Philippines in a positive degree. Fabricating a prototype unit of a working design of a water hyacinth fiber processor will produce weavable fibers with consistent quality at a high-speed. Pressing and cutting, into useful uniform strips or fibers for use in the handicrafts and furniture industry. These include the necessary design processes all the way to the testing methods of the machine, and the machine's effectiveness both in production rate and effectiveness.

There are several studies which focus on the design for an automated biofiber extractor machine. An article by [4], developed a machine for an on field hemp processor. The study produced a prototype which mechanically harvests raw hemp and then performs separation on the bast fiber and core fiber of the hemp through breaking and scutching. The fiber processing machine is designed for the use of dried water hyacinth fiber as the main feed.

2. Methodology

2.1Mechanical Design

This section will explain different parts, design and dimension used in the prototype of the machine. The motor that was used is a single phase 220V JSCC Motor 100YT200GV22 with 100GF20H Reduction Gearbox produced by Shantou Wingtai Equipment Corporation.

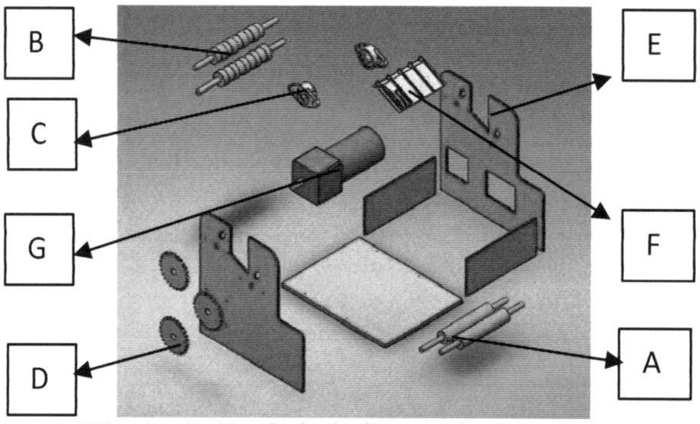

Figure 1. Exploded view of the Prototype

A- *Roller mechanism:* The roller press mechanism is based on the manual pressing machine used by the facility. Torque is the main design consideration for the entire roller press mechanism. Two pairs of rollers are used in the design. The first is used to feed the stalk into the machine and the second is used as a guide to pull the finished product out of the machine. Slotted rollers are used for alignment purposes. Using the average width of a water hyacinth stalk, it was determined that the length of the four slots is 23mm. Given the length of the slots, the spacers for the slots of the rollers should be 18mm. By using the torque of the motor, 32.67 and the rpm, which was determined to be 32.67, the power is found to be 0.193hp. These dimensions will be used to make the machine as cost effective as possible.

B- Springs: From the given parameters, the torque provided by the rollers is not enough to flatten the stalks. Springs are used to increase the force on the stalks to allow the hyacinth stalks to be flattened. RS Pro stainless-steel alloy compression spring measures 40.5 x 11.6 mm and has a spring rate of 7.87 N/mm. It provides a load of 169.99N. Bring the load total to 190.95N of force. A factor of safety of 1.881 was determined from the parameters of the simulation.

C- *Bearings:* The bearings selected were based on the available catalogue of PTI International, which was used in Guan Yiac Trading Corporation, where the flange bearing units were sourced from. Since the calculated shaft size of the roller press mechanism was 15 mm diameter UCFL 202- 03 was used for the machine.

D- *Chain drive mechanism:* The chain used in the drive system was based on the power of the chosen motor for the water hyacinth processor. Design of the chain drive system was based on [5]. The design horsepower was achieved by multiplying the service factor with the power of the motor. Chain No. 40 with a pitch of 0.5 inches was chosen as a suitable chain for the 200W motor. The output RPM and torque of the motor is 32 RPM at 42.8 N-m of torque. Considering that the RPM needed for the rollers is 50 RPM. Using the chain drive general rules it was determined that 17 teeth is the minimum amount needed unless the RPM is below 100 RPM. Excess teeth were used, since the output of the motor is below 100 RPM, to ensure no breakage will occur. The rollers will have smaller sprockets to step up the rpm to 50 RPM.

The chain drive system has a configuration of three sprockets. One main driver containing 27 teeth, and two 17 teeth driven sprockets, which are used to run the rollers. These were interconnected by a single chain. Since sag was observed, a tensioner was added. The center to center distance between the sprockets was computed to be 16.61 between sprocket 1 to 2, 11.73 for sprocket 2 to 3, and 15.78 between sprocket 3 to 1. The chain drive mechanism is shown below:

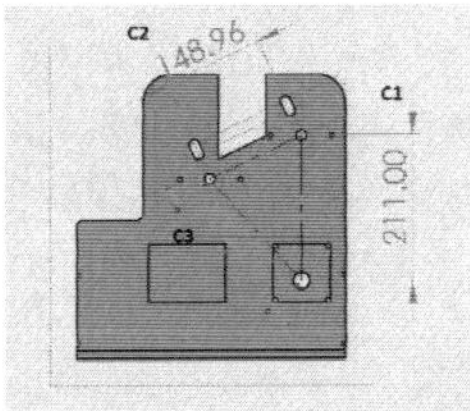

Figure 2. Chain drive mechanism

E- *Frame Dimension*: Three separate plates are used for the mechanism. The plates are separated because the machine has a modular design. 6mm mild steel plate due to its cost effectiveness and durability. The side plates are designed in such an angle to house the rollers with consideration for the comfort of the user. Having a modular design is advantageous because the cutting mechanism can be replaced whenever needed.

Shown below are the stresses experienced by the frame including the stress on the rolling mechanism slots.

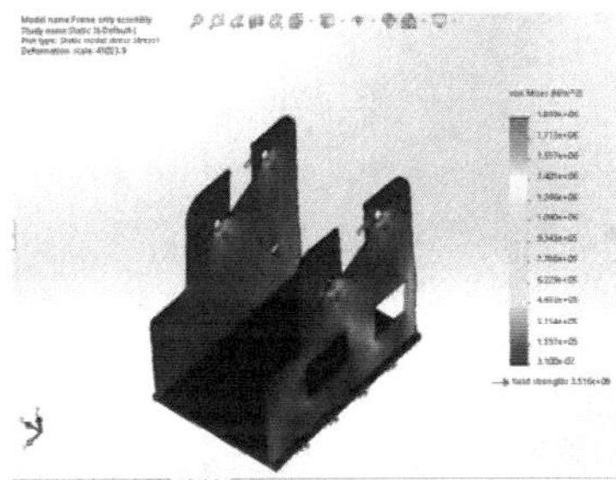

Figure 3: Simulation of the base frame

F- *Cutting Mechanism*: There are four available options for the cutting mechanism. These are: 0.5mm, 0.25mm, 0.75mm, and untrimmed. Each slot will hold a single option, these are not interchangeable. Since the mechanism has a modular design, the user may replace it with another cutting mechanism if the user wishes so. A hole will be included in the design of the mechanism where the trimmings of the stalk will fall through towards a bin. The trimmed stalk may then be used as a fertilizer enricher or a briquetting material. The first part of the cutting mechanism is a funnel guide which catches any pressed water hyacinth which comes out of the roller. Second, is a guide situated at the rightmost side of the slot, its main purpose is to ensure wall contact for stalk trimming consistency. Third, is the razor blade which comes into contact with the stalk, thus trimming it. Fourth, is a hole which catches the excess part of the stalk. The final part of the cutting mechanism is a guideway which guides the stalk to the secondary rollers, which introduces a pulling force which helps in the consistency of cutting. It is suggested that the thinnest end of the hyacinth may be inserted first so the thick end will not cause any problems for the machine. Shown in Figure 4 is the overview of one slot of the cutting mechanism.

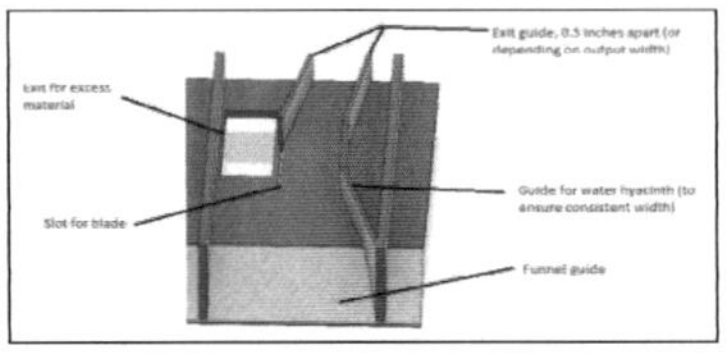

Figure 4. Cutting Mechanism

G- *Motor*: The motor must deliver high torque and low speed to ensure that the machine can move the heavy parts such as the roller press (with the springs) while at the same time delivering a reasonable speed which will fall within the range of the selected RPM. Too much speed may cause deterioration of hyacinth fibers upon contact with the cutting blade. Using these considerations, the researchers selected a Single Phase 220V JSCC Motor 100YT200GV22 with 100GF20H Reduction Gearbox for the mechanism.

Solid works and fusion360 are used for designing the machine. Solid works is a program used to design every component of the machine except for the cutter. On the other hand, the cutting mechanism was a collaboration design by the group since it has a complex design. After designing, the components were fabricated at U-Veex corporation and assembled there. CAD Model and Actual prototype shown in Figure 5 and 6

Figure 5. CAD Model **Figure 6**. Actual Prototype

2.2 Stalk sizes

The stalks have an average width of 18.4566mm and an average length of 773mm. From what was observed from the facility, it is seen that the works used stalks of various widths. In particular, the sizes 0.5mm, 0.75mm, and 0.25 mm are the most used. For this reason, the cutting mechanism uses these three widths for the trimming options.

2.3 Percent increase in output

The increase in output is measured and compared to the number of stalks pressed and cut manually in a process facility that provides livelihood programs to the residence of the area. The dried and cut stalks were observed and based on supervisor and worker's average output.

2.4 Consistency

Each stalk will be measured and any stalk that deviates more than 20% (i.e. deviates to 0.4 for the 0.5-inch cutter) will have to pass the ocular inspection of the facility manager to be deemed consistent. Every stalk that does not deviate to more than 20% will be deemed consistent automatically. Each of the three cutting sizes (0.25-inch, 0.5 inch, and 0.75 inch) will be tested with 100 stalks.

2.5Reliability

To measure the machine's reliability, the uptime, downtime, and number of failures occur during the whole operation. Uptime is referred to as the time it takes for the machine to run without failure while downtime is referred to as the time it takes to repair the machine when a failure occurs. The number of failures will be counted as we measure the total uptime and total downtime during the whole operation. The mean time to repair (MTTR) and mean time between repairs (MTBR) will be calculated in order to measure machine reliability, failure rate as well as availability.

3. Results and discussion

50 water hyacinth stalks were tested for their length and width to get the average dimensions of a stalk. The water hyacinth fibres have an average length of 769.22mm and a width of 18.46 mm. The said average values will be used as a reference for any design consideration which is dependent on the dimensions of a water hyacinth stalk. By adding allowances, a width of 23mm was used for the parameters for the slots of the cutting mechanism.

a. Consistency

The data table for consistency below shows the summary of the testing done for consistency.

Table 1. For 0.75-inch cutter

Total Number of Stalks	Deviated more than 20%		Did not deviated more than 20%
100	38		62
	Passed	Failed	
	25	13	

Table 2. For 0.5-inch cutter

Total Number of Stalks	Deviated more than 20%		Did not deviated more than 20%
100	24		76
	Passed	Failed	
	21	3	

Table 3. For 0.25-inch cutter

Total Number of Stalks	Deviated more than 20%		Did not deviated more than 20%
100	63		37
	Passed	Failed	
	12	51	

Out of the three slots, the 0.5-inch cutter shows the most consistency out of the stalks that was produced.

b. Percent increase in output.

The number of stalks produced by the manual method was determined by ocular inspection and solidified by the average of the worker's answers to the survey questions. However, the number of stalks produced by the manual plant will be based on the maximum yield in a day, which was determined to be 360 in a day's shift, calculated by multiplying the number of stalks produced manually per minute by the number of minutes worked by the staff, which is usually 90 minutes (plus 30 minutes of break to add up to two hours) assuming the workers experience no fatigue and that there is a steady supply of raw product. This will be called manual method (ideal) This will be divided by the number of maximum stalks produced by the machine in a single shift, which was determined to be 833, determined by operating the machine continuously for two hours (an average of 6 stalks per minute at 120 minutes).

To compute for the number of stalks produced in an hour, the value above should be multiplied by the average length of a stalk, which was determined to be 769 mm/stalk. This value is subtracted by 20% to account for any losses in non-ideal operation. This in turn would yield 784 stalks/shift. It is important to note that this yield can be multiplied by up to a factor of four if all slots are to be utilized. However, this will not be used by the researchers. This will be expressed as a percentage increase of yield which can be solved using the equation below.

It can therefore be concluded that the machine is able to increase the yield of processed stalks by up to 43.21% versus the manual method, ideally. This can be greater by utilizing more slots, and this does not include the fact that instead of merely being flattened in the manual process, the increase in yield by the automated process produces products which are already trimmed to consistent width.
One other method is to observe the number of stalks produced by the machine in a two-hour period, instead of using mathematical methods. The researchers timed the number of stalks produced in a two-hour period. The researchers were able to exhaust their supply of 400 stalks in one hour ten minutes, which by proportion is equal to 727 stalks fed.

c. Reliability

Table 4. Reliability

	0.75 inch	0.5 inch	0.25 inch
No. of stalks	40	45	40
No of failures	4	2	11
Downtime (s)	140	20	352
Uptime (s)	417	480	576
Total	557	500	928
MTTR (s)	35	10	32
MTBR (s)	104.25	240	52.36
Availability (%)	74.33	96	62.07
Failure rate	0.00718	0.00400	0.01185

During the reliability tests of the machine, several stalks were tested for each trimming size. Total uptime, downtime, and number of failures was counted during the tested operation. From the test results, the 0.5-inch trim option has achieved remarkable results with a 96% availability rating with a low failure rate and high MTBR. The 0.75-inch is acceptable, despite having more failures and less uptime. The 0.25-inch, however, has suffered poor results from the tests with an availability rating of only 62% with a high failure rate. This cutting width has an average reliability time or MTBR is only for 52 seconds running time which will affect the quality and consistency of the output stalks. For this case, the 0.25 inch cutting option will need major redesigns and planning in the future.

The results of the testing achieved acceptable results with regards to output rate, quality, and economic value. The manual process took longer due to the utilization of outdated tools and equipment used for pressing and trimming of the stalks. This in turn would affect the number of stalks produced as well as the products they are able to make such as baskets and placemats for every 2-hour shift.

This automatic platform offers a high rate of output due to eliminating manual processes which in turn significantly affects the feed rate and output rate of stalks of the machine. The machine was able to output almost twice as many stalks compared to what 2 workers can do using the manual process. The number of products that the workers were able to make were also doubled if they make use of the number of stalks they have produced on today's shift. As the ground survey was conducted the total number of workers inside the site were a minimum of 8 workers. The allocation of workers was stated that half of them would process the fibers while the rest would weave the stalks into products. It was additionally reported that the production of baskets and placemats per shift of eight workers were around 20 and 50, respectively. It is also important to note that the number of workers for one processing line would require two people. Thus, the manual processing station would require two processing lines with four workers, two workers for each line.

The implementation of the machine would also propose a different production strategy due to the difference in production output rate of processed stalks. The workers instead of being split in half to work on processing and the other in weaving, the workers would be more concentrated in weaving the processed stalks into products due to the sheer amount of processed stalks the machine is capable of producing. There would be only one worker out of eight of them that would be processing the stalks since the machine is capable of almost twice the output of two manual processors. The rest of the seven workers would be able to focus on weaving the stalks into products which would be able to produce more baskets, placemats, etc.

The workers' production time, length of breaks, and the time it takes to produce such products by the machine were taken into consideration when calculating for the number of products produced when using the automated processing machine.

d. Comparison of stalks produced

Comparing the stalks produced through manual cutting and through the automated method, it is seen that the stalks made through the automated method have more consistent dimensions as compared to the manual method. This is especially made apparent for the 0.5mm cutter. The comparison between the manual stalks and the automated stalks are shown below:

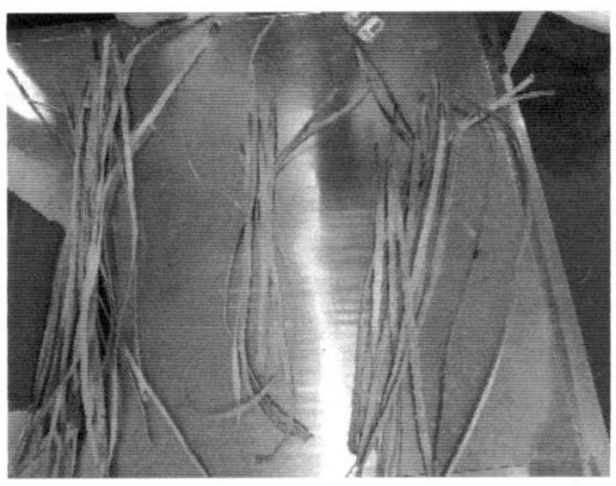

Figure 7: Manually produced stalks (left), stalks produced through the automated method (middle and right)

4. Conclusion

In this study a prototype machine was designed, fabricated, simulated, and tested to aid the local handicraft industry particularly the dried water hyacinth fibers by creating a low-cost, efficient, compact, user-friendly, and produce consistent processed dried stalk. The prototype machine was able to achieve its objective and yield an acceptable result by having fast feed rates and high output rate of processed stalks compared to manual processing methods and techniques the results obtained from the tests yielded a production rate approximately nine times more than 360 stalks per day of the manual production mode if all the slots are used. This compact machine might improve production capability in the water hyacinth handicraft and furniture industry.

5. References

[1] Sullivan P R and Wood R 2012 Water hyacinth (Eichhornia crassipes (Mart.) Solms) seed

[2] Gopal S 2018, January 23 Researchers innovate to make money out of water hyacinth Retrieved from https://india.mongabay.com/2018/01/researchers- innovate-to-generate-money-out-of-water-hyacinth/

[3] Comedis E, Ayran J, Camacho S, De Leon J and Segura R n d Modern eco-friendly containers: Transforming water lilies into proactive environmental product. *2017 DLSU Research Congress Proc.*

[4] Gratton J L 2002 Design and evaluation of an on field hem fibre processing machine Department of Biosystems Engineering University of Manitoba (0-612-76952-6)

[5] Renold J Roller Drive chain Selection and Engineering Information

[6] Guna V, Ilangovan M, Prasad M G and Reddy N 2017 Water hyacinth: A unique source for sustainable materials and products *ACS Sustainable Chem.Eng*, **5**
[7] Ighodalo O, Zoukumor K, Egbon C, Okoh S and Odu K 2011 Processing water hyacinth into biomass briquettes for cooking purposes *J Emerg Trends Eng Appl Sci (JETEAS)* **2** pp 305-307
[8] Keawmanee R 2015 Water hyacinth - The green potential *Thesis* (Rochester Institute)
[9] Ramirez N F, Hernandez S Y, Leon J C, Garcia R V, Lvova L D and Gonzalez G 2015 Composites from water hyacinth (Eichhornea crassipe) and polyester resin Fibers *and Polymers* **16** (1) pp 196–200
[10] Rezania S, Mohanadoss P, Din M F and Chelliapan S 2015 The diverse applications of water hyacinth with main focus on sustainable energy and production for new era: An overview *Renewable and Sustainable Energy Reviews* **41**(41), 943–
[11] Sisante J C 2008 Malabon turning pesky water lilies into export goods *GMA News Online* Retrieved from http://www.gmanetwork.com/news/news/nation/133485/
[12] Coffey S 2008 Water Hyacinth Furniture. Retrieved from apartmenttherapy.com/small-storage-4133
[13] Tacio H D 2009 Water hyacinth ecological value, environmental impacts. Retrieved August 18, 2009, from https://www.gaiadiscovery.com/nature- biodiversity/water-hyacinth-ecological-value- environmental-impacts.html
[14] Villanueva C 2017 How an entrepreneur turned a 'public enemy' into a money-making resource. Rappler. Retrieved from https://www.rappler.com/move- ph/177141- water-hyacinth-processing-community- livelihood

Forced and natural ventilation of a room with a combined solar chimney and Earth - to - Air Heat Exchanger system

V T Nguyen and Y Q Nguyen

Faculty of Engineering, Van Lang University, 45 Nguyen Khac Nhu St., Co Giang Ward, Dist. 1, Ho Chi Minh City, Viet Nam

Email: viet.nguyen@vlu.edu.vn, y.nq@vlu.edu.vn

Abstract. In this study, a passive ventilation and cooling technique which is widely used in sustainable building designs was investigated by a numerical simulation based on computational fluid dynamics (CFD). A system of a solar chimney (SC) and an Earth-to-Air Heat Exchanger (EAHE) is able to ensure thermal comfort in a house by generating an air flow which is induced by the thermal effect. The solar chimney drew ambient air (at 35°C) through the inlet of EAHE; the air flow was cooled by a lower temperature of the underground pipes (at 20°C) before circulating inside the house and leaving through the outlet of the solar chimney. Both the openings of the system were exposed to the atmosphere, and the atmospheric pressure was set at the inlet of the EAHE pipe and the outlet of the solar chimney. The performance of this ventilation and cooling system was analyzed under effects of design parameters, including the gap of the chimney, the diameter and length of the EAHE pipe as well as the solar heat flux applied on the wall of the air channel of the chimney. The results of the assessed parameters, such as temperature inside the house, show that the indoor temperature was reduced to 25°C and 26°C, respectively for the forced ventilation case of 0.2 m/s and natural ventilation case in the configuration with the EAHE pipe length of 11m. The coupled solar chimney and EAHE system can achieve the thermal comfort conditions for both cases of natural and forced ventilation.

Key words: solar chimney, EAHE, CFD, thermal comfort.

1. Introduction

As people have been spending more time indoors, for example about 90% in the US [1], energy consumption for HVAC systems in buildings has increased dramatically. In the USA, the energy for space cooling and heating takes 43% of the total energy consumed in buildings in 2007 [2]. That number in Australia was 42% [2].

To save energy in buildings, using renewable resources for space cooling and ventilation has been studied extensively recently [3-8]. Bansal et al. [3] proposed a solar chimney (SC) for natural ventilation of buildings. This device absorbs solar radiation and creates an air flow through its confined air channel for ventilating the connected building without fans. Furthermore, the air can be cooled prior to being supplied to the building by exchanging heat underground with an Earth – to – Air Heat Exchanger (EAHE) system [4-8]. As the soil temperature is sufficiently lower than the ambient air in summer [4,5], by bypassing the supply air through pipes buried underground, the air can be cooled.

Previous studies have shown that a coupled SC – EAHE system can achieve thermal comfort conditions for residential buildings [5,7] which are defined by the ASHRAE Standard 55 [9]. The research by Li et al. [5] demonstrated that the indoor air temperature and humidity were maintained at

21.3 – 25.1 °C and 50 – 78%, respectively. Elghamry and Hassan [6] reported that the SC – EAHE system in their experiment could reduce the air supply temperature up to 3.5 °C. The thermal comfort index PMV (Predicted Mean Vote) [9] in the study by Yu et al. [7] was within the acceptable range of ±1.0.

For thermal comfort conditions, the ASHRAE Standard 55 [9] also considers the distributions of air temperature and speed in the room. However, in those previous reports, only the average air temperature and humidity of the whole room were considered. In our previous report [10] we examined a coupled SC – EAHE system and focused on the natural ventilation flow induced by the solar chimney in the room. The results showed that an adequate ventilation rate was achieved with the proposed system. Air speed and temperature distributions inside the room were also maintained in thermal comfort ranges.

In this report, we examined the performance of the proposed coupled SC – EAHE system in our previous study [10] for the case of forced ventilation, where the air flow was forced through the system. In previous studies [5,7], the air speed and temperature only measured at one point in the room. This study considered the distributions of air speed and temperature inside the room at different elevations in the occupied zone of the room and then compared with those of the natural ventilation cases.

2. Problem formulation and numerical method

2.1. *System of a solar chimney and an Earth-to-Air Heat Exchanger*

As a passive cooling technique, a combination of a solar chimney and an Earth-to-Air Heat Exchanger ensures to provide a cooling air flow inside a house, as shown in figure 1a. A solar chimney is associated with the house and designed to absorb solar radiation. As the temperature of the chimney surface increases, the absorbed heat is transferred into the air in the channel of the chimney. The air inside the chimney is heated, the hot air rises, and an induced air flow is drawn through the EAHE inlet. Exploiting the potential difference between ambient and underground temperatures in hot climates, an EAHE cools the air flow during its passage through EAHE pipes. The cooling process takes place between the pipe surfaces and the air flow. On the assumption that the house is air-tight, only ambient air through the EAHE pipe circulates inside the house. Cooling the air flow before supplying into the house, a coupled SC-EAHE ensures the house is ventilated and cooled.

2.2. *Numerical model*

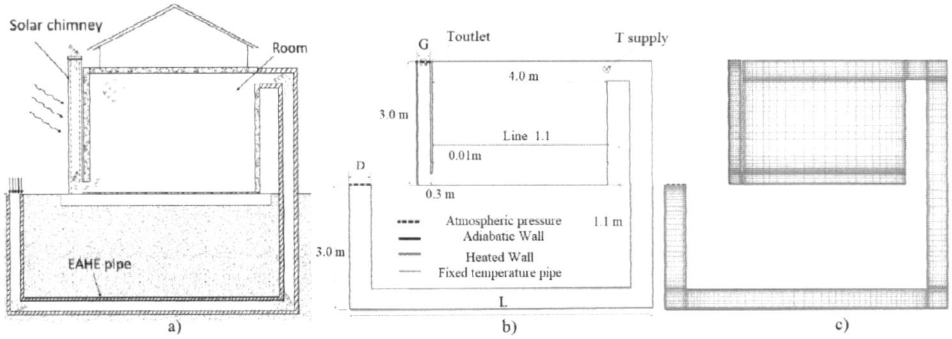

Figure 1. Schematic of SC-EAHE (a), computational domain (b), and mesh (c)

A CFD model was applied to investigate the behaviors of the air flow and heat transfer within the SC-EAHE system and the house. The setup was similar to those in previous researches [4,6,10]. A two-dimensional domain in the vertical plane was employed (figure 1b). Accordingly, the air flow and heat transfer were assumed to be steady and two- dimensional. Both modes of convection and radiation were integrated into the simulation of the heat transfer on the pipe and solar chimney surfaces. For simplicity, the process of heat conduction through the surfaces of the pipes was not directly simulated but modeled by a uniform temperature on the pipe surfaces. Similarly, a uniform heat flux was applied on the wall of the air channel of the chimney. The house walls were assumed to be adiabatic. A uniform vertical velocity profile was assumed to enter at the inlet of the EAHE whereas the air was assumed to leave the domain at the outlet of the solar chimney at atmospheric pressure.

Under the above assumptions, using the Reynolds – Averaged Navier Stokes equations [11,12], the problem was governed by the equations of conservation of the mass, momentum, and energy of the flow.

$$\frac{\partial u_j}{\partial x_j} = 0 \tag{1}$$

$$\frac{\partial (u_i u_j)}{\partial x_j} = -\frac{1}{\rho}\frac{\partial p}{\partial x_i} - g_i \beta (T - T_0) + \frac{\partial}{\partial x_j}\left(\nu \frac{\partial u_i}{\partial x_j} - \underline{u_i' u_j'}\right) \tag{2}$$

$$\frac{\partial (T u_j)}{\partial x_j} = \frac{\partial}{\partial x_j}\left(\frac{\nu}{Pr}\frac{\partial T}{\partial x_j} - \underline{T' u_j'}\right) \tag{3}$$

In this system of equations, u and T show time-averaged components of the velocity and temperature. u' and T' are the velocity and temperature fluctuations. __ represents the time – averaged operator. p, ν, ρ, and Pr are respectively the pressure, air kinematic viscosity, density, and the Prandtl number. g is the gravitational acceleration and β is the air thermal expansion coefficient. The Boussinesq approximation was applied to Equation (2). This approximation was used for the air density variation with temperature [12,13,14]. The turbulence stress $\underline{u_i' u_j'}$ in Equation (2) and the turbulent heat flux $\underline{T' u_j'}$ in Equation (3) were solved with the low – Reynolds number k–ε model, which is the standard k–ε model with modification for low -Reynolds number effects by a damping coefficient for the eddy viscosity. This model was also employed in previous studies of solar chimneys [11,14] because it allowed to obtain the fastest convergence and the best agreement with the experiment by Burek and Habed [15].

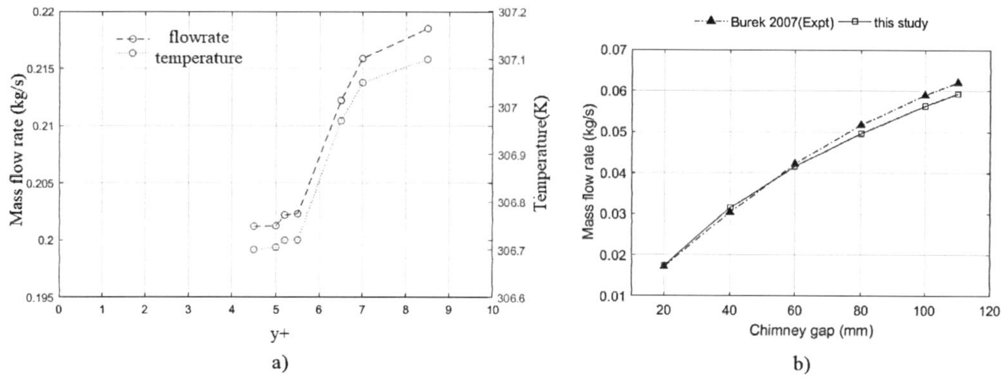

Figure 2. Examination of the mesh (a) and validation of the numerical model (b)

The Finite Volume Method on a structured rectangular mesh (figure 2c) was used to discretize the governing equations, Equation (1) to (3), using Ansys Fluent CFD code. The geometric discretization focused on the proximity of the walls. High mesh densities were applied near the walls where the variation of the flow variables was strong. To respond to reliability and quality issues for the mesh generation, the variation of the induced flow rate and the air temperature at the outlet of the chimney were examined when increasing the mesh density. The flow rate and temperature at the outlet were plotted versus y^+, the maximum non – dimensional distance of the first grid point from the surfaces of the pipe and the air channel. In the mesh independence test, the value of y^+ varied from 8.5 to 4.5, which corresponded to the discretization mesh in figure 1c was finer near the wall of the chimney and EAHE pipes. The differences between the air temperature and the induced flow rate computed with y^+ =4.5 and the ones computed with y^+ =5.0 were 0.5% and 0.01%, respectively. The corresponding differences between y^+ =5.0 and y^+ =5.5 were 1.55% and 0.11%. As a result, a mesh with $y^+ \leq 5$ was sufficient to give acceptable results for most simulations with different inlet velocities, as shown in figure 2a.

A validation of the above setups was verified with the data from the experiment by Burek and Habeb [15]. The flow rate at the outlet of the chimney was compared with the measured one in the case of the heat flux of 800 W/m^2. For the chimney gap in the range of 20 to 110 mm, the discrepancy between the predicted and measured results was less 5 %, as shown in figure 2b. As a result, this numerical model can be applied for this problem.

3. Results and discussions

The aim of this study is to investigate factors that affect the performance of the system SC-EAHE in forced ventilation compared to the natural case [10]. In the computational model, airflow at 35°C entering the EAHE was examined in natural ventilation regime and three different forced ventilation regimes in which the velocity of the airflow at the inlet of the EAHE pipe were 0.2, 0.4 and 0.8 m/s. The temperature of the pipe walls was set at 25°C. Although the underground temperature was about 20°C [4,5], a higher temperature was fixed for EAHE pipe due to the thermal resistance of the pipe thickness.

Parameters consisting of the induced air flow rate, air temperature supplied to the room, and the averaged values of the air speed and temperature along the line at the height of 1.1 m [9] were assessed with regard to the variation of four factors: the heat flux, the gap of the solar chimney G, the length L and diameter of the EAHE pipe D, as denoted in figure 1b.

3.1. *Effects of the applied heat flux on the system performance*

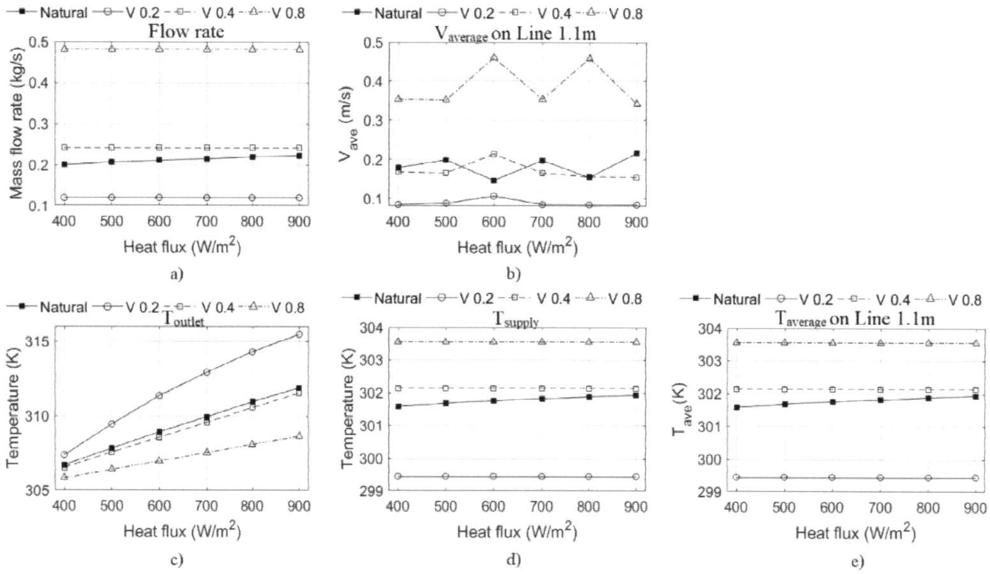

Figure 3. Performance of the system as the heat flux changed.

For the effects of the heat flux, a range from 400 to 900 W/m^2 was inspected when the inlet velocity changed from 0.2, 0.4 to 0.8 m/s. The setup was similar to those of the natural configuration [10]. In particular, G=0.3 m, D=0.5 m, and L=7.0 m. As the velocities were imposed at the EAHE inlet, the flow rates of the three flow regimes were proportional to their speeds in figure 3a. The flow rates in the three cases of forced ventilation were independent of the heat flux while there was an increase of 10% in the flow rate for the case of natural ventilation when the heat flux increased from 400 to 900 W/m^2[10]. An increase of about 30% was reported by Chen et al. [16], for a stand-alone chimney, when the heat flux varied from 400 to 600 W/m^2.

The number of times in one hour the air in the room, ACH, was calculated for both natural and forced ventilation cases. With the width of 4.0 m for the room and the width of 1.0 m for the chimney in figure

1a, the ACH was 7.5, 15, and 30, respectively, for the inlet velocity of 0.2, 0.4, and 0.8 m/s. For the natural ventilation case, the ACH was in the range of 12.5-14 [10]. Compared to the ACH of 3.0 for residential buildings, these values were far above [9].

In figure 3b, similarly to the flow rate, the air speed along the line at 1.1 m from the floor changes with the heat flux. For both natural and forced ventilation cases, the speeds were in the range from 0.1 to 0.45 m/s. It is within the comfort range [9].

The air temperature at the SC outlet, T_{outlet}, increases with the heat flux, as shown in figure 3c. The highest temperature at the outlet was recorded as the inlet velocity was the lowest, which was the case of the velocity of 0.2m/s. In contrast, the lowest T_{outlet} corresponds to the case of the velocity of 0.8 m/s. It is due to a large amount of the heat from the absorbed surface transferred into the air when the flow circulates slowly along the chimney. A similarity in the T_{outlet} results between the natural case and the case of 0.4 m/s velocity derives from the nearly 0.4 m/s air speed in the natural ventilation.

According to Li et al. [5], the air is cooled up to 10° K along a 57 m long EAHE pipe before supplying into the room. For such a 7 m long EAHE pipe, the reduction of the air temperature during its passage through EAHE pipes is expectedly lower in figure 5c. In contrast to a slight increase in T_{supply} for the natural ventilation, T_{supply} in forced ventilation cases is invariant within the range of the heat flux. Moreover, in figure 5d and 5e, the temperature along the line at 1.1 m from the floor is identical to the T_{supply}, which is due to no heat source existing inside the room, and the distribution of the temperature is uniform inside the room [10].

3.2. *Effects of the chimney gap on the system performance*

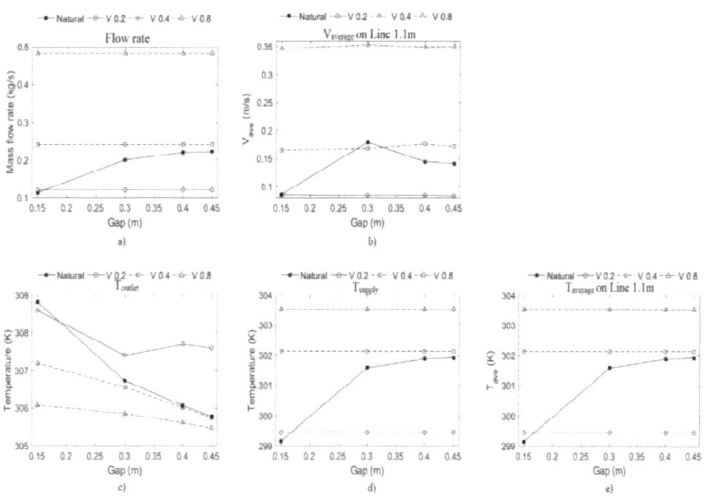

Figure 4. Performance of the system as the chimney gap changed.

The setup for the test on the chimney gap was as follows: the heat flux of 400 W/m^2, D= 0.5 m, and L= 0.7 m while the gap varied from 0.15 to 0.45 m. For a gap from 0.15 to 0.3 m, in figure 4a, the flow rate increases with the gap in the case of natural ventilation. For the gap above 0.3 m, the flow rate increases insignificantly, which is also reported by Khanal and Lei [17]. A slowdown in the increase in the flow rate is due to a reverse flow at the outlet of the chimney by examining the flow fields. In terms of the forced ventilation cases, the flow rate remains unchanged within the current range of the gap.

In figure 4b, the mean air speed at the height of 1.1 m increases with the gap from 0.15 to 0.3 m and slightly drops for the gap above 0.3 m in the case of natural ventilation. For the forced ventilation cases, the air speed is nearly unchanged although there is a small impact of the reverse flow. The air speeds satisfied comfort conditions [9]. The according ACH varies from 7.0 to 14.0 for natural setup while those for forced setup are invariant as 7.5, 15, and 30, respectively for velocity 0.2, 0.4, and 0.8 m/s.

For both natural and forced ventilation cases, T_{outlet} decreases with the gap, in figure 4c. For the inlet velocity of 0.8 m/s, although the temperature of the air supplied into the room is the highest, the temperature at the outlet is the lowest. It is because the time for the heat transfer into the air is reduced. Accordingly, a low velocity at the inlet results in a high temperature at the outlet.

The temperature along the line at the height of 1.1 m is similar to T_{supply}, as shown in figure 4d and 4e. In the forced ventilation cases, they are fixed when changing the gap. It is seen that their value depends on the air speed at the inlet. In the natural ventilation case, T_{supply} and the temperature at the height of 1.1 m increases with the gap.

3.3. *Effects of the EAHE pipe length on the system performance*

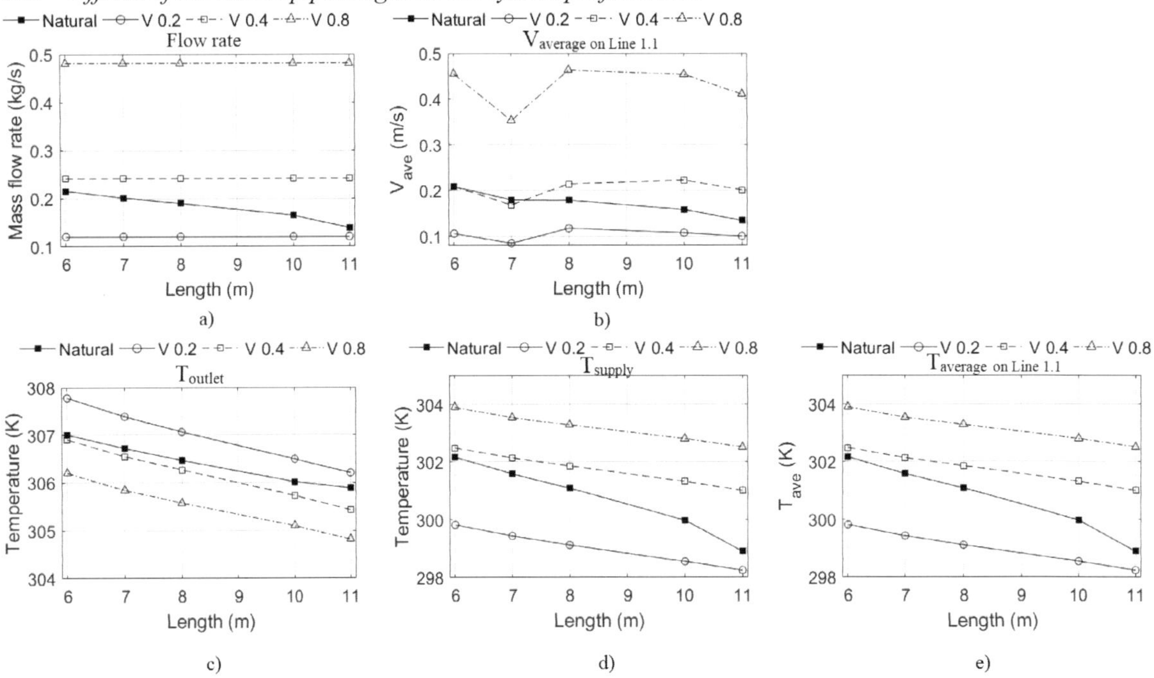

Figure 5. Performance of the system as the EAHE pipe length changed.

To examine the effects of the EAHE length, the test was used in the range of the length from 0.6 m to 11.0 m. Other dimensions were set as follows: G= 0.3 m, D= 0.5m, and the heat flux of 400 W/m^2. The flow resistance only influenced the flow rate in the natural ventilation case in which the flow rate was reduced when the length was extended (figure 5a). For the forced ventilation case, the flow rates are still invariant and proportional to the value of the inlet velocity. The distribution of the velocity fields at the height of 1.1 m tends to decrease in figure 5b. The temperatures at the outlet, the supply position, and line 1.1 decrease with the length, which is due to the heat exchanger within the EAHE pipes in a longer process when the pipe length increases, as shown in figure 5c, 5d, and 5e. The air speeds and the air temperature are suitable for comfort conditions.

3.4. *Effects of the EAHE pipe diameter on the system performance*

The test was set as follows: the heat flux of 400 W/m^2, L = 0.7 m, G= 0.3 m and the diameter D varies from 0.25 to 0.6 m. An increase in the flow rate with the diameter is reported in the work of Serageldin et al. [8]. This is also illustrated in figure 6a. The flow rate increases with the diameter for both natural and forced ventilation cases. It can be explained by the fact that the air flow passes through the EAHE pipes with less flow resistance when the diameter increases in the natural case. The ACH was calculated

IOP Conf. Series: Materials Science and Engineering 1109 (2021) 012009 doi:10.1088/1757-899X/1109/1/012009

and its range was from 4 to 14 and from 7.5 to 36, respectively, for the natural and forced ventilation cases.

Figure 6b shows the effects of the diameter on the air speeds. In figure 6c, 6d, and 6e, the temperature at the height of 1.1 m is similar to the T_{supply}. They increase with the pipe diameter and achieve a high value at high inlet speeds. However, lower inlet speeds result in higher temperatures at the outlet. The air speeds and the air temperatures satisfied the comfort conditions.

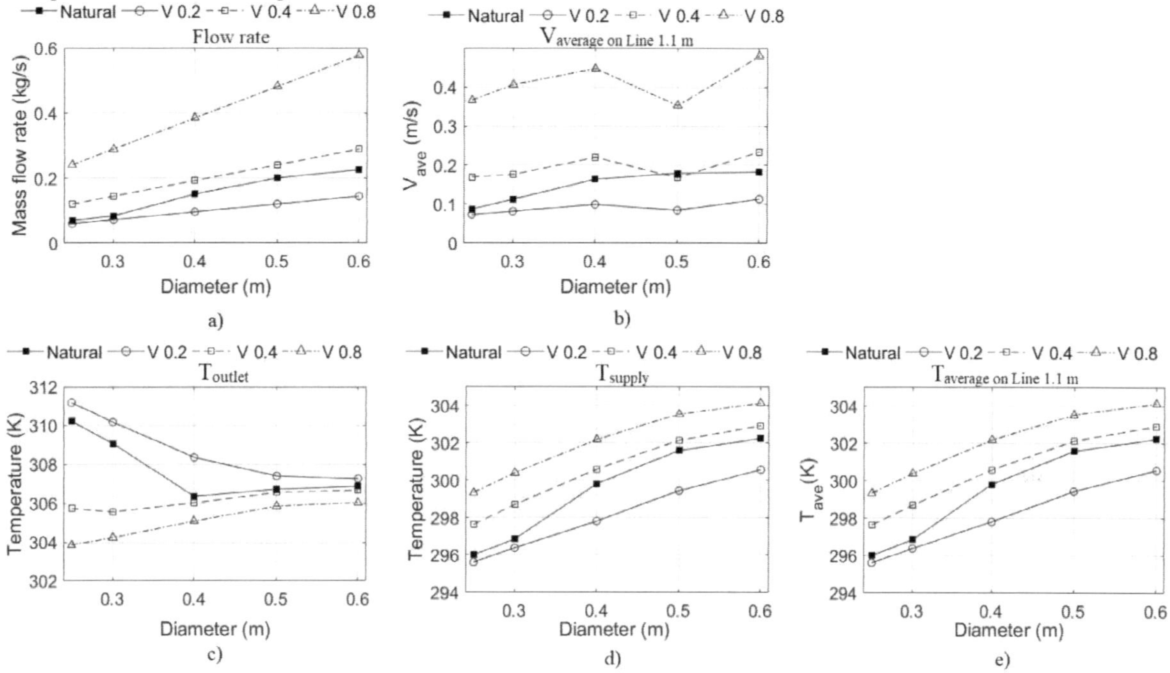

Figure 6. Performance of the system as the EAHE pipe diameter changed.

4. Conclusions

In this study, the configuration of the coupled solar chimney and Earth – to – Air Heat Exchanger system can satisfy the requirements for indoor thermal conditions by both natural and forced ventilation. The factors such as the heat flux, the gap of the solar chimney, the length and the diameter of the EAHE pipe were investigated. The results can be shown as follows:

For natural ventilation:

- The flow rate increases with the heat flux, the gap of the chimney, and the diameter of the EAHE pipe but decreases with the length of the EAHE pipe.
- The temperature supplied into the house slightly increases with heat flux, increases with the chimney gap, and the diameter of the EAHE pipe but decreases with the length of the EAHE pipe.
- The velocity at the height of 1.1 m increases with the heat flux and pipe diameter, decreases with the pipe length, increases with the chimney gap lower than 0.3 m but decreases with the chimney gap above 0.3 m.

For forced ventilation:

- The temperature supplied into the house is insensible to the heat flux, and the gap of the chimney increases with the diameter of the EAHE pipe but decreases with the length of the EAHE pipe.
- The velocity at the height of 1.1 m increases with the pipe diameter, decreases with the pipe length and remains unchanged compared with the heat flux and the chimney gap.

The air speeds and the temperature at the height of 1.1 m from the floor are within the comfort ranges. This work demonstrates an application of CFD in designing green buildings by considering both passive and active solutions for ventilation and cooling. In future studies, numerical methods or artificial intelligence can be used to optimize the distribution of the temperature in buildings.

References

[1] U.S. Environmental Protection Agency 1989 *Report to Congress on indoor air quality: Volume 2* (EPA/400/1-89/001C. Washington DC)

[2] Allouhi A, El Fouih Y, Kousksou T, Jamil A, Zeraouli Y and Mourad Y 2015 Energy consumption and efficiency in buildings: Current status and future trends *Journal of Cleaner Production* **109** 118-130.

[3] Bansal N K, Mathur R and Bhandari M S 1993 Solar chimney for enhanced stack ventilation *Building and Environment* **28** (3) 373–377

[4] Maerefat M and Haghighi A P 2010 Passive cooling of buildings by using integrated earth to air heat exchanger and solar chimney *Renewable Energy* **35** 2316–2324

[5] Li H, Yu Y, Niu F, Shafik M and Chen B 2014 Performance of a coupled cooling system with earth-to-air heat exchanger and solar chimney *Renewable Energy* **62** 468–477

[6] Elghamry R and Hassan H 2020 An experimental work on the impact of new combinations of solar chimney, photovoltaic and geothermal air tube on building cooling and ventilation *Solar Energy* **205** 142- 153

[7] Yu Y, Li H, Niu F and Yu D 2014 Investigation of a coupled geothermal cooling system with earth tube and solar chimney *Applied Energy* **114** 209–217

[8] Serageldin A A, Abdelrahman A K and Ookawara S 2018 Parametric study and optimization of a solar chimney passive ventilation system coupled with an earth-to-air heat exchanger *Sustainable Energy Technologies and Assessments* **30** 263–278

[9] ASHRAE 2010 ASHRAE Standard 55 – thermal environmental conditions for human occupancy,

[10] Nguyen V T, Nguyen Y Q and Huynh T N 2020 Natural ventilation and cooling of a house with a solar chimney coupled with an earth – to – air heat exchanger *The Int. Conf. on Modern Mechanics and Applications*

[11] Zamora B and Kaiser A S 2009 Optimum wall-to-wall spacing in solar chimney shaped channels in natural convection by numerical investigation *Applied Thermal Engineering* **29** (4) 762–769

[12] Kong J, Niu J and Lei C 2020 A CFD based approach for determining the optimum inclination angle of a roof-top solar chimney for building ventilation *Solar Energy* **198** 555–569

[13] Khanal R and Lei C 2014 An experimental investigation of an inclined passive wall solar chimney for natural ventilation *Solar Energy* **107** 461–474

[14] Zavala-Guillén I, Xamán J, Hernández-Pérez I, Hernández-Lopéz I, Gijón-Rivera M and Chávez Y 2018 Numerical study of the optimum width of 2a diurnal double air-channel solar chimney *Energy* **147** 403–417

[15] Burek S A M and Habeb A 2007 Air flow and thermal efficiency characteristics in solar chimneys and Trombe Walls *Energy and Buildings* **39** (2) 128–135

[16] Chen Z D, Bandopadhayay P, Halldorsson J, Byrjalsen C, Heiselberg P and Li Y 2003 An experimental investigation of a solar chimney model with uniform wall heat flux *Building and Environment* **38** (7) 893–906

[17] Khanal R and Lei C 2012 Flow reversal effects on buoyancy induced air flow in a solar chimney *Solar Energy* **86** (9) 2783–2794

IOP Conf. Series: Materials Science and Engineering 1109 (2021) 012010 doi:10.1088/1757-899X/1109/1/012010

Identification of key sugarcane harvester features using the Analytic Hierarchy Process

J A Gonzaga[1], E Dela Cruz[1], N S Lopez[1]

[1] Mechanical Engineering Department, De La Salle University, Manila, Philippines

Email: jeremias.gonzaga@dlsu.edu.ph

Abstract. The Philippine sugarcane industry is worth roughly 88 billion pesos and employs around 700,000 Filipinos. However, the competitiveness of the local industry ($0.17 per lb sugar) is currently under threat from international producers ($0.14 per lb sugar) because of the upcoming ASEAN 2015 integration. One of the major blockages to cost effectiveness is the low productivity of the local industry. This may be addressed with the use of a semi-automated mechanical harvester. As a preliminary study, the present work captures the expert preferences of various stakeholders in the selection of a semi-automated mechanical harvester. This is done using the Analytic Hierarchy Process (AHP) surveying tool. The study assumes the following important factors: a) Ease of use, b) Productivity, c) Adaptability to different terrains, d) with Lifter, and e) with Topper. Using a survey with a series of pairwise comparison questions, it is able to quantify even qualitative factors affecting decision making. The priorities of the surveyed stakeholders are averaged and discussed in the results section. Widely used in policy making and planning, the paper introduces the use of the AHP tool to commercial machinery design. The model may be expanded and improved to consider other key factors.

1. Introduction

The Philippine Sugarcane Industry is now facing a dilemma which has big opportunities and challenges ahead. Sugarcane is the fourth largest crop in the Philippines (after coconut, corn, and rice). It easily gives employment to around 700,000 Filipinos. The sugarcane industry contributes no less than P70 billion annually to Philippine economy [1]. In the Philippines, sugarcane is planted in about 422,500 hectares with a labor force of 62,000 farmers. The archipelago has 23 sugar producing provinces. As of Crop Year 2009-2010, 29 mills are operational and divided as follows: 6 mills in Luzon, 13 mills in Negros, 4 mills in Panay, 3 mills in Eastern Visayas and 3 mills in Mindanao [2]. In terms of the farm sizes, 75% of farms have sizes less than 5 hectares and another 11% have sizes of 5 to 10 hectares. Among the remaining farms, 11% have sizes 10 to 50 hectares, a mere 2% have sizes of 50 to 100 hectares, while only 1% has a size of over 100 hectares.

The problem arises during the harvesting season since most of the mechanical harvesters that are available are not suitable to the terrains in our country. Most of the advance designed harvesters are basically suited to developed countries which have wide and flat farm areas. In the Philippines manual harvesting of sugarcane is still very much in practice because of the adaptability issue. Manual harvesting is very tedious and labor-intensive. Because of this, the cost of production for sugarcane becomes high. The efficiency in the harvesting industry is measured by the number of crops one manages to harvest within a given span of time. An increase in efficiency in the harvesting process will lead to lower production costs which will eventually lead to a bigger profit margin. Aside from the harvesting problem, the Philippine Sugar Industry is also facing another challenge because tariff on imported sugar under *ASEAN Trade in Goods Agreement* (ATIGA) is now at 28% [1]. The ASEAN Free Trade Area (AFTA) will basically eliminate the tariff, making imported sugar cheaper compared to locally produced sugar. Because the cost of production in the Philippines is a little higher compared to the neighbouring competitors, possible entry of cheaper imported sugar to cater the needs of 96 million Filipinos is considered the biggest threat to the country's sugar producers. The cost of production of sugarcane in international markets is set at \$0.14, while the cost of production of sugarcane in the Philippines is at least \$0.17. Currently, the government has been imposing tariffs on imported sugar to protect the local producers. This inefficiency in harvesting contributes to further increase in costs of production. A solution to this would be to automate some parts of the harvesting process. According to [3], the Philippine agricultural sector's plan to sustain growth is to increase the farmers' access to technology.

In an interview with experts and officials from the Sugar Regulatory Administration (SRA), the authors learned that a mechanical harvester is almost always expected to have a lifter. The stalks sag a lot, especially due to the wet local climate and the characteristics of local sugarcane stalks (long and thin). Also, it has been reported that there are plenty of unused mechanical harvesters due to issues on adaptability to different terrains (i.e. sloped, rocky, or muddy). Lastly, having a topper would also be a huge help to the harvesters but the variability on the height of different sugarcane species makes this challenging. Having a mix of quantitative and qualitative factors makes it a challenging task to design an equipment that will have commercial success. The following is a case study demonstrating the capability of the Analytic Hierarchy Process (AHP) methodology to capture expert decision making with regards to machine design. For a designer, it is straightforward to produce cost effective and efficient machines, but it is quite a challenge to guarantee that the intended user will find value in it and actually use it. Capturing customer priorities using AHP before diving into the actual design process ensures that the designer fully understands the needs of the customer.

Examples of MCDM work conducted similarly for agriculture studies include using conflict analysis model to analyze crop choice of Philippine farmers [4]; using multi-attribute utility theory to study collaborative farming in Japan [5]; and using SWOT analysis and AHP to understand the decision-making of the local sugar industry in Algeria [6].

Using AHP, expert decision making of stakeholders may be captured. Both quantitative and qualitative factors may be quantified and converted into a single criterion, which will aid in the design of an appropriate harvester. The goal of this preliminary study is to describe a harvester design that would provide value to the local sugarcane industry. The paper is outlined as follows: a brief background on sugarcane and mechanical harvesters are discussed in the next section, to be followed by the methods, results, conclusions and acknowledgements in the proceeding sections.

2. Background: Sugarcane harvesting and mechanization

2.1 Sugarcane

Saccharum officinarum, the scientific name of sugarcane, is a member of the Poaceae or grass family. It is widely cultivated as it provides an approximate of 70% of the world's sugar. Sugarcane, also known as the noble cane, accumulates a significant amount of sucrose in its stalk, which is then harvested and processed to produce sugar crystals [7].

2.2 *Sugarcane in the Philippines*

The following sugar cane varieties are currently available in the Philippines: PHIL 99-1793, PHIL 97-2041, PHIL 97-1123, PHIL 97-0693, PHIL 94-0913, PHIL 93-3849, and PHIL 93-3155 [8]. Varieties are described by their stalk characteristics, stalk diameter, millable stalk, average yield, potential yield, and districts with high yield of a certain breed or variety. Indicated varieties are commercial hybrid cultivars; thus, were specially bred to maximize production.

Some common properties and characteristics of the sugar cane varieties in the Philippines are cylindrical, ovate, barrel and conoidal in shape; small flat, semi-bulging to flat and corky cracks/patches; bud grooves and no bud groove; dark yellow-purple stalks, green-purple stalks and brown-purple stalks; green –yellow and yellow-green internodes. The average diameter size, millable stalks and potential yield are 2.81cm, 2.31 stalks per stool at harvest and 2.34 LKg/TC; 132.68 TC/Ha, respectively.

Table 1. Levels of mechanization of various crops in the Philippines. [9]

Operation	Rice & Corn	Vegetables, Legumes & Root Crops	Coconut	Sugarcane	Fruits	Fiber Crops
Land Preparation	Intermediate to High	Low		Intermediate to High	Low	Low
Planting/Transplanting	Low	Low	Low	Intermediate to High	Low	Low
Crop Care Cultivating	Low	Low	Low	Low to High	Low	Low
Harvesting	Low	Low	Low	Low		Low
Milling/Village level processing	High	Low	Low	Low	Low	Low

2.3 *Mechanization and Sugarcane Harvesting Processes Locally*

Hand cutting is the most common harvesting method utilized throughout the world; however, more developed nations such as the USA, Thailand, and Brazil have adopted mechanical harvesting to increase the harvesting yield. After harvesting, the cane is loaded into a trailer or truck, where it is transported to a mill. Along with other local agricultural processes, the Philippines have low levels of harvesting mechanization for sugar cane crops. In Table 1, the levels of mechanization for various crops in the Philippines are presented.

The predominance of manual labor and lack of development in agricultural technology are the reasons for the low mechanization of the majority of agricultural operations. Table 1 above indicated the high level of mechanization of pre-harvesting sugarcane processes, which may be due to the presence of imported tractors utilized by large-scale farms; however, as stated by T.D. Lopez (personal communication, April 2016) in an online interview conducted by the authors, most Filipino farmers cannot afford these machines as the cheapest available in the market still exceeds Php 1.8 million. Furthermore, the lack of mechanization for sugarcane harvesting processes reduces the production yield as it cannot sustain the level of production yield that pre-harvesting processes possess. Reviews of many of the models that were developed at SASRI (South African Sugarcane Research Institute) on primarily the development of harvesting machines were taken down over a span of 20 years.

Fully automated sugarcane harvesters have limitations with regards to the terrain they are deployed on. In addition to their limitations, these harvesters are very expensive and can only be afforded by a few companies. Innovations have been made which led to the development of the semi-automated sugarcane harvester. Semi-automated sugarcane harvesters have addressed most of the requirements in the harvesting process such as proper consideration of the various slopes possibly encountered, the different types of cane that is harvested, as well as the topping requirements.

IOP Conf. Series: Materials Science and Engineering 1109 (2021) 012010 doi:10.1088/1757-899X/1109/1/012010

3 Methods

3.1 Analytic Hierarchy Process

The Analytic Hierarchy Process (AHP) methodology is used to design the survey questionnaire and to process the responses. AHP is a multi-criteria decision making method that was originally developed by Thomas Saaty [10]. It is a method to derive ratio scales from paired comparisons. The method is particularly useful when subjective opinions such as satisfaction and preference need to be quantified. The AHP methodology should capture how the decision makers in the sugarcane industry think, teach the authors how to prioritize over the numerous design considerations (e.g. ergonomics, machine performance and functions), and determine the most important factors contributing to the most effective design. All of these are necessary for the development of a potential commercial product. The methodology starts with a decision structure, which summarizes the factors that could affect the decision making of the stakeholders. The decision superstructure is shown on Figure 1 and is based on prior consultation with other stakeholders.

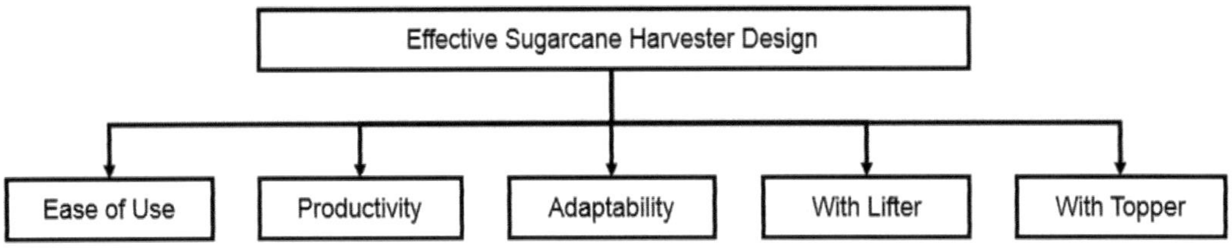

Figure 1. Decision superstructure based on preliminary consultation with stakeholders.

The survey questionnaire is designed as a series of pairwise comparisons between the different factors found in the superstructure. The comparisons require the stakeholder to identify which factor is more important and by how much (in a scale of 1 to 9, with 1 referring to equal importance and 9 referring to extreme importance). The responses are then converted into a matrix, as shown on Figure 2. If the factor on the vertical axis is deemed more important than the factor on the horizontal axis, x_{il} becomes the actual judgement value (from 1 to 9). On the other hand, if the factor on the horizontal axis is deemed more important than the factor on the vertical axis, x_{il} becomes the reciprocal of the actual judgement value. Upon completion of the matrix, the columns are normalized by dividing each term of the column to the sum of the column. After normalizing, the importance of each factor is called the priority vector and is determined by averaging each row (see Figure 3). The priority vectors of all respondents are aggregated by getting the arithmetic mean. This then becomes the importance rating of each factor. The readers are directed to Saaty [10] and Seyhan & Mehpare [11] for a more in-depth discussion on the AHP methodology.

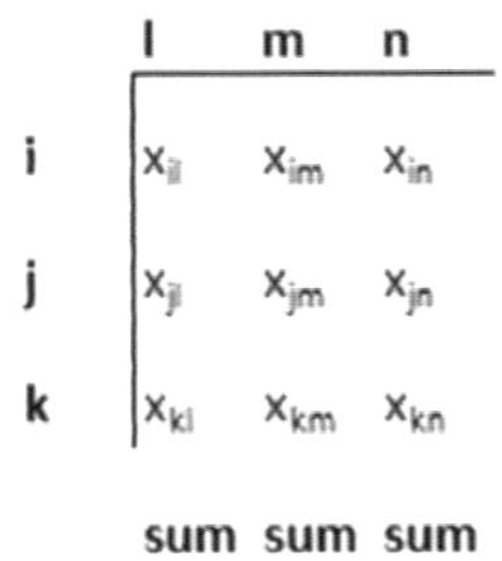

Figure 2. Matrix summarizing the survey responses.

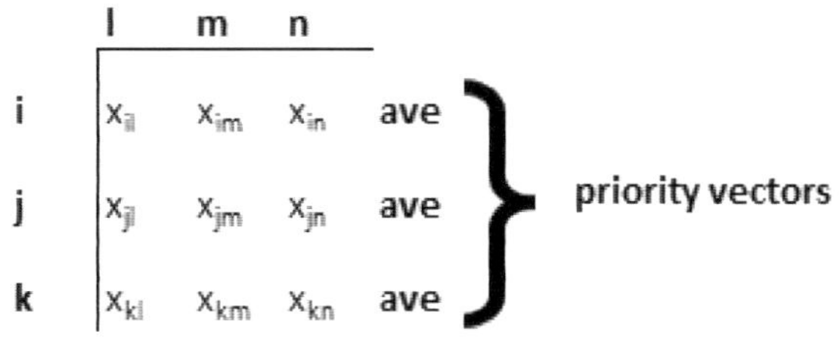

Figure 3. The average of each row is taken after normalizing each column. The average of each row is the priority vector of the particular factor.

3.2 Definition of factors

The following were the factors considered in the study and their descriptions.

3.2.1 Ease of Use. This refers to the user-friendliness of the machine. The machine should be simple to operate and should not require extensive training for the operator. The machine should also not cause body pain/strain on the operator.

3.2.2 Productivity. The machine should significantly reduce the harvesting time and manpower requirement.

3.2.3 Adaptability. The machine should be easily adaptable to most terrains.

3.2.4 With Lifter. The machine should come with a lifter to help straighten the canes before cutting.

3.2.5 With Topper. The machine should be equipped with a topper to cut.

The factors were shortlisted based on preceding interviews with stakeholders and literature review. A conscious effort was also done to ensure that they are simple enough for all kinds of respondents to understand.

3.3 Conduct of expert survey

The experts were a mix of planters, farm owners, engineers and technical service providers from the sugarcane industry. The survey was added as a side agenda of the recently concluded committee meeting on mechanical harvesting of the Sugar Regulatory Administration. The experts were presented with a 10-question survey. The questions were all pairwise comparisons of all considered factors. For each question, the expert had to select which of the two factors is more important than the other, and by how much degree (from equally important to extremely more important).

Table 2. Profiles of experts

	Age	Role	Had used mechanical harvesting before?	Location
Expert 1	51	Engineering	No	<no info>
Expert 2	56	Operations	Yes	Pampanga
Expert 3	46	Operations	No	Negros Occidental
Expert 4	<no info>	Engineering	No	Negros Occidental
Expert 5	39	Operations	No	Isabela

4 Results and discussion

The survey results validated the expert opinion of the interviewees from SRA. Figure 4 summarizes the results. The prioritization rating from the experts is translated into percent importance of each feature of the harvester.

It is definitely a requirement to have a lifter as 3 out of 5 experts selected it as their top priority. It is almost twice as important as all the other factors. Reiterating the point brought up by the experts from SRA, the wet local climate causes the stalks to sag a lot. This sagging affects ease and efficiency of harvesting a lot. A mechanical harvester is almost always expected to have a lifter to ensure the stalks are vertical when they are severed, otherwise, additional manpower will be required to assist in the lifting of the stalks. Adaptability is also the second most important factor. Likewise, 3 out of 5 respondents agreed that it should be the second priority. The variability of Philippine terrain makes it challenging to design a single machine that would be compatible to all conditions. Productivity, surprisingly, is not the main concern of the stakeholders. It is possible that this is a consequence of having more technical than business-minded respondents (2 from Engineering, 3 from Operations, 0 from Business Management). To recall, a possible reason as well is the fact that the major bottleneck in harvesting is actually the availability of trucks for hauling. Too much increase in productivity would be useless if the other problems (logistics) will not be addressed. Having a topper function was the second least important, while ease of use was ranked the last priority of the respondents. Not to be mistaken that farmer welfare is not important, this may imply that a sugarcane harvester has a naturally simple operation. There is practically no way to make it difficult to operate, and thus, designers can focus on other more important aspects.

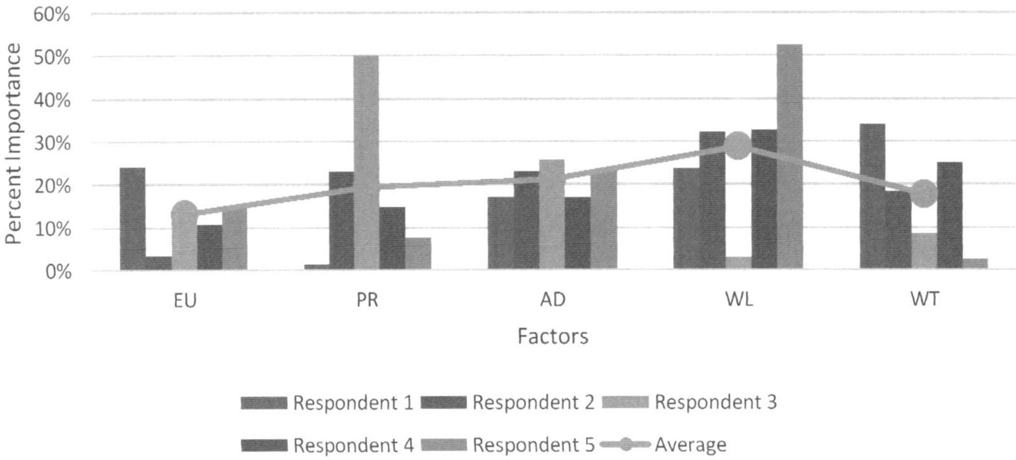

Figure 4. Summary of survey results. EU - Ease of Use; PR - Productivity; AD - Adaptability; WL - With Lifter; WT - With Topper.

5 Conclusion

The present study demonstrated that the AHP methodology can effectively capture expert opinion on decision making for machine design. The information derived from the interview with SRA officials agreed with the results of the AHP survey. The most important factor in a semi-automated mechanical sugarcane harvester is having a lifter (28.7%), followed by adaptability (21.2%), productivity (19.4%), having a topper (17.6%) and ease of use (13.1%). The model may be further extended to consider more factors. The next step would be to evaluate if such a design can be developed at an affordable cost.

Acknowledgement

The authors would like to acknowledge the University Research Coordination Office of De La Salle University for providing financial support to this study, and the Sugar Regulatory Administration for the invaluable support in the conduct of this study.

References

[1] Doronila A 22 December 2014 The Philippine Daily Inquirer, Philippine Sugar Millers Association, Inc. Retrieved March 18, 2016

[2] www.sra.gov.ph/the-philippine-sugarcane-industry-challenges-and-opportunities. Retrieved March 16, 2016

[3] Festejo-Abeleda M A 5 July 2017 *Philippine development plan 2017 to 2022 for agriculture, forestry and fishery sector: Sustaining inclusive economic growth.* Retrieved from Food and Fertilizer Technology Center for the Asian and Pacific Region: http://ap.fftc.agnet.org/ap_db.php?id=773

[4] Van Huylenbroeck G and Damasco-Tagarino D 1998 Analysing crop choice of Philippine vegetable farmers with multicriteria analysis *Journal of Multi-Criteria Decision Analysis,* 7, 160-168.

[5] Hayashida T, Nishizaki I, Ueda Y and Honda H 2012 Multi-criteria evaluation for collaborative circulating farming with collective operations between arable and cattle farmers *Journal of Multi-Criteria Decision Analysis.* Doi: 10.1002/mcda.1473.

[6] Achabou M A and Tozanli S 2009 The institutional environment in the strategic decision of the emerging companies: the case of the Algerian sugar industry *Journal of Multi-Criteria Decision Analysis,* 16, 139-148.

[7] Kew.org, (2015). Saccharumofficinarum (sugar cane) | Plants & Fungi At Kew. Available at: http://www.kew.org/science-conservation/plants-fungi/saccharum-officinarum-sugar-cane. Retrieved on March 27 2015.

[8] Sra.gov.ph, 'Sugarcane Varieties | SRA'. N.p., 2015. Web. 17 Apr. 2015

[9] Amongo R M, Amongo L D and Lorona M V 2011 Mechanizing Philippine Agriculture

[10] Saaty T L 1980 The analytic hierarchy process. McGraw-Hill, Inc, New York.

[11] Seyhan S and Mehpare T 2010 The analytic hierarchy process and analytic network process: an overview of applications *Management Decision* 48: 775–808.

Influence of outlet/inlet area ratio on performance of a vertical solar chimney for natural ventilation of buildings

Y Q Nguyen and T N Huynh

Faculty of Engineering, Van Lang University, 45 Nguyen Khac Nhu St., Co Giang Ward, Dist. 1, Ho Chi Minh City, Viet Nam
Email: y.nq@vlu.edu.vn

Abstract. In this study, effects of the contractions of the inlet and outlet areas of a vertical solar chimney were examined with a Computational Fluid Dynamics. The results show that the flow rate decreased with both the inlet and outlet areas. The outlet area influenced more than the inlet area did. The flow rate decreased linearly with the outlet area, however, was affected significantly only when the inlet area was below half of the air gap.

Keywords: Solar chimney, natural ventilation, inlet area, outlet area, CFD.

1. Introduction

Using renewable energy resources is a key factor of the green building rating systems. For example, in the LEED rating system of the U.S. Green Building Council, renewable energy credits up to 5 over a total of 110 points. One of the applications that can be based totally on renewable energy resources in buildings is the ventilation system. Natural ventilation utilizes natural driving forces for the air flow, such as effects due to external wind or thermal effects due to indoor – outdoor temperature difference [1].

Solar chimney is the common system based on the thermal effects for natural ventilation of buildings. The simple form of a solar chimney is a tube or pipe attached to a wall of a building. The tube or pipe absorbs solar radiation on its surfaces and heats the air inside its channel. As the air temperature rises, the stack effects can withdraw air from the building and discharge to the ambient. Accordingly, the building is ventilated naturally [2].

Among the design factors of a solar chimney, previous studies showed that the size and shape of the inlet of a solar chimney strongly influence its performance. The experimental data by Mathur et al. [3] and the numerical results by Zamora and Kaiser [4] indicated that the induced flow rate increased with the inlet size. However, effects of the inlet size ceased as it was above two times the air gap [4]. For a solar chimney attached to a wall of a building, increasing the inlet height from the floor resulted in lower induced flow rate [5]. Al-Kayiem et al. [6] tested different shapes of the inlet of a roof solar chimney. They reported that the flow rate was higher when the upper wall of the air channel was longer than the lower one.

Although most of the solar chimneys examined in the literature had equal areas for inlet and outlet, tests with different areas of the inlet and the outlet were also reported. Li et al. [7] conducted numerical simulations while Susanti et al. [8] took experiments for solar chimneys with different ratios of the areas of the outlet (A_{out}) and inlet (A_{in}). Both studies showed that the induced flow rate increased with A_{out}/A_{in}. The results by Susanti et al. [8] furthermore showed that the outlet area influenced more than the inlet did. For example, the case of $A_{out}/A_{in} = 312/14$ had higher flow rate than that of the case of $A_{out}/A_{in} = 14/312$. Similar conclusions were also seen in Al-Kayiem et al. [6].

In the above studies, effects of unequal area for the inlet and outlet were conducted for an inclined solar chimney [6, 8] or vertical solar chimney with different directions of the inlet and outlet [3, 4, 7]. It is practically interesting to test those effects for a vertical solar chimney with vertical inlet and outlet. In this study, we examined a vertical solar chimney with different areas for the inlet and outlet numerically. Performance of the chimney in terms of the induced flow patterns and induced flow rate was evaluated.

2. Description of the problem and the numerical model

2.1. Solar chimney

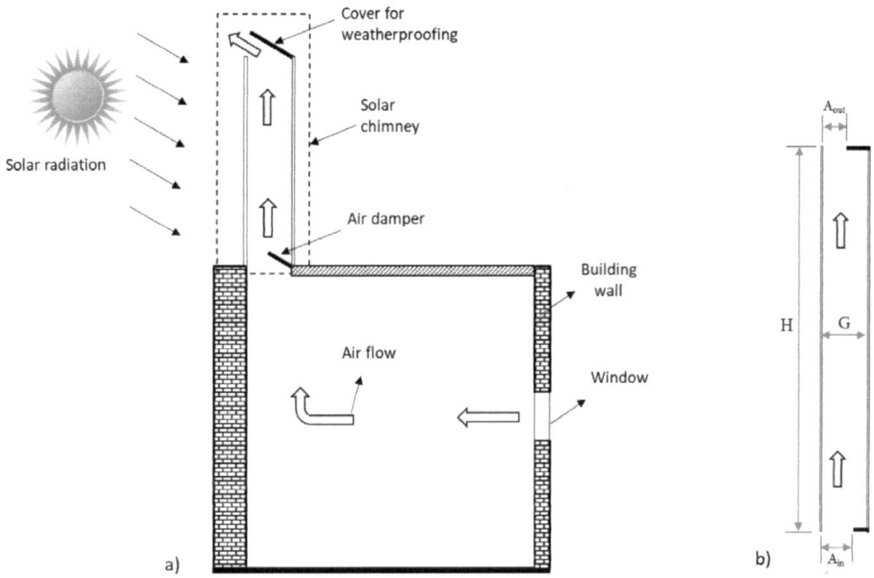

Figure 1. Schematic of a solar chimney integrated into a building (a) and its simple model (b).

A vertical solar chimney for ventilation of a building is depicted in figure 1a. The chimney is assumed on the roof of the building. Solar radiation is absorbed on the walls of the chimney. The absorbed heat is then transferred to the air in the channel. Warmed air rises up and induces an air flow through the building for ventilation.

In practical applications, an air damper for controlling the ventilation rate is placed at the inlet of the air channel, as sketched in figure 1a. At the outlet of the chimney, a cover for weatherproofing may also be required. With these additional elements, the inlet and outlet areas of the air channel are contracted and less than that of the air channel. Therefore, it is necessary to study the performance of a solar chimney when the air channel is contracted at the inlet and/or the outlet instead of a straight one.

A simple two – dimensional model of the solar chimney shown in figure 1a is presented in figure 1b. Major dimensions of the chimney are the height H and the gap G of the air channel. The inlet and outlet areas are denoted as A_{in} and A_{out}, respectively. For modeling heat transfer inside the air channel, for simplicity, the heat source is assumed inside the air channel. In figure 1b, the heat source can be on either the left or the right wall.

2.2. Governing equations

A two – dimensional, steady, incompressible, and turbulent flow in the vertical plane can be modeled with the Reynolds Averaged Navier – Stokes equations as follows.

$$\frac{\partial u_j}{\partial x_j} = 0 \tag{1}$$

$$\frac{\partial(u_i u_j)}{\partial x_j} = -\frac{1}{\rho}\frac{\partial p}{\partial x_i} - g_i\beta(T - T_0) + \frac{\partial}{\partial x_j}\left(\nu\frac{\partial u_i}{\partial x_j} - \underline{u_i' u_j'}\right) \tag{2}$$

$$\frac{\partial(T u_j)}{\partial x_j} = \frac{\partial}{\partial x_j}\left(\frac{\nu}{Pr}\frac{\partial T}{\partial x_j} - \underline{T' u_j'}\right) \tag{3}$$

where u and T stand for the time – averaged velocity and temperature; u' and T' are the fluctuating velocity and temperature; T_0 is the ambient temperature; p is the pressure; ν, ρ, β are the air kinematics viscosity, density, and thermal expansion coefficient; Pr and g are the Prandtl number and gravitational acceleration; $\underline{u_i' u_j'}$ and $\underline{T' u_j'}$ are the turbulence stress and turbulence heat flux which are solved with the RNG $k - \epsilon$ turbulence model.

2.3. *Numerical setup*

The above equations were discretized by the Finite Volume Method with the CFD (Computational Fluid Dynamics) code ANSYS Fluent (Academic version 2020R2). The computational domain and mesh are presented in figure 2.

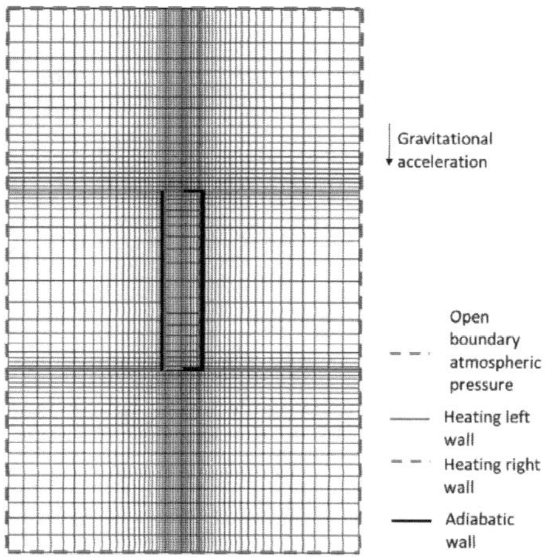

Figure 2. Computational domain and mesh.

The computational domain (figure 2) included both the air channel and ambient air. Gan [9] claimed that the extension outside the air channel is required to obtain reliable results. In figure 2, the extension was more than 10 times of the channel gap above the outlet and 5 times of the channel gap below the inlet and to the sides of the air channel.

The mesh was structured. It was finer near the solid surfaces. Mesh – refinement tests showed the numerical results changed below 1.0% when the non – dimensional distance y^+ of the first grid points near the solid surface was less than 1.5.

On the boundary of the domain, the pressure was assumed at the atmospheric value. All solid walls applied non – slip conditions. Air entered the domain at the ambient temperature which was fixed at 20°C in all tests. Inside the air channel, a uniform heat flux was applied on one wall while the opposite wall was able to receive radiative heat transfer from the other. Two cases of the heat source location were examined: heating the left or the right wall.

For the numerical setups, radiation between two walls in the air channel was calculated with the S2S radiation model in ANSYS Fluent. The SIMPLE method for the coupling between the continuity equation (equation (1)) and the momentum equation (equation (2)), and the PRESTO! method for interpolation of the pressure on the staggered mesh were selected. Second order upwind scheme for equations (2) and (3), and first order upwind scheme for the equations for k and ϵ were utilized.

2.4. *Validation*

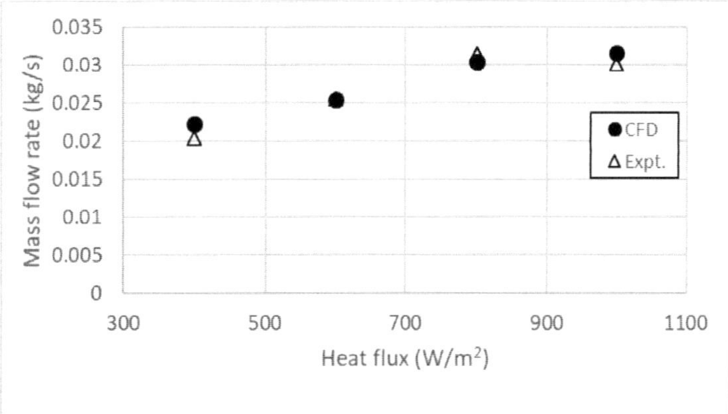

Figure 3. Induced flow rate obtained with the CFD model and the experiment [10].

Figure 3 shows the induced flow rate through a solar chimney obtained with the CFD model and the experiment by Burek and Habeb [10]. The chimney had a length of 1.025 m, a width of 1.0 m, and a gap of 40 mm. The heat flux changed in the range of 400 – 1000 W/m^2. The CFD model slightly under – estimated the flow rate at 400 W/m^2 and 1000 W/m^2 and over – estimated at 800 W/m^2. However, the maximum difference between them was only 9.2% at 400 W/m^2. This discrepancy should be within the uncertainties of the measurement. Therefore, the prediction of the CFD model was acceptably accurate.

3. **Results and Discussions**

In this section, the induced flow rate was compared among four cases:

- Without any area restriction at the inlet and outlet ($A_{out} = A_{in} = G, the\ base\ case$).
- With only restriction at the outlet ($0 < A_{out} \leq G, A_{in} = G$).
- With only restriction at the inlet ($A_{out} = G,\ 0 < A_{in} \leq G$).
- With restrictions at both outlet and inlet ($0 < A_{out} \leq G, 0 < A_{in} \leq G$).

In all tests, H and G were fixed to 1.0 m and 0.2 m, respectively. First, A_{out} and A_{in} were fixed to 50.0% of G, if any, while the heat flux changed from 250 to 1000 W/m^2. The heat source was either on the left or the right wall (see figure 2). Second, the heat flux was fixed to 500 W/m^2 while A_{out} and A_{in} changed from 0 to G.

3.1. *Comparison at different heat fluxes*

Figure 4 shows the induced flow rate obtained with four cases of the inlet and outlet areas at different heat fluxes. The heat source was on the left wall (LWH) of the right wall (RWH). In all cases, the flow rate increased with the heat flux. This trend has been widely reported in the literature [10 – 12] as the flow rate increased with the heat input in the air channel. The base case had the highest flow rate, as the flow did not experience any restriction of the flow area. Other cases always had lower flow rates.

Two interesting points are observed in figure 4. Firstly, with the same area of 0.5G, the case $A_{out} = 0.5G$ had much lower flow rates than those with the case $A_{in} = 0.5G$. For example, at 1000 W/m^2, the cases $A_{out} = 0.5G$ resulted in 0.0404 kg/s (LWH) while the case $A_{in} = 0.5G$ had 0.0623 kg/s (LWH), which was almost 50.0% higher. Therefore, the restriction of the outlet area influenced more than that of the inlet did. This point is further confirmed as the case $A_{out} = A_{in} = 0.5G$ and $A_{out} = 0.5G$ had similar flow rates. This observation was also reported by Susanti et al. [8] and Al-Kayiem et al. [6]. Secondly, between heating either side of the channel, LWH offered higher flow rate for the case $A_{out} =$

0.5G but lower flow rate for the case A_{in} = 0.5G. Particularly, the case A_{in} = 0.5G and RWH had a flow rate identical to that of the base case.

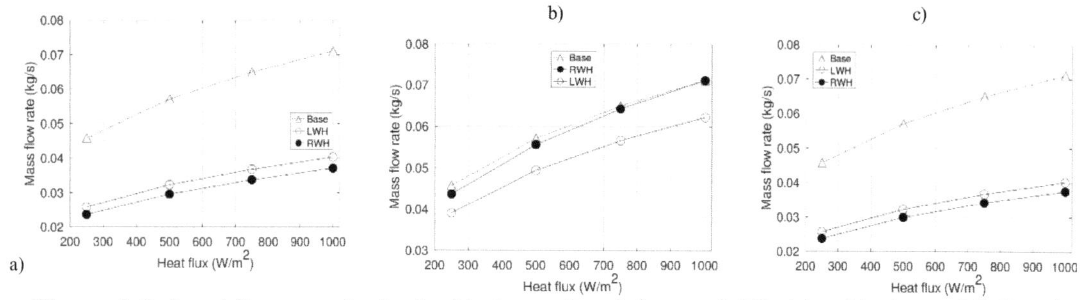

Figure 4. Induced flow rate obtained with $A_{in} = G$ and $A_{out} = 0.5G$ (a), with $A_{in} = 0.5\,G$ and $A_{out} = G$ (b), and $A_{out} = A_{in} = 0.5G$ (c), for heating the left (LWH) or the right (RHW) wall. The base case is for $A_{out} = A_{in} = G$.

Figure 5 displays the flow and temperature fields for the cases in figure 4. In the base case, the velocity and thermal boundary layers developed freely along walls of the air channel. When the outlet was contracted (figure 5b), the flow was obstructed. A separation zone appeared near the outlet which should increase the flow resistance; hence reduced the flow rate. The RWH had a separation zone larger than that of the LWH; hence lower flow rate, as seen in figure 4a. A separation zone near the inlet was also seen in figure 5c and caused the reduction of the flow rate compared to that of the base case. However, it is interesting to see that for RWH, the thermal boundary layer along the right wall in figure 5c was enhanced and yielded the flow field at the outlet similar to the base case. As a result, the flow rate in this case was close to that of the base case, as seen in figure 4a. When both openings were contracted (figure 5d), the flow fields near the outlet were identical to those in figure 5b; hence similar flow rates are seen between figures 4a and 4c.

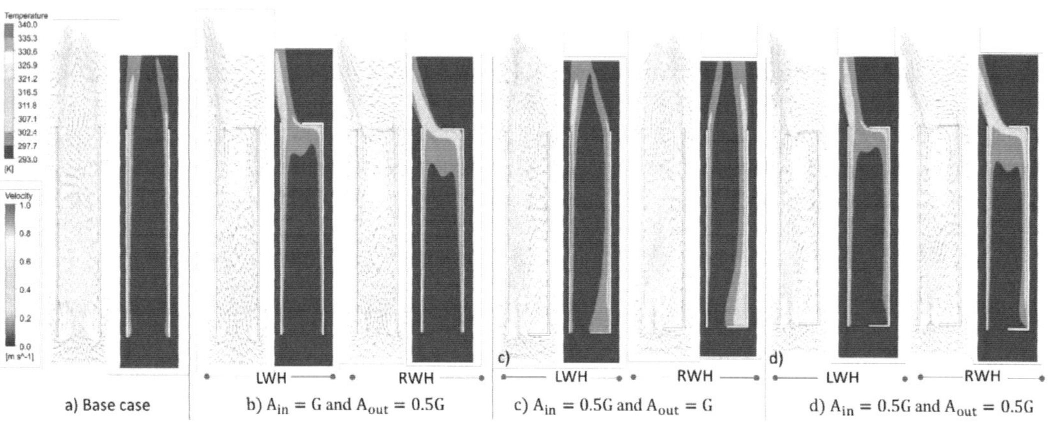

Figure 5. Flow (left) and temperature (right) fields obtained with four cases of the inlet and outlet areas.

3.2. *Comparison of different inlet and outlet areas*

3.2.1. Changing outlet and outlet areas Figure 6 shows the flow rate when one end of the channel was fully open while the other end was partially restricted. It is seen that the flow rate decreased with both A_{in} and A_{out}. The decrease rate was almost linearly for A_{out}. For A_{in}, the decrease was significant when A_{in} was smaller than 0.5G. RWH cases resulted in lower flow rate as A_{out} changed but higher flow rate as A_{in} changed. However, the maximum difference between the flow rates of LWH and RWH in figure 6 was only less than 12.0%.

The flow field in figures 5b and 5c may again help to explain the fact that A_{out} influences the flow rate more than A_{in} does. The separation zone near the outlet as A_{out} decreases (figure 5b) is larger than that near the inlet as A_{in} is contracted (figure 5c). Therefore, the flow with the contracted A_{out} is more obstructed and accordingly, the flow rate is less.

In figure 5b, between LWH and RWH, though the separation zones are similar, the main flow field near the heated wall for RWH may experience more resistance at the outlet due to sudden change of the flow path as A_{out} is contracted on the right side. Therefore, LWH offered slightly higher flow rate, as seen in figure 6a. In figure 5c, the thickness of the thermal and velocity boundary layers of the RWH case are more than those of the LWH. Accordingly, the flow rate of RWH is higher, as seen in figure 6b.

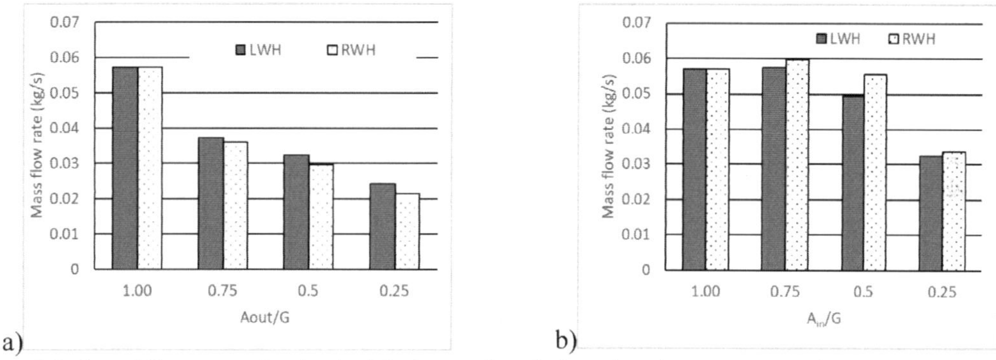

Figure 6. Induced flow rate obtained with $A_{in} = G$ and changing A_{out} (a) and with $A_{out} = G$ and changing A_{in} (b)

3.2.2. Changing the ratio of A_{out}/A_{in} In figure 7, the flow rate is plotted for different A_{out}/A_{in} ratios. Figure 7a shows the results for equal area cases. The flow rate also decreased as both areas were reduced, similar to figure 6a. Both cases of heating location resulted in identical flow rates, with the differences of within 11.0%. Figure 7b presents the flow rates for different areas. It is seen that for two cases $A_{out}/A_{in} = 0.75/0.5$ and $A_{out}/A_{in} = 0.5/0.75$, though the flow experienced the same areas of the two ends of the channel, the case of larger outlet ($A_{out}/A_{in} = 0.75/0.5$) had a higher flow rate, which is 17.5% for LWH and 6.7% for RWH. Similar points are also seen for two cases of $A_{out}/A_{in} = 0.75/0.25$ and $A_{out}/A_{in} = 0.25/0.75$, however, the differences are enhanced with 28.0% for LWH and 42.0% for RWH. Therefore, it is again confirmed that the outlet area influenced more than the inlet area did, as seen in figures 4 and 7.

a) 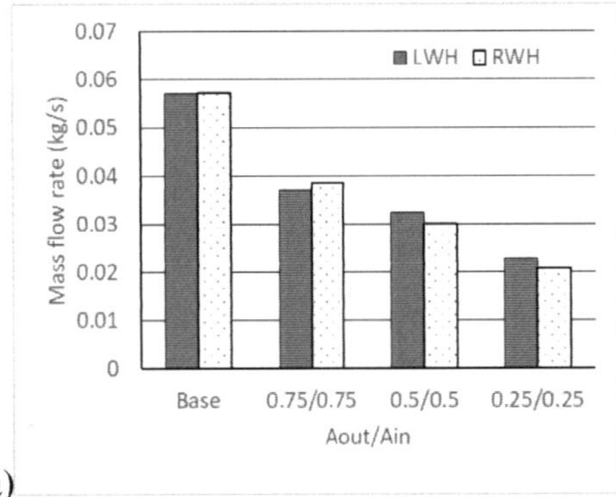

b) 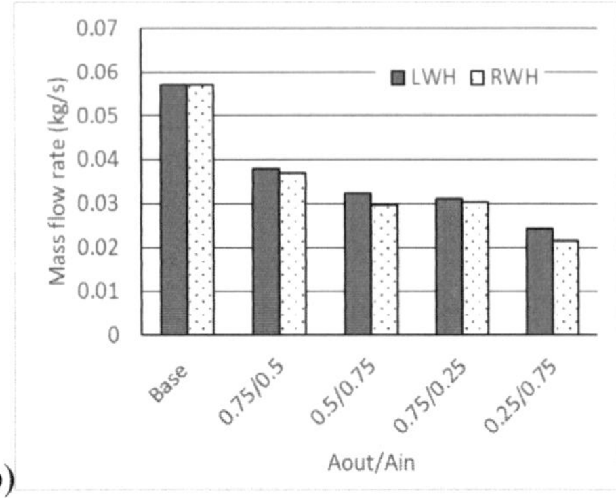

Figure 7. Induced flow rate obtained for different A_{out}/A_{in} ratios: a) Equal areas and b) Different areas.

4. Conclusions

From the results, the following points are observed:

- The flow rate decreased with both the inlet and outlet areas. While the flow rate decreased linearly with the outlet area alone, it was affected significantly when $A_{in} < 0.5G$.
- The outlet area influenced more than the inlet area did. The difference was significant as the areas decreased. As $A_{out} = 0.5G$, the flow rate was about 57.0% of that of the base case while $A_{in} = 0.5G$, the flow rate was about 87.0% of that of the base case (for LWH).
- When either the outlet or the inlet area was reduced, reducing A_{out} had higher flow rate with LWH but reducing A_{in} had higher flow rate with RWH.
- When both areas are reduced, both LWH and RWH had similar flow rates with the maximum difference was less than 11.0%.

References

[1] Awbi H 2003 *Ventilation of Buildings Second Edition* (Spon Press)

[2] Bansal N K, Mathur R and Bhandari M S 1993 Solar chimney for enhanced stack ventilation *Building and Environment* **28** (3) 373–377

[3] Mathur J, Bansal N K, Mathur S, Jain M and Anupma 2006 Experimental investigations on solar chimney for room ventilation *Solar Energy* **80** (8) 927–935

[4] Zamora B and Kaiser A S 2009 Optimum wall-to-wall spacing in solar chimney shaped channels in natural convection by numerical investigation *Applied Thermal Engineering* **29** (4) 762–769

[5] Shi L, Zhang G, Yang W, Huang D, Cheng X and Setunge S Determining the influencing factors on the performance of solar chimney in buildings *Renewable and Sustainable Energy Reviews* **88** (2018) 223–238.

[6] Al-Kayiem H H, Sreejaya K V and Chikere A O Experimental and numerical analysis of the influence of inlet configuration on the performance of a roof top solar chimney *Energy and Buildings* **159** (2018) 89–98.

[7] Li A, Jones P, Zhao P and Wang L 2004 Heat Transfer and Natural Ventilation Airflow Rates from Single-sided Heated Solar Chimney for Buildings *Journal of Asian Architecture and Building Engineering* **3** (2) 233-238

[8] Susanti L, Homma H, Matsumoto H, Suzuki Y and Shimizu M 2008 A laboratory experiment on natural ventilation through a roof cavity for reduction of solar heat gain *Energy and Buildings* **40** (12) 2196–2206

[9] Gan G 2010 Impact of computational domain on the prediction of buoyancy-driven ventilation cooling *Building and Environment* **45** (5) 1173–1183

[10] Burek S A M and Habeb A 2007 Air flow and thermal efficiency characteristics in solar chimneys

and Trombe Walls *Energy and Buildings* **39** (2) 128–135
[11] Chen Z D, Bandopadhayay P, Halldorsson J, Byrjalsen C, Heiselberg P and Li Y An experimental investigation of a solar chimney model with uniform wall heat flux *Building and Environment* **38** (7) (2003) 893–906.
[12] Shi L, Zhang G, Yang W, Huang D, Cheng X and Setunge S Determining the influencing factors on the performance of solar chimney in buildings *Renewable and Sustainable Energy Reviews* **88** (2018) 223–238.

IOP Conf. Series: Materials Science and Engineering 1109 (2021) 012012 doi:10.1088/1757-899X/1109/1/012012

Minimizing Dynamic Loading Resonance of Battery Pack Subjected to Road Profile Loads

Muhammad Farkhan Abdillah[1], Bentang Arief Budiman[1*], Ignatius Pulung Prasetio[1], and Poetro Lebdo Sambegoro[1]

[1]Faculty of Mechanical and Aerospace Engineering, Institut Teknologi Bandung, Ganesha street no. 10, Bandung, Indonesia

*bentang@ftmd.itb.ac.id

Abstract. One of the main problems in designing battery packing for electric vehicles is the appearance of deformation in the components due to vibration generated by road profile. The vibration brings the most significant effect on the battery packing due to low frequency and high amplitude. Resonance occurs at low frequency since most of the E-trike components have a low natural frequency. This should be avoided to increase the reliability of the battery pack due to mechanical loadings. Modal analysis, Quarter Car Model (QCM), and optimization were conducted in this work. Modal analysis is used to find the natural frequencies of the battery pack on several modes of vibrations. Simulink of the Quarter Car Model (QCM) produces the load from the road profiles based on ISO 8608. The load from the road profiles is used as the input in the Ansys 16.0 Harmonic Analysis. The output is the deformation frequency response for the bottom side case of the battery pack. To reduce the outcome of resonance, a damping mechanism is selected to reduce maximum amplitude from road profile at a resonance frequency. The simulation results show that the damping mechanism is the suitable technology to decrease the major amplitude of the deformations from road profile with reduction up to 54% lower than the previous model. The lower deformations of the battery pack component reduce the battery pack's possible accident due to vibration.

Keywords: Quarter car model, Battery packing, road profile loads

1. Introduction

The most challenging aspect of the battery pack design is to generate calculations around a new battery solution. There are a few simple calculations that can be made as a principal concept. Another thing is the design for the reliability and service of the battery pack. The safety and security of the battery pack should be the main concern from the design process. The design for reliability is a series of processes of failure mode effects (FMEA), test to failure (TTF), and accelerated life testing [1]. Correspondingly, quality and reliability are the clarification points to judge the battery pack's reliability level. For that reason, the engineers need to develop and design the battery pack that is influenced by the material, dimension, and function.

Some units can be viewed for the battery pack's reliability, which is related significantly to vibration, temperature, and pressure [2]. Vibration of the battery pack must be circumvented from the natural frequency of vehicle suspension system and sprung mass from 0 to 7 Hz, vehicle powertrain, the driveline, and the gearbox, from 7 to 20 Hz, and the vehicle chassis system from 20 to 40 Hz [3-4]. Lithium-ion batteries have to be utilized in safe and reliable operating conditions limited by temperature and voltage windows [3,5]. Battery temperature must be lower than 50°C [3]. Effective control and management of a battery management system maintain restriction conditions [3]. Marginal deviations such as exothermic reactions in producing toxic and flammable gas that can lead to the explosion can be triggered by the excessive heat build-up and physical damage of the battery packs [2].

The industry pushes battery pack technology to enhance more modern features such as high energy density. The lithium-ion battery pack's advantage, such as high energy density, generates challenges that must be faced. The battery pack tends to be sensitive to extreme conditions such as high temperature, overcharge, and short-circuit. The battery pack's standard components are combustible,e.g., plastic packing, separator, and electrolyte [6-7]. The abusive conditions destroy the battery pack's stabilization, which triggers chain reactions inside the battery and leads to thermal runaway. Besides the danger of abusive conditions, limited research has been investigated toward the battery pack's safety, mainly focusing on the fire-retardant battery pack components [8].

When the battery pack is damaged in the accident, there is a potential of the battery thermal runaway, fire, and explosion [9-11]. The conditions and prevention of the thermal overheating and electrical overcharging of the batteries are explained in the literature [12-13]. The literature review is mostly the innovation of rigid objects placed into the battery cells, modules, and packs. US Patent 7614469 also provides hollow structural members for battery packing that absorbs undesirable vibrations. Battery pack's structure through positioning elevation to avoid battery pack damage by a rear impact is discussed in US Patent 7070015. About 288 Li-ion pouch cells of Chevrolet Volt battery pack for 70% of the battery mass and around 55% of the pack volume are in the T-shaped steel vertical tray [3]. The battery pack is stranded over the chassis center under the passenger seats.

However, correspondingly, very little research discusses the relations of road profiles loads with velocity. Moreover, narrow findings have been covered about dynamic loadings of road profiles in correlation with battery pack resonance.This paper is aimed to investigate the optimization of the battery pack to reduce the resonance load from road profiles within different velocities. Road profiles are generated from ISO 8608 standard for different road roughness. A simple dynamic system, the quarter car model, is used to describe the loads from road profiles. The natural frequency of the battery pack is examined by simplifying geometry and utilizing finite element analysis. The road profile loads show maximum amplitude at the natural frequency of the battery pack. To minimize this load, the paper illustrates an optimization by using a rubber bolt as a damping mechanism for the battery pack.

2. Methodology

2.1. Road Profiles

It is possible to construct an artificial road profile using an equation and assuming a random phase angle φ with probabilistic distribution within the 0 until 2π range. The artificial profile $h(x)$ can be described as Equation (1) :

$$h(x) = \sum_{i=0}^{N} \sqrt{2.\Delta n.Gd\ (i.\Delta n)}\ cos(2\pi.i.\Delta n.x\ + \varphi\ ). \qquad (1)$$

The Matlab program is developed to give the output of the chart by Equation (1). The equation has the variable of L, N, and k that respectively are the length of the road simulation, the number of data points, and the value for ISO road classifications [14]. Equation (1) has the Fourier series function, and then the programs try to loop the function to store the data in variable h(x) that is the road function over the length of the road. The spatial frequency in the range between 1/L and N/L m^{-1}. Consequently, Equation (1) is programmed through Matlab R2018b of generating road profiles by assuming L is 2,500 m, and the number of the data points is 50.000. The constant ISO classes k of road profiles are determined to show roughness levels of road profiles. For each value of k, ranging from 3 to 9, road profiles are generated to see the different road roughness levels. This value represents the level of road surfaces causing high input force for different E-trike velocities, respectively, 40 km/h, 30 km/h, 20 km/h, and 10 km/h. Also, a graph of road displacements over time is made simultaneously in Simulink

Matlab R2018b for different road classes and time-variable that depends directly on the E-trike velocities.

2.2. Loads from road profiles

The road profile generates the E-trike oscillation, which depends on the road surface's displacement level. This randomness of road profile is described from Matlab R2018b program with representing equation that describes the roughness of road profiles. The dynamic vibrational load is not constant in time, but it depends on several factors such as vehicle mass, speed, and road surface randomness. For analytical purposes, Quarter Car Model (QCM) is generated to study E-trike's dynamic response [15]. Correspondingly, Figure 1 shows the QCM Model of a vehicle with a base movement of the road profiles of h with k is the equivalent stiffness of the E-trike, m denotes the mass of the battery pack, and c is the equivalent damping coefficient of the E-trike. The assumptions regarding the body of the E-trike is rigid and does not have relative displacement with the road surfaces. The assumption that the variable z is more than h is used in the equation of motion, which is shown by the following equation:

$$m\ddot{z} + c\,(\dot{z} - \dot{h}\,) + k\,(\,z - h) = 0 \tag{2}$$

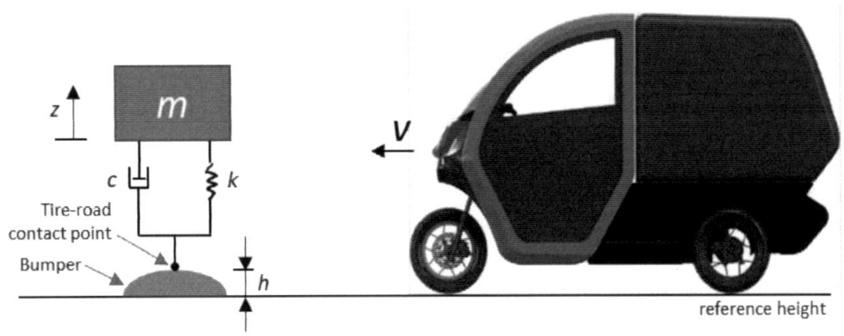

Figure 1. Quarter car model of the battery pack

2.3 E-trike

E-trike is a three-wheel vehicle to transport goods for medium ranges. The vehicle is suitable in Indonesia's traffic road since it has high mobility than the car and the truck. It is an electric vehicle that uses the electric motor as the driver. A battery pack is used to store the electrical power source. It consists of a connector, bracket, rubber, bolt, cell board, isolator, electric module, and 240 battery cells of Lithium Nickel Manganese Cobalt Oxide (NMC) battery (see Figure 2(a)). It is fixed by the bolt which is attached to the chassis. Battery Pack is mounted on the E-trike chassis below the driver, as shown in Figure 2(b). Two battery packs are used to drive SRM DC Electric Motor 5 kW and create the E-trike's maximum speed of 40 km/h [16].

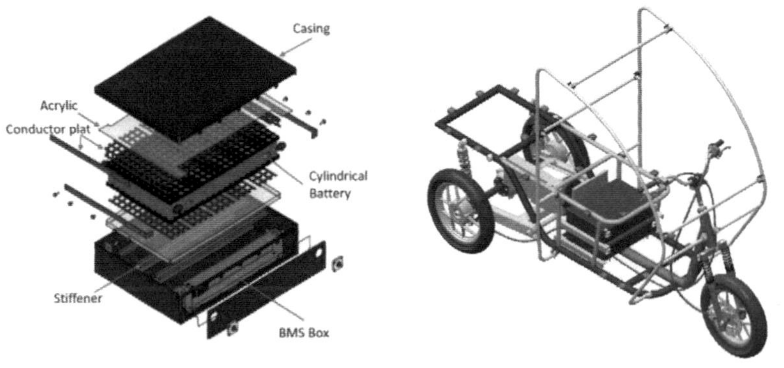

(a) (b)

Figure 2. (a) E-trike' battery pack and (b) E-trike structure with the battery pack

2.4 Modal Analysis and Harmonic Analysis of Existing Design

To get vibration response at a different location, harmonic analysis is set in the frequency range of 0 Hz until 100 Hz with the solution intervals of 200 segments. The output is the deformation graph at different locations, which are the bottom side. The vibration response does not work for all sides, but only one side that represents similar geometry. The setting in Ansys 16.0 starts with creating two sides as a fixed point and the other two as force excitations that are the load from road profiles. The reason is the usual possible movement, such as twisting from different existing edges. The force data is generated from Matlab R2018b Simulink, and the input of force in Ansys 16.0 is in the frequency domain. The harmonic analysis is specified into different E-trike velocities. It is possible to get the natural frequencies value from the graphs from the deformation frequency responses.

2.5 Optimization

The main issue in this paper is the resonance occurrences on the battery pack. The vibration load is produced by road profile displacements that excite force on the battery pack. The simplification of the battery pack is an important process for the analysis, and then the battery pack only remains the steel cases. The deformations of the battery pack cases may enhance the short circuit of the batteries due to the dynamic loadings from road profiles. To minimize the resonance damage, rubber bolts are simulated on the two small plates which is used to connect battery pack with the E-trike chassis in Figure 2(b). The small plates are located on the bottom side of the battery pack casing in Figure 2(a). where the load from road profiles occurs. The rubber bolt is meant to reduce the major amplitude of the deformation frequency responses from the load of road profiles.

3. Results and Discussion

3.1Results of Modal Analysis

Modal analysis results are natural frequency values on several modes depending on the settings that the user picks. There are four boundary conditions as fixed support at the four small plates. The shape of first mode vibration can occur since the motion of the battery is mostly in the gravity directions; thus, the top side has the greatest deformations. The maximum deformation on mode vibration I is 20 mm from Figure 3(a) to the inside of the battery pack on the top side. The rear side's deformations maximum value is 9 mm, which also exceeds allowable deformations of the battery pack. The least value of the deformations is 1 mm, which should be the maximum number. This error in real applications can damage the batteries. Figure 3(b) shows the deformation of mode vibrations 2 with the maximum values of 21 mm. The battery pack on the rear side only accepts maximum deformation of 1 mm on normal conditions based on the spacing between the cage and the batteries. Figure 3(c) shows the deformation caused by mode vibration 3, in which the direction is to the inside of the battery pack. Nevertheless, the simulations still can be used as information of deformation directions and natural frequency values for different modes [4].

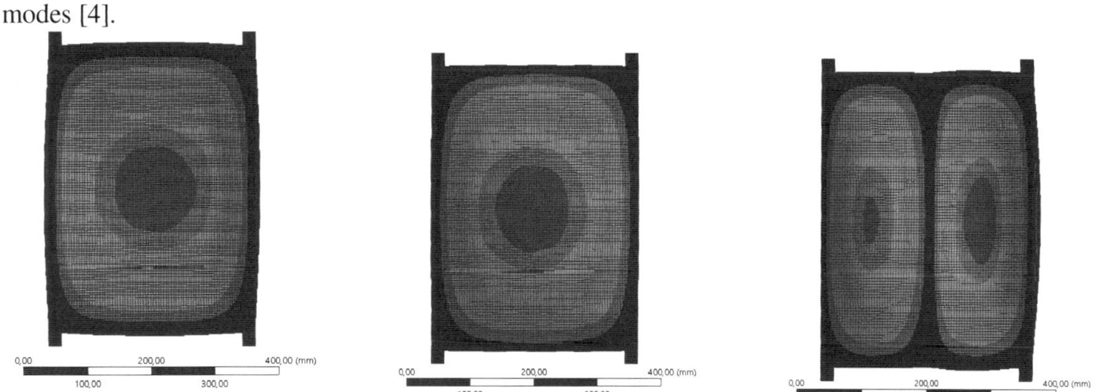

(a) (b) (c)

Figure 3. Several modes for different element sizes: (a) first mode; (b) second mode; (c) third mode

3.2 Road Profiles Graphs

Figure 4(a) shows road displacement from Matlab R2018b, which results from the ISO 8608 equation for 40 km/h. Also, Figure 4(b) illustrates road profiles for 30 km/h, and Figure 4(c) demonstrates road displacement for 20 km/h. Additionally, road displacement for the lowest velocity of 10 km/h is shown in Figure 4(d). The road class from the legend indicates lower limit from two consecutive road class. Equation (1) is used to generate road profiles over 2500 m length road over different velocities; 40 km/h, 30 km/h, 20 km/h, and 10 km/h [17]. The displacements are in meters for the four graphs below. Road class A-B has the least road displacements over time, meaning that it is the best quality road to give the E-trike the least force. Otherwise, road class G-H is the worst quality road since it has the displacements of 4 meters. Road profiles A-B has the averaged h = ± 5 mm, which is the best quality road condition. Road class B-C displacements are in the average of $h = \pm 10$ mm. Also, road class C-D has the average value of $h = \pm 15$ mm, and road class D-E has averaged road displacements of $h = \pm 300$ mm. The results agree with other research in [14]. Road Class F-G and G-H have the worst road class conditions. Figure 4(a) shows road class G-H of 4 meters displacement at about 20 seconds. This road class is usually located in rural areas. From the four graphs, it can be concluded that E-trike may only be used in city areas that have a road roughness of not less than 15 mm. Therefore, road class A-B, road class B-C, and road class C-D are the only road classes generated further in the harmonic analysis.

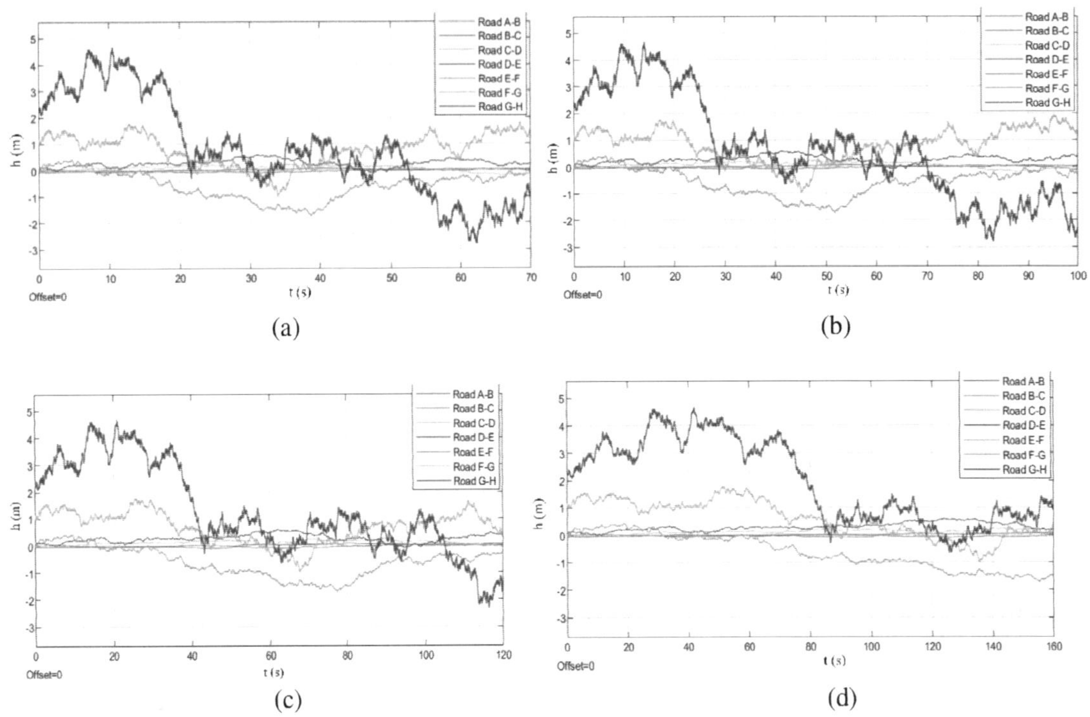

Figure 4. Road profiles (m) versus time (s) for: (a) 40 km/h, (b) 30 km/h, (c) 20 km/h, and (d) 10 km/h.

3.3 Force versus frequency graphs

The force versus time data from Simulink Matlab R2018b is used as the input for the Fast Fourier Transform (FFT) process. The signal filter that is used in the process is the Yulewalk with the matrix M = [1 1 0 0] and F = [0 0.2 0.2 1] for the 8th order. The frequency range is picked 0 Hz until 100 Hz since the maximum amplitude occurs below 100 Hz. The sampling frequency is 1000 Hz because the sampling frequency needs to be twice or larger than the maximum frequency in the analysis. The signal from Simulink Matlab R2018b is transferred to the low-pass filter before it is used as the input in the FFT. Figure 5 is the result of the force versus frequency data for different velocities.

Load by road profiles in 40 km/h shows that the maximum amplitude is 383,000 N^2 at the 15.32 Hz and 43.84 Hz frequencies. This indicates possible resonance at these frequencies. The force amplitude is decreasing highly at 32.72 Hz, and 62.81 Hz. Other velocities, however, show the maximum amplitude at these frequency ranges. Road displacements of the highest velocity have the least time spacing due to the highest velocity. This might alter the amplitude in Figure 5, which is larger than the maximum amplitude from the force on lower velocities.

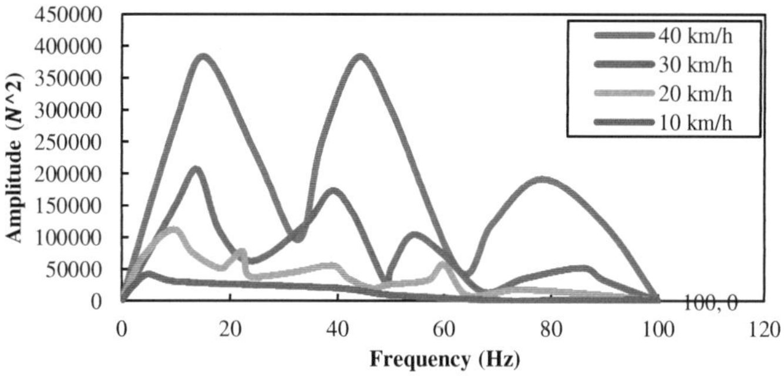

Figure 5. Force versus frequency for different velocities

3.4 Result of harmonic analysis

The frequency response graphs are used to find the resonance frequency of the battery pack under road profiles load. The reason is to choose optimized solutions regarding the value of natural frequency. The location of the peak amplitudes of deformation frequency response on every velocity in Figure 6(a) is at 10 Hz. The low-speed frequency response has the least value of maximum deformation within the range of 0 Hz until 100 Hz. Correspondingly, the deformation frequency response produces notable results of deformation under 20 Hz. By means that the frequency range of the battery pack under road profile loads only alters the natural frequency on the range of 0 Hz until 20 Hz. To validate the result, another harmonic analysis is generated to the normal stress frequency response of the bottom plate in Figure 6(b). The resonance frequency shows the same value of 10 Hz, meaning that 10 Hz is the valid natural frequency of the battery pack. The electric vehicle battery's natural frequency in urban driving response is at about 5 Hz [18]. The natural frequency of the dynamic load from road surfaces is about 10 Hz [18]. The natural frequency at all deformation frequency responses is at about 10 Hz, and the natural frequency value agrees with the results from previous research. However, the value of the maximum deformation is not rational, but it still gives a significant result of resonance frequency from the frequency response graphs.

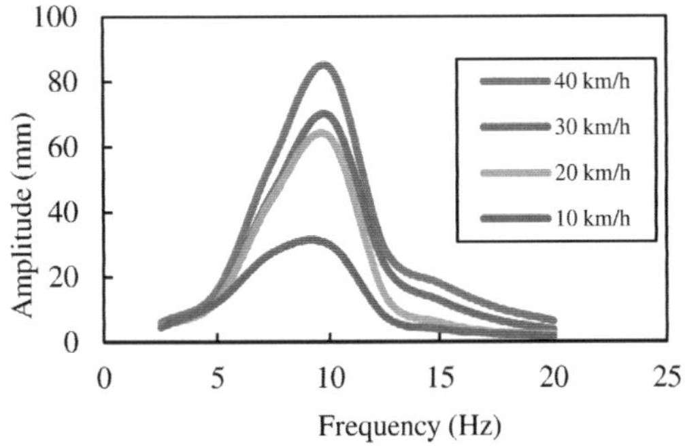

(a)

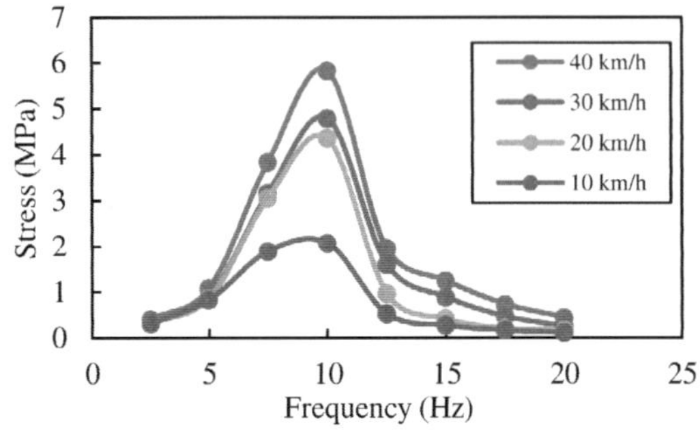

(b)

Figure 6. (a) Bottom side deformation frequency response for different velocities and (b) Bottom side normal stress frequency response.

3.5 Optimization by rubber bolts

The natural frequency results from modal analysis do not represent exact values. The different locations of boundary conditions alter the natural frequencies of the battery pack. These frequency locations can give excessive dynamic force to the battery pack due to the resonance phenomena. The E-trike motor driver is the SRM DC 5 kW motor, which has low maintenance costs [16]. The revolution speed of the motor shows the maximum values of 10,000 rpm. The motor is simulated between 2,000 rpm and 10,000 rpm, and the median value for 85% efficiency is about 2500 rpm. From this result, the critical frequencies from the motor speed are at 32.27 Hz, 41.67 Hz, and 166.67 Hz. By these values, the actions to minimize damage from the road profile loadings can be interpreted in two ways: shifting the natural frequencies higher than the motor's critical frequencies and minimizing the peak amplitude from the resonance phenomena. Since the natural frequency is at 10 Hz, it is chosen to minimize the peak amplitude by using a rubber bolt. The rubber bolt is mounted on the battery pack to see the difference result of peak amplitude.

The bottom side becomes the reference side to the analysis because it has similar shapes with the opposite side. The deformation direction of the bottom plate is on the z-axis that is parallel to the displacement of the road profiles. The harmonic analysis is used with the road profile loads by 10 km/h velocity. The reason is the data for four speeds represent similar graphs, and only one of the data should be analyzed as a source to show the optimized design result, which is 10 km/h. The optimized design by mounting rubber bolts, which the vibration characteristics are found in [19], on the connecting plates reduces the maximum amplitude of road profiles load. The damping coefficient of soft, medium, and hard rubber are respectively 103 Ns/m, 125 Ns/m, and 150 Ns/m [19]. The battery pack is simulated with The critical frequencies of the harmonic analysis without the damping mechanism are at 10 Hz. Figure 7 is the deformation frequency response for the soft, medium, and hard rubber damping mechanisms for 10 km/h. The natural frequency after mounting the rubber bolt is still at about 10 Hz. The maximum amplitude on the bottom side with soft rubber fastenings is reduced to 26.975 mm, which is 8% lower than the undamped battery pack. The natural frequency from medium rubber fastenings is at about 10 Hz. The peak amplitude on the bottom side is reduced to 19,47 mm, which is about 33% lower than the previous results. Similarly, the natural frequency from hard rubber is at about 10 Hz. The peak amplitude on the bottom side is reduced to 16.414 mm, which is about 45% lower than the previous results. It is proved that rubber acts to reduce the peak amplitude; following that, the hard rubber is the perfect type for reducing considerable maximum amplitude.

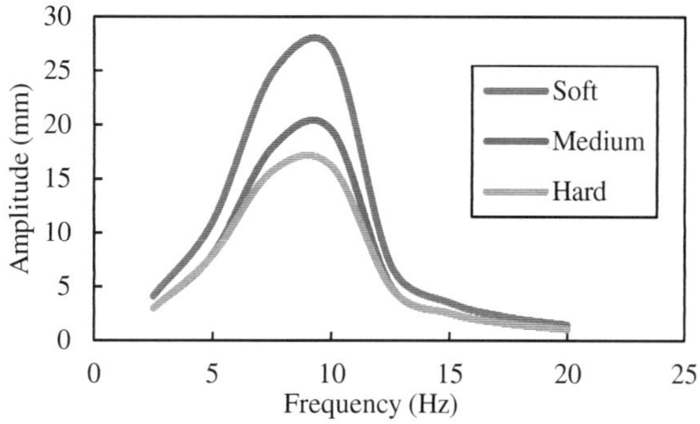

Figure 7. Rubber bolts assembling on the small plates of the battery pack for 10 km/h.

4. Conclusion

Road profiles are generated from the Equation (1) into road displacements to show the level of roughness for different class roads. Because E-trike is only used on the city roads, road class A-B, road class B-C, and road class C-D are used in the analysis. The results show that the displacement of the road profiles generates deformations to the battery pack as harmonic loadings from the QCM to show the loads from road profiles. Moreover, it is found that the E-trike accepts more load for greater velocities, and the frequency for each amplitude is also greater than fewer velocities. The harmonic analysis is generated to exhibit the resonance frequency from the battery pack that is given the road profiles in terms of twisting movement. The result shows that the original battery pack model shows the resonance frequency at 10 Hz. The maximum amplitude of deformation increases correspondingly with the expanding velocities. To minimize the resonance occurrence, the battery pack's proposed design consists of rubber fastenings attached to the bolt. The major amplitude of loads is decreased due to the vibration isolations. Rubber only acts as vibration isolation, which absorbs energy from road profile loads by internal frictions. The remaining energy is still transferred to the battery pack, which creates 55% of the original model's deformations. The simulation result is based on general calculations from the QCM model to harmonic analysis in FEM. Several assumptions are made to simplify the research. An experimental study should be examined further to get the accurate results of using rubber bolts to minimize resonance occurrences.

References

[1]. Warner, J.T., 2015. *The handbook of lithium-ion battery pack design: chemistry, components, types and terminology.* Elsevier.

[2]. Arora, S., Shen, W. and Kapoor, A., 2016. *Review of mechanical design and strategic placement technique of a robust battery pack for electric vehicles. Renewable and Sustainable Energy Reviews*, ***60***, pp 1319-1331.

[3]. Pesaran, A.A. and Keyser, M., 2001, January. Thermal characteristics of selected EV and HEV batteries. In *Sixteenth Annual Battery Conference on Applications and Advances. Proceedings of the Conference (Cat. No. 01TH8533)* pp 219-225. IEEE.

[4]. Arifurrahman, F., Indrawanto, I., Budiman, B.A., Sambegoro, P.L. and Santosa, S.P., 2018. Frame modal analysis for an electric three-wheel vehicle. In *MATEC Web of Conferences*, **197**, p. 08001. EDP Sciences.

[5]. Sambegoro, P.L., Budiman, B.A., Philander, E., Aziz, M., 2018, October. Dimensional and Parametric Study on Thermal Behaviour of Li-ion Batteries. In *5th International Conference on Electric Vehicular Technology (ICEVT)* pp 123-127. IEEE.

[6]. Karimi, G. and Li, X., 2013. Thermal management of lithium-ion batteries for electric vehicles. *International Journal of Energy Research*, ***37***(1), pp 13-24.

[7]. Doughty, D.H. and Roth, E.P., 2012. A general discussion of Li ion battery safety. *Electrochemical Society Interface*, ***21***(2), p 37.

[8]. Belov, D. and Yang, M.H., 2008. Failure mechanism of Li-ion battery at overcharge conditions. *Journal of Solid State Electrochemistry*, ***12***(7-8), pp 885-894.
[9]. Ouyang, D., Chen, M., Huang, Q., Weng, J., Wang, Z. and Wang, J., 2019. A review on the thermal hazards of the lithium-ion battery and the corresponding countermeasures. *Applied Sciences*, ***9***(12), p 2483.
[10]. Halimah, P.N., Rahardian, S. and Budiman, B.A., 2019. Battery Cells for Electric Vehicles. *International Journal of Sustainable Transportation Technology*, ***2***(2), pp 54-57..
[11]. Sambegoro, P.L., Budiman, B.A., Philander, E. and Aziz, M., 2018, October. Dimensional and Parametric Study on Thermal Behaviour of Li-ion Batteries. In *2018 5th International Conference on Electric Vehicular Technology (ICEVT)* pp 123-127. IEEE.
[12]. Zhu, J., Wierzbicki, T. and Li, W., 2018. A review of safety-focused mechanical modeling of commercial lithium-ion batteries. *Journal of Power Sources*, ***378***, pp 153-168.
[13]. Rodrigues, M.T.F., Babu, G., Gullapalli, H., Kalaga, K., Sayed, F.N., Kato, K., Joyner, J. and Ajayan, P.M., 2017. A materials perspective on Li-ion batteries at extreme temperatures. *nature energy*, *2*(8), pp 1-14.
[14]. ISO 8608 2016 Mechanical Vibrations - Road surface profile Data, *International Stand. Org.* **3**: **1-36**.
[15]. Agostinacchio, M., Ciampa, D. and Olita, S., 2014. The vibrations induced by surface irregularities in road pavements–a Matlab® approach. *European Transport Research Review*, *6*(3), pp 267-275.
[16]. Furqani, J. and Purwadi, A., 2018. Review of Switched Reluctance Motor Control for Acoustic Noise and Vibration Reduction. *International Journal of Sustainable Transportation Technology*, *1*(2), pp 59-62.
[17]. Marzuki, M.A.B., Azmi, M.F.M., Mohamed, A.R.N., Bakar, M.A.A. and Nawi, M.A.M., 2018. Pre-stress Modal Analysis of Space Frame Chassis Structure Using Finite Element Method. *International Journal of Engineering & Technology*, 7(4.36), pp 1545-1548.
[18]. Hooper, J 2016 *Understanding Vibration Frequencies Experienced by Electric Vehicle Batteries*. Master Thesis. (Conventry:University of Warwick) p 4.
[19]. Lépore Neto, F.P. and Santos, M.B.D., 2011. A Procedure for the Parametric Identification of Viscoelastic Dampers Accounting for Preload. *Journal of the Brazilian Society of Mechanical Sciences and Engineering*, *33*(3), pp 308-313.

Multidisciplinary design optimization for sustainable design using building information modeling

Le Hung Tien[1], Nguyen Van Tung[1]

[1]Faculty of Engineering, Van Lang University

tien.lh@vlu.edu.vn

Abstract. This paper illustrates the application of computational design using Rhinoceros software and Grasshopper plugin to investigate a real accommodation with respect to the solar radiation reduction. The paper also introduces the basic concepts of Building Information Modelling (BIM) utilization in construction. The multidisciplinary BIM model show that the Computable BIM Model can be used for Architecture, Structure, Mechanical – Electrical – Plumbing (MEP) to help architects, designers do sustainable design such as acoustic comfort, lighting design on effective cost manner with many rapid design options from the conceptual design phase, to facilitate wind load simulation, predict structural behaviour under fierce conditions, to estimate heat load, HVAC systems performance for energy saving and sustainable design.

1. Introduction

The climate change and the reduction of natural resources drive the world governments, local authorities and designers to act as quickly as possible. To respond to foreseen calamities and bad consequences due to climate change, many countries and organizations are taking their responsibilities. For example, the European Union committed to resolve this issue through the 2020 in 2020 initiative [1]. The "2020 in 2020" plans to reduce overall emissions to at least 20% below 1990 in 2020 as well as increase the rate of renewable energy in energy use to 20% by 2020.

Admittedly, Construction industry (Architecture Construction Engineering – AEC), the main body plays a major contribution to climate change. According to some recent studies, buildings consume 40% of global primary energy and contribute more than 30% of CO_2 emissions [2]. While Heating, Ventilation and Air Conditioning (HVAC) systems account for nearly 50% of energy consumption in buildings and account for 10% - 20% of total energy consumption [2].

For all these reasons, the new approach should be executed and implemented as soon as possible. Today, thanks to the recent developments in information technology, the construction industry is benefited with visual databases. The multidisciplinary BIM Model, fully equipped with almost needed data for various disciplines from architecture to structure and building mechanical – electrical systems (MEP).

2. Building Information Modelling background

2.1. *Building Information Model (BIM)*

Building Information Model (BIM) is a sharing model that reproduces building information, from project concepts to completing tasks and destroying phases of the entire project life cycle [2]. All project stakeholders can insert, update, extract, and modify information through the BIM information integration platform at different stages.

BIM, the latest technology trends in the construction industry, the progress in Information – Telecommunication Technology (ICT) recently leads to the advance in numerous applications in construction and Building Information Modeling (BIM) would be the remarkable result of ICT progress. In construction also, the demands for new cutting edge technologies like BIM are crucial for modern companies in order to keep the company remaining and staying competent.

Through the whole life cycle, BIM model is not only being used in building structure design, but also can be used for bidding and tendering management, construction management, cost calculation, equipment management, property management, etc., to realize dynamic sharing and reuse of building information data, In the continuous improvement of the construction of the entire life cycle of information at the same time the construction of the latter part of the management and maintenance. There are plenty of tools for energy simulation and analysis according to Tian et al. [20].

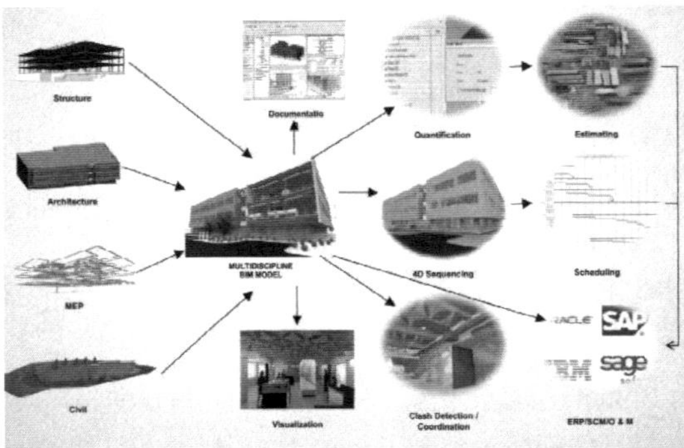

Figure 1. BIM Model for Multi disciplines (Adapted from Autodesk documents)

2.2. Sustainable design definition:

Sustainable Design is the intersection of sustainability and design innovation. Sustainable design is design that integrates, analyzes and optimizes environmental, social, and economic factors for the life of the product. Sustainable design is design that contributes to making the world a better place for future generations [2].

Figure 2. Sustainable Design Definition

Sustainable Design can deliver triple bottom benefits:

- **Economic Benefits:** Reduces the overall operating cost to the owner/occupant of the site or building, helps to improve productivity in employees.
- **Environmental Benefits:** Improve conditions of surrounding community and ecosystems, reduces overall cost to the owner/occupant.
- **Social Benefits:** Environmentally conscious construction practices will improve public health, safety and well-being.

Therefore, sustainable design is a desired goal and requirement for most construction projects.

3. Engineering design and computational models

Most of key BIM software providers such as Autodesk, Graphisoft, Nemetschek ...and a number of addins, plugins, software extensions developers provide all sphere solutions for sustainable design. There are many early works have been done to build the mathematical models [9] or using tools for building envelope optimization [3-5,7], cost or energy consumption minimization [5,8,10,14,15].

3.1. Analysis Tools for Sustainable Design

- **Conceptual Design Phase:** Autodesk Green Building Studio, Ecotect, IES VE Pro...

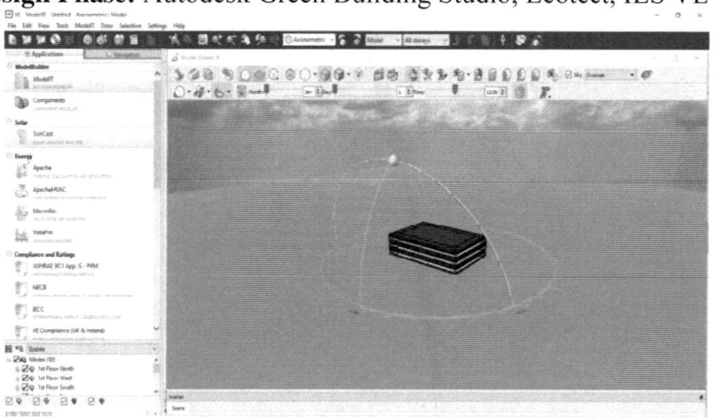

Figure 3. IES VE for conceptual analysis

- **Building Performance:** DesignBuilder, CFD Simulation, Simergy, Trace700...

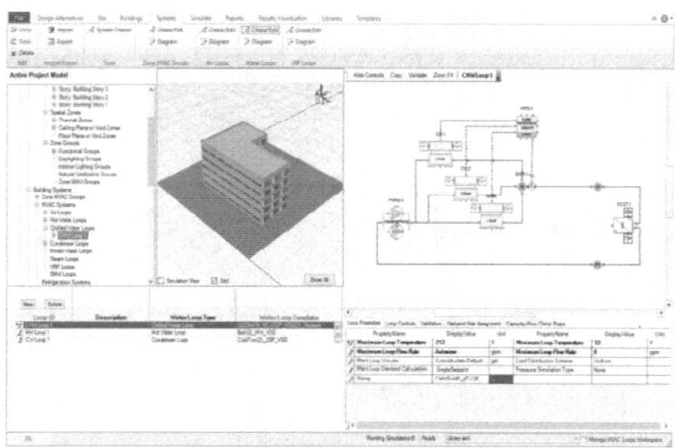

Figure 4. Simergy for MVAC operation analysis

- **Whole Building Energy Analysis:**
  - Solar Analysis: Light Analysis Extension for Revit, Ecotect...
  - Daylight Analysis: Light Analysis Extension for Revit, Ecotect
  - Thermal Analysis: Heat load calculation, Simergy, Trace700
- **Cost analysis:**
  - **OneClick LCA:** One Click LCA™ is an easy-to-use plugin that delivers an automated Life-Cycle Assessment and Life-Cycle Costing from Revit models.
  - **Tally:** Tally® is an Autodesk® Revit® application that allows architects and engineers to quantify the environmental impact of building materials for whole building analysis as well as comparative analyses of design options.

4. BIM Model for Multidisciplinary Design Optimization (MDO)

This session presents the method of finding the optimized form of the building facade to achieve optimal conditions for the visibility and the amount of Annual Sun Exposure (ASE) transmitted into the building.

NSGA-II algorithm (Non dominated Sorting Genetic Algorithm II) is used to solve the objective function. The algorithm is integrated into the Wallacei toolkit, a plugin for Rhinoceros software. The results are retrieved from the Ladybug toolkit on Grasshopper BIM platform. The Rhinoceros (Rhino) software was used instead of other design software such as AutoCAD, SolidWorks, ANSYS, Catia, etc…because Rhino software has Grasshopper plugin with many powerful features which are suitable for computational BIM for architecture while almost other software are for mechanical design.

The old Ngô Tất Tố Flat, located at 10°47'30.5"N 106°42'44.9"E, is investigated for effective façade design.

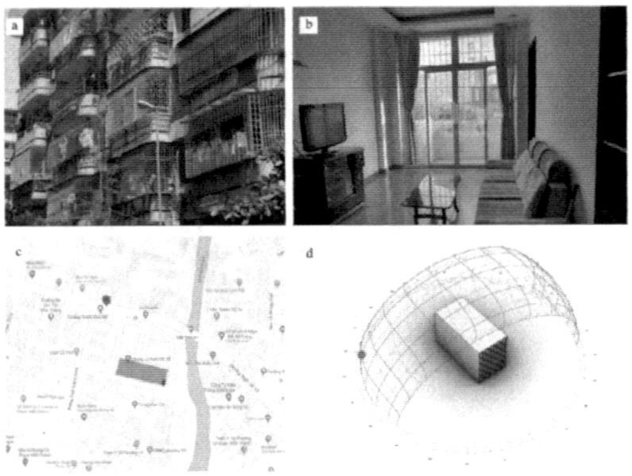

(a, b)Location, (c,d): Simulated room

Figure 5. BIM for Multidisciplinary Design Optimization (MDO)

Table 1. Simulation data

Name	Data
Length	5m
Width	2.5m
Height	3m
Area	12.5m^2
Operation hours	6AM -10PM
Function	Guest room + balcony

4.1. Weather data

Weather data plays a critical role in simulation. These simulations use the weather data from EnergyPlus, Department of Energy (DoE), USA for Ladybug toolkit.

Table 2. Design variable constraints

Design Variables	Range	Value
Number of Hexagon horizontal cells	4 … 10	6
Number of Hexagon vertical cells	4 … 10	6
Cell to Wall Ratio	0.40 … 0.80 (%)	40
Grid rotation angle in horizontal axis	-30 ... 30 (°)	60
Grid rotation angle in vertical axis	-30 ... 30 (°)	60

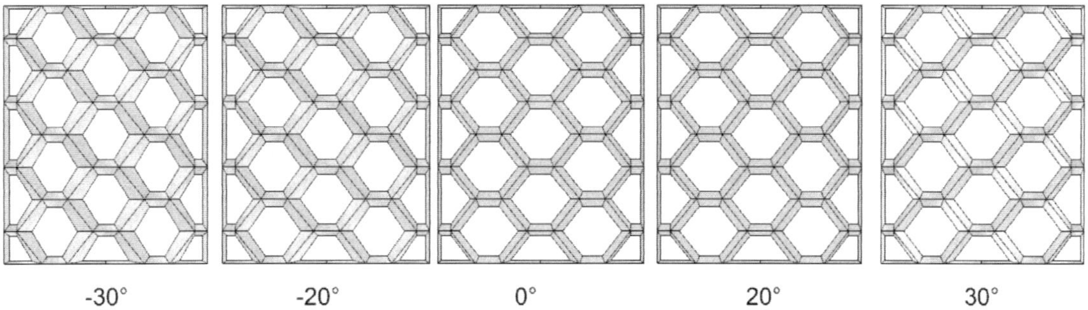

Figure 6. Façade components with different horizontal angle rotation

4.2. *Simulation results and discussions*

The objective function converges by 20 iterations with 10 options per iteration, finally 200 results analyzed to finalize the optimum solutions.

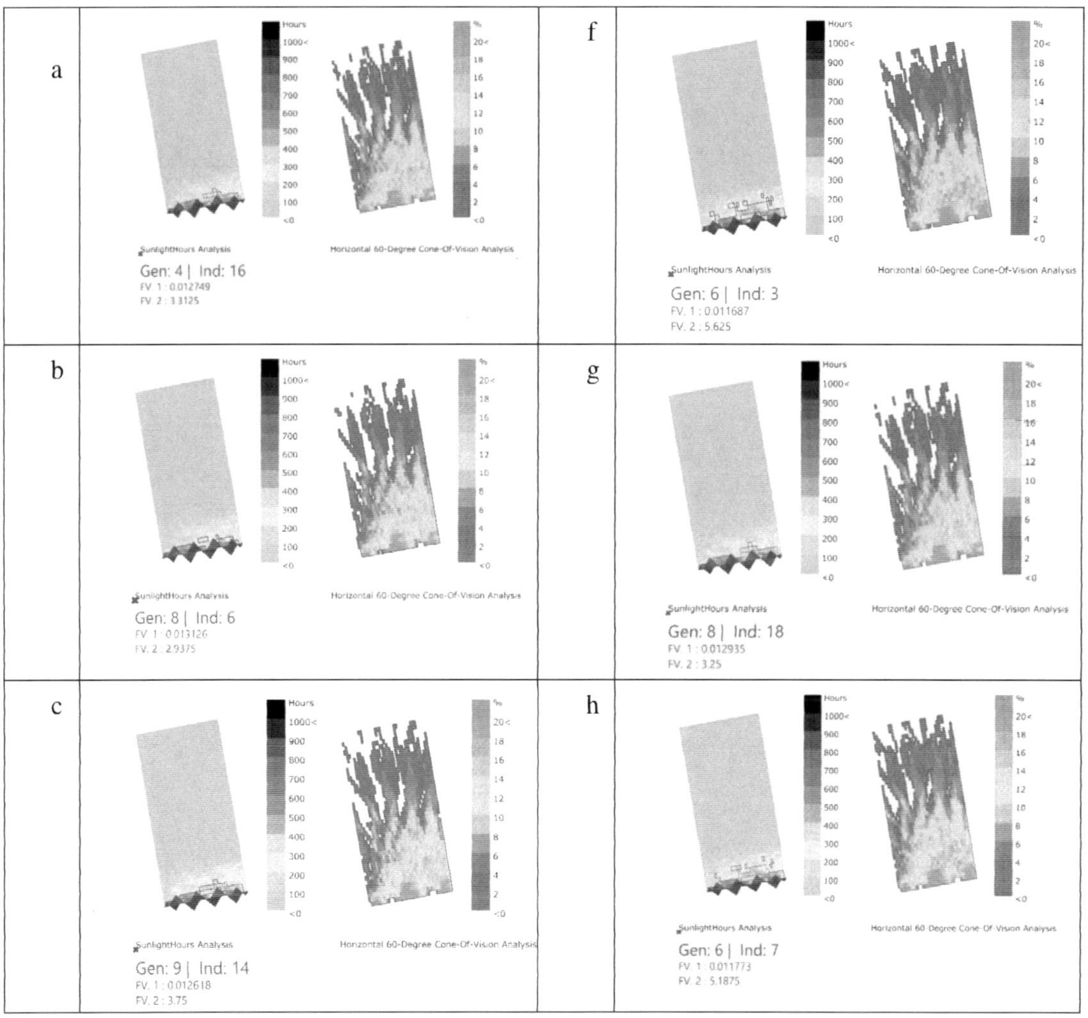

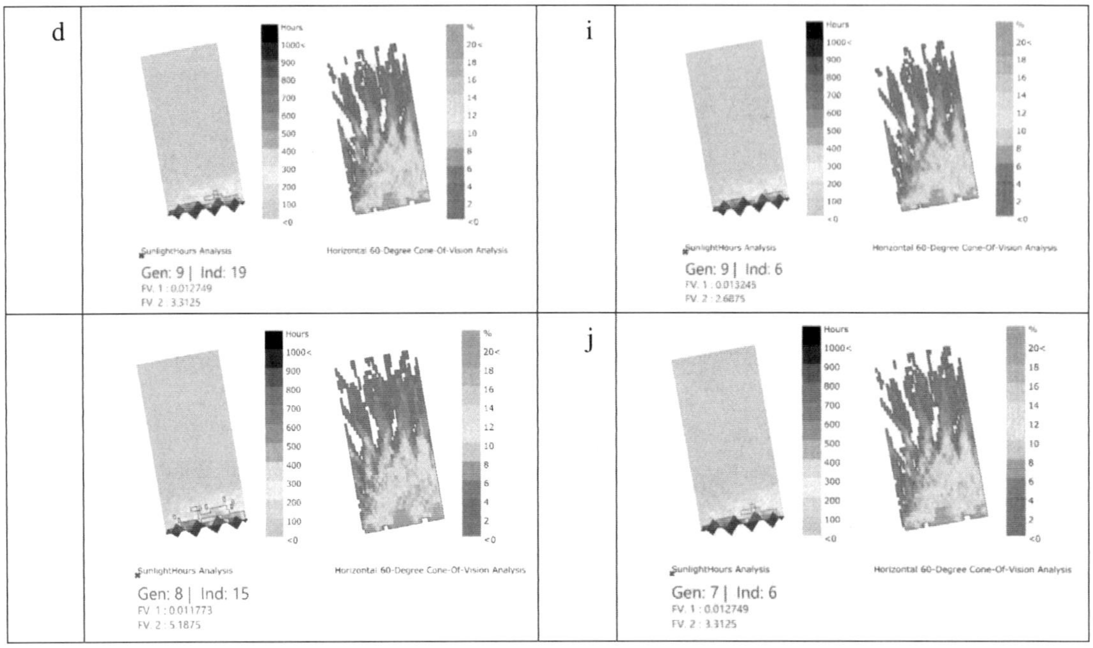

Figure 7. View analysis and Annual Sun Exposure results

Façade components simulated results with NSGA II are shown in figure 7 and table 3 in relevant scale respectively. All optimized analyses meet LEED v.4 with ASE <10% and View analysis in range 2% to 4%.

Table 3. Simulated data for objective function result comparison

	Option	**ASE (%)**	**View analysis >2%**
a	Option 1	3.3125	78.4%
b	Option 2	2.9375	76.2%
c	Option 3	3.75	79.2%
d	Option 4	3.3125	78.7%
e	Option 5	5.1875	79.9%
f	Option 6	5.625	85.6%
g	Option 7	3.25	77.5%
h	Option 8	5.1875	84.9%
i	Option 9	2.6875	75.5%
j	Option 10	3.3125	78.7%

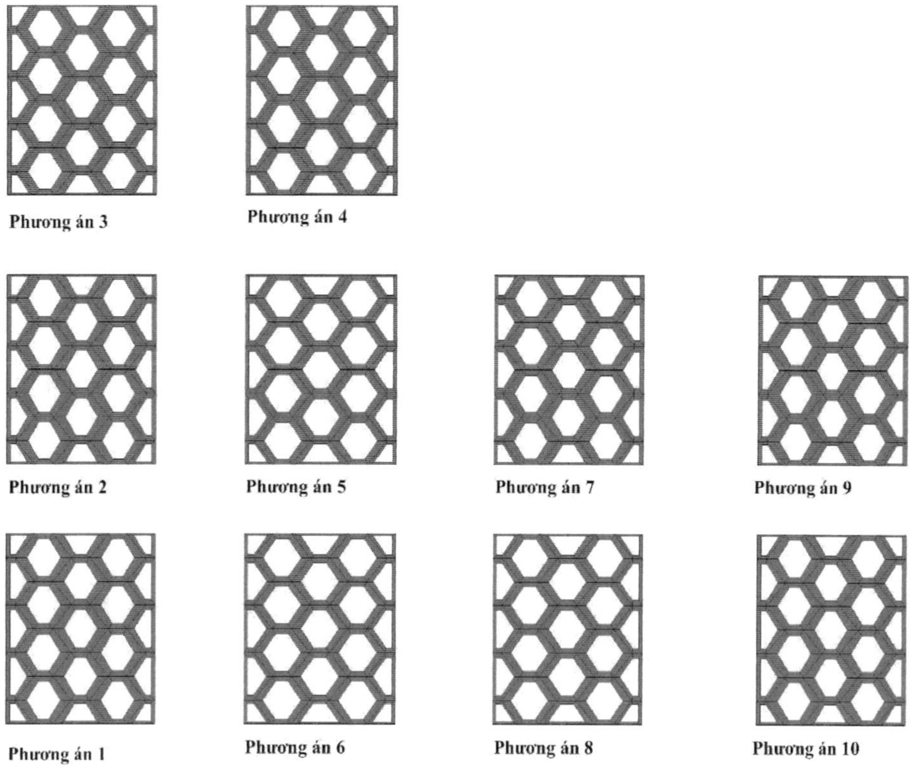

Figure 8. Form of selected façade components

5. Findings and Conclusions

Building Information Modelling received great concerns from largest firms, consultants and designers in the world. The multidisciplinary model can benefit all stakeholders from architectural, structural, mechanical designers, consultants and owners by all predictable outcomes on virtual models before the real buildings are built. The project stakeholders can estimate energy consumption, cost analysis, lifecycle cost with simulations done via virtual models.
This paper started the steps which show the solved optimization façade to reduce the solar radiation then energy consumption in general. The computable BIM models can help to perform energy analysis, cost analysis, wind load analysis…Those calculations were not able to be facilitated formerly.

This paper with all above mentioned software and tools shows sustainable design can be done easily for the overall optimization. The further works are carried on for Building Envelope Optimization and MVAC system optimization individually and finally the total solution reached using numerical methods like FEM and FVM.

References

[1] Architecture 2030 www.architecture2030.org
[2] Autodesk, "All documents for BIM and sustainable design".
[3] Caldas L G and Norford L K 2003 Genetic algorithms for optimization of building envelopes and the design and control of HVAC systems' *Journal of Solar Energy Engineering* **125**(3) p 343 doi: 10.1115/1.1591803.
[4] Cano E L *et al.* 2014 A strategic optimization model for energy systems planning *Energy and Buildings* (Elsevier B.V.) **81** pp 416–423. doi: 10.1016/j.enbuild.2014.06.030.
[5] Central P 2006 Building shape optimization using neural network and genetic algorithm

approach - Tags: NEURAL networks (Computer science) GENETIC algorithms'.
[6] Creese R C 2010 *Geometric Programming for Design and Cost Optimization (with illustrative case study problems and solutions), Second Edition, Synthesis Lectures on Engineering*. doi: 10.2200/S00314ED1V01Y201011ENG012.
[7] Garber B R 2009 Optimisation stories *Architectural Design* **79** (2) pp 6–13
[8] Grant E 1991 Constraint-based design by cost function optimization
[9] Hansen P, Lagouanelle J L and Messine F 2007 Comparison between Baumann and admissible simplex forms in interval analysis *Journal of Global Optimization* **37** (2) pp 215–228 doi: 10.1007/s10898-006-9045-9
[10] Kämpf J H and Robinson D 2010 Optimisation of building form for solar energy utilisation using constrained evolutionary algorithms *Energy and Buildings* (Elsevier B.V.) **42** (6) pp 807–814. doi: 10.1016/j.enbuild.2009.11.019.
[11] Ladenhauf D *et al.* 2016 Computational geometry in the context of building information modeling *Energy and Buildings* (Elsevier B.V.) **115** pp 78–84. doi: 10.1016/j.enbuild.2015.02.056.
[12] Lapinskiene V and Martinaitis V 2013 The framework of an optimization model for building envelope *Procedia Engineering* (Elsevier B.V.) **57** pp 670–677 doi: 10.1016/j.proeng.2013.04.085.
[13] Lartigue B, Lasternas B and Loftness V 2014 Multi-objective optimization of building envelope for energy consumption and daylight *Indoor and Built Environment* **23**(1) pp 70–80 doi: 10.1177/1420326X13480224.
[14] Marks W 1997 Multicriteria optimisation of shape of energy-saving buildings *Building and Environment* **32**(4) pp 331–339 doi: 10.1016/S0360-1323(96)00065-0.
[15] Marks W and Owczarek S 1997 Multicriteria optimisation of energy-saving buildings *Studies in Civil Engineering* **32**(4) pp 331–339.
[16] Okeil A 2010 A holistic approach to energy efficient building forms *Energy and Buildings* (Elsevier B.V.) **42**(9) pp 1437–1444. doi: 10.1016/j.enbuild.2010.03.013.
[17] Russell A D and Arlani G A (1981) Objective functions for optimal building design *Computer-Aided Design* **13**(6) pp 327–338. doi: 10.1016/0010-4485(81)90208-6.
[18] Shi X 2010 Performance-based and performance-driven architectural design and optimization *Frontiers of Architecture and Civil Engineering in China* **4**(4) pp 512–518 doi: 10.1007/s11709-010-0090-6
[19] United States Green Building Council *Building Impacts* presentation: *http://www.usgbc.org/DisplayPage.aspx?CMSPageID=1720*
[20] Tian Z C *et al.* 2015 Building energy optimization tools and their applicability in architectural conceptual design stage *Energy Procedia* (Elsevier B.V.) **78** pp 2572–2577 doi: 10.1016/j.egypro.2015.11.288
[21] Wang W, Rivard H and Zmeureanu R 2006 Floor shape optimization for green building design *Advanced Engineering Informatics* **20**(4) pp 363–378 doi: 10.1016/j.aei.2006.07.001

IOP Conf. Series: Materials Science and Engineering 1109 (2021) 012014 doi:10.1088/1757-899X/1109/1/012014

Numerical simulation of a solar chimney for natural ventilation of a building: Comparison of different computational domains

Trieu N Huynh and Y Q Nguyen

Faculty of Engineering, Van Lang University, 45 Nguyen Khac Nhu St., Co Giang Ward, Dist. 1, Ho Chi Minh City, Viet Nam

Email: y.nq@vlu.edu.vn

Abstract. Numerical simulations, particularly those based on the Computational Fluid Dynamics (CFD) Method, have been widely employed in engineering and design, such as design of green buildings. In this study, we investigated effects of the computational domain, a factor influencing the accuracy of a CFD simulation, of a solar chimney, a device for natural ventilation of buildings, on the predicted performance of a solar chimney attached to a building. Four different domains which, in turn, included both the chimney and the house, the chimney and the inlet length, the chimney with the horizontal inlet, and only the air channel of the chimney, were examined. The CFD model was based on the RANS equations and RNG $k-\epsilon$ turbulence model. The chimney had different heights, gaps, heat fluxes, and the location of the heat source in the air channel. The results show that the differences in the predicted flow rate and temperature rises through the chimney among the domains changed with the height, gap, and the location of the heat source. The simplest domain, which was a simple vertical rectangular channel, over – predicted the flow rate and under – predicted the temperature rise. The main cause of its worse performance is because of its inability to model the separation zones near the inlet of the air channel. Therefore, it is not recommended for high accuracy – simulations. Two other simpler domains, which included the inlet length and only the inlet, can be used when the required accuracy is within 10.0%. The full domain consisting of both the chimney and the building is preferred for the highest accuracy.

Keywords: solar chimney, natural ventilation, CFD, computational domain.

1. Introduction

Numerical simulations have been employed successfully in designing green buildings. They have been utilized for predicting wind load on buildings [1], designing natural ventilation solutions [2], or estimating energy consumption [3]. Among the natural ventilation solutions, solar chimneys have been examined by numerical simulations by several researchers [4–6].

Solar chimney refers to any mechanical system or building element which is based on thermal effects to induce airflow for ventilation or cooling of a building [7]. It absorbs solar radiation on a wall of an enclosed channel, which can be the cavity inside the building façade or a separate pipe or tube attached to a building. The absorbed heat warms the air in the channel and creates thermal effects to induce an air flow through the channel.

With a proper arrangement of the solar chimney, the induced air flow can ventilate or cool the building effectively. For example, it can provide sufficient ventilation rate [6] or thermal comfort conditions [8]. It can also save energy consumption of a building up to 50% [9].

For simulations of solar chimneys, the Computational Fluid Dynamics (CFD) method has been widely employed [4–6,10–13]. This technique is based on numerical approximations of the equations for conservation of mass, momentum, and energy to predict the flow and temperature patterns inside the solar chimneys and the connected buildings.

Among the factors influencing the accuracy of a CFD solution is the computational domain, which is the spatial volume or surface where the governing equations are approximated numerically. In the literature, different types of the domain have been employed for simulations of a solar chimney attached to a building. In some simulations [6,10], the domain consists of both the solar chimney and the building. In other cases [4,11], the domain only covers the solar chimney and air inlets on its side wall. A solar chimney is also simulated with a simple rectangular domain [12,13].

Gan [5] examined effects of the size of a computational domain for simulations of solar chimneys. He compared a simple rectangular domain covering only the air channel of a vertical solar chimney and an extended one which covered both the air channel and the surroundings of the solar chimney. Based on the results, he claimed that the simple domain is only suitable for solar chimneys with high ratios between the channel height and gap where the flow at the outlet of the air channel was not reversed.

As mentioned above, for simulations of a solar chimney, the employed domains are different for different studies where using simpler and smaller domains can save the computational cost. However, we have not seen a comparison among the results obtained with different domains for a solar chimney particularly for that connected to a building. Therefore, we examined effects of different types of the computational domain for simulation of the flow and temperature fields of a solar chimney attached to a building.

2. Method

2.1. *The problem*

Figure 1 describes a simple vertical solar chimney attached to a building in the vertical plane. The chimney comprises a cover plate placed outside a building wall to form a cavity, or channel, in the middle of the plate and the wall. The lower end of the channel is connected to the room while the upper end is open to the atmosphere. Solar radiation can be absorbed by the cover plate when the plate is opaque, or by the building wall, when the plate is transparent. The absorbed heat is transferred to the air in the channel. The air then rises due to thermal effects. Consequently, an air flow is induced through the room for ventilation.

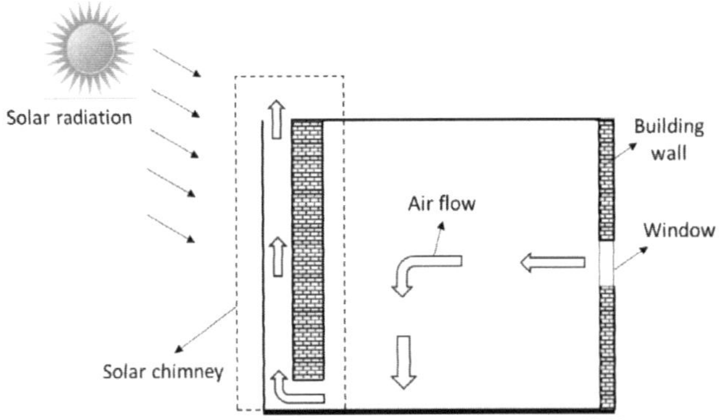

Figure 1. Solar chimney attached to a building.

In a typical solar chimney, heat exchange happens in the air channel and creates the thermal effect which is the main driving force of the flow. However, the results by Gan [5] shows that the induced flow is affected not only by the boundary conditions in the air channel but also the local dynamic conditions at the inlet and outlet of the air channel. Accordingly, for the problem considered in this

IOP Conf. Series: Materials Science and Engineering 1109 (2021) 012014 doi:10.1088/1757-899X/1109/1/012014

work, as seen in figure 1, the window and the inlet length from the room to the air channel should also affect the induced air flow in the air channel.

To examine effects of different parts of the system in figure 2 in the induced air flow, four different domains were considered, as presented in Figure 2. In figure 2a (domain SC+B), the domain covers both the solar chimney and the building. The domain is further extended beyond the window of the building to allow the air flow to adapt freely to the flow dynamics at the window. Air enters the domain on the open boundary (dashed lines) on the right side, receives heat in the air channel, and escapes through the upper end (outlet) of the air channel. In figure 2b (domain SC+L), the domain includes the air channel and the inlet length. Air enters the domain at the lower opening of the channel and leaves at the outlet. In figure 2c (domain SC+I), the domain is further simplified by excluding the inlet length from the domain in figure 2b. In figure 2d (domain SC), the domain is a simple rectangular channel covering only the channel height above the inlet. For consistent comparison among the domains, it is assumed that the air channel receives solar radiation along with the height above the inlet, i.e. the height h in figure 2a.

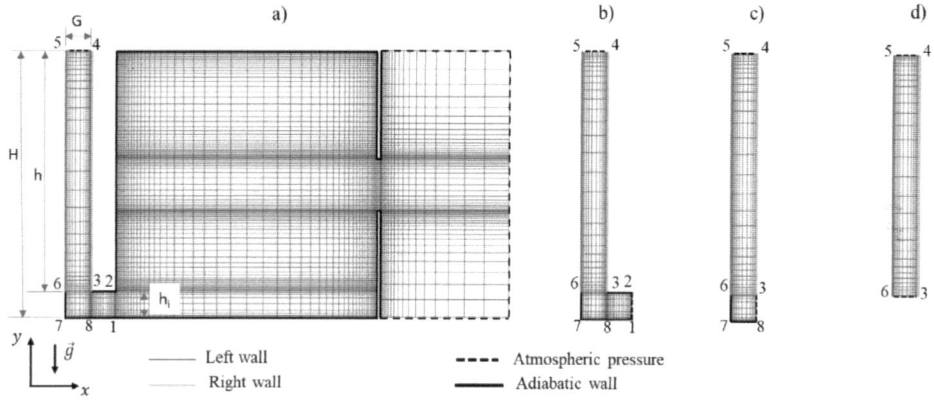

Figure 2. Computational domain including: a) Solar chimney and the building (SC+B), b) Air channel and the inlet length (SC+L), c) Air channel and horizontal inlet (SC+I), and d) Air channel only (SC)

2.2. *Numerical model*

The governing equations for the air flow and heat transfer were predicted by RANS (Reynolds Averaged Navier – Stokes equations) method in CFD. In these equations, time – averaged quantities for velocity, pressure, and temperature are described in equations (1) – (3).

$$\frac{\partial u_j}{\partial x_j} = 0 \tag{1}$$

$$\frac{\partial (u_i u_j)}{\partial x_j} = -\frac{1}{\rho}\frac{\partial p}{\partial x_i} - g_i\beta(T - T_0) + \frac{\partial}{\partial x_j}\left(\nu\frac{\partial u_i}{\partial x_j} - \underline{u_i' u_j'}\right) \tag{2}$$

$$\frac{\partial (T u_j)}{\partial x_j} = \frac{\partial}{\partial x_j}\left(\frac{\nu}{Pr}\frac{\partial T}{\partial x_j} - \underline{T' u_j'}\right) \tag{3}$$

With these equations, it is assumed that the flow and heat transfer are two dimensional and steady, and the air flow is incompressible. As the air in the channel is warmed, its temperature increases from T_0 at the inlet to T at a point in the channel. Variations of the air properties are described by the Boussinesq approximation in equation (2). Other quantities in equation (1) to (3) are as follows:

- u and u' are respectively the time – averaged and fluctuating velocity.
- T and T' are respectively the time – averaged and fluctuating temperature.
- p stands for pressure.
- ρ, ν, and β are the air density, kinematic viscosity, and thermal expansion, respectively.
- Pr and g are the Prandtl number and gravitational acceleration, respectively.

In equation (2) and (3), two turbulence quantities $\overline{u_i'u_j'}$ and $\overline{T'u_j'}$ were obtained with the RNG $k-\epsilon$ turbulence model, which has been used for simulations of solar chimneys in the literature [5,14]. Other turbulence models, such as the standard $k-\epsilon$, standard $k-\omega$, standard $k-\omega$ with modifications for low Reynolds number effects were also tested. The RNG $k-\epsilon$ offered the best convergence rate. Therefore, it was selected.

Equations (1) to (3) were numerically discretized by the Finite Volume Method on staggered rectangular meshes, as seen in figure 2, by the CFD code ANSYS Fluent (Academic version 2020 R2). The meshes were clustered near the solid surfaces. For all cases in figure 2, the mesh density was higher toward the solid walls. By changing the number of cells and the growth rate of the mesh cells from the solid surfaces, the induced flow rate through the chimney was compared. It was seen that the difference in the flow rate was less than 1.0% when the maximum non–dimensional distance of the first cell centre from the solid walls, or y^+, was less than 1.0. The according mesh resolution was more than 8000 cells in a solar chimney of H=1.0 m and G=0.1 m, and a heat flux of 1000 W/m^2. For all tests in this study, the mesh was carefully checked to satisfy that required maximum y^+.

Other settings of the numerical scheme included the SIMPLE method for the coupling between the continuity equation (equation (1)) and the momentum equation (equation (2)), the PRESTO! method for interpolation of the pressure on the staggered mesh, second order upwind scheme for equations (2) and (3), and first order upwind scheme for the equation for k and ϵ.

In figure 2, all open boundaries were applied with the atmospheric pressure. The ambient temperature T_0 was assigned at all inlets. All walls were assumed to be no – slip. As seen in figure 1, the heat source can be either on the left or the right wall of the channel. On the heated wall in the air channel, a uniform heat flux I was distributed. The opposite wall in the air channel received radiative heat transfer from the heated wall, which was described by the S2S model in ANSYS Fluent. Other walls were adiabatic. Details of the above numerical setup are similar to those in our previous works [15, 16].

The CFD model was validated against the experiment by Burek and Habeb [17] for a vertical solar chimney whose height was 1.025 m. The channel gap varied from 40 mm to 110 mm. A heat flux of 422 W/m^2 was applied on a wall of the air channel. The flow rate through the chimney obtained with the CFD model was compared with the measured one and plotted in figure 3. It is seen that the CFD model slightly over – predicted the flow rate. It may be due to heat loss through the walls of the air channel in the experiment that was not considered in the CFD model. However, the maximum discrepancy was less than 10.0% and should be within the measurement uncertainties.

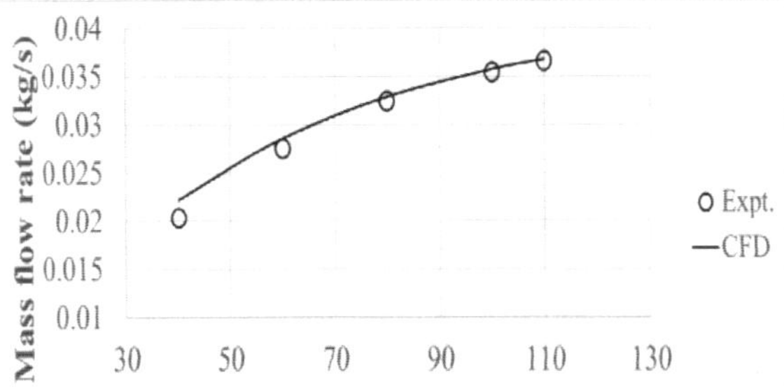

Figure 3. Computed (CFD) and measured (Expt.) flow rate through the solar chimney with a heat flux of 422 W/m^2 in the experiment by Burek and Habeb [17].

3. Results and Discussions

Comparisons of four computational domains shown in figure 2 are presented in this section in terms of the induced flow rate and temperature rise through the air channel. Changing factors included the chimney H, the heat flux applied on the heated wall, and the gap G of the air channel. The location of the heat source was either on the left (LWH) or the right wall (RWH) of the air channel.

3.1. *Induced flow rate*

The induced flow rate for different heights is presented in figure 4. Three heights of H=1.0 m, 2.0 m, and 3.0 m were tested. The heat flux varied from 200 to 1000 W/m^2. The air gap, G, and the inlet height, h_i, were fixed to 0.1 m in all tests.

For LWH (figure 4a), it is seen that at H=1.0m, all four domains offered flow rates similar to each other with the maximum difference of less than 5.0%. However, as the height increased further, at H=2.0 m and 3.0 m, the SC domain predicted the highest flow rate. Three other domains (SC+B, SC+I, and SC+L) had similar flow rates. At H= 3.0 m, the maximum difference between flow rates obtained with SC and with SC + B was up to 14.0% and the differences among three domains (SC+B, SC+I, and SC+L) were within 3.0%.

Similar trends were also seen for RWH in figure 4b. However, at H=1.0 m and 2.0 m, the differences in the flow rates among the domains were higher than those for LWH in figure 4a. The highest flow rate was with the SC+I at H=1.0m and with the SC at H=2.0 m. In both cases, the highest flow rate was 10.0% higher than that with the SC+B. At H=3.0 m, the SC also offered the highest flow rate, similar to figure 4a, and up to 13.0% higher than that with the SC+B.

From the observations for both LWH and RWH, it is concluded that the differences among the flow rates obtained with four domains depended on both the channel height and the location of the heat source for H=1.0 m and 2.0 m. However, at H=3.0 m, heating either side had very similar results.

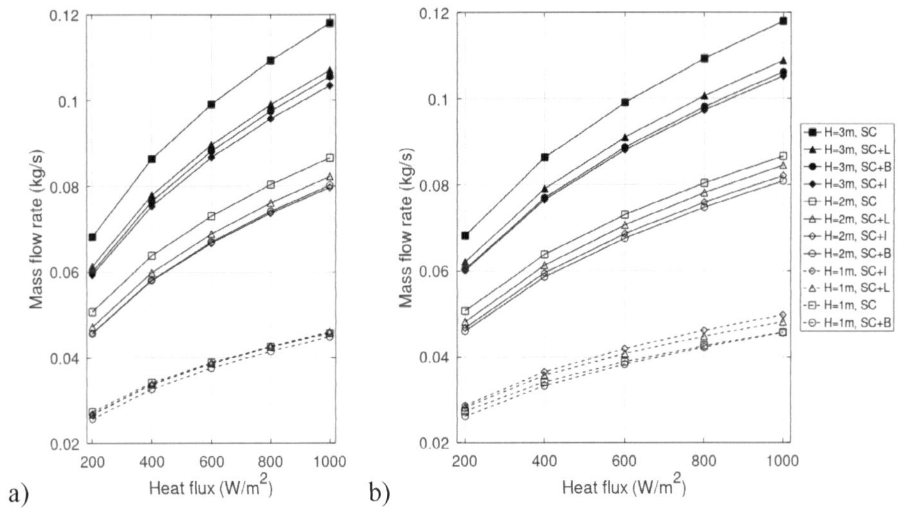

Figure 4. Induced mass flow rate at different heights for heating: a) Left wall, b) Right wall

Figure 5 compares the flow fields for H=3.0 m, G=0.1 m, and the heat flux of 600 W/m^2 on the left wall (LWH). The flow fields are seen to be different mainly near the inlet of the air channel. With SC+B, three separation zones existed and were marked as (i), (ii), and (iii) in figure 5a. Two zones (i) and (ii) also appeared in SC+L and SC+I domains (figures 5b and 5c) and were similar to those in the SC+B domain. It is seen that the size of the zone (iii) is smaller than those of the zones (i) and (ii). In contrast, the domain SC did not result in a separation zone at the inlet. By referring to the flow rates obtained with each domain in figure 4a for this case, it is expected that the reduction of the flow rates with SC+B, SC+L, and SC+I compared to those of the SC is due to the separation zones. These zones contracted the flow area near the inlet; hence increased the flow resistance, or pressure loss. As the separation zones in the domains SC+B, SC+L, and SC+I were identical, the induced flow rates of those domains should be also similar to each other, as seen in figure 4a.

In figure 6, the flow rates obtained with four domains at different gaps are presented for H=1.0 m and H=3.0 m. Similar trends are observed in both cases of heating. At H=1.0 m, the flow rates in all four domains are seen to be close to each other. The differences among them were within 5.0%. At H=3.0 m, at all gaps, three domains SC+B, SC+L, and SC+I had similar flow rates with the maximum difference of less than 5.0%. The flow rates with SC were higher than those with three other domains. The difference between the flow rates with SC and SC+B increased with the gap. That difference was about 2.0% at G=2.5 cm but was up to 12.0% at G=10 cm. Therefore, the difference was magnified as the gap increased.

Examination of the flow fields for the cases in figure 6 reveals that the separation zones near the inlet were similar to those in figure 5. However, their sizes increased with the gap and led to more flow resistance. Consequently, the difference in the flow rates obtained with the SC and the others were enhanced as the gap increased.

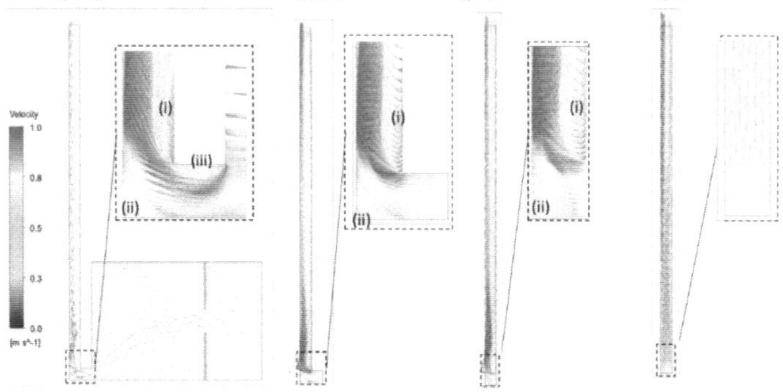

Figure 5. Flow fields for H=3.0 m, G=0.1 m, and the heat flux of 600 W/m^2 on the left wall (LWH).

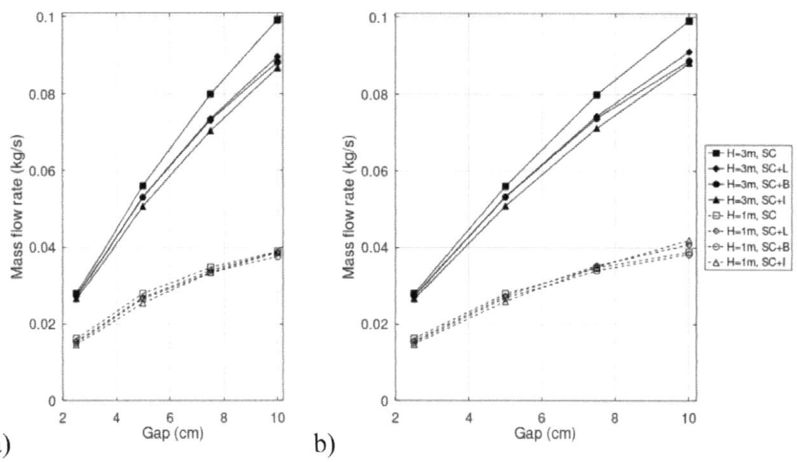

a) b)

Figure 6. Induced mass flow rate at different gaps and heights for heating left (a) and right (b) walls.

3.2. *Temperature rise*

Figure 7 shows the temperature difference between the inlet and the outlet of the air channel obtained with four domains for both cases of heating (H=1.0 m and 3.0 m, G=0.1 m). Similar trends and values for a given height are seen from figures 7a and 7b. In both cases of heating and both heights, the SC offered the lowest temperature rise while the SC+B offered the highest.

At H=3.0 m, three domains SC+B, SC+L, and SC+I had identical temperature rise which was up to 3.1 K higher than that of the SC. Figure 8 shows the temperature fields and confirms that difference. The temperature field near the outlet of SC is seen to be lower than those of the other cases, which are similar to each other.

The temperature fields near the inlet of the air channel may explain the difference in the temperature rise obtained with SC and with three other domains. For three other domains except for the SC, the separation zones contracted the flow field and enhanced the mixing and the expansion of the thermal boundary layers near both walls of the air channel. As a result, at the outlet, the temperature field is more uniform and the temperature rise was higher for three domains SC+B, SC+L, and SC+I.

Figure 7 shows that less difference in the temperature rise is seen among the domains for RWH. For example, at H=3.0 m, the difference between the SC and SC+B was 2.3 K for RWH but 3.1 K for LWH. It may be because when the right wall was heated, part of it was embedded in the separation zone (i) (figure 5), which should exchange heat worse than a straight wall did. Therefore, the effects of the separation zone (i) on enhancing the mixing of the temperature field should be less than those when the left wall was heated.

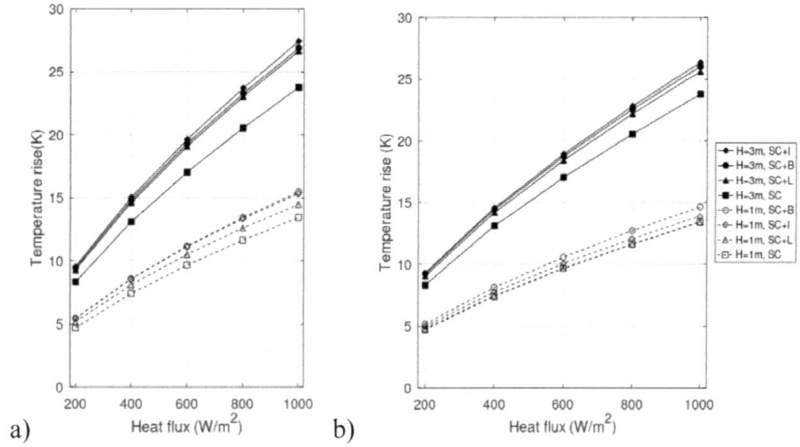

Figure 7. Temperature rise at different heights for heating left (a) and right (b) walls.

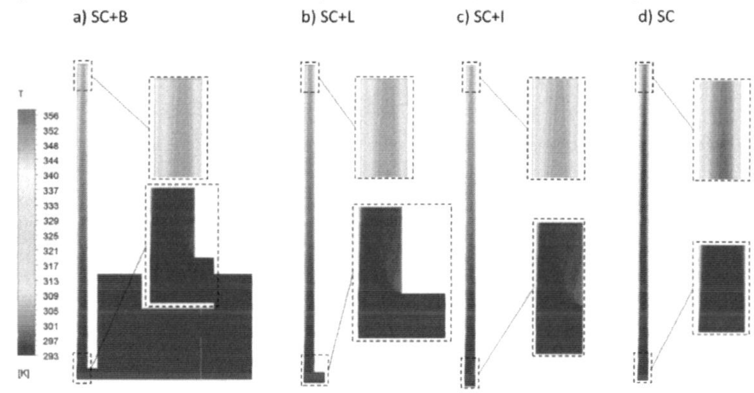

Figure 8. Temperature fields for H=3.0 m, G=0.1 m, and the heat flux of 600 W/m^2 on the left wall (LWH).

4. Conclusions

The results show that the differences in the induced flow rate and temperature rise through the solar chimney attached to a building simulated with four different domains changed with the height, gap, and location of the heat source, as follows:

- At H=1.0 m, all domains had similar flow rates with differences of within 5.0% for LWH but up to 10.0% for RWH.
- At H=2.0 m and 3.0 m, the SC always offered the highest flowrate while three other domains had similar flow rates of within 5.0% difference. The flow rate obtained with SC was up to 14.0% higher than those with the SC+B. LWH and RWH had similar performance.

The existence of the separation zones near the inlet in three domains SC+B, SC+L, and SC+I, which are similar to each other, may increase the flow resistance in those three domains and reduce their flow rate compared to that with the SC.

The SC always offered the lowest temperature rise while the SC+B offered the highest. The difference was up to 3.1 K at H=3.0 m for LWH and 2.3 K for RWH. Besides SC, three other domains had similar temperature rises with differences of within less than 1.0 K. The separation zones in three domains SC+B, SC+L, and SC+I may enhance the mixing of the temperature field and result in higher temperature rise in those three domains.

From the results, it is concluded that the SC domain should not be employed for this problem as it over – predicted the flow rate and under – predicted the temperature rise. The other two domains, i.e. SC+L and SC+I, may be used when the required accuracy is within 10.0%. For eliminating the effects of the domain on the results, the full domain, i.e. SC+B, should always be used.

The results from this study can serve as a reference for architects and building engineers using CFD for building design. Depending on the required accuracy of the solutions, a particular type of the computational domain can be referred. For example, for the first stage of the design, when testing different scenarios of a solar chimney, they may employ the domain SC for a quick comparison among the design scenarios. At later stages, such as for calculating energy and thermal comfort performance of the building, the full domain SC+B can be used for more accurate and detailed results.

References

[1] Thordal M S, Bennetsen J Chr and Koss H H H 2019 Review for practical application of CFD for the determination of wind load on high-rise buildings *Journal of Wind Engineering and Industrial Aerodynamics* **186** 155–168

[2] van Hooff T, Blocken B and Tominaga Y 2017 On the accuracy of CFD simulations of cross-ventilation flows for a generic isolated building: Comparison of RANS, LES and experiments *Building and Environment* **114** 148–165

[3] Al Touma A and Ouahrani D 2018 Performance assessment of evaporatively-cooled window driven by solar chimney in hot and humid climates *Solar Energy* **169** 187–195

[4] Zamora B and Kaiser A S 2009 Optimum wall-to-wall spacing in solar chimney shaped channels in natural convection by numerical investigation *Applied Thermal Engineering* **29** (4) 762–769

[5] Gan G 2010 Impact of computational domain on the prediction of buoyancy-driven ventilation cooling *Building and Environment* **45** (5) 1173–1183

[6] Shi L, Ziem A, Zhang G, Li J and Setunge S 2020 Solar chimney for a real building considering both energy-saving and fire safety – a case study *Energy and Buildings* **221** 110016

[7] Bansal N K, Mathur R and Bhandari M S 1993 Solar chimney for enhanced stack ventilation *Building and Environment* **28** (3) 373–377

[8] Tan A Y K and Wong N H 2012 Natural ventilation performance of classroom with solar chimney system *Energy and Buildings* **53** 19–27

[9] Miyazaki T, Akisawa A and Kashiwagi T 2006 The effects of solar chimneys on thermal load mitigation of office buildings under the Japanese climate *Renewable Energy* **31** (7) 987–1010

[10] Bassiouny R and Korah N S A 2009 Effect of solar chimney inclination angle on space flow pattern and ventilation rate *Energy and Buildings* **41** (2) 190–196

[11] Khanal R and Lei C 2012 Flow reversal effects on buoyancy induced air flow in a solar chimney *Solar Energy* **86** (9) 2783–2794

[12] Zavala-Guillén I, Xamán J, Hernández-Pérez I, Hernández-Lopéz I, Gijón-Rivera M and Chávez Y 2018 Numerical study of the optimum width of 2a diurnal double air-channel solar chimney *Energy* **147** 403–417

[13] Zhang T and Yang H 2019 Flow and heat transfer characteristics of natural convection in vertical air channels of double-skin solar façades *Applied Energy* **242** 107–120

[14] Zhang T, Zhao Y and Wang S 2014 Prediction of airflow rate through a ventilated wall module *Energy and Buildings* **82** 651–659

[15] Nguyen Y Q and Wells J C 2020 Effects of wall proximity on the airflow in a vertical solar chimney for natural ventilation of dwellings *Journal of Building Physics* **44** (3) 225–250

[16] Nguyen Y Q and Wells J C 2020 A numerical study on induced flowrate and thermal efficiency

of a solar chimney with horizontal absorber surface for ventilation of buildings *Journal of Building Engineering* **28** 101050
[17] Burek S A M and Habeb A 2007 Air flow and thermal efficiency characteristics in solar chimneys and Trombe Walls *Energy and Buildings* **39** (2) 128–135.

IOP Conf. Series: Materials Science and Engineering 1109 (2021) 012015 doi:10.1088/1757-899X/1109/1/012015

Performance evaluation and characterization of a locally designed water pump

A S Pascual[1], A L Fajardo[2], O F Zubia[2] and R A Luyun Jr.[2]

[1] Department of Science and Technology-Engineering Research and Development for Technology, Philippines
[2] College of Engineering and Agro-Industrial Technology (CEAT), UPLB, College, Laguna, Philippines

aspascual2@up.edu.ph

Abstract. *Voltron*, a locally designed water pump, is widely used among farmers in Central Luzon of the Philippines. The study aimed to evaluate the performance of the *Voltron* water pump. *Voltron* was tested at different pump shaft speed settings of 450, 600, and 700 rpm and with different blade (impeller vanes) numbers of 6, 8, 9, and 10 using V-belts and B-section pulleys transmission. The highest pump efficiency of 62.5% was observed using an 8-vane impeller under a 700-rpm setting. Currently, farmers utilized the 8-vane impeller under a 600-rpm setting. Statistical analysis showed that there is no significant difference between the pump efficiencies on the set pump shaft speeds, but there is a significant difference in the set number of vanes. On the other hand, the effects of pump shaft speed, the head, discharge, and fuel consumption increased as rotational speed was increased. The difference is around Php 17.61 per ha per day of operation favoring the usual setting upon the computation of the fuel cost. However, the 8-vane impeller under a 700-rpm setting will be more economical in the long run as it is determined to be more efficient than the normal setting being used by the farmers.

1. Introduction

Successful agriculture depends on the accessibility of water to farmers to increase cropping intensity and productivity. Aside from the seasonal rainwater, the utilization of a pumping system enables the farmers to draw water from existing open surface water resources (e.g., rivers, creeks, shallow groundwater) available within the area. The most common type of pump being utilized by farmers in the Philippines is the centrifugal pump.

In Vietnam and Thailand, farmers used an energy-efficient alternative to centrifugal pumps called Axial Flow Pumps or AFPs. According to the Cereal Systems Initiative for South Asia (CSISA) Factsheet (2014), AFPs can save more than 50% of the diesel/energy compared to the conventional centrifugal pumps when the lift is not more than 3 meters (9.84 feet). Since AFPs may only be used for lowland rice irrigation, local farmers in Region III developed a pump design that they claim to be as energy efficient as AFPs. The local pump design is called a Voltron pump as shown in Figure 1.

Figure 1. Field installation of the *Voltron* pump.

The Voltron pump uses an impeller in a volute casing submerged in water instead of a propeller as in AFPs. However, the Voltron pump can lift water with heads of more than 3 meters. The Voltron pump could be considered a centrifugal pump. A centrifugal pump is a rotating machine in which flow and pressure are generated dynamically. A centrifugal pump delivers useful energy to the fluid or "pumpage" largely through velocity changes that occur as this fluid flows through the impeller and the associated fixed passageways of the pump [2].

It is also a "rotodynamic" pump, namely, a radial-type centrifugal pump. The fluid enters the impeller axially through the inlet eye and is then forced to rotate by the impeller vanes. While rotating inside the impeller, the fluid moves outward, thus gaining an increase in pressure with a parallel increase in kinetic energy. The high velocity at the impeller exit is transformed into a pressure increase through the pump casing and discharge nozzle [3].

Just like many AFPs, the Voltron pumps are normally powered by a single-cylinder engine. Despite lacking technical information on pump performance, this peculiar pump setup is locally produced and continuously gaining popularity among local farmers in the Central Luzon of the Philippines.

This study aimed to determine the performance characteristics of the Voltron pump to validate the local farmers' claims on pump performance. The effect of varying the shaft speed and the number of impeller blades (vanes) on the pump's performance was likewise studied.

2. Materials and Methods

2.1. Preliminary Data Gathering

A survey was conducted to gather sufficient information about the operation of the *Voltron* pump in selected areas of Region III. The Voltron pumps have an overall length of at least 3.05 meters (10 feet). The usual number of impeller vanes was eight (8). A 6-inch reducer elbow (to 4-inch) at the discharge side was normally used. The elbow reducer can be rotated so that the discharge port can be directed as preferred. Farmers normally set the pump shaft speed between 300 and 600 rpm. The pulley combination varied per farmer, but most utilized a speed ratio of 1:2 (4-inch diameter pulley for the engine and an 8-inch diameter pulley for the pump). Double sheave pulleys were used to reduce belt slippage. A 5.5 kW (8 hp) diesel engine was commonly used primarily because diesel fuel is a lot cheaper than gasoline. Data gathered from these interviews served as the basis for setting the operational speed and number of vanes of the pump during testing.

2.2. Fabrication of Voltron pump

The Voltron pump (with a clutch system) used in the study was fabricated in a local shop in Brgy. Lagundi, Plaridel, Bulacan, Philippines. The fabricated Voltron water pump had a semi-open type of impeller made from MS plates and has an outside diameter of 29 cm (Figure 2). The volute casing and elbow reducer were constructed using gauge 16 GI sheets. The long shaft used was 1-inch cold-rolled steel and was supported by three pillow block bearings. A universal joint was used as coupling to facilitate a slight angle difference in elevation during pumping operation.

Shafting supports were made of GI pipes, flat bars, and angled bars. The engine base was constructed using 1.5-inch of steel angle bars. The pump has an overall 8-feet length as opposed to the usual 10-feet basically to fit in the testing facility's reservoir.

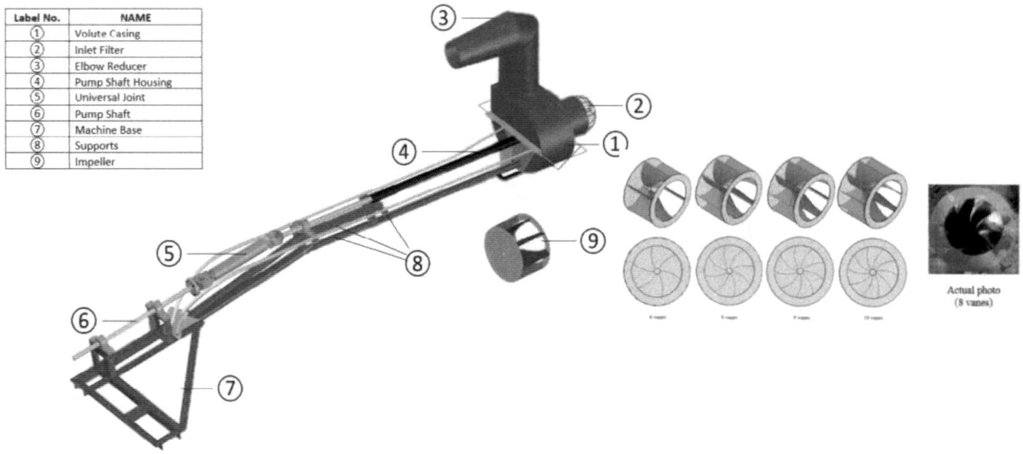

Figure 2. Detailed drawing of the *Voltron* pump with a different number of vane configurations.

2.3. *Method of test*

The pump was tested following PAES 115:2000 - Centrifugal, Mixed Flow and Axial Flow Water Pumps – Methods of Test at the Mechanical Testing Facility of the Agricultural Machinery Testing and Evaluation Center (AMTEC-UPLB), Los Baños, Laguna, Philippines. The schematic diagram of the test setup is shown in Figure 3.

A Yanmar TF 90 MLY diesel engine with 7.1 kW (9.5 hp) output power was used for this study to set an allowance for higher speed settings and possible higher power requirements. Based on the survey, the usual engine used by local farmers was 5.5 kW (8 hp). The said engine was mounted on a steel base, while the pump was submerged at a height from the water source.

The pump was connected to a torque transducer using 2B pulleys with diameters of 203 mm (8 inches) and 114 mm (4.5 inches) respectively and two pieces of B-69 V-belts. The other end of the torque transducer was driven by the diesel engine, which was also connected using similar belt-drive transmission. A 10-ml pipette attached to the engine's fuel tank and a stopwatch was used to measure the fuel consumption of the engine. The pressure gauge (head measurement) was only connected at the discharge side since the suction side of the pump was submerged underwater.

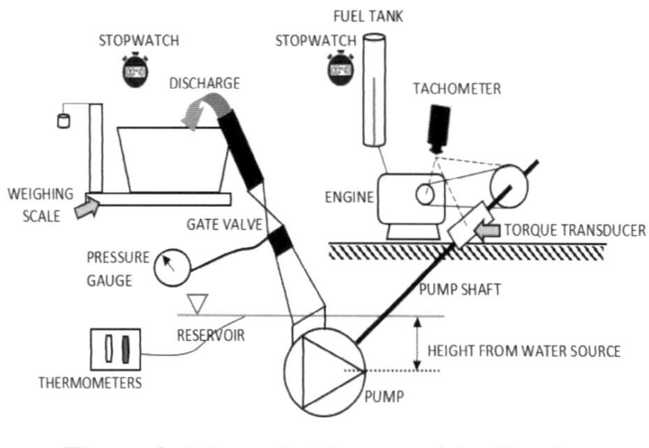

Figure 3. Schematic Diagram of the Test Set-up.

2.4. *Design of Experiment*

The experiment employed a Split Plot Design using a Completely Randomized Design. ANOVA was used to determine the interaction between pump shaft speed and the number of vanes. On the other hand, the speed for response variables such as pump efficiency, total head, water discharge, and fuel consumption were analyzed using the Statistical Tool for Agricultural Research (STAR) 2.0.1. When there were no significant differences found between the set interaction, the Least Significant Differences (LSD) Test was used at α=0.05. However, when significant effects were found on an interaction, means were separated using Tukey's Honest Significant Difference Test at α=0.05 with R version 3.6.1. The Shapiro-Wilk test and Bartlett's test were used to identify its normality and homogeneity, respectively.

The test was divided into two parts: (a) determination of the effect of shaft speed (450 ±10 rpm, 600 ±10 rpm, and 700 ±10 rpm) on the pump's performance, and (b) determination of the effect of the number of impeller vanes (6, 8, 9, and 10) on the pump's performance. Based on the interview, the usual speed for the pump ranges from 300 to 600 rpm while the usual blade number of the impeller is 8 vanes. The 300-rpm speed from the interview was not achieved in the study because its discharge was insufficient for data collection and analysis. On the other hand, 6, 9, and 10 impeller vanes were additionally chosen as the angles will be more exact in comparison with 7 vanes, which translated to easier fabrication.

Before starting the test, the height of the water level, the eye of the impeller, and gauge pressure was obtained with respect to the ground as references for computation. During testing, the following data were obtained and/or measured: 1) Level of water from the eye of the impeller, cm; 2) Pump shaft speed, rpm; 3) Torque transducer speed, rpm; 4) Engine speed, rpm; 5) Torque, kg-m; 6) Reading of pressure gauge on the discharge side, psi; 7) Time to consume 10 ml of fuel, s; 8) Water discharge, kg; 9) Time elapsed from initial to the final weight of water discharge, s.

Relative humidity and ambient temperature measurements were also gathered during the testing proper with the use of a hygrometer. The following pump performance information were also determined: 1) Total Head (m) vs Discharge Capacity (L/s); 2) Waterpower (kW) vs Discharge Capacity (L/s); 3) Pump Input Power based on shaft torque (kW) vs Discharge Capacity (L/s); 4) Fuel Consumption (L/hr) vs Discharge Capacity (L/s); 5) Pump Efficiency (%) vs Discharge Capacity (L/s). All equations and computations used were based on PAES 115:2000.

3. Results and Discussion

3.1. Pump characteristics curves

From the pump characteristic curves (Figures 4 to 7), the general trend observed was the indirect proportionality of the total head and discharge. The total head decreased as the discharge increased. Compared with most centrifugal pumps, the Voltron pump efficiency peak at a certain point then decreases after. However, the pump efficiency increases with head and discharge with 9 and 10 vanes at 700 rpm shaft speed settings. Also, waterpower and fuel consumption increased as head, discharge, and pumps shaft speed increased. Work done on the fluid and energy transfer (from mechanical power to fluid power) only occurs because of the vanes. Also, the pump performance characteristics and the overall efficiency depend mainly on the vane shape and number of vanes [3].

The highest pump efficiency obtained was 62.5% at 8 vanes with a 700-rpm setting (Figure 8). From the result of ANOVA in Table 1, the interaction of all sources of variation for pump efficiency and the head is highly significant. From the results of Tukey's HSD test based on pump shaft speed (Table 2), there is no significant difference between the pump efficiencies of the set pump shaft speeds at a 95% level of significance. However, there is a significant difference between pump efficiencies (except for 8 and 9 blades) at a 95% level of significance based on the number of vanes (Table 3).

Table 1. ANOVA table with pump efficiency as a response variable.

Sources of Variation	F value	P-Value
No. of vanes	119.06	0.0000^b
Pump shaft speed	13.49	0.0004^b
No. of vanes x Pump shaft speed	11.00	0.0001^b
CV(a)%=5.44 (b)%=9.82		

^{a} significant,
^{b} highly significant
^{c} not significant

IOP Conf. Series: Materials Science and Engineering 1109 (2021) 012015 doi:10.1088/1757-899X/1109/1/012015

Table 2. Effect of pump shaft speed on characteristics of the locally designed pump [d].

RPM	Pump Eff. (%)	Head (m)	Discharge (L/s)	Fuel Consumption (L/hr)
450	39.84 [a]	4.08 [c]	13.50 [c]	0.46 [c]
600	48.59 [a]	5.82 [b]	28.77 [b]	0.99 [b]
700	47.23 [a]	6.89 [a]	36.63 [a]	1.75 [a]

[d] Means in the same column followed by a common letter(s) are not significantly different at the 95% level by Tukey's HSD.

Table 3. Effect of the number of vanes on characteristics of the locally designed pump [e].

Blades	Pump Eff. (%)	Head (m)	Discharge (L/s)	Fuel Consumption (L/hr)
6	40.50 [ab]	5.17 [a]	25.95 [a]	1.11 [a]
8	53.56 [a]	5.99 [a]	27.32 [a]	1.06 [a]
9	51.83 [a]	5.97 [a]	25.84 [a]	1.11 [a]
10	35.00 [b]	5.27 [a]	26.08 [a]	0.98 [a]

[e] Means in the same column followed by a common letter(s) are not significantly different at the 95% level by Tukey's HSD.

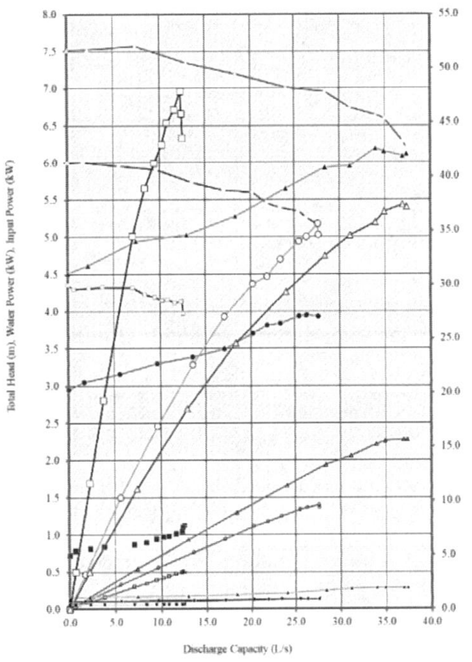

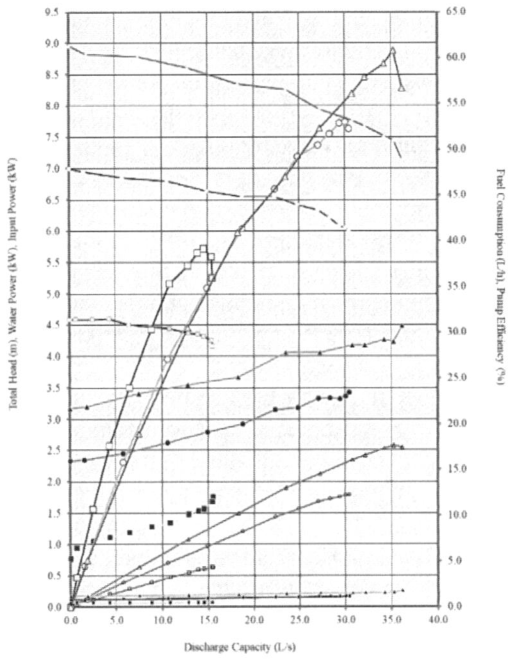

Figure 4. Pump characteristic curve of Voltron with 6 vanes at different pump shaft speed.

Figure 5. Pump characteristic curve of Voltron with 8 vanes at different pump shaft speed.

IOP Conf. Series: Materials Science and Engineering 1109 (2021) 012015 doi:10.1088/1757-899X/1109/1/012015

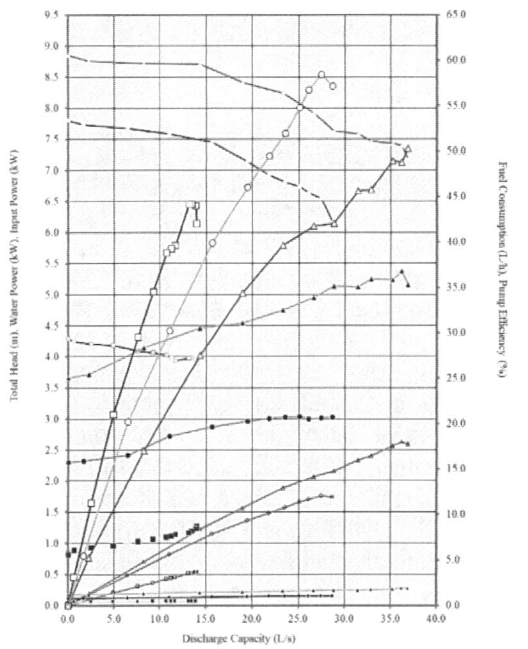

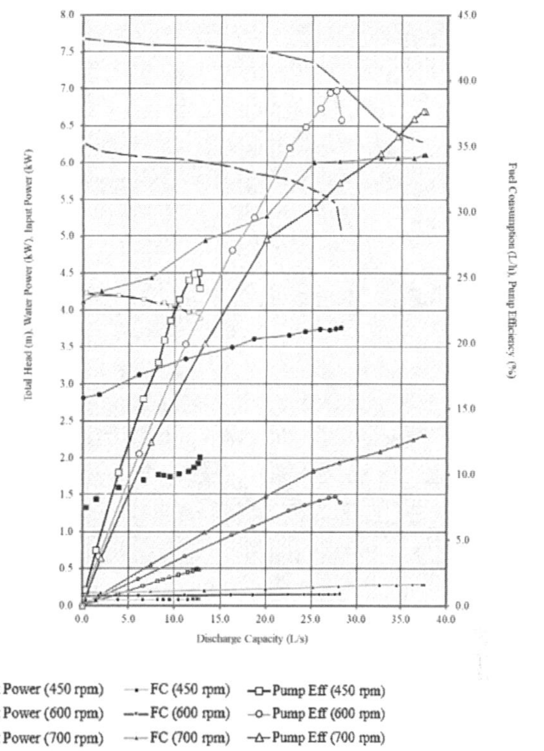

Figure 6. Pump characteristic curve of Voltron with 9 vanes at different pump shaft speed.

Figure 7. Pump characteristic curve of Voltron with 10 vanes at different pump shaft speed.

Plots in Figure 8 conformed with the findings of Chakraborty et al. [5]. Since 8 and 9 vanes have no significant difference, the trend for 450 rpm can be treated in a generally decreasing manner while the other two showed an increasing then decreasing behavior.

The result of other ANOVA on other response variables such as head, discharge, and fuel consumption manifested that the interaction between the number of vanes and pump shaft speed also have significant differences. As such, the data were further analyzed using Tukey's HSD.

Based on Tukey's HSD as presented in Tables 2 and 3, the head, discharge, and fuel consumption based on pump shaft speed have significant differences between each other. Meanwhile, they have no significant differences based on the number of vanes. As apparent in Figures 4 to 7, the total head, discharge, and fuel consumption increases as the pump shaft speed increases.

The highest total head obtained was 7.54 meters with 8 vanes at 700 rpm (Figure 5) while the highest discharge capacity of 37.47 L/s was obtained with 10 vanes at 700-rpm (Figure 7). Moreover, the highest fuel consumption obtained was 1.97 L/hr using 6 vanes at 700-rpm (Figure 4). From Figure 9, it could be observed that fuel consumption at 450 rpm and 600 rpm for all number of vanes is almost the same. On the other hand, the lowest pump input power (1.06 kW) was obtained using 6 vanes at 450 rpm setting while the highest (6.11 kW) was obtained using 6 vanes at 700 rpm (Figure 10). Meanwhile, the pump input power obtained at the farmer's setting of 8-vane, 600 rpm was 3.38 kW (Figure 10).

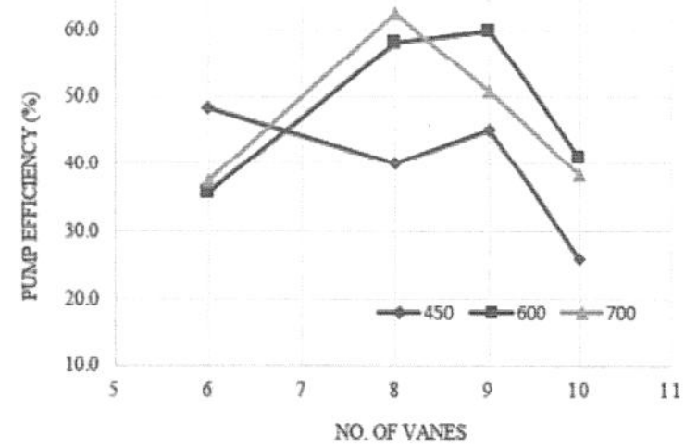

Figure 8. Pump efficiencies using the different number of vanes under different pump shaft speed settings.

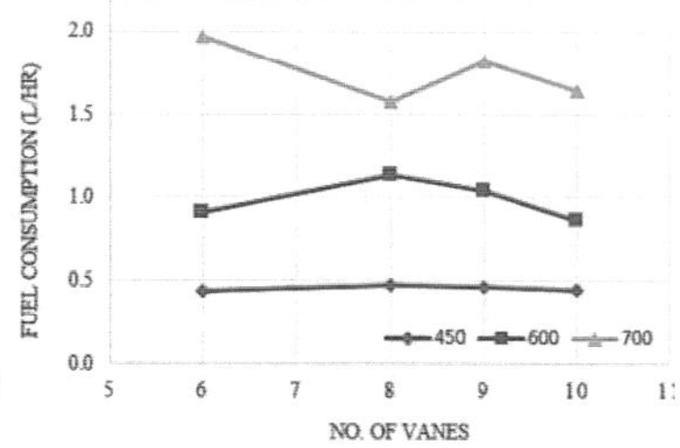

Figure 9. Fuel consumption using a different number of vanes under different pump shaft speed settings.

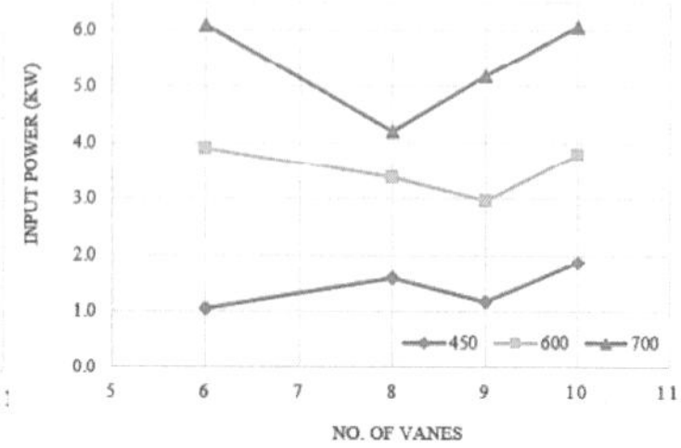

Figure 10. Pump input power using a different number of vanes under different pump shaft speed settings.

Overall, the behavior of the performance characteristics of this locally designed pump is similar as in the study of Chakraborty et al. except for the effect of increasing the number of vanes on the head. Congruent to the study of Liu et al., the study of Chakraborty et al. also concluded that increasing the number of vanes entailed increasing the head of the pump. The increasing head was due to increasing the number of vanes which increases the friction losses [7]. On the contrary, this effect was not present on the behavior of the locally designed water pump as the total head has no significant difference between the number of vanes. This may be because of the way it is operated locally –submerged underwater– unlike other centrifugal pumps being used by farmers that are operated above the water level and have a suction head.

3.2 *Fuel cost for operation and time to irrigate*

The highest pump efficiency of 62.5% was observed using an 8-vane impeller under a 700-rpm setting. It has a total head of 7.54 m, a discharge capacity of 35.60 L/s, fuel consumption of 1.58 L/hr, and an input power requirement of at least 4.22 kW (almost 6 hp).

Given one-hectare land with diversion water requirement (DWR) of 2 L/s per hectare and the performance data of this setting, the time and fuel cost for the operation were computed. With a discharge capacity of 35.60 L/s, the time to irrigate a hectare of land was estimated to be around 1.35 hours under lowland conditions. Assuming Php 45.00 per liter of diesel fuel, the said setting costs around Php 95.87 per hectare.

Since the farmers normally use the 8-vane impeller under 600 rpm setting as per the survey, the time and fuel cost for the operation under this setting were also computed and presented in Table 4. The 8-vane impeller under 600 rpm setting's time to irrigate a 1-hectare land was around 1.54 hrs under lowland conditions. Accordingly, the cost was around Php 78.26 per hectare under lowland conditions. These computations are only applicable for farm sites with elevation from a source of lower than 7.5 meters and 6.2 meters for the 700 rpm and 600 rpm settings, respectively. These are also applicable using the 4-inch diameter discharge as opposed to the original size of 6-inch diameter.

Table 4. Fuel cost for operation of the pump and the time to irrigate a hectare of land.

Setting	Discharge (L/s)	Time (hr)	Fuel consumptio n (L/hr)	Fuel consumed (L)	Price of fuel (Php/L) [f]	Fuel cost (Php)
8-vane, 700 rpm	35.60	1.35	1.58	2.1	45	95.87
8-vane, 600 rpm	31.19	1.54	1.13	1.7	45	78.26

[f] Assumption only

The difference in time for irrigation was around 11.4 minutes (in favor of the 8-vane, 700 rpm setting), while the difference in fuel cost was around Php 17.61 per hectare per day of operation (in favor of 8-vane, 600 rpm). Although the fuel cost per hectare per day of operation was less in the usual setting, farmers should be informed that it is more efficient to use the 8-vane, 700 rpm setting as it has the highest pump efficiency based on the results of this study. Its discharge is also higher than the usual setting, which will entail greater savings in the long run of the operation.

4. Conclusion

The highest pump efficiency of 62.5% was observed using an 8-vane impeller under a 700-rpm setting. Based on the results of Tukey's HSD test, this efficiency is significantly different from the other pump efficiencies.

Statistical analysis also showed that there is no significant difference between the pump efficiencies on the set speeds. However, there is a significant difference in the set number of vanes at a 95% level of significance. In contrast with the findings of Liu et al. and Chakraborty et al., the behavior of the performance characteristics varies according to the effect of the number of vanes. On the other hand, the head, discharge, and fuel consumption increased as rotational speed increased just as Chakraborty et al. argued.

Finally, assuming a diversion water requirement (DWR) of 2 L/s per hectare and a fuel price of Php 45.00 per liter, the difference in fuel cost between the 8-vane impeller under 700 rpm setting and the usual setting of 8-vane impeller under 600 rpm setting is around Php 17.61 per hectare per day of operation. Although the fuel cost is less in the usual setting, the 8-vane impeller under the 700 rpm setting is better in the long run as it is more efficient than the usual setting. Thus, the 8-vane impeller under the 700-rpm setting is recommended for farmers' use.

Acknowledgments

I wish to thank the following individuals and institutions and honor them for their invaluable contributions to this work. To DOST-SEI and ERDT for the opportunity of taking post-graduate studies; To AMTEC-UPLB's engineers and technicians for their unconditional support during the conduct of my study, most especially to Engr. Romy Eusebio for sharing his knowledge and expertise in testing the pump and analyzing the data; To Mr. Albert A. Tolentino and the farmers of Bulacan for sharing their experiences regarding the use and proper operation techniques of the locally designed water pump in my study; To Dr. Consorcia E. Reaño, my best friend Regine Olivas and Ma. Camille Gonzales for helping me with my Statistical Analysis; To my parents Angelita S. Pascual and Agapito T. Pascual, and siblings Angelo-Aga and Angeleo-Aga for their unsolicited support and love throughout my life; Finally, to God, all of these are for Your glory.

References

[1] *CSISA Factsheet* 2014 The axial-flow and mixed flow pumps for surface water irrigation drainage and fisheries 6–7 https://csisa.org/wp-content/uploads/sites/2/2014/06/Axial-Flow-Pump-Factsheet_Eng.pdf

[2] Cooper P 2008 *Section 2.1.i Centrifugal pump theory in: Pump Handbook, Fourth Edition* Igor J. Karassik, Joseph P. Messina, Paul Cooper, Charles C. Heald (eds.) (New York, New York: The McGRAW-HILL Companies, Inc.)

[3] Bad H M and Ahmed W H 2015 *Pumping machinery theory and practice* (West Sussex, United Kingdom: John Wiley & Sons Ltd)

[4] PAES 115-2000 Agricultural machinery – Centrifugal, mixed flow and axial flow water pumps – Methods of test *Philippine Agricultural Engineering Standards*

[5] Chakraborty S, Pandey K M and Roy B 2012 Numerical analysis on effects of blade number variations on performance of centrifugal pumps with various rotational speeds *International Journal of Current Engineering and Technology* vol **2** no 1 pp 143-152

[6] Liu H, Wang Y, Yuan S, Tan M and Wang K 2010 Effects of blade number on characteristics of centrifugal pumps *Chinese Journal of Mechanical Engineering* **23**(06) 742 https://doi.org/10.3901/CJME.2010.06.742

[7] Abo Elyamin G, Bassily M, Khalil K and Gomaa M 2019 Effect of impeller blades number on the performance of a centrifugal pump. *Alexandria Engineering Journal* (2019) **58** 39–48

Pipe sizing of district cooling distribution network including in energy transfer station

E B Lisaba[1], G L Augusto[1]

[1]Mechanical Engineering Department, De La Salle University, Manila, Philippines

edwin_lisaba@dlsu.edu.ph

Abstract. A district cooling system (DCS) distributes chilled water to different buildings connected to the main plant. The benefits of a DCS include greatly increasing energy efficiency and decreasing life cycle operation costs. The installation, maintenance, and design of a DCS can be expensive, however. Therefore, means of decreasing costs without cutting quality is necessary. Given this, a hypothetical DCS is presented with a constant temperature difference set-point of 9°C. The DCS was set to have a system diversity of about 80%. Hydraulic calculation will be used in the study in order to determine the smallest possible pipe sizes that would permit normal operation below allowable limits. The network is divided into 3 parts; namely the main line (ML), the plot take-off (PTO), and the plate-type heat exchangers (PHEs). The DCS will have a piping network connected to 12 energy transfer stations, each with 3 sets of PHEs. Initial guess values of pipe dimensions and friction factors will first be calculated using continuity and the Von Karman equation. A single-variable Newton-Raphson method is then used on the implicit Colebrook-White equation to estimate the actual friction factor for each pipe. Then, the calibrated pipe sizes will be determined by cross-referencing theoretical values with industry-standard sizes as specified in ASHRAE. In calculating the friction factor, it was found that the values increase as the given tonnage decreases, thus implying a decrease in pipe size as the chilled water flows toward the last building. The entire study was conducted on MATLAB. It must be noted that the study is only limited to the selection and design of a simple DCS.

1. Introduction

According to Dincer & Abu-Rayash [1], a district cooling system (DCS) distributes chilled water to different customer energy transfer stations through the use of a network of underground pipes. The installation of a district cooling system eliminates the need for separate buildings and rooms to have their own climate control devices such as air conditioners and heaters. Given this, a DCS helps save a significant amount from both initial and maintenance costs. DCS technology allows for an increase in energy efficiency by 40% and a decrease in lifecycle cost by 20%.

Krarti [2] states that a DCS can utilize three main technologies that will generate the chilled water necessary for distribution. First, electric chillers can be used since they have a lower global warming potential. The second option is through the use of free cooling, which involves using cold water from nearby bodies of water. Finally, absorption chillers can also be used by DCS from nearby sources of free heat, such as power plants and industrial works.

A district cooling system has three major components, namely the central plant, the network, and the energy transfer station (ETS). The central plant produces the chilled water through means of harvesting thermal energy using the technologies stated earlier. The water is then pumped through the network, or a series of pipes that connect the production facility to the customer ETS. Finally, the ETS, also called the customer-building interconnections, transfers heat to the network through the use of plate-type heat exchangers or PHEs [3]. It must be known that conventional methods of cooling, such as air conditioning units, constitute up to 50% to 70% of electricity demand within a building. The use of an ETS with secondary lines going throughout the building not only allows for a more environmentally friendly solution to large-scale climate control, but also for a more economic opportunity to reduce costs during its operating life cycle.

The projected friction factors present within the different sections of the network will be estimated using numerical optimization techniques. Previous studies have incorporated such techniques in distribution network design. For example, Augusto et al. [4] used exhaustive search methods in order to find the optimal pipe sizing dimensions that will yield the lowest overall cost. Another study by Liao [5] assigned confidence values to corresponding accuracies and uncertainties within a distribution system. Finally, in selecting the appropriate pipe sizes within a network, it was recommended by ASHRAE [6] to base design criteria off fluid velocity and pressure drop limits.

Given that the study will be performed on a hypothetical DCS, a change in temperature of about 9°C will be assumed between the chilled water return and supply lines. This is to follow the standards set by ASHRAE [7]. The ASHRAE Standard 90.1, published in 2016, dictates that the chilled-water cooling coils within the heat exchangers must allow for a minimum change in temperature of about 9°C between the supply and return lines. Any change in temperature that is lower than 9°C may suffer from "low delta-T syndrome." A study by Yan [8] has provided a mathematical explanation for this phenomenon. Put simply, lower values would result in chilled water systems to consume more energy, thus causing it to operate less efficiently. This can be the result of improper load distribution.

The installation of a DCS is an expensive venture. Mistakes can further amplify that cost in terms of wasted time and resources. Therefore, it is imperative to determine the necessary pipe dimensions given the physical limitations of the setup that would not only allow optimal running conditions, but to also ensure the lowest cost possible. This way, optimal pipe sizes can be pinpointed quickly through system simulation. A study by Walters [9] has shown that the costs for pump and pipe sizing can be cut by a substantial amount once numerical optimization methods have been introduced to fluid flow analysis. Numerical methods have proven to reduce costs without compromising system performance.

This study aims to create a program that will calculate the appropriate pipe sizing when designing a DCS for a system with twelve ETS. The program must first calculate the friction factor present in each pipe. This will be done using the Newton-Raphson Method (NRM) on the implicit Colebrook-White Equation as a function of the Reynolds Number (*Re*) and the pipe relative roughness ($\frac{\varepsilon}{D}$). Using the calculated friction factors, the calibrated main line pipe sizing will be determined using the pressure drop and velocity limit as reference. It must be stated that this study is limited to only selecting the cheapest possible pipe sizes. The costs of installation and maintenance are not covered nor considered. Furthermore, the study will be done on a DCS without secondary lines flowing throughout each building. The final pipe sizes will then be compared with the results obtained by Augusto et al.

2. **Methods and Data**

2.1. DCP Schematic

The district cooling plant (DCP) is composed of three different sections, namely the main line (ML), the plot take-off (PTO), and the plate-type heat exchanger (PHE). In this study, the appropriate pipe sizing for each section was determined. Figure 1 below shows the district cooling schematic to be used. It can be seen that there are twelve buildings, each with their respective energy transfer stations. The main line is represented by the solid line extending from the district cooling plant (DCP) to the rest of the buildings. Each building has a plot take-off line that connects them to the main line. Lastly, given that there is an ETS in each building, three PHEs connect it to the rest of the building. It must be noted that all three sections in the schematic were simplified, and that one line represents one pair of supply and return lines.

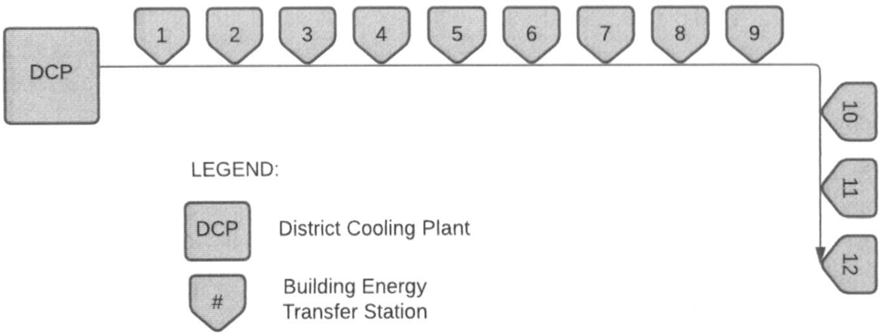

Figure 1: Basic schematic diagram of a DCP

2.2 Initial Pipe Dimensions

In calculating the initial pipe dimensions per sector, the flowrate ω passing through each pipe was determined. Eq. 1 below determines the flowrate in gallons per minute given the tons of refrigeration. It must be noted that the ΔC is the change in temperature between supply and return lines. This value was kept at a constant $9°C$ for all pipes in the system. After converting to cubic meters per second, the continuity equation, as shown in Eq. 2, was used to calculate the theoretical pipe size D_{theo} in terms of millimeters.

$$\omega = \frac{24TR}{1.8\Delta C} \tag{1}$$

$$D_{theo} = \sqrt{\frac{4\omega}{V\pi}} \tag{2}$$

The theoretical values were then cross-checked with possible pipe sizes within that range. The initial pipe sizing for each section was based on the smallest possible pipe sizes that could accommodate the velocity and pressure drop limits at particular sections in the DCS without failing. More specifically, the pressure drop within each pipe must not exceed $100\ Pa/m$. In Eq. 2, the velocity limit V within each pipe is different per section in the network. In the main line, the velocity limit was set to $3\ m/s$. The velocity limits at PTO and PHE were both set to $1.75\ m/s$. The values for these velocity limits, pipe lengths, and the tons of refrigeration per pipe were taken from the aforementioned study conducted by Augusto et al. Finally, the industry-standard pipe sizes were taken from the 2017 ASHRAE handbook [10].

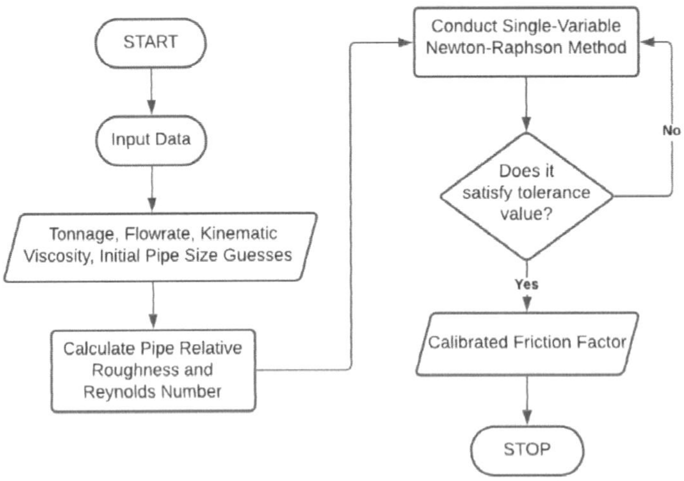

Figure 2: Friction factor calculation algorithm

2.2 Friction factor calculation

After acquiring the initial pipe sizes, the Newton-Raphson Method (NRM) was used to find the finalized friction factor for each section. Initial values were set first, and this was done by calculating the pipe relative roughness ($\frac{\varepsilon}{D}$) as shown in Eq. 3. The Von Karman Equation (Eq. 4), was then used to find the initial friction factor values. Eq. 5 then shows the Reynolds number calculation given a constant kinematic viscosity ν of $1.547E-6\ m^2/s$. It must be noted that the theoretical velocity V_{theo} was calculated using the theoretical diameter D_{theo}. The theoretical velocity is expressed in meters per second. Lastly, the inner diameter D_{in}, expressed in meters, was determined using the initial pipe size calculated earlier. The algorithm can be more easily seen in Figure 2.

$$\frac{\varepsilon}{D} = (6.571429E-6) \times \left(\frac{D}{7000}\right)^{-1} \quad (3)$$

$$\frac{1}{\sqrt{f_{initial}}} = 2\ log\ log\left(\frac{D}{\varepsilon}\right) + 1.14 \quad (4)$$

$$Re = \frac{V_{theo}D_{in}}{\nu} \quad (5)$$

$$\frac{1}{\sqrt{f}} = -2\ log\ log\left[\frac{\frac{\varepsilon}{D}}{3.7} + \frac{2.51}{Re\sqrt{f}}\right] \quad (6)$$

NRM was then used on the implicit Colebrook-White equation by following Eq. 7 below, where F is the original equation and dF is the derivative with respect to f. The process will start from the initial friction factor calculated earlier. The process will keep iterating until the relative approximate error is less than the tolerance value of 10^{-6}.

$$f_{new} = f_{old} - \frac{F}{dF} \quad (7)$$

2.3 Calibrated Pipe Sizing

Using the finalized friction factors, the calibrated pipe sizing was then acquired by following the same process used in finding the initial diameter. The finalized pipe sizes were determined by implementing the final friction factor in solving for the allowable velocity limit and pressure drop. The process will stop iterating once the smallest possible pipe size provides values that are within the acceptable range. The pressure drops ΔP, in terms of Pascals per meter, were calculated using the modified Darcy-Weisbach equation (Eq. 8) below. The variable V_{pipe} is the velocity within the pipe that is compared with the velocity limit and is expressed in meters per second. Similarly, D_{test} is the pipe size being tested expressed in meters. Figure 3 shows the algorithm for obtaining the calibrated pipe sizes.

$$\Delta P = \frac{f\rho}{2}\left(\frac{V_{pipe}^2}{D_{test}}\right) \quad (8)$$

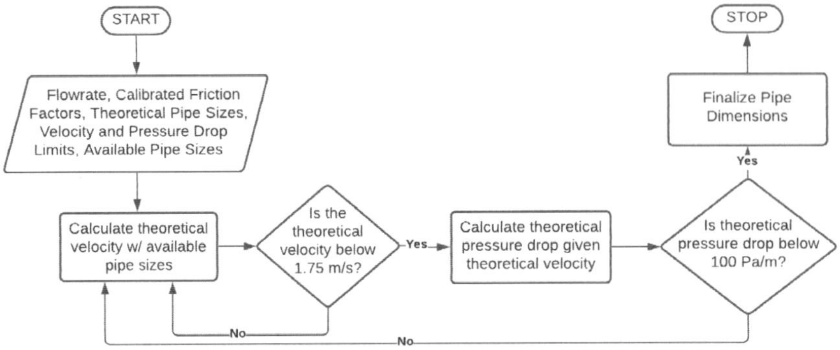

Figure 3: Calibrated pipe sizing

3. Results and Discussion

3.1 Estimated Friction Factors per Section

It must be noted that the entire study was done on MATLAB. The calculated friction factors for the main line, PTO, and PHE are shown in Tables 1a, 1b, and 1c below, along with the tons of refrigeration gathered from Augusto et al. Additionally, the study used a diversified flowrate of 80% given that the buildings will be assumed to not operate at full load at the same time. The system diversity of 80% applies to the ML, PTO, and PHE lines. Additionally, the flowrate going through one PHE is the PTO flowrate divided by 3. It can be seen that as the flowrate decreases, the friction factor increases. This corresponds to a decrease in the pipe size, which can be seen later on.

Table 1a: Friction Factor Values at Main Line

Pipe Designation	Tons of Refrigeration	Flowrate (gpm)	Initial Friction Factor	Calibrated Friction Factor
DCP – Tower 1	24,000	28444.44	0.01060	0.01204
Tower 1 – Tower 2	21,000	24888.89	0.01060	0.01219
Tower 2 – Tower 3	18,200	21570.37	0.01082	0.01235
Tower 3 – Tower 4	15,700	18607.41	0.01095	0.01252
Tower 4 – Tower 5	13,200	15644.44	0.01109	0.01272
Tower 5 – Tower 6	11,200	13274.07	0.01109	0.01293
Tower 6 – Tower 7	9,400	11140.74	0.01142	0.01314
Tower 7 – Tower 8	7,900	9362.96	0.01142	0.01337
Tower 8 – Tower 9	6,100	7229.63	0.01183	0.01369
Tower 9 – Tower 10	4,300	5096.30	0.01208	0.01418
Tower 10 – Tower 11	2,800	3318.52	0.01237	0.01483
Tower 11 – Tower 12	1,400	1659.26	0.01296	0.01599

Table 1b: Friction Factor Values at Plot Take-Off

Pipe Designation	Tons of Refrigeration	Flowrate (gpm)	Initial Friction Factor	Calibrated Friction Factor
ML – Tower 1	3,000	3555.56	0.01208	0.01479
ML – Tower 2	2,800	3318.52	0.01208	0.01492
ML – Tower 3	2,500	2962.96	0.01237	0.01504
ML – Tower 4	2,500	2962.96	0.01237	0.01504
ML – Tower 5	2,000	2370.37	0.01271	0.01536
ML – Tower 6	1,800	2133.33	0.01271	0.01556

ML – Tower 7	1,500	1777.78	0.01296	0.01585
ML – Tower 8	1,800	2133.33	0.01271	0.01556
ML – Tower 9	1,800	2133.33	0.01271	0.01556
ML – Tower 10	1,500	1777.78	0.01296	0.01585
ML – Tower 11	1,400	1659.26	0.01296	0.01599
ML – Tower 12	1,400	1659.26	0.01296	0.01599

Table 1c: Friction Factor Values at a Single Plate-Type Heat Exchanger

Pipe Designation	Tons of Refrigeration	Flowrate (gpm)	Initial Friction Factor	Calibrated Friction Factor
Tower 1 – PHE 1	1000	1185.19	0.01344	0.01654
Tower 2 – PHE 2	933.33	1106.17	0.01344	0.01669
Tower 3 – PHE 3	833.33	987.65	0.01344	0.01695
Tower 4 – PHE 4	833.33	987.65	0.01344	0.01695
Tower 5 – PHE 5	666.67	790.12	0.01343	0.01751
Tower 6 – PHE 6	600	711.11	0.01408	0.01749
Tower 7 – PHE 7	500	592.59	0.01408	0.01794
Tower 8 – PHE 8	600	711.11	0.01408	0.01748
Tower 9 – PHE 9	600	711.11	0.01408	0.01748
Tower 10 – PHE 10	500	592.59	0.01408	0.01794
Tower 11 – PHE 11	466.67	553.08	0.01408	0.01812
Tower 12 – PHE 12	466.67	553.08	0.01408	0.01812

3.2 Calibrated Pipe Sizes per Section

After acquiring the needed friction factor values, the calibrated pipe sizing was then determined. Tables 2a, 2b, and 2c below show the final pipe sizes needed for the DCP installation for each section. The calibrated pipe sizing was based on available industry-standard sizes according to ASHRAE. The final pipe sizes were then compared with the values obtained by Augusto et al.

Table 2a: Calibrated Pipe Sizes for the Main Line

Pipe Designation	Initial Pipe Size (mm)	Final Pipe Size (mm)	Augusto et al. (mm)
DCP – Tower 1	872.72	900	900
Tower 1 – Tower 2	816.35	900	900
Tower 2 – Tower 3	759.98	800	800
Tower 3 – Tower 4	705.86	750	750
Tower 4 – Tower 5	647.23	700	700
Tower 5 – Tower 6	596.18	600	700
Tower 6 – Tower 7	546.18	600	600
Tower 7 – Tower 8	500.71	600	600
Tower 8 – Tower 9	439.98	500	500
Tower 9 – Tower 10	369.40	450	450
Tower 10 – Tower 11	298.09	400	400
Tower 11 – Tower 12	210.78	300	300

Table 2b: Calibrated Pipe Sizes for the Plot Take-Off

Pipe Designation	Initial Pipe Size (mm)	Final Pipe Size (mm)	Augusto et al. (mm)
ML – Tower 1	403.99	450	450
ML – Tower 2	390.29	400	450
ML – Tower 3	368.79	400	400
ML – Tower 4	368.79	400	400
ML – Tower 5	329.86	350	350
ML – Tower 6	312.93	350	350

ML – Tower 7	285.66	300	300
ML – Tower 8	312.93	350	350
ML – Tower 9	312.93	350	350
ML – Tower 10	285.66	300	300
ML – Tower 11	275.98	300	300
ML – Tower 12	275.98	300	300

Table 2c: Calibrated Pipe Sizes for a Single Plate-Type Heat Exchanger

Pipe Designation	Initial Pipe Size (mm)	Final Pipe Size (mm)	Augusto et al. (mm)
Tower 1 – PHE 1	233.24	250	300
Tower 2 – PHE 2	225.32	250	250
Tower 3 – PHE 3	212.90	250	250
Tower 4 – PHE 4	212.90	250	250
Tower 5 – PHE 5	190.43	250	250
Tower 6 – PHE 6	180.66	200	250
Tower 7 – PHE 7	164.91	200	200
Tower 8 – PHE 8	180.66	200	250
Tower 9 – PHE 9	180.66	200	250
Tower 10 – PHE 10	164.91	200	200
Tower 11 – PHE 11	159.33	200	200
Tower 12 – PHE 12	159.33	200	200

3.3 Comparison of Data Values

The results show that there are slight differences between the calculated pipe sizes. It can also be seen that the values obtained seem to deviate even more as the water flows farther away from the plant. Additionally, the new values are one size smaller compared to those by Augusto et al. However, the majority of the data remains the same. A possibility of why the values do not match is because this study did not take into consideration the minor losses present in an actual network. The inputs that are common to both studies are only the flowrates through each building, system diversity, velocity limits, and water density. Furthermore, the code was written in such a way that it would terminate once it finds that both fluid velocity and pressure drop values are below their respective limits. Given that the minor losses were not considered, the friction factor within each pipe has deviated by a slight amount, thus causing discrepancies in the finalized pipe sizes.

4. **Conclusion**

This study was divided into two main parts. The first was to estimate the friction factor that exists in each pipe using the given tons of refrigeration. The second task was to determine the required pipe sizes using the calculated friction factors. It must be noted that the study was performed on a hypothetical DCS with a constant change in temperature of 9 degrees Celsius between the supply and return lines. In conducting the study:

The district cooling plant schematic outlines 12 buildings. Each of these buildings has one energy transfer station (ETS) that is connected to 3 plate-type heat exchangers (PHEs). Additionally, the system diversity was set to 80%.

The total tons of refrigeration outputted by the DCP was set to 24,000 TR. The flowrate going through each pipe, as well as the pressure drop and velocity limits, were taken from a study by Augusto et al. The pressure drop limit was set to $\mathbf{100}\ \boldsymbol{Pa/m}$. The velocity limit in the main line was set to $\mathbf{3}\ \boldsymbol{m/s}$, while those in the PTO and PHE were set to $\mathbf{1.75}\ \boldsymbol{m/s}$.

The initial values of the pipe size were found using the tonnage and theoretical flowrate. Similarly, initial values of pipe friction factor were found using the pipe relative roughness equation and the Von Karman Equation.

A single-variable Newton-Raphson Method was then used on the implicit Colebrook-White equation to find the finalized friction factors of each pipe. This was done using the initial friction factor values.

Using the ASHRAE 2017 Handbook as a guide, the calibrated pipe sizes were found. The algorithm was to use the theoretical pipe sizes as a guide to narrow down the range of potential sizes starting from the smallest to the largest. The final pipe dimensions were then selected once the fluid velocity and pressure drop are below the aforementioned limits.

The entirety of the study was conducted on MATLAB. Utilizing the smallest possible pipe diameters that permit allowable velocity and pressure values will ensure the lowest possible overall price in selecting the needed pipe sizes for a district cooling system.

References

[1] Dincer I and Abu-Rayash A 2020 Chapter 5 – Community energy systems *Energy Sustainability*, pp. 101-118.
[2] Krarti M 2018 Chapter 8 – Integrated design of energy efficient cities *Optimal Design and Retrofit of Energy Efficient Buildings, Communities, and Urban Centers*, pp. 471-545.
[3] Tredinnick S and Phetteplace G 2016 8 – District cooling, current status and future trends *Advanced District Heating and Cooling (DHC) Systems*, pp. 167-188.
[4] Augusto G L, Culaba A B and Tan R R 2013 Identification of design criteria for district cooling distribution network *Philippine Science Letters,* vol. 6, no. 2, pp. 182-197, October 2013.
[5] Liao H 2019 Review on distribution network optimization under uncertainty *Energies*, vol. 12, September 2019.
[6] *Fundamentals*, ASHRAE handbook 2013.
[7] *Energy standard for buildings except low-rise residential buildings,* ANSI/ASHRAE/IES standard 90.1-2016; 2016.
[8] Yan C 2018 Mathematical explanation and fault diagnosis on low delta-T syndrome in building chilled water systems *MDPI,* vol. 8, June 2018.
[9] Walters T 2002 Cutting costs in pump and pipe sizing *Chemical Processing,* pp. 35-38, June 2002.
[10] *Fundamentals,* ASHRAE handbook 2017.

Reliability based design of shaft for gearbox

N H Loc[1,2*],
[1]Faculty of Mechanical Engineering, Ho Chi Minh city University of Technology (HCMUT), Ho Chi Minh City, Vietnam
[2]VietNam National University of Ho Chi Minh City (VNUHCM), Ho Chi Minh City, Vietnam

*nhloc@hcmut.edu.vn

Abstract. In this article, the author presents research on the reliability-based design methods of mechanical systems, includes 3 stages: the first, the fault tree analysis (FTA) method is utilized for reliability analysis of single and double reducer into the logic diagram; the second that apply the probabilistic design methods, including moment matching method (MMM), first-order reliability method (FORM), and Monte Carlo simulation (MCS) for reliability-based design and to analyze the components of the mechanical transmission systems, and the last sensitivity analysis is performed in the post-design stage after a design solution is identified. The study also aims to set up RADME (Reliability-based analysis and design of the mechanical systems) computer programs applying the aforementioned methods. The reliability design problem was surveyed as reliability-based design optimization with one design variable.

Keywords: *Probabilistic design, Reliability-based design, Mechanical system, Shaft, Monte Carlo simulation, Sensitivity analysis, Gear reducer*

1. Introduction

In deterministic design, the variables are considered deterministic and calculated values are selected as the worse – case values or nominal values; however, by calculation, they are divided by safety factor. It is evident that the use of elements of safety may be either risky (under-designed) or conservative (over-designed). Overall, uncertainty exists in any engineering system. When it comes to designing complicated systems, a small variation of system input may cause a significant quality loss, and ignorance of such uncertainty may lead to catastrophic failure events. The probabilistic design includes design and analysis of mechanical transmission systems, reliability-based design optimization of machine structure, tolerance analysis, kinematics, and dynamics of the machine. This method ensures given reliability, safety, quality, and profitability of the product.

Three categories of methods exist for the task of uncertainty analysis. The first category is the sampling-based approach such as MCS [10-11]. The second category comprises approximation methods like first order (FORM) and second-order (SORM) reliability methods [1,2,7] The third category involves the response surface – using surrogate models to replace the performance function by Design of experiments[3,8].

2. Fundamentals of reliability-based design and analysis of shafts

Machine elements for speed reducer include gear drive, shaft, bearing, and housing. In the papers by [6,12] the details of reliability-based design and an analysis of gear drive were presented. The book by Nguyen (2016, 2018), the researchers have focused on the shaft, gear drive and bearing design in accordance with reliability.

2.1. Reliability-based design and analysis of shaft

The shafts must satisfy the condition of strength, stiffness, wear resistance, and vibration stability. Under strength conditions, for fast-rotating shafts, the fatigue limit is decisive. Calculating shaft reliability is essential because the load and fatigue limits have significant dispersion. In fact, fatigue damage accounts for 40 ÷ 50% of shaft failures. When abruptly overloaded, the shafts made of normalized, tempered steel can be impaired by large plastic deformation while those made of brittle and less malleable materials can be damaged by brittle destruction.

The characteristic of calculating fatigue strength is that the bending stress and torsional stress change according to different cycles. Specifically, bending stress alters in symmetrical alternating cycle while torsional stress changes according to the pulsating cycle. The load acting on the shaft is the force that exerts influence on the gear, belt, sprocket, coupling and we determine the reaction forces at the bearings. The force point depends on the bearing structure.

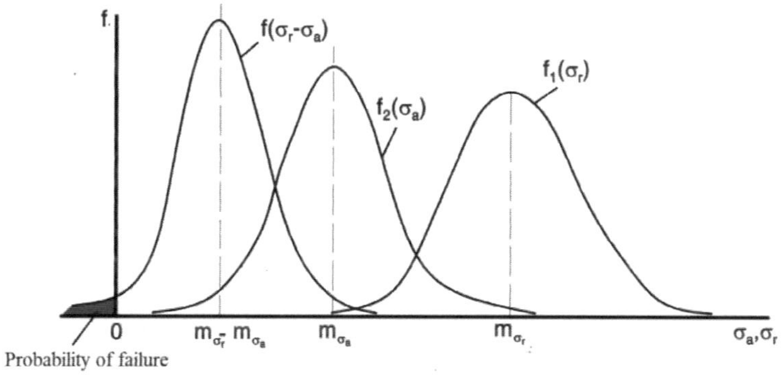

Figure 1. The probability density of the amplitudes of stress σ_a and the fatigue limits σ_r.

The curves of probability density of the amplitudes of stress σ_a and the fatigue limits σ_r are illustrated on the graph in Figure 1. Machine elements and mechanical structures are considered safe and reliable when limited value exceeds the calculating value. As can be seen, these curves intersect, which means that failure can occur.

The function g (σ_a, σ_r) = $\sigma_r - \sigma_a$ is called performance or limit – state function. When g (σ_a, σ_r) > 0, the shaft is considered safe, but when g (σ_a, σ_r) ≤ 0, the shaft is considered as unsafe.

$$m_g = m_{\sigma_r} + m_{\sigma_a}; \quad S_g = \sqrt{S_{\sigma_r}^2 + S_{\sigma_a}^2} \tag{1}$$

Equivalent bending stress of shaft with diameter d at gear position is determined by the formula:

IOP Conf. Series: Materials Science and Engineering 1109 (2021) 012017 doi:10.1088/1757-899X/1109/1/012017

$$\sigma = \frac{32M_{eq}}{\pi d^3} \tag{2}$$

For the drive shaft, there are eight cases of forces acting in a gear and shaft as shown in Figure 2. When calculating, the researchers proceed at the dangerous cross section of the shaft according to the maximum equivalent moment M_{eq}:

$$M_{eq} = \sqrt{M_X^2 + M_Y^2 + 0.75T^2} \tag{3}$$

Mean and standard deviation of the equivalent moment are calculated as follows:

$$\bar{M}_{eq} = \sqrt{\bar{M}_X^2 + \bar{M}_Y^2 + 0.75\bar{T}^2} \tag{4}$$

$$S_{M_{eq}} = \left(\frac{M_X^2 S_{M_X}^2 + M_Y^2 S_{M_Y}^2 + (0.75T)^2 S_T^2}{M_X^2 + M_Y^2 + 0.75T^2} \right)^{1/2} \tag{5}$$

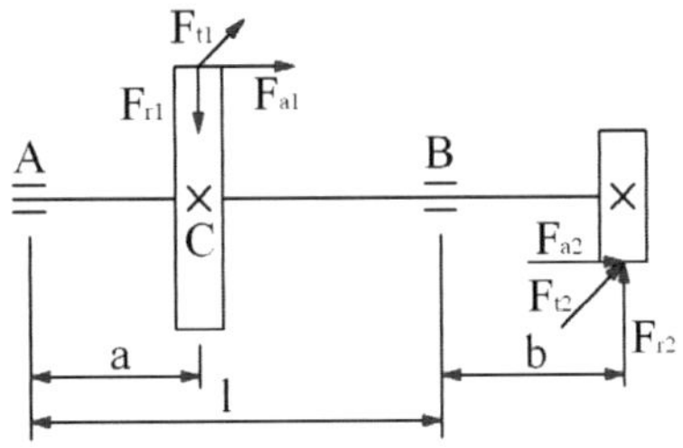

a) F_{r1} and F_{r2} in the opposite directions
F_{a1} and F_{a2} in the same direction

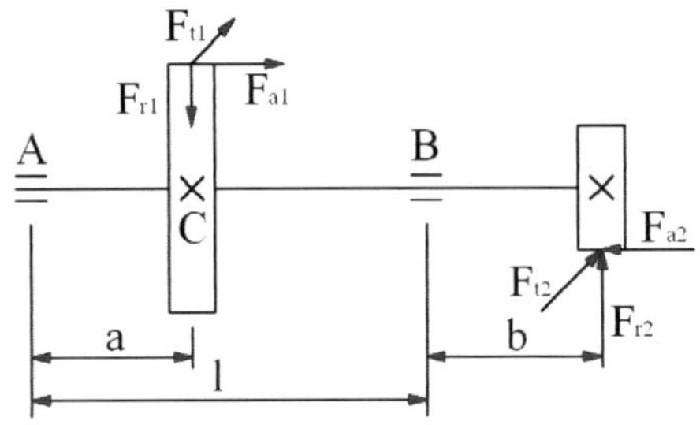

b) F_{r1} and F_{r2} in the opposite directions
F_{a1} and F_{a2} in the opposite directions

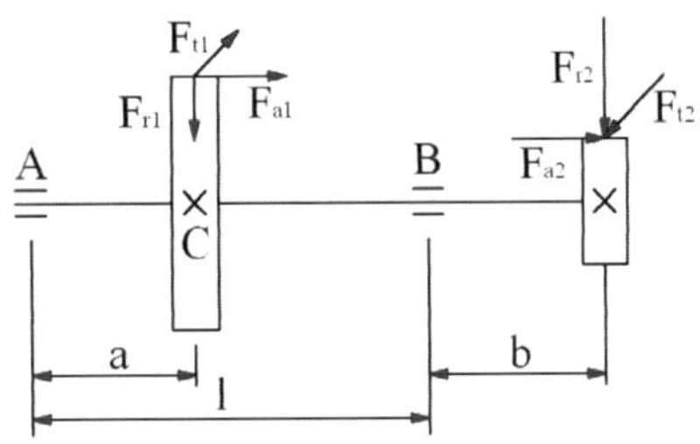

c) F_{r1} and F_{r2} in the same direction
F_{a1} and F_{a2} in the same direction

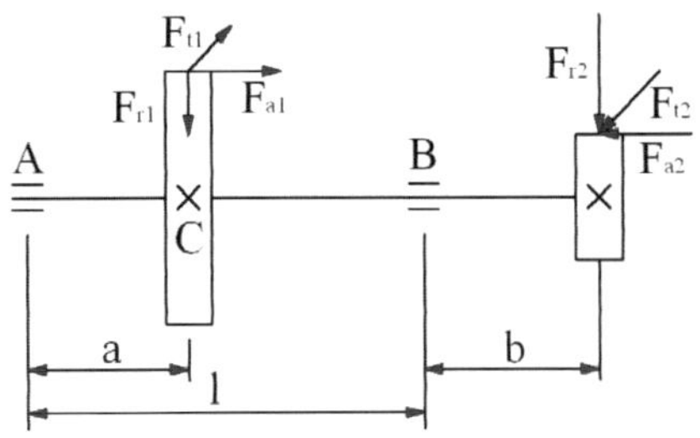

d) F_{r1} and F_{r2} in the same direction
F_{a1} and F_{a2} in the opposite directions

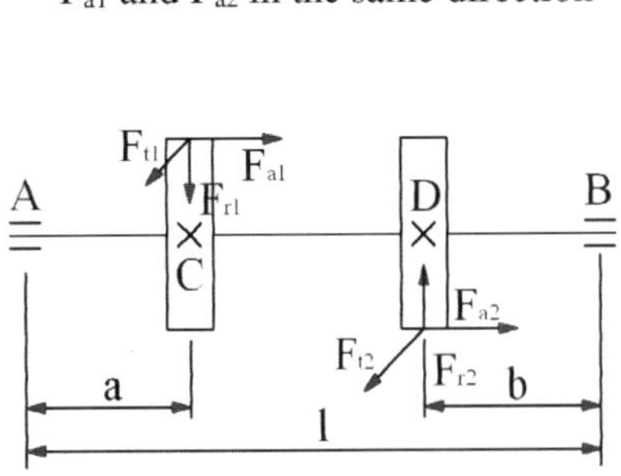

e) F_{r1} and F_{r2} in the opposite directions

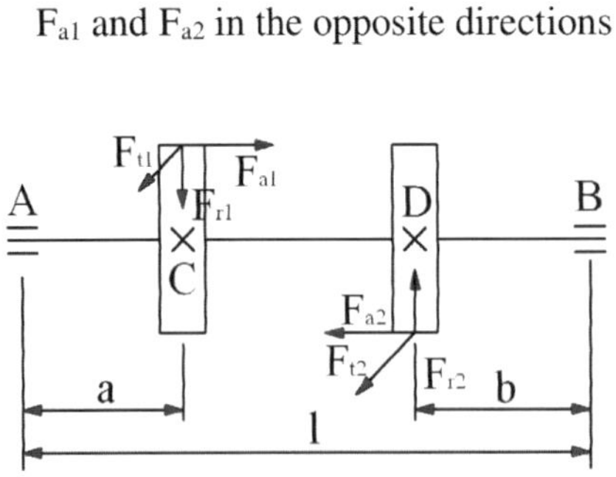

f) F_{r1} and F_{r2} in the opposite directions

F_{a1} and F_{a2} in the same direction F_{a1} and F_{a2} in the opposite directions

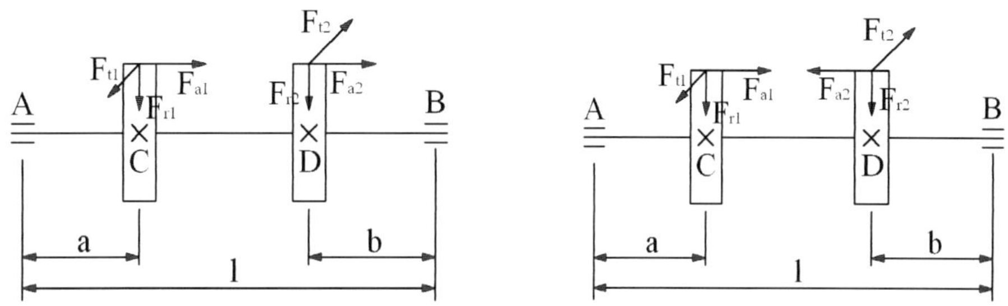

g) F_{r1} and F_{r2} in the same direction
F_{a1} and F_{a2} in the same direction

h) F_{r1} and F_{r2} in the same direction
F_{a1} and F_{a2} in the opposite directions

Figure 2. Forces acting in a gear and shaft for all cases.

Mean of the bending stress is determined by:

$$\bar{\sigma} = \frac{32\bar{M}_{eq}}{\pi \bar{d}^3} \tag{6}$$

Standard deviation of the bending stress is calculated by the formula:

$$S_\sigma = \left(\left(\frac{\partial \sigma}{\partial M_{eq}} \right)^2 S_{M_{eq}}^2 + \left(\frac{\partial \sigma}{\partial d} \right)^2 S_d^2 \right)^{1/2} = \left(\left(\frac{32}{\pi \bar{d}^3} \right)^2 S_{M_{eq}}^2 + \left(\frac{96M_{eq}}{\pi \bar{d}^4} \right)^2 S_d^2 \right)^{1/2} \tag{7}$$

The formula for moments, equivalent moments and standard deviations are listed in the Table 1.
The fatigue limit $\sigma_{\lim}$ of shaft material can be calculated as:

$$\sigma_{\lim} = \frac{\sigma_r \varepsilon_\sigma \beta}{K_\sigma} K_L \tag{8}$$

with σ_r - endurance limit for bending; ε_σ - scale factor taking into account the effect of the dimensions of the shaft; β - hardening factor; K_L - life coefficient; K_σ - effective stress concentration factors for bending.

Mean of the fatigue limit is: $\bar{\sigma}_{\lim} = \frac{\bar{\sigma}_r \bar{\varepsilon} \bar{\beta}}{\bar{K}_\sigma} \bar{K}_L$ (9)

The coefficient of variation of the fatigue limit is determined as:

$$v_{\sigma H \lim} = \sqrt{v_{\sigma_r}^2 + v_\varepsilon^2 + v_{K_\sigma}^2 + v_\beta^2 + v_{K_{HL}}^2} \tag{10}$$

Standard deviation is reckoned by $S_{\sigma_{\lim}} = v_{\sigma_{\lim}} \bar{\sigma}_{\lim}$ (11)

Reliability of shaft is determined by the quantile z_1:

$$z_1 = -\frac{\bar{\sigma}_{\lim} - \bar{\sigma}}{\sqrt{S_{\sigma_{\lim}}^2 + S_\sigma^2}} = -\frac{\bar{n} - 1}{\sqrt{\bar{n}^2 v_{\sigma_{\lim}}^2 + v_\sigma^2}} \tag{12}$$

with $\bar{n}$ – mean factor of safety which is determined by mean stress and fatigue limit $\bar{n} = \frac{\bar{\sigma}_{lim}}{\bar{\sigma}}$.

To solve a design problem, the mean of the shaft diameter $\bar{d}$ is found, then $\bar{n}$ and v_σ^2 are the functions of $\bar{d}$. Now the job is to replace the expression $\bar{n}$ and v_σ^2 into formula (12) and solving the equation, the researchers find $\bar{n}$, then $\bar{\sigma} = \frac{\bar{\sigma}_{lim}}{\bar{n}}$, from equation (6) $\bar{d}$ is found by the formula:

$$\bar{d} = 10\left(\frac{32\bar{M}_{eq}\bar{n}}{\pi\bar{\sigma}_{lim}}\right)^{1/3} \tag{13}$$

The fatigue limits of machine elements according to the bending and torsion stresses are interrelated and the common safety factor of $\bar{n}_\sigma$ is much smaller than that of $\bar{n}_\tau$. Therefore, the coefficient variation of fatigue limit can be taken from the coefficient variation of bending stress.

For stepped shaft, there is stress concentration distribution along the axial length. Usually a reliability assessment is conducted in a dangerous cross section. The reliability probability of the shaft is equal to the reliability probability of the all dangerous cross sections:

$$R_{shaft} = \prod_{1}^{n} R_i \tag{14}$$

2.2. *Sensitivity analysis*

Traditionally, sensitivity analysis is performed in the post-design stage after a design solution is identified. There is also great necessity for designers to conduct sensitivity analysis in the pre-design stage to gain valuable information about the model and its probabilistic behavior (Nam et al., 2006). This is especially important for models with high dimensions as well as for those with high nonlinearity, such as the invisible relationships between inputs and outputs.

The goal of pre-design sensitivity analysis is to reduce the size of the probabilistic design problem by eliminating insignificant random variables. Based on the ranking of all variables, unimportant design variables and noise parameters could be treated as deterministic variables and fixed as constants. In this case, at the initial design stage, a first-order response surface is constructed using all random variables and can be expressed as (x_i is coded design variable at the scale (-1, +1)):

$$y = b_0 + \sum_{i=1}^{k} b_i x_i \tag{15}$$

The global sensitivity index can be calculated as:

$$s_i = \frac{b_i}{\sum_{i=1}^{k} |b_i|} \tag{16}$$

The focus of the sensitivity analysis in the post – design stage is to answer the question of which random uncertainties should be further controlled to gain the largest improvement on the probabilistic performance of the response. The post–design sensitivity analysis is applied to prioritize available resources for variance reduction.

After the sensitivity analysis, with the reduced number of random variables, a high order response surface is constructed. Reliability based design and analysis problems are formulated in this high order response surface.

3. Results of reliability-based design and analysis of reducer shafts

Applying the aforesaid methods, the researchers could establish a RADME (Reliability-based analysis and design of the mechanical system) computer program which includes the three following parts: reliability-based design and analysis of the mechanical structure, transmission system and other.

In this section, RADME is employed to solve the reliability-based design and analysis of machine elements for speed reducer.

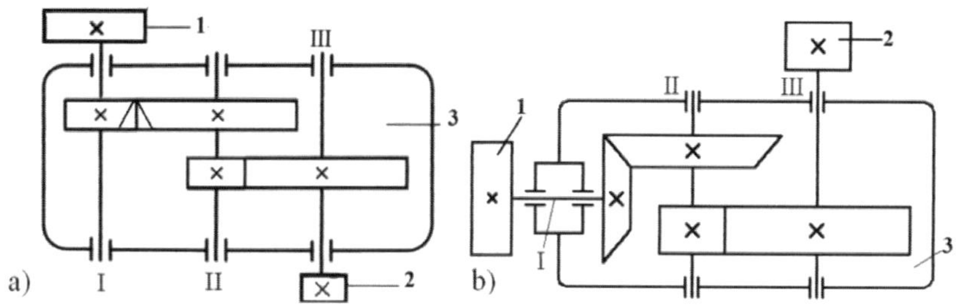

Figure 3. Spur double (a) and bevel-spur reducer (b): 1- Input shaft with rotating part; 2- Output shaft with rotating part; 3- Gear pair.

A popular speed reducer problem represents the design of the single reducer (Figure 3a) which is frequently used in many transmission systems such as light airplanes between the engine and propeller to allow each to rotate at its most efficient speed. As transmission power of reducer equals 7.5 kW, speed ratio u = 3.15 (Table 1), for double reducer $u_{12} = 3.15$, $u_{34} = 2.5$. For more speed ratio, double reducer is utilized (Figure 3b).

Table 1. Technical specifications for single and double reducer.

Shaft	I	II	III
Torsion moment, Nm	73.993	233.077	582.692
Speed, rpm	968	307.3	122.9

3.1. Fault tree analysis

Fault tree analysis (FTA) method is used for reliability analysis of single and double reducer into a logic diagrams in Figure 4.

Reliability R_1, R_2 of single and double reducer can be calculated as:

$$R_1 = R_{sf}^2 R_{gear} R^4_{bear} = R_{sf}^2 R_{cont} R_{bend} R^4_{bear} \quad (17)$$

$$R_2 = R_{sf}^3 R^2_{gear} R^6_{bear} = R_{sf}^3 R^2_{cont} R^2_{bend} R^6_{bear} \quad (18)$$

where R_{sf}, R_{cont}, R_{bend} and R_{bear} are respectively shaft, gear and bearing reliability with reliability of gear pair $R_{gear} = R_{cont}R_{bend}$ (contact and bending stress).

With the system reliability of 0.95, using fault-tree analysis and the distribution method of system reliability in mechanical design, the researchers have the reliability of each element in Table 2.

Table 2. Reliability of machine elements.

Elements	Gear	Shaft I	Shaft II	Bearing
Reliability R_l	0.999	0.999	0.999	0.988

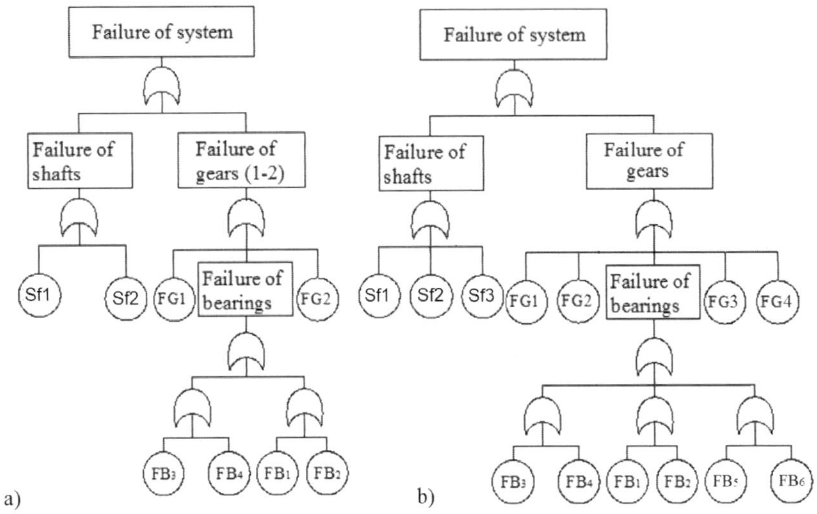

Figure 4. Fault – tree analysis of system: single reducer (a); spur double, bevel-spur reducer (b)

3.2. *Design and analysis of shaft*

Radial load in the end of input shaft is $F_{r2} = 800$ N in the opposite direction of load F_{r1}. Firstly the researchers design and analyse the shaft I. Input data are determined in Figure 5, reliability of shaft R = 0.999.
Using the RADME program, the researchers choose button *Position C* in *Case 1* (loads F_{r1} and F_{r2} contrariwise). Diameter of shaft is found at d = 26.744 mm with 10 random variables (Figure 5). The reliability based design optimization problem with one design variable is now solved using the proposed adaptive reduction of random variables and MCS d = 26.695mm. The researchers choose d = 28mm as standard diameter for shaft.

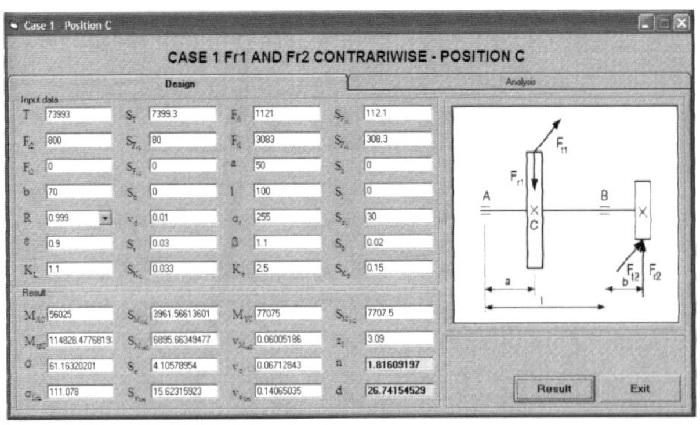

Figure 5. Dialog box for shaft design with 10 random variables

Following that, the reliability-based analysis of shaft I is performed with standard diameter d = 28mm. The calculation results from different methods are shown in Table 3.

Table 3. Reliability of shaft from different methods for single reducer .

Shaft	Diameter d, mm	Matching moment	FORM with MPP	MCS Method
Shaft I	28	0.9998527	0.999913	0.99990274

For bevel-spur reducer with transmission power of reducer equals 5.7 kW, torque moment in shaft I 90.1Nm, in shaft II 171.2Nm. Calculating, designing and analyzing the reliability of 3 shafts of a 2-speed bevel-spur reducer with the given data, we obtained the following results (Table 4).

Table 4. Reliability of 3 shaft from different methods for bevel-spur reducer.

Shaft	Diameter d, mm	FORM with MPP	MCS Method
Shaft I	31.5	0.9991257	0.99946
Shaft II	35	0.9990795	0.99997
Shaft III	42.5	0.9993105	0,99994

At the last design stage, we proceed to analyze the effect of random parameters on the shaft diameter d: a first order response surface is constructed, using all random variables. The first order response surface with 13 random variables can be expressed as (for shaft I):

$$y = 46.414795 - 1.902249x_1 - 1.900806x_2 - 7.100178x_3 - 4.956725x_4 - 0.365519x_5 - 0.097684x_6 - 0.097828x_7 + 35.651551x_8 + 10.101273x_9 + 6.060764x_{10} + 9.091146x_{11} - 1.818229x_{12} + 0.491370x_{13}$$

Sensitivity index: s[1] = -0.02237170; s[2] = -0.02386888; s[3] = -0.08915865; s[4] = -0.06224280; s[5] = -0.00458991; s[6] = -0.00122664; s[7] = -0.00122845; s[8] = 0.44768517; s[9] = 0.12684413; s[10] = 0.07610648; s[11] = 0.11415972; s[12] = -0.02283194; s[13] = 0.00617025.

If sensitivity index of specific variable is less than 1%, the variable is considered as determistic and is fixed at its mean value.

Using Minitab software, we have graphs of the dependence of shaft diameters on the change of random parameters. The regression equation is (Coefficients of regression equations b_o, b_1, b_{11} in Table 5):

$$d = b_0 + b_1 S_{\sigma r} + b_{11} S_{\sigma r}^{\ 2} \qquad (19)$$

Thereby we see that the standard deviation of the fatigue limit $S_{\sigma r}$ has the greatest influence. For example when the values of the fatigue limit standard deviation $S_{\sigma r}$ are from 2.55 to 51 MPa (corresponding to the coefficient of variation of fatigue limit $v_{\sigma r}$ is from 0.01 to 0.2), the size of diameter d increases significantly from 29.6276 to 38.0279 mm (Figure 6a). Dimensional accuracy has little effect on reliability. Hence in some cases this influence can be ignored.

Table 5. Coefficients of regression equations.

Diameter of shaft, mm	b_0	b_1	b_{11}	R-Sq, %
Shaft I d1	29.69	- 0.01517	0.003483	99.9
Shaft II d2	32.89	- 0.00925	0.003869	100.0
Shaft III d3	40.07	- 0.02131	0.004697	99.9

Similarly for shaft II and shaft III, the change of fatigue limit σ_r also has the biggest effect. With the above data, the shaft diameter III changes in turn according to the values from 32.8429 to 42.5373 MPa and the values for the diameter of the shaft III are from 39.9802 to 51.27 mm (Figure 6b).

When analyzing the reliability of shaft with two methods: FORM and MCS, are effective tools to give accurate results. However, if compared to the matching moment method, these two methods are much more complicated, large number of calculations, require loops and help with a computer, so when calculating the design of the shaft diameter, people often use the moment matching method.

For the FORM, convergence reliability index after 4 loops, results appear less than 1second, while with the MCS method, the number of trials must be up to 100 000 and it takes 175 seconds for accurate results. However, we find that the MCS method is more accurate than the FORM, especially when testing with extremely large numbers. Today, with the help of computers and software, the computation is no longer too difficult, the MCS method will be an effective tool to help designers get the exact calculation parameters.

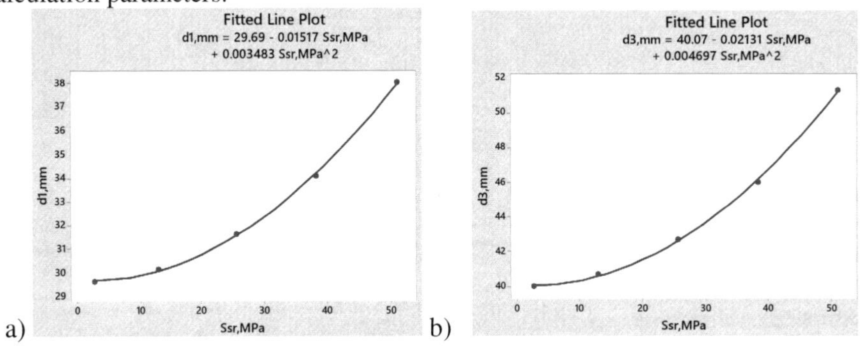

Figure 6. The dependence of shaft diameter I according to the change of standard deviation of the fatigue limit $S_{\sigma r}$.

Thanks to the values obtained from the reliability-based design and analysis of machine elements, the 3D models for single and double reducers can be built. The 3D modeling is illustrated in Figure 7.

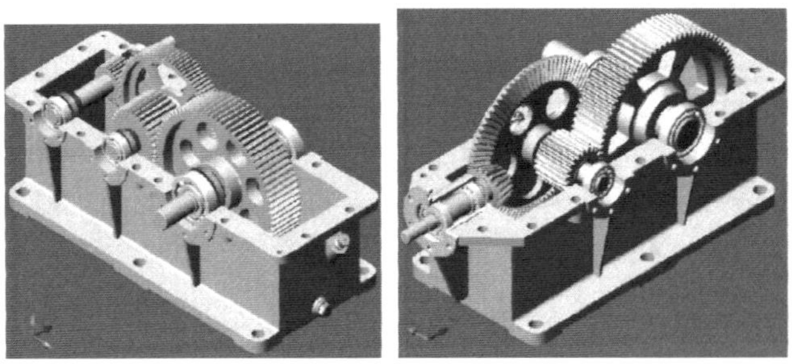

Figure 7. 3D modeling of single and double reducer.

4. Conclusions

In this paper, the researchers have presented the theoretical basis for design calculation and selection of the transmission shafts and gears with 3 stages. At the same time, the RADME program can be employed for reliability-based design and analysis of machine elements and structure:

- The probabilistic design method gives us more precise results, ensuring product quality, high reliability, and safety, depending on the uncertainty of random variables, and can apply RADME computation software in practical industrial design.
- The fault tree analysis (FTA) method is utilized for reliability analysis of single and double reducer into the logic diagrams. From the given reliability of a mechanical system, the researchers can distribute reasonable reliability for each machine element.
- The mechanical properties of a material (fatigue limit) have the highest effect on the reliability and design dimensions of the mechanical system with sensitivity index s[8]= 0.44768517. The lengths of shaft a, b, l, and diameter d have the smallest effect on the reliability. Therefore, with regard to reliability-based design and analysis of shaft, these dimensions a, b, d, l are unique (deterministic). The distribution of external forces dramatically impacts the reliability and design dimensions.
- The FORM, convergence reliability index after 4 loops, results appear less than 1 second and MCS - 175 seconds for accurate results. However, we find that the MCS method is more accurate than the FORM, especially when testing with extremely large numbers. With the help of RADME software, the computation is no longer too difficult, the MCS method will be an effective tool to help designers get the reliability testing of design parameters.
- In the future, we will complete the RADME software to automatically design all the mechanical drive system elements according to the reliability and apply the second-order (SORM) reliability methods.

Acknowledgement

We acknowledge the support of time and facilities from Ho Chi Minh City University of Technology (HCMUT), VNU-HCM for this study.

REFERENCES

[1] Choi, K. K. and Youn, B.D., Selecting Probabilistic Approaches for Reliability-Based Design Optimization (RBDO), *AIAA Journal*, **42**, No. 1 (2004), pp 124–131.

[2] Du, X. and Chen W. A most probable point-based method for efficient uncertainty analysis. *Design Manufacturing*, No. 4 (2001), pp 47-66.

[3] Harish, A, Reliability-Based Optimization: Formulations and Methodologies, *Doctoral thesis*, University of Notre Dam, (2004).

[4] Huibin, L. , Wei, Ch. and Agus Sudjianto, Probabilistic sensitivity analysis methods for design under uncertainty. *10th AIAA/ISSMO*, (2004).

[5] Nam, H. K. and Haoyu, W and Nestor, V. Q., Adaptive reduction of design variables using global sensitivity in reliability – based optimization, *International Journal of Reliability and Safety*, No. 1(1-2) (2006), pp.102-119.

[6] Nguyen, H. L. Reliability based design and analysis of gear drive, *Journal of Science and technology technical universities* (in Vietnamese). No. 05 (2006).

[7] Nguyen, H. L., Tran, V. T. and Pham, Q. T., Reliability–based analysis of machine structure using second-order reliability method. *Journal of Advanced Mechanical Design, Systems, and Manufacturing*, **13**, No.3 (2019).

[8] Nguyen, H. L. *Reliability based design and analysis of mechanical sysems*. Viet Nam National University of Hochiminh City Publishing house. 2016 (in Vietnamese)
[9] Nguyen, H. L.. Textbook Fundamentals of machine design. Viet Nam National University of Hochiminh City Publishing house. 2018 (In Vietnamese).
[10] Padmanabhan. D., Agarwal, H., Renaud, J. E., Batill, S. M, Monte Carlo Simulation in Reliability-Based Optimization Using Approximation Concepts, *Fourth International Symposium on Uncertainty Modeling and Analysis*, (2003).
[11] Qu, X. Reliability – based structural optimization using response surface approximations and probabilistic sufficiency factor. *Doctoral Thesis*, University of Florida (2004).
[12] Zhang, Y.M., Qiaoling, L., Bangchun, W., Practical reliability-based design of gear pairs, *Mechanism and Machine Theory*. Volume 38, Issue 12 (2003), pp. 1363-1370.

An innovative control system of a hybrid tankless water heaters using LabVIEW

R F Garcia[1,2] and A Chua[1*]

[1]Mechanical Engineering Department, De La Salle University, Manila, Philippines
[2]Engineering and Science Research Center, Polytechnic University of the Philippines, Sta. Mesa, Manila, Philippines

*rey_garcia@dlsu.edu.ph

Abstract. This paper presents an innovative control system of a hybrid tankless water heater that provides a rapid response to deliver hot water within a desired temperature range at lower energy consumption. The proposed tankless water heater provides a hybrid heating process that composes of a gas fired heater as primary heater and an electric heating system as secondary heater. LabVIEW designing software was used in designing the control system. A controller was designed using a PID controller which response with respect to given input parameter such as changes in water flow and the inlet temperature. Through alteration of heat source, this efficiently reduces the power consumption of the tankless water heater during low water demand, and eliminate overshoots. Simulation results demonstrated that the proposed system has performed efficiently in a range of water flow and temperature demand while reducing power consumption for about 30%.

1. Introduction

Majority of residential hot water production uses gas fired water heaters as it has the advantage of converting thermal energy to heat. Storage type or tank-based gas fired water heater is the most conventional water heater used for both industrial and residential application because of its capability to heat bulk of water at lower cost. Tank water heater works as simply as supplying cold water to the tank and injected at its bottom through a dip tube. As the water heats up, it naturally rises and is drawn off by a hot-water discharge pipe [1]. However, tank-based water heaters use a lot of gas for maintaining the desired water temperature in which it continuously fires the tank even if the demand of hot water is low. Tankless gas water heater (TGWH) gains the popularity as a replacement of tank-based heaters. It is the most efficient conventional method of generating heat from natural gas in a domestic hot water application [2]. At the rate of three gallons of hot water per minute, the TGWH has the capability to heat the water instantly compared to tank-based water heaters since a larger volume of water in a tank makes the tank-type heater to heat the water at a longer period of time [3]. One disadvantage of TGWH is the difficulty to produce stable hot water temperature as it is prone to large temperature swings. This is due to the changes in water flow, as it changes rapidly and unpredictable [4]. Majority of tankless water heater units are having fully modulating gas valves that can range from 10,000 BTU to 1,000,000 BTUs which limits the heating system at certain flow rate [5]. As the flow rate is inversely proportional to the outlet temperature for tankless water heaters, a very low flow rate may result overheating of water since the gas fired unit has a minimum flow limit. Contrarily, a temperature drops or a splash of cold-water occurred when water demand increases rapidly. On the other hand, an electric powered water heater gains the popularity as it can be easily controlled. Compared to gas fired heater, an electric water heater power consumption is controlled by the voltage applied to its heating element [6]. Recent work conducted by manufactures for controlling of TGWH output temperature were not been published as research paper, but resulted as patents instead such as the use of blending valve for combining overheated water and cold water to produce blended hot water at predetermined temperature [7], a heating system that has a multiple chamber connected in series [8]. Few research papers are can be found in the literature discussing techniques of TGWH temperature output control. One of those controlling techniques was proposed by Haissig and Woessner using An adaptive fuzzy control (AFC) algorithm to regulate the domestic hot water temperature and central heating water of a combi-boiler [9]. Henze et. al. develop and build an TGWH that utilize model predictive control (MPC) to minimized the outlet temperature error [10]. Costa et. al. make a model of a tankless gas water heater that composes of a tank acting as a thermal capacitance, a mixing valve for mixing tanked water and heated water, and a bypass valve that is connected to the outlet of heater connected to the output aims to eliminate the outlet temperature overshoot [4]. Furthermore, despite of giving an accurate outlet temperature control, a gas fired water heaters release CO2 emission from burning of natural gas that is needed to reduce or eliminate in the future works [11]. In Table 1, CO2 emissions of heating water using natural gas, and electric resistance water heaters was presented. This clearly shows that electric water heaters greatly contribute CH4 and CO2 emission when using natural gas or coal as source of electricity more than those of Gas water heaters. Even though the use of renewable sources of energy are being develop for water heating, its current efficiency were not enough to overcome the conventional gas fired heater [12-13].

Table 1. CO2 emissions of heating water using natural gas, and electric resistance water heaters.

Source of electricity	*GHG*	*Reference*	*Natural gas water heater*		*Electric water heater*	
			Storage	*Tankless*	*Storage*	*Tankless*
Coal	CH4	Alvarez et al [14]	0.82	0.6	0.2	0.19
	CO2		86.5	63.6	256	243
	CH4	Howarth et al [15]	1.38	1.02	0.2	0.19
	CO2		86.5	63.6	256	243
Natural Gas	CH4	Alvarez et al.	0.82	0.6	0.97	0.93
	CO2		86.5	63.6	125	119
	CH4	Howarth et al.	1.38	1.02	1.3	1.23
	CO2		86.5	63.6	125	119

Due to the stated problems with regards to environmental impact, power consumption, and systems controllability of water heating technology a hybrid heating system was proposed that is capable of delivering water within a desired temperature range at lower energy consumption and CO2 emissions. This paper presents an innovative control system of a Hybrid tankless water heater (HTWH). A gas fired water heater is used as a primary heating element for high flow water demand, while an electric water heater is used as a secondary heater for maintaining the water temperature in a buffer tank during low flow demand. To accurately measure flowrate in high temperature areas, the differential pressure sensor was calibrated using artificial neural network. Furthermore, the PID controller is used as the main controller for the both heaters and recirculating pumps. The simulation of the proposed model was made through LabView design software.

2. Modelling HTWH Components

Figure 1. illustrate the schematic diagram of the HTWH. Combination of two heating element is present in this design for heating different flow demand. A flow demand causes a fluid to enter the water heater at inlet and exit water heater at outlet. At point 1, the water inlet temperature was measure and sends the signal to the controller for heat control. From point 1, water enters directly to point 2 and at point 6 if bypass is desired. Flow meter used to measure the rate of water entering the primary heater, signal from flow meter was sent to the both controller of heater. When in moderate and high demand of water it sends signal to the primary heater to give enough amount of heat to reached the desired temperature at the outlet. A sudden drop of flowrate tends to increase the outlet temperature in conventional water heater, while flow surge decreases the temperature at the inlet. To resolve this issue, a thermal compensating differential pressure switch is used to detect surge of flow or sudden drop of flow between the point 6 and point 4, it signals the controller to turn on or off the primary heater or secondary heater. In a moderate demand of flow, secondary heater was off or at the lowest power setting to just maintained the heat at the buffer tank. When the demand of flow was high, both heater functions to generate give the desired temperature rapidly. And at when flow of water is too low, the primary heater was turned off while the secondary heater was on, also recirculation was added in the design to absorb the remaining heat from the primary heater, this would also resolve the problem of having cold sandwiches or the stack water that cooled down at the pipe while heater is not in used.

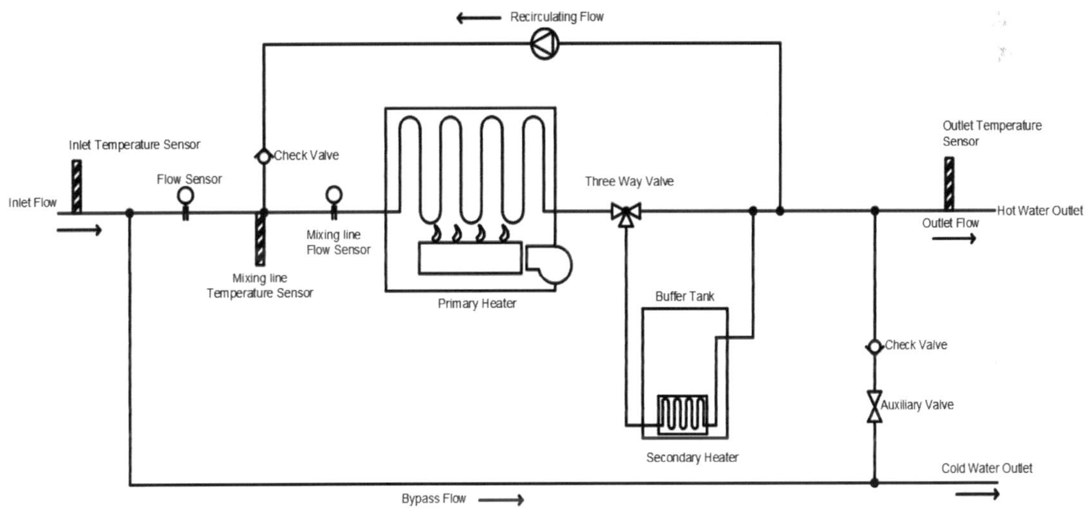

Figure 1. Schematic of a HTWH

To develop and simulate HTWH controller, a modelling methodology was established in which individual parts of each internal component are modelled, describing thermal, fluid and mechanical dynamics. Then models of each component are then interconnected which builds the whole HTWH system. For the fluid component, mass conservation law is applied to for controlled volume given by the equation

$$\dot{m}_{in} = \dot{m}_{out} \tag{1}$$

where $\dot{m}_{in}$ is the mass flow rate inlet and $\dot{m}_{out}$ is the mass flow rate outlet. Mass distribution inside the system where split into the three-way valve for the recirculation, bypass and buffer tank. The equation of mass flow rate split by the bypass line is written as

$$\dot{m}_{in} = \dot{m}_a + \dot{m}_b \tag{2}$$

where $\dot{m}_a$ and $\dot{m}_b$ are the mass going to heat cell and bypass line. A check valve is used to avoid counter flowing of water in the recirculation line. During overshoot the bypass line and recirculation line are open for tempering of water outlet, so the total mass entering the heat cell during overshoot is the addition of flow of recirculation and inlet flow deducting the bypass flowrate given by the equation

$$\dot{m}_h = \dot{m}_a + \dot{m}_c - \dot{m}_b \tag{3}$$

where $\dot{m}_c$ is the mass flow rate from the recirculation line.

2.1. Heat Cell

The thermal dynamics of the heat cell is based on the energy conservation equation for the control volume under analysis. Only the internal energy variations are considered, the kinetic and gravitational potential energy variations are assumed negligible and a constant mass inside each control volume is considered.

2.1.1. Primary Heat Cell. Primary heat cell mathematical model is derived from the energy conservation equation

$$Q = \dot{m}_h C_p (T_f - T_i) \tag{4}$$

Where Q is the heat needed to rise the temperature inlet, Ti into the final temperature, Tf at mass entering the heat cell, $\dot{m}_h$. Using the specific heat capacity of water, Cp. Figure 2. Illustrates the case structure of primary heat cell in LabView. Case structure is used as to turn on and off the heater whenever a boolean statement is true which is connected to the amount of flow at the inlet line.

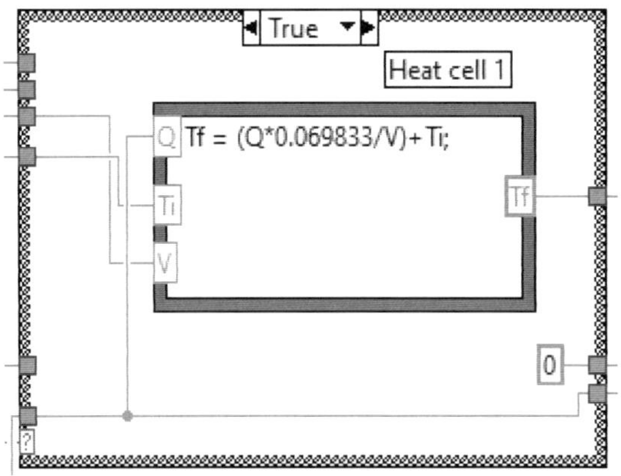

Figure 2. Case structure of Primary heat cell

2.1.2. Secondary Heat Cell. For Secondary Heat cell, uses the same energy equation as the primary heater. Secondary heater will turn on by a boolean statement connected into the flow sensor at the inlet line. For low demand of flow the boolean statement is false which turns on the secondary heater. The temperature inlet, T_i of the secondary heater are the total heat absorb by the recirculating line and the temperature of water inlet.

2.2. Recirculation and Bypass line

Bypass line was designed to split the water inlet to two part. At the end of the line cold water and hot water meets which tempers the outlet water at the desired water temperature. Bypass valve is either manual or automatically operated. The final temperature from the model of the bypass mixing valve is given by the equation

$$T_m = (m_b T_i + m_h T_f) / (m_b + m_h) \quad (5)$$

Where Tm is the temperature at the mixing valve, and Tf is the temperature from the outlet of the heater. Case structure of the mixing valve was shown in figure 3, it is turn on when the boolean statement is true which means the control valve for the bypass line was manually controlled, while false case structure where made by just connecting the input Tf into output Tm.

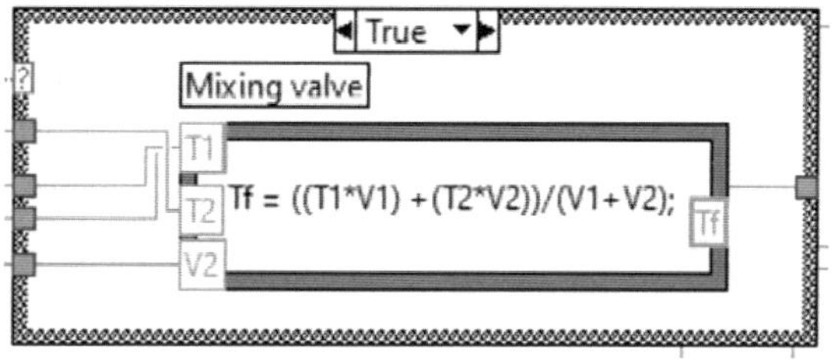

Figure 3. Case structure of mixing valve

Recirculation take effect when an overshoot was occurred or during low flow demand. Mixing of heated water and water inlet happened during recirculation, this for absorbing the excess heat from the primary boiler, or reusing the excess heat during overshoot. Case structured with formula node was used for another mixing valve using the equation 5.

2.3. Sensors

Sensors are one of the critical components of the HTWH. This this heating system two thermal sensors are used to monitor the temperature from the inlet, mixing and outlet of the system. each sensor sends signals to the controller for an accurate feedback. For measuring the flow and pressure of the system, three pressure sensors are used. Two sensors are used for measuring the flow rate of inlet and mixing valve while the other one is a Differential pressure switch sensor which has the capability to detect surge flow or flow drop. Each sensor sends signal to the main controller for the selection of which heater will turned on. A pressure sensor with thermal compensation is a must feature for the system since sensors has direct contact with the measuring fluid.

2.3.1. Flow Measurement. Flow measurement calculation was based on the differential pressure measured by the DP sensor. As shown in Fig. 4 the DP sensor is used in a straight pipe attached at the both pressure side separated by an orifice. For DP sensors placed at the mixing valve, a specially designed sensor was used to compensate the effect of high temperature. From the reading acquired by hall sensor the differential pressure will be computed. The formula used in this simulation for computing the flowrate by means of measured pressure difference is based on the ISO standards ISO 5167-1 and 5167-2 for measuring steady flow. Given by this formula

$$Q = C_d \sqrt{\frac{2\Delta P}{\rho}} \times \frac{A_2}{\sqrt{1-\left(\frac{A_2}{A_1}\right)^2}} \quad (6)$$

where Δp is the differential pressure between high pressure and low pressure side, ρ is the fluid density and C_d is the coeficient of discharge. The discharge coefficient C_c is given by empirical formulas, which depend on the ratio of the orifice bore and pipe inside diameter. Hence the density of the fluid varies as temperature increases due to thermal expansion of fluid. This may result to an additional error to the flowrate reading as in Equation 6 the density is required.To accurely measure the flowrate at different temperature the DIPPR105 equation 7 is used for computing the exact density of fluid [16]. The DIPPR105 equation is written as

$$\rho = \frac{A}{B^{1+\left(1-\frac{T}{C}\right)^{D}}} \tag{7}$$

where T is the temperature of water in Kelvin which has 273 K to 648 K temperature limit and A,B,C, and D are empirical constant with a value of 0.14395,0.0112,649.727, and 0.05107 consecutively.

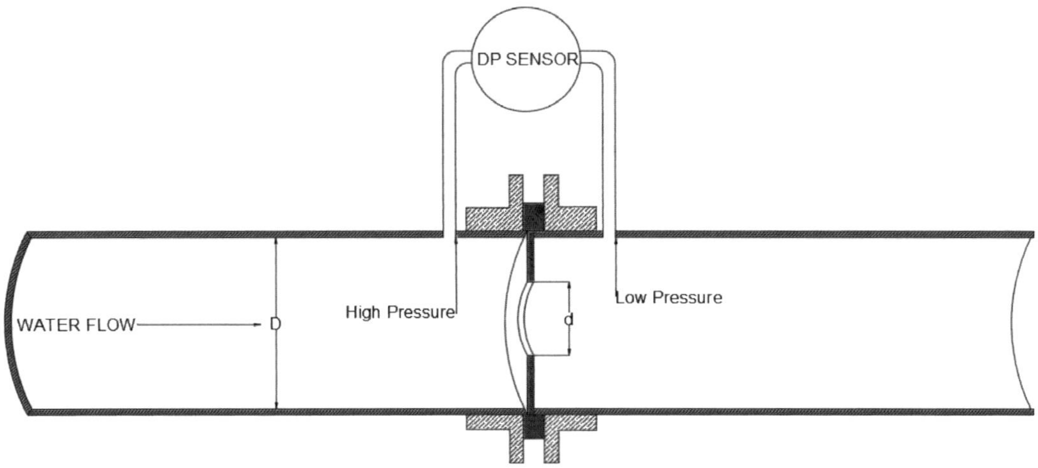

Figure 4. DP sensor using orifice to measure flowrate

2.3.2 DP sensor calibration. The calibration of pressure transmitter using ANN is shown in Fig. 5. In this work the training sets used to train the neural network is from the experimental data of Sinha and Mandal 2018 [17]. The desired value of the hall voltage was a result of the experiment at applied constant pressure at a room temperature of 27°C. The inputs to the neural network are the Hall voltage from hall sensor and varying temperature, and the ANN output is corrected output Hall voltage.

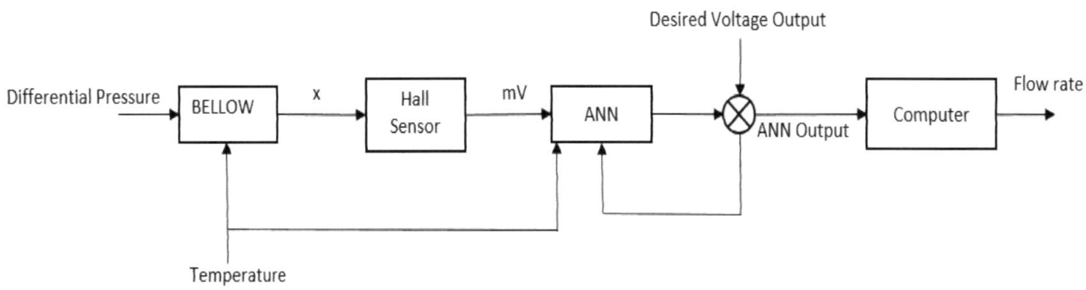

Figure 5. DP sensor using orifice to measure flowrate

ANN is trained in Neural Fitting tool in Matlab using Levenberg-Marquardt backpropagation algorithm to calibrate the sensor since the reading of DP sensor without calibration results an increasing error as temperature increases. The performance of calibration was shown in table 2. Wherein the DP sensor error was reduced from 3.51%-13.14% into 0.75% - 1.79%. This will give an accurate value for the controller the calculate the power needed to supply in order to meet the desired outlet temperature.

Table 2. Effect of High Temperature on DP sensor reading.

Temperature	*DP reading before calibration*	*DP reading after calibration*	*True Value*	*difference*	*% error before calibration*	*% error after calibration*
15°C	17.61	18.39	18.25	0.78	3.51%	0.75%
20°C	18.21	18.25	18.25	0.04	0.20%	0.02%
27°C	18.42	18.26	18.25	0.15	0.95%	0.14%
30°C	18.57	18.27	18.25	0.28	1.75%	0.23%
35°C	18.73	18.31	18.25	0.42	2.64%	0.32%
40°C	19.04	18.33	18.25	0.71	4.35%	0.45%
50°C	19.41	18.36	18.25	1.05	6.34%	0.60%
60°C	19.79	18.40	18.25	1.39	8.46%	0.84%
70°C	20.21	18.47	18.25	1.73	10.72%	1.23%
80°C	20.65	18.58	18.25	2.07	13.14%	1.79%

2.4 PID Controller

For the PID component, the feedback power is calculated based on the desired input temperature and output of process variable e.g outlet temperature. PID output was set at from 3000 to 60000 watts based on the maximum performance of the TGWH and 0 to 2000 watts base on the maximum power output of the EWH. PID gains were automatically tune from the LabView PID autotuning block.

3. Simulation

Individual components were modelled separate in LabView, by connecting each component models a HTWH formed and use in simulation. As illustrated in Figure 6. The HTWH model ware enclose by a while loop for an infinite simulation time frame. The volume flow rate was adjusted manually by adjusting the control knob during the run time. Adjusting know in the scale of 0-4 makes an opening of the valve which enables water to flow for a range of 0 – 10 liters per minute. The initial water temperature is a constant parameter in the simulation, it was set according to the average temperature of tap water in a tropical country that is 27 °C. The set-point temperature was set to be at 50 °C, above the average hot shower temperature. A monitoring panel was shown in figure 7. In the front panel the temperature of inlet and outlet was illustrated by red colored progress bar, a default numeric indicator of LabView for indicating temperature. Three knobs are place for the adjustment of heated water volume flowrate, bypass flow rate and desired temperature setting. Two led light was also place to indicate whether the heater is on or off, and which heater is running. And lastly a flow gauge was placed to monitor the volume of water flowing though the heater.

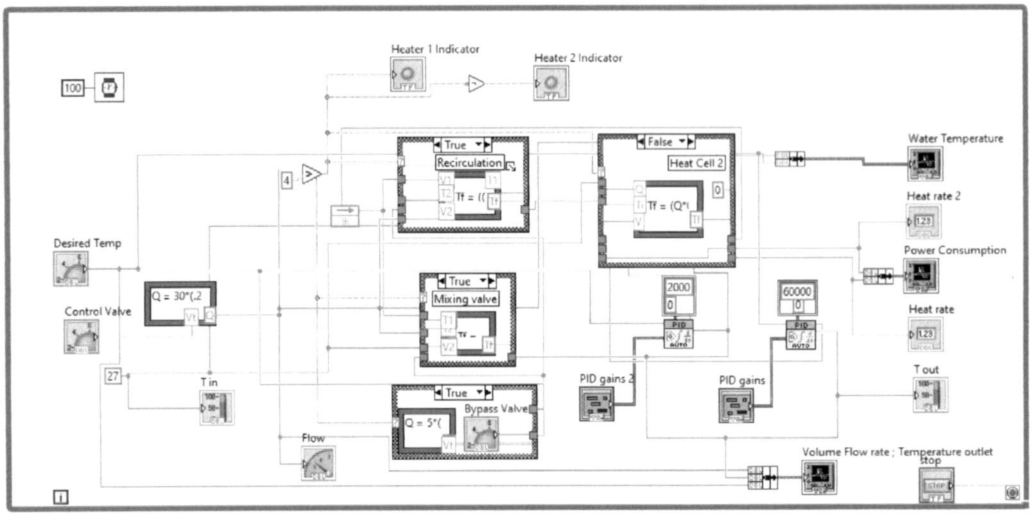

Figure 6. HTWT LabView block diagram

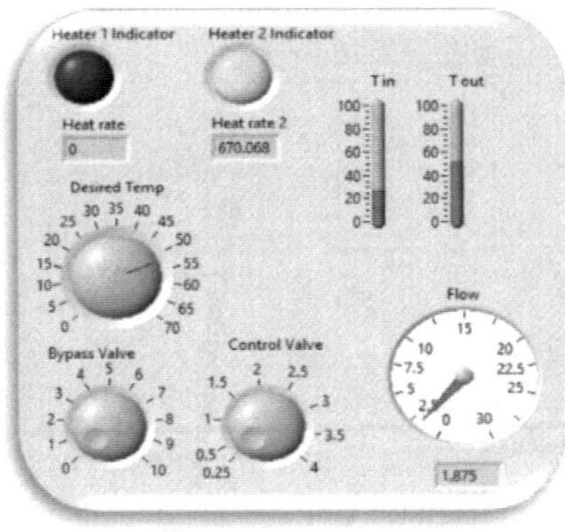

Figure 7. HTWT LabView front panel

4. Results and Discussion

Simulation results shows that the model response with less overshoots on the given flowrate changes as illustrated in Figure 8. Turning on of bypass valve during overshoots helps the system to reduce the high outlet temperature output rapidly, while recirculation of high temperature water throughout the system helps the secondary heater to maintain the desired temperature in a low flow demand scenario. The power consumption of the HTWH was shown in Figure 9. The red line indicated the power consumption of the primary heater while the blue line indicates the secondary heater consumption. It clearly shows that when the flowrate was below 5 lpm, the primary heaters shuts down then the secondary heater takes its place. As a result, it saves power for about 30% and lessen the amount of CO_2 emission from the primary heater.

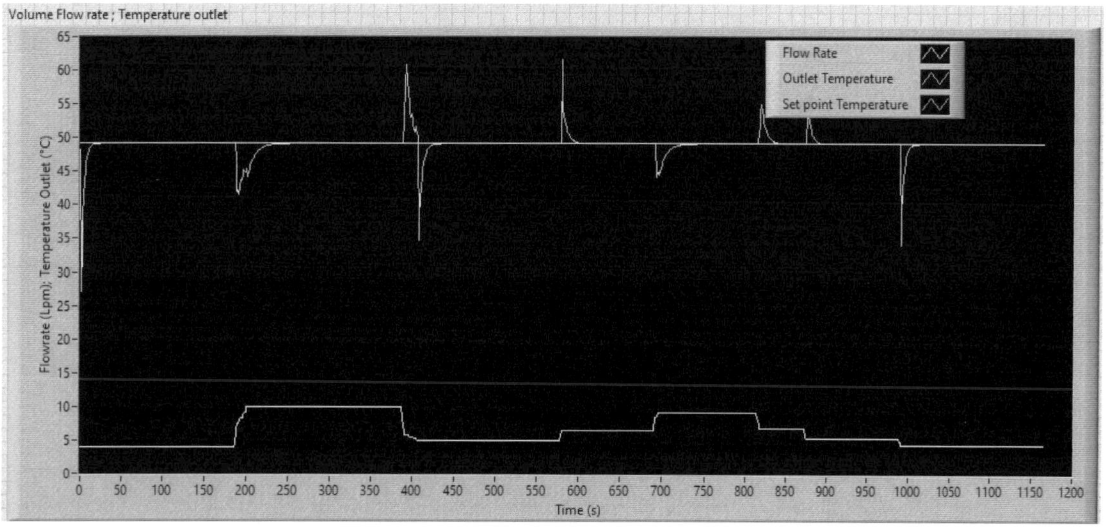

Figure 8. Simulated HTWH outlet temperature response to a water flow rate step.

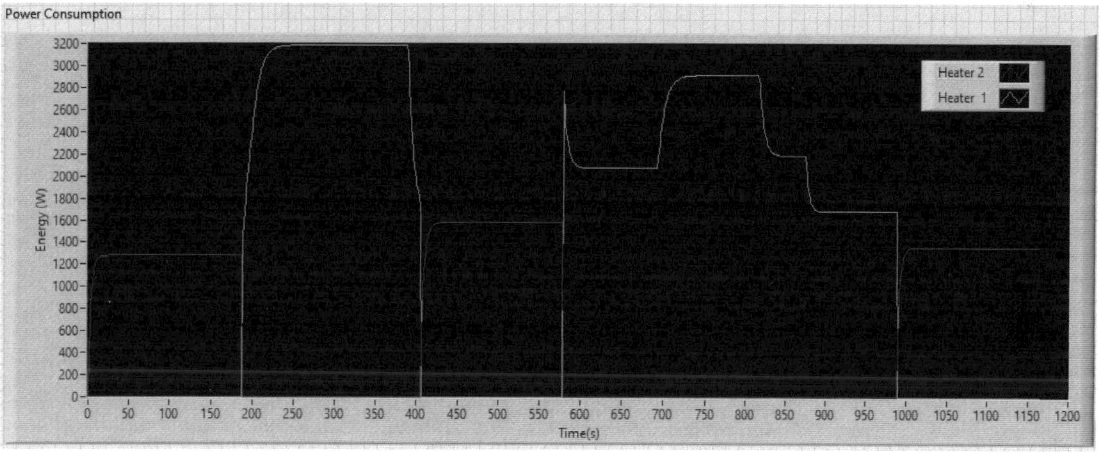

Figure 9. Simulated HTWH power consumption.

5. Conclusion

In this work, an innovative control system was developed for a Hybrid tankless water heater for the improvement of conventional tankless water heater. A hybrid heating system were composed of two heating cells specifically a gas fired and an electric operated heater that are controlled using PID controllers. Simulation results shows that the response of the model has able to give the desired water temperature at the different flowrates and responds within short period of time. Through recirculating flow and opening of bypass valve the overshoots were lessened. Moreover, the proposed control system has able to save energy by altering the use of heat cells in high and low flow demand for about 30% and lessens the CO_2 emission from the gas fired heater. For the future work of this research, the next task will be a research on an advanced controller which has able to predict the flow demand ahead of time for an advance heating method.

6. References

[1] O. Ibrahim, F. Fardoun, R. Younes, and H. Louahlia-Gualous, "Review of water-heating systems: General selection approach based on energy and environmental aspects," *Build. Environ.*, **72**, pp 259–286, 2014, doi: 10.1016/j.buildenv.2013.09.006.

[2] D. Bohac, M. S. Lobenstein, and T. Butcher, "Actual Savings and Performance of Natural Gas Tankless Water Heaters Prepared for Minnesota Office of Energy Security," *Cent. Energy Environ.*, 2010.

[3] A. F. Quintã, J. A. F. Ferreira, A. Ramos, N. A. D. Martins, and V. A. F. Costa, "Simulation models for tankless gas water heaters," *Appl. Therm. Eng.*, **148**, no. July 2018, pp 944–952, 2019, doi: 10.1016/j.applthermaleng.2018.11.095.

[4] V. Costa, J. Ferreira, and D. Guilherme, "Modeling and simulation of tankless gas water heaters to reduce temperature overshoots and undershoots," *12th Int. Conf. Heat Transf. Fluid Mech. Thermodyn. (HEFAT 2016)*, pp 1404–1409, 2016.

[5] I. Sridhar Deivasigamani, (Peoria, "CONTROL METHOD FOR A HYBRD TANKLESS," 2015.

[6] M. H. Nehrir and B. J. Lameres, "Multiple-block fuzzy logic-based electric water heater demand-side management strategy for leveling distribution feeder demand profile," *Electr. Power Syst. Res.*, **56**, no. 3, pp. 225–230, 2000, doi: 10.1016/S0378-7796(00)00124-3.

[7] W. Kubik, "Instantaneous Water Heater," US 7,140,378 B2, 2006.

[8] W. R. Sturm, "Fluid heating system," US 7, 206,506 B2, 2007.

[9] C. M. Haissig and M. Woessner, "An adaptive fuzzy algorithm for domestic hot water temperature control of a combi-boiler," *HVAC R Res.*, **6**, no. 2, pp 117–134, 2000, doi: 10.1080/10789669.2000.10391253.

[10] G. P. Henze, D. P. Yuill, and A. H. Coward, "Development of a model predictive controller for tankless water heaters," *HVAC R Res.*, **15**, no. 1, pp. 3–23, 2009, doi: 10.1080/10789669.2009.10390823.

[11] B. Hong and R. W. Howarth, "Greenhouse gas emissions from domestic hot water: Heat pumps compared to most commonly used systems," *Energy Sci. Eng.*, **4**, no. 2, pp 123–133, 2016, doi: 10.1002/ese3.112.

[12] C. S. Jwo, Z. J. Chien, Y. L. Chen, and C. C. Chien, "Development of a wind directly forced heat pump and its efficiency analysis," *Int. J. Photoenergy*, vol. 2013, pp. 1–8, 2013, doi: 10.1155/2013/862547.

[13] R. Dhinakaran, R. Muraliraja, R. Elansezhian, S. Baskar, S. Satish, and V. S. Shaisundaram, "Utilization of solar resource using phase change material assisted solar water heater and the influence of nano filler," *Mater. Today Proc.*, no. xxxx, 2020, doi: 10.1016/j.matpr.2020.06.460.

[14] R. A. Alvarez, S. W. Pacala, J. J. Winebrake, W. L. Chameides, and S. P. Hamburg, "Greater focus needed on methane leakage from natural gas infrastructure," *Proc. Natl. Acad. Sci. U. S. A.*, **109**, no. 17, pp 6435–6440, 2012, doi: 10.1073/pnas.1202407109.

[15] R. W. Howarth, R. Santoro, and A. Ingraffea, "Methane and the greenhouse-gas footprint of natural gas from shale formations," *Clim. Change*, **106**, no. 4, pp 679–690, 2011, doi: 10.1007/s10584-011-0061-5.

[16] Dortmund Data Bank Software & Separation Technology, "Pure Component Equations," *Fitting Pure Compon. Equations*, pp 3–26, 2014.

[17] S. Sinha and N. Mandal, "Design of a Smart Pressure Transmitter and Its Temperature Compensation Using Artificial Neural Network," *J. Control. Autom. Electr. Syst.*, 2018, doi: 10.1007/s40313-018-00430-1.

Analysis of the effects of geometry on the press fit application in automotive power modules

A T Ubando[1,2], J Gonzaga[1,2], E Arriola[1,2], R L Moran[1,2], N R E Lim[1,2], J P Mercado[1,2], A Conversion[1,2], D Belarmino[1,2]

[1] Mechanical Engineering Department, De La Salle University, 2401 Taft Avenue, 1004 Manila, Philippines
[2] Thermomechanical Analysis Laboratory, De La Salle University – Manila, Laguna Campus, LTI Spine Road, Laguna Blvd, Biñan, Laguna, Philippines

aristotle.ubando@dlsu.edu.ph

Abstract. Semiconductors and electronics have been found to have an increasing use in automobile design. One key component of an automobile is the power module, which is a high voltage component that finds itself in extreme operating conditions. Due to this condition, the power module demands for highly effective terminal connections that can withstand the extreme operating condition specifically on the terminals that will connect the power module to the automobile system. Press fit technology has been a solution for cheap and reliable methods of creating interconnections since the 1980s; however, necessary innovations exist for more environmental design, and increase demand in quality and quantity. Press fits are simple, force-fitting connectors that allow permanent bonds without solder or adhesive bonding and a shortened process time. This research presents a unique methodology on the evaluation of the effects of the geometry to the insertion force and retention force using finite element analysis. The result showed that the geometry of the pin in a press fit pin configuration has a significant effect on the insertion force, retention force, as well as in the equivalent stress.

1. Introduction

As electronics are now fundamental to the modern automobile, semiconductor packages are put under harsh operating conditions of the vehicle and local climate. Furthermore, miniaturization and the demand for greater functionality drive the demand for denser interconnections in electronics. The power module is a primary component, exposed to 150 Celsius or higher ambient operating temperature and high vibrations forces [1]. This all generates harsh thermal cycling for their interconnection. A white paper by Lynch & Longford [1] outlines the ambient temperatures of hood and near engine temperature requirements for power module assemblies.

Press fit technology is a 'force fitting' interconnection solution that has been in use in automobile power modules and electronics since the 1980s. 'Force fitting' means that two or more components are pressed into place and locked together via the plastic deformation that occurs as components are pressed together [2]. Press fit pins were originally in use for telecommunication electronics in the 1960s[3] and were only introduced later to automotive design.

Table 1. OEM automotive temperature classes by [1] which highlighted the under hood and near engine ambient temperature ranges.

Temperature Class	Temperature Range		SAE Temperature Classifications	
			Class	**Ambient Temp.**
	Low	**High**	1	-40 to +85°C
1		+105°C	2	-40 to +100°C
2	-40 °C	+125°C	3	-40 to +125°C
3		+150°C	4	-40 to +150°C
4		+175°C	5	-40 to +175°C

While our research concentrates on the compliant press fit pin inserted in a solid press-in zone, another type exists that is a solid pin for a compliant press-in zone [4]. Compliant pins enjoy more favored use compared to solid pins, due to lower insertion force, elimination of damage to the plated through holes (PTH) insertion zone, and more reliable signal [5]. The compliant, press fit pins gain the best economy compared to most electronic joints and possess an ease in assembly, repair, and interchangeability. Removing the soldering process eliminates the thermal cycling, contaminant and heat flux, and environmental impact associated with it [6]. These designs are fitted into the PTH of a PCB, which vary in sizes up to 0.8 mm. The FR-4 plate finishes of a PCB are most commonly Tin, Silver, Gold, and OSP, and have a thickness of 1.57 mm or higher. In the press-in zones, plating can be reduced to 1 micron, while the allowable contact resistance in any connection may only go as high as 0.5 mW [4].

A white paper by Lynch & Longford [1] presents a press fit design in 0.64 mm and 0.8 mm thickness. This experimented with CuSn5 (15% IACS), 7025 CuNiSi (40% IACS), and two special brass alloys with 30% and 80% IACS; the interchanging of materials allows the design to survive in either 125 or 150 Celsius operating conditions.

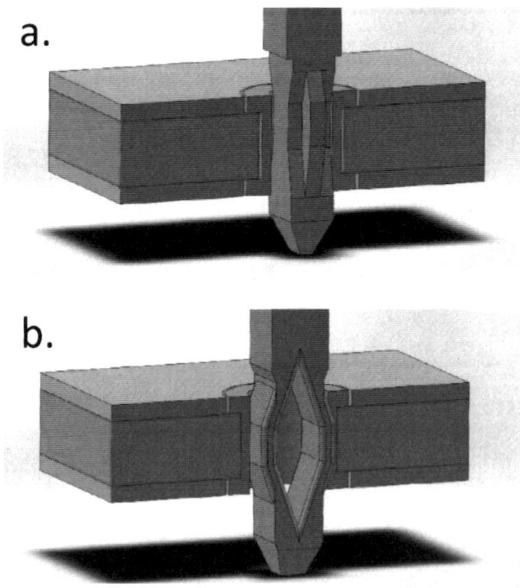

Figure 1. (a.) solid press-fit pin with compliant PTH versus (b.) compliant press-fit pin with solid PTH [6]

High contact pressure from macroscopic 'squeezing' forces are concentrated in localized areas, allowing press-fit pins to deform into the press-in zones. Recrystallizing of free electrons occur on the contact metal surfaces, creating a cold weld phenomenon and increasing bonding forces. This phenomenon is more prominent in Tin plated surfaces, hence the use of Tin-lead (SnPb) surface finishes for the compliant pins and PTH [6]. Tin-lead finishes are also able to effectively reduce contact resistance down to 0.1 mW while increasing mechanical stability. These innovations were able to reduce the necessary retention force below 20 Newtons [5]. However, the push for lead-free alternatives in electronic components poses a challenge in creating metal finishes that can possess a similar efficiency.

Chou and Hilty [6] conducted a design of experiment (DOE) study to reduce the retention force required for lead-free, compliant press fit pins. The research found that the hole size has a linear inverse relationship with insertion force while finish of the compliant, press-fit pins has a direct relationship with retention force. The minimum radius a metal contact is able to carry current is greater than 10 micrometers [2], meaning that bumps in metal surfaces will vary in their ability to carry a current, affecting overall resistance of a metal to metal contact. The retention force, typically between -20 and 40 Newtons, is important for the stability of the press-fit pin while insertion force is necessary to guide pin application. [1, 6]

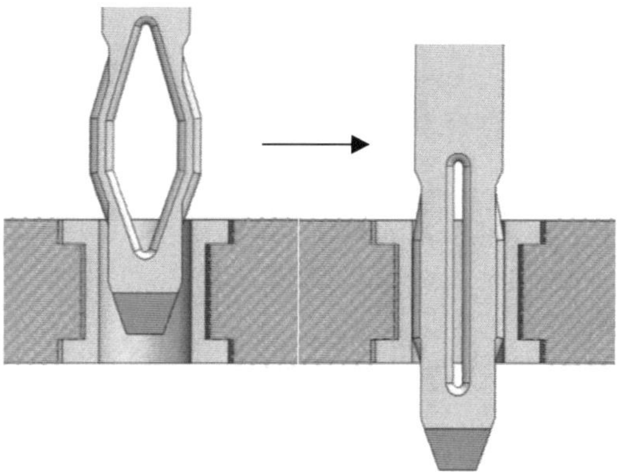

Figure 2. Insertion of compliant press-fit pin into solid PTH [1]

Corrosion, organic wastes, and contamination leads to contact issues into the press-fit pin technology. Most problems arise from corrosion; fusing surfaces of noble metals are known to reduce corrosive effects. Cracks and signal disruption will still occur due to deformation in the material. Stotze et al. [2] explores non-destructive reliability tests and analysis of press-fit technology, for power modules in climactic conditions. The author also explores the fritting phenomenon, where corrosion build up is burnt away by low voltage. Finite Element Analysis (FEA) was used by Fellner et al.[7] to explore the stability and connection reliability of molded interconnection devices (MID) press-fit pins for 3-dimensional PCB substrates. Their model was able to describe the behavior of the coefficient of friction, assisted with empirical data. Their model predicts force-displacement curves, retention forces, and mechanical and thermal stresses of the pins.

This paper uses FEA to create a model to predict necessary press-in forces and contact/sliding frictional forces that affect retention forces in a compliant, press fit pin.

2. Methodology

The determination of the different forces and stresses in a press fit connection was conducted using finite element analysis. Finite element analysis (FEA) is used to determine the distribution of stress in a complex geometry [8]. As discussed by Ogrodnik [9], finite element analysis is fully accepted for simulation of experimentations. The predictions result of FEA for strains, warpage, stress, etc. are widely acceptable as predictions of reality. The application of FEA is suitable for drop test analysis due to its superior capability of analyzing diversified problems with complicated geometry [10]. FEA utilizes the piecewise approximation wherein the geometry is divided into smaller body (discrete subregions) called finite elements. Each of these elements is represented by a set of nodes which are controlled by a set of partial differential equations with respect to its physical disciplines [10]. The whole analysis is carried out flowing the methodological framework of the study as shown in **Figure 3**.

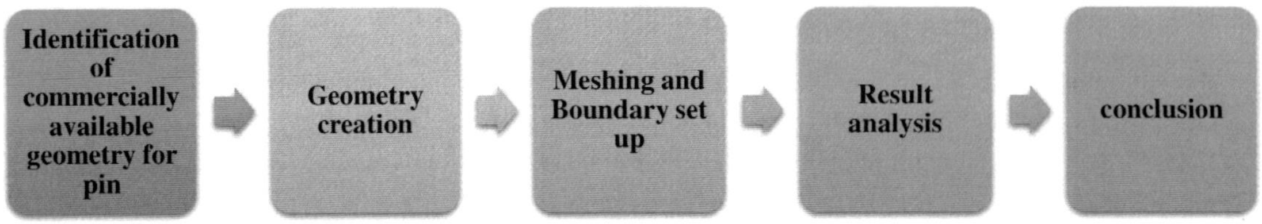

Figure 3. Methodological Framework

The geometry evaluated came from the commercially available pins [11]. The geometry selected was created using Solidworks 2020 under the license of De La Salle University. The mesh was set to fine setting for a more accurate result. Specifically, meshing focuses on the refinement around the contact area between the pin and the plated through hole. A refined mesh strategy, together with a target minimum number of elements within the targeted contact region, was applied to ensure the accuracy of results. The simulation platform used in the analysis was ANSYS Workbench 2020 R1 licensed under the De La Salle University in the Thermomechanical Analysis Laboratory while the finite element analysis was based on the study of Wu et al. [12]. The computer utilized for the simulation is equipped with Xeon Gold Processor with 20 cores and 128 GB RAM.

Table 2. Different geometry used in the simulation.

Type of geometry	Illustration
Eye of Needle	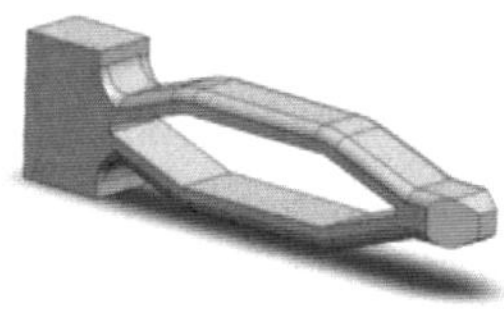
M-Shape	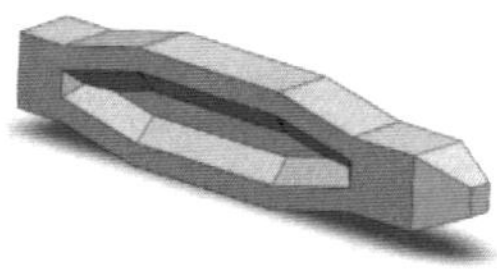
Multi-Spring	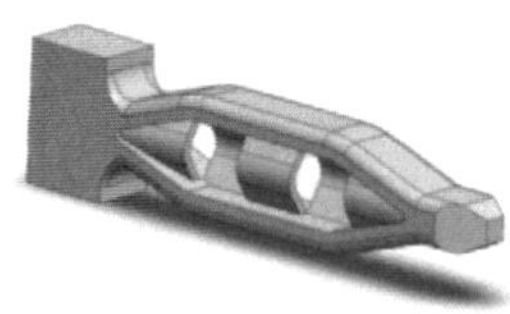
Barrel	

The main assumption on the simulation was the starting point which was the pin was already inserted in the platted through hole. The analysis starts on the pull out of the pin in the platted through hole. The retention force, insertion force, and equivalent force will be determined and evaluated on each geometry. Different types of geometry used in the evaluation were shown in **Table 2**. A frictional contact condition was applied, where the target was the barrel PTH. A coefficient of friction of 0.2 was applied for these contact conditions.

The thickness of the pin used in all geometry is 25 μm while the barrel has the thickness of 1.6 mm. The properties of the materials used in the simulation was adopted from the experimentation of Tohmyoh et al [13] which is phosphor for the pins and copper for the barrel. **Table 3** summarized the material properties of both components.

Table 3. Material properties of the components used in the simulation.

Property	Pins (Phosphor)	Barrel (Copper)
Yield stress (MPa)	731	305
Poisson's ratio	0.30	0.35
Young's modulus (GPa)	77.5	21.0
Hardening modules (MPa)	103	200

A mesh independence test was conducted to control the quality and efficiency of the simulation in consideration of the computational time and accuracy of the mesh are both satisfactory. The simulation time took approximately 40 mins per geometry.

The model was configured to identify the effects of the different geometry on the insertion force, retention force, and in the equivalent stress in the press fit set up. Simulation overview is illustrated in **Figure 4**.

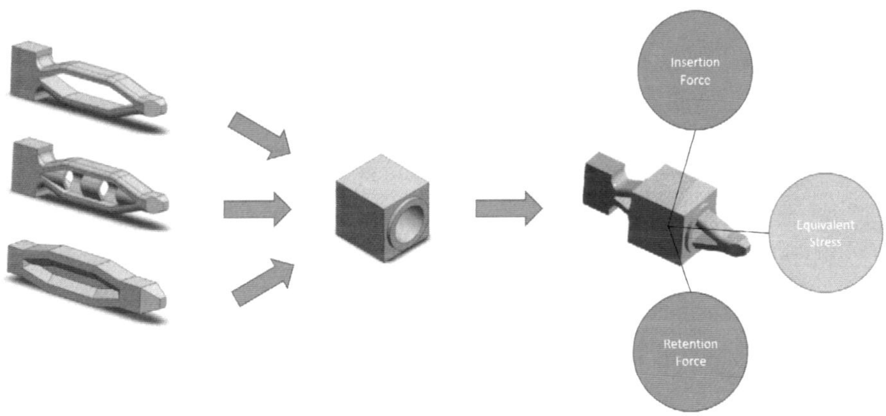

Figure 4. Simulation overview

Each pin geometry was initially inserted in the barrel then the simulation model will identify the insertion force, retention force, and equivalent stress of the press fit set up to evaluate the effect of the geometry while keeping the materials constant and coefficient of friction at 0.2.

3. Results and discussion

The simulation results showed significant impact of the geometry with respect to the insertion force, retention force, and on the equivalent force of the press fit configuration. The simulation output indicates that the multi-spring pin needed the highest force (41.1 N) in terms of insertion while the eye of the needle geometry requires the least force (6.91 N) (**Figure 5**). This pertains to the force required to assemble the press fir configuration [14]. With the results obtained, the barrel assumes the frictional force and the pressure on the press fit configuration which leads to a three-dimensional stress state occurrence on the barrel. During the insertion, the stress due to compression exists on the radial direction of the barrel, while the tensile stress occurs in an axial direction on the circumference. Considering that the copper barrel has high ductility, the von mises stress can be used as the indicator for parametrical study on failure analysis.

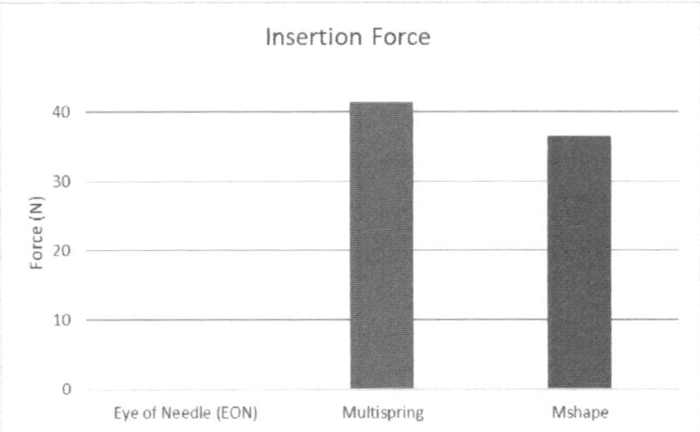

Figure 5. Insertion force comparison of the three geometry

The retention force can be defined as the maximum load exerted in the pull-out simulation. This force is governed by the normal force acting in between the through hole and the pin as well as the coefficient of friction acting on the contact region [15]. The results have shown that the M-Shape pin holds the highest retention force with 257.3 newtons compared to Multi-spring (93.3 N) and Eye of needle geometry (10.7 N) (**Figure 6**).

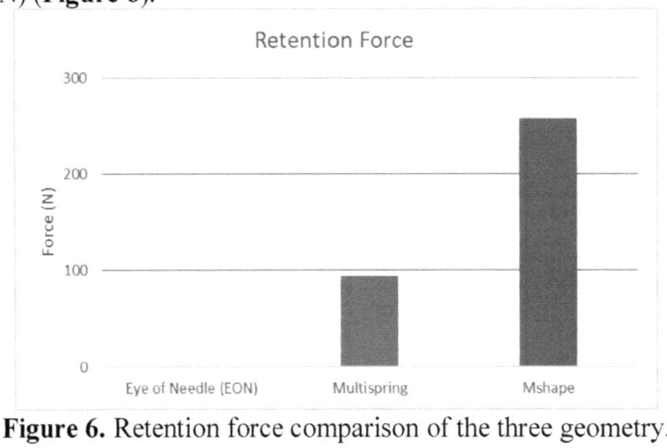

Figure 6. Retention force comparison of the three geometry.

And lastly, the M-shape yielded the most equivalent stress with 370 MPa followed by the multi-spring (138.3 MPa) and eye of the needle (48.1 MPa) (**Figure 7**). High retention force is preferred on the application of press fit. This force ensures the reliability of the connections of the semiconductor devices on the electronic control units which are crucial in automobile operation.

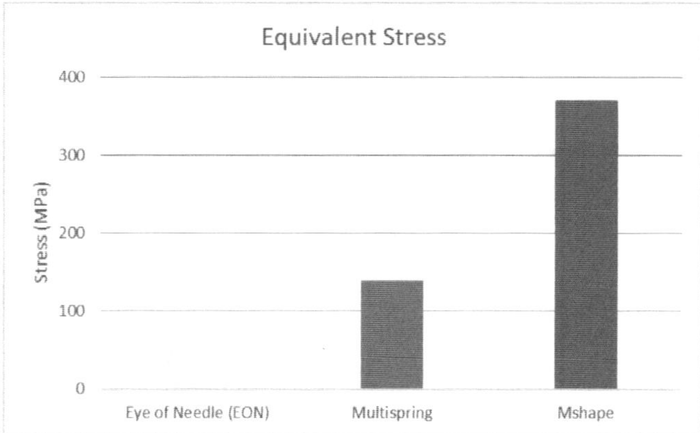

Figure 7. Equivalent stress comparison of the three geometry.

4. Conclusion

The study focuses on the investigation of the geometry of the pins in the press fit configuration and how it affects the forces involved such as insertion force, retention force and equivalent stress. The results had shown that the geometry of the pin in a press fit pin configuration has a significant effect on the insertion force, retention force, as well as in the equivalent stress. It is also noteworthy that the insertion force does not have a direct relationship with the retention force. It can be observed in the case of the multi-spring and in the M-shape pin where the multi-spring resulted into a 12% higher insertion force but yielded on a significantly 93% lower retention force compared to the M-Shape. In this study, out of the three geometries, the M-shape pin showed the best parameter with lower insertion force needed yet, it has the highest retention force for the press fit pin configuration.

The authors recommend further investigation on the relationship of the insertion force and retention force as well as the factors that affect these forces such as contact surface area and temperature.

Acknowledgment

The authors would like to acknowledge the Philippine's Department of Science and Technology with the Collaborative Research and Development to Leverage Philippine Economy (DOST-CRADLE) program project no. 08897 under the Philippine Council for Industry, Energy and Emerging Technology Research and Development (PCIEERD) project and Department of Science and Technology - Engineering Research and Development for Technology (DOST-ERDT) for the assistance and funding of this research.

References

[1] Lynch J and Longford A, 2009 Adapting press-fit connection technology for electronic modules in harsh environments (Intterplex Industries, Inc.)
[2] Stolze T, Thoben M, Koch M and Severin R 2015 Reliability of PressFIT connections (Infineon Technologies, Warstein)
[3] Corman N, Myers M and Copper C 2003 Friction behavior of press-fit applications: test apparatus and methodology *Forty-Ninth IEEE Holm Conf. on Electrical Contacts* (Harrisburg, PA, USA, 2003: Tyco Electronics Technology)
[4] "Press-Fit Connections for the Automotive Industry (Catalog 1654403-1)." (TE Connectivity Germany) (accessed.
[5] Goel R 1981 An analysis of press-fit technology *Electronic Components Conference* (Atlanta Georgia)
[6] Chou G J S and Hilty R D 2003 Effects of lead-free surface finishes on press-fit connections *The IPC Annual Meeting 2003* (Minneapolis, MN, USA)

[7] Fellner T, Zukowski E and Wilde J 2009 Modelling of process and reliability of press-fit interconnections *Int. Conf on Thermal, Mechanical and Multiphysics Simulation and Experiments in Micro-Electronics and Micro-Systems* (Freiburg, Germany)

[8] Mills N J 2007 Chapter 6 - Finite element modelling of foam deformation *Polymer Foams Handbook* (N. J. Mills Ed. Oxford: Butterworth-Heinemann) pp 115-145.

[9] Ogrodnik P 2020 Chapter 10 - Evaluation (validation and verification) *Medical Device Design (Second Edition)* (P. Ogrodnik Ed.: Academic Press) pp 317-375.

[10] Bi Z 2018 Chapter 1 - Overview of finite element analysis," in *Finite Element Analysis Applications* (Z. Bi Ed.: Academic Press) pp 1-29.

[11] F. C. LTD. Appearances and Shapes of Press-Fit Terminals https://www.finecs.co.jp/en/knowledge/press_fit.html?fbclid=IwAR28fMPQDxmEFchu5aRomzABcIeLcZ7MTrUM8xxRkP9HBYZKpXIBb5qgkk (accessed December 1, 2020).

[12] Wu B-H *et al.* 2018 Study on the contact performance of electronic EON connectors under axial vibration *IEEE Transactions on Components, Packaging and Manufacturing Technology* **8**(12) pp 2090-2097

[13] Tohmyoh H, Yamanobe K, Saka M, Utsunomiya J, Nakamura T and Nakano Y 2008 Analysis of solderless press-fit interconnections during the assembly process *Journal of Electronic Packaging* **130**(3)

[14] Kim S-Y, He B, Shim C-S and Kim D 2013 An experimental and numerical study on the interference-fit pin installation process for cross-ply glass fiber reinforced plastics (GFRP)," *Composites Part B: Engineering* **54** pp 153-162 2013/11/01/ 2013, doi: https://doi.org/10.1016/j.compositesb.2013.05.006.

[15] Palaniappan A, Li L and Lee T-K 2018 Impact of press-fit connector pin microstructure elastic response to PCB through-hole Cu wall interface long-term contact reliability *2018 IEEE 20th Electronics Packaging Technology Conference (EPTC)* (IEEE) pp 890-893.

Balancing control for single-wheel unicycle Robot using the Sliding mode controller

P L Nguyen[1,4*], T P Nguyen[2] and T M Ngo[3]

[1] Faculty of Engineering, Van Lang University, Ho Chi Minh City, Vietnam

[2]Institute of Engineering, HUTECH University, Ho Chi Minh City, Vietnam

[3]Refrigeration Division, Tan Son Nhat International Airport, Ho Chi Minh City, Vietnam

*luan.np@vlu.edu.vn

Abstract. In recent years, sliding mode control has been disseminated in the industry, including nonlinear control, and is considered as one of the simplest methods for nonlinear system control, including uncertainly model-based systems that are affected by external magnetic interference or parameter conversion. This paper shows the method of sliding mode control for a single-wheel unicycle-like mobile Robot. First, the Robot kinematics model was established by using a self-balancing system based on the balancing control principle in two models; flywheel inverted pendulum. In order to simplify the complicated problem with high nonlinearity, this model was balanced in each direction. Then the system model was emulated, showing the result that the separation sliding mode control worked properly with a large angle of deflection (0.17 rad), and the wheel rolled slowly with the relatively low speed (~2.1 rad/s), while the flywheel rotated slowly with high-speed of rod control (~28 rad/s).

1. Introduction

Along with the development of science and technology, today robots have the ability to replace humans working in hazardous environments, in manufacturing, or serving people for entertainment needs or problems in daily life day… Thanks to the continuous development of technology, robots have been built to serve many different purposes. In particular, the development of automatic balancing systems is also very strong. Over the years it has become a familiar topic in the teaching environment of engineering and automation. More and more products apply the theory of balance for fast, compact movement in short distances.

Balancing of a Two-Wheel Unicycle Robot system is an application developed from the self-balancing reverse pendulum system. For a self-balancing single-wheel unicycle Robot system, it is more complicated, it is a problem of combining the balance of the inverted pendulum and the balance of the handwheel, balancing the unicycle Robot in two directions. This is a complex system with high nonlinearity and instability easily. Issues related to this system including controller design, signal-processing filter design to filter measurement signals... are complex problems in the control field. Currently, there are many products applying the equilibrium principle using various algorithms such as PID, Fuzzy fuzzy logic, LQR... [1-4]. In this paper, a self-balancing single-wheel unicycle Robot system using a sliding controller is presented.

The objective of this paper is to build a simulation model for a self-balancing single-wheel Unicycle-like Robot system using the sliding control technique. The mathematical model of the system is divided into two models, one in the form of an inverted pendulum (Pitch angle), the other in the form of a flywheel pendulum (Roll angle) from which to design the sliding and forward controller model simulation on Matlab simulink software to check the accuracy as well as see the system's characteristics.

2. Unicycle Robot modeling

The Unicycle Robot system includes a ground contact wheel to move, the robot body, turntable 1 and reaction wheel 2 to change the direction of the robot and keep the robot in balance.

OXYZ is the ground coordinate system; $O_bx_by_bz_b$ is the coordinate system attached to the robot; α, β, γ are Pitch, Roll, Yaw angles respectively; θ, ϕ, ψ are the rotation angle of wheel, balance flywheel and directional swing wheel, respectively.

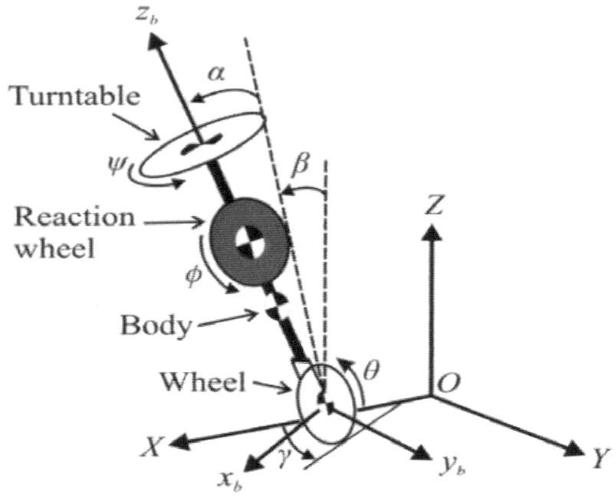

Figure 1. Model of single-wheeled Unicycle Robot in space

The self-balancing single-wheel system is a balancing vehicle system based on the principle of balance control with two models: an inverted pendulum and a flywheel pendulum. We can divide the model into three parts: the body, the upper part is a rotating flywheel that can rotate around the shaft to create a moment of inertia to keep the vehicle balanced in a left-right direction according to the flywheel pendulum balance principle, the lower part is the wheel that can move back and forth to keep the vehicle balanced in the front-back direction according to the reverse pendulum balance principle. This is a relatively complex model with high nonlinearity, so to simplify the problem, we divide the system into two small models, that is, we balance the model in each direction according to the above equilibrium principles [5]. And combining the two models after splitting requires that in each model (inverted pendulum and flywheel pendulum), The system must have stable equilibrium in the small region adjacent to the equilibrium point to ensure that the changes of each model have little effect on each other.

The model parameters are shown in Table 1.

Notation	Unit	Interpretation
$\alpha\ \beta\ \gamma$	rad	Pitch, Roll, Yaw angle deviation respectively.
$\theta\ \phi\ \psi$	rad	Rotation angle position of wheel, balance flywheel, turntable.
$m_w\ m_b\ m_d\ m_f$	kg	Respectively the mass of the wheel, robot frame, balance flywheel, and directional flywheel.
$\tau_w\ \tau_d\ \tau_f$	N.m	In turn is the controlling force acting on the wheel, balance flywheel, and directional flywheel.
$I_{wx}\ I_{wy}\ I_{wz}$	Kg.m2	The inertial moment of the wheel on the three axis x, y, z axis respectively.
$I_{bx}\ I_{by}\ I_{bz}$	Kg.m2	The inertial moment of the robot frame follows the x, y, z axis respectively.
$I_{dx}\ I_{dy}\ I_{dz}$	Kg.m2	The balance flywheel moment of inertia according to the x, y, and z axis respectively.
$I_{fx}\ I_{fy}\ I_{fz}$	Kg.m2	The handwheel moment of the directional flywheel in the direction of the x, y, and z axis respectively.
l	m	Distance from wheel to rotating flywheel
l_r	m	Wheel radius.

l_{wd}	m	Distance from balance flywheel to rotating directional flywheel.

The dynamic equation of the unicycle built from the Euler-Lagrange equation is as follows:

$$\frac{d}{dt}\left[\frac{\partial L}{\partial \dot{q}}\right]-\frac{\partial L}{\partial q}=Q \qquad (1)$$

In which:

T is the kinetic energy of the whole system,

V is the potential energy of the system,

$q=[\alpha \quad \beta \quad \gamma \quad \phi \quad \theta \quad \psi]^T$, $Q=[0 \quad 0 \quad 0 \quad \tau_d \quad \tau_w \quad \tau_f]^T$

We have the kinetic equations of the self-balancing unicycle on the Pitch, Roll and Yaw axis.

a. Axis Pitch.

$$(I_{by}+I_{dy}+I_{fy}+\frac{l^2}{4}(\mathrm{m}_b+4\mathrm{m}_f)+l_{wd}^2 m_d)\ddot{\alpha}+(\mathrm{l}_{wd}\,m_d+\frac{l_{mb}}{2}+lm_f)l_r\ddot{\theta}\cos\alpha\text{-}(\mathrm{l}_{wd}\,m_d+\frac{l_{mb}}{2}+lm_f)g\sin\alpha=0 \qquad (2)$$

$$l_r(\mathrm{l}_{wd}\,m_d+\frac{l_{mb}}{2}+lm_f)\ddot{\alpha}\cos\alpha-ll_r(m_d+\frac{\mathrm{m}_b}{2}+m_f)\dot{\alpha}^2\sin\alpha+(I_{wy}+l_r^{\,2}(m_b+m_d+m_f+m_w))\ddot{\theta}=\tau_w \qquad (3)$$

b. Axis Roll.

$$I_{dx}\ddot{\phi}+((I_{bx}+I_{dx}+I_{fx}+I_{wx}+l_{wd}^2 m_d+l^2 m_f)+\frac{m_b}{4}(l+2l_r)^2+2l_r(l_{wd}m_d+lm_f)+l_r^{\,2}(m_d+m_f+m_w))\ddot{\beta}$$

$$-(l_r(m_d+m_f+m_w)+\frac{m_b}{2}(l+2l_r)+l_{wd}m_d+lm_f)g\sin\beta=0 \qquad (4)$$

$$I_{dx}\ddot{\phi}+I_{dx}\ddot{\beta}=\tau_d \qquad (5)$$

c. Axis Yaw.

$$I_{fz}\ddot{\psi}+(I_{wz}+I_{dz}+I_{bz}+I_{fz})\ddot{\gamma}=0 \qquad (6)$$

$$I_{fz}\ddot{\psi}+I_{fz}\ddot{\gamma}=\tau_f \qquad (7)$$

3. Unicycle Robot controller design using the Sliding mode controller

A diagram of the sliding control system block is shown in figure 2.

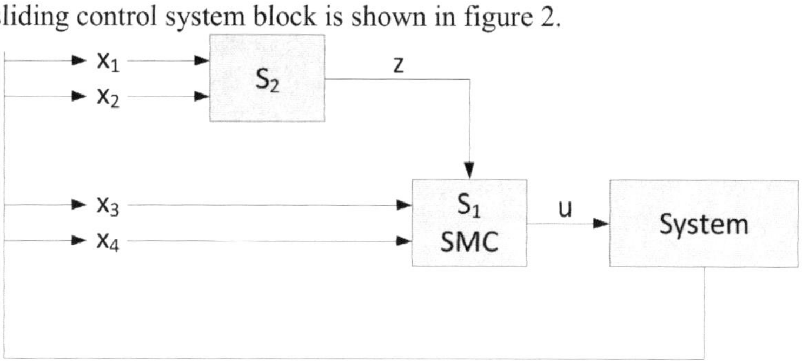

Figure 2. Block diagram representing the decoupled sliding mode control

$$s_1(x) = \dot{e}_1 + \lambda_1 e_1 = x_2 - x_{2d} + \lambda_1(x_1 - x_{1d} - z) = 0$$
$$s_2(x) = \dot{e}_2 + \lambda_1 e_2 = x_4 - x_{4d} + \lambda_2(x_3 - x_{3d}) = 0 \quad (8)$$

Compared with the classic slide controller, we find that the control target has been changed.

$$x_1 = x_{1d} + z$$
$$x_2 = x_{2d}$$
$$x_3 = x_{3d}$$
$$x_4 = x_{4d} \quad (9)$$

Similar to the classic sliding controller we have the following control signals.

$$u = u_1 = \hat{u}_1 - K_1 sat\left(s_1(x) b_1(x) / \phi_1\right); K_1, \phi_1 > 0 \quad (10)$$

With

$$\hat{u}_1 = -b_1^{-1}(x)\left(f_1(x) + \lambda_1 \dot{e}_1\right) \quad (11)$$

Output z is the value

$$z = K_2 sat\left(s_2(x) / \phi_2\right); K_2, \phi_2 > 0 \quad (12)$$

The controller works in the following sequence:

When $S_2 \neq 0$ therefore $z \neq 0$ The requirement for the controller is generate signals u reduce deviation S_2 with value 0. when the $S_2 \to 0$, follow $z \to 0$, and $s_1 \to 0$ thus the system control target can be achieved.

4. Results and discussion

From the mathematical model describing the system found above, the author of the paper has designed the model simulating it on Matlab Simulink with the actual parameters and system models as shown in Figures 3 and 4.

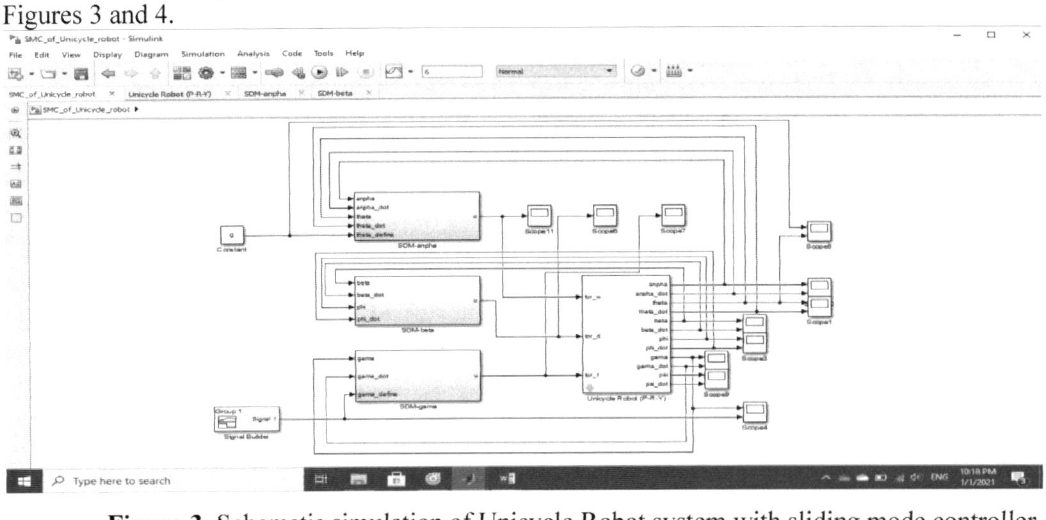

Figure 3. Schematic simulation of Unicycle Robot system with sliding mode controller

IOP Conf. Series: Materials Science and Engineering 1109 (2021) 012020 doi:10.1088/1757-899X/1109/1/012020

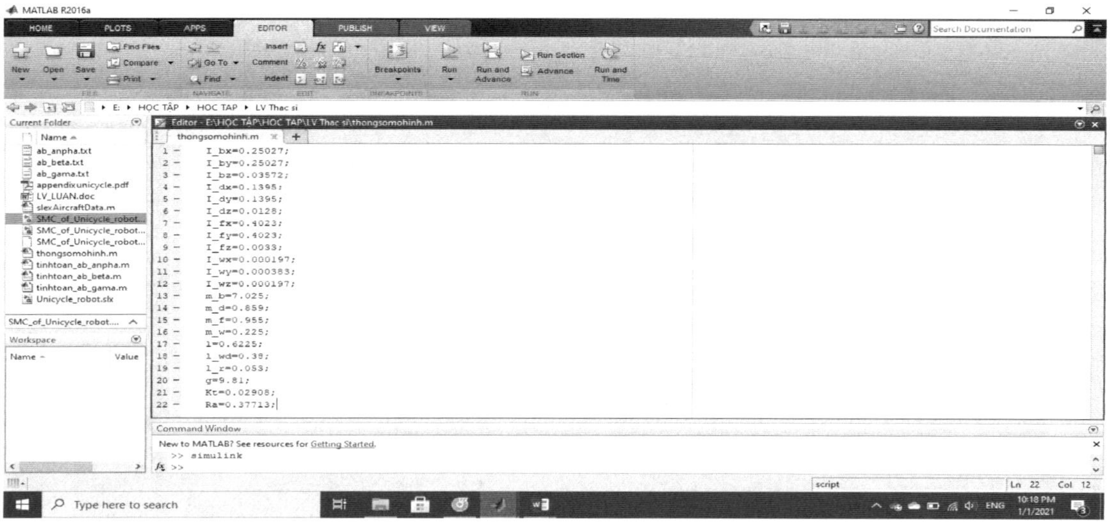

Figure 4. The actual configuration of the system

4.1. Simulate math model without control effect

Response results:

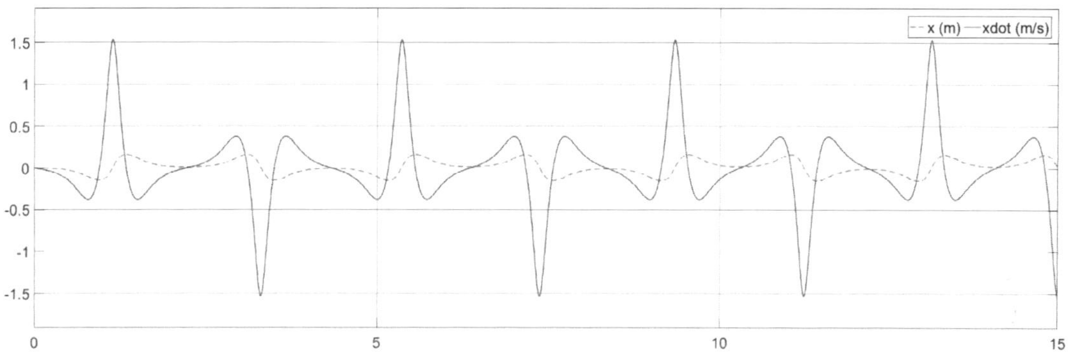

Figure 5. Robot corner speed and position response without control signal

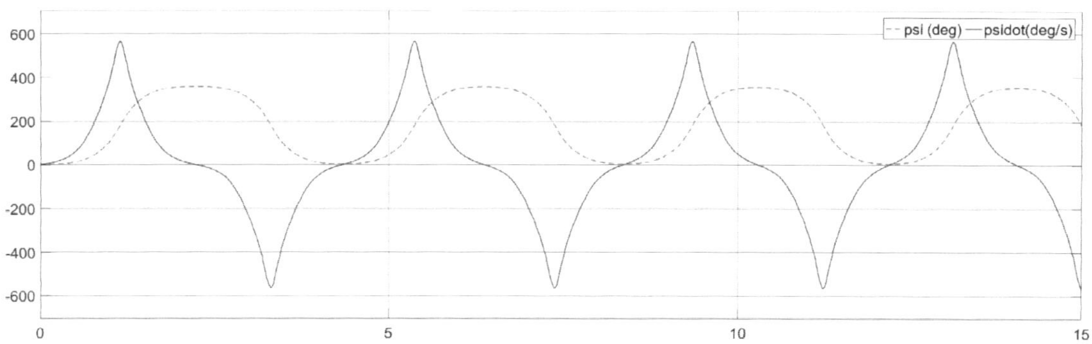

Figure 6. Response and PITCH angular velocity without control signal

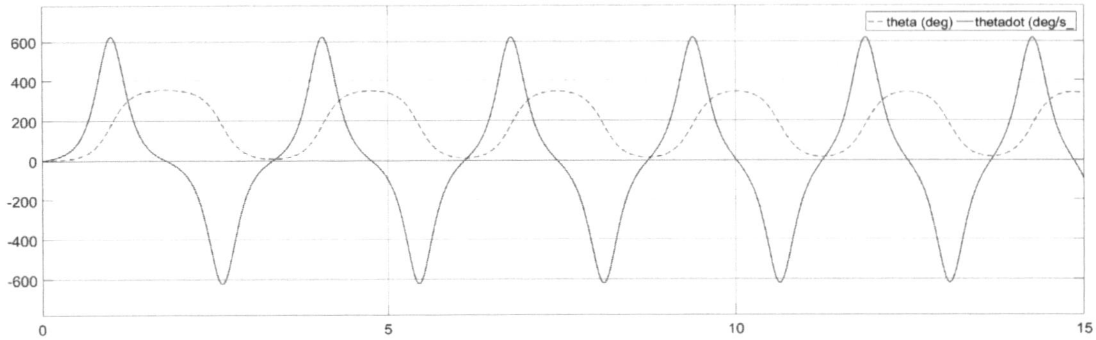

Figure 7. Response and ROLL angular velocity without control signal

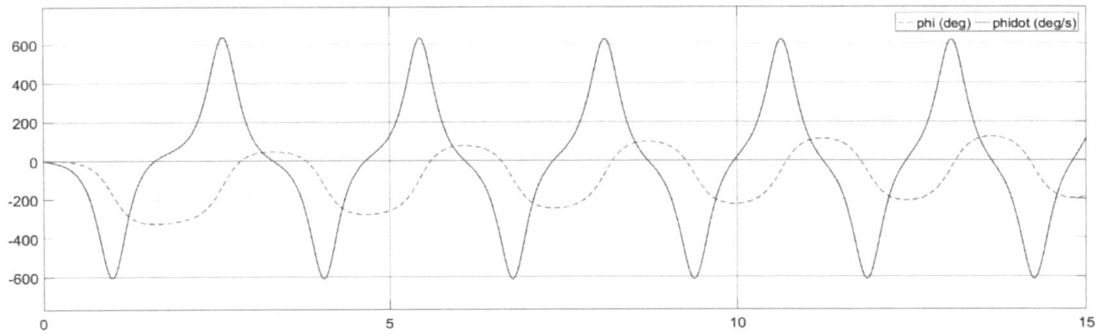

Figure 8. The flywheel position and speed response when not operating

From the above graphs, we see: when there is no signal to control and ignore the friction at the rotation axis as well as the air friction, if the system deviates from the initial angle 0 it will vibrate like a pendulum conditioning around the 180^0 equilibrium point.

4.2. System simulation with decoupled sliding mode control

Output response when using decoupled sliding mode control

With the following initial conditions:

$$\begin{bmatrix}\alpha & \beta & \gamma & \dot{\alpha} & \dot{\beta} & \dot{\gamma}\end{bmatrix}=\begin{bmatrix}0.17 & 0.17 & 0.17 & 0 & 0 & 0\end{bmatrix}$$

$$\begin{bmatrix}\theta & \phi & \psi & \dot{\theta} & \dot{\phi} & \dot{\psi}\end{bmatrix}=\begin{bmatrix}0 & 0 & 0 & 0 & 0 & 0\end{bmatrix}$$

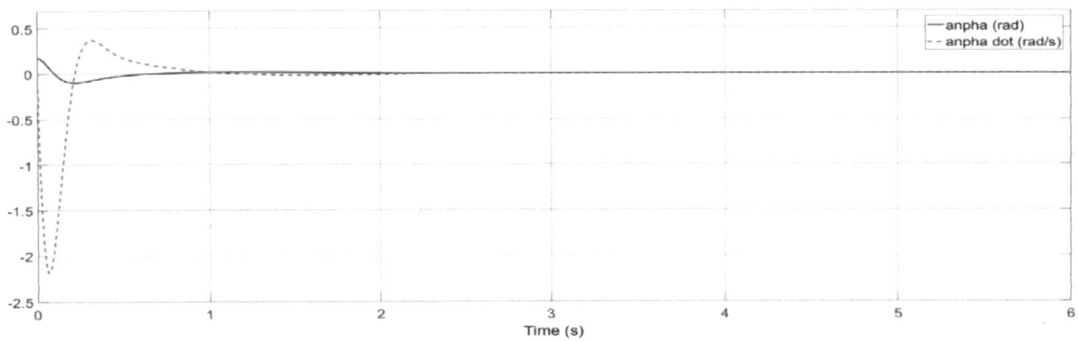

Figure 8. PITCH angle response and PITCH angular velocity with decoupled sliding mode control

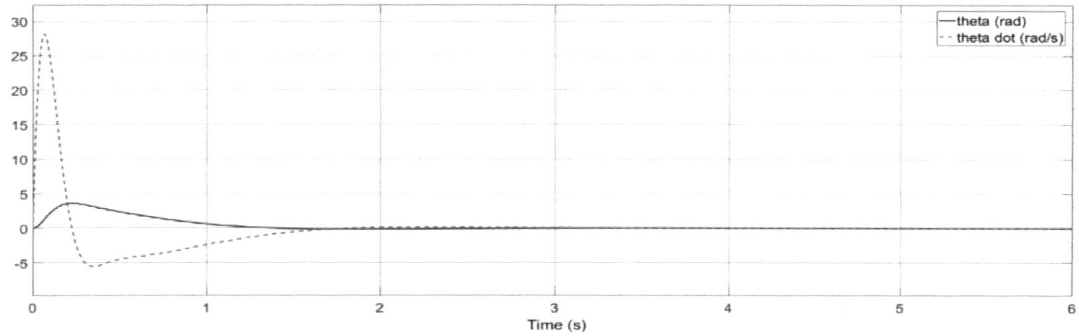

Figure 10. Response to wheel position and speed with decoupled sliding mode control.

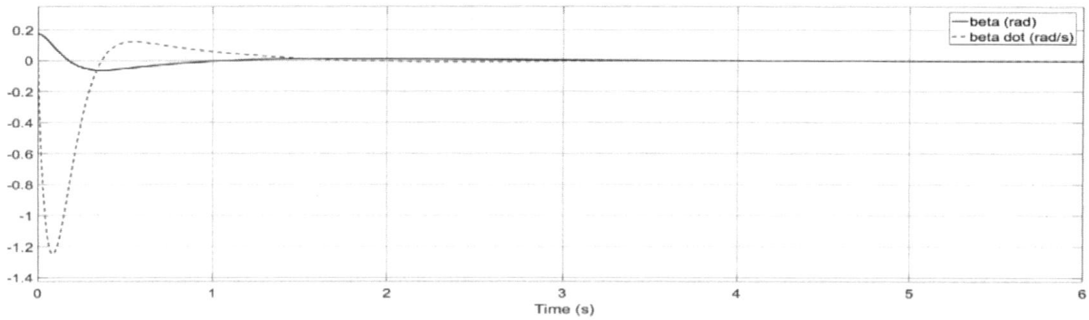

Figure 11. Response of deflection angle and ROLL angle velocity with decoupled sliding mode control.

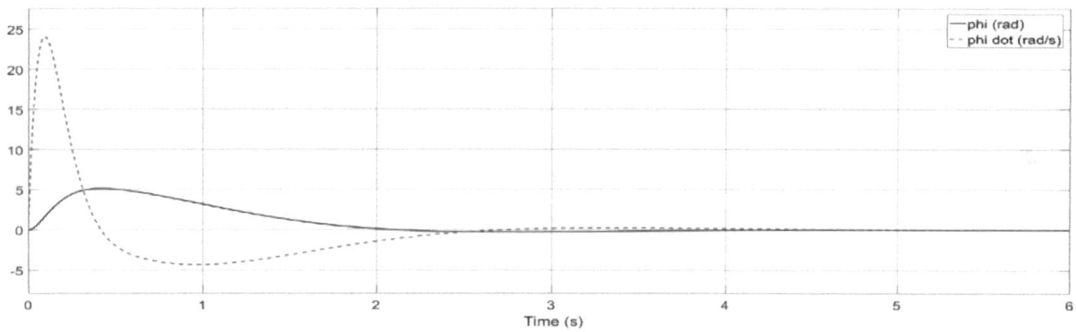

Figure 12. Flywheel position and speed response with decoupled sliding mode control

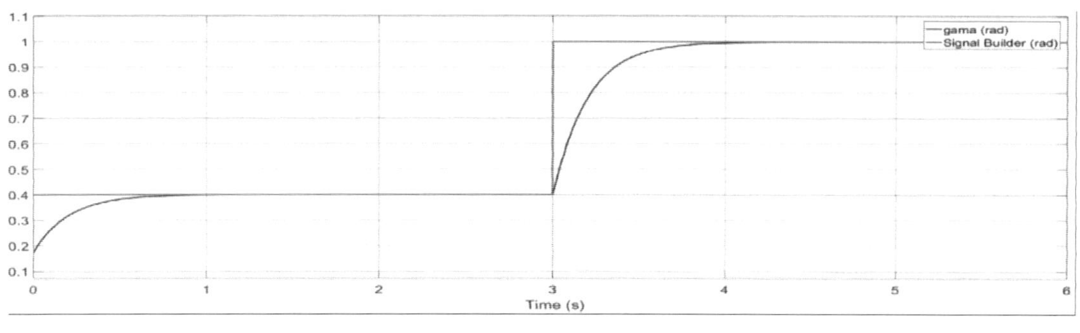

Figure 13. Response to YAW rotation angle with a given value

The simulation results show that the decoupled sliding mode control works well with a large initial deviation angle (0.17rad):

- The pitch and roll angle responds to 0 quite quickly (~ 1s)
- The wheel spins quite a bit and the wheel speed is relatively low (~ 2.1 rad / s).
- The flywheel is also quite low and control speed should be high (~ 28 rad / s).

The paper explored the single-wheel robot, modeled the system, and designed a decoupled slide controller for the balance holding system. The simulation results on Matlab Simulink show the efficiency of the decoupled sliding mode control

References

[1] Jae-Won An, Min-Gyu Kim and Jangmyung Lee, *"Control of a Unicycle Robot using a Non-model based Controller"*, Journal of Institute of Control, Robotics an d Systems (2014).

[2] Seong I. Han and Jang M. *Lee "Balancing and Velocity Control of a Unicycle Robot Based on the Dynamic Model"*, IEEE Transactions on Industrial Electronics January 2015.

[3] Xiao-gang Ruan and Yu-feng Wang, *"The Modelling and Control of Flywheel Inverted Pendulum System*", Artificial Intelligence and Robot Institute Beijing University of Technology Beijing, 100124, China.

[4] Zhao jie and Ren Sijing, *"Sliding Mode Control of Inverted Pendulum Based on State Observer"*, 2016.

[5] M.J. Mahmoodabadi, S. Momennejad, A. Bagher, *"Online optimal decoupled sliding mode control based on moving least squares and particle swarm optimization"*, 2014.

IOP Conf. Series: Materials Science and Engineering 1109 (2021) 012021 doi:10.1088/1757-899X/1109/1/012021

Design, Manufacture, and Testing of a Low-Cost Force Platform with 3-Axis Load Cell

F Ferryanto*, **M M Akbar, S Wicaksono, A I Mahyuddin**

Mechanical Design Research Group
Faculty of Mechanical and Aerospace Engineering
Institut Teknologi Bandung

*ferryanto@ftmd.itb.ac.id

Abstract. Force platform is an instrument to measure Ground Reaction Force (GRF) and locate Center of Pressure (CoP) in motion analysis. Although it is very useful, the cost of the commercially available force platform is relatively high. Therefore, the objective of this work is to develop a low-cost force platform with 3-axis load cell. The development was initiated with Design Requirement and Objective (DR&O) of the force platform. To measure the force, four 3-axis loadcells were placed in each corner of the force platform to measure a maximum force of 1500 N. The dimension of the aluminum 5058 force platform was 500 x 500 mm. After the manufacturing process, the force platform was calibrated to obtain 12 correction factors for the four loadcells in the three-axes. It was obtained that the RMSE of the correction factors was 0.87% in z direction and lower than 2% for x and y direction of GRF. The CoP was also tested to ensure the functionality of the developed force platform. The RMSE errors for the CoP were 5 mm and 5.9 mm in x and y coordinates, respectively. With relatively small RMSE, the force platform could be used to obtain GRF and locate the CoP in gait analysis.

Keywords: *low-cost, force platform, gait analysis, ground reaction force, center of pressure*

1. Introduction

Human daily activities, such as walking or running, have been researched for many purposes for the last decades. Numerous approaches have been made to analyze the living-being movement and apply the results for robotic technology, increase athletic performance, or even diagnose medical injuries. In the movement analysis, a force platform is mostly used to obtain biomechanical parameters, such as Ground Reaction Force (GRF) and Center of Pressure (CoP). The force platform uses a load cell or piezoelectric as its primary sensor to measure GRF and determine the location of CoP that is required for the study. The capacity of the force platform is based on the sensor capability to measure the magnitude and direction of the force. Single-axis load cell gives only the vertical force data. Additionally, the 3-axis load cell is able to monitor force applied in the medio-lateral (F_x), anterior-posterior (F_y), vertical (F_z) direction, and force moment relative to the GRF, M_x, M_y, and M_z [1,2].

Kistler, Bertec, and AMTI are prominent manufacturers that provide force plates with a wide variety of specifications. These commercial force plates price is relatively high, with an average cost of US$10,000-US$20,000 (140-280 million rupiahs), several even reach the price of US$62,000 (900

IOP Conf. Series: Materials Science and Engineering 1109 (2021) 012021 doi:10.1088/1757-899X/1109/1/012021

million rupiahs) [2]. Vernier may provide a low-priced force platform (US$400). Nevertheless, the specification is significantly inferior to that of Kistler's or AMTI's force plates that are capable of measuring 3-axis GRF [3].

There were already several researches in developing low-cost force platforms, such as the research conducted by Silvia et al. [2] and Jaafar et al. [4]. In [2], although the production cost of the force platform was one-tenth of the commercial force platform, it was still not affordable. Moreover, the developed force platform still needed final validation and comparation with the commercial force platform. Another low-cost force platform was developed by Jaafar et al. that could only measure GRF in one dimension [4]. Although, it had large capacity, the force platform resulted in unacceptable error in measuring small GRF.

Because of the costly commercial force platforms and limitations of previous researches, developing a low-cost force platform that could measure GRF in three directions and CoP become the objective of the present work. The current work was initiated by developing an affordable 3-axis load cell by Wicaksono et al. [5]. Then, the 3-axis load cell was used as the force transducer in this research. Subsequently, an affordable force platform was designed and manufactured.

2. Design Requirement and Objective

The first step in designing a low-cost force platform was to determine the design requirement and objective (DR&O). The DR&O consists of demand and wish criteria for the designed force platform. The criteria in the DR&O became the guidance in developing the force platform. Table 1 summarized the DR&O in the present work.

Table 1 Design Requirement and Objective of Designed Force Platform

Category	Requirement	Demand/Wish
Cost	1. Force platform cost is relatively more affordable than commercial force platform	Demand
Performance	2. Force platform must be able to measure ground reaction force	Demand
	3. Force platform must be capable of locating the center of pressure	Demand
	4. The maximum load is 1500N	Demand
	5. User-friendly data acquisition (Force plate to PC)	Wish
	6. Force platform is easy to assemble and works fixed or portable	Wish
Physique	7. Maximum overall force platform is less than 30kg	Demand
	8. The proportional dimension of the force platform	Wish

As shown in Table 1, one of the demand criteria is the force platform in this work should be affordable. Furthermore, the maximum load of the force platform is to be 1500 N since the average human weight would approximately be 57.5 – 80.7 kg [6] and the ground reaction force (GRF) is up to 1.3 times of weight in walking movement [7].

3. Design of Force Platform

The force platform consists of three essential components, i.e. lower platform, upper platform and force transducer. The lower platform was used to hold the force transducer that are capable to measure forces in three independent directions. In the present work, the load cell developed in previous research was used as the force transducer of the force platform [5]. In total, there are four load cells that were located in each corner of the platform to enhance the accuracy of the force platform. The load cells were bolted to both lower and upper platform. Hence, the forces from the upper platform could be transferred to the load cells.

The dimension of the force platform is determined by the step length of human gait since a ground reaction force is generally measured per foot. This length of step may be estimated at half of the stride length. The spatio-temporal parameters of gait consist of walking speed, stride length, cadence, and cycle time, and is presented in Table 2 for Indonesian subject in the 18 – 49 age group. From Table 2, it may be seen that the average stride length of Indonesian men and women are 1.2 meters and 1.02 meter, respectively. Hence, the force platform dimensions will be 500 x 500 mm, as may be seen in Figure 1. The thickness of the upper and lower platform was 10 mm and 6 mm, respectively.

Table 2 Spatio-temporal gait parameters of subject in 18–49 age-group in Indonesia [8]

Variables	Male		Female	
	Mean	SD	Mean	SD
Walking Speed	1.09	0.11	1.02	0.12
Stride length (m)	1.20	0.08	1.11	0.10
Cadence	109.29	7.84	110.36	9.78
Cycle time (s)	1.10	0.13	1.10	0.14

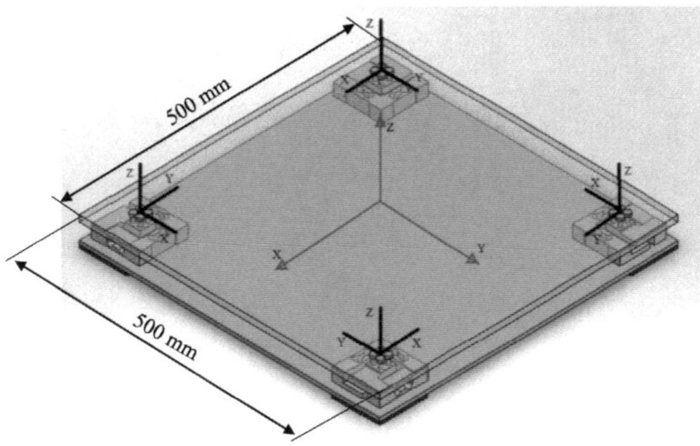

Figure 1 The isometric view and dimension of force platform

Based on DR&O, the plate should be able to withstand force of 1500 N, but it should be rigid and weighs less than 30 kg. This indicates the material should have low density but high stiffness. There were several materials that fulfilled the criteria, such as stainless steel and aluminum. However, stainless steel has higher density than the aluminum. Therefore, aluminum was chosen as the material for the lower and upper platform. Three types of aluminum were considered, and their mechanical properties and prices are compared. The results are tabulated in Table 3. It may be seen that, aluminum 5058 is the lightest (lowest density) and the cheapest of the three types. While aluminum 7075 has the highest modulus of elasticity. Hence, both aluminum 5058 and 7075 were chosen as the candidates for the platform material.

Table 3 Aluminum mechanical properties [9]

Parameter	Aluminum 5058	Aluminum 6061	Aluminum 7075
Density (g/cc)	2.66	2.7	2.81
Tensile yield Strength (MPa)	228	276	503
Ultimate tensile stress (MPa)	317	310	572
Modulus of Elasticity (GPa)	70.3	68.9	71.7
Elongation at Break	16%	12%	11%
Poisson's Ratio	0.33	0.33	0.33
Price	3rd (cheap)	2nd	1st (expensive)

The next step in designing the force platform was to assess the deformation due to the load from the human body on the upper platform. The upper platform should have insignificant deformation since it will affect the movement analysis. Finite element method by ANSYS was used to evaluate the deformation of the upper platform when it was loaded with a force of 2000 N. Figure 2 shows the result of deformation analysis for both aluminum 5058 and 7075. Since they have similar deformation, aluminum 5058 was selected as the material for the lower and upper platforms due to its lower density and price. Then, by using the dimension and material of the platform, it is calculated that the total mass of the designed force platform was 13.81 kg, included the four loadcells as the force transducer.

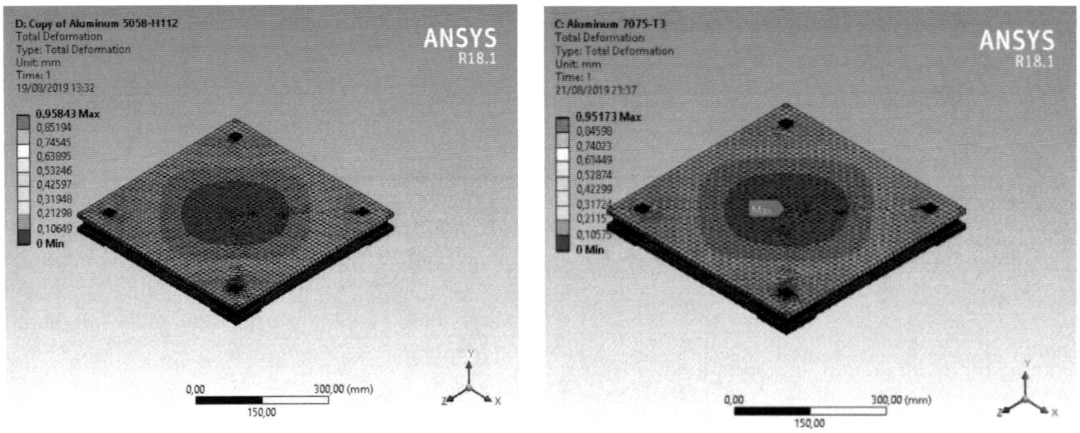

(a) (b)
Figure 2 Deformation of force platform with (a) aluminum 5058, (b) aluminum 7075

4. Manufacture of the Force Platform

Referring to the force platform design, manufacture begins with aluminum 5058 procurement with a dimension of 500 mm × 500 mm with 6 mm and 10 mm thickness. Since each load cell was mounted to the upper and lower platform by four bolts, there should be four holes in each corner of lower and upper platform. In total, 16 holes were drilled in the platform. The final assembly of the force platform was presented in Figure 3. The total cost for the present force platform was around US$1,400.

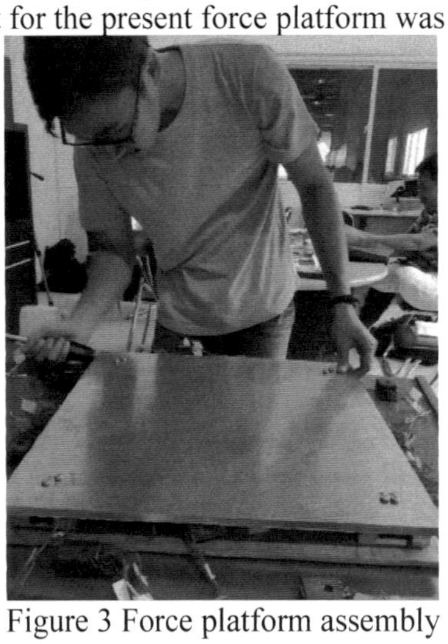
Figure 3 Force platform assembly

5. Calibration and Testing Result

The final assembly of the force platform was then calibrated so that it could measure the ground reaction force (GRF) accurately. When the force platform was loaded, the signal from the load cells were acquired by NI 9237 DAQ with the help from SignalExpress software. SignalExpress is a data-logging software from National Instrument and it can acquire signals from many channels simultaneously.

The signal measured by each load cell ($f_{i,j}$) however was not the same with the actual force loaded to the force platform. Hence, there should be a correction factor $C_{i,j}$ for the measured signal by i^{th} - load cell in j^{th} - direction. Since there were four load cells in the force plate, the variable i was numbered from 1 to 4. Moreover, the variable j was numbered 1, 2, 3 for x, y, and z direction, respectively.

The ground reaction force in each direction should be the sum of corrected signal from each load cell in the same direction. Therefore, the relationship between signal from load cell $f_{i,j}$ and actual force F_j was depicted in Equation (1) below.

$$F_j = \sum_{i=1}^{4} C_{i,j} f_{i,j}$$

As implied in the Equation (1), there are 12 correction factors to be found before the force platform could be utilized. So, there should be at least 4 experiments for each direction with known $f_{i,j}$ and F_j to calculate the value of all the correction factor. However, larger number of experiments will increase the accuracy of the correction factors.

The calibration method in z direction was to apply a static human body weight on the force platform, as illustrated in Figure 4(a). The weight was measured using a scale with a precision of 0.05 kg, or about 0.5 N, and it is assumed as an actual force in z direction. In the calibration process, the measurement of load cell signal was conducted 12 times to obtain correction factor $C_{i,3}$ accurately with variation of load and position. The load was varied from 46.75 kg, 49.3 kg, 64.25 kg, 73.35 kg, 86.4 kg, and 99.15 kg. After estimating the correction factor for each load cell and direction, root mean square error (RMSE) was evaluated to assess the accuracy of the ground reaction force measurement by the force platform. The RMSE was calculated from the difference between the actual and measured GRF. The RMSE of GRF in z direction was 5.22 N. This error was acceptable since it is 0.87% of the average force loaded to the force platform, i.e. 600 N.

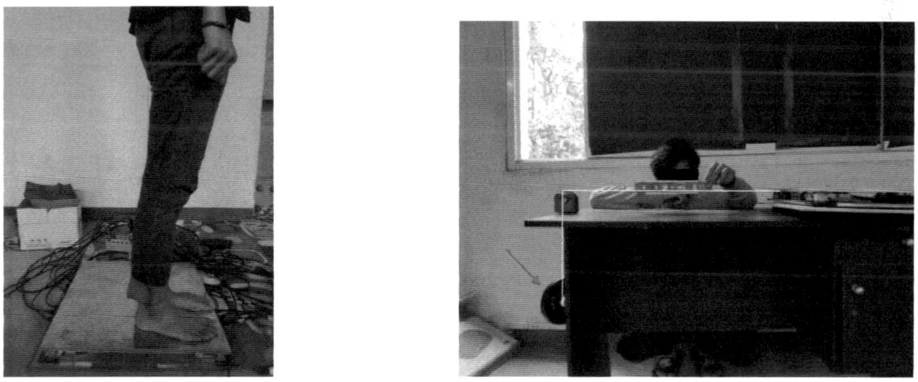

(a) (b)

Figure 4 Force platform calibration in (a) z direction, (b) x and y direction

The principle of calibration in the x and y direction is roughly the same with calibration in z-direction. However, the arrangement of force platform and its loading should be modified. The loading of the force platform in x and y direction was conducted by a string that was guided to z-direction by a pulley, as can be seen in Figure 4(b). The actual load used was a dead weight, which varies from 1.25 kg, 2.5 kg, 3.75 kg, and 5 kg. The experiments were done 16 times for each x and y direction to obtain accurate correction factor. From the experiments conducted, it was obtained that the RMSE of the force platform was 0.69 N and 1.27 N, respectively. The RMSE in both directions was tolerable since it was less than 2% compared to the average loading given to the force platform in x and y direction.

After obtaining the correction factors accurately, the next step was to check the force platform performance on locating the Center of Pressure (CoP). The center of pressure (CoP) represents a point of application of the ground reaction force. In checking the CoP, the force platform was loaded by a pointed stick on the pre-determined location. Then, the calculated CoP was compared to the location of the loading in the term of RMSE. The RMSE for locating the CoP in this experiment was 5 mm and 5.9 mm in x and y coordinate, respectively. This number is acceptable because when compared to the dimensions of the force platform and the surface area of the foot, the error value is relatively small.

Since the RMSE of GRF and CoP testing was relatively small, the force platform testing could be continued to measure GRF and CoP of human gait. Two participants with body weigh 715.1 N and 614.1 N walked normally and set one foot, either right or left, on the top surface of the force platform. The data acquisition system recorded the GRF and CoP. Moreover, the data will be processed by the MATLAB program. Figure 5 shows the result of GRF and CoP for one of the participants. Note that the x and y direction in the Figure 5(b) is the mediolateral and anteroposterior direction, respectively. As compared to the GRF and CoP in the Whittle [10], the GRF and CoP obtained from the developed force platform was not much different with the references. Therefore, it could be said that the present work in developing the affordable force platform was successful.

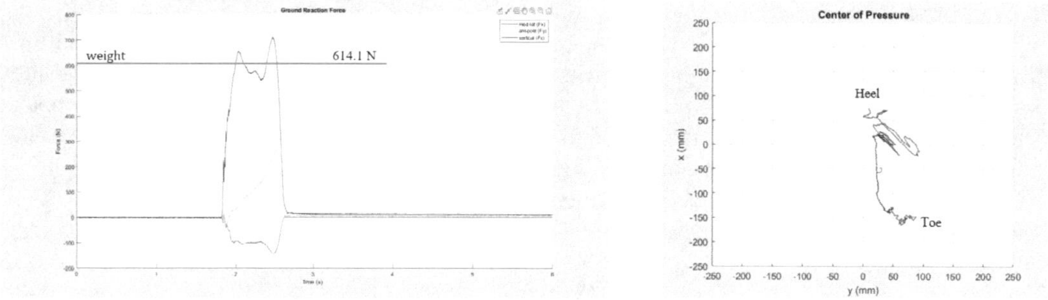

(a) (b)

Figure 5 (a) GRF and (b) CoP measured from the developed force platform

6. Conclusion

In the present work, a low-cost force platform that could measure GRF in three direction and CoP have been successfully developed with the cost only around US$1,400. The force platform used the four pre-developed 3-axis load cells in each corner as the force transducer. The total capacity of the force platform was 1500 N. The dimension of the force platform was 500 x 500 mm with thickness 5 mm and 7 mm for lower and upper platform, respectively. The lower and upper platform was made from aluminum 5058 since this material is inexpensive and rigid enough to support the maximum vertical GRF. After the manufacturing process, the force platform was calibrated to obtain correction factor for each load cell in x, y, and z direction. Then, the force platform was evaluated, and it was acquired that the RMSE of GRF in vertical direction was 0.87%. Moreover, the RMSE of GRF for x and y direction was lower than 2%. In addition, the force platform was also investigated to measure the CoP and its RMSE was 5 mm and 5.9 mm in x and y coordinate, respectively. With relatively small RMSE, the force platform could be used to obtain GRF and CoP of gait analysis.

Acknowledgement

We gratefully acknowledge *Program Penelitian, Pengabdian kepada Masyarakat, dan Inovasi* (P3MI) 2020, Institut Teknologi Bandung and *Penelitian Unggulan Perguruan* Tinggi (PUPT) 2020 scheme research grant from Ministry of Research Technology and Higher Education in supplying the financial support.

References

[1] Robertson DG, Caldwell GE, Hamill J, Kamen G, Whittlesey S. *Research methods in biomechanics*. Human kinetics; 2013 Nov 1.

[2] Silva MG, Moreira PV, Rocha HM. *Development of a low cost force platform for biomechanical parameters analysis*. Research on Biomedical Engineering. 2017 Sep;33(3):259-68.

[3] Product Category [Internet]. Vernier. [cited 2020 Nov 30]. Available from: https://www.vernier.com/product-category/

[4] Ridwan M, Lan MF, Fauzi M, Weeratunga K. *Development and validation of a force platform to measure ground reaction forces of national athletes*. In2017 12th IEEE Conference on Industrial Electronics and Applications (ICIEA) 2017 Jun 18 (pp. 95-98). IEEE.

[5] Wicaksono S, Ariana GP, Ferryanto F, Wibowo A, Mahyuddin AI. *Design optimization, manufacturing, and testing of affordable 3-axis load cell for reliable force plate component*. InAIP Conference Proceedings 2019 Dec 10 (Vol. 2187, No. 1, p. 050008). AIP Publishing LLC.

[6] Walpole SC, Prieto-Merino D, Edwards P, Cleland J, Stevens G, Roberts I. *The weight of nations: an estimation of adult human biomass. BMC public health.* 2012 Dec 1;12(1):439.

[7] Robertson DG, Caldwell GE, Hamill J, Kamen G, Whittlesey S. *Research methods in biomechanics*. Human kinetics; 2013 Nov 1.

[8] Mahyuddin AI, Mihradi S, Dirgantara T, Moeliono M, Prabowo T. *Development of Indonesian gait database using 2D optical motion analyzer system.* ASEAN Engineering Journal. 2012;2(2):62-72.

[9] A. S. Material, Aluminum Alloy, *ASM Aerospace Specification Material*, [Online]. (Available: http://asm.matweb.com. Accessed July 2019)

[10] Whittle MW. *Gait analysis: an introduction*. Butterworth-Heinemann; 2014 May 12.

Development of a local unmanned aerial vehicle (UAV) pesticide sprayer for rice production system in the Philippines

E P De Padua[1], R C Amongo[2], E P Quilloy[2], D C Suministrado[2], J C Elauria[3]

[1]Department of Agricultural and Biosystems Engineering, Visayas State University, Visca, Baybay City, Leyte, Philippines
[2]Agricultural Machinery Division, Institute of Agricultural Engineering, University of the Philippines – Los Baños, Laguna, Philippines
[3]Agricultural and Bio-processing Division, Institute of Agricultural Engineering, University of the Philippines – Los Baños, Laguna, Philippines

eldon.depadua@vsu.edu.ph

Abstract. Currently, commercially available UAV Pesticide Sprayers in the Philippines are very expensive. Therefore, this study aimed to develop a locally made automated UAV Pesticide Sprayer for rice production system that is much cheaper than the current price in the market. Moreover, the One-liter Capacity UAV Pesticide Sprayer has an autopilot flight control system-allowing the pilot to fly the UAV without directly maneuvering the controls. Pixhawk 2 Cube and Arduino Uno were used as flight and spray controllers, respectively, while Mission Planner 1.3.68 was used as the UAV sprayer's ground control station. The result from the UAV sprayer's field performance test revealed that its effective field capacity was 750 m^2 per 10 minutes, and its application rate was 3.2 liters per 1,000 m^2 of the field. Meanwhile, it was observed that its battery's life was too short to cover at least a tenth of a hectare. Nonetheless, it was found that relative to the tank capacity and field capacity of the UAV Pesticide Sprayer, the one that was developed in this study was the cheapest, compared to the current price of UAV Sprayers in the market.

1. Introduction

Aerial robotics is one of the most anticipated research areas in agriculture today [1]. It is considered a breakthrough technology with very high potential in precision agriculture [2]. It has even attracted a great deal of scientific interest due to sensor electronics' advances and the emergence of fast and compact microcontrollers and lightweight high-capacity batteries [3]. Particularly, Unmanned Aerial Vehicles (UAVs) have become popular in the field of agriculture (i.e., seeding, spraying, and mapping) due to their high capability [4]. In pesticide spraying, the use of UAV is promising. It keeps the operators safe from possible crash accidents [5] and even protects them from direct exposure to pesticides [6], which is a major concern in agricultural practice [7]. However, despite its potential benefits, the current market price of UAV sprayers is quite expensive.

In the Philippines, the use of UAV sprayer is quite challenging. Aside from the fact that the country's level of mechanization is low, particularly in crop care and cultivation for rice and corn [8], the commercially available UAV Pesticide Sprayers in the country are very expensive. Thus, this study aimed to develop a local version of UAV Pesticide Sprayer for rice production systems, which is much cheaper than the commercially available ones.

2. Methodology

2.1. Design Concept

The schematic diagram shown in figure 1 generalizes the mechanism of the entire UAV Pesticide Sprayer system. The designed and assembled UAV Pesticide Sprayer has an autopilot flight control system, which allowed the pilot to fly the UAV without directly maneuvering the controls. Besides, Mission Planner 1.3.68 was used as the ground control station of the UAV sprayer. The UAV can be triggered to turn on/off, takeoff or land, change UAV's flight mode (i.e., Autopilot, Position Hold, Altitude Hold), and return to launch turn on/off the sprayer's pump through the Radio Transceiver (Remote Control). Meanwhile, a liquid level sensor was installed on the pesticide sprayer assembly of the UAV to monitor the tank capacity. When the sensor detects during the spraying operation that the tank has no more fluid, it would automatically stop, and a buzzer would automatically make a long beep sound. The long beep sound serves as a signal to the operator to trigger the UAV Pesticide Sprayer to "return to launch" for a refill.

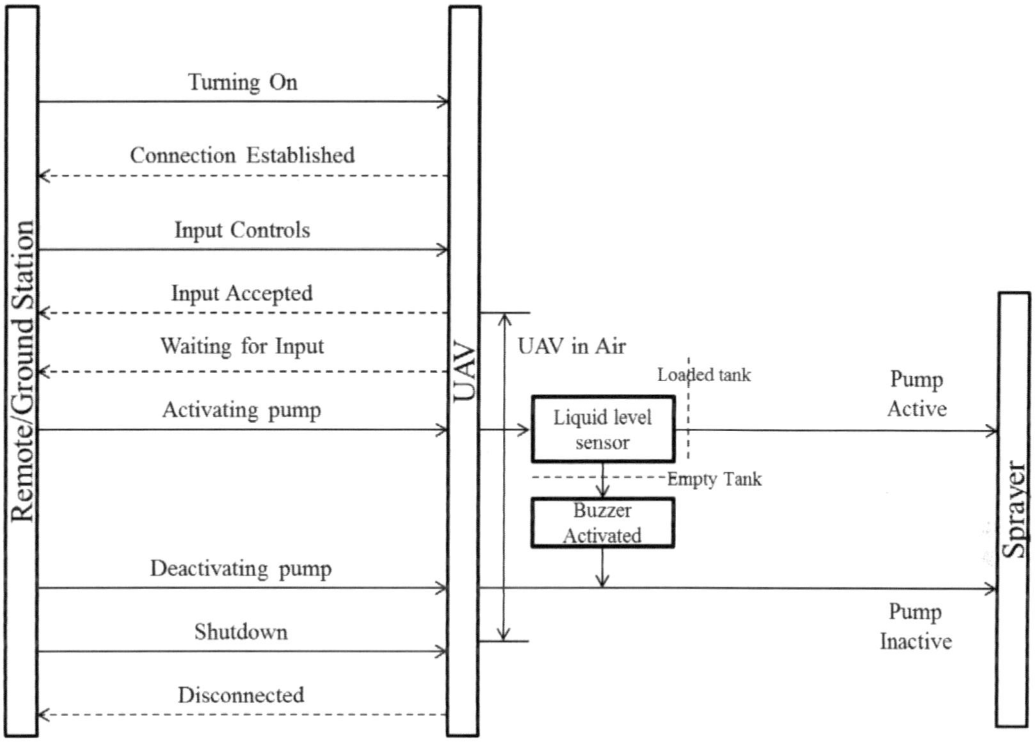

Figure 1. The UAV Pesticide Sprayer's Schematic Diagram showing how the sprayer is activated or inactivated.

2.2. Field Performance Test

2.2.1. General Field Performance Test Parameters. The field performance test of the UAV Pesticide Sprayer was based on PAES 113:2000. However, in testing the swath, some modifications in the test method were done so it would be appropriate with the mechanism of the UAV sprayer. Meanwhile, a mission in the Mission Planner was prepared for each trial. The mission was set to 2.5 m/s forward speed, 3 m operating altitude, 5 m altitude during the takeoff and landing, 2 m spacing between lines (rows), and a working area that measures approximately 500 m^2. The mission was implemented three times with the UAV pesticide sprayer at full tank capacity. Subsequently, the following field performance test parameters were gathered: actual operating time, unproductive time (refilling time, adjustment time, time to takeoff, travel time to the first waypoint, and time to return to launch), effective field capacity, effective working width, forward speed, and application rate of the UAV pesticide sprayer. Moreover, effective field capacity and application rate of the UAV Pesticide Sprayer were computed using the following formulae:

2.2.1.1. Effective Field Capacity (EFC). The UAV Pesticide Sprayer's actual capacity to cover a specific working area per total time (that it is committed to doing the field operation).

$$EFC = SWE_{ff} = \frac{A}{t_e + t_u} \quad (1)$$

Where EFC is the effective field capacity (ha/h), S is the UAV's forward speed, W is the swath, E_{ff} is the UAV's field efficiency, A is the actual area covered (ha), t_e is the effective UAV sprayer operating time, performing its full width of action (h), t_u is the unproductive time including refilling time, adjustment time, takeoff time, travel time to first waypoint, and return time to launched point (h).

2.2.1.2. Application Rate (AR). The amount/volume of pesticide solution applied to a particular working area.

$$AR = \frac{Qxt_{actual}}{A} \quad (2)$$

Where AR is the application rate (L/ha), Q is the nozzle discharge (L/h), A is the actual area covered (ha), t_e is the effective UAV sprayer operating time, performing its full width of action (h).

2.2.2. Swath at Hover. The UAV Pesticide Sprayer's swath was determined by hovering the UAV two meters above the dry concrete pavement (nozzle tip to ground). The UAV Pesticide Sprayer position was held, and the sprayer was turned on until the wetted area of the spray became visible on the concrete pavement. The longest diameter of the wetted area (the length perpendicular to the forward direction of the UAV Pesticide Sprayer) was recorded as swath. This method was done three times for validation.

2.2.3. Statistical Method for Data Analysis. Descriptive statistics was used in the analysis of data. Specifically, the standard deviation, z-score, and coefficient of variation were determined to test the dispersion and precision of the gathered data.

2.2.4. Software Development. The algorithm used to activate the spraying activity of the UAV Pesticide Sprayer is presented in figure 2. The algorithm determines when the user should start refilling the tank or turning off the pump and when the spraying would start or end. This algorithm was materialized using a liquid level sensor, a pump, and a radio transceiver (RC Controller). The algorithm was adopted during the laboratory and actual field testing of the UAV Pesticide Sprayer.

2.2.5. Simple Cost Analysis. The most commonly used and commercially available imported UAV Pesticide Sprayers in the Philippine market were compared to the developed local UAV Pesticide Sprayer. Each UAV sprayer's cost was scaled-down relative to its tank capacity and field capacity, respectively, to compare their costs. Moreover, the following assumptions and considerations were observed: (1) Other things being equal, only UAV Sprayer's tank capacity and field capacity affect its cost, (2) UAV sprayer's cost is proportional to its tank capacity and field capacity, (3) UAV Sprayers must have undergone AMTEC Test or have been used for research activities, and (4) Information about UAV Sprayer's cost must be from an authorized distributor or the end-user.

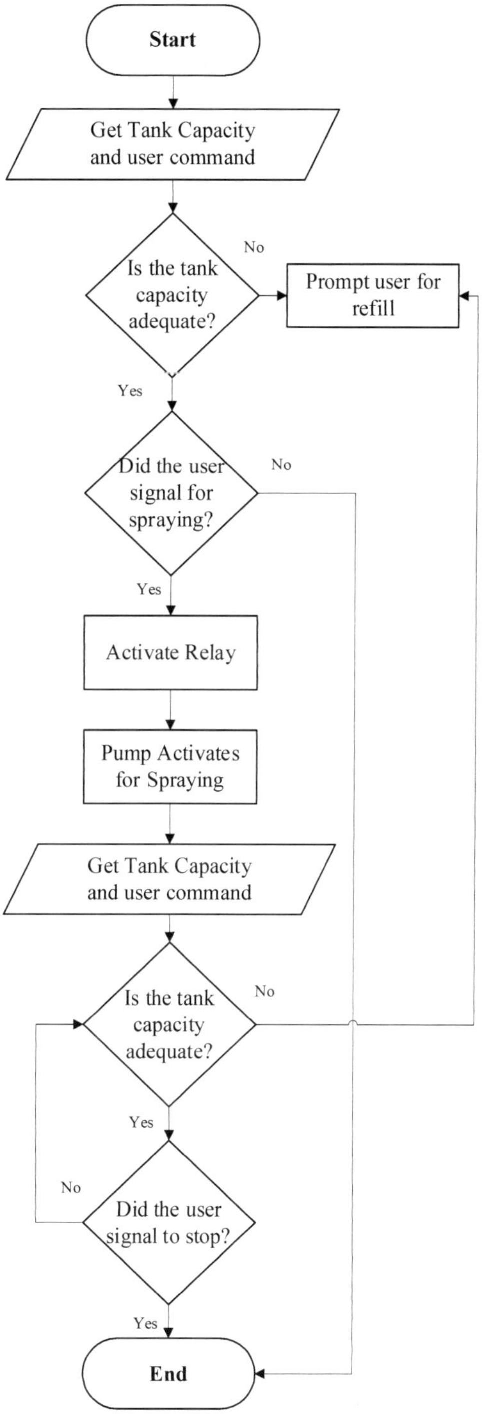

Figure 2. Flow chart showing the rules for activating the UAV spraying activity.

3. Results and Discussion

3.2. General Specifications

The locally designed UAV Pesticide Sprayer (figure 3) has a dimension of 60 cm X 60 cm X 48 cm with props. It weighed 3.815 kg with no load and 4.875 kg with a load of one liter of fluid.

Figure 3. The locally developed Unmanned Aerial Vehicle (UAV) pesticide sprayer

The tank was designed for a one-liter capacity with a working pressure of 414 kPa. Furthermore, a single head single hole nozzle with a solid cone spray pattern was used.

3.3. Field Performance Test

3.3.1. Swath. The actual swath (swath when the propellers were working) was determined before the UAV pesticide sprayer was sent to do the spraying mission. Table 1 presents the obtained values of the different parameters used in the study. The average swath of 2.27 m was obtained as the UAV Pesticide Sprayer hovered 2.46 m above the ground. A minimal dispersion of data for the swath was observed as it only had a 1.17% coefficient of variation. Also, a wider swath was observed in the actual field test (2.27 m) in comparison to the swath obtained in the laboratory test (1.94 m). The propeller's air blown effect and the increase in altitude of the UAV sprayer while hovering during the test caused the spreading out of the spray as it hit the ground. Further investigation on this observation should be done to know the effect of the propeller's air blown on the swath of spray. Also, no extreme value in swath was observed during the test.

Table 1. Swath at hover.

Trial	Altitude (m)	Swath (m)
1	2.44	2.30 (1.13)
2	2.49	2.26 (-0.38)
3	2.46	2.25 (-0.76)
Average	2.46	2.27
CV	-	1.17%

Note: Values in parentheses are the standard deviations of each data from the mean (z-score).

3.3.2. UAV Pesticide Sprayer Field Performance Test Results. In evaluating the UAV Pesticide Sprayer's field performance, the mission was prepared in the Mission Planner and was used in three flights. Each flight was considered a trial; hence, three trials were conducted for the actual field testing. Also, the actual sampling area was 527 m^2. This area was used to solve for the effective field capacity. The summary of the gathered data is presented in Table 2. In general, the UAV Pesticide Sprayer performed its spraying operation for 107.10 seconds (1 minute and 47.10 seconds).

Moreover, the UAV sprayer's computed effective field capacity was 0.45 ha/h or 750 m^2 per 10 minutes. The small value obtained for the effective field capacity was basically due to the battery's short life- it barely covered at least a tenth of a hectare. Perhaps, the option to use a battery with a higher milliAmp-hour (mAh) rating could help resolve the problem. However, replacing the current battery with a higher mAh-rating would mean additional weight to UAV's payload, thus resizing and re-matching the corresponding UAV components, i.e., DC motors propellers. Further analysis and optimization of the UAV sprayer's battery will be conducted.

The decrease in speed was also assumed to affect the UAV sprayer's effective field capacity since field capacity is also a function of UAV's forward speed (see equation 1). The slightly unstable altitude of the UAV as it completed the mission was suspected of causing the fluctuation in UAV's forward speed. Nevertheless, the UAV sprayer's unstable altitude needs to be investigated further since the researchers hypothesized that the fluctuation was mainly due to the gradually decreasing payload of the drone as it sprayed out the fluid inside the tank and the excessive vibration of the DC pump used. The calibration of the flight controller's barometer might also be a contributor; however, it was assumed that the flight controller used was in good condition, and its sensors are working well since it was new.

Furthermore, a huge dispersion of data was observed in the recorded actual operating time of the UAV Pesticide Sprayer. During the second trial of the sprayer, the UAV Pesticide Sprayer was interrupted during its mission because the pump stopped spraying due to loosened wires. The sprayer was then triggered to "Return to Launch" to fix the loosened connections. After then, the UAV sprayer was resumed to its last waypoint (the point where it last sprayed). However, due to human error, the UAV sprayer was resumed to a different waypoint. Instead of resuming the mission at point 25, the mission was resumed at point 35 (the mission had 45 overall waypoints). As a result, two rows were missed, and the operating time of the UAV sprayer was shortened. Despite this error, the mission was still considered a trial since the two important parameters to compute the UAV sprayer's effective field capacity were attained – the Covered Area and Total Time (productive time and unproductive time).

Table 2. Field performance test result based on PAES 113:2000.

Particulars	Test Number			Average
	1	2	3	
Actual operating time, s	100.49	77.72	143.10	107.10 (30.99%)
Unproductive time, s	334.62	329.03	290.64	318.10 (7.53%)
Effective field capacity, ha/h	0.44	0.47	0.44	0.45 (3.84%)
Effective working width, m	2.27	2.27	2.27	2.27
Forward speed, m/s	1.95	2.18	2.41	2.18 (10.51%)
Application rate, L/ha	35.70	32.07	28.92	32.23 (10.52%)

Note: Values in parentheses are the coefficient of variation of each parameter.

On the other hand, a similar error happened during the third trial except that the UAV sprayer was resumed to point 17 instead of point 27. Consequently, two rows overlapped, causing a longer operating time and extended covered area. Notwithstanding, the mission was considered since it would offset the missed rows' equivalent field capacity from the previous error. Subsequently, small variation was found in the effective field capacity, forward speed, and unproductive time of the sprayer. Meanwhile, the UAV sprayer's application rate was 32.23 L/ha or 3.2 L per 1, 000 m^2.

3.3.3. Simple Cost Analysis. Table 3 shows the costs of the different UAV Pesticide Sprayer models from foreign manufacturers and the cost of the locally developed UAV Pesticide Sprayer. For this study's purpose, only the cost of each UAV Sprayer relative to its tank capacity and field capacity were determined to compare the cost of each unit and to know whether the developed UAV Pesticide Sprayer in the study is cheaper or not. AGILA Drone D10 and AGILA Drone M12 of Eagle Brothers were used as the basis for comparison on this study since only these UAV sprayers have undergone AMTEC (Agricultural Machinery Testing and Evaluation Center) Testing as of June 2020. AMTEC is the premiere and reference testing center of agricultural and fisheries machinery in the Philippines.

Moreover, DJI Agras MG-1 was also included since it has been used by the Philippine Rice Research Institute (PhilRice) and International Rice Research Institute (IRRI) in their research and extension activities [9]. The tabulated costs of commercialized UAV sprayers are their approximated values based on personal interviews with UAV sprayer operators/evaluators from AMTEC and IRRI and personal interviews with drone sprayer distributors in the country. After considering the above-mentioned factors and assumptions, the result showed that as the UAV Pesticide Sprayers work with the same tank capacity and field capacity, the locally developed is the cheapest among the drone sprayers being compared.

Table 3. Cost analysis for the local UAV Pesticide Sprayer versus imported UAV Sprayer models.

Model	Tank Capacity (L)	Field Capacity (ha/hr)	Cost (PHP)	Cost per Liter (Php/L)	Cost per ha/hr, (Php/ha/hr)
Local UAV Pesticide Sprayer	1	0.45	63,143.64	63,143.64	140,319.20
DJI Agras MG-1	10	3.6	1,000,000.00	100,000.00	277,777.78
AGILA Drone D10	10	4	720,000.00	72,000.00	180,000.00
AGILA Drone M12	12	5	980,000.00	98,000.00	196,000.00

4. Summary and Conclusion

In summary, the UAV Pesticide Sprayer weighs 3.815 kg and has a one-liter tank capacity. The performance test results showed that the local UAV pesticide sprayer's computed effective field capacity was 750 m^2 per 10 minutes, and its application rate was 32.23 liters per hectare or 3.2 liters per 1,000 m^2 field. It was also found that relative to the tank capacity and field capacity of the UAV Pesticide Sprayer, the locally developed was the cheapest. Hence, the study's objective to develop a local and cheap version of UAV Pesticide Sprayer for rice production in the Philippines was achieved.

5. Recommendation

Further studies regarding the optimization of the UAV sprayer's design must be done to ensure its safety, durability, and reliability. Subsequently, cost analysis relative to UAV Pesticide Sprayer's other capabilities should be done.

References

[1] Mogili U M and Deepak B B V L 2018 Review on application of drone systems in precision agriculture. *Proc. Computer Science* (Elsevier) (133) 502–503

[2] Abdullahi H, Mahieddine F and Sheriff R 2015 Technology impact on agricultural productivity: A review of precision agriculture using unmanned aerial vehicles *Wireless and Satellite Systems 2015* (USA: Springer International Publishing) 388–400

[3] Boursianis A, Papadopoulou M, Diamantoulakis P, Liopa-Tsakalidi A, Barouchas P, Salahas G, Karagiannidis G, Wan S and Goudos S 2020 Internet of things (IoT) and agricultural unmanned aerial vehicles (UAVs) in smart farming: A comprehensive review *Internet of Things* **9** p 2

[4] Gaponov I and Razinkova A 2012 Quadcopter design and implementation as a multidisciplinary engineering course. *Proc. of IEEE Int. Conf. on Teaching, Assessment, and Learning for Engineering 2012* (USA: IEEE/Curan Associates, Inc.) p 1

[5] Kim J, Kim S, Ju C and Son H 2019 Unmanned aerial vehicles in agriculture: A review of perspective of platform, control, and applications *IEEE Access* **17** (105) 100-112

[6] Faiçal B, Freitas H, Gomes P, Mano L, Pessin G, De Carvalho A, Krishnamachari B and Ueyama J 2017 An adaptive approach for UAV-based pesticide spraying in dynamic environments *Computers and Electronics in Agriculture* **138** 210-23

[7] Wang A, Han Y, Li X, Andaloro J, Chen P, Hoffmann W, Han X, Chen S and Lana Y 2020 Field evaluation of spray drift and environmental impact using an agricultural unmanned aerial vehicle (UAV) sprayer *Science of the Total Environment* **737** 2

[8] Desale R, Chougule A, Choudhari M, Borhade V and Teli S 2019 Unmanned aerial vehicle for pesticides spraying. *Int. J. for Sci. and Adv. Rsch.in Tech.* **5** 79-82
[9] Cao L , Cao C, Wang Y, Li X, Zhou Z, Li F, Yan X and Huang Q 2017 Visual determination of potential dermal and inhalation exposure using Allura Red as an environmentally friendly pesticide surrogate *ACS Sustain. Chem. Eng.* **5** 3882
[10] Suministrado D 2013 Status of agricultural mechanization in the philippines. *Reg. For. on Sus. Agr. Mech. in Asia and the Pacific* Quingdao, China
[11] Adriano L 2019 Autopilot agricultural drone sprayer tested in Ilocos Norte *Philippine News Agency* 14 March 2019
[12] Arcalas J 2019 PHL farms to take flight in the future as government supports drone use *BusinessMirror* 14 February 2019

IOP Conf. Series: Materials Science and Engineering 1109 (2021) 012023 doi:10.1088/1757-899X/1109/1/012023

Drop test analysis of ball grid array package using finite element method

J Gonzaga[1,2], A T Ubando[1,2], E Arriola[1,2], R L Moran[1,2], N R E Lim[1,2], J P Mercado[1,2], A Conversion[1,2], D Belarmino[1,2]

[1] Mechanical Engineering Department, De La Salle University, 2401 Taft Avenue, 1004 Manila, Philippines
[2] Thermomechanical Analysis Laboratory, De La Salle University – Manila: Laguna Campus, LTI Spine Road, Laguna Blvd, Biñan, Laguna 4024, Philippines

aristotle.ubando@dlsu.edu.ph

Abstract. Proper design of semiconductors is the most essential step in ensuring electronic product reliability during manufacturing. The adoption of design optimization approach enhances the performance and reliability of the semiconductor package, hence, minimizes product failure. Semiconductor reliability especially in the application to automobiles is very crucial. High product failure tolerance in semiconductors in automobiles is required due to problems such as defects and malfunctions that are directly linked to casualty accidents. Such high tolerance requires keen quality control up to the semiconductor solder component level. This study aims to do a sensitivity analysis of the solder balls' material properties and its effect on the stresses experienced by the semiconductor component with respect to its reliability using drop test. A drop test analysis was simulated wherein a shorter distance of 5 mm between the package and platform was implemented to make the simulation time faster. The density and modulus of elasticity of the solder ball material were identified as the independent factors while the stresses experienced by the BGA package during the drop test is the dependent variable. The results have shown that the maximum stress for all runs was found in the same area in the package. Changing the modulus of elasticity showed greater effect on the impact stress compared to the alteration of material density.

1. Introduction

Failure test analysis is a crucial step in the manufacturing of semiconductor packages to test its reliability. Crack propagation analysis, warpage analysis, and drop test are some of the examples of quality control measured to ensure product quality. This test analysis normally requires actual products from the manufacturing department that will be subjected to experimental analysis. Hence, it is expected to consume resources and time that can affect the manufacturing of products. Simulations of failure tests analysis can be introduced to the system to significantly reduce the enormous amount of physical testing while offering time efficiency to the manufacturing sector [1].

Semiconductor design is one of the considered critical measures for ensuring the reliability and lifespan of semiconductor packages [2]. Common cause of failure is generated by shared causes according to Sakurahara et al [3]. These causes had been quantified and modeled based on the study of Mi et al [2]. In association with the different system analysis for reliability, importance of component analysis is beneficial for system design, system control, and improvement of reliability [4]. This highlights the importance of design optimization on the research and development portion of the product reliability test. According to Dunn et al. [5], the design optimization will aid the analysis by the identification of a system design that will produce a result for which the error is minimized when measuring a set of samples. This typically looks for objective functions that have the partial derivatives desired to be zero at value.

Semiconductors are considered to be one of the key components of an automobile. It plays an important role in modern automotive specifically on its electronic control units [6]. These controls provide stability of operations to the consumer, and fuel efficiency. These control units are typically used in automobile and space equipment applications [7]. These controls may be subject to free fall due to automobiles moving when in operation. In such cases, the mechanical and electrical functionality of these controls can be unreliable. Drop test analysis is a test that identifies the mechanical reliability of electronic components in drop. Using finite element analysis on simulation models will significantly reduce the control unit's reliability risk in the design optimization stage. Automotive semiconductors often require higher levels of requirements on humidity, temperature, and most importantly, in tolerance rate when it comes to tolerance failure compared to general semiconductors [8]. The high tolerance is required due to problems such as defects and malfunctions that are directly linked to casualty accidents. Safety and reliability are essential requirements on the specification of automotive semiconductors. According to Chae and Kim [6], reliability refers to the magnitude where the product operates with no failure accounted for at a specific period while safety is the degree of minimizing the effect of product failures to consumers. Such high tolerance requires keen quality control up to the semiconductor solder component level. Ball Grid Array (BGA) is a popular technology used on semiconductor applications due to its dense interconnectivity and high lead count [9]. Its' characteristic of self-centering has given its advantage on semiconductor applications. These are easy to install because it has no leads, but it still suffices the high thermoelectrical requirements of a semiconductor. BGA packages have high transistor counts but it has small footprint however, the technology is known to manifest cracking of solder joints between interconnections due to hard drops [10].

There are studies available on the drop test method to analyze the impacted state and expected vibration in automobiles. This particular test was run by Shirangi et al. [11] on Electronic Control Units (ECU) or known as the computer boxes of the automobiles. The experimentation set up and procedure can accommodate varying weight of the semiconductors while monitoring the speed of the test subject with the use of an on-board sensor. The actual results were the basis of the simulation model using ANSYS for validation.

A study made by Balakrishnan et al. [7] comparing the effectiveness of finite element analysis from drop test. The study had concluded the finite element analysis is suitable for capturing the dynamic behavior of the drop test analysis. Material properties of solder balls are to be varied due to the limited number of material compatible for semiconductor solders. To date, there are no direct relevant studies for the stress analysis with respect to drop tests.

In line with this, the study aims to perform a sensitivity analysis of the solder balls' material properties and its effect on the stresses experienced by the semiconductor package with respect to its reliability on drop test.

2. Materials and Methods

A BGA assembly was employed for this analysis, the specifications are derived from the Intel databook on BGA packaging [12]. The BGA assembly consists of several components such as the die chip that is the heart of the semiconductor device, molding compound that protects the package, substrate, solder balls, and PCB that provides connectivity. An illustration of a BGA package together with its major components can be seen in Figure 1.

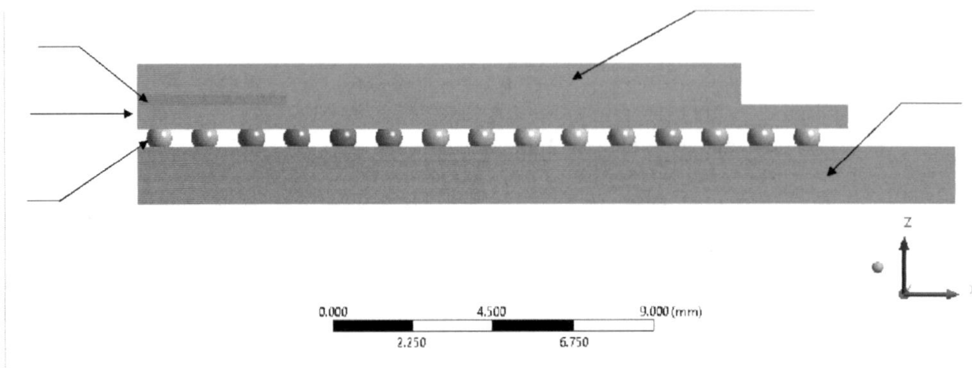

Figure 1. Illustration of Ball Grid Array assembly and components

The symmetric nature of the BGA allowed further simplifications to the model such as incorporating a quarter model for the simulation in order to reduce the total number of nodes and elements needed in the analysis [13][14]. The model utilizes a die thickness of 0.27 mm and 1.16-mm-thick EMC. The ball grid array package that is used for solder ball attachment has the dimension of 40 x 40 x 1.7 mm. A symmetrical boundary constraint was applied in the model. It is noteworthy that the expanded center of the geometry model was fixed with dz = 0 to hinder the rigid body motion while the boundary of dx = dy = 0 was set due to the symmetry planes.

In this study, the following material properties of the BGA package derived from various literature [9][10] were applied and is summarized in Table 1.

Table 1. Material properties for the drop test simulation analysis

Component	Modulus of Elasticity, E (MPa)	Poisson's ratio, v	Density, ρ (kg·m^{-3})
EMC	17000	0.25	1900
Die Chip	170000	0.28	2300
Substrate	11600	0.42	3950
PCB	17200	0.42	1850
Structural Steel	200000	0.30	7850

2.1. *Meshing*

The general meshing applied to the model focused on the application of second order hexahedral elements to yield as few elements as possible [14][15]. Refined meshing was applied towards the solder balls to ensure the accuracy of stress calculations around the area of concern. A total of 65000 nodes and 226000 elements were produced in the mesh generation. Through a mesh independence test, these numbers of nodes and elements were identified. The selected elements of these mesh settings were found to be excellent in both the orthogonal quality and skewness criteria for the finite element analysis [9].

2.2. *Drop test analysis*

For the analysis setup, the BGA assembly is to be dropped from a height of 1 m towards a structural steel platform. However, to speed up the analysis, the semiconductor package is moved to about 0.5 mm from the structural steel platform. Standard earth gravity load was applied towards the BGA assembly while the bottom surface of the platform was fixed. By reducing the distance between the BGA and the platform, the time frame involved in the analysis is also reduced; hence, the simulation time is shorter.

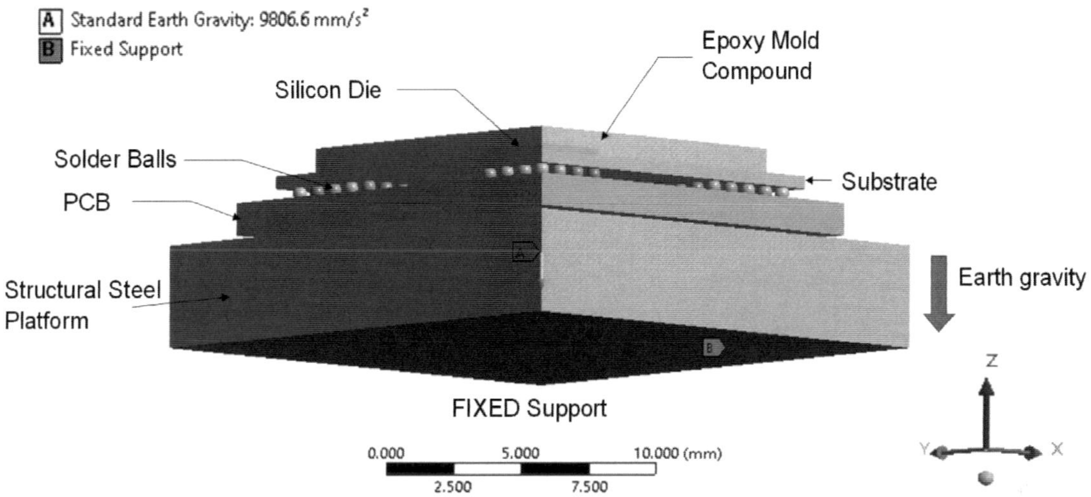

Figure 2. Drop test simulation setup

Initial conditions were also applied in the model such as the instantaneous velocity the package experiences at about 50 mm from the platform. This can be computed using free fall equations and would yield to a velocity of 4.428 m/s. Consequently, for this analysis the end time was calculated to be 0.00015 s. Under these loads and constraints, the package is simulated to be free falling from a height of 1 m with an initial velocity of zero.

2.3. *Design of experiment*

The reliability of a BGA package is highly dependent on the state of the solder balls, as such, this study explores the variations in the material properties of the solder balls to be used in the BGA assembly and evaluates it in terms of a standard drop test. The density and modulus of elasticity of the solder is identified as the independent variables while the von-Mises stress during impact is recorded as the dependent variable. In order to determine the effects of these material properties on the stress experienced by the package, particularly in the solder balls, a two-factor factorial design is implemented where the extreme "-" and "+" conditions from the nominal value are subject to the design. A -10% of the nominal value was set for the "-" condition while a +10% of the nominal value was set for the "+" condition. Table 2 summarizes the simulation runs for this study.

This design ensured an adequate spread of data points, while gaining an accurate correlation between the targeted material properties versus the resulting impact stress. This allowed the researchers to utilize a lower density of data points.

3. **Results and discussion**

The maximum impact stress has been consistently located in the solder balls for all runs. Figure 3 illustrates the impact stress experienced by the BGA package in the various runs.

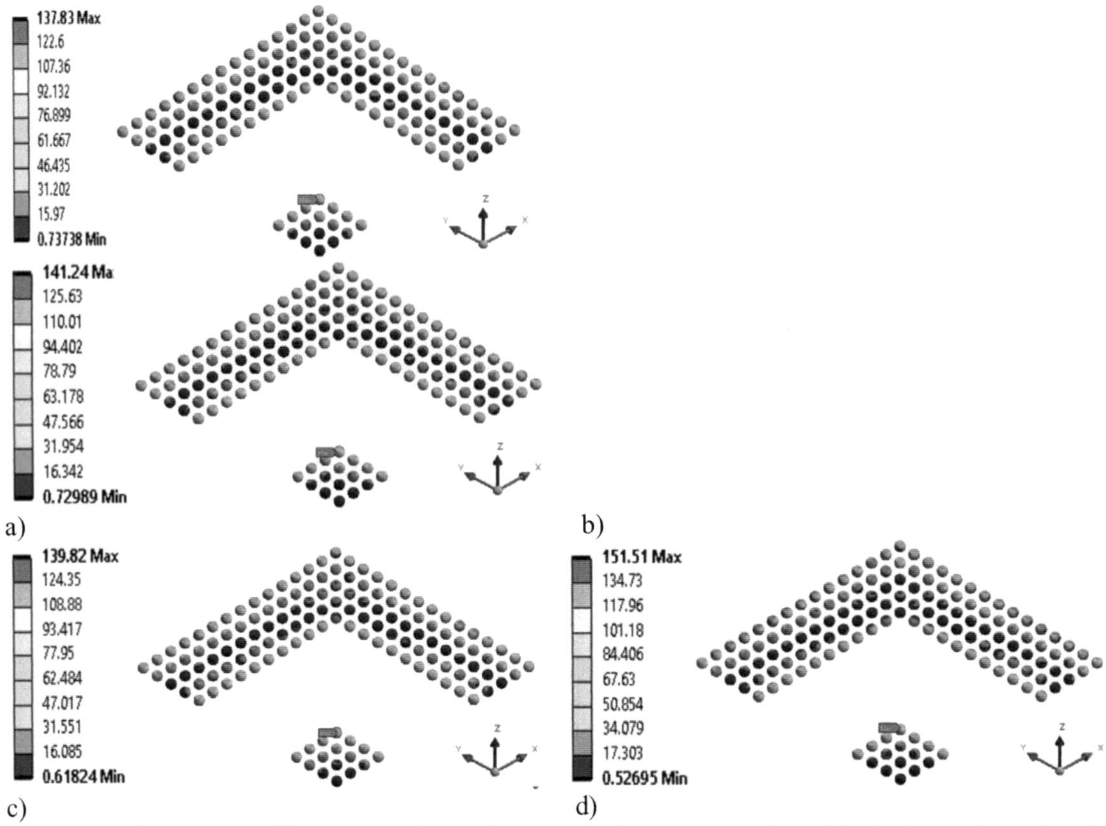

Figure 3. Stress profile showing the resulting equivalent stress on the solder ball area for a) Run 1 (-,-), b) Run 2 (-,+), c) Run 3 (+,-), and d) Run 4 (+,+). All values in MPa.

The tabulated results can be seen in the last column of Table 2.

Table 2. Equivalent Stress experienced in the solder ball area

Run	ρ (kg·m^{-3})	E (MPa)	Impact Stress (MPa)
1 (-,-)	6624	31896	137.83
2 (-,+)	6624	38984	141.24
3 (+,-)	8096	31896	139.82
4 (+,+)	8096	38984	151.51

The results were further subjected to statistical tests to determine the significance and influence of the identified variables towards the stress experienced in the solder balls. A multivariate regression analysis was applied and it was determined that the overall p-value of the fit model based on the gathered simulation data is 0.39 which is greater than 0.05. This implies that there is not enough statistical evidence to support the claim that varying the density or modulus of elasticity can significantly affect the stress experience in the ball solder area of a BGA package when subjected to a drop test. Upon further inspection, the parameter estimate of density is 0.38 which is slightly higher than the modulus of elasticity which has a p-value of 0.32. This suggests that based on the available simulation data on this scenario, the modulus of elasticity has a slightly higher influence over the impact stress in the package than the density parameter.

4. Summary and Conclusion

For a BGA package, one of its major design points to be considered is its solder balls. Majority of its functionality is highly dependent on the connectivity and state of the solder balls attached to it. As such, it is important that these components survive the operating conditions where the package will be subjected to. In this paper, the idea of conducting a sensitivity analysis on the material properties of the solder ball is done to determine its influence over the stresses that the package experiences. A drop test analysis was simulated wherein a shorter distance of 5mm between the package and platform was implemented to make the simulation time faster. The density and modulus of elasticity of the solder ball material was identified as the independent factors while the stresses experienced by the BGA package during the drop test is the dependent variable. A sensitivity analysis was conducted where a two-factor design was developed to determine the significance of altering the identified parameters over the stress in the package. The maximum stress for all runs is found in the same area in the package. Moreover, the gathered simulation data suggests that there is no significant evidence to imply that varying the material properties of the solder ball would affect the stresses it experiences. However, it would be interesting to further evaluate the other components of the package and identify the significant factors that might contribute to reliability concerns from a drop test.

Acknowledgments

The authors would like to thank the DOST-ERDT consortium for its support towards their studies. The authors also wish to express their gratitude towards the PHILIPPINE COUNCIL FOR INDUSTRY, ENERGY, AND EMERGING TECHNOLOGY RESEARCH AND DEVELOPMENT (DOST-PCIEERD) with Project No. 8897, 2020 for funding this work.

References

[1] Cai Y, Wang G, Li G and Wang H 2015 A high performance crashworthiness simulation system based on GPU *Advances in Engineering Software,* vol. 86, pp. 29-38, 2015/08/01/ 2015, doi: https://doi.org/10.1016/j.advengsoft.2015.04.003.

[2] Mi J, Beer M, Li Y-F, Broggi M and Cheng Y 2020 Reliability and importance analysis of uncertain system with common cause failures based on survival signature *Reliability Engineering & System Safety* **201** p 106988 2020/09/01/ 2020, doi: https://doi.org/10.1016/j.ress.2020.106988.

[3] Sakurahara T, Schumock G, Reihani S, Kee E and Mohaghegh Z 2019 Simulation-informed probabilistic methodology for common cause failure analysis *Reliability Engineering & System Safety* **185** pp 84-99

[4] Kuo W and Zhu X 2012 *Importance measures in reliability, risk, and optimization: principles and applications* (John Wiley & Sons)

[5] Dunn W L and Shultis J K 2012 7 - Inverse Monte Carlo *Exploring Monte Carlo Methods* (W. L. Dunn and J. K. Shultis Eds. Amsterdam: Elsevier) pp 171-202.

[6] Chae S Y and Kim W J 2012 Plan for reaction of automotive ECU software and SOC for supporting ISO26262 *Korea Information Processing Society Review*

[7] Balakrishnan K, Sharma A and Ali R 2017 Comparison of explicit and implicit finite element methods and its effectiveness for drop Test of electronic control unit *Procedia Engineering,* **173** pp 424-431 2017/01/01/ 2017 doi: https://doi.org/10.1016/j.proeng.2016.12.042.

[8] Yun G, Jung H-W and Park S 2018 Prediction of field failure rate using data mining in the automotive semiconductor *Procedia Computer Science* **139** pp 512-520 2018/01/01/ doi: https://doi.org/10.1016/j.procs.2018.10.242.

[9] Lim N R E G, Ubando A T, Gonzaga J A and Dimagiba R R N 2020 Finite element analysis on the factors affecting die crack propagation in BGA under thermo-mechanical loading *Engineering Failure Analysis* **116** doi: 10.1016/j.engfailanal.2020.104717.

[10] Birzer C, Rakow B, Steiner R and Walter J 2005 Drop test reliability improvement of lead-free

fine pitch BGA using different solder ball composition *2005 7th Electronic Packaging Technology Conference* **1** p 7 doi: 10.1109/EPTC.2005.1614403.

[11] Shirangi M H, Simo G T, Wang Z, Unnikrishnan R and Heinrich T 2014 A novel drop test methodology for highly stressed interconnects in automotive electronic control units *2014 IEEE 64th Electronic Components and Technology Conf. (ECTC)* pp 108-113 doi: 10.1109/ECTC.2014.6897274.

[12] Intel 2000 2000 Packaging Databook 14-1 *Ball Grid Array (BGA) Packaging* (Intel)

[13] Xiuyun Hao, Liancheng Qin, Daoguo Yang and Shilong Liu 2003 Thermal-mechanical stress and fatigue failure analysis of a PBGA *Fifth International Conf. on Electronic Packaging Technology Proceedings* (ICEPT2003., Shanghai, China) pp 438-442.

[14] Lim N R E, Arriola E, Moran R L, Mercado J P, Dimagiba R, Gonzaga J and Ubando A 2019 Material properties sensitivity analysis based on thermomechanical analysis on warpage for ball grid array semiconductor packaging *2019 IEEE 11th Int. Conf. on Humanoid, Nanotechnology, Information Technology, Communication and Control, Environment, and Management (HNICEM)*

[15] Moran R L, Arriola E, Lim N R E, Mercado J P, Dimagiba R, Gonzaga J and Ubando A 2019 Reduction of IC package warpage through finite element analysis and direct optimization 2019 *IEEE 11th Int. Conf. on Humanoid, Nanotechnology, Information Technology, Communication and Control, Environment, and Management (HNICEM)*

Evaluation of action sport camera optical motion capture system for 3D gait analysis

N Rizaldy, F Ferryanto*, A Sugiharto & A I Mahyuddin

Mechanical Design Research Group
Faculty of Mechanical and Aerospace Engineering, Institut Teknologi Bandung
Jalan Ganesha 10, Bandung 40132, Indonesia

*ferryanto@ftmd.itb.ac.id

Abstract. The high price of motion capture systems on the market has been a hindrance for many hospitals, health agencies and rehabilitation centers to acquire them. Due to its significance in gait analysis, the ITB Biomechanics Research Team has developed affordable motion capture systems since 2008. The advent of Action Sport Cameras (ASC) provide an affordable and portable alternative to the available motion capture systems. Hence, this work investigated the relative accuracy of an ASC based motion capture system. The well-established Vicon V8 with 10 cameras system is utilized in this study to evaluate the proposed ASC Mocap system. Simultaneous gait data acquisition is necessary to ensure valid comparison. Evaluation of the ASC based system was carried out by comparing the spatio-temporal and kinematics parameters obtained with that obtained with Vicon. Based on the parameters obtained, the 4 GoPro Hero 5 cameras system developed by the ITB Biomechanics Team's is deemed to be suitable to be used in Gait Analysis, especially the spatio-temporal parameters that closely match the Vicon results. For the kinematic parameters, the results obtained while qualitatively similar, in many cases differ in magnitudes. This problem is currently is still under investigation and will be reported in future works.

Keywords: *Motion capture system, gait parameters, 3D, action sport camera, Vicon, spatio-temporal, kinematics*

1. Introduction

Gait is a personal characteristic of walking, a motion that is crucial in daily life of a person. Walking involves a complex motion to maintain balance and coordinate muscles to propel the body forward in stride. Since walking is an acquired skill it takes time for gait to mature. However, there are many possibilities that gait could deviate from normal walking. It is important to understand gait abnormality that a patient walk could be observed and recorded to obtain related gait parameters. Normal gait requires that many systems, including strength, sensation and coordination, function in an integrated fashion. Many common problems in the nervous system and musculoskeletal system will show up in the way a person walks [1].

Objective to measure or quantify a patient gait is commonly achieved by using optical motion capture (mocap) systems. Motion capture (mocap) is the process of recording the movement of objects in the form of objects or humans. This mocap technology originally came from the development of life sciences (biology) which was used to analyze gait [2]. Motion capture consists of a part that marks and tracks the body and a part that converts that information into data that is useful for research and applications. The accuracy and precision obtained is determined by how the data is collected. This is because each motion capture system has different capabilities and limitations with regards to the accuracy and flexibility of data. Many motion capture systems are designed for indoor use, but several are effective and suitable for outdoor operation [3-6].

Today the process of recording the movement of these objects is widely and commercially used by VFX studios [7-8], sports therapists [9], neuroscientists, clinical specialists [10], biomechanists [11], validation and control of computer vision and robotics [2,12]. There are several motion capture systems already in circulation, such as Qualysis, Vicon, Motion Analysis, Xsens, and OptiTrack [13]. Which needs fixed setup, usually in a laboratory environment. Among commercial mocap systems, Vicon has higher resolution and has been certified according to standard that covers design, production and support for motion capture systems and includes software development for motion capture systems, measurement and analysis of three-dimensional structures [14-15]. Vicon has been taken as the gold standard in the mocap [16].

In efforts to provide affordable systems, ITB Biomechanics Research Team due to the needs in their investigations of gait and walking motion, have developed several optical motion capture systems. Initially with a single home-video camera for 2D gait analysis, and later 3D walking motion systems using two cameras [17-18]. On the other hand, the advent of Action Sport Cameras (ASC) and their affordability provide an alternative to the commercially available mocap systems. This work investigates the relative accuracy of an ASC based motion capture system. The proposed system consists of 4 (four) *GoPro Hero 5* cameras and related data acquisition components. The well-established 10 cameras Vicon V8 is utilized in this study to evaluate the proposed ASC mocap system. Simultaneous gait data acquisition is necessary to ensure valid comparison. Evaluation of the ASC based system was carried out by comparing the spatio-temporal and kinematics parameters obtained to those of Vicon.

2. Methodology

Determination of the scale ratio needs to be performed in order to obtain comparable results for the two systems. Also, to enable result comparison, the data collection needs to be synchronized and taken at the same rate. To that end, the camera capture speed were set at 120 fps for both ASC mocap and Vicon systems. The data comparisons included one walking cycle for the left leg and the right leg obtained from a simultaneous data collection.

In the present work, due to its portability, the ASC mocap system was set-up at the Laboratory of Work System Design and Ergonomics, Faculty of Industrial Technology, Institut Teknologi Bandung, where the Vicon system is located. This would enable simultaneous data collection. The ASC based system set-up was determined by ensuring that the selected positions of the action cameras allow for recording of all markers on a subject while in walking motion at certain intervals. The direction of the black arrow in Figure 1 shows the direction of motion of the subject during the data collection.

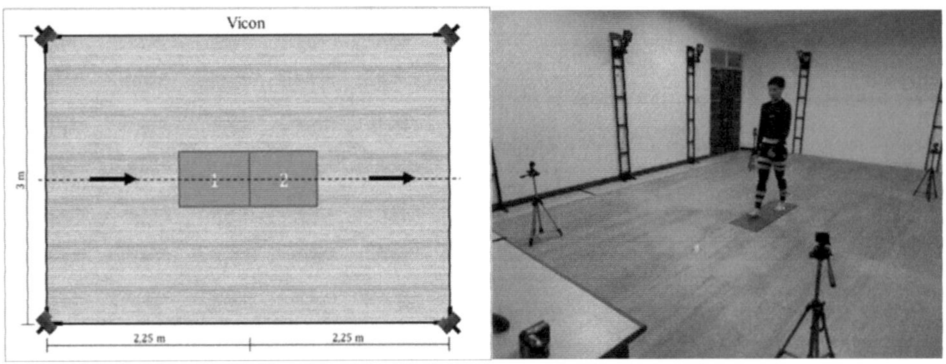

Figure 1 Experiment set-up

The subject selected has a normal BMI, and does not have physical disabilities that could alter the gait or walking motion. Next, the subject's anthropometric measurements were taken to obtain necessary data input for both Mocap systems.

Marker placements on the subject were based on specific marker placement configurations for each system. Each active markers were placed according to ASC based system configurations and each passive markers were placed according to plug in gait marker placements set by Vicon. Markers placement on subject are shown in Figure 2. Modified markers were used as solution for handling markers that have overlapping position due to different type of markers and different marker configuration (Figure 3).

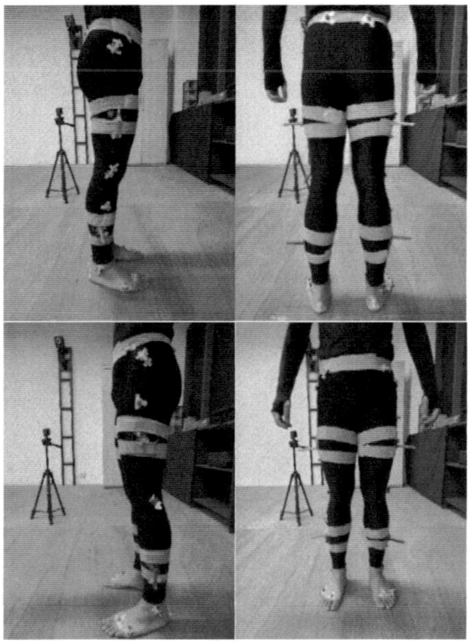

Figure 2 Marker placement on subject

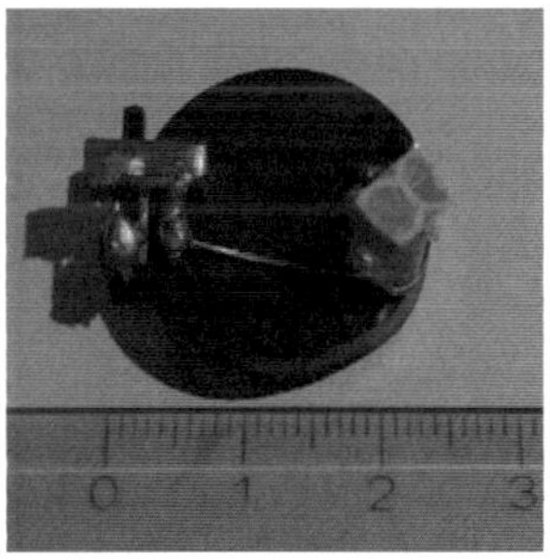

Figure 3 Modified Marker

Before doing a trial or taking walking motion data on the subject. First, the calibration was carried out on the two existing motion analysis systems. The ASC Mocap system calibrates cameras by moving a 2-dimensional object and using Zhang camera calibration technique [19]. While for Vicon its active wand was employed to calibrate the cameras.

Before taking the data, the subject was asked to do walking exercises on the existing set-up. This is done so that the subject could walk at his preferred normal walking speed, and would not be hindered d by the set-up or the equipment and markers that are attached to the subject. Data collection for the two motion analysis systems was then carried out simultaneously.

Determination of the global reference axis of the motion analysis system is needed to synchronize the origin of the coordinates systems used by the two systems. The global reference axis used by the author as a guide in data collection is the global reference axis of the Vicon motion analysis system. This reference axis has its origin at the middle of the two force-plates in the data collection set-up, as shown in Figure 4.

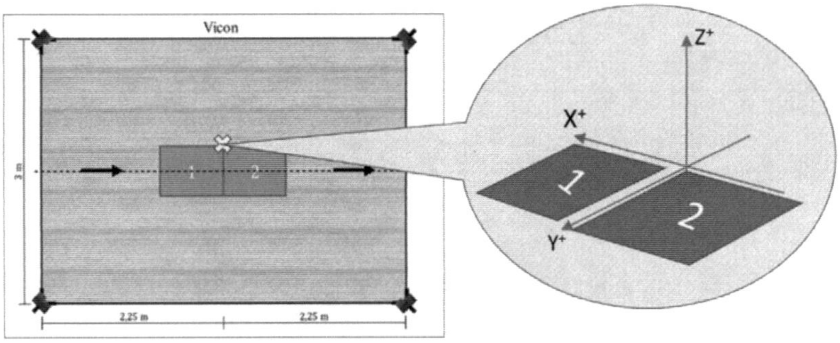

Figure 4 Illustration of the global reference axis of the motion analysis systems

Data processing from the results of data collection of the two motion analysis systems is carried out separately. Vicon has an established data processing system. While the ASC based mocap system employs the developed motion analysis system data processing software with an application interface as shown in Figure 5.

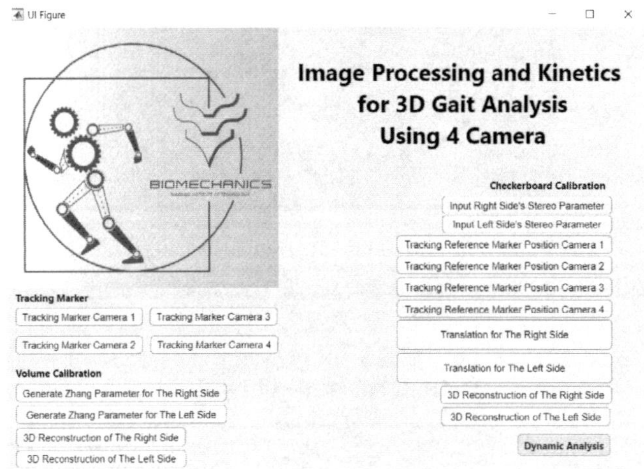

Figure 5 Motion analysis application interface of ASC mocap

Comparison of the kinematic parameters of the subject are focused on the sagittal plane. because in this field changes in the anatomical angle indicate the dominant visible and most significant human walking characteristics.

3. Results and Analysis

The subject selected was a 22-year old male, weighs 69.5 kg, and a height of 1.73 m with a BMI of 23.25. It was confirmed beforehand the absence of any disease or any dysfunction that could affect the gait motion. After the BMI calculation and clarification are done, the subject was deemed to have meet the requirements for data collection as a normal gait subject. The resulting gait spatio-temporal parameters and kinematic joint angles obtained by the two systems are tabulated in Table 1.

IOP Conf. Series: Materials Science and Engineering 1109 (2021) 012024 doi:10.1088/1757-899X/1109/1/012024

Table 1 Spatio-temporal gait parameters

Variables	Data	
	Vicon	ASC
Walking Speed (m/s)	1.04	1.08
Stride length (m)	1.24	1.27
Cadence (step/ min)	101.41	101.41
Cycle time (s)	1.18	1.18

Table 1 shows that the spatio-temporal data obtained in both systems are in agreement. The differences between the two systems are in the speed of the walking and the stride length. The difference that occurs between the spatio-temporal parameter results of the walking motion analysis system of the ASC-based mocap has an error of 3.09% on the spatio-temporal parameter data of the Vicon walking motion analysis system used as a reference. The error that occurs is relatively small when compared to the reference, so it may be stated that the 3D walking motion analysis system of the ITB Biomechanics Team has succeeded in obtaining the spatio-temporal parameter of human walking motion well.

The joint angular kinematics are joint angles in the sagittal plane, i.e. hip, knee, and ankle angles for both right and left foot. The comparison of results obtained by ASC mocap to that of Vicon are presented in Figures 6 – 8, along with gait reference database [20].

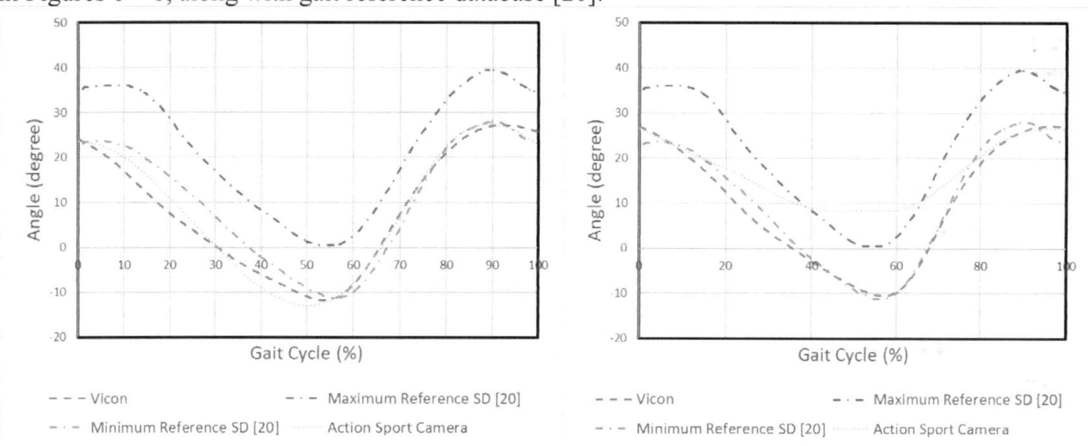

Figure 6 Left Hip Angle (left) and Right Hip Angle (right)

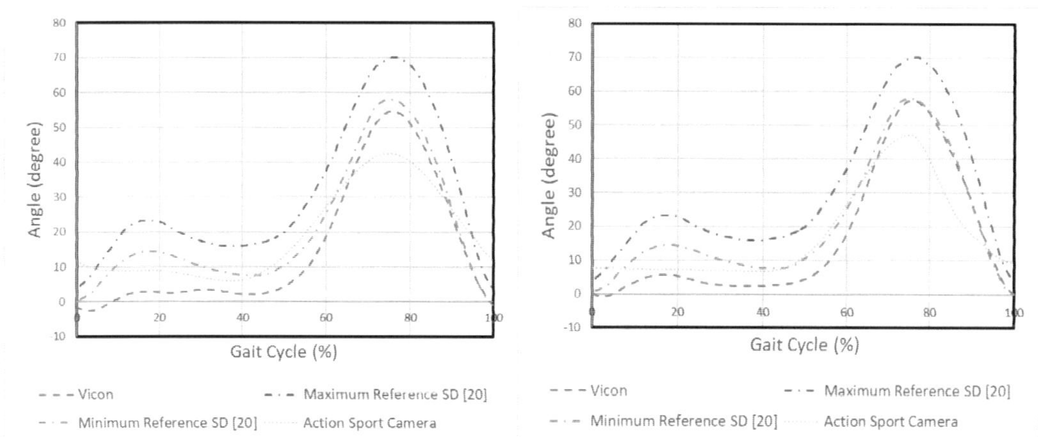

Figure 7 Left Knee Angle (left) and Right Knee Angle (right)

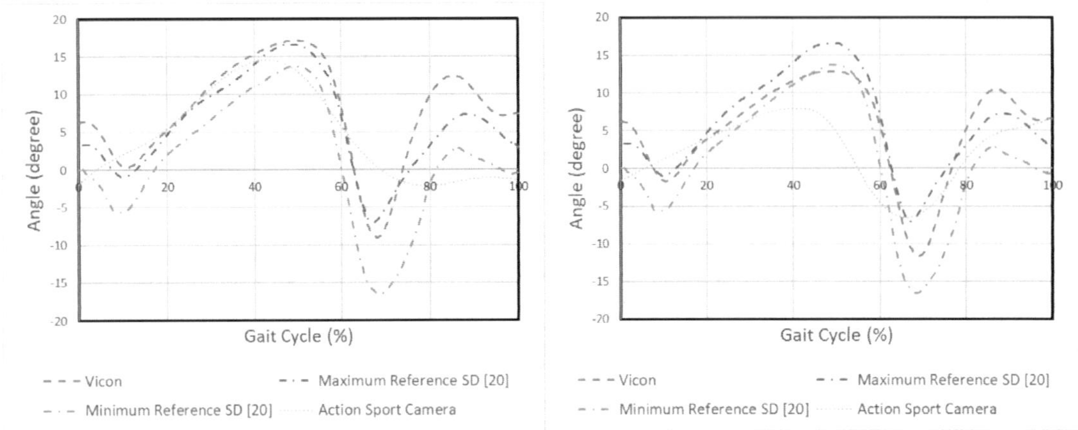

Figure 8 Left Ankle Angle (left) and Right Ankle Angle (right)

From the graphs obtained, the results of the processing of the Vicon mocap kinetics parameters can be used as a reference because it has a trend and the result timing that resembles the existing reference walking motion database. The results of the action camera mocap kinematics parameter data processing show data that have a trend and timing that is almost similar to the results of the Vicon mocap on the right hip angle, left hip angle, right knee and left knee of the subject although there are still differences in amplitude in some areas. The very significant difference in the results of the kinematic parameter data between the action camera mocap and the Vicon mocap is in the results of the angles of the subject's right and left ankles. The results obtained have relatively different trends / patterns and timing.

The difference in the results of the joint angle obtained is due to differences in human body modelling. The human body modeling of Vicon mocap system and action camera mocap system use three markers for the reconstruction of body segments such as calves and thighs. The modeling of the human body between the two systems differs in the foot reconstruction section where the Vicon mocap uses three markers that form a triangle for foot segment reconstruction while in the action camera mocap uses two markers forming a line pattern for foot segment reconstruction. This is the underlying cause of the considerable difference between the results of kinematics data processing of the right and left ankle angle of the action camera mocap and the Vicon mocap. Figures 9 (a) and (b) show the configuration of the human body modeling on the action camera mocap and Vicon.

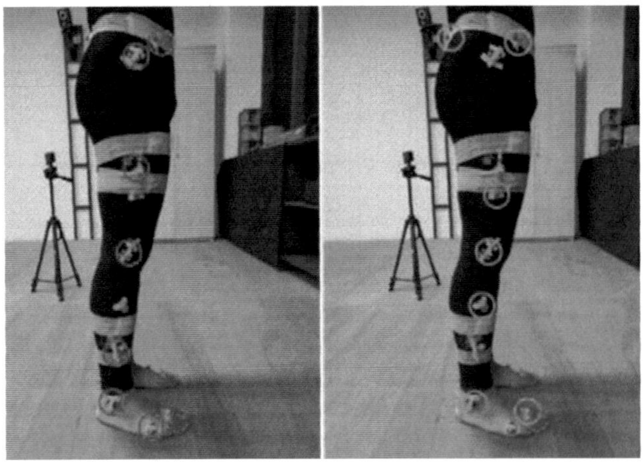

(a) (b)

Figure 9 Human body modelling configuration (a) Action sport camera and (b) Vicon

4. Conclusion

After taking data and processing data for both mocap, it was found that the walking motion obtained was a normal walking motion. It is indicated that the results of the joint angle Vicon from the subject's walking motion are within the standard deviation range of the 3-dimensional walking motion data reference, so that Vicon can also be used as a reference in comparison.

Using Vicon as a reference in the analysis of existing walking motion results in an error of 3.09% in the calculation of each walking speed and stride length for ASC-based mocap system spatio-temporal parameters.

With this small error, it can be said that the ASC-based mocap yield accurate spatio-temporal gait parameters. However, for the calculation of joint angles, the Vicon mocap and ASC-based system have a fairly good value in calculating the hip angle and knee angle although there are still differences in amplitude and trend of the same walking motion. Calculation of ankle angle for both mocap has far enough results. This is presumably due to differences in human body modeling. From this, it may be concluded that the ASC mocap is still not very accurate in calculating the joint angles.

In future works, the ITB Biomechanics human body modeling mocap will be readjusted and several correction factors in the calculation of amplitude and trend joint angle.

Acknowledgement
We gratefully acknowledge *Penelitian Unggulan Perguruan* Tinggi (PUPT) 2020 scheme research grant from Ministry of Research Technology and Higher Education in supplying the financial support.

References

[1] Jameson JL. *Harrison's principles of internal medicine*. McGraw-Hill Education,; 2018.
[2] *What is Motion Capture?: What Can I Use Motion Capture For?* [Internet]. Vicon. 2020 [cited 2020 Nov 30]. Available from: https://www.vicon.com/about-us/what-is-motion-capture/
[3] Moeslund TB, Hilton A, Krüger V. *A survey of advances in vision-based human motion capture and analysis.* Computer vision and image understanding. 2006 Nov 1;104(2-3):90-126.
[4] Bernardina GR, Cerveri P, Barros RM, Marins JC, Silvatti AP. *In-air versus underwater comparison of 3D reconstruction accuracy using action sport cameras*. Journal of Biomechanics. 2017 Jan 25;51:77-82.
[5] Begon M, Colloud F, Fohanno V, Bahuaud P, Monnet T. *Computation of the 3D kinematics in a global frame over a 40 m-long pathway using a rolling motion analysis system*. Journal of Biomechanics. 2009 Dec 11;42(16):2649-53.
[6] Zefinetti FC, Vitali A, Regazzoni D, Colombo G. *Goalkeeper's Performances Assessed with Action Cameras Based Mocap System*. In International Conference on Applied Human Factors and Ergonomics 2020 Jul 16 pp 259-266. Springer, Cham.
[7] Pullen K, Bregler C. *Motion capture assisted animation: Texturing and synthesis*. In Proceedings of the 29th annual conference on Computer graphics and interactive techniques 2002 Jul 1 pp 501-508.
[8] Zordan VB, Hodgins JK. *Motion capture-driven simulations that hit and react*. In Proceedings of the 2002 ACM SIGGRAPH/Eurographics symposium on Computer animation 2002 Jul 21 pp 89-96.
[9] Bates NA, Ford KR, Myer GD, Hewett TE. *Impact differences in ground reaction force and center of mass between the first and second landing phases of a drop vertical jump and their implications for injury risk assessment.* Journal of Biomechanics. 2013 Apr 26;46(7):1237-41.
[10] Gefen A, Megido-Ravid M, Itzchak Y, Arcan M. *Biomechanical analysis of the three-dimensional foot structure during gait: a basic tool for clinical applications.* J. Biomech. Eng.. 2000 Dec 1;122(6):630-9.

[11] Safayi S, Jeffery ND, Shivapour SK, Zamanighomi M, Zylstra TJ, Bratsch-Prince J, Wilson S, Reddy CG, Fredericks DC, Gillies GT, Howard III MA. Kinematic analysis of the gait of adult sheep during treadmill locomotion: parameter values, allowable total error, and potential for use in evaluating spinal cord injury. Journal of the Neurological Sciences. 2015 Nov 15;358(1-2):107-12.
[12] Ott C, Lee D, Nakamura Y. Motion capture based human motion recognition and imitation by direct marker control. In Humanoids 2008-8th IEEE-RAS International Conference on Humanoid Robots 2008 Dec 1 (pp. 399-405). IEEE.
[13] Bernardina GR, Monnet T, Pinto HT, de Barros RM, Cerveri P, Silvatti AP. Are Action Sport Cameras Accurate Enough for 3D Motion Analysis? A Comparison With a Commercial Motion Capture System. Journal of Applied Biomechanics. 2019 Feb 1;35(1).
[14] Medical devices directive [Internet]. Medical devices directive - Nexus 2.9 Documentation - Vicon Documentation. [cited 2020 Nov 30]. Available from: https://docs.vicon.com/display/Nexus29/Medicaldevicesdirective
[15] Vicon Motion Systems Limited. Vicon Vantage Reference [Internet]. USA: VESA; April 2016 [cited 2020 Nov 30]. 216 p. Available from https://www.prophysics.ch/wp-content/uploads/2017/06/ViconVantageReference.pdf
[16] Lamine H, Bennour S, Laribi M, Romdhane L, Zaghloul S. Evaluation of calibrated kinect gait kinematics using a vicon motion capture system. Comput. Methods Biomech. Biomed. Eng. 2017 Oct 30;20:111-2.
[17] Mihradi S, Dirgantara T, Mahyuddin AI. Tracking of Markers for 2D and 3D gait analysis using home video cameras. International Journal of E-Health and Medical Communications (IJEHMC). 2013 Jul 1;4(3):36-52.
[18] Mihradi S, Dirgantara T, Mahyuddin AI. Development of an optical motion-capture system for 3D gait analysis. in 2011 2nd International Conference on Instrumentation, Communications, Information Technology, and Biomedical Engineering 2011 Nov 8 (pp. 391-394). IEEE.
[19] Zhang Z. A flexible new technique for camera calibration. IEEE Transactions on pattern analysis and machine intelligence. 2000 Nov;22(11):1330-4.
[20] Whittle MW. Gait analysis: an introduction. Butterworth-Heinemann; 2014 May 12.

IOP Conf. Series: Materials Science and Engineering 1109 (2021) 012025 doi:10.1088/1757-899X/1109/1/012025

Evaluation of the integration of recycling unit in an iron manufacturing plant

E Arriola[1,2], I H V Gue[1,3], A T Ubando[1,2,4], R R Tan[4,5]

[1] Mechanical Engineering Department, De La Salle University, 2401 Taft Avenue, 1004 Manila, Philippines
[2] Thermomechanical Analysis Laboratory, De La Salle University – Manila: Laguna Campus, LTI Spine Road, Laguna Blvd, Biñan, Laguna 4024, Philippines
[3] Mechanical Engineering Department, FEU – Institute of Technology, 839 P. Paredes Street, 1015 Manila, Philippines
[4] Center for Engineering and Sustainable Development Research, De La Salle University, 2401 Taft Avenue, 1004 Manila, Philippines
[5] Chemical Engineering Department, De La Salle University, 2401 Taft Avenue, 1004 Manila, Philippines

aristotle.ubando@dlsu.edu.ph

Abstract. Manufacturing of new steel from virgin iron ore is not only an energy intensive process but also generates a considerable amount of annual carbon footprint with respect to the world's total carbon generation. It is actually considered as the highest carbon footprint generated among the heavy industries. Considering these characteristics of steel making, recycling technologies from scrap metal had been an area of interest in the circular economy framework. Production of new steel from scrap metals only requires as little as 10% of the energy used in the production of steel from virgin materials. This study aims to integrate a scrap metal recycling unit in an existing iron manufacturing plant using an optimization model that evaluates the carbon footprint. LINGO is used for the optimization model. The results have shown that carbon emission was reduced by 10.35% and significantly reduced one of the raw materials, coal, by 35.44%. The total power consumption was also reduced by 15.42%.

1. Introduction

The steelmaking industry is notably connected in the development of the global economy. As majority of the world's economic sector requires steel, economic growth is reliant on the industry's production. The industry, however, is known to be resource-intensive and energy intensive. This is in addition to the logistical challenges between resource-rich countries and steel producing countries as sea trade becomes a necessary step [7]. The report of the International Energy Agency determined that the current steelmaking production results to 6.7% of the world's annual carbon footprint, ranking the industry as the highest among 'heavy industries' [9]. The same report also ranked the industry as the largest consumer of coal as an energy source. Considering its demand will increase three folds in the next 30 years, sustainable production practices must be implemented to ensure a sustainable future.

Although steel is a recyclable material, current production and consumption patterns have yet fully optimized this characteristic. Secondary steel production has significant advantages over primary steel production. Mousa et al. [15] showed that secondary steel requires approximately half of the required energy than that of primary steel. Although there is a larger energy requirement, primary steelmaking still accounts for three quarters of the world's production. Shifting current practices towards cleaner production is necessary. Ma et al. [12] indicated that the steelmaking industry will need to shift towards a new sustainability paradigm, Circular Economy (CE), to maintain its development and improve its environmental impact.

The concept of CE promotes resource efficiency, waste minimization, and prolonged product life. Its application is not constrained to a specific scale as it is applicable to companies, industrial parks, and cities [10]. CE is a solution for synchronizing business models and sustainable development goals [6]. Towards the Sustainable Development Goals set by the United Nations, CE possesses the capacity in resolving three of its goals, namely, SDG-7 (Affordable and Clean Energy), SDG-8 (Decent Work and Economic Growth), and SDG-13 (Climate Action). Research works estimate that a CE transition for most economic systems will improve annual GDP by 2%, increase employment rate by 1.6% and reduce carbon emission by 24.6% [1].

Implementing CE for the manufacturing sector requires coordination from the political institutions and industry corporations [11]. A necessary step for implementation is the adoption of performance metrics to assess system performance against circularity. Notably, existing research works have designed a variety of indicators; of which, each one has its own specific applications [17]. Additionally, target performances can be set in realizing circularity [14]. Implementing these concepts among steelmaking industries will enable movement towards its sustainability.

As the use of indicators will result in an intricate reevaluation, the design process needs to be systematic. Mathematical programming possesses a systematic approach for industrial system design. Existing works have shown its potential in the design of bioenergy systems with negative carbon emission, as shown in Ubando et al. [18], and of trigeneration plants with product price change, as shown in Gue et al. [8].

This method is applied for steelmaking plants as well. For carbon reduction, existing works have integrated the plant with a polygeneration system, as shown in Ghanbari et al. [4], and with carbon capture and utilization, as shown in Ghanbari et al. [5]. Other works have applied the concept in biomass systems. Ubando & Chen [19,20] designed it with a biomass-based polygeneration system. Of which, Ubando et al. [19,20] extended its application with the consideration of negative carbon footprint. Their model reached an optimized result of 2.7 million tons of carbon dioxide reduction with USD 6.91 billion profit per year.

Given that the Ellen MacArthur Foundation [2] distinguished CE frameworks as biological and technical cycles, a technical regenerative cycle is needed to sustain the resource requirement of the steelmaking industry. Ohno et al. [16] designed a linear programming model to determine optimum flow of scrap metals from end-of-life vehicles in the Japan economy. Meanwhile, Majji & Kesavarao [13] used a chance-constrained approach to determine optimal scrap metal mixing. These studies, however, did not consider indicators and targets as parameters in the design.

CE transition of steelmaking entails the inclusion of indicators and targets in the design process. Consideration of these parameters to existing system designs is challenging and may entail non-optimal choices. Therefore, this work presents a mathematical programming approach in integrating circularity targets in the design of a steelmaking plant. The developed model can help future research works in the design of a circular steelmaking plant. Additionally, findings of this work illustrate the potential of adopting circularity in the industrial system.

Steel and iron industries provide the basic material for industrialization and economic development. However, this is an energy-intensive industry, thus it needs a large number of resources and it has a large amount of carbon footprint. In line with this, the study aims to evaluate using fuzzy optimization the integration of a recycling unit to the iron manufacturing plant and how it will minimize its carbon footprint as well as its energy consumption due to the production of new steel through recycling requires only 10 times less energy compared to steel produces from virgin iron ore [21].

2. Methodology

A fuzzy linear programming model was developed to evaluate the effect of the attachment of a recycling plant to the iron manufacturing plant. The model was basically composed of the following objective function:

$$Maximize\ \lambda_o + \sum_{q=1}^{q} \frac{\lambda_q}{M} \qquad \forall q = 1, 2, 3 \tag{1}$$

s.t.

$$\lambda_o \leq \lambda_q \qquad \forall q = 1, 2, 3 \tag{2}$$

Where.

λ_o = overall membership level
λ_1 = satisfaction level for the product demand
λ_2 = satisfaction level for the annualized profit goal
λ_3 = satisfaction level for the carbon footprint
M = Large scalar number

The 2nd term of Equation 1 ensures the Pareto optimality of the solution. As per the aggregation of max-min, the λ_o should be less than or equal to the individual goals, λ_i level of fulfillment (Equation 2). Equation 3 represents linear energy and material balance of the product output y of the iron recycling plant integrated in the main iron manufacturing plant.

$$\sum_{j=1}^{j} A_{ij} x_j = y \qquad \forall j \tag{3}$$

Where.

A = process matrix of the iron manufacturing plant integrated with recycling unit
x = Process scaling vector
y = production level of the integrated system.

The A matrix includes the five-vector process of the main iron manufacturing plant and the single process vector of the added iron recycling unit. The model assumes that all process vectors in the A matrix are in steady state. The process scaling vector x simultaneously determines the capacity of the recycling plant and the optimum capacity of the main iron manufacturing plant. The product output y is modelled with the triangular fuzzy goal algorithm. The constraints and profit goals of the iron manufacturing plant are defined in Equation 4.

$$AP \geq AP^L + \lambda 2(AP^U - -AP^L), \tag{4}$$

$$AP = \sum_{k=1}^{k} EF_k \qquad \forall k = 1,2 \tag{5}$$

$$EF_1 = HP^T y, \tag{6}$$

$$EF_2 = AFCC^T x_n \qquad \forall n = 6,7,8,9,10 \tag{7}$$

Where.

AP = Annualized profit of the plant

AP^L = Lower bounding limit of AP
AP^U = Upper bounding limit of AP
EF = Economic factors
EF_1 = Annualized gross profit of the plant
EF_2 = Capital cost of the plant
P^T = Transposed price for each stream
CC^T = Transposed capita cost per process unit
H = Total plant hours in a year
AF = annualizing factor
N = specific process unit in the system

The carbon footprint goal is also considered in the model. Equation 8 describes how the model accounts for the carbon footprint of the integrated manufacturing plant. Equation 10 utilizes the linear minimization membership function to minimize the carbon footprint of the plant.

$$z \leq z^U + \lambda_3(z^L - z^U) \tag{8}$$
$$z = \sum_{r=1}^{r} Zr \quad \forall r = 1,2 \tag{9}$$
$$z_1 = b^T x \tag{10}$$
$$z_2 = c^T y \tag{11}$$

Where.

z = Plant's carbon footprint
z^L = carbon footprint lower limit
z^U = carbon footprint upper limit
z_r = different types of carbon footprint
b^T = transposed carbon emission vector per process
c^T = transposed carbon emission vector per product

And lastly, Equation 12 shows the process scaling vector x of the main iron manufacturing plant operating at an optimal capacity range. Equation 13 is created to ensure that the overall satisfaction level falls within 0 to 1.

$$x^L \leq x_n \leq x^U \quad \forall n = 1,2,3,4,5 \tag{12}$$
$$0 \leq \lambda_o \leq 1 \tag{13}$$

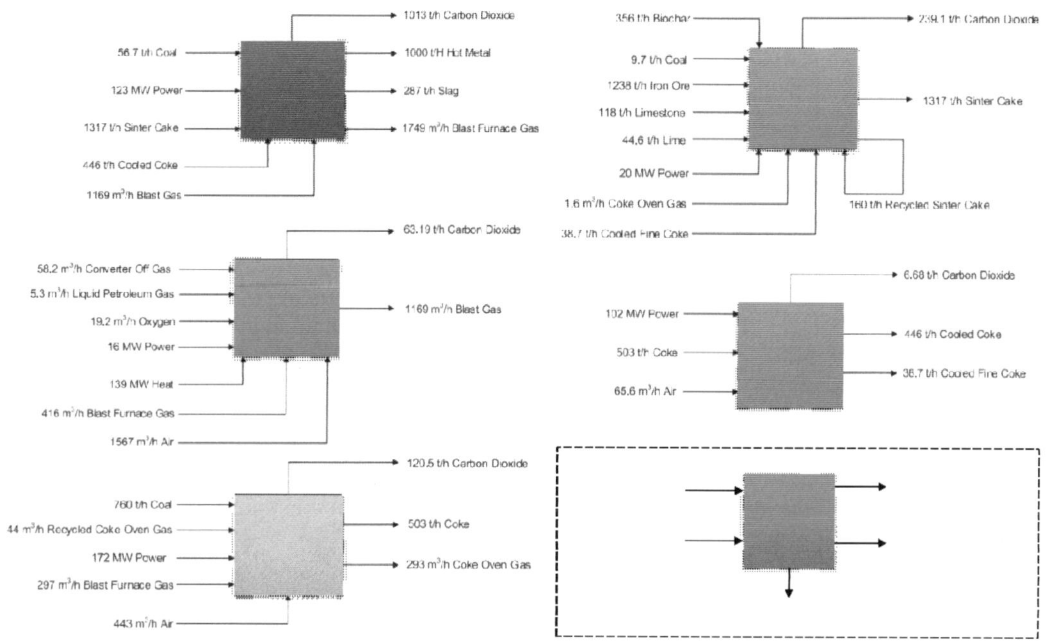

Figure 1. Material and Energy balance for the Iron plant and Recycle Unit.

The model was developed using a laptop computer powered by Intel core i7-6700HQ at 2.60 GHz with 8 GB of RAM. An educational licensed Lingo 18.0 aided with Microsoft Excel was utilized to solve the developed model. LINDO (Linear, Interactive, and Discrete Optimizer) were the software developers of Lingo which offers solvers and different programming models that have a wide variety of applications. Lingo basically aids the user to identify the best solution in an optimization problem considering well-defined constraints with the most efficient time. This program has the ability to read and write data on to an Excel spreadsheet and Microsoft access.

The materials and energy balance of the main iron manufacturing plant (**Figure 1**) was adopted from the study of Ubando et al. [22]. Majority of the prices of the product stream of the iron manufacturing plant integrated with a recycling plant were adopted from the study of Ubando et al. [22] and Ghanbari et al [5] in consideration of a conventional steelmaking process.

The optimization model considered an iron manufacturing plant which includes the 5 main processing units: blast furnace, sinter machine, hot stove, coke dry quenching, and coke oven. These process units are represented in different blocks in Figure 1 with their respective input-output material and energy balance. These assumptions allow the model to treat the operational capacity linearly. The blast furnace produces pig iron as the main product with by-products of slag and blast furnace. The blast gas was generated by the hot stove through heating and mixing the blast furnace gas with other raw materials. The coke oven on the other hand carbonized the coal through heat treatment to produce coke for the blast furnace. Before the utilization of hot coke in the blast furnace, it will be cooled first using an inert gas in the coke dry quenching. And finally, the sinter machine produces the sinter cake by mixing various raw materials with solid fuels.

The recycle unit considered to be integrated to the existing iron plant includes separation processes that will sort the scrap iron to the light and heavy residues from the scrap metal. This integration will lower the

IOP Conf. Series: Materials Science and Engineering 1109 (2021) 012025 doi:10.1088/1757-899X/1109/1/012025

carbon emission of the iron manufacturing plant as well as lowering the needed raw material requirements to obtain the optimal plant capacity.

3. Results and Discussion

The product stream generated by the optimization model is shown in Table 1. The streams that are significantly affected by the integration of the recycling unit to the iron manufacturing plant are the consumption of coal and power. While it is noteworthy that the carbon dioxide production is reduced from 1067.58 ton/hr to 962.70 after the integration of the recycling unit. The resulting carbon emission compared to the study conducted by Ubando et al. (2020) is 23% lower. Hence, an iron manufacturing plant integrated with a recycling unit has a lower carbon footprint compared to the iron manufacturing plant integrated with a biomass-based poly-generation system.

Table 1. Optimal streams for iron manufacturing plant with recycling unit and without recycling unit.

Process Unit	With Recycling Unit	Without Recycling Unit
Coal	-661.12	-945.92
Converter Off Gas	-46.56	-46.56
Liquified Petroleum Gas	-4.24	-4.24
Oxygen	-15.36	-15.36
Air	-1660.48	-1660.49
Scrap Metal	-320.00	0
Iron Ore	-660.40	-990.40
Limestone	-94.40	-94.40
Lime	-35.68	-35.68
Recycled Coke Oven Gas	-35.20	-35.20
Recycled Sinter Cake	-128.00	-128.00
Power	303.20	353.87
Heat	-111.20	-111.20
Hot Metal	800.00	800.00
Slag	229.60	229.60
Blast Furnace Gas	828.80	828.80
Sinter Cake	240.00	240.00
Coke Oven Gas	233.12	233.12
Scrap Iron	0.00	0.00
Cooled Coke	0.00	0.00
Blast Gas	0.00	0.00
Light Scrap Metal Residue	23.47	0
Heavy Scrap Metal Residue	3.20	0
Coke	0.00	0.00
Cooled Fine Coke	0.00	0.00
Carbon Dioxide	962.70	1067.58

The coal consumption was reduced (Figure 2) due to the integration of the recycling unit. The raw materials used in the production of iron was reduced and replaced with the scrap iron produced by the recycling unit. The scrap iron had lowered the requirements of the manufacturing plant in terms of raw materials (iron ore) that cascaded its effect to the lowering of heating requirement of the process hence lowering the coal and power requirement of the production. This also affects the carbon footprint generation of the whole production due to lesser fresh materials being used in the production.

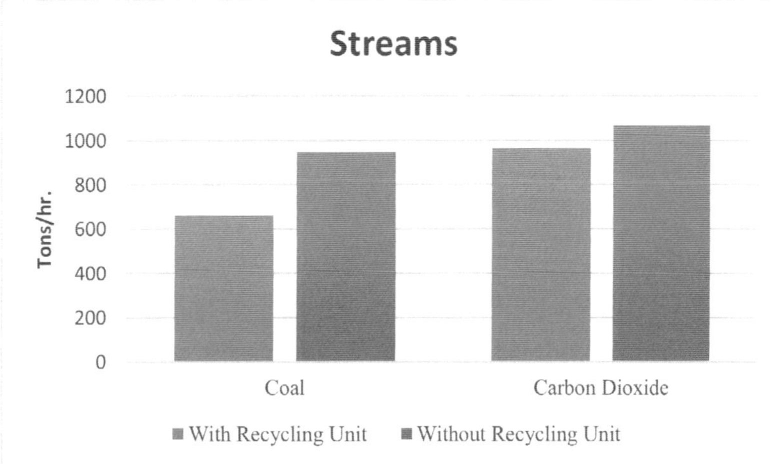

Figure 2. Coal and Carbon dioxide streams after the integration of recycling unit.

4. Conclusion

In the study, an iron manufacturing plant with an integrated recycling unit was evaluated. The generated model to optimize the iron manufacturing plant indicated that the carbon emission of the plant decreased by 10.35% when the recycling unit was introduced. It was also observed that the coal requirement as a raw material was also reduced significantly by 35.44%. The model had identified that the optimal operational capacity of the iron production plant integrated with a recycling unit is at 80% capacity. The total power consumed in the production was also decreased by 15.44% however a detailed analysis on energy consumption and can be conducted in the future studies specifically on the reduction of energy requirement on each process unit after the incorporation of the recycling unit. The cumulative energy demand for energy environment consideration can also be considered for further analysis of the model. This can also include the embodied carbon emissions especially on the consumption of coal in the iron manufacturing plant.

Acknowledgment

The authors would like to acknowledge the Philippine's Department of Science and Technology – Engineering Research and Development for Technology (DOST-ERDT) for the scholarship given to one of the authors of this study

References.

[1] Aguilar-Hernandez G A, Dias Rodrigues J F and Tukker A 2021 Macroeconomic, social and environmental impacts of a circular economy up to 2050: A meta-analysis of prospective studies *J Clean Prod* **278** 123421 doi:10.1016/j.jclepro.2020.123421.

[2] Ellen MacArthur Foundation 2013 *Towards the circular economy: economic business rationale for an accelerated transition* Ellen MacArthur Foundation, United Kingdom

[3] Geng Y, Sarkis J and Bleischwitz R 2019 How to globalize the circular economy Nature **565** doi:10.1038/d41586-019-00017-z.

[4] Ghanbari H, Saxén H and Grossmann I E 2013 Optimal design and operation of a steel plant integrated with a polygeneration system *AIChE J* **59** 3659–70 doi:10.1002/aic.14098.

[5] Ghanbari H, Helle M and Saxén H 2015 Optimization of an integrated steel plant with carbon

capturing and utilization processes *IFAC-PapersOnLine* **48** 12–7 doi:10.1016/j.ifacol.2015.10.069.
[6] Ghisellini P, Cialani C and Ulgiati S 2016 A review on circular economy: the expected transition to a balanced interplay of environmental and economic systems *J Clean Prod* **114** 11–32 doi:10.1016/j.jclepro.2015.09.007.
[7] González I H and Kamiński J 2011 The iron and steel industry: a global market perspective. *Gospodarka Surowcami Mineralnymi - Miner Resour Manag* **27**, 5-28
[8] Gue I H V, Ubando A T, Aviso K B and Tan R R 2019 Optimal design of a trigeneration plant using fuzzy linear programming with global sensitivity analysis on product price uncertainty. *Energy Procedia* **158** 2176–81 doi:10.1016/j.egypro.2019.01.617.
[9] IEA 2020 *Iron and Steel Technology Roadmap* IEA, Paris
[10] Kirchherr J, Reike D and Hekkert M 2017 Conceptualizing the circular economy: An analysis of 114 definitions *Resour Conserv Recycl* **127** 221–32 doi:10.1016/j.resconrec.2017.09.005.
[11] Lieder M and Rashid A 2016 Towards circular economy implementation: a comprehensive review in context of manufacturing industry *J Clean Prod* **115** 36–51 doi:10.1016/j.jclepro.2015.12.042.
[12] Ma S, Wen Z, Chen J and Wen Z 2014 Mode of circular economy in China's iron and steel industry: a case study in Wu'an city *J Clean Prod* 2014 **64** 505–12 doi:10.1016/j.jclepro.2013.10.008.
[13] Majji R and Kesavarao V V S 2019 Optimization of steel making with scrap mixing. *Int J Mech Eng Technol* **10**(9) 100–111.
[14] Morseletto P 2020 Targets for a circular economy. *Resour Conserv Recycl* 2020 **153** 104553 doi:10.1016/j.resconrec.2019.104553.
[15] Mousa E, Wang C, Riesbeck J and Larsson M 2016 Biomass applications in iron and steel industry: an overview of challenges and opportunities. *Renew Sustain Energy Rev* **65** 1247–66 doi:10.1016/j.rser.2016.07.061.
[16] Ohno H, Matsubae K, Nakajima K, Kondo Y, Nakamura S, Fukushima Y and Nagasaka T 2017 Optimal recycling of steel scrap and alloying elements: Input-Output based Linear Programming Method with its application to end-of-life vehicles in Japan. *Environ Sci Technol* **51**:13086–94. doi:10.1021/acs.est.7b04477.
[17] Saidani M, Yannou B, Leroy Y, Cluzel F and Kendall A 2019 A taxonomy of circular economy indicators *J Clean Prod* **207** 542–59 doi:10.1016/j.jclepro.2018.10.014.
[18] Ubando A T, Culaba A B, Aviso K B, Ng D K S and Tan R R 2014 Fuzzy mixed-integer linear programming model for optimizing a multi-functional bioenergy system with biochar production for negative carbon emissions *Clean Technol Environ Policy* **16** 1537–49 doi:10.1007/s10098-014-0721-z.
[19] Ubando A T and Chen W-H 2019 Multi-objective design of a biomass-based polygeneration for iron and steel manufacturing *IOP Conf Ser Earth Environ Sci* **268** doi:10.1088/1755-1315/268/1/012085.
[20] Ubando A T, Chen W-H, Tan R R and Naqvi SR 2019 Optimal integration of a biomass-based polygeneration system in an iron production plant for negative carbon emissions *Int J Energy Res* 1–17. doi:10.1002/er.4902.
[21] Harvey L D D 2020 Iron and steel recycling: Review, conceptual model, irreducible mining requirements, and energy implications. *Renewable and Sustainable Energy Reviews*, 110553.
[22] Ubando A T, Chen W-H, Tan R R and Naqvi S R 2020 Optimal integration of a biomass-based polygeneration system in an iron production plant for negative carbon emissions. *International Journal of Energy Research, 44*(12), 9350-9366.

Exploring simplification methods in reducing simulation time for drop test analysis

A T Ubando[1,2], J Gonzaga[1,2], N R E Lim[1,2], J P Mercado[1,2], E Arriola[1,2], R L Moran[1,2], A Conversion[1,2], D Belarmino[1,2]

[1] Mechanical Engineering Department, De La Salle University, 2401 Taft Avenue, 1004 Manila, Philippine
[2] Thermomechanical Analysis Laboratory, De La Salle University – Manila: Laguna Campus, LTI Spine Road, Laguna Blvd, Biñan, Laguna, Philippines

aristotle.ubando@dlsu.edu.ph

Abstract. – Automotive electronics can produce an unprecedented level of passenger comfort and optimized performance of engine and components. However, the increasing intricacy of these electronic devices makes them more susceptible to damage due to thermal loads, mechanical drops, and engine vibrations. While physical handling of automotive components is comparatively rare, the weight and complexity of specific components such as the ECU, power module, and sensor modules, makes any drop functionally fatal. Simulation approach is considered to be essential in solving these problems by minimizing costs and effort. This paper attempts to solve the issue of mechanical drops using an explicit dynamics finite element analysis solver. Moreover, simplification methods are applied to further minimize solution costs.

1. Introduction

Automotive electronics have become increasingly useful in optimizing the performance of a vehicle's drive and engine power, while providing unprecedented passenger comfort. Keeping up with a bench-marked quality, or eventually surpassing it is necessary for these companies and manufacturers to remain competitive in their industry. Thus, semiconductor devices and electronics are key in their success. Electrical components need to work after exposure to harsh environmental conditions of the automotive housing, and mechanical conditions or issues due to power supply.

Portable electronics are a subject of most drop tests due to their decreasing size [1], while automotive electronic components will rarely be ever handled in their life cycle. Save for installation or replacement, mishandling and dropping these devices should be rare. However, certain electronics such as the Electronic Control Units (ECUs) and power modules are heavy components with complex intricacy. Furthermore, components such as camera or proximity sensors are attached together with heavier components on the PCB structure, making drops more severe and functionality fatal to the electronic systems.

The semiconductor package must protect a device from external factors that could degrade the performance of a semiconductor chip. Increased complexity and functionality coupled with the sleekness of design has translated into more interconnections, increasing intricacy and vulnerability [2]. Manufacturers of integrated circuit (IC) components need to guarantee the reliability of their products under mechanical drop or shocks, utilizing a standard drop test. The standard method of drop testing for

IOP Conf. Series: Materials Science and Engineering 1109 (2021) 012026 doi:10.1088/1757-899X/1109/1/012026

semiconductor devices since 2005 is the JESD22-B111, formulated by the Joint Electron Device Engineering Council (JEDEC) [3]. The procedure and test rig are a board level drop test meant to evaluate and compare drop performance of surface mount electronic components for handheld electronic product applications [4]. Acceleration devices quantify the force and speed of a drop. This test rig best assesses designs sensitive to excessive flexure of a circuit board. The JESD22-B111 is a standardized procedure that can reproduce results of similar and normally observe situations. The test rig is a reportedly excellent basis for tests of automotive electronic control units [5].

The packaging technology being tested, the Ball Grid Array (BGA), is very popular in mobile devices for its dense interconnectivity and high lead count [6]. It is known to be self-centering and possesses an ease of installment, due to having no leads, while fulfilling high performance and high thermoelectrical requirements. The BGA package has a small footprint yet high transistor count [7]. However, this technology is known to experience solder joint cracking between the interconnections and PCB material due to hard drops [1].

Shirangi et al. [8] formulates a novel drop test method for automotive Electronic Control Units (ECU), specifically to test high-shock conditions and expected intense vibration in automobiles. Their test rig and procedure allow the testers to vary the weight of the semiconductor device while monitoring speed of a test subject using an on-board accelerometer. By confirming their findings via FEA, they find the data reliable. Such research also opens the possibility of applying structural Finite Element Analysis (FEA) simulations to replicate the standard JESD22-B111.

Research by Balakrishnan et al. [9] presents a comparison of a set of different explicit and implicit impact simulations using the ANSYS LS-DYNA solver, in simulating a standard, 1-meter drop of an ECU. The research finds explicit/implicit transient methods to best simulate the drop as a dynamic system; however, at high computational cost and time. An explicit method is able to capture strain and damage due to impact force on the drop at relatively lower time and computational costs. It is noted by the research of Balakrishnan et al. that FEA analysis needs to strike a balance between convergence time, computational costs, and accuracy of results. Chung and Kwak [6] present a comparative study between an FEA model using a detailed geometry of a BGA package and a simplified version of the geometry using beam elements. Their research used a drop test rig following the JESD22-B111a standards to confirm their results. The experiments were able to make a fair trade between accuracy and computational time and costs, creating accurate results at a shortened time frame.

This study aims to explore additional methods that can significantly reduce computational time while maintaining accuracy for an explicit dynamics analysis of a BGA drop test.

2. Materials and Methods

The material assignment, properties, and geometry used in this study are detailed in this section. Material properties for the analysis were derived from papers that have utilized BGA in FEA analysis [10,11] and were summarized in Table 1.

Table 1. Material properties for the drop test simulation analysis

Component	ρ (kg·m^{-3})	E (MPa)	ν
Die Chip	2300	170000	0.28
EMC	1900	17000	0.25
Substrate	3950	11600	0.42
Lead Free Solder	7360	35440	0.36

PCB	1850	17200	0.42
Structural Steel	7850	200000	0.30

The geometry construction was based on a generic BGA model [7] using SolidWorks 2019. Explicit Dynamics analysis system from ANSYS Workbench 2020 R1 was utilized for the drop test analysis via the research license of De La Salle University with supervision from the Thermomechanical Analysis Laboratory. The simulation was done using a workstation with Intel Xeon Gold with 20 cores and 128 GB RAM.

To alleviate the load in the analysis, a quarter model is employed as shown in **Figure 1.**, this is made possible due to the symmetric nature of the package. This is an established practice to improve simulation time without compromising its accuracy [12].

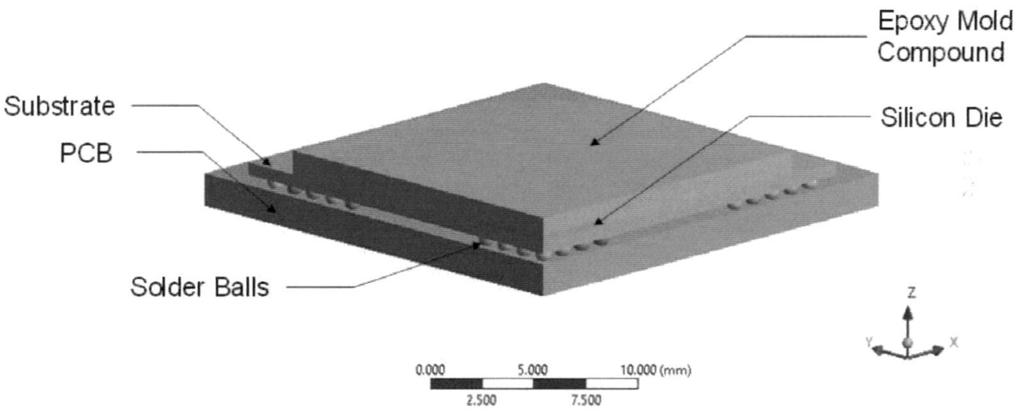

Figure 1. Quarter geometry used for Ball Grid Array assembly

2.1. Finite Element Analysis and explicit dynamics analysis

Finite Element Analysis (FEA) has been increasingly applied in manufacturing and production of several industries [13]. It is widely used as an alternative to costly design evaluation and prototype testing. In FEA, several factors and parameters can be assessed under defined boundary conditions. For the analysis in this study, a drop test is simulated which is a standard test conducted to ensure that the product condition is maintained from manufacturing to the end consumers. In such advanced analysis, simulation software such as ANSYS explicit dynamics can provide adequate tools necessary to arrive at accurate results with regards to capturing the physical phenomenon that the model undergoes through. The explicit dynamics analysis system is designed to predict large material deformations, and stresses and failure in the interactions between multiple bodies due to high impact loading.

2.2. Drop test analysis

In this study, the drop test is conducted by dropping the BGA assembly from a 1 m height to a structural steel platform. However, in order to simplify the simulation and improve the calculation time a variable simulated height is used that would reduce the end time analysis required, hence reducing solution time as can be seen in **Figure 2.**

In order to properly simulate the drop test conditions of a 1 m drop height, initial conditions were determined to be applied to the model. Kinematic free fall equations were employed to determine the instantaneous velocity of the BGA at a defined distance towards the structural steel platform assuming it was initially dropped at a height of 1 m with 0 initial velocity.

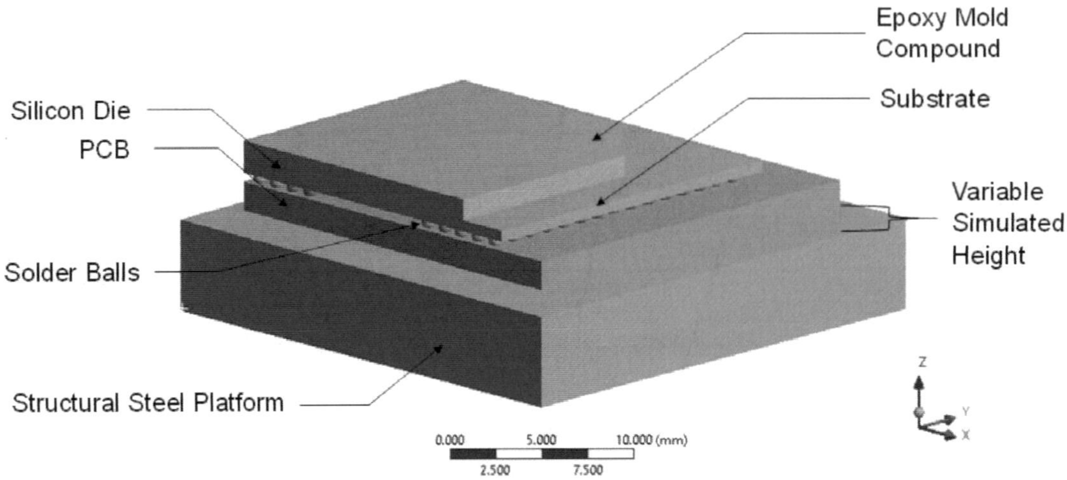

Figure 2. Drop test setup

$$h = \left(\frac{1}{2}\right) g t^2 \quad (1)$$

$$v = v_o + gt \quad (2)$$

The height difference can be determined by subtracting the variable simulated height from the original 1 m drop height. Placing it in (1) will yield the time it takes for the object to reach the desired variable height. Substituting this time in (2) will give the resulting instantaneous velocity or the initial velocity input needed for the analysis, assuming v_o to be zero.

Another necessary input for the drop test analysis is the end time step. This defines the time frame where the analysis is being evaluated. It is important to determine the optimal end time that would yield the desired result from the drop test without overestimating it as this also directly affects the solution time. In this study, the end time is determined by subtracting the time determined in (1) from the time it takes for the model to hit the ground assuming it is dropped at a 1 m height, which is ~0.4516 s. Standard earth gravity was applied in the overall BGA assembly to simulate the dropping motion aside from the set initial velocity. Moreover, a fixed support is added in the bottom surface of the structural platform to counteract the dropping of the semiconductor package.

2.3. *Meshing and Shared topology*

The general meshing algorithm is applied in the model. Hexagonal elements of second degree are prioritized in order to minimize the number of elements generated while a more refined meshing is applied towards the solder balls as this is the area of concern for this analysis. The meshing configuration is illustrated in **Figure 3.**

The factor influence of shared topology is also explored in this study. By applying shared topology, components that are meant to have a bonded condition will be sharing nodes; hence the total number of nodes and elements should be reduced.

2.4. *Design of experiment*

A full factorial design will be applied in this study considering two independent factors (variable simulated height and shared topology) in order to identify how each factor affects the dependent factors (simulation time and accuracy). Three variable simulated heights were identified, primarily 0.5mm, 1mm, and 5mm. An "N" in the shared topology column indicates that no shared topology feature was applied while a "Y" indicates that the run has applied shared topology. A summary of the identified factors is shown in **Table 2**.

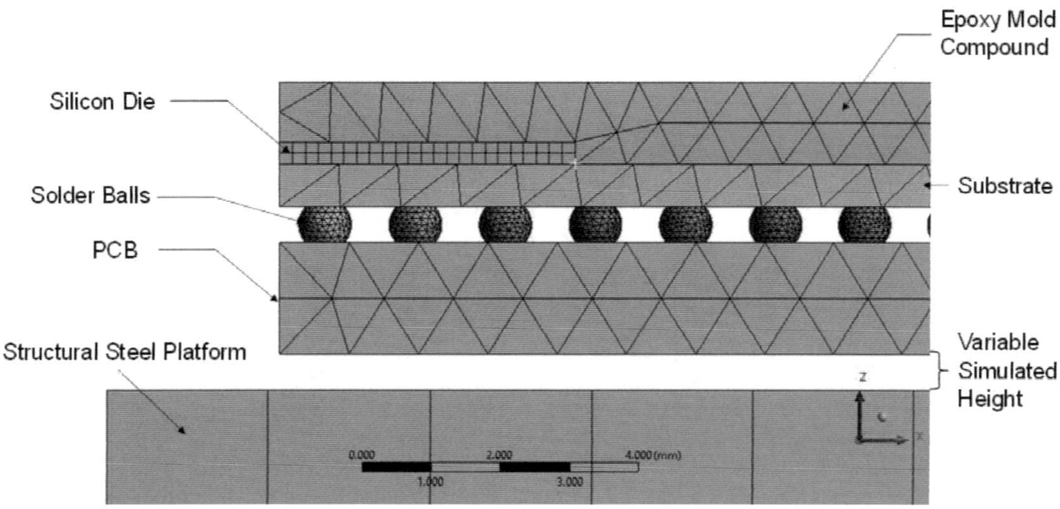

Figure 3. Meshing of model

Table 2. Full factorial design of experiment

Variable Simulated Height (mm)	Shared Topology (Y/N)	Simulation Time (hrs)	Impact Stress (MPa)
0.5	Y	-	-
1.0	Y	-	-
5.0	Y	-	-
0.5	N	-	-
1.0	N	-	-
5.0	N	-	-

3. Results and discussion

This section presents the findings of the simulation runs.

3.1. *Meshing and Shared topology*

The following table summarizes the number of nodes and elements generated with and without the application of shared topology.

Table 3. Node and element reduction after shared topology

	Without Shared Topology	With Shared Topology	Percent Decrease (%)
Number of Nodes	90854	65016	28.44
Number of Elements	396086	226448	42.83

By applying shared topology in the model setup, the total number of nodes and elements have been significantly reduced. In principle of FEA, the lesser nodes/elements would require less computing power and yield faster computational time. Nonetheless, it is important to also consider maintaining the accuracy of the solution even after reducing the number of nodes/elements in the model.

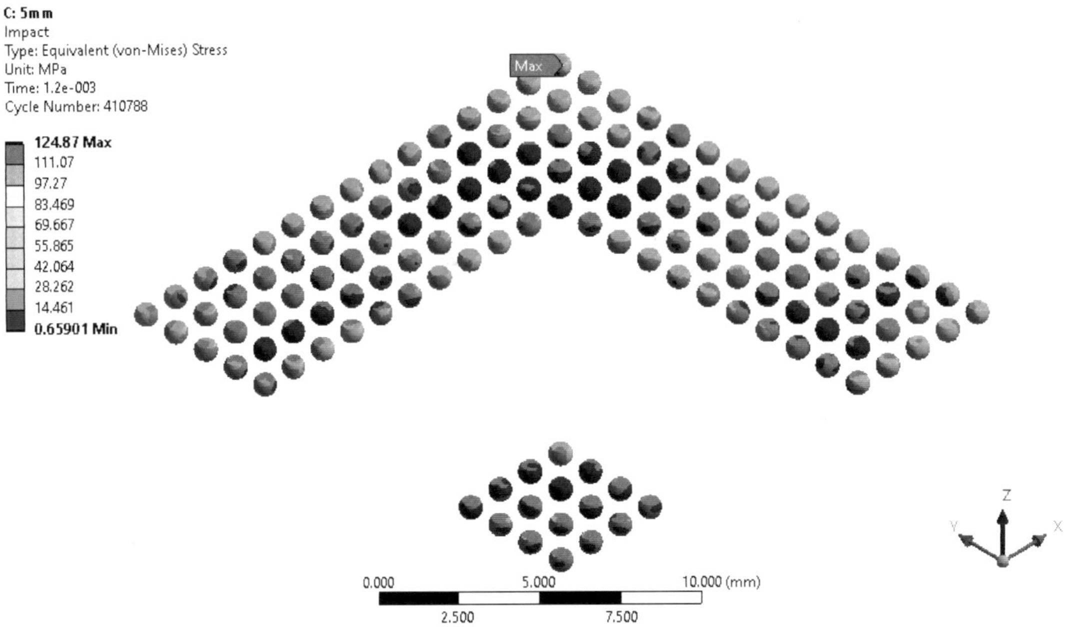

Figure 4. Stress profile of solder balls for 5mm variable simulated height and shared topology

3.2. *Full factorial design of experiment*

The summary of resulting simulation time needed, and impact stress calculated in each setup is shown in **Table 4**.

Table 4. Full factorial design of experiment

Variable Simulated Height (mm)	Shared Topology (Y/N)	Simulation Time (hrs)	Impact Stress (MPa)
0.5	Y	1.7	123.52
1.0	Y	2.8	125.54
5.0	Y	9.3	124.87
0.5	N	3.2	123.71
1.0	N	4.1	125.48
5.0	N	12.9	124.69

A multiple regression analysis was initially made in order to determine the parameter estimate or influence of the two factors: variable simulated height and shared topology on the simulation time. The overall model fit has an R square of 0.983 and a p-value of 0.002 indicating that there is at least one factor that influences the simulation time. The calculated p-value for the variable simulated height is 0.001 while the p-value for the shared topology factor is 0.038, both of which are less than the alpha 0.05 implying that both factors affect the simulation time.

Another regression analysis was made with regards to the impact stress calculated in each run. However, the overall p-value of the model fit is 0.922 which is greater than 0.05. This suggests that there is no significant difference in varying the factors in terms of the calculated impact stress.

4. Summary and Conclusion

In this study, the main objective was to determine methods that can further simplify FEA, specifically explicit dynamic drop test analysis. A standard drop test was simulated for a BGA assembly by dropping the package at a 1 m height from a structural steel platform. For this analysis, two factors were explored in minimizing the solution time while maintaining the accuracy of the solution. A variable simulated height is introduced where the model is made close to the structural steel platform and by applying initial conditions such as initial velocity, it would simulate a 1 m free fall drop. Next, model simplification is also done by applying shared topology. This method connects the nodes of shared surfaces; thereby reducing the total number of nodes and elements.

Significant reduction in the nodes and elements was observed after applying shared topology. A full factorial design of experiment was generated in order to determine the influence of the variable simulated height and shared topology towards simulation time and accuracy of solution. After conducting the six runs, statistical tests were made and it was found out that both factors affect the simulation time with both p-values being greater than 0.05. At the same time, both factors do not significantly affect the accuracy of results as the calculated p-value for the overall fit model is 0.92, which is greater than 0.05. Hence, it would be preferable to have a setup with the least simulation time since the results are not significantly affected. In conclusion, simulating a variable simulated height instead of actual 1 m drop and applying shared topology can significantly reduce the simulation time without affecting the accuracy of the analysis. Further analysis that employs drop tests can apply these simplification methods that reduce simulation time while maintaining solution accuracy.

Acknowledgments

The authors wish to acknowledge everyone who made this possible, specifically DOST-PCIEERD for their continued support towards the improvement of R&D in the country, the Thermomechanical Analysis Laboratory with its support from project grant under DOST CRADLE: 8897, for the guidance in the conduct of this study. Moreover, the authors would also want to extend their gratitude towards the DOST-ERDT consortium for funding their graduate studies.

References

[1] Birzer C, Rakow B, Steiner R and Walter J 2005 Drop test reliability improvement of lead-free fine pitch BGA using different solder bali composition *Electronics Packaging Technology Conf.*

[2] Lim N, Arriola E, Moran R, Mercado J P, Dimagiba R, Gonzaga J and Ubando A 2019 Material properties sensitivity analysis based on thermomechanical analysis on warpage for ball grid array semiconductor packaging *2019 IEEE 11th Int. Conf. on Humanoid, Nanotechnology, Information Technology, Communication and Control, Environment, and Management, Ilocos*

[3] Joint Electron Device Engineering Council (JEDEC) 2016 Board level drop test method of components for handheld electronic products *JEDEC*

[4] Krist M, Pan J, Farris A and Vickers N 2009 Drop impact dynamic response study of JEDEC JESD22-B111 test board (Cal Poly State University, San Luis Obispo, CA)

[5] Jiang D-S, Tzeng Y L, Wang Y-P and Hsiao C 2006 Board level drop test and simulation of CSP for handheld application *2006 7th Int. Conf. on Electronics Packaging Technology, Shanghai*

[6] Chung S and Kwak J B 2020 Comparative study on reliability and advanced numerical analysis of BGA subjected to product-level drop impact test for portable electronics *Electronics* **9**(1515)

[7] Intel, "2000 Packaging Databook 14-1," in Ball Grid Array (BGA) Packaging, Intel, 2000.
[8] Shirangi M, Tsebo S G, Wang Z, Unnikrishnan R and Heinrich T 2014 A novel drop test methodology for highly stressed interconnects in automotive electronic control unit *2014 Electronic Components & Technology Conference, Orlando*
[9] Balakrishnan K, Sharma A and Ali R 2017 Comparison of explicit and implicit FEA methods and its effectiveness for drop test of ECU *Procedia Engineering* (173) pp 424-431
[10] Lim N R E G, Ubando A T, Gonzaga J A and Dimagiba R R N 2020 Finite element analysis on the factors affecting die crack propagation in BGA under thermo-mechanical loading. Engineering Failure Analysis *Engineering Failure Analysis*, 104717. doi:10.1016/j.engfailanal.2020.104717
[11] Liu J J, Berg H, Wen Y, Mulgaonker S, Bowlby R and Mawer A 1995 Plastic ball grid array (PBGA) overview *Materials Chemistry and Physics* 40(4) 236–244 [Online]. Available: https://doi.org/10.1016/0254-0584(95)80004-2
[12] Zhang X, Wang Y, Im J and Ho P S 2012 Chip–package interaction and reliability improvement by structure optimization for ultralow-k interconnects in flip-chip packages *IEEE Transactions on Device and Materials Reliability* **12**(2) pp 462-469
[13] Kostandyan E E and Sørensen J D 2011 Reliability assessment of solder joints in power electronic modules by crack damage model for wind turbine applications *Energies* **4**(12)

IOP Conf. Series: Materials Science and Engineering 1109 (2021) 012027 doi:10.1088/1757-899X/1109/1/012027

Finite element analysis of frictional coefficients in press fit pins

J Gonzaga[1,2], A T Ubando[1,2], E Arriola[1,2], R L Moran[1,2], N R E Lim[1,2], J P Mercado[1,2], A Conversion[1,2], D Belarmino[1,2]

[1] Mechanical Engineering Department, De La Salle University, 2401 Taft Avenue, 1004 Manila, Philippines
[2] Thermomechanical Analysis Laboratory, De La Salle University – Manila: Laguna Campus, LTI Spine Road, Laguna Blvd, Biñan, Laguna 4024, Philippines

aristotle.ubando@dlsu.edu.ph

Abstract. The automotive industry places electronic components in vastly different conditions compared to that for residential purposes. Thus, there are certain components that have found an essential use in this industry. One such component is the press fit pin. The press fit pin is seen as an alternative to soldering because it is cheaper to implement and more reliable at times. All technologies however, come with disadvantages. One of the common issues encountered by press fit pins are the excessive stresses on both the pin and the hole during insertion which tend to cause deformation or cracks in the substrate material. Because of this, the main topics of research on press fit pins are most commonly on the insertion forces and retention forces present. These forces occur on the interfaces between the pin and the hole where it is inserted. Therefore, there have been several studies on the friction between the two components. The studies however, only explored the frictional coefficient of their present system which reduces its applicability on other designs. This work seeks to present an in-depth finite element analysis on the effects of varying frictional coefficients would have on the equivalent stress and retention forces present on the press fit pins.

1. Introduction

Electronics are becoming more frequently used in automobiles and other industrial machines. Most components and features of the modern vehicles are controlled by electronic components. Automobiles are mainly used for transportation wherein they undergo many different conditions such as violent weather and rough roads. Because of these, the electronic components within them are also subjected to the same conditions. Components within the engine compartment tended to experience temperatures of 125 °C to -40°C in the winter [1]. These are greatly differed from the normal operating range of conventional household appliances. There are also global movements, such as the Restriction of Hazardous Substances Directive of the European Union, aimed at removing hazardous materials from electronic devices. Due to these, the design of electronic devices has been constantly adapting to the various conditions that they will be subject to. The elimination of lead, which was one of the most used elements in solder, became a priority in the electronics industry. This led to the increase in interest in alternatives to solder [2]. One of the measures to adjust to the conditions is the development of the press fit pin.

Press fit connections are also deemed as the successor of soldering as they also have the added benefit of making the manufacturing processes more reliable. This is because they remove the need for the high temperatures involved in the soldering process. The press fit technology creates connections between components electrically and mechanically through a pin which is inserted through a hole where the elastic deformation between the components holds it in place [3]. There are two main types of press fit pins, the active pin, and the compliant pin. Active pins are usually designed as rigid connectors usually with straight edges to facilitate the insertion. Compliant pins on the other hand, are made to deform when inserted into the hole. The most common pin design is the Eye of Needle (EON) as shown in Figure 1.

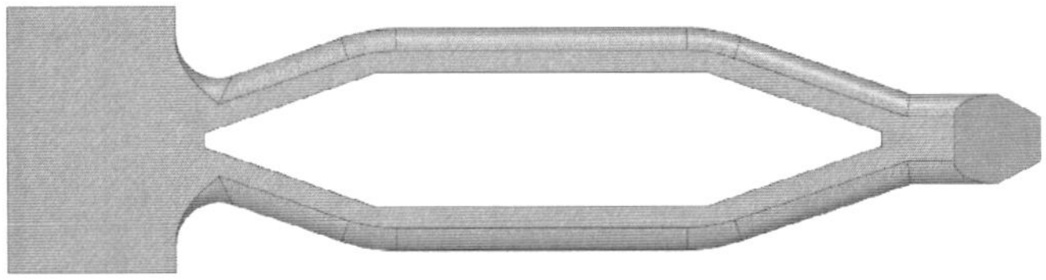

Figure 1. Illustration of an Eye of Needle press fit connector inserted into a hole.

Though this technology could overcome the problems of soldering, excessive deformation sometimes occurs in the assembly which causes failures [4]. To minimize this, there have been numerous studies on the different parameters of the press fit connections such as insertion force, retention force, pull out force, contact force. Majority of the studies on press fit pins are based on physical experiments though several studies have also involved finite element analysis [4-7] to gain a more in-depth view of the system.

The use of press fit connectors in the automotive industry subjects them to high temperatures, vibration, and corrosion. The various aspects of the press fit pin's performance such as its reliability [7], [8], degradation [3], thermal cycling performance [3-5], frictional performance [2,6,9,10], creep [5], and whisker formation[11-13].

Of the studies on the frictional performance of the press fit pin connectors only one used FEA [6], and they concentrated on determining the actual frictional coefficient of the system. Though it is important to understand the frictional coefficient of exiting systems, this does not give information that could be adapted to other press fit pin designs. The current work seeks to present the effects varying frictional coefficients of press fit pins on the retention force and equivalent stresses for application in future designs.

2. Methods

This study used Finite Element Analysis (FEA) to investigate the effects of varying Coefficients of Friction (COF) on the retention force of EON press fit connectors. This current study chose to use the EON design because it is the most used type of press fit connector. This is also to enable the findings from this study to be more applicable to most of the press fit pins. The FEA software used in this study was ANSYS 2020 R1. The simulations were run on an HP Z4 workstation with a single Intel Xeon Gold 6230 CPU and 128GB of RAM.

FEA is a numerical procedure that divides objects into elements connected by nodes. The elements are commonly in the shape of triangles but may also be composed of other shapes. The number of elements in a simulation varies depending on the size of the object to be analyzed and the resolution needed. These elements and nodes are interconnected in what is called the mesh of the model. A solver would then compute for the values of the parameters of interest of each node of the element. When all the information of the values on each node are combined, they would produce the complete picture of the effects that forces, temperature, gravity, or other conditions have on the object or system. Because of this, FEA is especially suited to investigate the forces and deformation on mechanical systems such as press fit connectors.

The press fit pin is to deform when inserted. This behavior could be defined by the following equation:

$$\delta = \frac{PL}{AE} \quad (1)$$

Were δ is the deformation, P is the force acting on the pin, L is the original length, A is the area, and E is for the elastic coefficient.

Another parameter to be analyzed are the stresses created by the contact of the pin to the walls of the hole. This could generally be described by:

$$\sigma = \frac{F_s}{A} \quad (2)$$

Where σ is the stress produced and F_s the force acting on the body.

One of the main parameters to be analyzed in this study is the retention force which is created due to the frictional forces acting on the pin and hole. It is given by:

$$F_f = F_n * \mu \quad (3)$$

Where F_f is the frictional force, F_n is the normal force, and μ is the COF. It was applied to each element and computer relative to its location on the model. (Comment 1.d.)

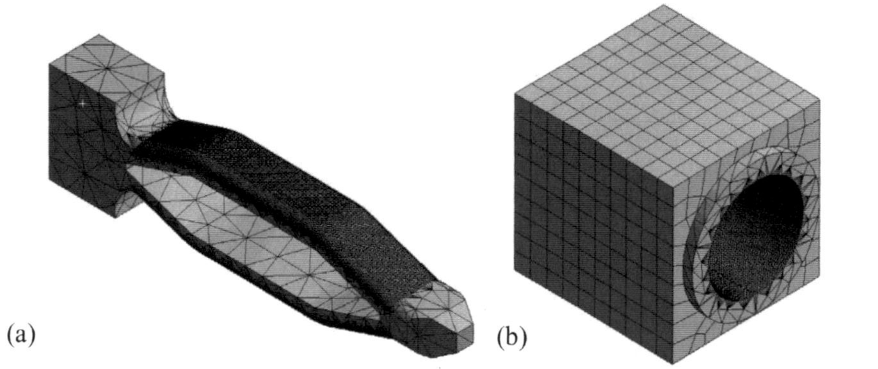

Figure 2. (a) Image of the Pin with its mesh; (b) Image of the hole with its mesh.

This study used the built in ANSYS meshing software to create the mesh of the pin and hole. The sizing of the mesh at the interfaces between the two parts is set at 0.02mm to accurately model the interference between the two which is also set at the same value. The rest of the mesh was automatically generated by the meshing software. The resulting model was comprised of 108,571 elements with

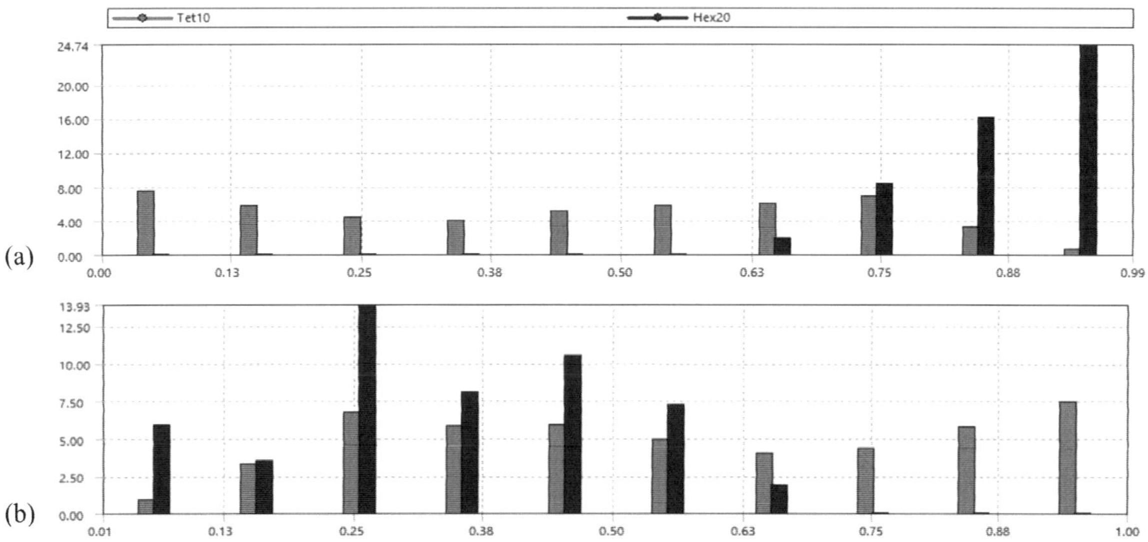

Figure 3. (a) Graph of the orthogonal quality of the mesh; (b) Graph of the skewness of the mesh.

195,720 nodes as seen in Figure 2. The mesh had mostly Tet10 and Hex20 elements. Figure 3 shows the orthogonal quality and skewness of the mesh on the x-axes and the percentage of elements in the y-axis. For the orthogonal quality, the closer its value to one, the better; for the skewness, the closer to zero the better.

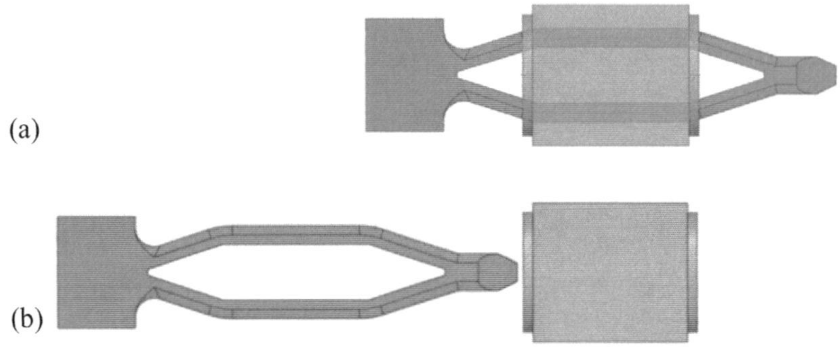

Figure 4. (a) The initial position of the system; (b) The final position of the system.

Fixed supports were added to the outer sides of the box of the hole to prevent movement. A frictionless support was also added to the faces of the hole near the sharp end of the pin to prevent it from being pulled along with the pin. From its initial position, the pin was displaced 2.75 mm until it was fully out of the box as seen in Figure 4. Frictional contacts were set between the surface of the pin and the inside of the hole. The pin had a length of 4.8mm and was based on default specifications. In addition, the width of the pin was 1.2 mm, diameter of the hole to be inserted into is 0.96 mm and the sides of the box are 1.5 mm.

The main assumption made in the study was that the pin was already inserted at the start of the analysis and was to be pulled out as seen in Figure 4. The stresses and forces within the system would then be collected then analyzed.

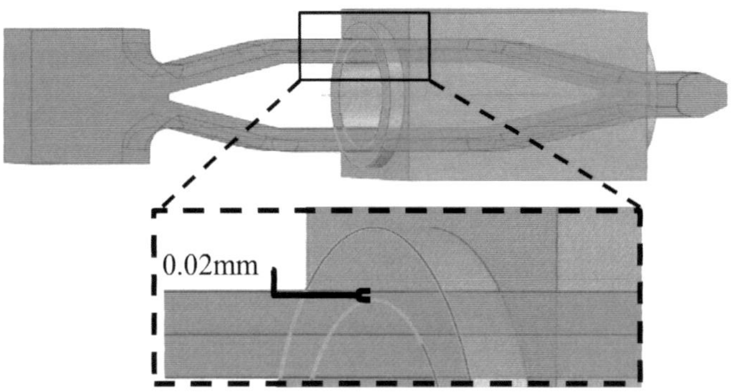

Figure 5. Enlarged view of the interference between the press fit pin and the hole.

Press fit connectors are held in place due to the elastic deformation of either the pin or the hole. To account for this in the simulation, the width of the pin was made 0.02mm wider than the radius of the cylindrical hole that it is inserted in as seen in Figure 5. The analysis type used in the simulation was static structural then the simulation was run with two steps. Since
it was assumed that the pin was already inserted at the start of the simulation, during the first step, the pin could deform to produce the normal force between it and the hole. During the following step, the pin will be pulled out to determine the retention force. The geometry, materials, and displacement of the pin and hole are held constant while the COF is varied. The values of COF used in the study ranged from 0.2 to 0.4 with intervals of 0.05 for each run for a total of five simulation runs. The values for the frictional coefficient was based on the range of results from the study of Tohmyoh et al. [6]. The maximum stresses occurring on the system would then be recorded. The materials used in the study are summarized in Table 1 and Table 2. It is to be noted that the PCB Substrate material is orthographic thus it has different properties in different directions as seen in Table 2. X, Y, and Z in Table 2 refer to the axes, which define the properties the material has along them. The same applies to XY, YZ, XZ which refer to the planes where the properties are defined. The mesh of both the pin and the hole were made finer only in the areas that are in contact with each other. This is to increase the accuracy of the simulations while minimizing the computational time.

Table 1. Material properties of Cu and CuSn5.

	Hole(Cu)	**Pin(CuSn5)**
Young's Modulus (MPa)	40200	118000
Poisson's Ratio	0.35	0.3
Bulk Modulus (MPa)	44667	98333
Shear Modulus (MPa)	14889	45385

Table 2. Material properties of the PCB substrate.

PCB substrate	XY	YZ	XZ	X	Y	Z
Young's Modulus (MPa)				17000	7450	17000
Poisson's Ratio	0.13	0.13	0.42			
Shear Modulus (MPa)	3000	3000	2400			

3. Results and Discussion

The five simulation runs approximately ran for 30-40 mins each. The results of the simulation show a relatively large stress on the first COF setting with an equivalent stress of around 48 MPa. The succeeding three COF settings had values close to each other with a trend of increasing retention force. There was another jump in the value of equivalent stress at 0.4 COF. This shows that there is a nonlinear relationship between equivalent stress and COF. The retention force, however, behaves as expected wherein it increases as the COF also increases. Figure 6 shows the COF on the x-axis and the values of equivalent stress and retention force on the y-axis. The visualization allows for analysis of the relationship of the variables used in the system. Table 3 displays the actual values obtained from the simulation. The desirable characteristics for press fit connections are low stresses and high retention forces. As can be seen from the results, the desirable range of COF lies between 0.2 and 0.4.

Table 3. The results of the simulation.

Coefficient of Friction	Equivalent Stress (MPa)	Retention Force (N)
0.20	48.07	10.67
0.25	34.27	13.16
0.30	34.71	15.58
0.35	34.80	17.98
0.40	40.50	20.35

The is normally not adjustable variable since it largely depends on the materials used as coatings on the pin and hole. Knowing the desirable COF would however, aid designers during the material selection process since the COF of materials is usually available from suppliers. Apparatus such as that described in [9] would make the determination of COF of materials faster and more efficient. The results of this study could be used in conjunction with the findings from [14] for a complete materials selection process that takes into account friction and the thermomechanical properties in the design of semiconductor devices.

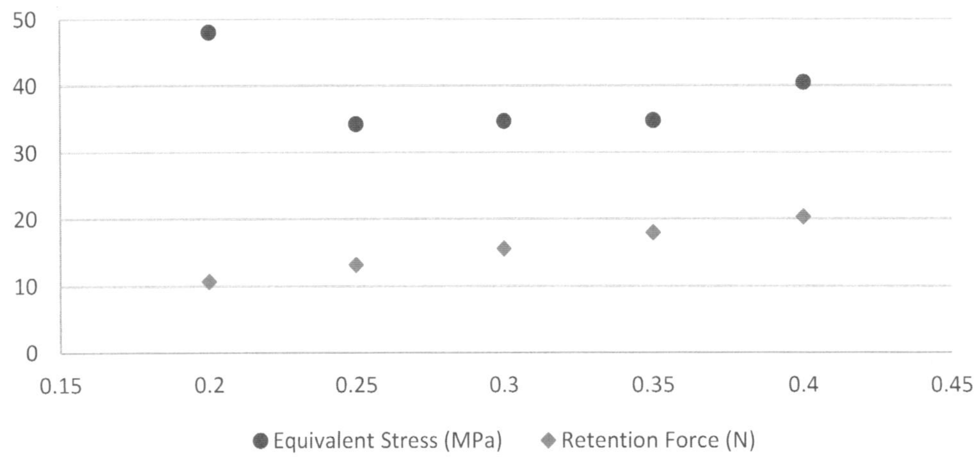

Figure 6 . Graph of the relationship between forces and their equivalent stresses

4. Conclusion

The simulation focuses on effects of the coefficient of friction (COF) to the press fit configuration. The simulation results had shown that there is an increasing trend between the COF and the retention force. Based on the data gathered, COF had significantly affected the retention force of the pin, as the COF increases, the resulting retention force also increases. It is also noteworthy that the equivalent stress does not behave linearly with the COF. The change in equivalent stress value for the COF of 0.2 to 0.25 was significantly greater than that of 0.25 to 0.3. Though the change in retention force between the COFs had less variations. A more in-depth simulation will be done to further analyse the trend of the equivalent stress with respect to the COF. Based on the trend obtained from the result, the COF value can be considered as a parameter for the design of a pin in a press fit configuration to increase or decrease the retention force based on the configuration requirements. Future studies will explore the stresses and retention at varying normal forces and COFs. The insertion force will also be investigated. Detailed experimentation and modelling are recommended by the author to further analyse the correlation of different parameters that are related in press fit. These significant findings can aid in the design guidelines of a press fit configuration.

Acknowledgements
The authors want to recognize the Engineering Research and Development for Technology program of the Department of Science and Technology for supporting the authors in the pursuit of their graduate degrees.
This work is/was primarily funded by the Philippine Council for Industry, Energy and Emerging Technology Research and Development (DOST-PCIEERD) with Project No. 08897, 2020.

References

[1] Kato M 2014 Historical overview of electronics and automobiles: Breakthroughs and innovation by electronics and electrical technology *Encycl. Automot. Eng.* (Ic) pp 1–13 doi: 10.1002/9781118354179.auto212.

[2] Nomura Y, Saitoh Y, Furukawa K, Minami Y, Horiuchi K and Hattori Y 2006 Press-fit connector for automobile ECUs *Electrical Contacts - 2006. Proceedings of the 52nd IEEE Holm Conf. on Electrical Contacts* pp 211–216, doi: 10.1109/HOLM.2006.284089.

[3] Palaniappan A, Li L and Lee T K 2018 Impact of press-fit connector pin microstructure elastic response to PCB through-hole Cu wall interface long-term contact reliability *2018 IEEE 20th Electronics Packaging Technology Conf., EPTC 2018* pp 890–893, doi: 10.1109/EPTC.2018.8654324.

[4] Riedel G J, Spindler C, Alaperä I, Schmidt R and Held M 2012 Current capability of press-fit contacts of IGBT modules *IET Conf. Publ.* **2012**(592) CP 2012 doi:10.1049/cp.2012.0338.

[5] Fellner T, Zukowski E and Wilde J 2009 Modelling of process and reliability of press-fit interconnections *2009 10th Int. Conf. Therm. Mech. Multi-Physics Simul. Exp. Microelectron. Microsystems, EuroSimE 2009* pp 1–7 doi: 10.1109/ESIME.2009.4938502.

[6] Tohmyoh H, Yamanobe K, Saka M, Utsunomiya J, Nakamura T and Nakano Y 2006 Determination of friction coefficient of a press-fit pin in thin plating *JSME Int. Journal, Ser. A Solid Mech. Mater. Eng.* **49**(3) pp 363–369, 2006, doi: 10.1299/jsmea.49.363.

[7] Wu B H *et al.*, 2018 Study on the contact performance of electronic EON connectors under axial vibration *IEEE Trans. Components, Packag. Manuf. Technol* **8**(12) pp 2090–2097 doi: 10.1109/TCPMT.2018.2800117.

[8] Tamburro P J 1975 Reliability of press-fit pins in printed wiring boards doi: 10.1109/EIC.1975.7458482.

[9] Corman N, Myers M and Copper C 2003 Friction behavior of press-fit applications: Test apparatus and methodology *Electr. Contacts, Proc. Annu. Holm Conf. Electr. Contacts* pp 38–44

[10] Lewis S, Bekele A, Tike T and Bastin V 2016 Characterization study of strain generated during press fit connector insertion in printed circuit board assembly *Proceedings of SMTA Int.* **IC** pp 25–29

[11] Crandall E *et al.* 2017 New LITESURF plating for the mitigation of whisker risks in press-fit applications *IEEE Holm Conf. on Electrical Contacts* **7**(685)

[12] He F *et al.* 2019 Press-fit whisker and its growth contributing factors-part i *China Semiconductor Technology International Conference 2019, CSTIC 2019* **201** pp 1–4, doi: 10.1109/CSTIC.2019.8755785.

[13] Liang J, Morand G, Dariavach N and Shangguan D 2006 In-situ observation of nucleation and growth of whiskers on Sn-Pb HASL coating layer inside plated through holes with press-fit pins *International Symp. and Exhibition on Advanced Packaging Materials, Processes, Properties and Interfaces* pp 1–7

[14] Lim N R E *et al.* 2019 Material properties sensitivity analysis based on thermomechanical analysis on warpage for ball grid array semiconductor packaging *2019 IEEE 11th Int. Conf. Humanoid, Nanotechnology, Inf. Technol. Commun. Control. Environ. Manag. HNICEM 2019* pp 0–4 doi: 10.1109/HNICEM48295.2019.9073437.

HALE UAV ITB perpetual flight

Mochammad Agoes Moelyadi[1]), Muhammad Fikri Zulkarnain[2]
Flight Physics Research Group
Faculty of Mechanical and Aerospace Engineering
Institut Teknologi Bandung
Jl. Ganesha 10 Bandung 40132
Indonesia

[1]moelyadi@ae.itb.ac.id
[2]zulkarnain.fikri@gmail.com

Abstract. The challenge developing High Altitude Long Endurance (HALE) UAV is to obtain the ability endurance of more than 24 hours (perpetual flight). The ability to fly for a long time needs to consider: high aerodynamic efficiency, high motor-prop efficiency, hybrid battery-solar panel energy source, light weight structures, efficient control, and optimum flight mission. This paper discusses about strategy, method and requirements to calculate perpetual flight mission for the case of HALE UAV ITB.

1. Introduction

Development of a High Altitude and Long Endurance (HALE) UAV may align to current issues such as regional border, disaster, and foreign areas. Various HALE has been developed as seen Figure 2. One of them is Zephyr UAV [1-3]. The development of Zephyr UAV starts at 2003 with only 15 minutes' endurance and climb to 27,000 feet, to set a record with Zephyr-s at 2018 flying for 26 days with recorded highest altitude at 74,000 feet.

Atlantik Solar is a small HALE platform, which record 3 days endurance flying at constant altitude of 2,000 feet [4]. Solar Impulse in an example of high altitude long endurance manned aircraft, which has a record of 5 days of flight with the highest altitude of 28,000 feet [5]. Korean Aerospace Research Institute has developed a series of HALE UAV [6], with the latest design EAV-3 which has reached 18,000 km by using own design propeller at 2018 [7]. ApusDuo is a HALE UAV developed by UAVOS, which has multi-boom, tandem wing configuration [8-10].

The development of High-Altitude Long Endurance UAV in ITB is a part of the ITB UAV Development, as shown in Figure 1, with a mission to develop a UAV platform that can use for surveillance and controlled the territory of the Republic of Indonesia. HALE ITB has started to be developed since 2016, and this development includes design, prototyping and structural and flight testing. The performance of HALE UAV ITB previous iteration with MTOW of 20 kg and a wing span of 12 meters has been calculated by Hakim [11]. It has a mission to fly to 20,000 feet for only 20 hours. The current iteration of HALE UAV ITB (Version 4) is designed to have higher aerodynamic characteristics, higher operational capabilities, while also has better structure efficiency compared to previous versions. The newest iteration of HALE UAV ITB has 50 kg MTOW with a mission to fly at more than 50,000 feet and fly for more than 24 hours.

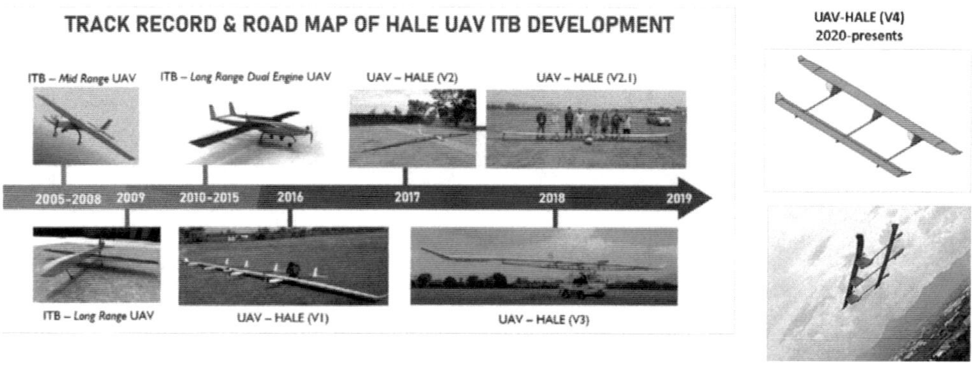

Figure 1. HALE UAV ITB Development Timeline

The challenge of developing HALE UAV is to design an aircraft which can reach a very high altitude and obtain long endurance by flying at that altitude. Fulfillment of those two requirements may need to consider several aspects: optimum aerodynamics, optimum motor and propeller combination, optimum battery and solar panel, optimizing flight control and automation, minimize weight, and optimize mission profile. In the case of HALE UAV ITB, minimum energy concept is used, of which total energy required during flight must be minimized.

Very high aerodynamic efficiency is usually achieved by using a very high aspect ratio wing, which resulted in a very minimum induced drag. Motor and propeller design is very important for both climb phase and cruising phase. Usually, this is achieved by developing a propeller specially designed for high altitude flight, as used by Zephyr [12]and EAV-3 [7].

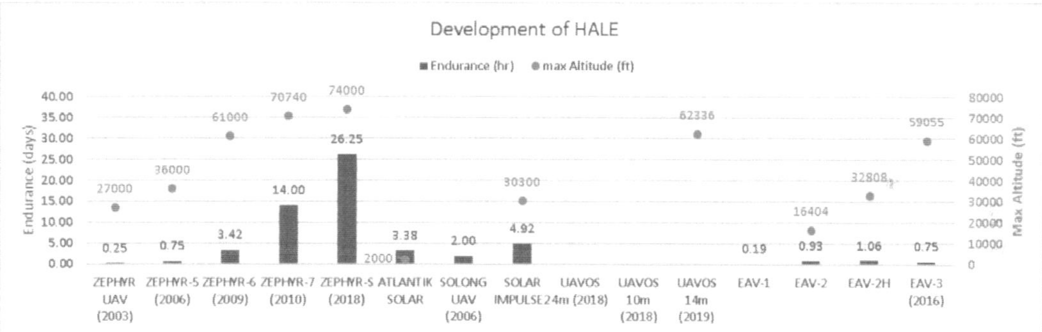

Figure 2. Max Altitude and Endurance Performance of Various HALE Aircraft.

To achieve long endurance, HALE UAV used battery as energy storage for night flight, and solar panel to charge the battery during the day. If possible, the battery with high energy density and solar panel with high efficiency should be used.

Airframe weight needs to be as low as possible, to increase battery weight and reduce total weight. Usually there are two strategies, by using relatively low load factor design as used by Zephyr and EAV-3, or by using flexible wing as used by UAVOS [13].

Final aspect to be considered is the mission profile. The most basic one is climb and then cruise at a constant altitude, which is used by Atlantik Solar [4]. However, as observed from the larger HALE UAV, as shown in Figure 3., there are problem with using this mission profile, especially due the weight of battery [10]. Thus, to reduce the weight of battery, at night, the aircraft will descend into lower altitude, as used by Zephyr [3], Solar Impulse [5], and UAVOS [10]. After the night flight, the aircraft will climb back into its intended altitude to continue cruising during the day.

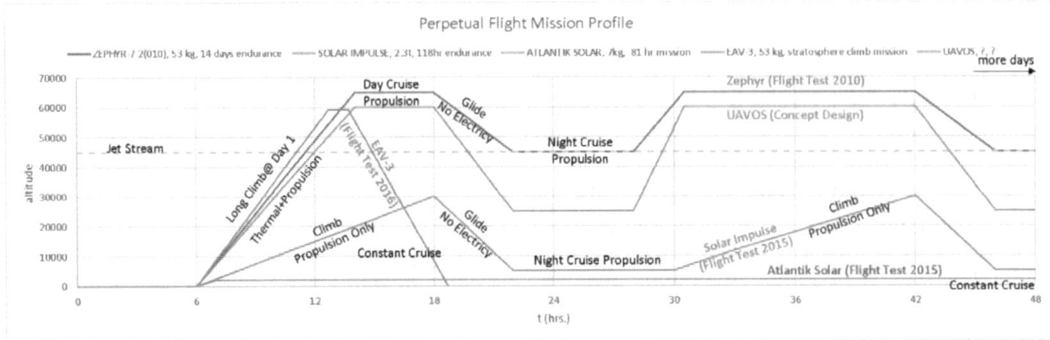

Figure 3. Mission Profile of several HALE UAV

To make sure HALE ITB V4 can fulfill its mission, a new analysis must be conducted to obtain feasibility analysis for HALE UAV ITB performance. There are three mission profile to be analyzed: 20-hour flight at 20,000 feet; more than 24-hour flight at 20,000 feet; and more than 24-hour flight at altitude of at least 50,000 feet.

2. Theory and Methodology

The performance calculated will be the optimum climb and cruise condition, battery weight, and solar panel area, which fulfill those mission. Method of energetic simulation, which simulate energy obtained and energy used during flight, is used to validate the calculation. The methodology is shown in Figure 4. Several variables such as aerodynamic efficiency, propeller efficiency, airframe weight, battery density, solar panel efficiency, etc., will be varied, to obtain the requirement for each mission profile. Spreadsheet is used to calculate aircraft performance and calculate energetic simulation.

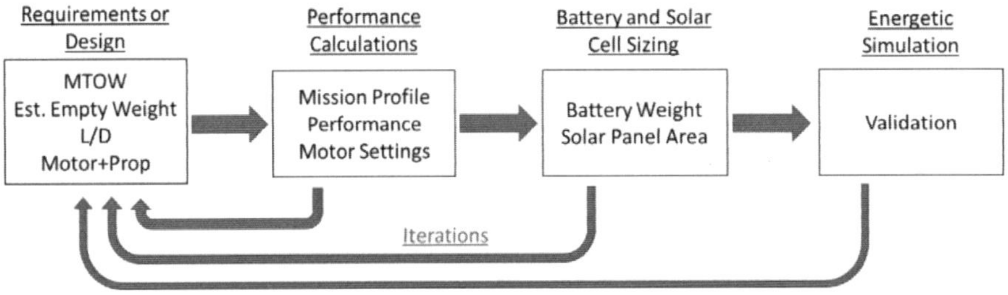

Figure 4. Methodology for energetic simulations

2.1 *Aircraft Performance Calculation*

Aircraft is assumed to four phases of flight: climb, day cruise, glide, and night cruise. Climb, cruise, and glide performance use equations from Reference [14]. Motor and propeller performance will use NACA propeller design from reference [15] to match motor and propeller from T-motor [16]. This is done as T-motor specifications are only provided for sea level static condition, while aircraft performance calculation needs to consider speed and density. It is found that combination of motor and propeller is close to a 14 degrees' pitch NACA propeller. The main output of aircraft performance calculation is to obtain throttle and power required during each flight phase, which will be used in battery, solar panel, and energetic simulation calculations.

There are four different motors considered: MN1005, U12, U13, and U13II. Two diameter of propeller are considered, 30 inches and 32 inches. It is assumed that the lowest throttle settings possible is 40% for MN1005 and 50% for others, as the data provided in motor specifications. The motor specifications used are maximum power, maximum thrust, and maximum rpm, while their variation at various throttle is assumed to be similar which is shown in Figure 5.

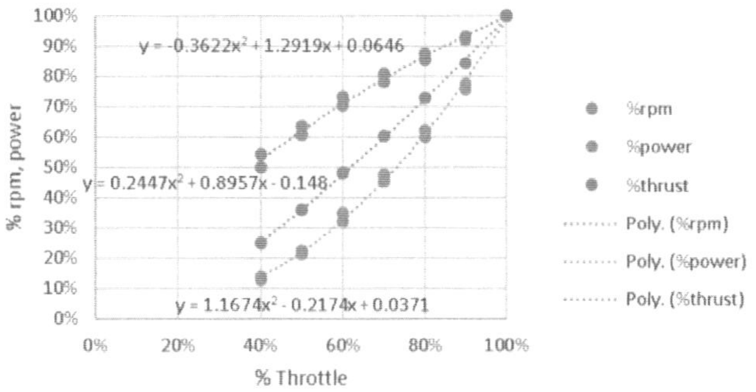

Figure 5. Typical motor performance.

2.2 Battery and Solar Panel Sizing

Solar cell needs to be able to provide energy for the entire flight, while the battery is used during night. From aircraft performance calculations, energy required can be calculated. Total energy required divided by total energy produced by solar cell will yields solar cell area. Total energy required during night and two hours during day will be used to obtain the battery size. The battery size and solar panel area is then iterated during energetic simulation to obtain the most optimum design.

2.3 Energetic Simulation

Energetic simulation is used as a method of validation for solar panel and battery sizing, as well as aicraft performance and weight sizing, as has been done in reference [4]. As shown in equation 1, this method calculates the rate of battery energy change as a power expended and gained throughout the flight. By using discretization, battery energy state as a function of time can be obtained by multiplying power expended and gained throughout the flight with Δt, as shown in equation 1. As the calculated method is not very detailed, the value of Δt may not need to be small and the value of half an hour is deemed to be accurate enough for the calculation.

$$\frac{dE_{battery}}{dt} = P_{solar} - P_{out} \tag{1}$$

$$E_{battery}(t) = \big(P_{solar}(t) - P_{out}(t)\big)\Delta t \tag{2}$$

Power expended (P_{out}) was calculated previously during performance calculation and is varied according to the aircraft flight phase and the mission profile. An additional constant power of 10 W is added to simulate other system such as flight control and actuator. Charging power of the solar cell (P_{solar}) is obtained by multiplying solar cell area with irradiance. The irradiance model, as a function of time, used a modified version from reference [17] [18]. The irradiance model is assumed to be 12 hours and start at 6 in the morning to better fit condition in Indonesia. Maximum irradiation is assumed to be the efficiency of solar cell.

3. Results and Discussions

To obtain the minimum energy, calculating the energy and power required for every altitude is done. Climb time, total climb energy, and cruise power for altitude 20,000 and 50,000 ft is shown in Figure 6. At an altitude of 20,000 feet, the most efficient for climbing is MN1005x4 engine. However, four MN1005x4 engines have higher required cruise power compared to three engine MN1005. This, as well as for U13, and U13II is due to low altitude and assumed lower limit of engine throttle, the aircraft cruise at angle of attack lower than the optimum angle of attack. The next optimum engine configuration is MN1005x3 which has lower required climb and cruise energy. This is due to higher number of motor needs lower throttle for each motor, while for electric motor observed, the motor has higher efficiency at lower throttle.

For altitude of 50,000 feet, only three combination of engine is capable to climb to this altitude, which are MN1005x4, U13x3, and U13IIx2. For climbing, the best configuration is U13IIx2, which has the lowest energy required and climb time. Lower climb time is very important because HALE UAV needs to be able to escape from lower altitude quickly to avoid more turbulent air. However, for cruise performance, the best option is using MN1005x4. However, it should be noted that has worse climb performance, and four engines may lead to complexity is system design compared to only two engines.

Variation of MTOW, aerodynamic efficiency, and propeller efficiency and its effects on energy consumption is also studied. As seen in Figure 7, the increase in MTOW resulting in the increase of energy required both during cruise and during climb.

Variation of aerodynamic efficiency and propeller efficiency is shown in Figure 8. The aerodynamic efficiency is represented by maximum L/D, while the propeller efficiency is represented by NACA equivalent propeller pitch. The better the efficiency, power required and energy required can be reduced. The aerodynamic efficiency has greater effects for cruise performance, while propeller efficiency is having greater effects for climb performance. One of the findings during calculation is that the typical propeller efficiency encountered is only 0.2 to 0.4 during climb and 0.6-0.7 during cruise, which is relatively low.

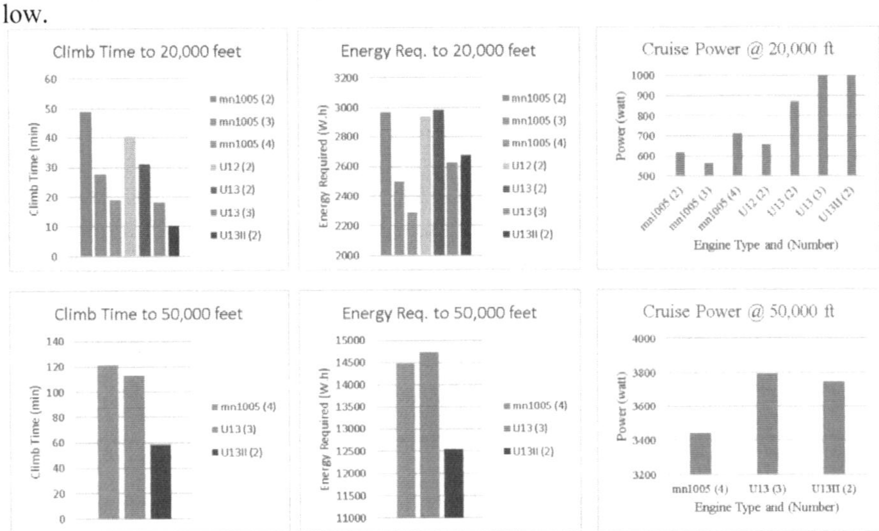

Figure 6. Effect of Variation on engine combination on performance.

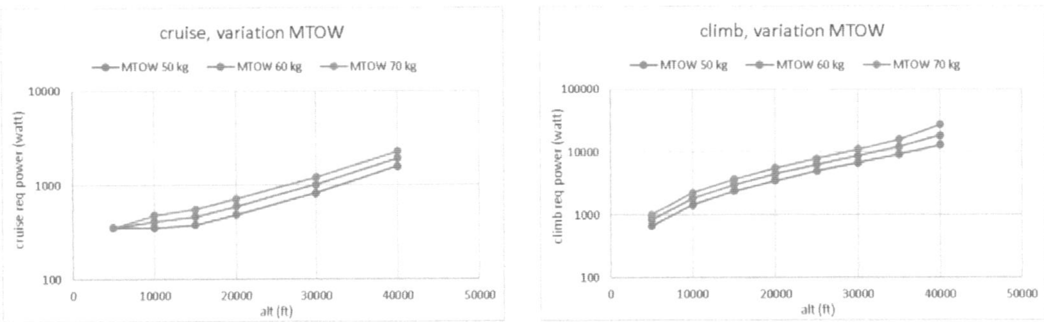

Figure 7. Effect of variation on MTOW, 2xMN1005, L/D 27, 10deg prop.

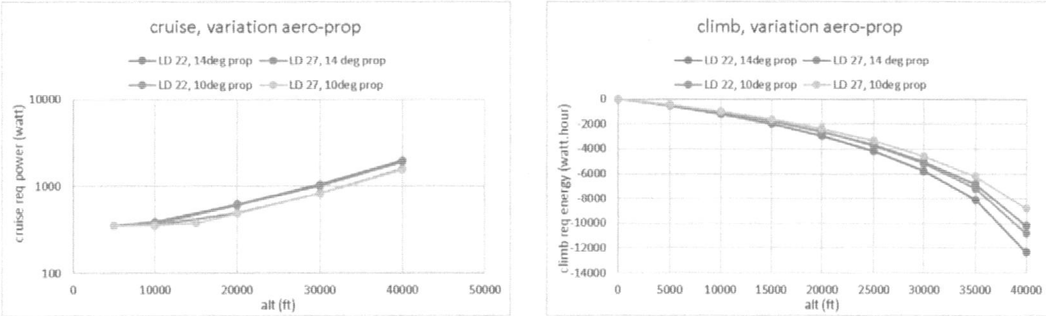

Figure 8. Effect of variation on aero and propeller efficiency.

3.1 Perpetual Flight

The main result for energetic simulation are motor-propeller configuration, aerodynamics efficiency, airframe weight requirements, battery requirements, and solar panel requirements, which fulfil the three mission case previously discussed. These results are shown in Table 1. As shown in Table 1, the MTOW is also increased due to increase reserve weight for airframe.

The mission profile in Figure 9 can be divided into two parts. The first part is for the first 24 hour of the flight, which includes takeoff and climb from sea level to desired altitude, day cruise at desired altitude, descending from desired altitude to lower altitude

Table 1. Results of Requirements for three mission case

	Case 1	Case 2	Case 3
Max altitude	20,000 feet	20,000 feet	50,000 feet
Night altitude	5,000 feet	5,000 feet	25,000 feet
Endurance	21 hour	more than 24 hour	more than 24 hour
MTOW	50 kg	60 kg	70 kg
Wing Loading	2.1 kg/m^2	2.1 kg/m^2	2.1 kg/m^2
Power Loading	2.2 kg/hp	2.2 kg/hp	1.4 kg/hp
Propeller	30x10,5" (off the shelf)	32", higher efficiency	32", higher efficiency
Max L/D	22	27	33
Airframe Weight	60%	50%	30%
Battery Weight	10.3 kg	18 kg	29 kg
Battery Energy Density	265 Wh/kg	265 Wh/kg	350 Wh/kg
Solar Cell Weight	7.7 kg	8.4 kg	16.2 kg
Solar Cell Area	11 m^2	14 m^2	27 m^2
Solar Cell Efficiency	14%	21%	27%
Solar Cell Area Density	0.7 kg/m^2	0.6 kg/m^2	0.6 kg/m^2

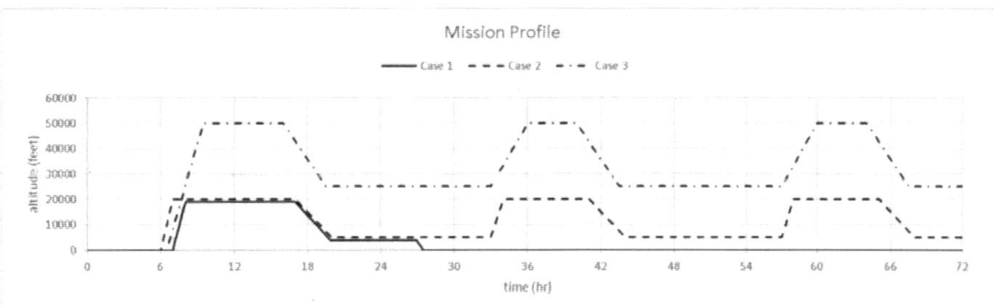

Figure 9. Mission Profile for the three mission case

The mission profile for Case 1, in shown in Figure 9 s climb to and cruise at 20,000 feet for 9 hours, and then glide to 5,000 feet and cruise for 6 hours,. The climb started later at 7 o'clock to take advantage of larger energy absorption from solar power, which also decrease battery weight. The results of energetic simulation is shown in Figure 10.

The mission profile for Case 2 is similar from Case 1, at which at day the aircraft cruise at 20,000 feet while at night the aircraft descend to 5,000 feet. For day 1, the aircraft can climb earlier than Case 1 at 6 o'clock or earlier, as the battery requirement, as seen in Case 2 energetic simulation results in Figure 11, is defined by the low altitude night cruise. The following day, the aircraft start to climb from 5,000 feet to 20,000 feet at 9 o'clock.

For Case 3, the climb starts at 6.30 AM for day 1 and 9.00 AM for the following day, to lower the battery requirement. The aircraft cruise at 50,000 feet during day and descend to 25,000 feet at night. Due to the larger difference in day-night cruise and higher altitude, the total time of high altitude cruise is only lower compared to Case 2. The energetic simulation for Case 3 is shown in Figure 12.

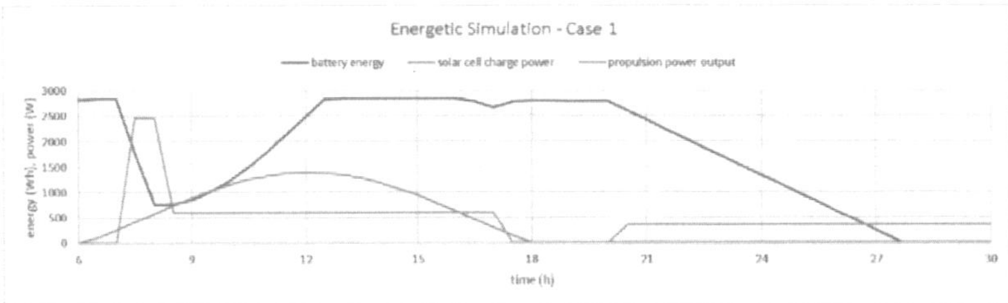

Figure 10. Energetic Simulation Results – Case 1

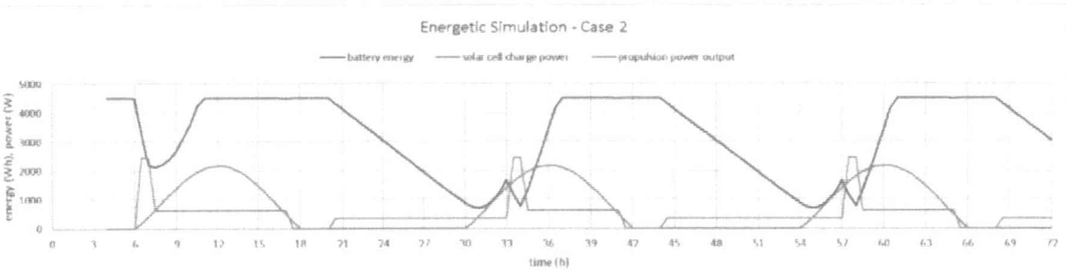

Figure 11. Energetic Simulation Results – Case 2

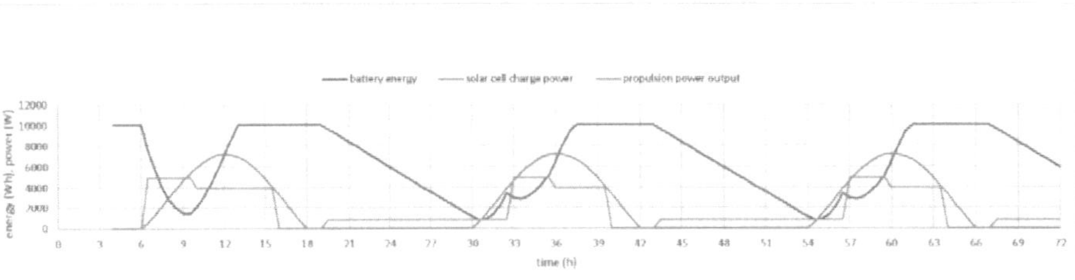

Figure 12. Energetic Simulation Results – Case 3

4. Conclusion

Study has been conducted for three case: Case 1 – altitude 20,000 feet and 21hr endurance; Case 2 – altitude 20,000 feet and endurance at least 24 hours; and Case 3 – altitude of at least 50,000 feet and endurance at least 24 hours. The performance has been calculated resulted in power and energy requirement for flight for each case. A design requirement including MTOW, aerodynamic, motor-propeller combination, airframe weight, battery requirement, and solar panel requirements has been developed. Case 1 to 3 has more stringent requirement due to the requirement for higher endurance and higher altitude. The mission profile to fulfil mission Case 1, 2 and 3 has also been developed, parallel to the design requirements. For battery saving, the aircraft fly at lower altitude at night compared to day cruise. Case 1 and 2, night altitude is at 5,000 feet while Case 3's night altitude is 25,000 feet.

For future works, a more detailed calculation needs to be done, which may include maneuver, environment, and electrical efficiency. The more detailed calculation may need a more sophisticated program or code to run, such as using python. A study on the aircraft variable included in the calculation may need to be conducted. This includes: aerodynamic characteristics; propulsion technology; construction and structure; battery, solar panel, and energy management; and also operational and control aspect. Test and experiment may need to be made to further deepen the knowledge for those aspects.

References

[1] *Jane's 2017 High-flying bird: Zephyr remains in the vanguard of solar-powered flight*
[2] Rapinett A 2009 *Zephyr: A High Altitude Long Endurance Unmanned Air Vehicle* (Surrey: University of Surrey Dissertation)
[3] Thisdell D 2018 Airbus sets flight endurance record with Zephyr UAV. [Online] August 8, 2018. [Cited: August 26, 2020.] https://www.flightglobal.com/airbus-sets-flight-endurance-record-with-zephyr-uav/129186.article.
[4] Oettershagen P et al 2015 A solar-powered hand-launchable UAV for low-altitude multi-day continuous flight. *IEEE International Conference on Robotics and Automation*
[5] Solar Impulse Foundation Our Adventure. [Online] https://aroundtheworld.solarimpulse.com/adventure.
[6] Hwang S J, Kim S G and Lee Y G 2016 Developing High Altitude Long Endurance (HALE) Solar-powered Unmanned Aerial Vehicle (AUV) *Journal of Aerospace System Engineering* **10** pp 59-65
[7] Park D, et al 2018 Design and Performance Evaluation of Propeller for Solar-Powered High-Altitude Long-Endurance Unmanned Aerial Vehicle *International Journal of Aerospace Engineering* **2018**
[8] UAVOS UAVOS ApusDuo AIRCRAFT COMPLETED FLIGHT TESTS. [Online] https://www.uavos.com/about-us/press-releases?id=120.
[9] UAVOS New UAVOS HAPS ApusDuo Variant Successfully Completed Test Flight. [Online] https://www.uavos.com/new-uavos-haps-apusduo-variant-successfully-completed-test-flight.
[10] UAVOS THE EXPERIENCE OF UAVOS IN THE CREATION OF HAPS. [Online] https://www.uavos.com/the-experience-of-uavos-in-the-creation-of-haps.
[11] Hakim M L, et al 2018 Solar Cell to Support Perpetual Flight of High Altitude Long Endurance UAV ITB *Journal of Physics: Conf. Series* **1005** 012038.
[12] de Mattos B S, Secco N R, and Salles, E F 2013 Optimal Design of a High-Altitude Solar Powered Unmanned Airplane *J. Aerosp. Technol. Manag., São José dos Campos* **5** pp 349-361
[13] Стратилатов A, et al 2020 Method for controlling an aircraft and aircraft (variants) WO2020056481A1 Worldwide
[14] Raymer D P 1992 *Aircraft Design: A Conceptual Approach* (Washington D.C. : AIAA)
[15] McCormick B W 1979 *The analysis of propellers including interaction effects* (NASA)
[16] T-Motor *T-Motor Store* https://store-en.tmotor.com/.
[17] Duffie J A 2013 *Solar Engineering of Thermal Processes* (John Wiley & Sons)
[18] Noth A 2008 *Design of Solar Powered Airplanes for Continuous Flight* (Zurich : ETH Zurich)

IOP Conf. Series: Materials Science and Engineering 1109 (2021) 012029 doi:10.1088/1757-899X/1109/1/012029

Influence of solar panel on wing aerodynamic characteristics of HALE UAV

Nu'man Amri Maliky[1], Mochammad Agoes Moelyadi[1], and Ema Amalia[1]

[1]Mechanical and Aerospace Engineering, Institute of Technology Bandung

maliky.numan@gmail.com, moelyadi@ftmd.itb.ac.id

Abstract. HALE UAV needs solar energy to maintain its flight in the day and night. The solar panel located on the upper surface may potentially affect aerodynamic characteristics of the HALE wing. The solar panel generates a heat resulted from solar radiation absorption and surface roughness which becomes parameters to be studied. To identify the phenomena of interaction of flow motion and the heat as well as the roughness, dimensionless numbers such as Reynolds Number (Re), Nusselt Number (Nu), and geometry reference (Ks/C) are used. The simulations for flows at the numbers are conducted using CFD method based on the solution of RANS equations. As a result, the heat transfer gives significant change in flow density near the surface which causes buoyancy effect and the change of flow velocity profile, while the roughness influences the flow characteristics disturbing the flow to become turbulent. They increase aerodynamic drag of 2.6 % and aerodynamic efficiency in the amount of 5%.

Keywords: HALE UAV, solar cell, heating, roughness, aerodynamic characteristics, computational fluid dynamics

1. Background

Most HALE (High Altitude Long Endurance) Unmanned Aerial Vehicles have been developed with various missions such as surveillance, monitoring, mapping, or pseudo-satellite UAV. To fulfill the mission, HALE UAVs have optimal flight conditions at high altitudes, around 60,000 ft above sea level, with long duration flight, more than one day as mentioned by Dalamagkidis [1-2]. Under these flight conditions the UAV needs a lot of energy which are fulfilled by installing solar panels on its wings. The HALE flight operating speed is low around 25 m/s (20 KEAS). Figure 1 shows Solar panels mounted on the upper wing surface of HALE UAV which change airflow around the wing, in turn, influence on aerodynamic performance of the HALE UAV.

There are several factors that the solar panel potentially affect the aerodynamic performance of the HALE UAV namely heat transfer and surface roughness of the solar panel. In addition, the HALE with a long wingspan and low airspeed at high altitude generates low Reynolds number flow over the wing. This may cause airflow in the HALE wing to be easily disturbed, Winslow [3], and the characteristics of low Reynolds numbers flow have a high tendency to occur separations due to external interference.

Figure 1. Example of HALE UAV: UAVOS *[4]*

Solar panel may absorb the radiation exposure from a sunlight emitted from the sun to produce electricity. Then, this absorption emit heat on the surface and interacts with air flow through the solar surface changes aerodynamic characteristics of the HALE UAV wing. Research conducted by Kumar [5] and Samiee [6] explains that the heat transfer that occurs on an airfoil affected the aerodynamic performance of the airfoil.

In addition, the solar panel surfaces have roughness that may alter laminar air flow to become turbulent. This causes the frictional forces that occur on the surface increases. This problem is closely related to airflow at a low Reynolds number where the location of the transition point occur earlier. Experiments carried out by Chakroun [7] prove that roughness on the wing surface will affect its aerodynamic performance, especially the addition of a large drag.

2. Basic Theory

2.1 Low Reynolds Number Flight

Air flow at low Reynolds numbers has special characters which dominate by viscous effect on flow field near the surface. This viscosity produces a friction affecting aerodynamic performance of a body. There is a condition called no-slip condition where flow velocity at the surface is equal to zero and the velocity change gradually in a normal direction with respect to the surface and meets uniform velocity at outer region. A boundary region where a dominant viscosity effect inside is called as the boundary layer as sketched for the boundary layer on an airfoil in Figure 2. The boundary layer also has several regions which are divided based on flow characteristics. The region having regular flow streamlines and calm properties are called laminar occurred in the starting of the boundary layer, while the region with irregular and convoluted air streamlines are called turbulent. Both of flow regions have different forms of velocity profiles. In a fluid flow, these two flow regions may occur simultaneously. The connection between them is a transition region.

Alsahlani [2] states that there are several factors to be considered in the aerodynamic design of the HALE UAV. The operations of the HALE at very high altitude with low velocity flight may occur at a small Reynolds number which may produce a great impact on aerodynamic design of the HALE UAV. At that altitude, the atmosphere density is low, to keep the flight position at a given speed the UAV should be designed to have airfoil with high lift coefficient at low Reynolds number. In addition, airfoil characteristics at low Reynolds numbers are strongly influenced by laminar flow separation. [3]

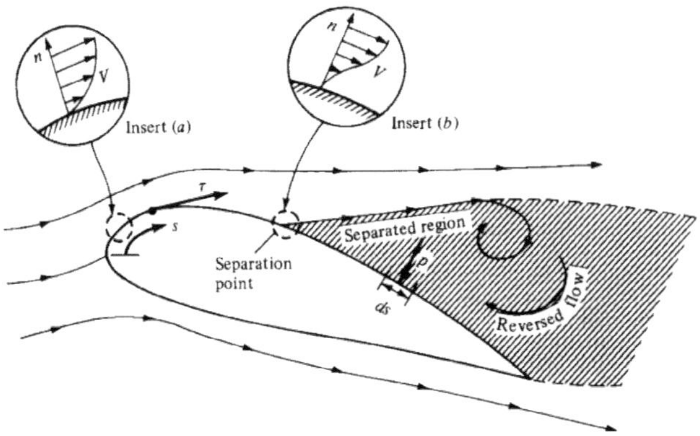

Figure 2. Effects of Viscosity on Airfoil: Boundary Layer and Separation *[8]*

There are several factors of flow transition including velocity and turbulence of freestream flow, adverse pressure gradient, roughness and heat transfer. These factors may give alteration of fluid properties inside the boundary layer such as density, temperature and viscosity. Fluid density and viscosity can be changed due to temperature, as governed by the ideal gas law in Equation 1, and Sutherland's law in Equation 2. [8]

$$p = \rho RT \tag{1}$$

$$\frac{\mu}{\mu_0} = \left(\frac{T}{T_0}\right)^{\frac{3}{2}} \frac{T_0+110}{T+110} \tag{2}$$

The frictional force on the airfoil surface depend on viscosity and fluid velocity gradients in the boundary layer in the normal direction of the surface as written in Equation 3. [8]

$$\tau_{wall} = \mu(T)\left(\frac{\partial u}{\partial n}\right)_{wall} \tag{3}$$

The fluid properties in the flow field around an airfoil may be obtained by solving Navier-Stokes equations which consists of continuity equation, momentum equations and energy equation as given in Equations 4 to 6, respectively [8].

$$\nabla \cdot V = 0 \tag{4}$$

$$\rho \frac{\partial V}{\partial t} + \rho(V \cdot \nabla)V = -\nabla p + \rho g + \nabla \cdot \tau \tag{5}$$

$$\rho \frac{\partial h_t}{\partial t} - \frac{\partial p}{\partial t} + \nabla \cdot (pV) = \rho \dot{q}_{rad} + \nabla \cdot \dot{q} + \nabla \cdot (V \cdot \tau) + \nabla \cdot (\lambda \nabla T) \tag{6}$$

Where $V, \rho, p, T, \tau, g, h_t, \dot{q}$ are velocity, density, pressure, temperature, shear stress, gravity, total enthalpy and heat transfer rate, respectively.

2.2 Computational Fluid Dynamics Simulation

The simulations are conducted by Computational Fluid Dynamics method solving the governing equations on spatial discrete points in a computational domain. The computation uses a computer to do iterative numerical process of the discrete equation either in explicit direct solution or implicit matrix solution. The boundary conditions are required in the simulation to obtain more realistic solution. The boundary conditions may be applied on body surface, incoming and outcoming flows as well as farfield boundary. In addition, the solution of flowfield can be extracted from values of fluid properties at every nodal in the computational domain. Figure 3 shows the steps that used for CFD simulation. [9]

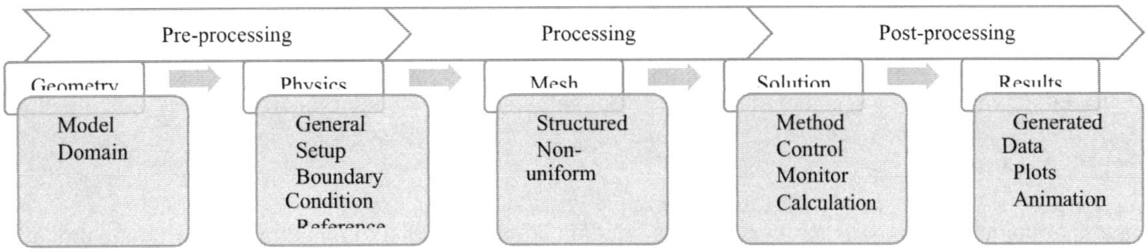

Figure 3. CFD Simulation Flowchart

The boundary layer on the CFD can be captured by adjusting the minimum meshing height on the airfoil surface. The height needs to be adjusted to the y^+ approach which is closely related to u^+. Equations 7 to 9 set the values of u^+ and y^+. Small y^+ (below 5) is needed to capture the boundary layer. [10]

$$u^+ = \frac{u}{u_\tau} \tag{7}$$

$$y^+ = \frac{\rho u_\tau y}{\mu} \tag{8}$$

$$u_\tau = \sqrt{\frac{\tau_{wall}}{\rho}} \tag{9}$$

The shear stress and friction coefficient uses the Equations 10 and 11 respectively, [11].

$$\tau_{wall} = C_f \frac{1}{2} \rho U_\infty^2 \tag{10}$$

$$C_f = (2(Re_x) - 0.65)^{-2.3} \tag{11}$$

The study is conducted by using ANSYS Fluent software to model the heat transfer on the wing surface. The heat flux or the wall temperature is required as an input for boundary conditions of the system. The roughness is modelled by sand-grain parameter uses conservation values. The values of the roughness for this study are given in Table 1. [12]

Table 1. Surface Roughness Conversion Parameter Value.

Roughness Parameter	**R_a**	**R_{RMS}**	**R_{zd}**
Estimated Sand-Grain Roughness, ε	$\varepsilon = 5.863\ R_a$	$\varepsilon = 3.100\ R_{RMS}$	$\varepsilon = 0.978\ R_{zd}$

2.3 HALE UAV Design

The specification of the wing design of the HALE UAV is used to be analyzed are listed in Table 2. The airfoil on HALE UAVs uses K3311 airfoil. This airfoil has the shape as shown in Figure 4.

Figure 4. K3311 Airfoil

Table 2. Analyzed Aircraft Design's Dimension.

Parameter	Value	Unit
MTOW	50.00	Kg
Chord	0.85	M
Wing Span	15.00	M
Wing Area	12.75	m^2
Taper Ratio	1.00	-
Criuse Speed (target)	25.00	m/s
Cruise Altitude (target)	18.00	Km
Cruise Altitude (target)	60.00	Kft

3. Modelling

The analysis is conducted for cruise conditions with variation of parameters adjusted for the altitude of the flight. The cruise conditions are on altitude 60,000 ft with angle of attack of two degrees. The air properties used for the analysis are shown in Table 3. The variation parameters are the heat flux (W/m^2) and sand-grain roughness (K_s/c). The extreme values of the parameters, used to show the design point, is the maximum heat absorption according to Wasfi [13] and roughness of the solar cell [14].

Table 3. Air Properties on Altitude 60,000 ft

Altitude	Pressure (Pa)	Density (kg/m3)	Temperature (K)	Reynolds Number
60,000 ft	7171.64	0.115318	216.65	15×10^4

Figures 4 and 5 show the wing mesh formed with a C-type topology of multi-block consisting of sub-domains for the wing with the computational domain size of 5 times of the wing chord for radius of the freestream inlet and 15 times of wing chord for the downstream. The mesh density and shape are adjusted according to the needs of the flow physics. Around the surface of the wing, the finer mesh is made to capture the viscous flow on the airfoil.

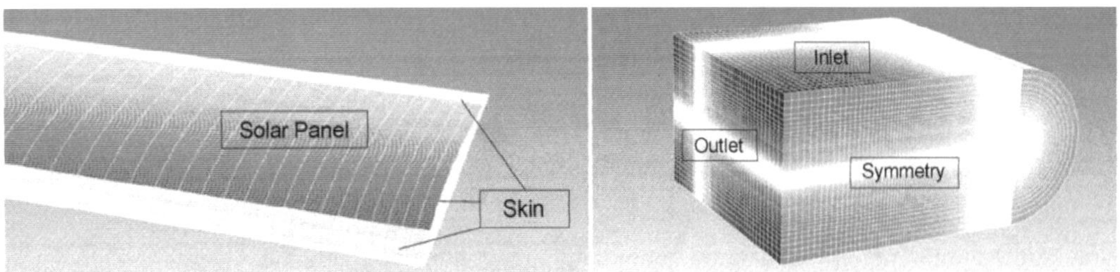

Figure 5. Mesh for HALE UAV Wing

Table 4. Properties for Boundary Conditions on the Mesh.

Parameter	Value	Unit
Inlet	Velocity Inlet	front, top, bottom, and right wing farfield
Solar Panel	Wall	upper wing surface at 0.1-0.9 chord
Skin	Wall	the entire wing surface is outside the definition of the solar panel
Symmetry	Symmetry	surface on the wing root
Outlet	Pressure Outlet	farfield the back of the wing

4. Simulation Results and Analysis

The validation is done by comparing the simulation results of K3311 airfoil with experiments taken from the reference book "Summary of Low-Speed Airfoil Data" by Michael S. Selig [15]. Figures 6 and 7 show lift and drag coefficients of HALE wing at zero angle of attack at various Reynolds numbers. From this comparison data, the simulation yields small value of error below than 5%. It explains that the simulation can capture effect of low Reynolds Number well.

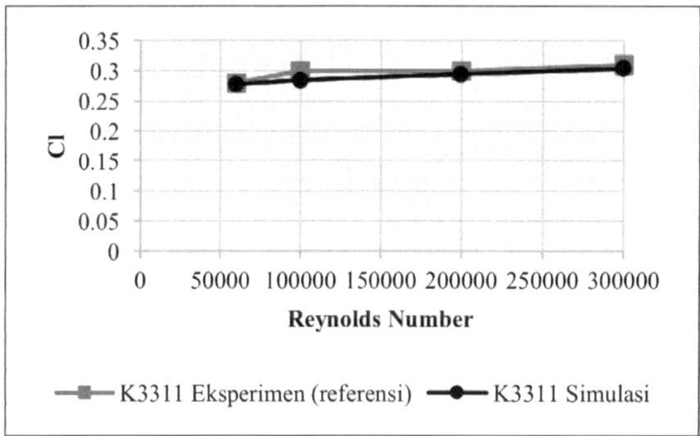

(a)

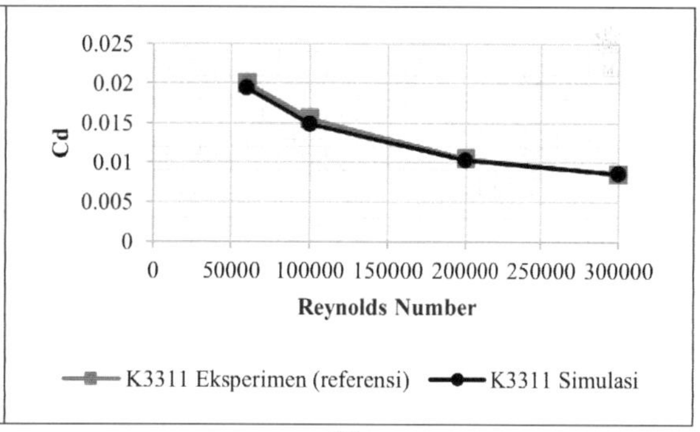

(b)

Figure 6. Comparison of simulation and experiment for wing with Airfoil K3311: (a) lift coefficient and (b) drag coefficient

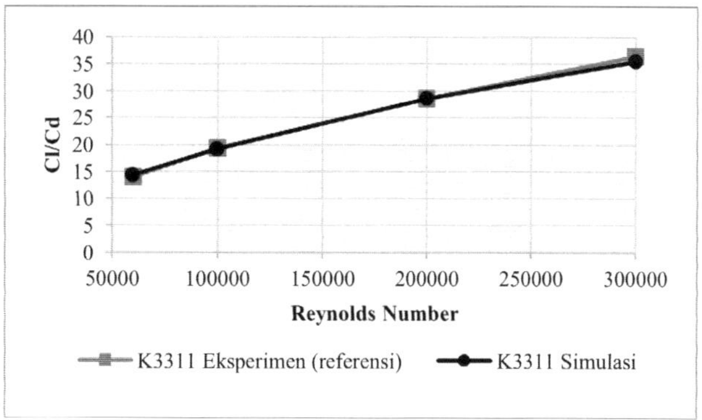

Figure 7. Comparison of simulation and experiment for wing with Airfoil K3311: aerodynamic efficiency

Figure 8 and 9 shows the effect of the combination of heat flux and roughness on drag and aerodynamic efficiency, respectively. The values of design point at heat flux of 1000 W/m^2 is also provided in the graphs.

The simulation results for off-design condition are performed at lower and higher heat flux and higher sand grain size. The decrease in drag can be obtained by reducing heat flux at lower sand-grain size less than 0.02 %. While, for higher heat flux above 1000 W/m^2, the drag increases and becomes linearly higher increases with larger sand-grain size. In addition, aerodynamic efficiency of the HALE UAV wing becomes higher with reducing heat flux at lower sand-grain size less than 0.02 %. But the reduction of aerodynamic efficiency occurs when the heat flux increases. This reduction in efficiency will result in an increase in the power needed by the HALE UAV.

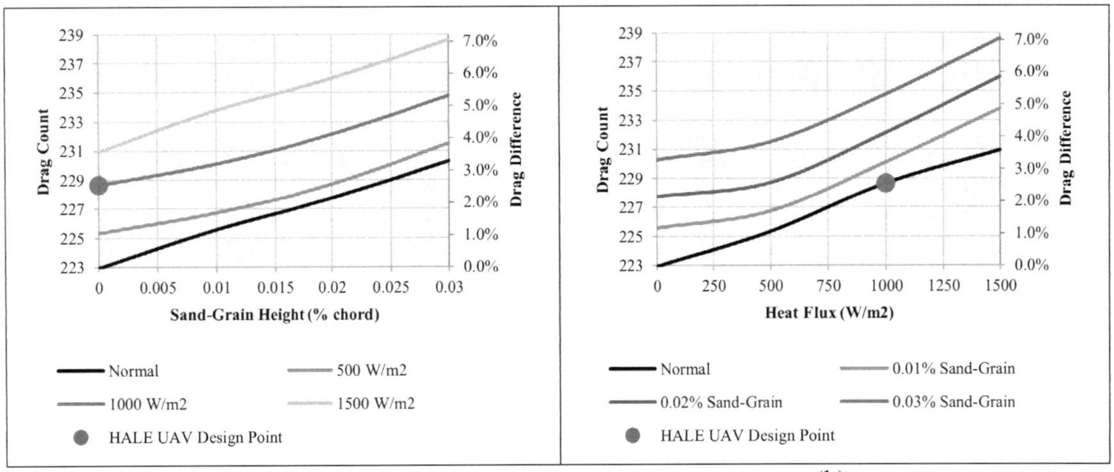

(a) (b)

Figure 8. C$_D$ Graph UAV HALE Wings and Extreme Design Points

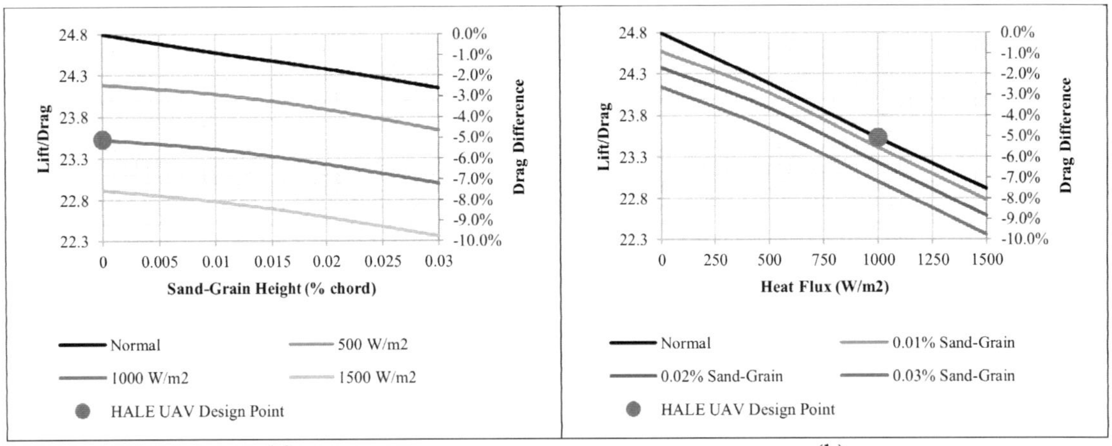

(a) (b)

Figure 9. L/D Graph UAV HALE Wings and Extreme Design Points

Figure 10 depicts the effect of heating and roughness on friction and turbulence intensity. The effect of heat flux of the solar panel gives slightly increase in local friction coefficient and turbulent intensity in the position of 10-90% of the wing chord. While, the effect of roughness gives higher increases in local friction coefficient and turbulent intensity in the position of 10-90% of the wing chord. The transition region moves forward due to the roughness.

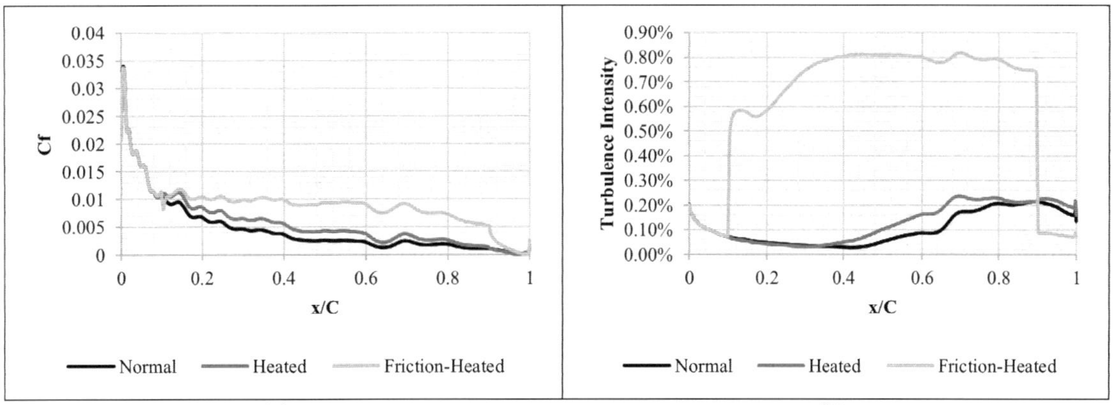

(a) (b)

Figure 10. Effect of Heating and Roughness on Friction and Turbulence Intensity

Table 5. Comparison of Normal and Extreme Values on HALE Solar Panels.

Parameters	Normal	Extreme Values
Sand-Grain Height	222.79 nm	703.56 nm
Heat Flux	0 W/m^2	1000 W/m^2
Drag Count	222.934	228.649
Drag Addition	-	2.6%
Lift to Drag Ratio	24.7977	23.5242
Decreasing Aerodynamic Efficiency (L/D)	-	5.1%

These simulations conclude that the effects of heat flux and roughness due to the installation of solar panels on aerodynamic characteristics of the HALE UAV is quite significant. The drag of the HALE wing increases by 2.6% and the decreased aerodynamic efficiency by 5.1% on the extreme possibility. With these results, we can be sure that the HALE UAV wing is quite safe from the influence of solar panels installed on the surface of the wing.

This extreme point is likely to occur during the day, where the sun is shining with the maximum heat at very high altitudes. Based on the previous analysis, the highest possible heat flux that occurs when this condition can reach 1000 W/m^2. In fact, the heat flux varies depending on the altitude of flight, latitude and longitude, season, and time (morning, afternoon, or evening). By analyzing the extreme condition, the possibilities of the effect of the flux and roughness on aerodynamic characteristics of the HALE wing can be ascertained to be smaller.

5. Conclusion

Based on the simulation results, it can be concluded that the installation of the solar panel as HALE wing surface gives quite significant influence the HALE UAV aerodynamic performance due to heat transfer and surface roughness. The decrease in drag can be obtained by reducing heat flux at lower sand-grain size less than 0.02 %. For higher heat flux above 1000 W/m^2, the drag becomes linearly higher increases with larger sand-grain size. Aerodynamic efficiency of the HALE UAV wing increases with reducing heat flux at lower sand-grain size less than 0.02 %, vise versa. The effect of heat flux of the solar panel gives slightly increase in local friction coefficient and turbulent intensity in the position of 10-90% of the wing chord. While, the roughness affects higher increases in local friction coefficient and turbulent intensity in the position of 10-90% of the wing chord. In extreme conditions, where the HALE UAV is operating on its main mission during the day, the influence has quite significant effect on the flow characteristics caused the change of aerodynamics performance in less than 5%.

References

[1] Dalamagkidis K 2015 Classification of UAVs *Handbook of Unmanned Aerial Vehicles* (Munchen, Springer Science+Business Media Dordrecht)
[2] Alsahlani A A M 2017 Design of a swept-wing high-altitude long-endurance unmanned air vehicle (HALE UAV) (Machester: Engineering Research Centre, University of Salford)
[3] Winslow J 2018 Basic understanding of airfoil characteristics at low Reynolds numbers *Journal of Aircraft* vol 55 no 3 pp 1050-1061
[4] Stratsilatau A 2018 Solar power moves aircraft and industry into the future [Online]. Available: https://www.uavos.com/solar-power-moves-aircraft-and-industry-into-the-future. [Accessed 14 November 2020].
[5] Kumar A V M 2019 Effect of temperature on the aerodynamics of airfoil in low Reynolds number flow (Maryland: American Institute of Physics)
[6] Samiee A 2018 Improvement of airfoils aerodynamic efficiency by thermal camber phenomenon at low Reynolds number *Journal of Aerospace Technology and Management* vol 10
[7] Chakroun W 2004 Effect of surface roughness on the aerodynamic (Tokyo, Japan: Japan Associati for Wind Engineering)
[8] Anderson J D 2001 Fundamentals of Aerodynamics 3rd Edition (New York: McGraw-Hill).
[9] Anderson J D 1995 Computational fluid dynamics: The basics with applications (New York: McGraw-Hill Science)
[10] jousefm, "What is y+ (yplus)?," SimScale, May 2018. [Online]. Available: https://www.simscale.com/forum/t/what-is-y-yplus/82394. [Accessed 21 04 2020].
[11] Schlichting H 1979 Boundary layer theory (New York: Springer)
[12] Adams T 2012 A simple algorithm to relate measured surface roughness to equivalent sand-grain roughness *International Journal of Mechanical Engineering and Mechatronics* vol 1 no 1 p 66
[13] Wasfi M 2011 Solar energy and photovoltaic systems (New Jersey: IEEE)
[14] American Elements, "Silicon Single Crystal," American Elements, [Online]. Available: https://www.americanelements.com/silicon-single-crystal-7440-21-3. [Accessed 3 March 2020
[15] Selig M S 1995 Summary of low-speed airfoil data (Virginia: SoarTech)

Interval forecasting model for time series based on the fuzzy clustering technique

T Vovan[1] and D Phamtoan[2,3,4*]

[1]College of Nature Sciences, Can Tho University
[2]University of Science, Ho Chi Minh City, Vietnam
[3]Vietnam National University, Ho Chi Minh City, Vietnam
[4]Faculty of Engineering, Van Lang University, Ho Chi Minh City, Vietnam

*dinh.pt@vlu.edu.vn

Abstract. This paper proposes the forecasting model for the fuzzy time series based on the improvement of the background data and fuzzy relationship (IFTC). This algorithm is built based on the fuzzy cluster analysis which the suitable number of clusters for series is considered. The problem of interpolating data according to fuzzy relationships of time series in the trapezoidal fuzzy number is also established. The proposed model is illustrated step by step by a numerical example and effectively implemented by the Matlab procedure. The IFCT has advantages in comparing to other models via the several indexes such as the MAE, MAPE and MSE with the Enrollment dataset.

1. Introduction

Forecasting is the process of making predictions based on historical data, knowledge and experience of the related problems. Because of its important role in many fields, forecasting has been paying much attention by scientists. It is an important science basis for projects, policies and appropriate development strategy. Thus, forecasting is always interested in managers and scientists. Regarding data, time series is popular and has great demand on forecasting in reality. For this data, the two main models used for forecasting are regression and time series. The regression model has conditional constraints on data that are difficult to satisfy in reality, so it has the disadvantage in many cases [5,8]. The time series model (TS) is considered to have many more advantages, so it is used very popularly today, Many scientists utilized the TS models as Auto regression (AR), Autoregressive Intergrated Moving Average (ARIMA) to apply in Economy, environment and hydrology [7,13,17].

For the mechanical and manufacturing engineering field, there are some outstanding studies as Fang-mei tseng et al. [21] proposed a procedure of fuzzy seasonal time series and apply this method to forecasting the production value of the mechanical industry in Taiwan. This method includes interval models with interval parameters and provides the possibility distribution of future value. From the results of practical application to the mechanical industry, it can be shown that this method makes good forecasts. Further, this method makes it possible for decision makers to forecast the possible situations based on fewer observations than the SARIMA model and has the basis of pre-procedure for fuzzy time series. Trendafilova [22] considered some possibilities to use pure time series analysis for damage diagnosis in vibrating structures. The author introduced the basics of the state space methodology and discussed a number of possible methods to extract damage sensitive features from the state space representation of the attractor of a vibrating system. In similar direction, Irina Trendafilova and Emil Manoach [20] introduced two viable VHM (Vibration health monitoring) methods that use large amplitude vibrations and are based on nonlinear time series analysis and suggested explore some changes in the state space geometry/distribution of the structural dynamic response with damage and their use for damage detection purposes. However, to build a best TS model, the time series must stop and its error must be the white noise. These conditions are considered the challenge that are difficult to satisfy in reality. Therefore, in many cases, the forecasting results are poor when using TS models. Although, many authors as Abreue et al. [2], Ghosh, Chowdhury and Prajneshu [12], Tai [19], Yu and Huarng [23] have improved the previous model, they still face many difficulties.

One of the research directions that has been interested by many statisticians is defuzzification for the original data to get the relationship of the elements in series. Song and Chissom [18] are pioneers in this direction. The TS models normally include three phases: (i) determine the background set from the original data, divide the interval for the background set, and find the number of elements for each interval; (ii) build the fuzzy relationship, and (iii) interpolate and predict the time series. For (i), many authors have used the minimum and maximum values of the original data to define the background set [7,8]. Moreover, Huarng [13], Huarng and Yu [23] proposed the two new techniques for determining intervals of the background set based on the mean and distribution of the entire series. Another way to build the background set use the clustering method. This method is a new research direction at present. For (ii), some researchers have performed this problem, for instance, Song and Chissom [18] used the matrix operations and Chen [8] utilized the fuzzy relationship. Meanwhile, Aladag [3] used the Neural Network to define the fuzzy relationship. For (iii), most of study applied the centroid method in order to perform this phase [8,13,20].

This paper contributes to the three stages (i),(ii) and (iii) for the TS model. For (i), after normalizing the data, the model find the number of groups that belong to each centroid of the time series. These centroids are calculated according to the algorithm to determine the suitable number of clusters (SNC). For (ii), based on the FCM, we build the algorithm to find the fuzzy relationship between the centroid and each value of the time series by interval after that perform the process of interpolating according to research of Liu Hao-Tien [16]. This is the contribution to stage (iii). Combining all the enhancements, this article proposes an interpolating model for time series that is better than the existing models based on the benchmark dataset.

The remainder of this paper is structured as follows: Section 2 considers some conceptions about time series, fuzzy number related to the proposed model. Section 3 presents the step by step of the developed algorithm. Section 4 illustrates the proposed algorithm and compares to other ones based on the benchmark dataset. The final section is the conclusion.

2. Some conception

Definition 1. Give U is universe set, $U = \{u_1, u_2, ...u_n\}$. Fuzzy set A of U is defined as follows:

$$A = \{\mu_A(u_1)/u_1, \mu_A(u_2)/u_2, ..., \mu_A(u_n)/u_n\},$$

where μ_A is the membership function A, $\mu_A : U \to [0,1]$. $\mu_A(u_i)$ indicates the grade of membership of u_i in A, $\mu_A(u_i) \in [0,1]$, $1 \le i \le n$.

Definition 2. Suppose $F(t)$ is computed from $F(t-1)$, the fuzzy logical relationship between $F(t)$ and $F(t-1)$ is illustrated by the following formula:

$$F(t) = F(t-1) * R(t, t-1),$$

where, * is is an arithmetic operator, $R(t, t-1)$ is the fuzzy logical relationship. If $F(t-1) = A_{t-1}$ and $F(t) = A_t$ then the fuzzy logical relationship is signed as $A_{t-1} \to A_t$.

Definition 3. Given a series of original data $\{X_i\}$ and the predictive value $\{\hat{X}_i\}$, $i = 1, 2, \ldots, n$. Then, we have some parameters as follows :

- Mean squared error: $MSE = \frac{1}{n}\sum_{i=1}^{n}\left(\hat{X}_i - X_i\right)^2$.
- Mean absolute error: $MAE = \frac{1}{n}\sum_{i=1}^{n}\left(\frac{|\hat{X}_i - X_i|}{X_i}\right)$.
- Mean absolute percentage error: $MAPE = \frac{1}{n}\sum_{i=1}^{n}\left(\frac{|\hat{X}_i - X_i|}{X_i}.100\right)$.

The value of these parameters are much smaller, then the proposed model is the best.

Defnition 4: A trapezoidal fuzzy number A* can be defined as follows: A*=(a,b,c,d)

$$\mu_{A*}(x) = \begin{cases} 0, x < a \ or \ x > d \\ \frac{x-a}{b-a}, a \le x \le b \\ 1, \ b \le x \le c \\ \frac{d-x}{d-c}, c \le x \le d \end{cases}$$

3. The proposed algorithm

Let $X_i, i = \overline{1, N}$ is time series. The proposed fuzzy time series model (FTSF) includes the seven steps as follows:

Step 1. Normalize the time series to scale 100:

$$Y_i = 100X_i / \max\{X_i\}, i = 1, 2, \ldots, N.$$

IOP Conf. Series: Materials Science and Engineering 1109 (2021) 012030 doi:10.1088/1757-899X/1109/1/012030

Step 2. Determine the suitable number of clusters for the universal set U. This problem is performed by the SNC (Suitable number of clusters) algorithm

Step 2.1. When $t=0$ (the number of current iterations is zero), initialize $V^{(0)}=\left\{V_1^{(0)},V_2^{(0)},...V_N^{(0)}\right\}=X=\left\{Y_1,Y_2,...Y_N\right\}$, $\varepsilon>0$ is a very small number.

Step 2.2. Update the sequence of centroids by the formula below:

$$V_i^{(t+1)}=\frac{\sum_{j=1}^{N} f\left(V_i^{(t)},V_j^{(t)}\right).V_i^{(t)}}{\sum_{j=1}^{N} f\left(V_i^{(t)},V_j^{(t)}\right)},i=1,...,N,$$

$$\text{where } f\left(V_i^{(t)},V_j^{(t)}\right)=\begin{cases} e^{\left(-\frac{d}{\lambda}\right)} & \text{if } d\left(V_i^{(t)},V_j^{(t)}\right)\le c_s \\ 0 & \text{if } d\left(V_i^{(t)},V_j^{(t)}\right)>c_s \end{cases},$$

with $d(V_i^{(t)},V_j^{(t)})$ is Euclidean distance between $V_i^{(t)}$ and $V_j^{(t)}$. $c_s=\frac{1}{n(n-1)}\sum_{i<j}d\left(V_i,V_j\right)$ is the average of the similar index of cluster of all pairs of data and $\lambda=\frac{d}{r}$. In fact, the window size λ determines the number of clusters in the time series, if $\lambda\to 0$, each element will belong to their own cluster, by contrast, all elements of the time series will belong to one cluster ($\lambda\to\infty$). The value of λ depends on constant r and c_s. In the numerical example of this articles, we take λ = d/5.

Step 2.3. Repeat Step 2 until $\max_i c\left(V_i^{(t+1)},V_i^{(t)}\right)<\varepsilon$.

In this algorithm, after each iteration, each $V_i^{(t)}$ converges to the centroid of cluster containing it. This process stops when the variation of all $V_i^{(t)}$ through two of adjacent iterations are less than ε. When ε is large, the algorithm will stop faster but the number of cluster can be unsuitable In this article, we choose $\varepsilon=0.0001$ for illustrative example.

Step 3. Determine the specific elements in each cluster by the Fuzzy Cluster Analysis (FCA) algorithm with four steps as follows:

Step 3.1. Divide Y_i, i=1,N into k intervals $C_1,C_2,...,C_k$ randomly. Establish the initial partition matrix $U^{(0)}=[\mu_{ij}]_{k\times n}$, with $\mu_{ij}=1$ if the jth element belongs to the c_i and $\mu_{ij}=0$ for otherwise.

Step 3.2. Find the representative element v_i for each population by the formula below:

$$v_i=\left(\sum_j^n \mu_{ij}{}^2 Y_j\right)/\left(\sum_j^n \mu_{ij}{}^2\right),$$

where $1\le i\le k$, μ_{ij} is the probability of jth element assigned to C_i.

Step 3.3. Update the new partition matrix $U^{(1)}$ by the following principle:

$$\mu_{ij}^{(1)} = \frac{1}{\sum_{l=1}^{k}[d(v_i,Y_j)/d(v_l,Y_j)]^2},$$

if $d(v_i,Y_j)>0$ for all $i=1,2,...,k$ and $\mu_{ij}^{(1)}=0$ for otherwise ($d(v_i,Y_j)$ is the Euclidean distance from v_i to Y_j).

Step 3.4. Repeat Step 3.2 and Step 3.3 until the following condition is satisfied:

$$\max_{ij}\left(\left|\mu_{ij}^{(t+1)}-\mu_{ij}^{(t)}\right|\right)<\varepsilon.$$

When Step 3 stops, we will be obtained a matrix of size $(k\times n)$. In this matrix, we have the sum of each column that always equal 1 ($\sum_{j}^{k}\mu_{ij}=1$). If $\max\{\mu_{ij}\}=\mu_{im}, 1\le m\le k$, then the element $Y_i, 1\le i\le n$ is put in cluster $C_m, 1\le m\le k$.

Step 4. Utilize $c_i=v_i, i=1,2,\ldots,k$ from step 3 and determine the universal set U as follows:

$$U=[c_1-\frac{c_2-c_1}{2},c_k+\frac{c_k-c_{k-1}}{2}].$$

Divide the universal set U into k intervals:

$$U_1=[c_1-\frac{c_2-c_1}{2},\frac{c_1+c_2}{2})=[D_1;D_2);\ U_i=[\frac{c_{i-1}+c_i}{2},\frac{c_{i+1}+c_i}{2})=[D_i;D_{i+1}),\ i=\overline{2,k-1}$$

$$U_k=[\frac{c_{k-1}+c_k}{2},c_k+\frac{c_k-c_{k-1}}{2}]=[D_{k-1};D_k)$$

Step 5. Calculate the different mean of intervals by the following equation:

$\bar{D}=\frac{1}{k}\sum_{i=1}^{k}(D_i-D_{i-1})$. With k intervals (k-fuzzy numbers) and $A_1,A_2,...,A_k$ can be defined as follows:

$$A_1=[D_1-\bar{D},D_1,D_2,D_2+\bar{D});$$
$$A_2=[D_2-\bar{D},D_2,D_3,D_3+\bar{D});$$
$$\ldots$$
$$A_k=[D_{k-1}-\bar{D},D_{k-1},D_k,D_k+\bar{D});$$

where $A_i, i=\overline{1,k}$ is the fuzzy numbers.

Step 6. If $Y_j\in U_i, 1\le i\le k, 1\le j\le N$ then $Y_j\to U_i$ and the fuzzy logical relationship of k intervals is $U_i(Y_i<Y_l)\to U_l$.

- If the fuzzy logical relationship group of U_i is empty ($U_i\to\varnothing$) then $FI=A_i$.
- If the fuzzy logical relationship group of U_i is one-to-one then $FI=A_{i-1}$.
- If the fuzzy logical relationship group of U_i is one-to-many,

  $$U_i\to U_i,U_i\to U_a,...,U_i\to U_b, 1\le i,a,b\le k$$

  then $FI=(A_i+A_a+A_b)/(i+a+b)$.

Step 7. Calculate the forecasting interval for time series by the following equation

$$FY = FI * \max(X_i)/100,$$

where, $FI = [FI_{1,j}; FI_{2,j}]$ and $\begin{cases} FI_{1,j} = (D_i - D_{i-1}) * \alpha + D_{i-1} \\ FI_{2,j} = D_k - (D_k - D_{k-1}) * \alpha \end{cases}$, α is the degree of confidence of forecasted intervals.

4. The numerical example

In this section, we use the Enrollment data of Alabama University (ACD) to illustrate the proposed algorithm. This data is used in many researches as Song and Chisson [18], Chen [8], Huarng [13] Singh [17] and it is shown in the Table 1.

Table 1. Original and standardized data and the convergence value of clusters in the Enrollment data

Year	X_i	Y_i	$V^{(18)}$	Cluster	Year	X_i	Y_i	$V^{(18)}$	Cluster
1971	13055	67.513	67.513	C_1	1982	15433	79.811	79.964	C_5
1972	13563	70.140	70.140	C_2	1983	15497	80.142	79.964	C_5
1973	13867	71.712	71.712	C_3	1984	15145	78.321	78.372	C_{11}
1974	14696	75.999	75.999	C_4	1985	15163	78.414	78.372	C_{11}
1975	15460	79.950	79.964	C_5	1986	15984	82.660	82.621	C_{12}
1976	15311	79.180	79.198	C_6	1987	16859	87.185	87.181	C_9
1977	15603	80.690	80.631	C_7	1988	18150	93.862	93.862	C_{13}
1978	15861	82.024	82.064	C_8	1989	18970	98.102	97.859	C_{14}
1979	16807	86.916	87.181	C_9	1990	19328	99.953	99.977	C_{15}
1980	16919	87.495	87.181	C_9	1991	19337	100.000	99.977	C_{15}
1981	16388	84.749	84.749	C_{10}	1992	18876	97.616	97.859	C_{14}
1972	13563	70.140	70.140	C_2					

The steps of the proposed algorithm are presented as follows:

Step 1: Standardize the time series on the scale 100, we have the result presented at column Y_i of Table 1.

Step 2: Determine the number of clusters for time series by the SNC algorithm. This result is shown by Figure 1.

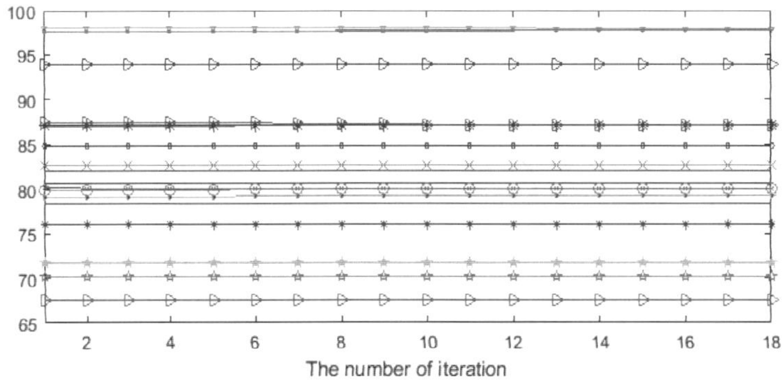

Figure 1. The convergence of time series after 18 iterations

Figure 1 shows that Enrollment data converge into 15 clusters. The numeral result obtained in Table 1.

Step 3: For 15 clusters obtained in Step 2, we determine the centroid points of each cluster according to the SNC algorithm. The outcome is shown in Table 2.

Table 2. The value of centroid clusters

Cluster	Centroid (v_i)	Cluster	Centroid (v_i)
C_1	67.513	C_9	87.181
C_2	70.140	C_{10}	84.749
C3	71.712	C_{11}	78.372
C_4	75.999	C_{12}	82.621
C_5	79.964	C_{13}	93.862
C_6	79.198	C_{14}	97.859
C_7	80.631	C_{15}	99.977
C_8	82.064		

Table 2 shows the value of 15 representing clusters for time series.

Step 4. Divide the universal set U into 15 intervals as follows:

U_1=[66.1995; 68.8266],U_2=[68.8266; 70.9262],U_3=[70.9262; 73.8558],
U_4=[73.8558; 77.9819],U_5=[77.9819; 79.5810],U_6=[79.5810;79.9142],
U_7=[79.9142; 81.3472],U_8=[81.3472; 84.6222],U_9=[84.6222; 85.9651],
U_{10}=[85.9651; 81.5605],U_{11}=[81.5605; 80.4961],U_{12}=[80.4961; 88.2410],
U_{13}=[88.2410;95.8604],U_{14}=[70.9262; 73.8558],U_{15}=[95.8604; 98.9180]

Step 5. Calculate $\overline{D} = 2,3224$ and A_i, i = 1,..,15, we have the Table 3:

Table 3. The computing result of $\overline{D}$

A_1	63.8771	66.1995	68.8266	71.1490
A_2	66.5042	68.8266	70.9262	73.2486
A_3	68.6038	70.9262	73.8558	76.1782
A_4	71.5334	73.8558	77.1855	79.5079
A_5	74.8631	77.1855	78.7846	81.1070
A_6	76.4622	78.7846	79.5810	81.9034
A_7	77.2586	79.5810	80.2976	82.6200
A_8	77.9752	80.2976	81.3472	83.6696
A_9	79.0249	81.3472	82.3421	84.6645
A_{10}	80.0197	82.3421	83.6850	86.0074
A_{11}	81.3626	83.6850	85.9651	88.2875
A_{12}	83.6427	85.9651	90.5211	92.8435
A_{13}	88.1987	90.5211	95.8604	98.1828
A_{14}	93.5380	95.8604	98.9180	101.2404
A_{15}	96.5956	98.9180	101.0354	103.3578

Step 6. Establish the fuzzy logical relationship as follows:

Table 4. The fuzzy logical relationship of elements

1971	Y_1	U_1	--	1982	Y_{12}	U_7	$U_{11} \rightarrow U_7$
1972	Y_2	U_2	$U_1 \rightarrow U_2$	1983	Y_{13}	U_8	$U_7 \rightarrow U_8$
1973	Y_3	U_3	$U_2 \rightarrow U_3$	1984	Y_{14}	U_5	$U_8 \rightarrow U_5$
1974	Y_4	U_4	$U_3 \rightarrow U_4$	1985	Y_{15}	U_5	$U_5 \rightarrow U_5$
1975	Y_5	U_7	$U_4 \rightarrow U_7$	1986	Y_{16}	U_{10}	$U_5 \rightarrow U_{10}$
1976	Y_6	U_6	$U_7 \rightarrow U_6$	1987	Y_{17}	U_{12}	$U_{10} \rightarrow U_{12}$
1977	Y_7	U_8	$U_6 \rightarrow U_8$	1988	Y_{18}	U_{13}	$U_{12} \rightarrow U_{13}$
1978	Y_8	U_9	$U_8 \rightarrow U_9$	1989	Y_{19}	U_{14}	$U_{13} \rightarrow U_{14}$
1979	Y_9	U_{12}	$U_9 \rightarrow U_{12}$	1990	Y_{20}	U_{15}	$U_{14} \rightarrow U_{15}$
1980	Y_{10}	U_{12}	$U_{12} \rightarrow U_{12}$	1991	Y_{21}	U_{15}	$U_{15} \rightarrow U_{15}$
1981	Y_{11}	U_{11}	$U_{12} \rightarrow U_{11}$	1992	Y_{22}	U_{14}	$U_{15} \rightarrow U_{14}$

Find solution of the fuzzy logical relationship.

Table 5. The result of fuzzy numbers

Fuzzy numbers	Fuzzy numbers
$U_1 \rightarrow U_2$	$U_9 \rightarrow U_{12}$
$U_2 \rightarrow U_3$	$U_{10} \rightarrow U_{12}$
$U_3 \rightarrow U_4$	$U_{11} \rightarrow U_7$
$U_4 \rightarrow U_7$	$U_{12} \rightarrow U_{12}; U_{12} \rightarrow U_{11}; U_{12} \rightarrow U_{13}$
$U_5 \rightarrow U_5; U_5 \rightarrow U_{10}$	$U_{13} \rightarrow U_{14}$
$U_6 \rightarrow U_8$	$U_{14} \rightarrow U_{15}$
$U_7 \rightarrow U_6; U_7 \rightarrow U_8$	$U_{15} \rightarrow U_{15}; U_{15} \rightarrow U_{14}$
$U_8 \rightarrow U_9; U_8 \rightarrow U_5;$	

Calculate the value of the FI

Table 6. The outcome of the FI

Year	Interval 1	Interval 2	Year	Interval 1	Interval 2
1971	[63.877, 66.200]	[68.827, 71.149]	1982	[77.259, 79.581]	[80.298, 82.620]
1972	[66.504, 68.827]	[70.926, 73.249]	1983	[77.219, 79.541]	[80.464, 82.787]
1973	[68.604, 70.926]	[73.856, 76.178]	1984	[76.944, 79.266]	[80.563, 82.886]
1974	[71.533, 73.856]	[77.186, 79.508]	1985	[77.441, 79.764]	[81.235, 83.557]
1975	[77.259, 79.581]	[80.298, 82.620]	1986	[77.441, 79.764]	[81.235, 83.557]
1976	[77.219, 79.541]	[80.464, 82.787]	1987	[83.643, 85.965]	[90.521, 92.844]
1977	[77.975, 80.298]	[81.347, 83.670]	1988	[84.401, 86.724]	[90.782, 93.105]
1978	[76.944, 79.266]	[80.563, 82.886]	1989	[93.538, 95.860]	[98.918, 101.240]
1979	[83.643, 85.965]	[90.521, 92.844]	1990	[96.596, 98.918]	[101.035, 103.358]
1980	[84.401, 86.724]	[90.782, 93.105]	1991	[95.067, 97.389]	[99.977, 102.299]

1981	[84.401, 86.724]	[90.782, 93.105]

Step 7. The time series is predicted by the intervals with 80% of the degree of confidence that is shown in Table 7 below.

Table 7. The outcome of forecasting according to interval

Year	FY	Year	FY
1971	[12711.183, 13398.817]	1982	[15298.755, 15616.957]
1972	[13219.185, 13804.816]	1983	[15291.044, 15649.161]
1973	[13625.184, 14371.315]	1984	[15237.922, 15668.353]
1974	[14191.683, 15015.177]	1985	[15334.111, 15798.189]
1975	[15298.755, 15616.957]	1986	[15334.111, 15798.189]
1976	[15291.044, 15649.161]	1987	[16533.252, 17593.884]
1977	[15437.324, 15819.934]	1988	[16679.951, 17644.370]
1978	[15237.922, 15668.353]	1989	[18446.710, 19217.593]
1979	[16533.252, 17593.884]	1990	[19037.960, 19627.040]
1980	[16679.951, 17644.370]	1991	[18742.335, 19422.316]
1981	[16679.951, 17644.370]		

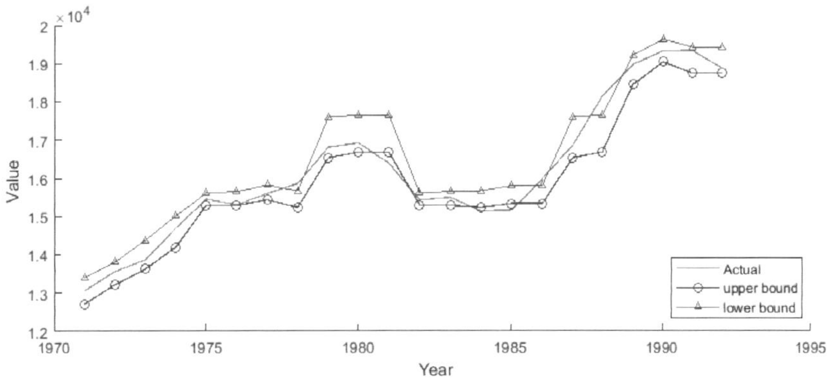

Figure 2. The line graph presents the comparison between actual outcome and upper and lower bound of forecasting result.

Figure 2 shows time series predicted based on the confidence level is 80%. Furthermore, they also present the fluctuating level of original data according to upper and lower bound.

Compare to the forecasting result of some method, we have the Table 8:

Table 8. The result of algorithms for the Enrollment dataset

Criteria	Chen (1996) [8]	Huarng (2001) [13]	Abbasov (2003) [1]	Lee (2004) [15]	Chen (2004) [7]
MAE	502.38	299.15	479.57	296.15	293.45
MAPE	3.08	2.45	2.87	2.69	1.76
MSE	413,980.98	226,611	342,326	255,227	138,366.80
Criteria	**Singh (2007) [17]**	**Yu (2010) [23]**	**Khashei (2011) [14]**	**Chen (2013) [9]**	**Ghosh (2015) [12]**
MAE	254.16	216.50	211.12	314.34	298.68
MAPE	1.53	2.15	2.12	2.17	1.82
MSE	95,305	47,231.03	31,021	41,235	186,421

Criteria	Egrioglu (2016) [10]	Bindu (2016) [4]	Yusuf (2017) [24]	Tai (2018) [19]	Proposed
MAE	192.15	285.28	182.51	168.84	132.73
MAPE	1.83	1.65	1.62	1.02	1.04
MSE	34,280	174,390.90	31,752	28,525.00	26,461

According to Table 6, we can see that the proposed model has the better outcome than others because it has the lowest parameters.

5. Conclusion

The article has presented a fuzzy forecasting model for time series, a common data type in reality. Based on many important improvements from the unsupervised learning algorithm, the existing fuzzy time series, and the trapezoidal fuzzy numbers, we propose the new interval forecasting model. In addition, this study has applied well machine learning technique in training for fuzzy time series. This is a new research direction in the current digital age. The proposed model has performed better than the existing ones based on the Enrollment dataset. With this new model, in the next time, we will continue to improve the steps to divide the domain time series and the training data based on improving the machine learning technique, and apply to many important forecasts to reality.

Acknowledgments

For Dinh PhamToan, this work was funded by Vingroup Joint Stock Company and supported by the Domestic Master/ PhD Scholarship Programme of Vingroup Innovation Foundation (VINIF), Vingroup Big Data Institute (VINBIGDATA)

References

[1] Abbasov A and Mamedova M 2003 Application of fuzzy time series to population forecasting. *Vienna University of Technology* **1** pp 545-552

[2] Abreue P H, Silva D C, Mendes-Moreira J, Reis L P and Garganta J 2013 Using multivariate adaptive regression splines in the construction of simulated soccer team's behavior models. *International Journal of Computational Intelligence Systems* **6** pp 893-910

[3] Aladag S, Aladag C H, Mentes T and Egrioglu E 2012 A new seasonal fuzzy time series method based on the multiplicative neuron model and SARIMA, *Hacettepe Journal of Mathematics and Statistics* **41** pp 145-163

[4] Bindu G, Rohit G 2016 Enhanced accuracy of fuzzy time series model using ordered weighted aggregation. *Appl. Soft Comput* **48** pp 265-280

[5] Box G E P and Jenkins G M 1970 *Time series analysis: Forecasting and control, Holden-Day, San Fancisco* 546 pages

[6] Chen J and Hung W 2015 An automatic clustering algorithm for probability density functions, J. Stat. *Comput. Simul* **48** pp 3047-3063

[7] Chen S M and Hsu C C 2004 A new method to forecast enrollments using fuzzy time series, *International Journal of Applied Science and Engineering* **2** pp 234–244

[8] Chen S M 1996 Forecasting enrollments based on fuzzy time series, *Fuzzy sets and systems* **81** pp 311–319

[9] Chen S M, Kao P Y 2013 TAIEX forecasting based on fuzzy time series, Particle swarm optimization techniques and support vector machines, *Inf. Sci.* **247** pp 62-71

[10] Egrioglu S, Bas E, Aladag C H, Yolcu U 2016 Probabilistic fuzzy time series method based on artificial neural network, *American Journal of Intelligent Systems* **62** pp 42-47

[11] Eren B, Vedide R, Uslu U and Erol E 2014 A modied genetic algorithm for forecasting fuzzy time series, *Applied Intelligence* **41** pp 453–463

[12] Ghosh H, Chowdhury S, Prajneshu S 2015 An improved fuzzy time-series method of forecasting

based on L-R fuzzy sets and its application, *Journal of Applied Statistics* **43** pp 1128–1139
[13] Huarng K 2001 Heuristic models of fuzzy time series for forecasting, *Fuzzy sets and systems* **123** pp 369–386
[14] Khashei M, Bijari M, Hejazi C S R 2011 An extended fuzzy artificial neural networks model for time series forecasting, *Iranian Journal of Fuzzy Systems* **3** pp 45-66
[15] Lee H S, Chou M T 2004 Fuzzy forecasting based on fuzzy time series, *International Journal of Computer Mathematics* **81** pp 781-789
[16] Liu Hao-Tien 2007 An improved fuzzy time series forecasting method using trapezoidal fuzzy numbers, *Fuzzy Optimization and Decision Making* **6** pp 63-80
[17] Singh S R 2007 A simple method of forecasting based on fuzzy time series, *Applied mathematics and computation* **186** pp 330–339
[18] Song Q and Chissom B S 1993 Fuzzy time series and its models, *Fuzzy sets and systems* **54** pp 269-277
[19] Tai Vovan 2018 An improved fuzzy time series forecasting model using variations of data, *Fuzzy Optimization and Decision Making* **18** pp 151-173
[20] Trendafilova I and Manoach E 2008 Vibration-based damage detection in plates by using time series analysis, *Mechanical Systems and Signal Processing* **22** pp 1092-1106
[21] Tseng F M, Tzeng G H and Yu H C 1999 Fuzzy seasonal time series for forecasting the production value of the mechanical industry in Taiwan, *Technological Forecasting and Social Change* **60** pp 263-273
[22] Trendafilova I 2006 Vibration-based damage detection in structures using time series analysis, *Proceedings of the Institution of Mechanical Engineers, Part C: Journal of Mechanical Engineering Science* **220** pp 261-272
[23] Yu H K, Huarng K 2010 A neural network-based fuzzy time series model to improve forecasting, *Expert Systems with Applications* **37** pp 3366-3372
[24] Yusuf S M, Mohammad A, Hamisu A A 2017 A novel two-factor high order fuzzy time series with applications to temperature and futures exchange forecasting, *Nigerian Journal of Technology* **36** pp 1124-1134

Nanotechnology in the Philippines: Development of framework for technology adoption

N R E Lim[1], G N C Santos[2], A T Ubando[1], and A B Culaba[1]

[1]Mechanical Engineering Department, De La Salle University – Manila, Manila, Philippines

[1]Physics Department, De La Salle University – Manila, Manila, Philippines

nino_lim@dlsu.edu.ph

Abstract. Groundbreaking advancements have driven significant industries to adopt and take advantage of the key technologies in order to thrive in these adjustments coming from the Fourth Industrial Revolution (FIRe). However, with the threat of the Coronavirus disease 2019 (COVID-19) causing catastrophic setback to the society and economy, scientific endeavors have shifted towards finding a solution to this pandemic. It is high time to evaluate how the current technology from the FIRe can help a country, such as the Philippines, survive such outbreaks and unforeseen circumstances. Advancements in the field of nanotechnology aim to develop smaller and inexpensive testers, and overall newer and more effective medical solutions. However, the extent on how these technologies can be realized depends on the interplay between the country's policies, and the strong link between industry and academe. This paper raises policy recommendations from synthesizing the trends in nanotechnology and forecasting how it develops so as the country can readily adapt. Moreover, this paper identifies the barriers and constraints, as well as the factors necessary for the technology adoption in the country.

1. Introduction

With the promise of economic growth, and comfort of living, most technological innovations have seen productivity over the last decades [1,2]. These developments have been identified as significant factors in prosper and survival for any firm, company, or country. [3-5]. As such, important attention must be placed upon emerging technologies to be able to anticipate the evolution and future applications of the technology [6]. Technological innovation systems have been developed to address the importance of several relevant actors such as firms, government, and academe, and the interactions surrounding the technology [7,8]. And with the advent of the Fourth Industrial Revolution (FIRe), several disruptive technologies have developed that substantially affected the productivity, and operational efficiency of industries [9]. Among these disruptive technologies, nanotechnology is acknowledged as one of the major worldwide technology research initiatives of the 21st century [10,11].

Nanotechnology, as defined by the U.S. National Nanotechnology Initiative (NNI), deals with the applications of particles within the size of 1 to 100 nm [12]. Additionally, the development, and manufacturing of materials with structures, fibers, or platelets smaller than 100 nm involves nanotechnology [13]. This includes the discovery of novel properties and material behavior in the nanoscale, as well as research and development of devices, methodologies, and system frameworks to gather data on a nanomaterial [14]. Nanotechnology serves as a key to a promising future, paving the way for more advances in the fields of engineering, materials science, electronics, and other disciplines such as in energy, biomedicine, food, and environment [15]. Moreover, it can help in understanding nature, creating advanced manufacturing equipment, developing innovative medical treatments, and even disturb societal norms [16]. The nanoscale threshold is where the behavior of atoms or molecules are a function of the composition, structure, and size. It is within this level that fundamental properties are explored, and miniscule changes in the composition can produce novel properties that would considerably revolutionize the areas where the technology is applied such as in engineering, industry, safety and security [15].

The Philippines and the world have seen the rapid growth of technological advancements amidst the waves from the FIRe. Noticeable changes are observed in terms of economy and increase in industrial productivity; various business frameworks have been established to adapt to these shifts and thrive in the years to come. However, the scope of how to harness the advantages from these emerging technologies depend on the country's ability to project the developments and adapt accordingly with the various disruptions brought about by the technologies [17]. The adaptation process would entail careful analysis of technological trends, observing how such advancements affect the stability of the society and economy. Additionally, this would require proper preparations and implementation from the stakeholders such as the government, research facilities, and industries.

The potential of nanotechnology has been acknowledged by the Philippines and as such was able to develop a technological roadmap on how the technology will develop and affect the lives of the Filipinos. That was 10 years ago, the previous decade focused the resources of the country on developing the 4 identified prioritization areas of nanotechnology which are: (1) energy, (2) food and agriculture, (3) environment, and (4) development of nanocomposite materials from local minerals [18]. However, with the presence of the COVID-19 pandemic that hinders the development of most industries, these previously drafted roadmaps and technological adoption plans are being impeded. The presence of the pandemic has and will continuously negatively affect the country's economic performance and production capacity due to the disruption in the global supply chain and increase in mortality rate [19]. In line with this, the paper aims to raise policy recommendations for the development and adoption of nanotechnology in the Philippines that would anticipate the technology trends and identify how nanotechnology can help to combat pandemic.

The rest of the paper is constructed as follows. The Section 2 discussed the ongoing technological trends in the field of nanotechnology applications in the global and local scale. In Section 3, the several factors, barriers, and constraints for technology adoption were identified and briefly discussed, while Section 4 discussed the impacts of nanotechnology in other fields. Finally, in Section 5 laid out the policy recommendations that would aid in the adoption of nanotechnology in the country.

2. Trends in Nanotechnology Application

2.1. Global trends

Dealing with materials at the nanoscale (10-9), the study of tiny materials has huge impacts on several applications. The field of nanotechnology spreads its application in various fields; as the nanoscale control of structure, size, and composition have proved useful in terms of food and agriculture [20], biomedical applications [21], and drug delivery [22].

A SCOPUS search on nanotechnology trends over the last 5 years (2015-2020) was made and has resulted in over 1 402 papers, upon which 715 are journal articles, and 687 are review papers. The resulting papers scoped to different fields such as Materials Science, Engineering, Chemistry, Biochemistry, Chemical Engineering, Pharmacology, Physics, Medicine, Environmental Science, and Energy. Fig. 1 illustrates the correlation among the major keywords and topics with regards to nanotechnology trends. The figure is generated by calculating the amount of times the topics or keyword was mentioned or discussed in the identified papers. The size of the font and nodule indicates the relevance of the topic with respect to the network established on nanotechnology trends. From this figure, four main clusters were identified: (1) disease and drug delivery, (2) sensors, and biosensors, (3) computational simulation of nanomaterials, and (4) experimentation on nanomaterial properties.

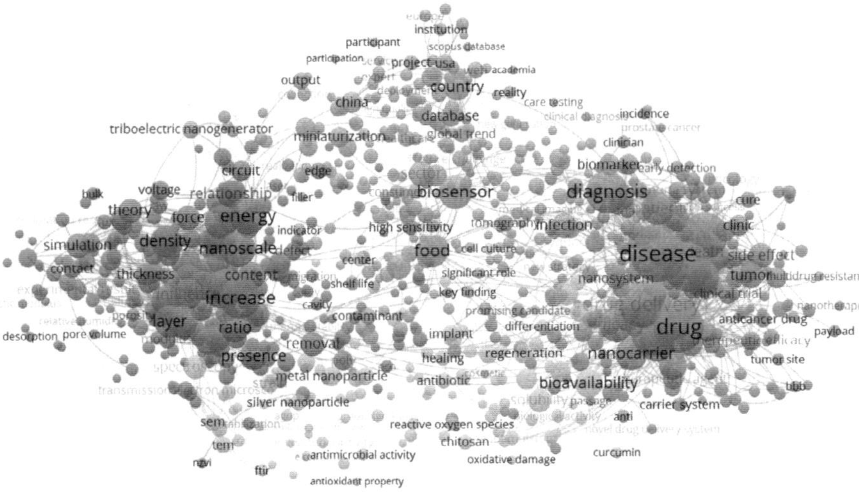

Figure 1. Keywords and topics on nanotechnology trends from 2015 to 2020.
Source: Adapted from VOSViewer

Medical applications of nanomaterials are one of the well researched fields of nanotechnology. Most of the studies in this cluster focus on drug delivery; particularly utilizing nanomaterials in chemotherapies that guide the treatment to identified cancer cells, directing the removal of tumors in tissues or organs, improving the effectiveness of radiation-based therapy [23-25], and the use of nanocarriers in treatment of Central Nervous System disease [22]. The cluster on sensors, and biosensors primarily focuses on the aspect of detection. Nanotechnology equipment and methodology are developed to detect pathogenic bacteria [26]. Gold nanoparticles are used for medical imaging and treatment management for pancreatic cancer [27]. In addition, scaffoldings for skin tissue are developed by fabricating nanofibers [28]. The cluster on computational nanotechnology delves on discovering new materials and designing it to have suitable properties for the intended applications. Research on custom tailoring the material properties of graphene for various applications [29] and designing superhydrophobic nano-based coatings [30] are some of the studies being explored in this cluster. On the other hand, the experimental side of nanotechnology deals with varying the structures and composition of materials at the nanoscale then finding new purposes and applications such as in epoxy nanocomposites [31], nano-biopesticides [32], and nanomaterials as antimicrobial agents for food safety [33]. Identifying the main clusters can help in determining and predicting the direction of the development in the field of nanotechnology.

2.2. *Local trends*

In 2009, the Department of Science and Technology (DOST) of the Philippines formally acknowledged the field of nanotechnology as one of the priorities for Research and Development (R&D). As such, 14 local scientists were gathered to develop a roadmap for nanotechnology in the Philippines that identified major areas of development that received attention and funding [18].

The applications of nanostructured solar energy devices were pursued in a collaborative effort from the top universities in the country and DOST. Solar cells were developed with embedded nanoparticles and graphene. Nanosensors that utilizes nanoparticles for environmental assessment and damage control were also researched. In the field of agriculture, Mesoporous Silicate Nanoparticles (MSNs) were used for enabling gene transformation in the plant and animal cells, several pesticides and fertilizers with nanoparticles were developed, and applications of nano-based sensors were made for precision agriculture [34]. Several funded projects funded by the DOST, aimed to utilize nanomaterials produced from the naturally-occurring minerals such as development of nanocomposites using recycled polycarbonate and modified montmorillonite, synthesis of strong plastics from the introduction of carbon nanotubes, and development of halloysite-epoxy hybrid nanocomposite as molding compound in semiconductor industry applications [18].

StatNano, a statistics gateway founded to monitor the latest developments in the field of nanotechnology has reported that the Philippines currently has established 6 companies that have developed 23 products in the fields of automotive, medicine, construction, textile, and home appliance. In terms of research, the field of nanotechnology made up 4.53% of the total ISI-indexed articles of the country, and a total of 0.50% out of the total research articles published in 2019 [35]. With the low numbers in terms of research, for a field that is recognized by the country to be a billion dollar industry globally [34]; more attention and resources should be allocated to harness the advantages coming from this emerging technology.

In the government's effort to attain world-class nanotechnology laboratories in the country, the Nanolab was established to provide technical services for the different local industries, and to develop and implement research initiatives in the field of nanotechnology [36]. Several experts consider nanotechnology to be one of the important disruptive technologies, capable of incorporating the technology in early detection and prevention, as well as to enhance diagnostic and treatment capabilities [34]. The future aspects of nanotechnology are deemed bright with the attention that the government has placed early on to this technology. The top efforts for research in the next years are expected to focus on bioactive nanostructures for drug delivery, improved imaging techniques, filtration using nanomaterials, and applications for nanoparticles and nanocomposites that enhance the functionality of natural materials such as in silica, limestone, and clay [36]. The projected advancements on these researches would also imply that the spread of nanotechnology in the curriculum is necessary to produce essential human resources, and procure the appropriate equipment to upgrade existing laboratories focusing on nano research.

2.3. Role of nanotechnology in the pandemic
The presence of COVID-19 pandemic has caused devastating effects to the economy and society, as well as affecting productivity in various scientific fields and programs [37]. This provided opportunities for emerging technologies to be integrated in the current battle against the pandemic [38]. Nanotechnology-based approaches are one of the potential directions in combatting the pandemic. The potential advantages coming from the nature of the size and morphology of the nanoparticles, facilitate the drug delivery precision to minimize threats coming from the immune system response [39].

The nano-based approach towards infectious diseases have been previously studied. In a paper by Kerry, et al. [40], the different applications on viral sensing and antimicrobial properties of identified nanoparticles were presented. Additionally, major antiviral approaches that utilizes nano-based technology have been presented to aid in strategizing possible methodologies in combating pandemics. The review also provided some insights with regards to how the development of nanoparticles would not be enough until the pathogenesis of a virus is understood.

Nanotechnology has been taking its course and realizing its role in this pandemic; with its main role of developing means of virus detection, reliable personal protection equipment (PPE), and better medical response. Nanosensors were being developed that would enable to detect viral infection even before symptoms start to appear. An Israeli company, Scentech Medical have started testing the breath analyzer that was developed through the use of nanoparticles and aims to be able to diagnose in 10 seconds [41]. Attention is also being placed on improving the efficiency of the PPE such as the incorporation of nanofibers in the filters being used on the masks. Furthermore, research on the various antiviral properties of nanoparticles are being made.

3. Nanotechnology Adoption in the Philippines
Through the collective effort of researchers, engineers, and scientists worldwide, the development of a technology continues to move forward. Potential benefits and drawbacks may arise as the technology matures. And in order to fully harness the advantages from an emerging technology, proper actions

must be taken by the stakeholders to facilitate the entry of such technology to the society and market. For groundbreaking technologies such as nanotechnologies, several factors are identified that would positively influence the adoption of this technology in a country [42].

First factor to consider is the benefits that the technology brings to the country. What are the advantages gained once the technology is spread in the market? Secondly, gauge the usefulness or utility that the benefits provide; identify the practical use and application of the technology. Next, determine the needs of the country and how can the entering technology address it. Another factor that influences the entry of a technology to the country would be the maturity level and how readily available the current state of technology is. It is important for the technology to reach a maturity level that would allow easier adoption to the country. Lastly, the realness of technology and how it is to be communicated to a layman. Understanding the basic concepts of the technology would lead to the acceptance of the technology.

This section identifies the important factors to drive the adoption of nanotechnology in the Philippines and the barriers that are to be encountered to achieve the objective of harnessing the technological benefits nanotechnology provides.

3.1. Facilitating Factors

Nanotechnology has continued to prosper from the support of public and private sectors of different countries. The increase in efforts for adoption of its technology has also resulted in multiple commercialization failures [43]. As such, research frameworks and technological innovation systems are being developed to identify the key factors in nanotechnology adoption.

One of the major applications of nanotechnology involves sensors, and biosensors. In line with this, the potential growth of nanotechnology as a field and industry relies on the expansion of the electronics and opto-electronics industry [44]. The development of advanced electronic devices is in conjunction with the discoveries in the field of nanotechnology. As the semiconductor and electronics further improve, so is the involvement of nanotechnology; it would not take long for the public to acknowledge the benefits of nanotechnology and welcome its adoption to the country.

Ever since the initial applications of nanotechnology were realized, one of the concerning aspects are the environmental hazards the technology might bring forward. As such, it is important to also acknowledge and focus on the eco-friendly aspects of nanotechnologies to ease the adoption of the technology [45]. With the special properties being presented by nanomaterials, this leads to interesting applications that promises environmental protection as opposed to the current materials being used. The development of stronger and more durable materials would lead to the increase the life of a product, this translates to less additional constructions and maintenance. Moreover, nanomaterials are being used in building insulations to optimize the energy being used.

As one of the emerging technologies that has the potential to disrupt the daily activities in several industries and economic activities of countries, it is important that attention and support is provided for the field of nanotechnology. An increase in the gross domestic expenditure on research and development (GERD) and income level to promote corporate nanotech patent applications would play a substantial factor in the engagement activities in nanotechnology, as well as its adoption in the country [46].

3.2. Barriers and constraints for adoption

Aside from identifying the facilitating factors that would aid in the ease of adoption of nanotechnology in the Philippines, it is also important to determine the barriers and constraints that impedes, or even inhibits the entry of the technology in the country. As part of the technologies that have emerged in the FIRe, nanotechnology is considered to be in its early stages, and to be part in that stage there are several barriers that might appear when the technology is to be adopted.

Technology transfer has been a major hurdle in any technological field that aims to be adopted and developed. For nanotechnology, it is estimated to be at least 3 years for the research to be commercialized [47]. This presents multiple financial risks and is not ideal for investors when a technology takes that long to be commercialized. In turn, this results in less attention and support to be placed in the field; ultimately delaying its progress for adoption.

Another problem with a technology in the early stage is that it progresses quickly. When several advances are being made, most of the equipment and devices become outdated. Moreover, research in nanotechnology requires proper infrastructures that are expensive such as laboratories and equipment to innovate further in the field.

In terms of the business sector, small businesses do not have the capacity to be able to access necessary equipment to adopt nanotechnology in their operations. Additionally, the public perception towards nanotechnology is that it is unsafe [47]. The involvement of the citizens and public approval play an important role towards adoption of nanotechnology in the country. Lastly, the lack of experts on the field of nanotechnology is negatively affecting its acceptance.

3.3. Philippines – Nanotech policy

In the 2018 Asia Nano Forum, the Philippines presented the status of nanotechnology in the country, the identified factors, and constraints for the technology transfer and commercialization [48].

The Philippine nanotech policy acknowledged that the adoption of nanotechnology in the country is highly dependent on funding from government agencies, government research institutions, and other research organizations. Identified barriers and constraints for adoption include the amount of time required for the realization of the R&D in the field of nanotechnology. In addition, most equipment required for nanotech research are costly making the investment towards the R&D to be extremely capital intensive. The local scene must also be wary towards the fast-paced developments in the global dimension of nanotechnology, efforts would be futile when more efficient methods or techniques are already developed in other countries. Lastly, the nanosafety issues from R&D to the end-product phase raises concern. As an emerging technology, there are uncertainties in terms of the long-term effect that the nanomaterials might bring towards human health and environment.

On the other hand, the nanotech policy recognizes the legal framework that would facilitate the technology transfer and commercialization of nanotechnology in the Philippines. Several policies are identified to help in this effort such as the Technology Transfer Act for ease of adoption in the market, Guidelines on Intellectual Property Validation to protect the scientists and researchers, Commercialization of Information Sharing to capitalize on important findings, and Technology Transfer Protocols for the systematic transfer of knowledge involve in the manufacturing, or development of products. Furthermore, under the nanotech policy, efforts are being made to market and promote nanotechnology towards the public and the industry. Small businesses and start-ups will be given access to technology experts to help commercialize the nanotechnologies and will also be given access to government funded laboratories.

4. Impacts of Nanotechnology

Aside from the analysis of how the technology has developed and where its future direction lies, it is also important in the development of policy recommendations to assess the impacts of the technology in aspects that would determine its viability for adoption. The development of nanotechnology would entail some influence on the different sectors such as economy, environment, and society.

4.1. Economic Impacts

Investments made towards R&D of a technology spreads throughout the entire economy of the country [49]. The development of nano-based products and processes is intensive for human and capital resources. The innovation cycles are fast-paced, and a knowledgeable workforce is required; several public and private funding were drawn by the promise of nanotechnology as a disruptive technology that will alter the current livelihood and operations in the world [50]. Billions of dollars of cumulative global R&D investments towards nanotechnology were estimated [51]. The United States alone has invested over $27 billion towards nanotechnology research since 2001 [52]. As a maturing technology, it would require several years before the economic benefit of nanotechnology to be realized. However, promise from the investment is shown as revenue generation in the European nanomaterials market has reached $2.5 billion in 2015 with a projection of $9.1 billion by 2022 [53].

Additionally, the development of the field would yield to growth in job opportunities [54]. As more nanomaterials are being discovered and advanced processes are developed, this would require human resources to cater the needs for the advancements. The economic impact of nanotechnology in a country can also be viewed through a time lens [55]. Partnerships and linkages are one of the main benefits in the short-term for the development of nanotechnology in a country. Several research facilities and instruments would be acquired, and capacity building is of utmost urgency. In the medium-term once nanotechnologies have planted their roots; supply-chain structure would be established, skilled graduates are generated, the R&D cycle would be further enhanced, and commercialization of the nano-based technologies would be improved. This would lead to harnessing the economic benefits of nanotechnology in the long-term such as in the effect on the Gross Domestic Product (GDP) and faster rate of Return on Investment (ROI).

4.2. Environment impacts

The effect and impact of nanotechnology towards living organisms and the ecosystem has been researched since the emergent of the technology. Particular concerns and drawbacks from the technology have been raised early on such as the uncertainties and lack of knowledge with regards to the long-term effects, and the so-called nanopollution [54]. Undesirable health or environmental outcomes from an emerging technology leads to backlash and criticism from the public and key stakeholders, that would eventually lead to stricter regulations [56]. Major environmental concerns are being heavily investigated such as the effects of nanoparticles towards its surroundings. Engineered nanoparticles such as silver negatively impacts the quality as the nanoparticle binds with the solid particles of soil causing degradation of soil fertility [57]. Proper life cycle analysis must be made in the development and production of nanotechnology products to determine the amount of energy, water, and toxic chemicals involved. In addition, the production of carbon nanofibers contributes significantly greater towards global warming and ozone layer depletion than the production of conventional materials [58].

On the other hand, the positive economic benefits of nanotechnology include the potential to develop materials that are more ecofriendly than the current raw materials being used in the construction industry today [59]. The development of nanomaterials can boost the energy and resource efficiency in the chemical industry. Additionally, in terms of energy storage, a nanotechnology-based approach is being employed to optimize energy production that would significantly help in combating climate change [60]. Nanomaterials are also being developed as filters for wastewater clean-up and detection of chemical and biological agents in the air and soil [61].

4.3. Societal impacts

One of the alarming impacts that nanotechnology brings to society are the health-related concerns. The absorbance of nanoparticles in the bodily fluids or tissues of the body present some risks. Developed nanoparticles with particularly low solubility that were transported to a living organism could accumulate in the system for a long time [54]. Government funded nanotechnology research is also interested in the growth of the technology towards military applications. Applications such as atomically precise manufacturing are one of the interests in the development of nanorobotics and weapons for military affairs and surveillance [62]. Other societal impacts for the growing nanotechnology involve the domination of big companies with regards to the Intellectual Properties (IP) claims that are applied and patented [63]. This gives the big companies power to control the rate of advancement and adoption of nanotechnology in a country. The advancement of nanotechnology also impacts the demand for scientists, engineers, and technicians. Moreover, technologies would affect the capabilities of the government and important agencies in terms of citizen tracking and surveillance [64].

5. Policy Recommendations

The analysis of development and growth of nanotechnology, as well as its future directions is essential in determining potential policy recommendations in the adoption of nanotechnology in the Philippines. Moreover, the identification of facilitating factors, and barriers for adoption helped to focus on the drivers and address the constraints in the acceptance of the technology. In this paper, a framework was developed and proposed to identify and assess the roles of key actors in the development of nanotechnology in the country as can be seen in Fig. 2. The main advancements in the field of nanotechnology has shown correlating characteristics to a General Purpose Technology (GPT) [65]. A GPT exhibits its potential from identifying and connecting key firms, industries, and other stakeholders across different sectors [66]. As such, the proposed framework aims to recognize important connections between varying sectors and how the interplay among them can benefit the advancement of nanotechnology. Key players were identified to be the academe, industry, and the government regulating bodies. Government policies play a crucial role in regulating the behavior and developments in the field, it would also control the flow of support and attention towards grants and funding supports. The regulating bodies are responsible for assessing the government policies and revise it in accordance with the progress in the maturity of nanotechnology. Moreover, a team of nanotechnology experts should be made that would coordinate and regulate the funding supports, as well as serve as the linkage between government and industry. Funding agencies from both the public and private sector would provide support to promising academic institutions, and research laboratories engaging in nanotech research; in addition, efforts towards conferences, workshops, and training to promote nanotechnology acceptance will be made. The financial support will also spur the formation of Research Innovation Hubs and Centers of Excellence for nanotechnology, that gives access for Micro, Small, and Medium Enterprises (MSME), and start-ups with adequate support from experts and access to relevant facilities. The academe side would focus on capacity building catering to the technological advancements and industrial needs. Knowledge generation would be the main driving force in this framework; by addressing the societal needs in terms of another outbreak or disaster, the research direction would be shifted accordingly coming from this framework. The advances in the field of nanotechnology coming from the knowledge of societal needs would challenge the current policies; the regulating bodies will formulate more appropriate policies and regulations to address the societal demands. By directing the efforts towards knowledge generation and listening to societal needs, nanotechnology in the Philippines is set to significantly aid in the country's progress.

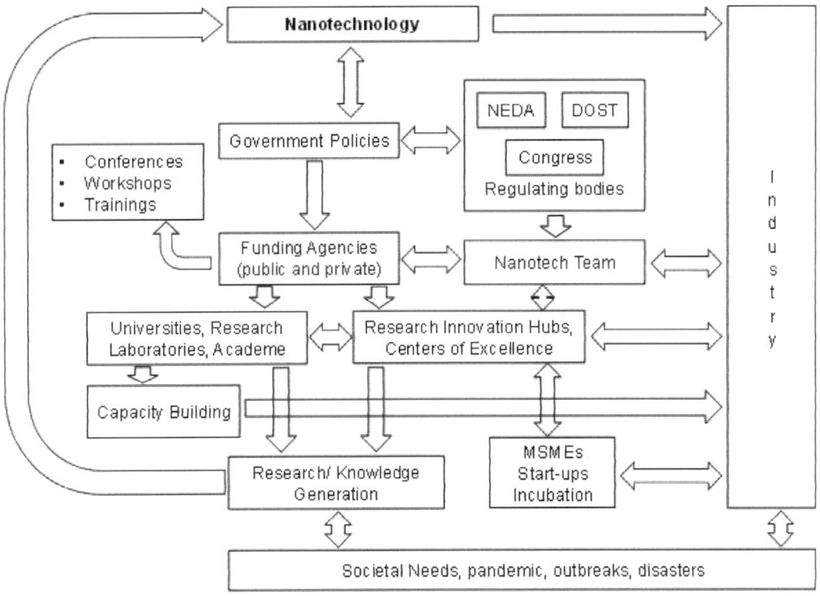

Figure 2. Proposed framework for nanotechnology assessment in the Philippines

5.1. *Labor and employment*

The development of the field of nanotechnology opens doors for more job opportunities. When a certain process is developed or discovered, skill specific jobs are required for the industry. The implementation of proper technological transfer protocols and information towards the society would yield high returns to labor. Labor policies on promotion of employment and human resource development, as well as livelihood and entrepreneurship are important. Opportunities for all to venture and succeed in the field of nanotechnology is essential in its spread and acceptance in the country. Proper PPE must be always provided for employees that are exposed to harmful substances in the event of production and development of nanomaterials. Furthermore, measures for cases of pandemic or outbreaks should be in place to ensure the employment of workers.

5.2. Capacity building

In the path of advancement for nanotechnology, it would require human capital. The progress of nanotechnology needs the help of more hands and brains, as such this policy recommendation recognizes the importance of human resources and aims to provide proper support towards the force behind the development of nanotechnology. Proper support towards building programs that are fair to the skilled individuals and will enable their passion in exploring nanotechnology in this country must be provided. The lack of experts would severely factor in the reason as to why nanotechnology adoption in the country would take a longer time.

Government initiatives such as the Balik Scientist Program (BSP), that recognizes Filipino experts that have trained abroad to come back to the country and share their expertise, would be great in reinforcing the available human resources of the country. The knowledge transfer from specialists that have practiced their profession outside of the country is an essential help in guiding the current state of Science and Technology in the Philippines, as well as sharing insights on how other countries have tackled and adopted the emerging technologies.

On the other hand, the academe should also play a role in developing quality education that equips students with expertise and knowledge through technical training. Alignment to industrial needs would be beneficial overall.

5.3. Government policy and laws

A positive perception towards nanotechnology by the public will yield benefits in the adoption and progress of the field in the country. Consequently, public acceptance and decisions are highly affected by the level of trust placed upon the government, academe, and industries [67]. The interplay between the stakeholders defines the role of government to take care of policies to protect the society from potential risks [68]. Conversely, the researchers and industries manage the risks associated with nanotechnology. The relationship between government and its citizens remains crucial as the residents that trust the government to prioritize the interest of the public would be generally welcoming towards the technological campaigns by the government such as in the introduction of nanotechnology for future product applications [69].

In addition, policies towards fair opportunities on exploring nanotechnology should be the one of the main focus, such as providing necessary support and equipment access to small scale businesses, incentives for industries, and safety nets for investors. Moreover, support should also be provided in the early stages of manufacturing of nanomaterials as these are costly and a significant factor to the success of the technological adoption. Lastly, proper implementation from the responsible organizations must be done to ensure that these countermeasures will help in the progress of nanotechnology in the country.

5.4. Research development and innovation

Support should be given to research in nanotechnology, particularly in the field of microfabrication research where most advances are being researched and developed. The local trend of development of nanocomposites from naturally occurring minerals must be funded, as well as research on drug delivery, filtration systems, and medical imaging techniques.

Competitive grants will also spark the interest of further research in the field, as would lead to obtaining necessary equipment for research. The development of innovation hubs with regards to specific research categories on nanotechnology would boost its advancement. In line with the established government programs, the Niche Centers in the Regions for Research and Development

(NiCeR) would capacitate nanotechnology related research through substantial funding and would be a significant help in obtaining necessary infrastructure for the development of nanotechnology in the country. Lastly, open access to available nanotechnology research and findings. The availability of modern results would enhance the rate of progress in terms of research; additionally, the public would be given access to valuable data and would lead to further appreciation heading to the faster adoption and acceptance of technology in the country.

5.5. Policy interventions summary

From the identified barriers and constraints for technology adoption in Section III, and the proposed policy recommendations in Section V, a summary of how the policy recommendations can address the determined barriers and constraints is shown in Figure 3.

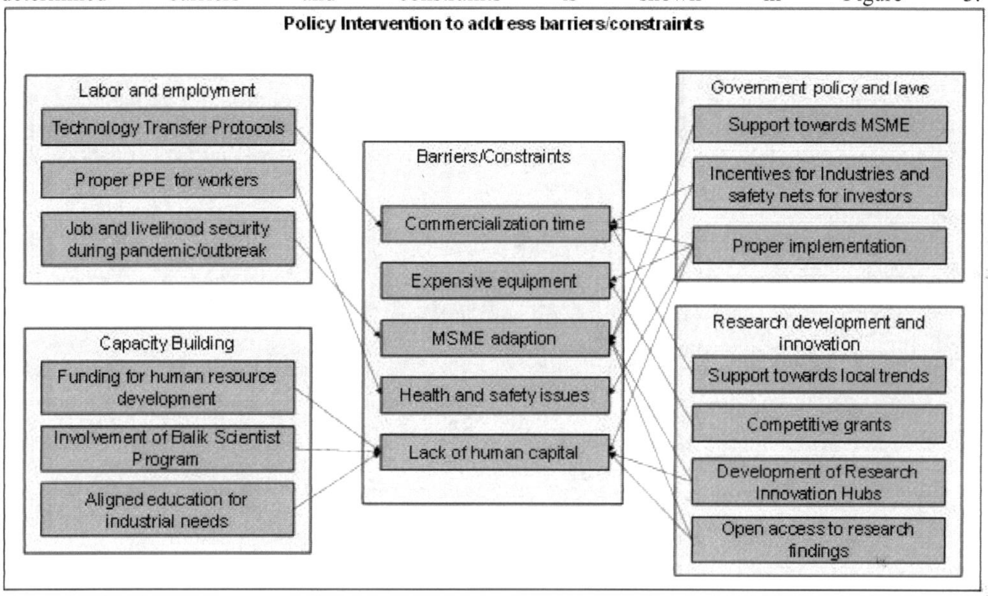

Figure 3. Proposed policy interventions to address barriers and constraints for nanotechnology adoption in the Philippines

6. Conclusions

Nanotechnology is one of the emerging technologies that has the potential to disrupt the daily activities of various sectors at a global scale. Interesting trends on how the technology has evolved can be observed. Research on the nanoscale applications have caused uproars in huge industries such as in automotive, construction, and healthcare.

The potential of nanotechnology has been acknowledged by the Philippines, as such a technology roadmap was established 10 years ago to guide the development of nanotechnology in the country. However, the presence of the COVID-19 pandemic has drastically affected the timeline of most technological roadmaps. Several barriers and constraints for the adoption of nanotechnology in the Philippines were identified. On the other hand, the facilitating factors that would enable the acceptance of the technology in the country was also explored. Moreover, the impacts brought about by the advancements in nanotechnology towards major sectors such as the economy, environment, and society were also discussed. From the gathered information on the trends, factors for adoption and impacts of the technology, policy recommendations were raised to help establish the roots of nanotechnology in the country and support it in its technological maturity.

A framework was developed and proposed that takes into consideration the major stakeholders and the relationship amongst them that would enable the flourish of nanotechnology in the country. The knowledge driven framework aims to cope up with societal needs or in cases of pandemic/outbreaks. Furthermore, the policy recommendations aimed to address the identified constraints for technological adoption of nanotechnology in the country.

While policy recommendations are important to help the governing bodies in tackling the topic on technology adoption. It would be ineffective if there is no established trust between the stakeholders. Citizens would be scared to accept new technology if the government is not confident in the performance of that technology or due to flaws in the information relay between the researching bodies and common folk. On the other hand, when strong linkages are established between the key actors such as in academe and industry, a projection of increased productivity will be seen in the field of nanotechnology. It is critical to capitalize and invest on emerging technologies and its maturity level should match the capabilities of the country to minimize the potential risks. Furthermore, it is important to raise awareness towards the public and clearly define the advantages and disadvantages of nanotechnology to ease its acceptance in the society. It has been six decades since Richard Feynman proposed the idea of manipulating things on a small scale, now the technology previously thought of is taking shape and is preparing to take over the operations of the world. It is now within the realm of finding how to fully harness the technology through proper deliberation and policy making of the stakeholders can a country like Philippines move forward.

7. References

[1] Bagheri M, Mitchelmore S, Bamiatzi V and Nikolopoulos K 2018 Internationalization orientation in SMEs: The mediating role of technological innovation *Journal of International Management* [Online]. Available: doi:10.1016/j.intman.2018.08.002 [Accessed 26-Sep-2020]

[2] Van Leeuwen G and Klomp L 2006 On the contribution of innovation to multi-factor productivity growth. *Economics of Innovation and New Technology* 15(4-5), 367–390 [Online]. Available: doi:10.1080/10438590500512927 [Accessed 26-Sep-2020]

[3] Bianchi C, Glavas C and Mathews S 2016 SME international performance in Latin America: The role of entrepreneurial and technological capabilities *In Business Association of Latin American Studies (BALAS) Annual Conf.*

[4] Ganotakis P and Love J H 2012 *The Innovation Value Chain in New Technology Based Firms: Evidence from the UK*, J. Prod. Innov. Manag. 29, 839–860

[5] Schumpeter J A 1934 *The Theory of Economic Development: an Inquiry into Profits, Capital, Credit, Interest, and the Business Cycle* Transaction Publishers

[6] Berg S, Wustmans M and Bröring S 2018 Identifying first signals of emerging dominance in a technological innovation system: A novel approach based on patents *Technological Forecasting and Social Change* [Online]. Available: doi:10.1016/j.techfore.2018.07.046

[7] Nevzorova T and Karakaya E 2020 Explaining the drivers of technological innovation systems: The case of biogas technologies in mature markets *Journal of Cleaner Production* [Online]. Available: doi:10.1016/j.jclepro.2020.120819

[8] Bach H, Bergek A, Bjørgum Ø, et al. 2020 Implementing maritime battery-electric and hydrogen solutions: A technological innovation systems analysis *Transportation Research Part D: Transport and Environment, Volume 87* [Online]. Available: https://doi.org/10.1016/j.trd.2020.102492.

[9] Pacchini A P T, Lucato W C, Facchini F and Mummolo G 2019 The degree of readiness for the implementation of Industry 4.0. *Computers in Industry,* 113 [Online]. Available: doi:10.1016/j.compind.2019.103125

[10] Mangematin V and Walsh S 2012 The future of nanotechnologies *Technovation*, 32(3-4), 157–160 [Online]. Available: doi:10.1016/j.technovation.2012.01.003

[11] Robinson D K 2009 Co-evolutionary scenarios: an application to prospecting futures of the responsible development of nanotechnology, *Technol. Forecast. Soc. Chang.* 76, 1222–1239

[12] National Nanotechnology Initiative. What it is and how it works|Nano, [Online]. Available: https://www.nano.gov/nanotech-101/what

[13] Genet C, Errabi K and Gauthier C 2012 Which model of technology transfer for nanotechnology? A comparison with biotech and microelectronics. *Technovation*, 32(3-4), 205–215 [Online]. Available: doi:10.1016/j.technovation.2011.10.007
[14] Kostoff R N, Koytcheff R G and Lau C G Y 2007 Global nanotechnology research literature overview. *Technological Forecasting and Social Change*, 74(9), 1733–1747, 2007. [Online]. Available: doi:10.1016/j.techfore.2007.04.004
[15] Chen H and Roco M C 2009 *Mapping Nanotechnology Innovations and Knowledge—Global and Longitudinal Patent and Literature Analysis* Springer Science + Business Media, New York
[16] Roco M C 2007 "National Nanotechnology Initiative - Past, Present, Future," in Handbook on Nanoscience, Engineering and Technology, 2nd ed., Taylor and Francis, 3.1-3.26, 2007.
[17] Dadios E P, Culaba A B, Albert J R G, et al 2018 Preparing the philippines for the fourth industrial revolution: A scoping study *Working Papers* id:12901, eSocialSciences. PIDS Discussion Paper Series 2018-11
[18] Dayrit F M 2010 Nanotechnology Roadmap for the Philippines: An overview, Tech Monitor
[19] Sy C, Bernardo E, Miguel A, et al. 2020 Policy development for pandemic response using system dynamics: a case study on COVID-19 *Process Integration and Optimization for Sustainability* [Online]. Available: doi:10.1007/s41660-020-00130-x
[20] He X, Deng H and Hwang H 2018 The current application of nanotechnology in food and agriculture *Journal of Food and Drug Analysis* [Online]. Available: doi:10.1016/j.jfda.2018.12.002
[21] Leso V, Fontana L and Iavicoli I 2019 Biomedical nanotechnology: Occupational views, *Nano Today*, 24, 10-14 [Online]. Available: https://doi.org/10.1016/j.nantod.2018.11.002.
[22] Saeedi M, Eslamifar M, Khezri K and Dizaj S M 2019 Applications of nanotechnology in drug delivery to the central nervous system *Biomedicine & Pharmacotherapy*, 111, 666–675 [Online]. Available: doi:10.1016/j.biopha.2018.12.133
[23] Ding S-S, He L, Bian X-W and Tian G 2020 Metal-organic frameworks-based nanozymes for combined cancer therapy *Nano Today*, 35
[24] Jin J-O, Kim G, Hwang J, Han K H, Kwak M and Lee P C W 2020 Nucleic acid nanotechnology for cancer treatment *Biochimica et Biophysica Acta - Reviews on Cancer*, 1874 (1)
[25] Xiong Y, Wang Y and Tiruthani K 2019 Tumor immune microenvironment and nano-immunotherapeutics in colorectal cancer *Nanomedicine: Nanotechnology, Biology, and Medicine*, 21
[26] Alafeef M, Moitra P and Pan D 2020 Nano-enabled sensing approaches for pathogenic bacterial detection *Biosensors and Bioelectronics*, 165
[27] Banstola A, Emami F, Jeong J-H and Yook S 2018 Current applications of gold nanoparticles for medical imaging and as treatment agents for managing pancreatic cancer *Macromolecular Research*, 26 (11), pp. 955-964
[28] Keirouz A, Chung M, Kwon J, Fortunato G and Radacsi N 2020 2D and 3D electrospinning technologies for the fabrication of nanofibrous scaffolds for skin tissue engineering: A review *Wiley Interdisciplinary Reviews: Nanomedicine and Nanobiotechnology*, 12 (4)
[29] Stergiou A, Cantón-Vitoria R, Psarrou M N, Economopoulos S P and Tagmatarchis N 2020 Functionalized graphene and targeted applications – Highlighting the road from chemistry to applications. *Progress in Materials Science*, 114
[30] Nguyen-Tri P, Tran H N, Plamondon C O, Tuduri L, et al. 2019 Recent progress in the preparation, properties and applications of superhydrophobic nano-based coatings and surfaces: A review *Progress in Organic Coatings*, 132, pp. 235-256
[31] Ahmadi Z 2019 Epoxy in nanotechnology: A short review *Progress in Organic Coatings*, 132, 445–448 [Online]. Available: doi:10.1016/j.porgcoat.2019.04.003
[32] Kremer R J 2019 Bioherbicides and nanotechnology: Current status and future trends *Nano-Biopesticides Today and Future Perspectives*, 353–366 [Online]. Available: doi:10.1016/b978-0-12-815829-6.00015-2
[33] Bahrami A, Delshadi R, Jafari S M and Williams L 2019 Nanoencapsulated nisin: An engineered natural antimicrobial system for the food industry *Trends in Food Science & Technology* [Online]. Available: doi:10.1016/j.tifs.2019.10.002
[34] Department of Science and Technology (DOST), DOST outlines nanotechnology roadmap, https://www.dost.gov.ph/knowledge-resources/news/38-2009-news/366-dost-outlines-nanotechnology-roadmap.html

[35] StatNano, Philippines. https://statnano.com/country/philippines
[36] Department of Science and Technology (DOST), Making the Country Competitive through Nanotechnology, One Expert, In the spotlight, STII-Bridge, https://news.oneexpert.gov.ph/stii-bridge/making-country-competitive-nanotechnology/
[37] Kumar A, Gupta P K, and Srivastava A 2020 A review of modern technologies for tackling COVID-19 pandemic *Diabetes & Metabolic Syndrome: Clinical Research & Reviews*, 14(4), 569–573 [Online]. Available: doi:10.1016/j.dsx.2020.05.008
[38] Mbunge E 2020 Integrating emerging technologies into COVID-19 contact tracing: Opportunities, challenges and pitfalls *Diabetes & Metabolic Syndrome: Clinical Research & Reviews* [Online]. Available: doi:10.1016/j.dsx.2020.08.029
[39] Bhavana V, Thakor P, Singh S B and Mehra N K 2020 COVID-19: Pathophysiology, treatment options, nanotechnology approaches, and research agenda to combating the SARS-CoV2 pandemic *Life Sciences* [Online]. Available: doi:10.1016/j.lfs.2020.118336
[40] Kerry R G, Malik S, Redda Y T, Sahoo S, Patra J K and Majhi S 2019 Nano-based approach to combat emerging viral (NIPAH virus) infection *Nanomedicine: Nanotechnology, Biology and Medicine* [Online]. Available: doi:10.1016/j.nano.2019.03.004
[41] Solomon S 2020 Aiming for 10-second result, breathalyzer detector for COVID begins early tests, *The Times of Israel* https://www.timesofisrael.com/israeli-startups-breathalyzer-detector-for-covid-19-begins-early-tests/
[42] Gupta N, Fischer A R H, van der Lans I A and Frewer L J 2012 Factors influencing societal response of nanotechnology: an expert stakeholder analysis *Journal of Nanoparticle Research*, 14(5) [Online]. Available: doi:10.1007/s11051-012-0857-x
[43] Lee C-J, Lee S, Jhon M S and Shin J 2013 Factors influencing nanotechnology commercialization: an empirical analysis of nanotechnology firms in South Korea *Journal of Nanoparticle Research*, 15(2) [Online]. Available: doi:10.1007/s11051-013-1444-5
[44] M.J. Pitkethly, Nanomaterials—the driving force. Matter Today 7(12):20–29, 2004
[45] Andersen M 2011 Silent innovation: corporate strategizing in early nanotechnology evolution. *J Technol Trans* 36(6):680–696
[46] Shapira P, Youtie J and Kay L 2011 National innovation systems and the globalization of nanotechnology innovation *J Technol Trans* 36(6):587–604
[47] McNeil R D, Lowe J, Mastroianni T, Cronin J and Ferk D 2007 Barriers to nanotechnology commercialization *A final report prepared for US Department of Commerce*. The University of Illinois at Springfield [Online]. Available: http://www.ntis.gov/pdf/Report-BarriersNanotechnologyCommercialization. Pdf
[48] Asia Nano Forum, Annual Report 2018. “Philippines – Nanotech Policy”, 2018. [Online]. Available: https://www.asia-anf.org/wp-content/uploads/2018/12/AnnualReport2018_1215_Final.pdf
[49] Doran J and Ryan G 2019 Does nanotechnology research generate an innovation premium over other types of research? Evidence from Ireland. *Technology in Society* [Online]. Available: doi:10.1016/j.techsoc.2019.101183
[50] Wonglimpiyarat J 2005 The nano-revolution of schumpeter's Kondratieff cycle *Technovation* 25, 1349–1354, 2005.
[51] Sargent Jr. J F 2013 The National Nanotechnology Initiative: Overview, Reauthorization, and Appropriations Issues, DTIC Document
[52] National Nanotechnology Initiative, National Nanotechnology Initiative (NNI) Budget, 2019. [Online]. Available: https://www.nano.gov/about-nni/what/funding
[53] Allied Market Research, Europe Nanomaterials Market by Type of Material, by End User – Opportunity Analysis and Industry Forecast, 2014–2022, 2016.
[54] Purohit R, Mittal A, Dalela S, Warudkar V, Purohit K and Purohit S 2017 Social, environmental and ethical impacts of nanotechnology *Materials Today: Proceedings*, 4(4), 5461-5467
[55] Baucher M A, Scott R, Cannizzaro C, Standridge S, Nesbitt E and Fadel T 2013 *Symp. on Assessing the Economic Impact of Nanotechnology: Synthesis Report*
[56] Köhler A R and Som C 2014 Risk preventative innovation strategies for emerging technologies the cases of nano-textiles and smart textiles *Technovation* 34, 420–430
[57] Cattaneo A G, Gornati R, Sabbioni E, Chiriva-Internati M, Cobos E, Jenkins M R and

Bernardini G 2010 Nanotechnology and human health: risks and benefits *Journal of Applied Toxicology*, 30: 730–744

[58] Khanna V, Bakshi B R and Lee L J 2008 Carbon nanofiber production *Journal of Industrial Ecology*, 12(3), 394–410 [Online]. Available: doi:10.1111/j.1530-9290.2008.00052.x

[59] Pacheco-Torgal F 2019 Introduction to nanotechnology in eco-efficient construction *Nanotechnology in Eco-Efficient Construction*, 1–9 [Online]. Available: doi:10.1016/b978-0-08-102641-0.00001-3

[60] Murphy P, Munshi D, Kurian P A, Lakhtakia A, Bartlett R V and Chakraborty M 2017 Nanotechnology, Society, and Environment . Reference Module in Materials Science and Materials Engineering [Online]. Available: doi:10.1016/b978-0-12-803581-8.10311-x

[61] Momba M N B, Baloyi L, Mpenyana-Monyatsi L and Kamika I 2017Nanotechnology-based filters for cost-effective drinking water purification in developing countries *Water Purification*, 169–208 [Online]. Available: doi:10.1016/b978-0-12-804300-4.00005-8

[62] Umbrello S and Baum S D 2018 Evaluating future nanotechnology: The net societal impacts of atomically precise manufacturing. *Futures*, 100, 63–73 [Online]. Available: doi:10.1016/j.futures.2018.04.007

[63] Genieser L and Gollin M 2007 Intellectual property issues in nanotechnology *Journal of Commercial Biotechnology*, 13(3), 195–198 [Online]. Available: doi:10.1057/palgrave.jcb.3050052

[64] Soltani A M and Pouypouy H 2019 Standardization and regulations of nanotechnology and recent government policies across the world on nanomaterials. *Advances in Phytonanotechnology*, 419–446 [Online]. Available: doi:10.1016/b978-0-12-815322-2.00020-1

[65] Shea C M, Grinde R and Elmslie B 2011 Nanotechnology as general-purpose technology: Empirical evidence and implications *Technol. Anal. Strateg. Manag.* 23, 175–192

[66] Bresnahan T F and Trajtenberg M 1995 General purpose technologies 'engines of growth'? *J. Econom.* 65, 83–108, 1995.

[67] Chen M F, Lin Y P and Cheng T J 2013 Public attitudes toward nanotechnology applications in Taiwan, *Technovation* 33, 88–96 [Online]. Available: https://doi.org/10.1016/j.technovation.2012.11.008.

[68] Kamarulzaman N A, Lee K E, Siow K S and Mokhtar M 2020 Public benefit and risk perceptions of nanotechnology development: Psychological and sociological aspects. *Technology in Society*, 62 [Online]. Available: doi:10.1016/j.techsoc.2020.101329

[69] Brown J and Kuzma J 2013 Hungry for information: public attitudes toward food nanotechnology and labeling *Rev. Pol. Res.* 30, 512–548 Reference lists

Acknowledgments

The first author would like to acknowledge the Engineering Research and Development for Technology (ERDT) consortium and the Department of Science and Technology – Science Education Institute (DOST-SEI) for supporting the author's academic endeavour in pursuing Doctor of Philosophy in Mechanical Engineering degree.

Noise filtering algorithm for control of motion systems

Hnin Lae Wah, Aung Myo Thant Sin

Department of Mechatronic Engineering, Yangon Technological University
Yangon, Myanmar

hninlae.gtc@gmail.com

Abstract. Measurement noise and external disruptions are problematic in real world applications and every control system design. These unwanted disturbances affect the measured output signal and cause the system to provide an inaccurate output. We need to eliminate these noises for stabilizing of the control system design. In this paper, we propose how to eliminate effectively the noises in motion control systems using the Kalman filter. We investigate practically the Kalman filtering algorithm based on the measurements generated by the encoder sensor fixed on DC motor. The Kalman algorithm is used for noise filtering because it is recursive algorithm and it can handle the noises that occur when the sensor is used to read data. The performance of KF method is enhanced by adjusting its parameters according to the innovation requirements. In the experimental results, the velocity estimation of the Finite-difference (FD) method is demonstrated to compare with the velocity estimation of the Kalman filtering (KF) method.

Keywords - Kalman filter, finite-difference, motion control, Process Noise Covariance, Measurement Noise Covariance

1. Introduction

Nowadays in many applications such as robot manipulator, hundreds of assorted areas, including all navigation forms (aviation, land and sea), moving object tracking, control motors, controlled motion systems, channel tracking in communications, traffic monitoring, fuzzy logic and neural network training, the detection of underground radioactivity and others control, we need to know the precise information and data as position, velocity, and acceleration. However, it is a challenge to obtain precise information in practice. Sensors and GPS are commercially available and are used to monitor the process and to obtain the information, and data, this data from monitoring system is important for decision or control system. However, in practice, the massive data and signals from the sensors and GPS are inaccurate, unsatisfactory and never available in the pure or noiseless form and it is a challenging task to obtain precise information. Every signal faces some distortion due to background noises. Hence, every data and signal must be handled with noise reduction tools. There are available several noise filtering algorithms. However, many of the algorithms have two major weaknesses. Firstly, some algorithms are genuinely complex and hard to implement real world events. Secondly, some algorithms filter the noises or unwanted parts of a signal, but can change the characteristics of the original signal in the process.

One of the most widely applied algorithms in real-time noise filtering and reduction is the Kalman filtering algorithm that can be utilized for stochastic estimation. Kalman filter is a common algorithm to estimate the system state from noisy data and signals, first released in 1960 [1]. Kalman filter has ideal accuracy in linear estimation, so it has been widely accepted and has received attention since it has been theorized and utilized in different areas [2]. Kalman filter uses a set Mathematical equation for predicting the actual value from the noisy measured information and estimates recursively the state of a dynamic system with two diverse models (i.e. dynamic model and observation model) [3]. While the dynamic system model describes the behaviour of the state vector, the observation model establishes the relationship between measurements and the state vector. Both models are related with statistical properties to portray the exactness of the models. The Kalman filter performs iteratively and recursively some set of mathematical equation on considering that the noise is Gaussian [4].

The performance of the KF method is limited by the numerical accuracy of the calculations and the common problem associated with Kalman filter is how to choose the appropriate process noise covariance and measurement noise variance. Proper set up of the noise variances makes a converged filter [5]. No appropriate set up of the noise variances may obviously decrease Kalman filter's performance and even diverge the filter [6]. Almost all the previous studies, the noise variances are kept constant during the estimation. In the Kalman filtering algorithm, the gain K is computed the model parameters and the noise covariance of process and measurement. So the process and measurement noise variances are usually measured priorly during filter operation. Most publications on noise filtering present simulation results to demonstrate the performance of the KF method as there are not known experimental results .In this paper, we try to gather a deeper insight on how the noise variances affect the filtering results and we test practically the performance of Kalman filter for motion control of a shaft of DC motor which is fixed by encoder sensor. The incremental shaft encoder is one of the most commonly used sensors for obtaining position measurements in electric motors [7]. The results in this paper are experimented, rather than simulated. The performance of the KF method has been evaluated experimentally using a DC motor and encoder sensor fixed on the motor that provides the position measurement only.

This paper is organized as follows: section II reviews the basics of the Kalman filter when the measurement and process noise variances are priorly needed and then implements the system and observation models with transition matrices and Kalman filter equation. Section III implements a Kalman filter to estimate the state of the motor shaft by removing the periodical disturbances introduced by using the position measurement only from the incremental shaft encoder sensor. In section IV we evaluate the experimented data from section III and the estimation results are presented. The experimental results demonstrate the performance of the Kalman filter in motion controlled system. Section V makes the concluding statements.

2. Background Knowledge of Kalman filter

In order to filter the noises or unwanted parts of uncertainty input or noisy data, Kalman filtering algorithm is widely used for real time data processing and is a recursive estimator means that No need to store all past measurements and doesn't need to reprocess all data each time step and only the current measurement data and the estimated state of the previous time step are required. The system state estimation provided by the Kalman filter can minimize the mean square estimation error within a given range of observations. The Kalman filter estimates the state of a dynamic system, based on two different models, namely the dynamic system model and observation model, that tend to be more precise that than the measurements alone. For using a Kalman filter, the appropriate system model must first be selected. After that, it needs observation model to get a sequence of noisy observations.

2.1. System Model and Observation Model

To estimate the state of the system, various models can be used depending on the filter considered problems, such as second order and third order kinematic models. Using basic equations of velocity, distance and time, the position of a moving object at time t is at x location, then at time $t+1$ the object will be at location $x + \left((t + 1) - t\right) * \dot{x}$, where $\dot{x}$ is velocity of moving object. Moving in further, acceleration value is available and can add it to above equation and update new position of a moving object to $x + \left((t + 1) - t\right)\dot{x} + \frac{1}{2}\ddot{x}\left((t + 1) - t\right)^2$. And then jerk, snap, etc. are available in further movement. Formulizing above equations, current state of a moving object consists of its current position, velocity, acceleration, jerk, etc. as a vector X. So vector X will have elements which are

$x, \dot{x}, \ddot{x}, \dddot{x}$,....representing position, velocity, acceleration, jerk,...with change in time as *delta_t= (t + 1)- t.* Using Taylor series and derivatives of position vector for motion paths with respect to time, the following equations for tracking motion paths become:

$$Position\ ; \quad x = x_0 + \dot{x}_0\,\Delta t + \frac{1}{2}\ddot{x}_0\Delta t^2 + \frac{1}{6}\dddot{x}_0\,\Delta t^3 + \frac{1}{24}\ddddot{x}_0\,\Delta t^4 + \frac{1}{120}\,\ddot{\ddot{\dot{x}}}_0\,\Delta t^5 + \cdots \tag{1}$$

$$Velocity;\ \dot{x} = \frac{dx}{dt} = \dot{x}_0 + \ddot{x}_0\Delta t + \frac{1}{2}\dddot{x}_0\,\Delta t^2 + \frac{1}{6}\ddddot{x}_0\,\Delta t^3 + \frac{1}{24}\,\ddot{\ddot{\dot{x}}}_0\,\Delta t^4 + \cdots \tag{2}$$

$$Acceleration;\ \ddot{x} = \frac{d\dot{x}}{dt} = \ddot{x}_0 + \dddot{x}_0\,\Delta t + \frac{1}{2}\ddddot{x}_0\,\Delta t^2 + \frac{1}{6}\,\ddot{\ddot{\dot{x}}}_0\,\Delta t^3 + \cdots \tag{3}$$

$$Jerk;\ \dddot{x} = \frac{d\ddot{x}}{dt} = \dddot{x}_0 + \ddddot{x}_0\,\Delta t + \frac{1}{2}\,\ddot{\ddot{\dot{x}}}_0\,\Delta t^2 + \cdots \tag{4}$$

For second-order kinematic model, the position and velocity of the moving paths are expressed by the linear state space $\begin{bmatrix} x \\ \dot{x} \end{bmatrix}$. The position, velocity and acceleration of the moving objects are described by the linear state space $\begin{bmatrix} x \\ \dot{x} \\ \ddot{x} \end{bmatrix}$ for third-order and $\begin{bmatrix} x \\ \dot{x} \\ \ddot{x} \\ \dddot{x} \end{bmatrix}$ for fourth-order model respectively. Calculation with matrix operation is more effective than solving individual equation.

2.1.1. For second-order kinematic model. The position and velocity equations of the moving object 'in equation (1) and (2)' are used with the first and second derivatives being velocity and acceleration. So,

$$\begin{bmatrix} position \\ velocity \end{bmatrix} = \begin{bmatrix} x_k \\ \dot{x}_k \end{bmatrix} = \begin{bmatrix} x_0 + \dot{x}_0\Delta t + \frac{1}{2}\ddot{x}\,\Delta t^2 \\ \dot{x}_0 + \ddot{x}\Delta t \end{bmatrix}$$

$$\begin{bmatrix} x_k \\ \dot{x}_k \end{bmatrix} = \begin{bmatrix} 1 & \Delta t \\ 0 & 1 \end{bmatrix} \begin{bmatrix} x_0 \\ \dot{x}_0 \end{bmatrix} + \begin{bmatrix} \frac{1}{2}\Delta t^2 \\ \Delta t \end{bmatrix} [\ddot{x}] \tag{5}$$

Comparing with 'equation (5)' and Newton's laws of motion concluding that $X_k = AX_0 + Bu$, will get X_k, A and B

$$X_k = \begin{bmatrix} x_k \\ \dot{x}_k \end{bmatrix},\ A = \begin{bmatrix} 1 & \Delta t \\ 0 & 1 \end{bmatrix} \text{ and } B = \begin{bmatrix} \frac{1}{2}\Delta t^2 \\ \Delta t \end{bmatrix}$$

The initial state value of the moving objects with perfect precision is

$$\begin{bmatrix} x_0 \\ \dot{x}_0 \end{bmatrix} = \begin{bmatrix} 0 \\ 0 \end{bmatrix} \tag{6}$$

where x_0 and $\dot{x}_0$ are the initial position and velocity. In the real world, the velocity will be disturbed by noise caused by wind gusts, potholes and other terrible realities. The noise of velocity is a random variable that changes with time. Therefore, a more realistic velocity equation would be:

$$\text{Velocity;}\quad \dot{x}_k = \dot{x}_0 + \ddot{x}\Delta t + v_k,$$

A comparable condition can be determined for the position:

$$\text{Position;} \quad x_k = x_0 + \dot{x}_0 \Delta t + \frac{1}{2}\ddot{x}\Delta t^2 + p_k$$

where v_k is the velocity noise and p_k is the position noise.

The system and observation models were set up agreeing to the kinematic movement characteristics within the normal scenarios ('see in equation (7) and (8)'). The observation model is established by a measurement system, which can be represented by a linear equation in the following form,

$$X_k = A_k X_{k-1} + B_k u_k + w_k \,, \tag{7}$$

$$z_k = H x_k + v_k \,, \tag{8}$$

where $A \in \Re^{n \times n}$ is the state transition matrix related to the previous state X_{k-1}, $B \in \Re^{n \times 1}$ is the control input matrix that relates the control input $u_k \in \Re^1$ and the subscripts k and $k-1$ is the time index. The random variables (w_k and v_k) represent the process and measurement noises (due to such things as instrumentation errors) respectively, which are assumed to be zero mean Gaussian white noise with normal probability distributions, (i.e. $w_k \sim N(0, Q_k)$ and $v_k \sim N(0, R_k)$, where Q_k and R_k are the corresponding covariance matrices). z_k is the observation or measurement made at time k, x_k is the state at time k and H is the observation model which maps the state space into the observed space. The vector X_k contains all of the information about the present state of the system, but cannot measure X_k directly [8]. We measure z_k instead of X_k , which is a function of X_k that is corrupted by the noise v_k. Using z_k to obtain an estimate of X_k , but we don't have to extract information from the face value of z_k, because it will be destroyed by noise. We can utilize the information that it provides to a certain extent, but we cannot completely believe it. The first thing we have to do is to calculate the difference between the measurement state z_k and the model state X_k, this is also called the innovation:

$$y_k = z_k - H X_k \tag{9}$$

The innovation, y_k is defined as the difference between the measurement (observation) z_k and its prediction X_k using the information available at time k. The Kalman filter contains the propagation of state estimation and the error covariance matrices at one time step to next one. X_k state variable was having 2 variables (position and velocity) so its covariance matrix is a *2 x 2* matrix.
Predicted Uncertainty or Process Covariance matrix:

$$P_k = A_k P_{k-1} A_k^T + Q_k \tag{10}$$

where P_{k-1} is previous estimated covariance matrix or state uncertainty. If the *initial values of position and velocity are known completely*, the error covariance matrix should be initialized with zero covariance matrix:

$$P_{k-1} = \begin{bmatrix} 0 & 0 \\ 0 & 0 \end{bmatrix} \tag{11}$$

where $Q_k = B_k B_k^T \sigma^2 = \begin{bmatrix} \frac{1}{2}\Delta t^2 \\ \Delta t \end{bmatrix} \begin{bmatrix} \frac{1}{2}\Delta t^2 & \Delta t \end{bmatrix} \sigma^2$

$$Q_k = \begin{bmatrix} \frac{1}{4}\Delta t^4 & \frac{1}{2}\Delta t^3 \\ \frac{1}{2}\Delta t^3 & \Delta t^2 \end{bmatrix} \sigma^2 \tag{12}$$

where σ^2 the process model .

2.1.2. For third-order kinematic model. The position, velocity and acceleration equations of the moving object 'in equation (1), (2) and (3)' are used with the first, second and third derivatives being velocity, acceleration and jerk. So,

$$\begin{bmatrix} position \\ velocity \\ acceleration \end{bmatrix} = \begin{bmatrix} x_k \\ \dot{x}_k \\ \ddot{x}_k \end{bmatrix} = \begin{bmatrix} x_0 + \dot{x}_0\Delta t + {}^1\!/_2\, \ddot{x}_0\, \Delta t^2 + {}^1\!/_6\, \dddot{x}_0\, \Delta t^3 \\ \dot{x}_0 + \ddot{x}_0\Delta t + {}^1\!/_2\, \dddot{x}_0\, \Delta t^2 \\ \ddot{x}_0 + \dddot{x}_0\, \Delta t \end{bmatrix}$$

$$\begin{bmatrix} x_k \\ \dot{x}_k \\ \ddot{x}_k \end{bmatrix} = \begin{bmatrix} 1 & \Delta t & {\Delta t^2}/_2 \\ 0 & 1 & \Delta t \\ 0 & 0 & 1 \end{bmatrix} \begin{bmatrix} x_0 \\ \dot{x}_0 \\ \ddot{x}_0 \end{bmatrix} + \begin{bmatrix} {\Delta t^3}/_6 \\ {\Delta t^2}/_2 \\ \Delta t \end{bmatrix} [\,\dddot{x}_0\,] \tag{13}$$

Comparing with 'equation (13)' and Newton's laws of motion concluding that $X_k = AX_0 + Bu$;

$$X_k = \begin{bmatrix} x_k \\ \dot{x}_k \\ \ddot{x}_k \end{bmatrix}, A = \begin{bmatrix} 1 & \Delta t & {\Delta t^2}/_2 \\ 0 & 1 & \Delta t \\ 0 & 0 & 1 \end{bmatrix} \text{ and } B = \begin{bmatrix} {\Delta t^3}/_6 \\ {\Delta t^2}/_2 \\ \Delta t \end{bmatrix}$$

$$Q_k = \sigma^2 \int_0^T G_k G_k^T = \sigma^2 \int_0^T \begin{bmatrix} {\Delta t^2}/_2 \\ \Delta t \\ 1 \end{bmatrix} \begin{bmatrix} {\Delta t^2}/_2 & \Delta t & 1 \end{bmatrix} dt$$

$$Q_k = \sigma^2 \int_0^T \begin{bmatrix} {\Delta t^4}/_4 & {\Delta t^3}/_2 & {\Delta t^2}/_2 \\ {\Delta t^3}/_2 & \Delta t^2 & \Delta t \\ {\Delta t^2}/_2 & \Delta t & 1 \end{bmatrix} dt = \sigma^2 \begin{bmatrix} {\Delta t^5}/_{20} & {\Delta t^4}/_8 & {\Delta t^3}/_6 \\ {\Delta t^4}/_8 & {\Delta t^3}/_3 & {\Delta t^2}/_2 \\ {\Delta t^3}/_6 & {\Delta t^2}/_2 & \Delta t \end{bmatrix} \tag{14}$$

2.1.3 For fourth-order kinematic model. The position, velocity, acceleration and jerk equations of the moving object 'in equation (1), (2), (3) and (4)' are used with the first, second, third and fourth derivatives being velocity, acceleration, jerk and snap. So,

$$\begin{bmatrix} position \\ velocity \\ acceleration \\ jerk \end{bmatrix} = \begin{bmatrix} x_k \\ \dot{x}_k \\ \ddot{x}_k \\ \dddot{x}_k \end{bmatrix} = \begin{bmatrix} x_0 + \dot{x}_0\Delta t + \frac{1}{2}\ddot{x}_0\, \Delta t^2 + \frac{1}{6}\dddot{x}_0\, \Delta t^3 + \frac{1}{24}\ddddot{x}_k\, \Delta t^4 \\ \dot{x}_0 + \ddot{x}_0\Delta t + \frac{1}{2}\dddot{x}_0\, \Delta t^2 + \frac{1}{6}\, s_0\Delta t^3 \\ \ddot{x}_0 + \dddot{x}_0\, \Delta t + \frac{1}{2}\, s_0\Delta t^2 \\ \frac{1}{2}\, s_0\Delta t^2 \end{bmatrix}$$

$$\begin{bmatrix} x_k \\ \dot{x}_k \\ \ddot{x}_k \\ \dddot{x}_k \end{bmatrix} = \begin{bmatrix} 1 & \Delta t & {\Delta t^2}/_2 & {\Delta t^3}/_6 \\ 0 & 1 & \Delta t & {\Delta t^2}/_2 \\ 0 & 0 & 1 & \Delta t \\ 0 & 0 & 0 & 1 \end{bmatrix} \begin{bmatrix} x_0 \\ \dot{x}_0 \\ \ddot{x}_0 \\ \dddot{x}_0 \end{bmatrix} + \begin{bmatrix} {\Delta t^4}/_{24} \\ {\Delta t^3}/_6 \\ {\Delta t^2}/_2 \\ \Delta t \end{bmatrix} [\,\ddddot{x}_0\,] \tag{15}$$

Comparing with 'equation (15)' and Newton's laws of motion concluding that $X_k = AX_0 + Bu$;

$$X_k = \begin{bmatrix} x_k \\ \dot{x}_k \\ \ddot{x}_k \\ \dddot{x}_k \end{bmatrix}, A = \begin{bmatrix} 1 & \Delta t & \Delta t^2/2 & \Delta t^3/6 \\ 0 & 1 & \Delta t & \Delta t^2/2 \\ 0 & 0 & 1 & \Delta t \\ 0 & 0 & 0 & 1 \end{bmatrix} \text{ and } B = \begin{bmatrix} \Delta t^4/24 \\ \Delta t^3/6 \\ \Delta t^2/2 \\ \Delta t \end{bmatrix}$$

$$Q_k = \sigma^2 \int_0^T G_k G_k^T = \sigma^2 \int_0^T \begin{bmatrix} \Delta t^3/6 \\ \Delta t^2/2 \\ \Delta t \\ 1 \end{bmatrix} \begin{bmatrix} \Delta t^3/6 & \Delta t^2/2 & \Delta t & 1 \end{bmatrix} dt$$

$$Q_k = \sigma^2 \int_0^T \begin{bmatrix} \Delta t^6/36 & \Delta t^5/12 & \Delta t^4/6 & \Delta t^3/6 \\ \Delta t^5/12 & \Delta t^4/4 & \Delta t^3/2 & \Delta t^2/2 \\ \Delta t^4/6 & \Delta t^3/2 & \Delta t^2 & \Delta t \\ \Delta t^3/6 & \Delta t^2/2 & \Delta t & 1 \end{bmatrix} dt = \sigma^2 \begin{bmatrix} \Delta t^7/252 & \Delta t^6/72 & \Delta t^5/30 & \Delta t^4/24 \\ \Delta t^6/72 & \Delta t^5/20 & \Delta t^4/8 & \Delta t^3/6 \\ \Delta t^5/30 & \Delta t^4/8 & \Delta t^3/3 & \Delta t^2/2 \\ \Delta t^4/24 & \Delta t^3/6 & \Delta t^2/2 & \Delta t \end{bmatrix} \quad (16)$$

2.2. *Kalman Gain*

Kalman gain is basically used to decide update value for moving object's state and is just a simple percentage formula. Gain value is to indicate how much we trust innovation and compare uncertainty in prediction and measurement methods by calculating percentage of uncertainty in system model value to that of total uncertainty.

$$Kalman\ Gain\ (K_k) = \frac{Uncertainty\ in\ model's\ state}{(Uncertainty\ in\ model's\ state\ + Uncertainty\ in\ observation)} = \frac{(P_k * H^T)}{((H * P_k * H^T) + R)} \quad (17)$$

Where uncertainty in predicting state is matrix P_k *(2x2 matrix)* for second-order model and that measurement noise Covariance matrix or uncertainty matrix in measured value is matrix *R(1x1 matrix)*. Both of them are of different size. Hence, we use *H(1x2)* and transpose of *H(2x1)* matrices to convert *P* matrix to correct size or dimensions. If system model value and its uncertainty *(X, P)*, measure value and its uncertainty *(z, R)* and Kalman gain values have in hand, it is enough to correct the estimated state $\hat{X}$ and its corresponding covariance matrix $\hat{P}$. The hat over the $\hat{X}$ means that it is the state estimation [9]. Kalman filter estimates a state iteratively from one time step to the next step using the above equations into two groups: prediction and correction [10]. An implementation flowchart of the KF algorithm is summarized in 'figure 3'.

Kalman filter can adjust its gains according to innovation requirements. However, it has large computation complexity due to its multiple state variables and has difficulty in calculating the matrices. Kalman gain K_k is computed by the given model parameters and noise covariances of the process and the measurement respectively. To determine K_k, the noise covariance matrices Q_k and R are priorly needed to estimate from the observation data. The process noise covariance matrix Q_k might change each time step according to Q_k equation. So the measurement noise variance R is estimated priorly during filter operation.

3. Implementation of Kalman Filter to Motion control

In order to survey the Kalman filter's performance on real world signal and data, the filter is evaluated experimentally using a digital position encoder fixed on the 12V brushed DC motor with a 100:1 gear ratio, controlled by the Arduino IDE from a PC computer with a sampling period of $1\mu s$. For this experimental setup shows in 'figure 1', a sinusoidal signal with amplitude of 50deg and *r(t)= Amplitude * sin(2 * pi * f * time),* and an encoder which can provide 64 counts per revolution of the motor shaft are integrated into the experimental process. The position readings from the encoder are acquired at fixed time intervals each Δt = 1 µs, it counts the number of pulses from the encoder sensor during a fixed interval of time. If the fixed time interval is too long, the encoder will not sense any new data until the next consecutive time.

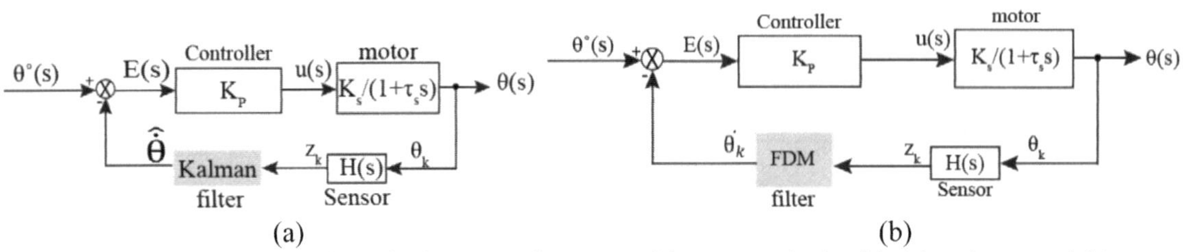

Figure 1. Block diagram of the velocity control system of the motor shaft with (a) Kalman and (b) FDM algorithms.

The position generated from the incremental encoder is highly affected by the quantization noises and Gaussian noises. In order to see the noise behavior and resolution of the encoder sensor, the output signal (z_k) obtained from the encoder sensor and its uncertainty of the measurement is evaluated with a sampling period of time. Experimental results express the change of the velocity estimations by using Kalman filtering. For this investigation, finite difference method (FDM) is integrated into the experimental system for comparison with Kalman filtering method. 'Figure 1' shows the block diagram sketches of the velocity control system of the motor shaft with (a) Kalman and (b) FDM algorithms.

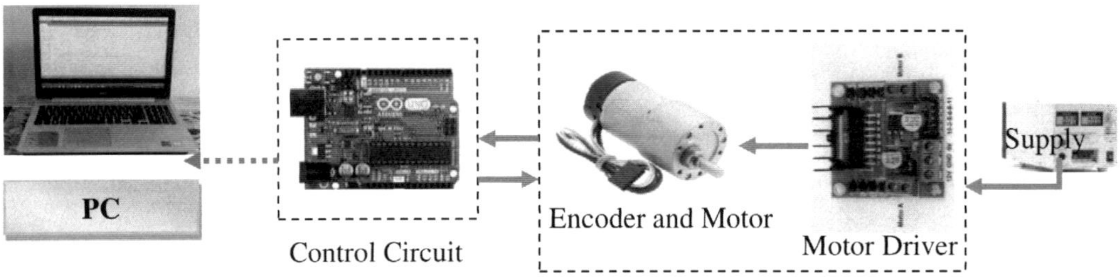

Figure 2. Experimental setup for Noise Filtering.

The control algorithm needs to estimate the velocity of the motor shaft. Note that the velocity is not measured. In order to solve this problem, one can use the Kalman filter with the encoder sensor for position measurement (see in 'figure 1(a)'). The Kalman model can be used to describe the relationship between two measurements: a measurement (collected in the vector X) and a second measurement (collected in the vector z_k). Then, the goal of the Kalman filter is to perform the "best" measurement fusion between the primary and secondary measurements. Otherwise, if we need to estimate the rate of change of position with respect to time in this case, we can use the finite difference method to calculate the approximate value of the speed. Three types of the finite difference method are basically considered: forward, backward, and central finite differences. The backward finite difference is applied to compare with KF. The formula for the backward finite difference is:

$$\dot{\theta} = \frac{\theta_t - \theta_{t-1}}{T^2} \tag{18}$$

where $\dot{\theta}$ is the output velocity estimation of backward FD method, θ_t represents the angle of the motor shaft at current time t, θ_{t-1} is the angle of the motor shaft at previous time t-1 and T^2 is the sample time period.

In off-line sample measurements, determining the measurement noise covariance R is usually performed with the help of VAR( ) function in excel and the computed variance will be used as the constant value of R in the Kalman process. However, the measurement error does not remain constant in on-line measurements. So, the measurement noise is needed to adjust adaptively through experimentation to obtain desired filter performance during filtering operation. The position readings from the encoder are stored in a data file for analysis using MATLAB. Users need to set the initial values $\hat{X}_0$ and P_0 for KF in the initialization step. Different from the most publications which keep Q_k and R constant. In this paper, the Q_k of the KF is iteratively estimated and updated during each correction step and R is adaptively adjusted through the experiments.

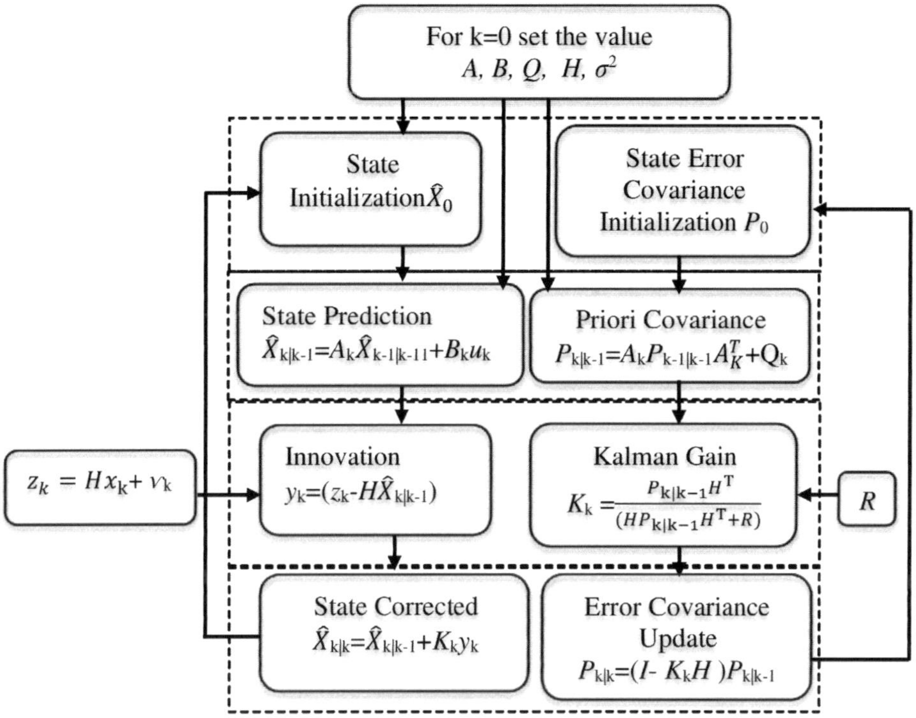

Figure 3. Structure of Kalman filter.

4. Experimental Results

In this paper, to investigate the Kalman filter performance with second-order model, the noisy position data from the digital encoder sensor is filtered with the Kalman filtering (KF) method and Finite Difference (FD) method used for comparison with KF. The velocity estimation results of the KF and FD methods are demonstrated using the experimental output data of a DC motor's shaft angle. 'Figure 4(a) and 4(b)' show the output velocity estimation of FD filtering method using 0.125Hz and 1Hz and then 'figure 4(c)' gives the velocity estimation of KF method using 0.125Hz. Firstly, both filtering methods produce the noisy velocity estimation as shown in 'figure 4' in which red lines represent the desired velocity signal, and the blue lines represent FD and KF velocity estimations recorded from the experiments. As above-explained, Kalman filter can adjust its gains according to innovation requirements so we can adjust its measurement noise covariance, but we cannot adjust FD filtering as needed.

Figures from 5(a) to (c)' give the output velocity estimation with 0.125Hz under the different measurement noise covariance R value and we can see the performance of Kalman filter from the experimental results. According to the experimental evaluation, 'figure 5 (a)' show noisy velocity estimation using small magnitude of R=0.01 and 'figure 5 (c)' has large overshooting and a significant signal time delay at a large magnitude of R=10. The action of this filter eliminates the noise in the choice of appropriate measurement noise variance R=1.5 (see blue 'figure 5 (b)'). And then these experiments repeat at frequency changing by adjusting the value of R shown in 'figure (6)'. As you can see, 'figure 6 (a)' has noisy output signal and 'figure 6 (c)' produces a little overshoot and delay output signal. In 'figure 6 (b), more smooth velocity output is produced by using the value of R=1.

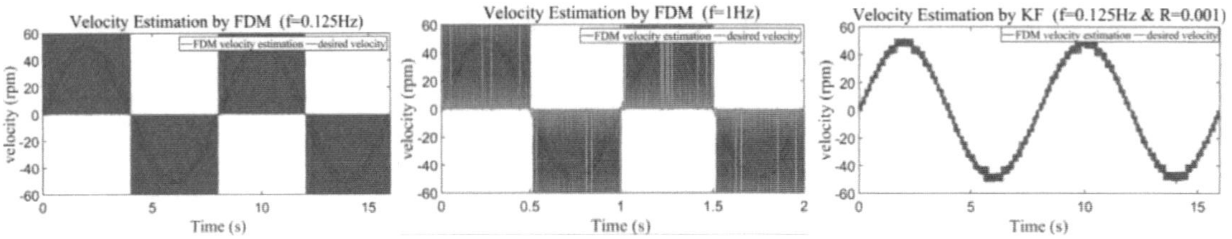

(a) (b) (c)

Figure 4. (a) Velocity estimation result of FD with 0.125Hz, (b) FD result with 1Hz and (c) Velocity estimation result of KF with f= 0.5Hz and R=0.00001 by using second-order model.

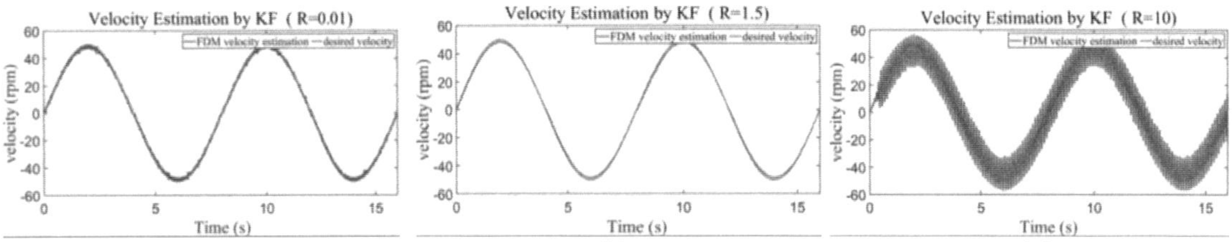

(a) (b) (c)

Figure 5. Velocity estimation of KF, filter performance evaluated on sinusoidal motions with 0.125Hz and changing magnitude of R (a) R=0.01, (b) R=1.5 and (c) R=10.

According to the above-mentioned experimental analysis and test, it may be seen that the FDM method cannot eliminate the noises of input signals and it produces a noisy output estimation at any frequency. The Kalman filtering algorithm produces a smooth filter result, but it requires the appropriate value of measurement noise covariance and, may get a more accurate performance if the measurement noise covariance is adaptively adjusted in real time. When the measurement of noise value is inappropriate, the output result will be far from the desired value. The produced estimation should be smooth without a large delay time to avoid instabilities. As it can be experimented, we should increase the magnitude of R if the estimation can't keep smooth performance without overshoot and reduce the magnitude of R if the user seems to be overshooting. The value of R should be increased as much as possible to obtain a properly smooth signal without a great filter delay. In order to keep the filter responsive to the changes of frequency changes, the value of R should be reduced if the filter cannot follow the measured position signal.

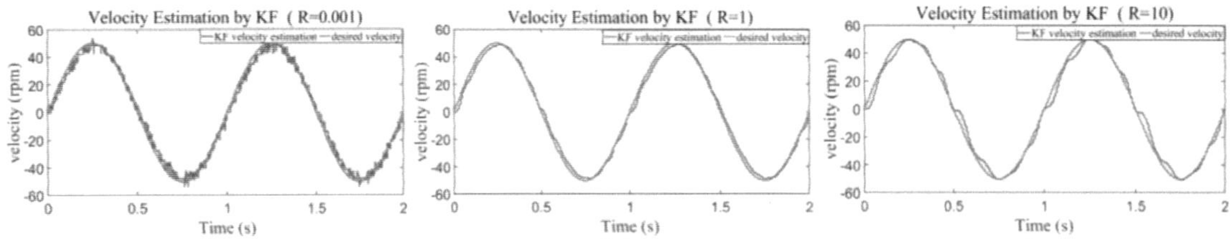

(a) (b) (c)

Figure 6. Velocity estimation of KF, filter performance evaluated on sinusoidal motions with 1Hz and changing magnitude of R (a) R=0.001, (b) R=1 and (c) R=10.

When the system model and measurement relationship are perfect, the Kalman filter can provide the best (minimum variance, unbiased) state estimation for noisy measurements at any frequencies. When they are not perfect, the filter can incorrectly estimate the state, and the estimation will diverge. It can be seen that the filtering with KF has more precision and less noise than that by differentiation method. The velocity estimations provided by FDM (see blue line in 'figure 4(a) and 4(b)') is not as good as velocity estimations obtained from the KF method (see blue line in 'figure 5(b) and 6(b)').

5. Conclusion

In this paper, the performance of the Kalman filtering algorithm has been investigated based on real-time experimental measurements in a motion control system. Regarding this, the velocity estimation of the KF method has been analyzed in various frequency ranges using the incremental encoder. According to this investigation, the accuracy of the KF method is limited by the numerical precision of the calculations and by the adaptive choosing of the measurement noise covariance. In order to obtain the smooth and optimal state estimation results from Kalman filter without delay time, its measurement noise covariance R is needed to adjust adaptively through the experimentation during filter operation. To maintain the filter responds to the changes of frequencies, we should adaptively adjust the measurement covariance. We should increase the magnitude of R if the estimation can't keep a smooth estimation without a large delay time, and reduce the magnitude of R if the user seems to be overshoot. The measurement noise covariance R is an significant influence and important parameter for the Kalman filter.

References

[1] Kalman R E 1960 A new approach to linear filtering and prediction problems *J. Basic Eng. (ASME)* vol 82D pp 35–45
[2] Bar-Shalom Y, Li X R, and Kirubarajan T 2001 *Estimation with Application to Tracking and Navigation: Theory Algorithm and Software* (New York: Wiley)
[3] Tripathi R P, Ghosh S and Chandle J O 2016 Tracking of object using optimal adaptive Kalman filter *2nd IEEE International Conference on Engineering and Technology*
[4] Kleinbaur R 2004 Kalman filtering with implementation of MATLAB pp 5-12
[5] Almagbile A, Wang J and Ding W 2010 Evaluating the performances of adaptive Kalman filter methods in GPS/INS integration *Journal of Global Positioning Systems* vol 9 no 1 pp 33-40
[6] Mohamed A and Schwarz K 1999 Adaptive Kalman filtering for INS/GPS *Journal of Geodesy* vol 73 pp 193-203
[7] Rodriguez-Maldonado J 2019 Estimation of angular velocity and acceleration with Kalman fillter, based on position measurement only *Measurement* pp 130-136
[8] Thombre D V, Nirmal J H and Lekha D 2009 Human detection and tracking using image segmentation and Kalman filter *International Conference on Intelligent Agent & Multi-Agent Systems* Chennai pp 1-5
[9] Welch G and Bishop G 2001 An introduction to the Kalman filter *SIGGRAPH* vol 7 no 1 Citeseer
[10] Kirchhoff J and von Stryk O 2018 Velocity estimation for ultralightweight tendon-driven series elastic robots *IEEE Robotics and Automation Letters* vol 3 no 2 pp 664-671

Object detection for drones on Raspberry Pi potentials and challenges

Tran Quang Khoi[1*], Nguyen Anh Quang[2], Ngo Khanh Hieu[1,3]

[1] Department of Aerospace Engineering, HCMUT, VNU-HCM
[2] Technical R&D Manager at SII Canada
[3] VNU-HCM Key Laboratory for Internal Combustion Engine

[*]khoi.tq.aero@gmail.com

Abstract. The paper presents preliminary research results about implementing an object detection program on a Single Board Computer. These results are used later to develop applications for drones. The object identification program is developed in Python using the TensorFlow library. The authors have succeeded in implementing and testing this object identification module using the artificial neural network model SSDMobileNet V2 on the Raspberry Pi 3B+. The results in this paper demonstrate the potential of this module for further development in the future. Based on the simulation and real-world results, the authors showed that a good outcome is achievable with limited resources for the AI module. Along with a high-precision object detection feature, this module can also estimate the distance and velocity of the "human" object with good accuracy. Besides, the paper also proposes several solutions to increase the performance and most importantly, the real-time feature of the developed module.

Index Terms— Object detection, Raspberry Pi, SSDMobileNet, Computer vision, Artificial neural network, Convolutional Neural Network.

1. Introduction

Unmanned Aerial Vehicle (UAV) is one of the recent rising research trends, in particular the topics related to Artificial Intelligence (AI)/Machine Learning (ML). The applications of the UAV + AI/DL is being investigated and developed for different industries such as art, transportation, security, etc. For the next generation of UAVs, one of the basic requirements is the ability to localize and identify the interested objects, which is fundamental for more advanced features such as real-time decision-making and high-precision indoor/outdoor navigation.

Even though object localization algorithms can offer a good accuracy, as a trade-off, they usually require powerful computational hardware. This is a real challenge for UAV designers and developers because of the constraints related to Size, Weight, and Power for these extra payloads. In the recent years, with the outstanding development in the computer sciences and other related domains, a new solution known as Single Board Computer (SBC) has been introduced. Thanks to its lightweight yet powerful processing capability (sometimes even with an external on-board graphic processing unit), the single-board computers (computer boards) have gradually overcome the above problem, becoming the best solution for both commercial and research-based UAV. Along with the new advances in hardware, the development in the field of AI/Machine Learning in recent years has also produced more lightweight, faster, and more reliable neural network algorithms and models, reducing the requirements for the computational resources (memory, clock rate, parallel computation, etc.).

Thanks to the above factors, the implementation of an on-board real-time object detection module using neural networks has become feasible. Some examples for these developments and the related researches can be referred in [1-2]. In this paper, the authors present their recent work on implementing an object detection program capable of estimating the distance and velocity of objects, then performing experiments to determine the applicability of this technology to small drones. The core hardware of this work is a cost-of-the-shelf yet popular Raspberry Pi 3B+ [3], and the AI module is implemented using Tensorflow library developed by Google [4]. Figure 1 below shows an example for the outputs of our work.

The next sections of this paper are structured as follows: In section 2 we make an overview about the computer vision and artificial neural network; Section 3 will show the implementation of the object detection module; Section 4 presents the experiments to determine performance of the integrated solution, along with the results of these tests. As described in detail in this Section, the authors have successfully implemented the program on the Raspberry Pi 3B+. The distance estimation and velocity estimation ability of the program give a good result with low errors (see Fig. 1). After that, Section 5 presents the potentials and challenges for the further research continuing this work in the real-time aspect of the module. Finally, Section 6 will present different future works based on the obtained results presented in this paper.

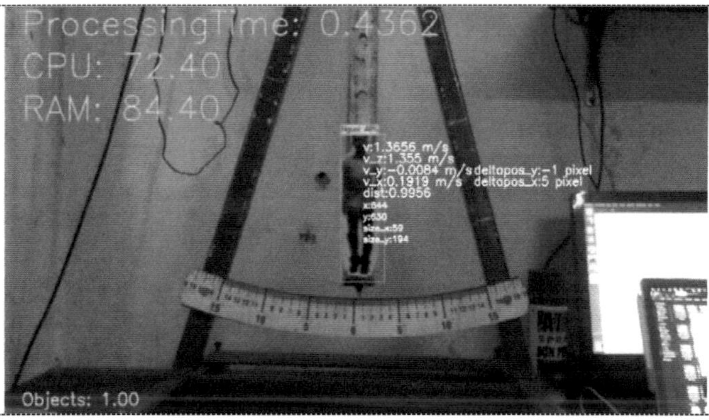

Figure 1. Object detection and distance estimation, velocity estimation results.

2. Background

2.1. Artificial neural network and object detection

The artificial neural network (or neural network for short) is a mathematical model that was introduced in 1943 to simulate the way in which human biological neurons work. The neural network is made up of basic units, i.e. neurons, which is in fact a group of different basic operations such as Multiplication, Addition, and linear/non-linear Activation A pioneer in the study of neurons is logic unit model made by Warren McCulloch and Walter Pitts in 1943 and the perception model introduced by Frank Rosenblatt in 1957 [5]. Although inspired by biological neurons, artificial neurons and artificial neural networks are completely different in nature. Human brain has about 100 billion neurons working in parallel meanwhile artificial neural networks are essentially just mathematical functions that are executed sequentially by computers. Studies of artificial neural networks favour math and technology rather than biology, although there are some mathematical models for ideas are made by observing how the brain works (see Fig. 2).

Mathematical representation of the neuron output:

IOP Conf. Series: Materials Science and Engineering 1109 (2021) 012033 doi:10.1088/1757-899X/1109/1/012033

$$y_k = \varphi(s_k) = \varphi(\sum_{j=0}^{m} \mathrm{w}_{kj} x_j + \mathrm{b}_j) \quad (1)$$

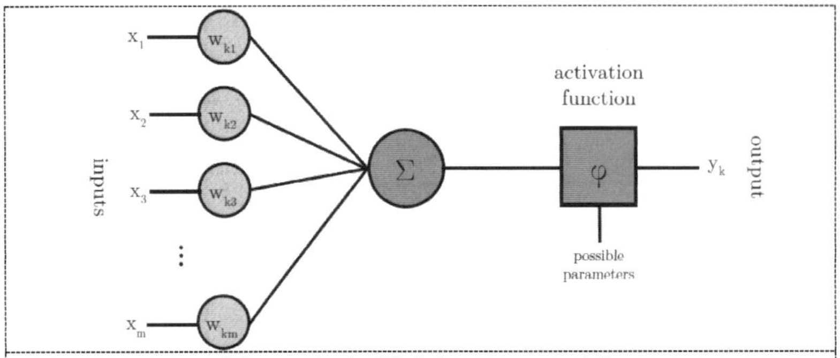

Figure 2. Artificial neuron model.

The training of artificial neural networks means finding the appropriate set of weights and deviation coefficients for each neuron in the network so that the final output is as closed as possible to the input references (i.e. expectation of the trainer), hence minimizing the objective function (cost function). To train neural networks we use back-propagation techniques in conjunction with optimization algorithms such as Gradient Descent [6].

To solve object detection and identification problems in computer vision, we use a convolutional neural network. The convolutional neural network has a different architecture than a normal neural network. The latter transforms input through a series of hidden layers. Each layer is a set of neurons and fully connected with the neurons on the previous layers. The final layer will represent the prediction of the network. Meanwhile, the convolution neural network (CNN) has neurons that are not fully connected to each neuron of the next layer, but only to a small region (some neurons). This section of the CNN is called the feature extractor, its function is to detect and extract different features in the image. Finally, a fully-connected flatten output layer will show the prediction result [7].

From the fundamental convolutional neural network AlexNet in 2012, there have been different convolutional neural network architectures developed so far (for example SSD and YOLO as shown in Fig. 3). Each network architecture has its own advantages in processing speed or resource consumption or accuracy (see Fig. 4).

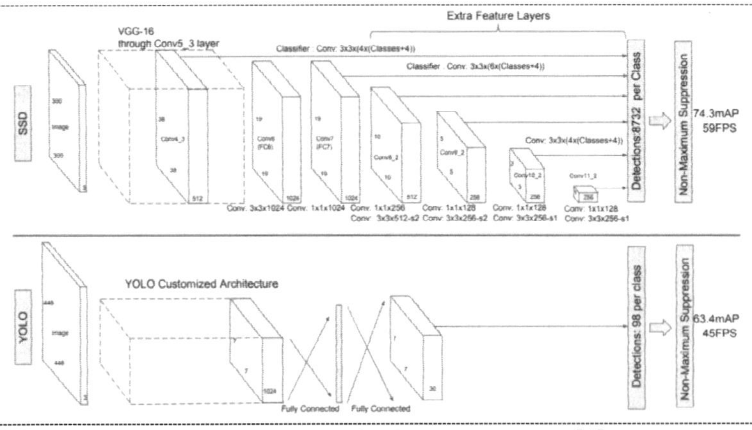

Figure 3. SSD vs YOLO meta-structure [1] .

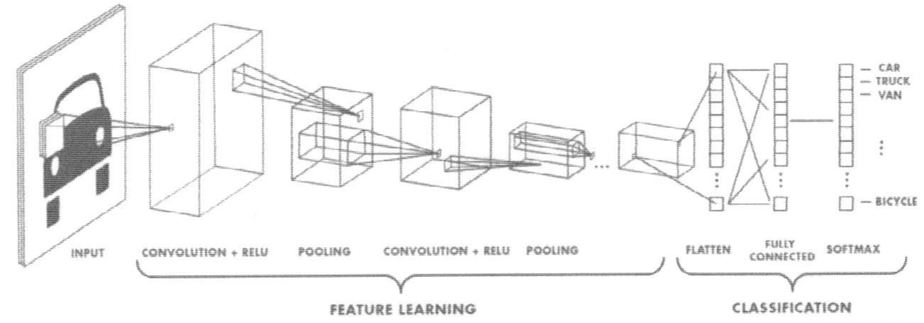

Figure 4. CNN structure for Object detection [2] .

2.2. *Distance and velocity measurement*

The distance estimation from the camera to the subject in the image is based on the image's formation taken through thin lenses (Fig. 5):

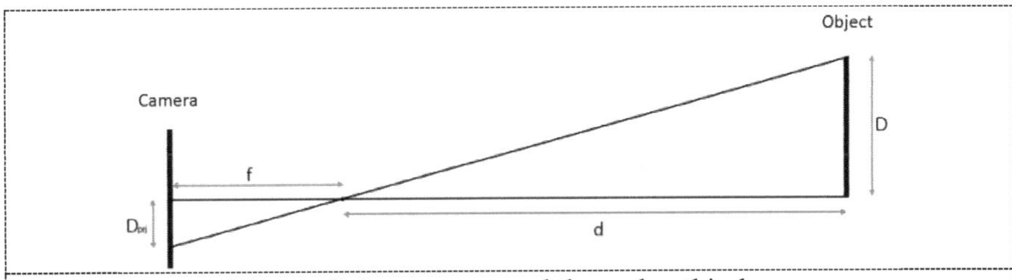

Figure 5. Image captured through a thin lens.

Where:

- f: camera lens focal (mm)
- d: distance between camera and object (mm)
- D: size of object (mm)
- D_{prj}: size of object in image (pixel)

The relationship between quantities is as follows:

$$d(mm) = f(mm).\frac{D}{D_{prj}} \quad (2)$$

$$D_{prj}(mm) = D_{prj}(pixel) * pixel\_to\_mm\_ratio \quad (3)$$

$$pixel\_to\_mm\_ratio = D_{prj}(pixel)/D_{mm} \quad (4)$$

In order to calculate the velocity of an object, it is necessary to obtain at least 2 frames has the same object and follow the steps (see Fig. 6):

a. Determine the position of the object in the image and find the box center coordinates in the first frame.
b. Determine the position of the object in the image and find the box center coordinates in the second frame.
c. Calculate the differences in position of box center between 2 frames.
d. Calculate the time difference between 2 frames.
e. Calculate the velocity by taking the ratio of the change in position of box center and time difference between 2 frames.

Figure 6. Bounding box center position change between 2 video frames

Note: No filters are applied to predict the velocity and position of the bounding box center due to research purpose only.

2.3. *Raspberry Pi 3B+*

For this work, we focus on the Raspberry Pi (RPi) 3B+, the latest generation of single-board computers of the Raspberry Pi Foundation in the UK at that time. Nowadays, there are thousands of diverse applications based on Raspberry Pi, ranging from Internet of Things to robotics and Industry 4.0 [8][9]. Some important parameters of the RPi 3B+ can be found in Table 1 below.

Table 1. Raspberry Pi 3B+ configuration

CPU	Broadcom BCM2837B0 quad-core A53 (ARMv8) 64-bit @ 1.4GHz
RAM	1GB LPDDR2 (900 MHz)
Connection options	Wifi 802.11 n, Bluetooth 4.1, 40-pin GPIO
Power consumption	800 mA, 5 V (4 W)
Total weight	50 gr
Graphic card	None, but support OpenGL for advanced image processing feature.

As can be seen from the Table above, RPi 3B+ is very energy-efficient and lightweight. Furthermore, thanks to the new powerful ARM processor, RPi 3B+ becomes a good candidate for the onboard processor for UAV applications.

3. Implementation

3.1. *TensorFlow API*

TensorFlow Object Detection API is designed specifically to build, train, and deploy different object detection models based on image databases such as COCO, Kitti, Open Images, AVA v2.1 and iNaturalist Species Detection. Each of these models has been pre-trained by computer scientists at Google for various artificial neural network meta-structures such as RCNN, SSD, Faster RCNN, SSDLite, etc. Each neural network meta-structure has different characteristics and depending on the needs, users can choose the appropriate model to use [10]. In addition, TensorFlow also allows users to train their own object detection neural network (from scratch or based on the pre-trained model) to suit each industrial specification. This API provides functions so that users can create their own database and conduct neural network training. However, this job is very time consuming and needs computers with powerful Graphical Processing Unit (GPU) to train or to fine-tune the neural networks.

3.2. COCO dataset

COCO (Common Object In Context) developed by Microsoft is a fairly large database with 330,000 images in 91 categories of common objects. Over 200,000 of them are labeled and ready for artificial neural networks training [3]. In particular, the most photographed subjects are people (Fig. 7).

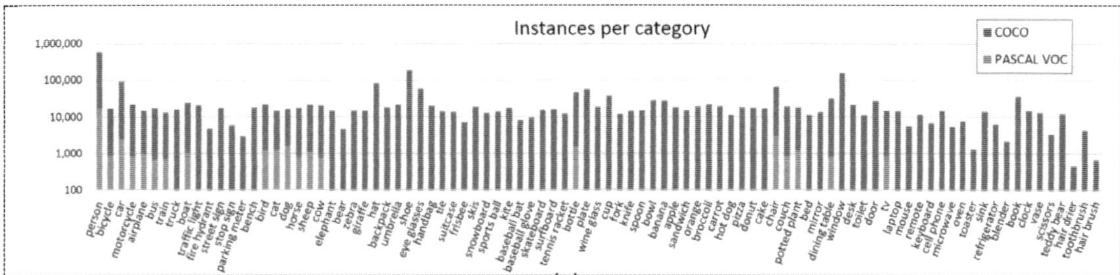

Figure 7. Number of images in each category of COCO [3]

3.3. Choose artificial neural network model

Since the algorithm must be low in resource consumption, the author prioritized the processing time of TensorFlow's existing artificial neural network models. As illustrated in Fig. 8, comparison results show that SSDlite_MobileNet_V2 model (SSDlite convolutional neural network architecture using MobileNet feature extractor version 2) has best performance (0.81) with the output is a bounding box, this model is chosen to implement in the object detection module presented in this paper.

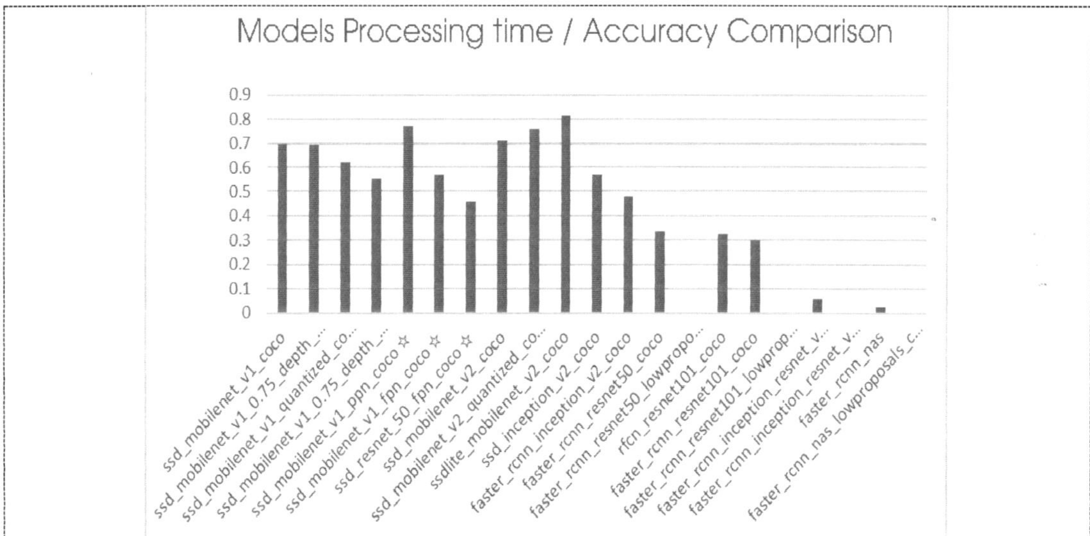

Figure 8. CNN models comparison

4. Experiment

To investigate the program's performance on object detection accuracy, distance estimation accuracy and velocity estimation accuracy, we perform 2 types of test: static image and video.

The experimental flow chart is as follow:

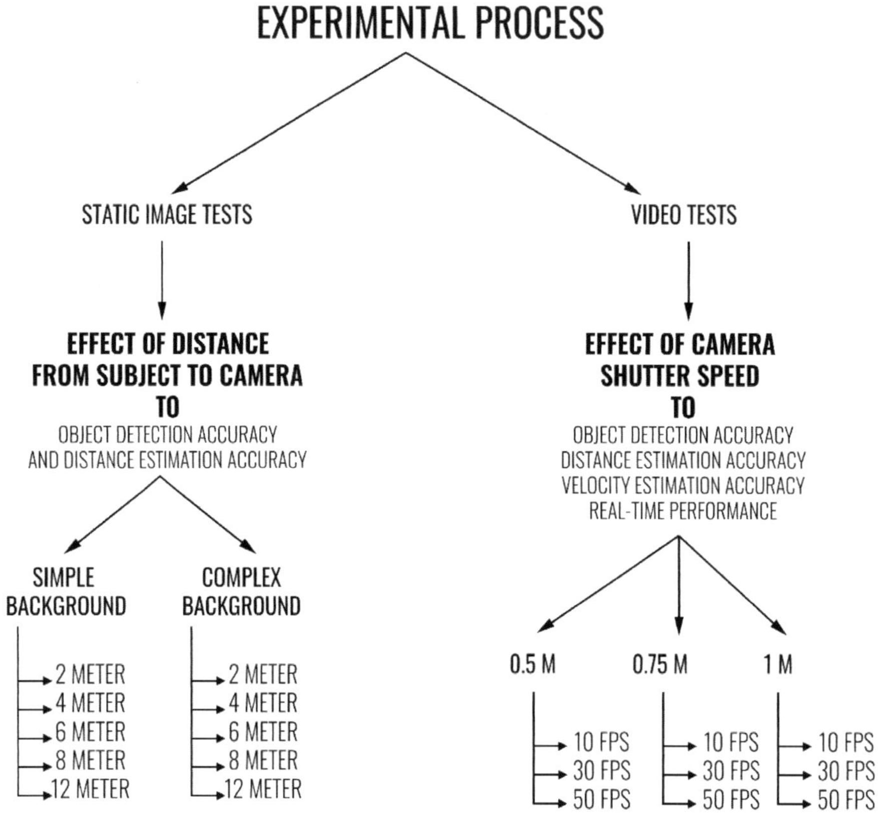

4.1. Static image tests

This test focuses on evaluating the ability of the program to detect objects for still images in order to obtain the optimal distance between the camera and the object where the program gives the best results (lowest error in both vertical and horizontal direction).

Experiment process: (see Fig. 9 – 11)

a. Take photos of a person at distances d = 666f (H 2 m), 1,333f (H 4 m), 2,000f (H 6 m), 2,666f (H 8 m), 4,000f (H 12 m).
b. At each distance, take pictures at horizontal positions equal to 0.5 m from the center to the left and right.
c. Calibrate the program's parameters at a distance of d = 666f (H 2 m) so that the distance estimation of the program is best.
d. Repeat step c for the remaining cases.
e. Make conclusions about the effects of object position in horizontal and distance to program's detection accuracy.
f. Make conclusions about the program's distance estimation accuracy at different distances.

The subject is a human photo (Fig.9). A simple grey background (Fig.10) is generated using Photoshop to compare the performance of the program with real life background (Fig .11) (complex background).

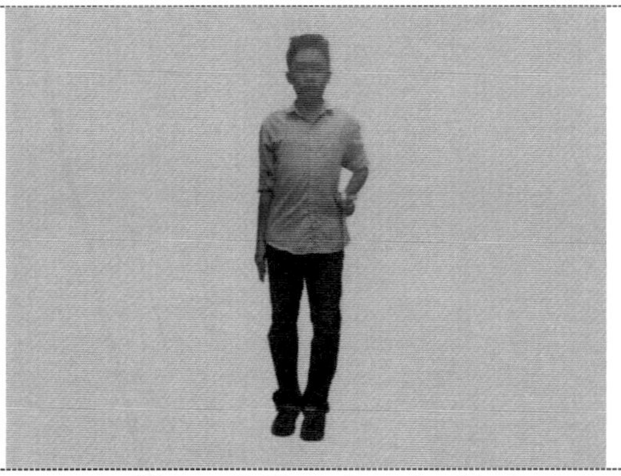

Figure 9. Experiment subject

Figure 10. Subject in complex background

Figure 11. Subject in a simple background

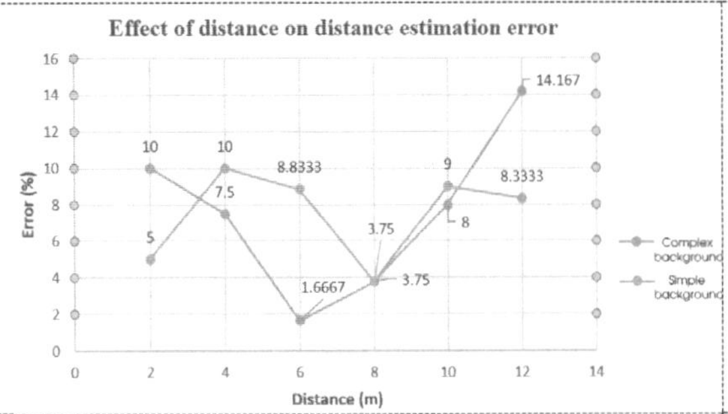

Figure 12. Effect of distance on distance estimation error

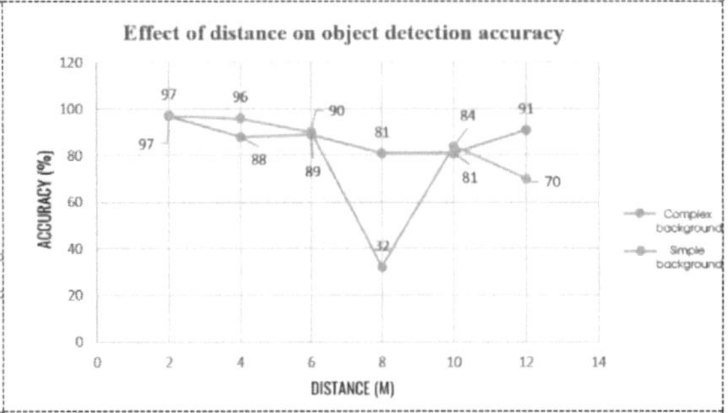

Figure 13. Effect of distance on object detection accuracy

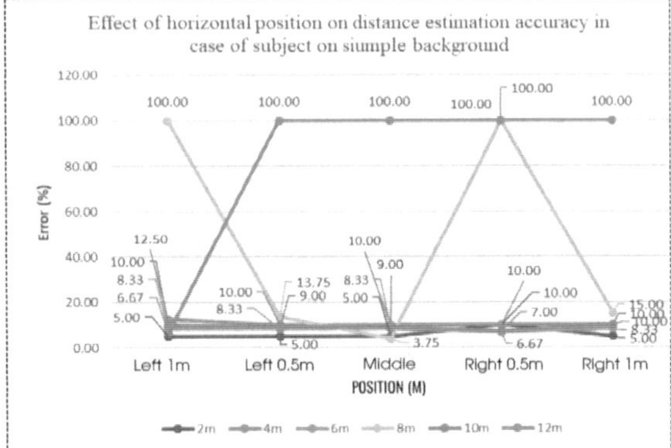

Figure 14. Effect of horizontal position on distance estimation accuracy in case of subject on simple background

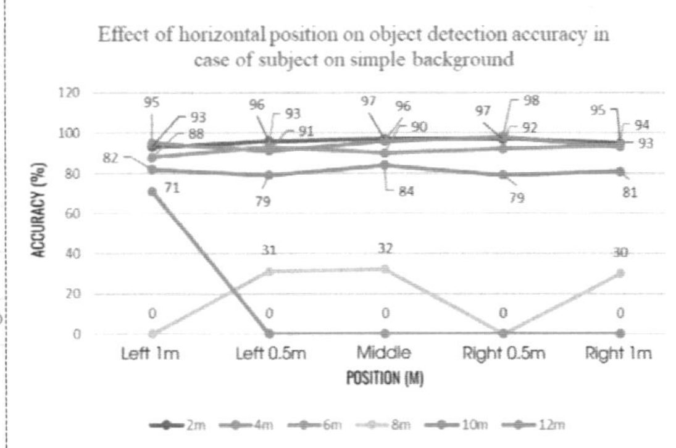

Figure 15. Effect of horizontal position on object detection accuracy in case of subject on simple background

5. Results

The program can identify objects up to 12 m (Fig.13). At a longer distance, i.e. smaller object; the precision (detection and labeling) of the program degrades gradually. The distance estimation accuracy is relatively good in the range of 2 m - 6 m (error not exceeding 10%).

Figure 12 show the differences in horizontal axis has a small impact on the accuracy of object detection and distance estimation. The effective range of the program is from 4 m to 6 m. The maximum detection distance of the program is 12 m. And the minimum estimated distance of the program is 2 m.

Figure 14 and 15 show that in effective range of the program (4m – 6m), object's horizontal position has little effect on program's accuracy.

5.1. Video tests

This experiment aims at evaluating the object detection ability of the program to determine the effect of distance and camera shutter speed to the accuracy of the recognition, distance and velocity estimation of the program as well as the program real-time performance. The test object (human image - size 19 cm) is mounted on a single pendulum which oscillates with a magnitude of 5 degrees, the pendulum length (to the midpoint of the object) is 0.52 m. Friction bypassed, parameters are calculated as follows (Fig. 16):

- The oscillation cycle: $$T = \frac{2\pi}{\omega} = 2\pi\sqrt{\frac{l}{g}} = 1,4466(s) \quad (5)$$

- Maximum speed at equilibrium point: $v_{max} = \omega S_0 = \omega \alpha_0 l = 0{,}1971(\frac{m}{s})$ (6)

The testing system is placed at different distances (corresponding to the distance of 4 m, 6 m, 8 m in real world) and oscillates (maximum velocity at the equilibrium position equal to 1.58 m/s in real world). The author recorded the system movement for 1 minute at each distance and each shutter speed value, extracted the image from the video recorded. The extracted images are fed to the program to evaluate: accuracy of identification, distance and velocity estimated compared to real world (Fig. 17).

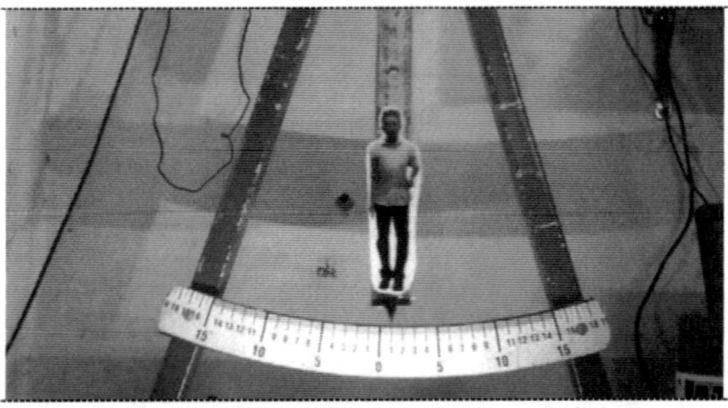

Figure 16. Experiment single pendulum

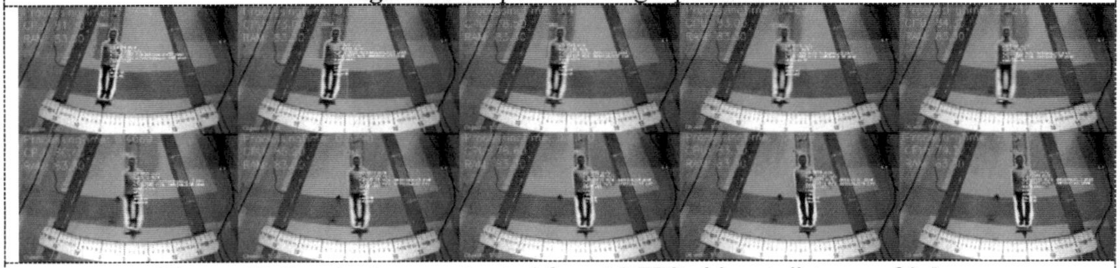

Figure 17. Result photos extracted from 10 FPS video at distance of 0.5 m

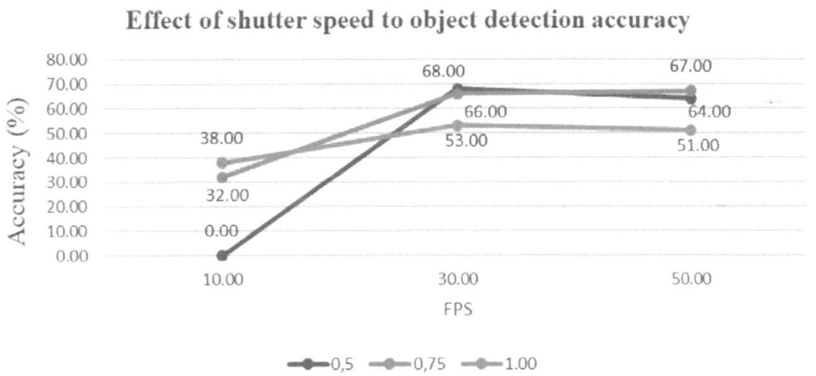

Figure 18. Effect of shutter speed to object detection accuracy

Figure 18 shows that at low shutter speed, the program's object detection accuracy is lower. This is due to image blur effect on a moving object. At the shutter speed of 1/30 second or faster, the program's accuracy nearly the same.

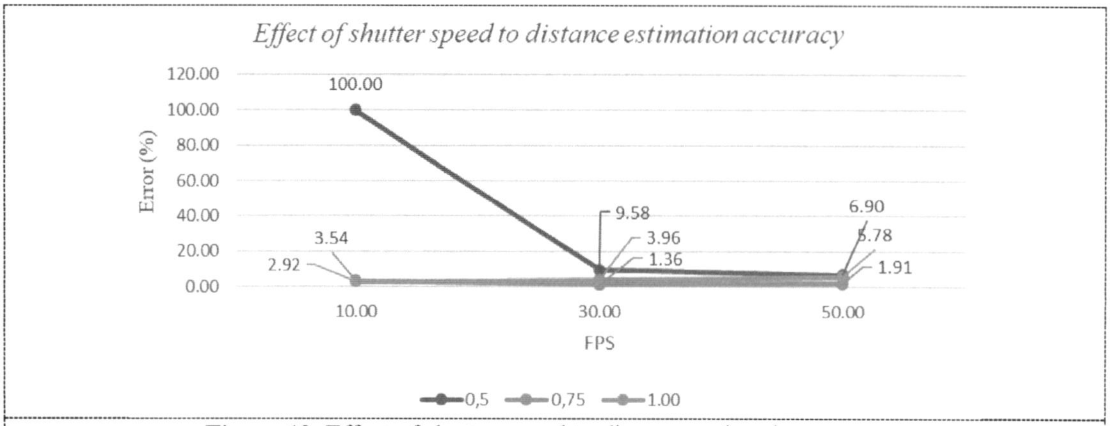

Figure 19. Effect of shutter speed to distance estimation accuracy

Figure 19 shows that the shutter speed has little effect on distance estimation accuracy thanks to the method to estimate the distance is based on object recognition. As long as the program recognizes the object, it will estimate the distance based on known sizes of the object.

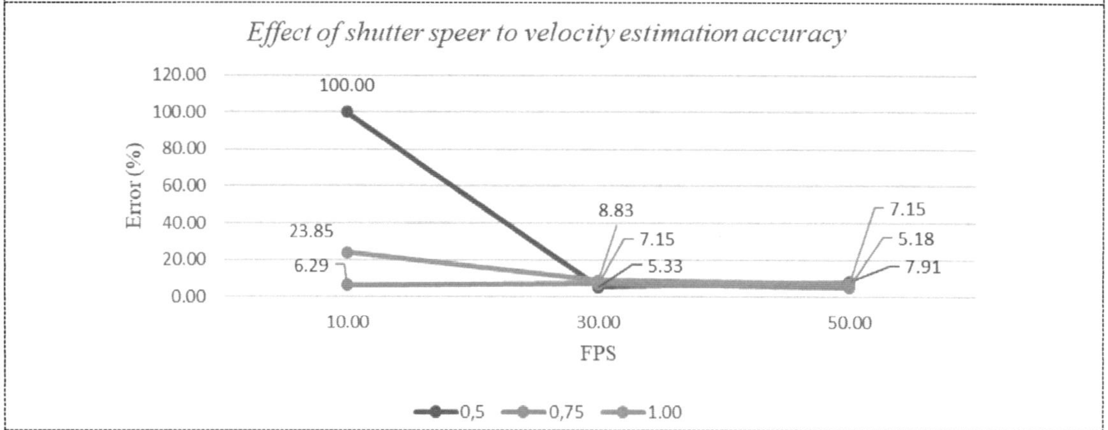

Figure 20. Effect of shutter speed to velocity estimation accuracy

Figure 20 shows that velocity estimation error reaches maximum value at the longest distance for a shutter speed of 1/10 second. For shutter speeds of 1/30 second, the error does not exceed 9% at all distances. For shutter speeds of 1/50 second, the error does not exceed 8% at all distances. When increasing the shutter speed, the velocity estimation error tends to decrease.

The estimated error of the program is within an acceptable range (not exceeding 10%) at shutter speeds of 1/30 second and 1/50 second. Shutter speed of 1/10 second gives a large velocity estimation error (nearly 24%) and in some cases it is not possible to identify the object, so the distance and velocity cannot be estimated.

5.2. Real-time performance

The processing time of the program determines the capability to use in real time. Real-time is defined as the ability to process all images that the camera obtains in a shorter or equal amount of time that the camera needs to capture the images. For example, a camera can obtain 10 frames per second (10 FPS), the processing time of the program for each image must be less than or equal to 0.1 second to be considered real-time.

The real-time performance is calculated by taking 1 second divided by the total time the program processes all frames obtained from the camera in 1 second.

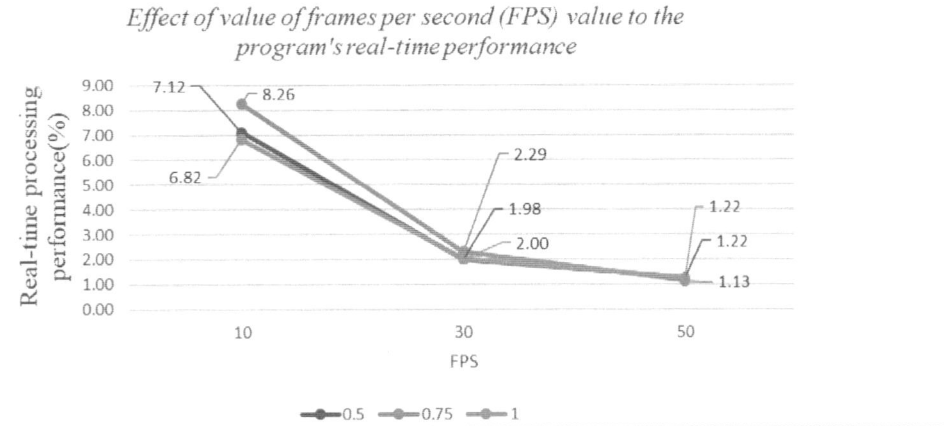

Figure 21. Effect of value of frames per second (FPS) obtained by the camera to the program's real-time performance

Results: (see Fig. 21)

FPS (frames per second) value has a strong influence on the program's real-time performance. As FPS increases, the real-time performance of the program decreases, because the program needs to handle more tasks.

At the lowest FPS (10 fps), the system takes approximately 14 seconds to process 1 second of video. This is due to the hardware limitations of Raspberry pi 3B + computers.

Similarly, at 30 fps, the program's real-time processing capability only reached 2.29% of requirement. This shows that the program cannot be used to handle situations that require rapid response.

The program can only handle a maximum of about 0.71 FPS. The real-time response of the program decreases linearly as the FPS value of the camera increases.

6. Conclusions

The program can run on Raspberry Pi 3B+ smoothly without any problem and shows good accuracy in object detection and identification for both still images and videos. The estimated distance and velocity accuracy is also quite good, which can be used in different features such as monitoring/avoiding obstacles.

Real time performance is bad even with one of the simplest and well-known object detection models. This is a real challenge because in many situations, the drone must act quickly and need fast response from the program to make decisions.

7. Future work

Based in the experience and obtained results in this study, some of the important future work are: (1) Improve the algorithm, using more compact and low computational resource neural network models to increase the program's real-time performance, e.g TensorflowLite and quantized model; (2) Develop the ability to track and estimate the velocity and distance of many objects simultaneously; (3) Make use of graphic card processing power on single-board advance computers like NVIDIA's Jetson Nano to improve real-time performance; and (4) Integrate the object detection program to a fully operational obstacle avoidance system for UAVs, especially quadrotors.

References

[1] Liu, Wei, et al, "SSD: Single shot multibox detector," in *European conference on computer vision.*, 2016.

[2] Mathworks.com, "MathWorks.com," MathWork, [Online]. Available: https://it.mathworks.com/discovery/convolutional-neural-network-matlab.html.

[3] Schnerr, G. H. and Sauer, J., "Physical and Numerical Modeling of Unsteady Cavitation

Dynamics," in *International Conference on Multiphase Flow*, New Orleans, USA, 2001.
[4] S. Boyce, "AI-on-raspberry-pi-with-the-intel-neural-compute-stick," 31/1/2019. [Online]. Available: https://hackaday.com/2019/01/31/ai-on-raspberry-pi-with-the-intel-neural-compute-stick/..
[5] "Raspberrypi," Raspberry Pi Foundation, 2019. [Online]. Available: https://www.raspberrypi.org/..
[6] Wikipedia, "TensorFlow," Google, 2019. [Online]. Available: https://vi.wikipedia.org/wiki/TensorFlow..
[7] Nicu Sebe; Ira Cohen; Ashutosh Garg; Thomas S. Huang, Machine Learning in Computer Vision, Springer Science & Business Media, 2005..
[8] Long. L. N., Gupta. A, "Scalable massively parallel artificial neural networks," Journal of Aerospace Computing, Information, and Communication, pp. 3 - 15, 2008..
[9] Wikipedia, "Backpropagation," [Online]. Available: https://en.wikipedia.org/wiki/Backpropagation..
[10] viblo.asia, "Mạng Nơ-ron tích chập," [Online]. Available: https://viblo.asia/p/mang-no-ron-tich chap-p1-DZrGNNjPGVB..
[11] A. Krizhevsky, I. Sutskever, G. E. Hinton, "ImageNet Classification with Deep Convolutional Neural Networks," 2012. [Online]. Available: https://www.cs.toronto.edu/~fritz/absps/imagenet.pdf. [Accessed 3 3 2019]..
[12] Mlab.vn, "Raspberry-pi-la-gi," [Online]. Available: http://mlab.vn/11025-raspberry-pi-la-gi gioi-thieu-cac-ung-dung-cua-raspberry-pi-3.html..
[13] Reglisse44, "Instructables.com - The drone pi," [Online]. Available: https://www.instructables.com/id/The-Drone-Pi/..
[14] "Tensor FLow Object Detection Model Zoo," Google, 2019. [Online]. Available: https://github.com/tensorflow/models/blob/master/research/object_detection/g3doc/detection_model_zoo.md.
[15] "TensorFlow-Object-Detection-on-the-Raspberry-Pi," 25/2/2019. [Online]. Available: https://github.com/EdjeElectronics/TensorFlow-Object-Detection-on-the-Raspberry-Pi..
[16] ""TensorFlow-Object-Detection-on-the-Raspberry-Pi," 25/2/2019. [Online]. Available: https://github.com/EdjeElectronics/TensorFlow-Object-Detection-on-the-Raspberry-Pi.".

Sliding speed effect in press-fit connection using Finite Element Analysis

A T Ubando[1,2], J Gonzaga[1,2], A Conversion[1,2], D Belarmino[1,2], E Arriola[1,2], N R E Lim[1,2], R L Moran[1,2], J P Mercado[1,2]

[1] Mechanical Engineering Department, De La Salle University, 2401 Taft Avenue, 1004 Manila, Philippines
[2] Thermomechanical Analysis Laboratory, De La Salle University – Manila: Laguna Campus, LTI Spine Road, Laguna Blvd, Biñan, Laguna 4024, Philippines

aristotle.ubando@dlsu.edu.ph

Abstract. Solderless press-fit contacts in power module packages are now gaining more attention in the automotive industry especially in hybrid electric vehicles because of the Restriction of Hazardous Substances (RoHS) directive to eliminate lead-based connections. The production of press-fit assembly is also faster, more convenient, safer, and cheaper than soldered connections. In this paper, the effect of insertion speed of pin to the press-in force as well as the pull-out speed to the retention force was investigated using finite element analysis. An eye-of-the-needle (EON) shape compliant pin with general dimensions was adapted and a transient structural analysis in ANSYS mechanical was used to add and vary the speed of the press-fit pin geometry. Four simulations were conducted with different speeds. The results obtained by FEA show that the speed when kept at constant has no significant effect both on the press-in and retention force. However, when the speed changes due to acceleration, the press-in and pull-out force are both affected.

1. Introduction

Press-fit technology is now commonly used in electronic packages because of its connection reliability [1]. It also becomes even more popular in the power modules that are used for power conversion in the area of renewable energies and automotive industries especially in hybrid electric vehicles which are subject to various outdoor conditions [2]. Press-fit contacts were first introduced in the telecommunication industry in the late 1960's [3] but are now gaining more attention because of the Restriction of Hazardous Substances or RoHS directive to eliminate lead-based connections in electronic package especially in the automotive industry as the demand for solderless connections in hybrid power trains continues to increase.[4].

A press-fit contact is achieved when a pin is inserted into a plated through hole (PTH) of the printed circuit board (PCB). **Figure 1** shows a typical Eye-of-the-Needle (EON) press-fit pin with its initial and final position on the PCB. The compliant pin (the elastic portion) deforms during the insertion process and creates a contact force against the hole. Since the press-fit pin has a slightly larger cross-sectional dimension than the hole, interference occurs and generates a gas-tight metal to metal contact between the press-fit pin and the plated through hole [5]. When compared to a soldered connection, a solderless press-fit assembly process is faster, more convenient, and safer [1]. In addition, since a press-fit connection is a solderless process, this reduces the production cost because the soldering materials and equipment like robots are eliminated and no additional cleaning step for soldering flux leftovers [6]. This makes a solderless press-fit pin connection cheaper than a soldered connection.

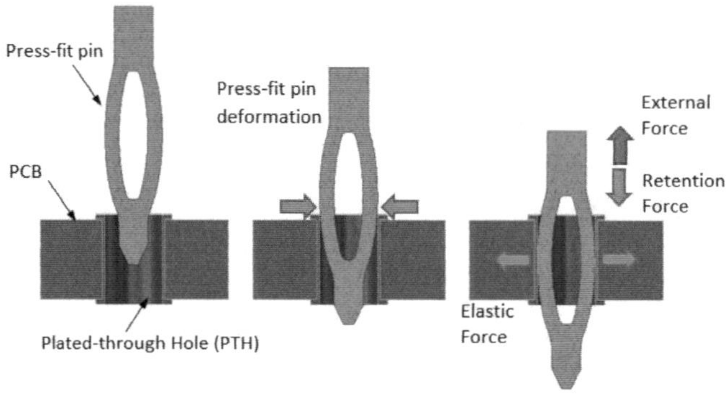

Figure 1. Press-fit process [5].

One specification of the press-fit process that needs to conform with the IEC 60352-5 standard is the press-in and retention force. According to Ring and Alt [4], these forces were affected mainly by these physical parameters, namely, the shape of the pin, the material of pin (core material and plating), diameter and homogeneity of PCB plated through hole (PTH), thickness of PCB and amount of laminate layers, material and the thickness of PTH plating, and press-in speed and temperature. In this study, a finite element method (FEM) is employed to study the effect of insertion and pull-out speed on the press-in and retention force of the press-fit assembly, respectively.

Finite element method has already been used by several researchers to study the behavior of the press-fit pin and PCB assembly. Yang et al. [1] developed a 3D finite element model to simulate the insertion process of press-fit assembly especially to study by simulation the possible failures of the press-fit pin during the laser soldering process. Riedel et al. [2] also used a 3D FEM as an additional tool to validate the measurement of electric current distribution within the pin-via-trace geometry. Another 3D FEA was performed by Ring and Alt [4] to examine the surface strain gradient on the PCB during the press-fit insertion process and was compared to two different sensor technologies. While Tohmyoh et al. [7] used a 3D FEA with an experimental set-up to determine the coefficient of friction of the compliant pins in the plated through holes (PTH) of PCB. Moreover, a very recent paper by Wang and Xu [8] exhaustedly used FE simulation to obtain and analyze the mechanical behaviors (which include mechanical stress, thermal stress, and contact force) of a typical eye-of-the-needle (EON) press-fit pin. However, no prior works yet have been published on developing a FE model to investigate the effect of sliding velocity on the press-in and retention force of press-fit pin.

This study aims to determine if any changes on the sliding speed during the press-in and pull-out process could significantly affect the magnitude of press-in and pull-out force of the press-fit pin, respectively using finite element analysis.

2. Finite Element Analysis

The basic principle of press-fit contact is that a terminal pin is inserted into the through hole of a printed circuit board (PCB). There are two types of press-fit pin terminals; the compliant pin which has the elastic behavior and the solid pin which is the terminal at the other end of the pin and is usually the one attached to the power module side [9]. In this paper, only the compliant pin having the elastic behavior is the focus and the press-fit pin being defined here.

Since the press-fit pins are injected into the PCB hole using assembly jigs, the amount of force to press-in the pins should be limited only to a certain degree to achieve reliable connections. The maximum force needed to press-in the pin into the plated through hole of PCB is called the press-in force, while the minimum force needed to pull-out the press-fit pin from the hole is called the retention force. Both these two forces should be within the standard value specified by the IEC 60352-5 [10]. The said standard also specified a recommended sliding speed during the press-fit assembly as the speed may also influence the quality of press-fit connections [5].

When installing press-fit connectors, press jigs or servo systems are designed for precision control of speed, position, and pressing forces [11]. Tomhyoh et al. [7, 12] used a constant speed 0f 0.833 mm/s to analyze the press-fit connection. While Ring and Alt [4] used a constant speed of 12 mm/min or 0.2 mm/s in recording the press-in force. However, no prior study yet recording the press-in force and the retention force while varying the sliding speed. Moreover, the recommended sliding speed by IEC 60532-5 standard is a range from 25 mm/min to 50 mm/min (0.417 mm/s to 0.833 mm/s) [5]. To verify the influence of sliding speed on the press-in and retention force, a finite element analysis of the press-fit assembly was modelled in this study.

Figure 2 illustrates the applied and reaction forces during contact of pin and plated through hole surfaces. During the press-in or insertion process, a load F is applied in y – direction. The plated hole in PCB is deformed and exerted a force to oppose the applied force by the press-fit pin. Using Newton's second law of motion, the summation of forces in y-axis is given by

$$\sum F_y = ma$$

$$F - R_y - \mu P = ma$$

rearranging, gets

$$F = R_y + \mu P + ma \quad (1)$$

where F is the load applied to insert the press-fit pin, R_y is the y-component of reaction force in contact region, P is the normal force working on the pin in x-axis with μ as the assumed friction coefficient, m is the mass, and a is the acceleration of the pin in y-direction.

Due to the slight deformation in plated through hole during assembly, R_y and P are given by

$$R_y = \sum sin\alpha.|f| \quad (2)$$

and

$$P = \sum cos\alpha.|f| \quad (3)$$

where the symbol $\sum$ is just referring to the summation of all nodes contribution on the contacting surface [12]. Notice that from equation (1), the required press-in force F increases when the acceleration a increases.

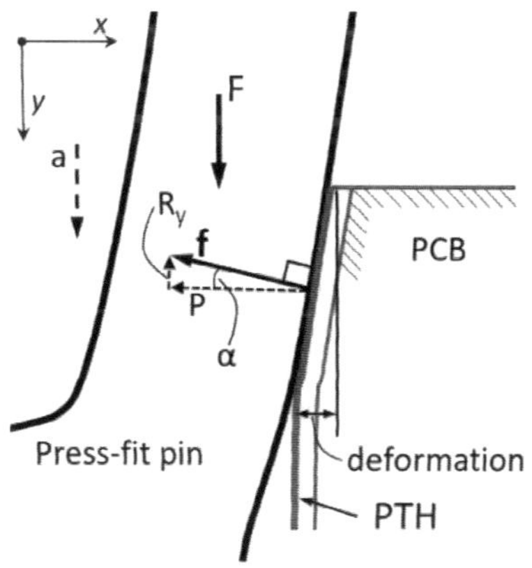

Figure 2. Free body diagram of press-fit contact region.

Moreover, when the acceleration is zero (a = 0), that is when the velocity is constant, the last term in equation (1) will be zero. That is,

$$F = R_y + \mu P \quad (4)$$

Hence, as long as the velocity is constant throughout the insertion process, its magnitude does not affect the magnitude of press-in force.

For the pull-out force, the direction of load F would be in opposite. Thus,

$$F = -R_y + \mu P + ma \quad (5)$$

A transient structural analysis in ANSYS Workbench was implemented on the FE model of the press-fit assembly to validate how the sliding speed influences the press-in and retention force. The geometric model of the press-fit assembly is shown in **Figure 3**. The dimensions of the EON compliant pin were taken from [8]. The 1.6 mm thickness of the PCB is a standard thickness and the hole dimensions, as well as the plated through hole were selected as general dimensions.

The parts of the press-fit model are made of three different materials, phosphor bronze, copper, and epoxy resin [7]. Their mechanical properties are listed in **Table 1**. Since some areas on the contact surface during the press-fit process undergo plastic deformations, a plastic model is adopted which is a Bilinear Isotropic Hardening model [13].

Table 2 summarizes the contact definition of the press-fit contact during the press-in and pull-out process. The assumed value of the friction coefficient was 0.2. Some values used for friction coefficient may be more than 0.2 [8], but to avoid convergence issues and no need to increase the mesh quality, a value not more than 0.2 was selected. The contact between the PCB hole and the PTH surface was bonded. Moreover, **Figure 4** shows the load and boundary conditions of the model. A symmetry region was added to lower the number of mesh elements and so the computational time. The total number of nodes was 51577 with 14786 elements. Finer mesh elements were added to the contact surfaces with a minimum element size of 0.03 mm. **Figure 5** shows the meshing of the model.

Table 1. Mechanical properties of three parts of press-fit assembly [8].

Material	Young's modulus (MPa)	Poisson's ratio	Yield strength (MPa)	Tangential modulus (MPa)
EON compliant pin (phosphor bronze)	77500	0.3	731	103
PTH (copper)	21000	0.3	305	200
PCB (epoxy resin)	10800	0.3	151	7930

Table 2. Contact definition.

Contact model	Surface to surface
Type	Frictional
Coefficient of friction	0.2
Offset	Set manually on the geometry
Behavior and formulation	Program controlled
Contact element	PTH contact surface
Target element	Compliant pin contact surface

Frictionless supports were added as constraints to the four sides of the PCB. Another frictionless support was added to the bottom of the PCB by which the reaction force exerted by the press-fit pin was taken. There was no fixed support added since the frictionless supports kept the PCB from moving already. Frictionless supports were also added to the symmetry plane to restrict it from deforming in that region. For the press-in process, the compliant pin was inserted into the PCB by displacing the pin 2.5 mm from its initial position. While for the pull-out process, the pin was ejected from the PCB and displaced to its original position. Finally, a velocity was added in the direction of the compliant pin.

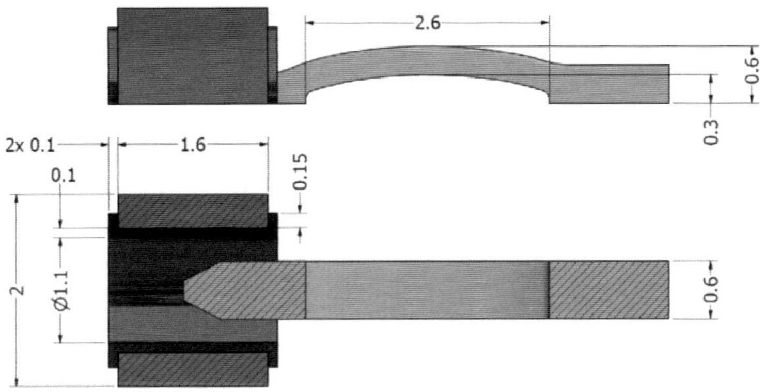

Figure 3. Press-fit geometric model.

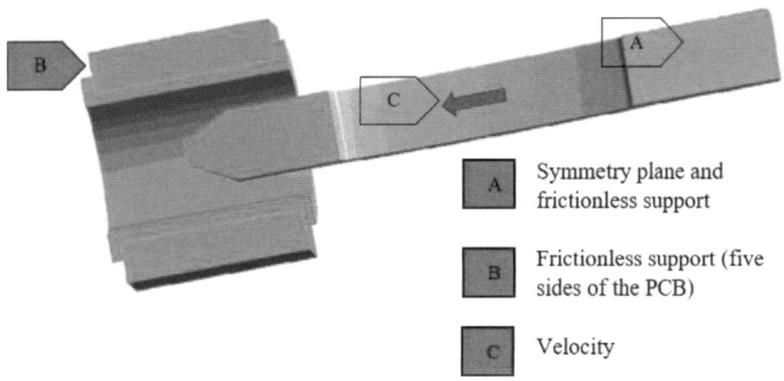

Figure 4. Load and boundary conditions.

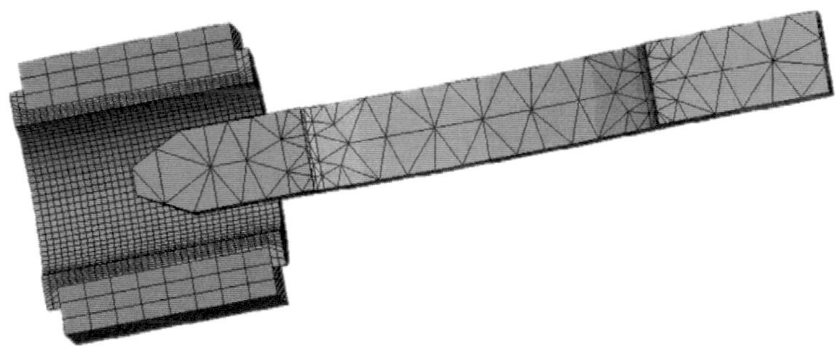

Figure 5. Mesh model.

3. Results and discussion

The FE simulation of the press-fit assembly is based on the parameters shown in table 3. Press-in and retention forces were obtained by getting the component of reaction force in the direction of pin's displacement at the frictionless support in the bottom surface of the PCB.

The sliding speed, both for the press-in and pull-out process, was varied accordingly. For the first three simulations, the speeds were kept at constant throughout the simulation. The lowest speed tested was 0.2 mm/s to account for the speeds slightly lower than the recommended speed by IEC 60352-5 standard. Next, was the 20 mm/s speed which is far higher than the recommended speed. The obtained forces between 0.2 and 20 mm/s speed did not significantly differ from each other so another significantly higher speed was tested which is 50 mm/s to confirm the results. The force vs. displacement curves are shown in **Figure 6**. The press-in and retention force curves of the first three simulations were almost identical except for the slight difference between their maximum retention forces. The obtained maximum press-in force in this study was 15% higher than the experimental result by Wang and Xu [8]. While the maximum pull-out force was 20% lower than the experimental value.

Table 3. Maximum press-in and retention force obtained from FEA.

Maximum force	Sliding speed			
	0.2 mm/s	20 mm/s	50 mm/s	(0.5 – 50) mm/s
Press-in force (N)	56.18	56.181	56.181	54.938
Retention force (N)	30.932	30.15	31.261	31.815

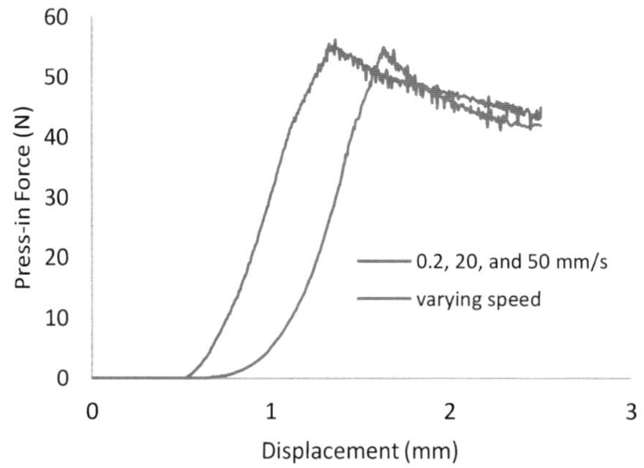

(a)

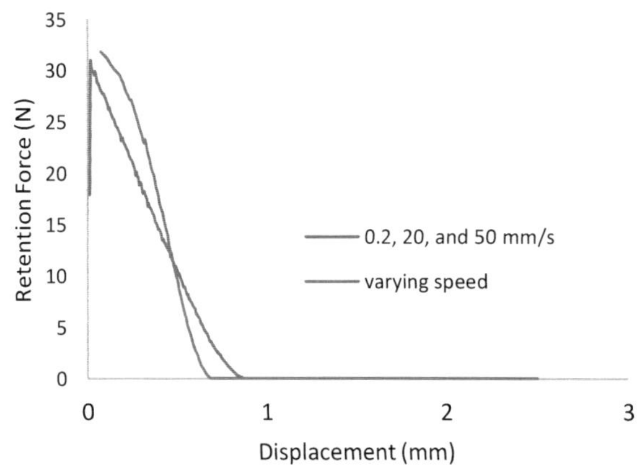

(b)

Figure 6. Relationship between (a) press-in force and displacement (b) retention force and displacement.

To see the outcome when the process is not at a constant speed, the fourth case was added where the speed was varied from 0.5 mm/s to 50 mm/s. **Figure 7** shows the velocity plot for this simulation. This time, acceleration or deceleration was happening. The contact of the pin and PTH surface begins at around 0.5 mm from its initial position so an acceleration was added after the 0.5 mm displacement to see if the acceleration affects the press-in force. Then, it decelerates before a full stop.

For the pull-out speed, an acceleration was added from 0 to 0.5 mm displacement and then a deceleration for the rest. The results show that the force-displacement curves of both the press-in and pull-out process were affected when an acceleration is added. The maximum values of forces were also slightly affected by the addition of acceleration. It can be observed in **Figure 6b** that the pull-out force from 0 to 0.5 mm displacement is higher compared to the curve with constant speed, and then lower when deceleration begins. For the press-in force, an offset can be noticed from the curve with constant speed in **Figure 6a**. The maximum press-in force in the case with acceleration was lower than the maximum press-in force with constant speed.

Moreover, **Figure 8** shows a sample of the final position of the press-fit assembly obtained from the simulation displaying the deformations. It shows the deformations on the barrel's surface and on the compliant pin. During the contact, only the corners of the compliant pin are in contact with the PTH and not the whole surface of the pin as can be noticed from the top view of the figure. Thus, an electrical resistance may increase due to the presence of contact gap. Adding larger fillet radius on the pin's corner may eliminate this gap and may improve the connection reliability.

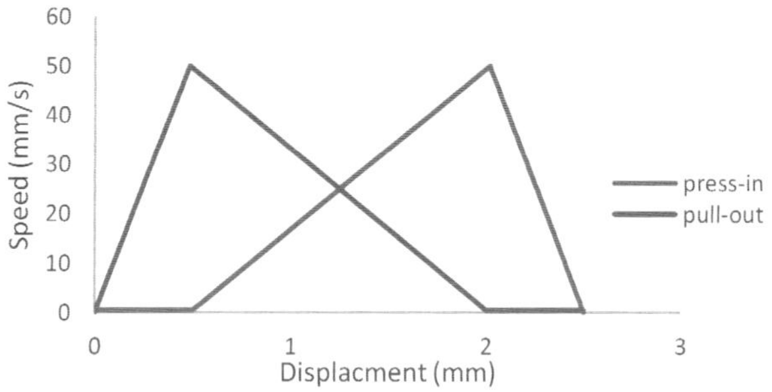

Figure 7. Varying the speed (0 – 50 mm/s) for the fourth simulation.

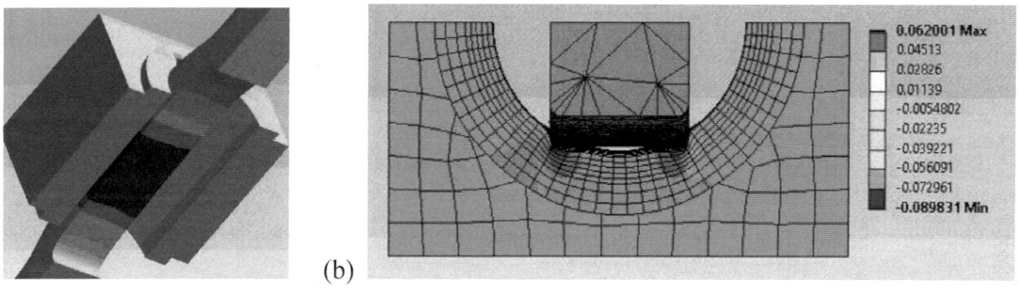

(a) (b)

Figure 8. Final position of press-fit assembly with the deformations in (a) isometric and (b) top view.

4. Conclusion

The effect of the sliding speed on the press-in forces and retention force was studied based on finite element analysis. The simulation was tested in different constant sliding speeds, 0.2 mm/s, 20 mm/s, and 50 mm/s. A simulation also having a changing speed from 0.5 to 50 mm/s instead of a constant speed was conducted. The force vs. displacement curves were plotted and the maximum forces were recorded. The maximum press-in force at constant sliding speed was 56.18 N but when acceleration was added, the maximum press-in force was 54.94 N. For pull-out force, the maximum force was 30.15 N and 31.82 N for constant speed and for the case with acceleration, respectively. Based on the FEA results and as predicted theoretically, it can be concluded that the acceleration affects the press-in and pull-out force of the press-fit pin but not when the sliding speed is kept at constant. Moreover, any changes on the speed during press-in assembly may affect the reliability of connections such as when bending of press-fit pins occurs due to misalignment and may provide higher stresses on contact surfaces which may be considered for further studies.

Acknowledgment
The authors would like to acknowledge the Philippine's Department of Science and Technology with the Collaborative Research and Development to Leverage Philippine Economy (DOST-CRADLE) program under the Philippine Council for Industry, Energy and Emerging Technology Research and Development (PCIEERD) with project No. 8897, 2020, for the assistance and funding this research. The authors would also like to recognize the DOST-ERDT consortium for the scholarship program awarded to the authors in taking up Graduate study.

References
[1] Yang E, Qian R, Yao A and Liu Y 2019 The Process parameter simulation of press-fit pin in a power module *2019 20th International Conference on Electronic Packaging Technology, ICEPT 2019*
[2] Riedel G J, Spindler C, Alaperä I, Schmidt R and Held M 2012 Current capability of press-fit contacts of IGBT modules *IET Conference Publications* **2012** pp 10–15
[3] Corman N, Myers M and Copper C 2003 Friction behavior of press-fit applications: Test apparatus and methodology *Electr. Contacts, Proc. Annu. Holm Conf. Electr. Contacts* pp 38–44
[4] Ring K and Schreier-Alt T 2009 Reliability of press fit contacts and adjoining SMD components on printed circuit boards *Electrical Contacts, Proceedings of the Annual Holm Conference on Electrical Contacts* (IEEE) pp 223–231
[5] Anon 2019 Mounting Instructions for PIM Modules (Q0, Q1, Q2, F1, F2) *Semicond.* pp 1–17
[6] Peter S 2011 Press-fit pin enables solder-less assembly *Power Electron. Technol.* **37**
[7] Tohmyoh H, Yamanobe K, Saka M, Utsunomiya J, Nakamura T and Nakano Y 2008 Analysis of solderless press-fit interconnections during the assembly process *J. Electron. Packag. Trans. ASME* **130** pp 0310071–6
[8] Wang R and Xu L 2021 Mechanical behavior investigation of press-fit connector based on finite element simulation and its reliability evaluation *Microelectron. Reliab.* **116** 114010
[9] Anon 1995 Press-Fit Connections for the Automotive Industry pp 1–18
[10] Crandall E R, Schabert F, Schmidt H, Bleicher M, Fili T, Fischer W E M Von, Borhauer C, Thoss S, Villarreal J and Kerckhof B 2017 New LITESURF plating for the mitigation of whisker risks in press-fit applications *Electrical Contacts, Proceedings of the Annual Holm Conference on Electrical Contacts* **7** pp 132–138
[11] Parenti D. M J 2003 Validating press-fit connector installation *Circuits Assem.* **14** pp 26–29
[12] Tohmyoh H, Yamanobe K, Saka M, Utsunomiya J, Nakamura T and Nakano Y 2006 Determination of friction coefficient of a press-fit pin in thin plating *JSME Int. Journal, Ser. A Solid Mech. Mater. Eng.* **49** 363–369
[13] You B, Lou Z, Luo Y, Xu Y and Wang X 2015 Prediction of Pressing Quality for Press-Fit Assembly Based on Press-Fit Curve and Maximum Press-Mounting Force *Int. J. Aerosp. Eng.* **2015**
[14] Lim N R E, Arriola E, Moran R L, Mercado J P, Dimagiba R, Gonzaga J and Ubando A 2019 Material Properties Sensitivity Analysis Based on Thermomechanical Analysis on Warpage for Ball Grid Array Semiconductor Packaging *2019 IEEE 11th International Conference on Humanoid, Nanotechnology, Information Technology, Communication and Control, Environment, and Management, HNICEM 2019* pp 0–4
[15] Moran R L, Arriola E, Emil Lim N R, Patrick Mercado J, Dimagiba R, Gonzaga J and Ubando A 2019 Reduction of IC Package Warpage through Finite Element Analysis and Direct Optimization *2019 IEEE 11th International Conference on Humanoid, Nanotechnology, Information Technology, Communication and Control, Environment, and Management, HNICEM 2019* pp 0–5

Study on replication of a nonlinear dynamical system's trajectory using a machine learning technique

Tri Quoc Truong[1]

[1]Faculty of Engineering, Van Lang University, 45 Nguyen Khac Nhu street, Co Giang ward, District 1, Ho Chi Minh city, Vietnam

tri.truong@vlu.edu.vn

Abstract. Nonlinear system exhibits various solution orbits depending on varying parameters. It is important to detect the system's behavior. In some cases, however, the mathematical modeling of the dynamic is completely unknown. By using a recent advance in the Machine Learning technique named Reservoir Computing, we replicate the solution orbits based only on data collected along with time evolution. We numerically confirm the effectiveness of Reservoir Computing in time series prediction.

1. Introduction

Recently, Machine Learning (ML) method, typically Deep Learning (DL) techniques have attracted intensive research interest in the nonlinear system [1-9]. By using limited data such as solutions of Ordinary Differential Equations (ODEs) or Partial Differential Equations (PDEs) as the input, the ML model is trained to estimate or replicate the evolution of the system. At the output, we then received an approximate solution's trajectory that is close to the real one. ML technique has shown great success in the task of nonlinear problems such as prediction of time-series evolution [3-5], chaos controlling [6, 7], or synchronization [8].

In principle, the study of a nonlinear dynamical system is often analyzed on a system that is known before. In many cases, nonlinear phenomena cannot be modeled in the accurate mathematical equation. Many researchers have proposed various ML techniques which are able to accomplish the desired purpose basing only on observed data. Among the ML methods, Reservoir Computing (RC) has been adopted to replicate the dynamic's trajectory. RC is a kind of Recurrent Neural Network (RNN) which is first employed by Jaeger et al. in [9]. The authors implemented such a model to learn the mechanism of biological brains. This method gained computational cost in the task of predicting chaotic time series over previous techniques. Further, it also showed the potential of engineering application in the communication field. In the same manner, Pathak et al. improved a procedure for calculating the Lyapunov exponent by using data observed in the Lorenz system and the Kuramoto-Sivashinsky system [4]. The proposed method was successful in computing the positive and zero Lyapunov exponents with great accuracy. However, it failed to predict the time-evolution for long iteration. Further, such a system is quite difficult to analyze higher dimensional dynamics.

Herein, we consider replicating the trajectory solution in a coupled piecewise-constant oscillators (PWCOs) by using the RC method. To the best of our knowledge, this is the first study of application of the ML technique on PWCOs. The detailed discussion of PWCOs is introduced in Ref. [10], the governing equation of which has a piecewise constant. Hence, it is relatively easy to obtain the rigorous solution. Thus, we expect that PWCOs is a great system to apply the ML method to the nonlinear dynamical systems with good accuracy. First, by varying the system parameter, we observe the periodic trajectory with period one and period two. Then, based on RC, we replicate the two above trajectories. The numerical results show the excellent prediction of the trajectory.

2. Piecewise-constant oscillators

Circuit model of piecewise-constant oscillators (PWCOs) is illustrated in Figure 1. The circuit consists of following elements: three capacitors (C_1 to C_3), and two hysteresis elements. The voltages across the capacitors is denoted as v_1 to v_3. The hysteresis behaves as shown in Figure 2. In case of the solution on the upper branch of $H(v) = I_h$, v is increasing. As $v = V_{th}$ the solution jumps to the lower branch of $H(v) = -I_h$. Further, the solution on the lower branch jumps to the upper branch when v decreases and reaches $v = -V_{th}$. By repeating such that behavior, $H(v)$ takes only to constant values I_h and I_h. By using Kirchhoff's law, the circuit equation is written as follows:

$$\frac{C_1C_2 + C_1C_3 + C_2C_3}{C_1C_2 + C_1C_3 + C_2C_3}\frac{dv_1}{dt} = \frac{C_2 + C_3}{C_3}H_1(v_1) + H_2(v_2)\ \frac{C_1C_2 + C_1C_3 + C_2C_3}{C_1C_2 + C_1C_3 + C_2C_3}\frac{dv_2}{dt}$$
$$= \frac{C_1 + C_3}{C_3}H_1(v_1) + H_1(v_1) \qquad (1)$$

Via rescaling,

$$v_1 = V_{th1}x, v_2 = V_{th2}y, t = \gamma\tau\ h_x(x)I_{h1} = H_1(V_{th1}x), h_y(y)I_{h2} = H_2(V_{th2}y)\ \gamma = \frac{V_{th1}(C_1C_2+C_1C_3+C_2C_3)}{I_{h1}C_3}\frac{I_{h2}}{I_{h1}} = \gamma_2, \frac{V_{th1}}{V_{th2}} = \alpha, \frac{C_2+C_3}{C_3} = \gamma_1, \frac{I_{h2}}{I_{h1}}\frac{V_{th1}}{V_{th2}}\frac{C_1+C_3}{C_3} = \beta, \qquad (2)$$

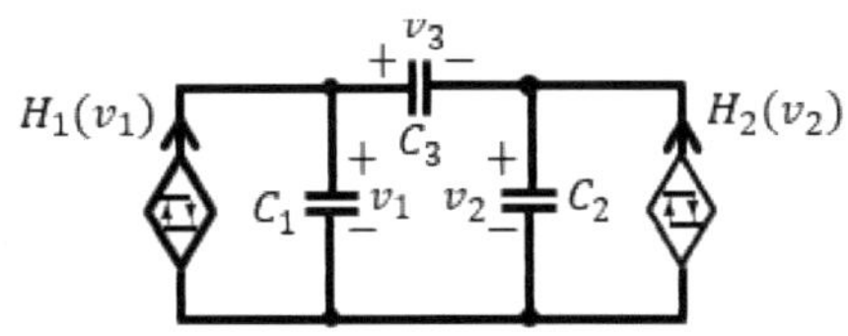

Figure 1. Circuit model of the PWCOs.

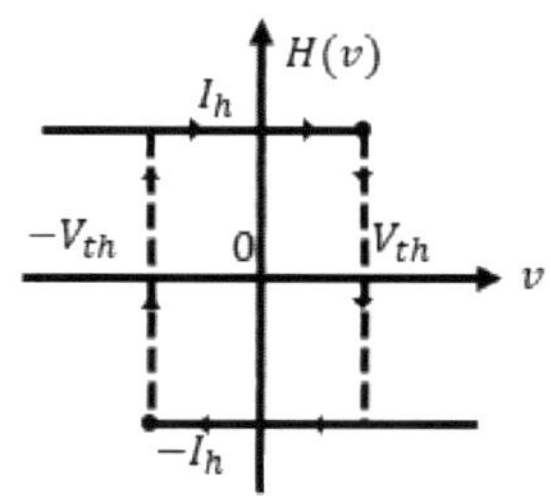

Figure 2. Behavior of a hysteresis element.

The circuit's dynamic is rewritten as following equation:

$$\dot{x} = \gamma_1 h_1(x) + \gamma_2 h_2(y)\ \dot{y} = \alpha h_1(x) + \beta h_2(y) \qquad (3)$$

where $\gamma_1, \gamma_2, \alpha, and\ \beta$ are system parameters. In this study, we set $\alpha = 1, \gamma_1 = 2.5$ and $\gamma_2 = 1.2$. $h_1(x)$ and $h_2(y)$ is a normalized hysteresis loop which behaves the same manner in Fig. 2. By varying the parameter β, the system obtains different trajectories in two dimensions. For simplicity, in this study we only focus on the trajectories depicted in time evolution of variable x as shown in Figure 3 and Figure 4. In these figures, the horizontal axis represents the time-series evolution. After the long transient time ($\tau = 200$), the variable x that expresses the value of the current v_1 is drawn in the vertical axis. Owning to the affected by the hysteresis element $h_1(x)$ and $h_2(y)$, the behavior of the system is repeated in a range of -1 to 1.

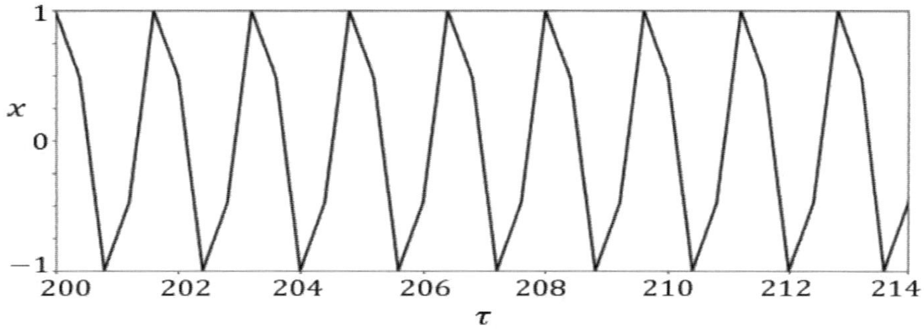

Figure 3. Periodic trajectory with period one ($\beta = 2.5$).

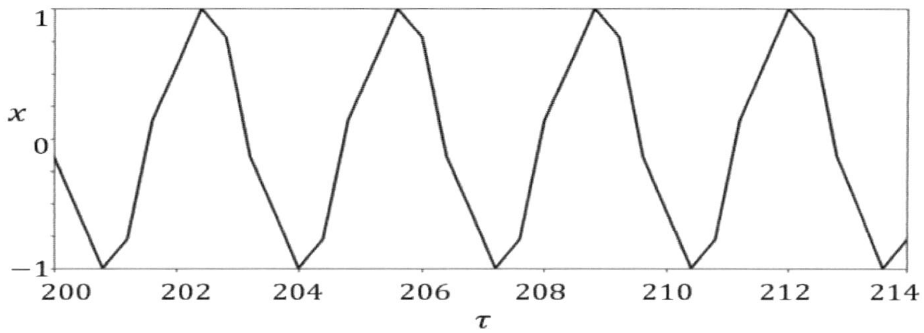

Figure 4. Periodic trajectory with period two ($\beta = 7.1$).

3. Trajectory replication in the piecewise-constant oscillators

3.1. *Reservoir Computing*

The general Reservoir Computing (RC) structure is illustrated in Figure 5. We assume that RC comprises five components: input $x(t)$, Input-to-Reservoir block, Reservoir, Reservoir-to-Output block and output $\hat{y}(t)$. The input $x(t)$ is fed into the network through the Input-to-Reservoir block. In this block, $W_{in}x(t)$ is combined with reservoir state $r(t)$ to generate the Reservoir output $r(t + \Delta t)$, which is expressed as follows:

$$r(t + \Delta t) = tanh[Ar(t) + W_{in}x(t)]$$

where A, and W_{in} is an adjacency matrix and a linear input weight matrix respectively. We note that the adjacency matrix and input weight matrix is initially randomly obtained.

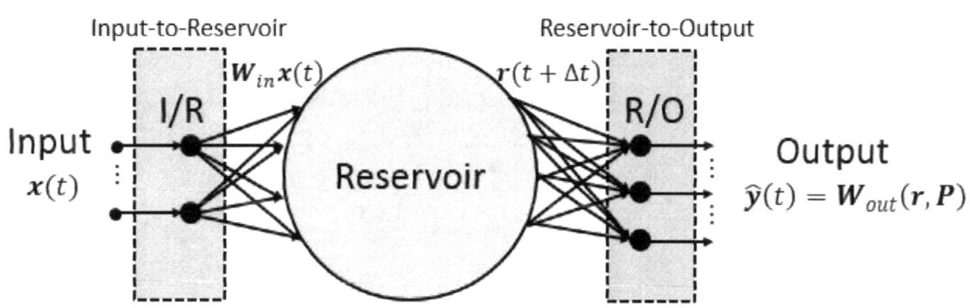

Figure 5. The general structure of Reservoir Computing.

The reservoir state $r(t+\Delta t)$ is combined with adjustable parameters denoted by matrix P through the Reservoir-to-Output block as shown in Figure 5. In addition, W_{out} connects the vector $r(t+\Delta t)$ to the output $\hat{y}(t)$, which is given by

$$\hat{y}(t) = W_{out}(r(t+\Delta t), P)$$

As the output, we received the predicted value $\hat{y}(t)$, which is expected to close to the real value $y(t)$ as much as possible. In the numerical implementation, we used limited data containing both $x(t)$ and $y(t)$, called a dataset as the input, the Reservoir Computing is trained to learn the dynamical system's characteristics.

3.2. *Results*

In order to collect the dataset, which is used for training model, we solve the Eq. 3 then obtain 3000 values containing both x and y after the transient time ($\tau = 200$). In this study, the dataset is divided into 2 parts: 70% for training the model, called train set and 30% for testing the model, called test set. Table 1 shows the first 10 values of x in the dataset. They are plotted in Figure 6. Wherein, the horizontal axis and vertical axis expresses the iteration number and the value of variable respectively. Here, the dash line is the solution trajectory. The solid point denotes the boundary. At the first iteration ($n = 1$), let $x_1 = -0.48$ be the point on the trajectory where the solution starts. At the second iteration ($n = 2$), the trajectory reaches the next boundary at $x_2 = 1$. In the same manner, the system repeats the behavior. We can observe that for only 5 iterations, the system shows again a periodic solution.

Table 1. Example of first 10 values of x in the dataset.

Iteration	1	2	3	4	5	6	7	8	9	10
x	−0.48	1.00	0.48	−1.00	−0.48	1.00	0.48	−1.00	−0.48	1.00

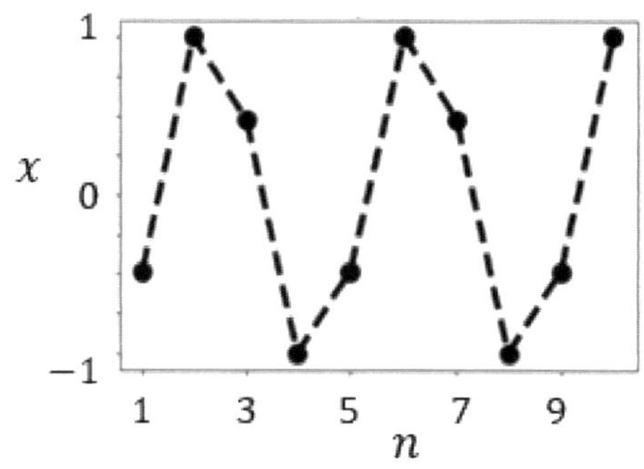

Figure 6. The time-series of x for first 10 iterations.

The above collected data is fed into the network. As mentioned in the Introduction section, the RC is a type of RNN which can take the information of prior variables to influence the current variables and the next variables. Thus, the RC is suited to learn the structure of nonlinear dynamical system. After training, the prediction result in the range of time $200 \leq \tau \leq 214$ is shown in Figure 7 ($\beta = 2.5$) and Figure 8 ($\beta = 7.1$). The predicted trajectory and the real trajectory are marked in the red color, and the black color respectively. According to the results, we can observe that the predicted trajectory is suited to the real one.

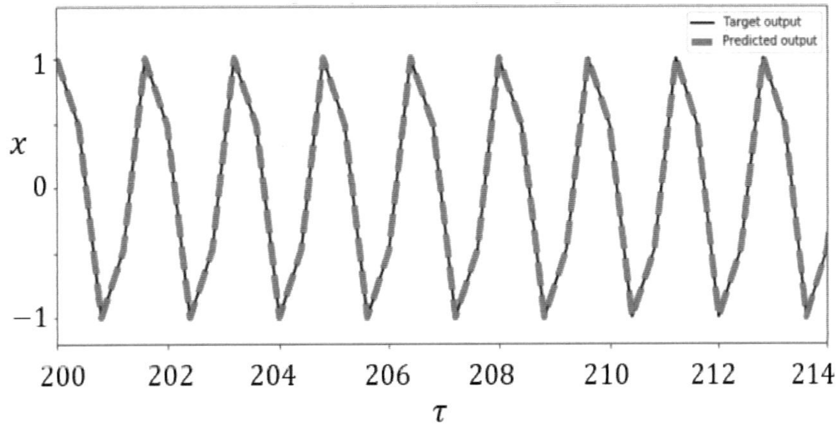

Figure 7. The time series prediction (red line) and the real trajectory (black line) of the piecewise-constant oscillators ($\beta = 2.5$).

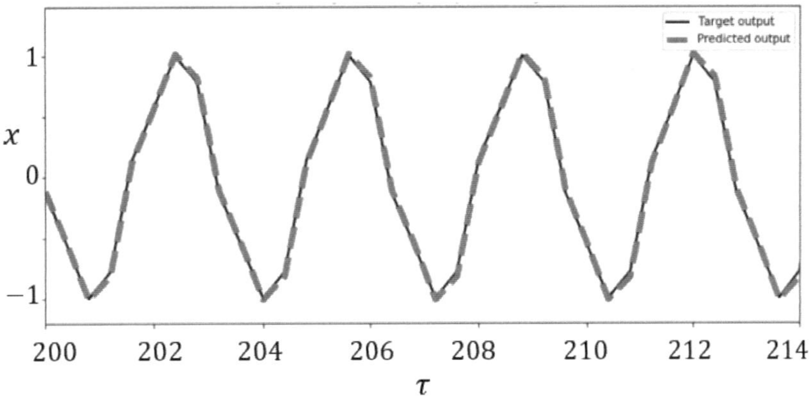

Figure 8. The time series prediction (red line) and the real trajectory (black line) of the piecewise-constant oscillators ($\beta = 7.1$).

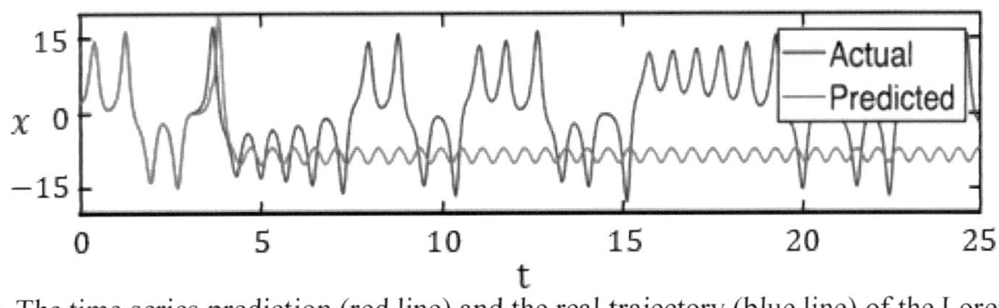

Figure 9. The time series prediction (red line) and the real trajectory (blue line) of the Lorenz system for $0 < t \leq 25$ in [4].

In Figure 9, by using the RC method, the time evolution of solutions of the Lorenz system is simulated by Pathak J et al. in [4]. The correct time series prediction is observed in short time about $t \cong 3$ and then is deviated from the real trajectory. By contrast, the PWCOs shows a great prediction in the longer time of 13. We find that PWCOs have some advantages when applying the ML techniques, typically RC. First, the solution can be derived explicitly in each piecewise-constant region and it is also connected explicitly at the boundaries. Its simplicity of PWCOs structure enables the ML model to be learned better. Hence, the model will demand the less data to get a great time-series prediction. Second, for more difficult ML tasks, such as applying in the Deep Learning model which demands huge number of data, the PWCOs is a great choice for collecting the big data of solution due to its high speed of calculation. Moreover, because the PWCOs has been studied in a systematic procedure, we suggest that it is easy to apply the ML method in the higher-dimensional system.

4. Conclusion

In this study, we have employed a famous ML method known as Reservoir Computing to reconstruct the trajectory of the coupled PWCOs. The time-series prediction observed in this system shows a great accuracy as comparing with the Lorenz system. We suggest that the PWCOs is the suitable system to apply the ML technique in the nonlinear dynamic for following reasons.

(i) Its structure is quite simple.
(ii) The solution is derived explicitly with a high speed of computation.
(iii) The dynamic is studied in a systematic procedure.

In future work, we consider the application task of trained PWCOs, especially in the synchronization of two independent oscillators.

References

[1] LeCun Y, Bengio Y and Hinton G 2015 Deep learning *Nature* **521**(7553) p 436.
[2] Wu Z, and Christofides P D 2019 Economic machine-learning-based predictive control of nonlinear systems *Mathematics* **7(6)** 494.
[3] Jin X, Shao J, Zhang X, An W and Malekian R 2016 Modeling of nonlinear system based on deep learning framework *Nonlinear Dynamics* **84**(3) 1327.
[4] Pathak J, Lu Z, Hunt BR, Girvan M and Ott E 2017 Using machine learning to replicate chaotic attractors and calculate Lyapunov exponents from data *Chaos: An Interdisciplinary Journal of Nonlinear Science* **27**(12) 121102.
[5] Pathak J, Hunt B, Girvan M, Lu Z and Ott E 2018 Model-free prediction of large spatiotemporally chaotic systems from data: A reservoir computing approach *Physical review letters* **120**(2) 024102
[6] Bucci M.A, Semeraro O, Allauzen A, Wisniewski G, Cordier L, and Mathelin L 2019 Control of chaotic systems by deep reinforcement learning *Proceedings of the Royal Society A* **475**(2231) 20190351.
[7] Zhu Q, Ma H, and Lin W 2019 Detecting unstable periodic orbits based only on time series: When adaptive delayed feedback control meets reservoir computing *Chaos: An Interdisciplinary Journal of Nonlinear Science* **29**(9) 093125.
[8] Weng T, Yang H, Gu C, Zhang J, and Small M 2019 Synchronization of chaotic systems and their machine-learning models *Physical Review E* **99**(4) 042203.
[9] Jaeger H, and Haas H 2004 Harnessing nonlinearity: Predicting chaotic systems and saving energy in wireless communication *Science* **304**(5667) 78.
[10] Truong T.Q, Tsubone T, Sekikawa M, and Inaba N 2020 Border-collision bifurcations and Arnol'd tongues in two coupled piecewise-constant oscillators *Physica D: Nonlinear Phenomena* **401** 132148.

Using ANFIS technique in prediction of reverse flow in a solar chimney

Minh-Thu T Huynh, Thinh N Doan and Y Q Nguyen

Faculty of Engineering, Van Lang University,
45 Nguyen Khac Nhu St., Co Giang Ward, Dist. 1, Ho Chi Minh City, Viet Nam

Email: y.nq@vlu.edu.vn

Abstract. In solar chimney, in order to save resources in Computational Fluid Dynamics (CFD), this study aims to combine Adaptive Neuro-Fuzzy Inference System (ANFIS) to predict reverse flow and investigate the ventilation performance. The inputs to the model are heat flux in the range of 200-1000W/m^2, gap of 0.04-0.25m and height of 1-1.4m, while outputs are penetration depth and mass flow rate. The results of R^2 and RMSE are reasonable for training and testing data, hence the ANFIS model is validated.

Keywords: solar chimney, CFD, machine learning, ANFIS, reverse flow

1. Introduction

Passive ventilation for buildings using solar chimney has been studied since 1993 by Bansal et al. [1]. With a well-designed solar chimney, it is said that the air changes per hour (ACH) even meet the standard recommended by ASHRAE [2]. In addition, indoor temperature can be reduced up to 3.5°C from an experimental study of Chungloo and Limmeechokchai [3]. In a study by Miyazaki et al. [4], a Tokyo's office building can save up to 50% of the annual energy required for ventilation.

For solar chimney principle, due to higher temperature air heated by solar energy makes air flow out of the channel; hence, induces natural ventilation. However, Zamora and Kaiser [5] have detected the sudden change of the flow pattern due to the development of reverse flow regions, as they said it should be an optimal inter-plate spacing that maximizes the induced mass flow rate. Reverse flow in the solar chimney was also examined quantitatively by calculating its penetration depth by Khanal and Lei [6]. They have proposed the inclined wall design in order to enhance ventilation performance. Reverse flow in a solar chimney normally occurs at the outlet and in reality, it is quite complicated as it depends on the weather condition, height, and width or gap of the solar chimney. Since it causes a decrease in ventilation, Nguyen [7] has tried to suppress the reverse flow, which, by rearranging the heat transfer surface on the opposite side in the upper and lower halves of the solar chimney.

Machine Learning (ML) has been applied to many fields in order to solve complicated problems in Fluid Dynamics. As in the study of evaporator performance in Organic Rankine Cycle, due to its high nonlinear behavior and high thermal inertia, Enayatollahi et al [8] has utilized machine learning to solve such a complicated problem. Pourtousi et al [9] have also combined CFD and ANFIS for the bubble column hydrodynamics study. They mentioned that using both experiment and numerical methods in the field are still difficult due to cost and time. Another study of Beigzadeh [10] is to investigate the heat transfer performance in flat and discontinuous fins as it was complex. Computational Fluid Dynamics is utilized to provide data for Adaptive Neuro-Fuzzy to model the performance.

One of the popular Machine learning techniques is ANFIS, which stands for Adaptive Neuro-Fuzzy Inference System. It was a class of adaptive networks that incorporate both neural networks and fuzzy logic principles and first introduced by Jang [13]. ANFIS has two popular types: Mamdami [14] and Sugeno [15]. Since ANFIS combines two advantages of fuzzy logic and neural networks, it can solve many kinds of complicated nonlinear problems in acceptable time and does not require expert knowledge.

In this work, we propose to implement an adaptive neuro-fuzzy inference strategy (ANFIS) with sugeno type to predict the penetrating depth and mass flow rate of solar chimney for different scenarios. CFD models were built to provide data for training and testing the ANFIS model.

2. Modeling setup

2.1. Modeling in ANSYS

The solar chimney model in this study is in the vertical plane with channel gap G and height H, as sketched in figure 1. A uniform heat flux I is applied on one wall. The opposite wall is considered as adiabatic with no-slip conditions. Air at the inlet and outlet of the channel are set as atmospheric pressure. Flow is assumed steady and incompressible. The conservation equations of mass, momentum and energy are described by RANS (Reynolds Averaged Navier – Stokes equations) method in CFD on a staggered grid using a finite volume solver ANSYS Fluent Academic version 2020 R2. For solving the turbulence quantities, the RNG k-ϵ turbulence model is used.

In the numerical setups, as the momentum and mass conservation equations were solved simultaneously, the SIMPLE numerical scheme was selected for handling the coupling. In addition, as the staggered grid system was employed, the pressure at the cell center was interpolated to the cell faces with the PRESTO! Method. Other discretization schemes of the equations included the second-order upwind scheme for momentum equation and energy conservation equation, and first-order upwind scheme for the equations for k and ϵ are applied.

The above setups have been tested intensively and compared with experimental data. Details can be seen in our previous studies [11][12].

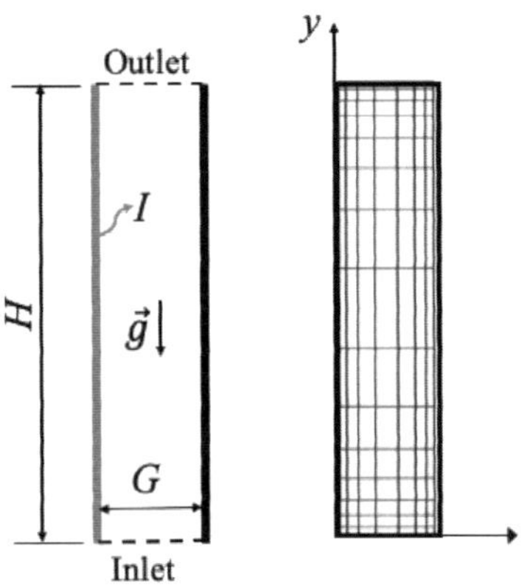

Figure 1. Solar chimney model in ANSYS: the computational domain (left) and mesh (right).

2.2. Modeling in ANFIS

The typical ANFIS structure is sketched in figure 2. It is assumed with two inputs and one output. This study is based on the model developed by Jang [13].

The membership function (MF) in the current study is selected as Gaussian functions:

$$\mu_i = e^{\frac{-(x-c_i)^2}{2\sigma_i^2}} \quad (1)$$

where σ_i and c_i with $i = (1, 2)$ are the fuzzy set parameters.

To evaluate the accuracy of produced outputs, two statistical indices of R^2 and RMSE are used.

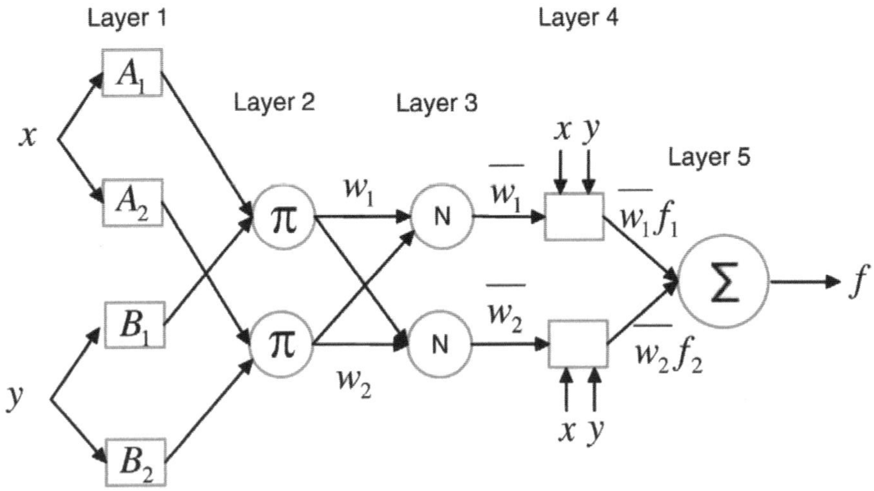

Figure 2. Typical ANFIS architecture.

3. Results and discussions

3.1. Reverse flow from the CFD model

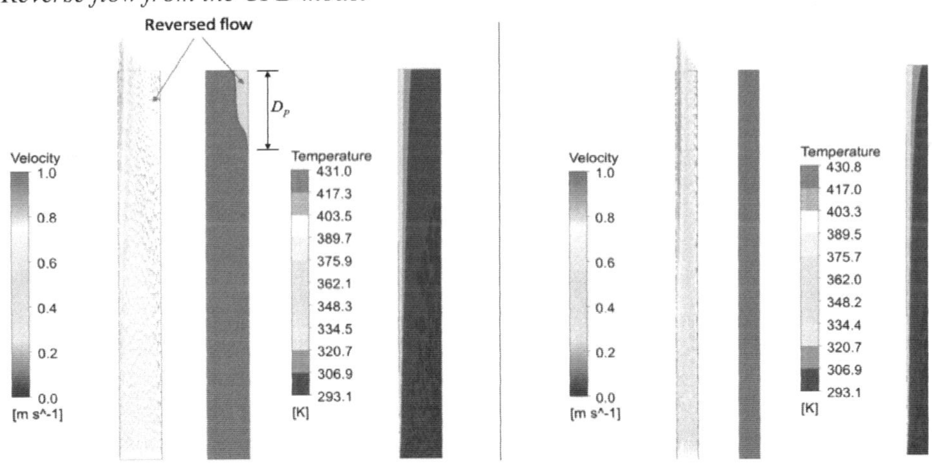

(a) H=1m, I=1000W/m^2, G=0.1m, reverse flow. (b) H=1m, I=1000W/m^2, G=0.05m, no reverse flow.

Figure 3. Examples of flow fields obtained with the CFD model: velocity vector and contour, and temperature fields.

Figure 3 shows examples of the flow field obtained with the CFD model for the solar chimney with H=1.0 m and I=1000W/m^2. The gap changed from G=0.1 m (figure 3a) to 0.05 m (figure 3b). The heat source was on the left wall of the air channel. It is seen that at the lower gap (G=0.05 m, figure 3b), the thermal layer developed along the heated surface and induced an air flow in the whole channel gap. At the larger gap (G=0.1 m, figure 3a), the thermal layer was close to the heated surface and did not expand to the center of the channel. At the outlet of the channel, the right side of the channel was not heated. A reverse flow appears near the right wall at the outlet. The area of the reverse flow is indicated by the green region in the middle image of figure 3a.

The reverse flow in figure 3a contracted the flow area at the outlet and increase flow resistance. As a result, the induced air flow rate was influenced. Figure 4 shows the induced flow rate and the penetration depth, D_p (see figure 3a), of the reverse flow for a solar chimney with H=1.1 m and I=400 W/m^2. A reverse flow existed for G>0.04 m. D_p increased with the gap. As the reverse flow region was still small, at G=0.04 m and 0.08 m, the flow rate continued to increase with the gap. However, as the gap was above 0.08 m, the reverse flow region exceedingly increased and resulted in a reduction of the induced flow rate. At G=0.24 m, the flow rate was almost similar to that at G=0.04 m.

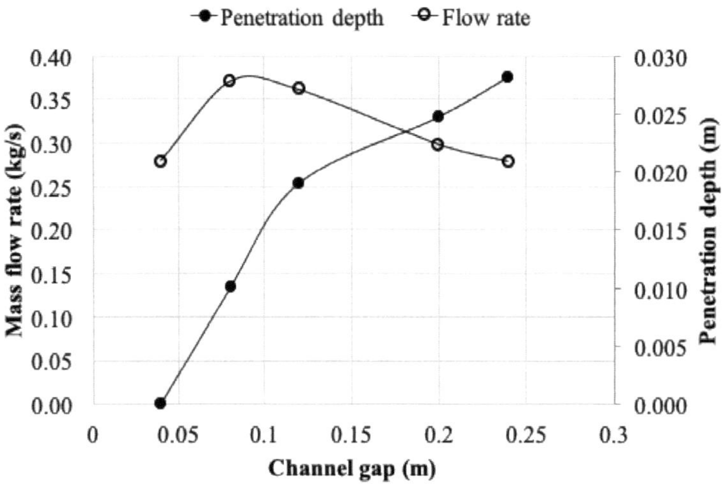

Figure 4. Mass flow rate and penetration depth change with channel gap at I=400 W/m^2, H=1.1 m.

3.2. *ANFIS model training*

As aimed to predict D_p and the mass flow rate with the ANFIS model, data were generated with the CFD model with different G, H, and I to the penetration depth D_p, if exists, and mass flow rate $\dot{m}$. The range of parameters considered in this study are 0.04-0.25 m for gap width, 1-1.4 m for solar chimney height and up to 200-1000 W/m^2 for heat flux. Total number dataset is 112, in which 80% is used for training models and the rest 20% is for testing. All dataset is shuffled randomly.

In this ANFIS model, 2 Gaussian membership functions are used for I and H; while 5 Gaussian membership functions for G. The fuzzy set parameters of each membership function are in table 1. After training, respectively for D_p and $\dot{m}$, the results give RMSE of 5.4E-05 and 4.5E-06; and R^2 of 0.99998 and 0.99996. Those results are quite reasonable for the outputs in this study.

Table 1. ANFIS model inputs.

Input	MF	Type of MF	σ	c
I (W/m^2)	in1mf1	Gaussianmf	3.82E+02	2.00E+02
	in1mf2	Gaussianmf	3.82E+02	1.10E+03
G (m)	in2mf1	Gaussianmf	2.23E-02	4.00E-02

	in2mf2	Gaussianmf	2.23E-02	9.25E-02
	in2mf3	Gaussianmf	2.23E-02	1.45E-01
	in2mf4	Gaussianmf	2.23E-02	1.98E-01
	in2mf5	Gaussianmf	2.23E-02	2.50E-01
H (m)	in3mf1	Gaussianmf	1.70E-01	1.00E+00
	in3mf2	Gaussianmf	1.70E-01	1.40E+00

3.3. *ANFIS model validation*

In order to validate the above ANFIS model, as mentioned above, the rest 20% of the data set is estimated using the above training model. Figure 5 shows the comparison between CFD results and those from the ANFIS model for penetration depth and mass flow rate for different solar chimney gaps.

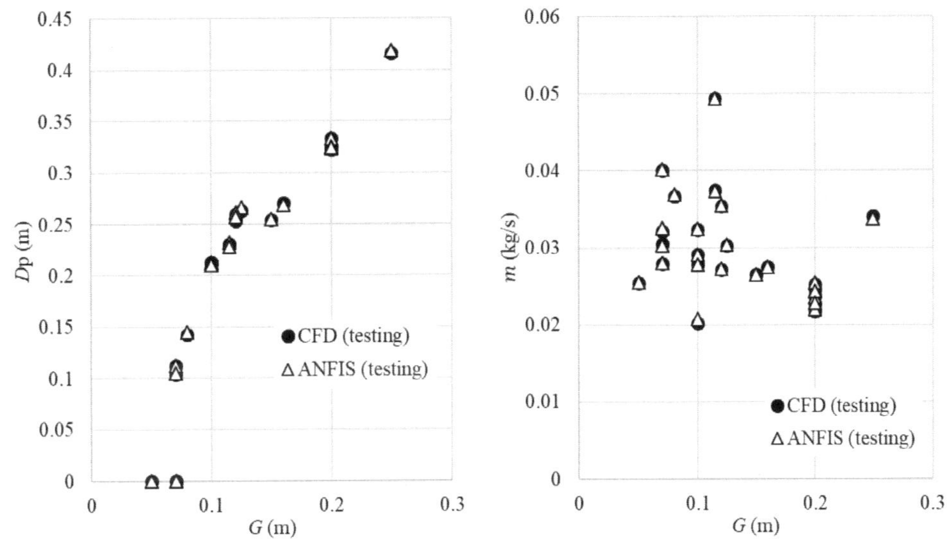

Figure 5. CFD vs. ANFIS for depth penetrating and mass flow rate.

From the result, respectively for D_p and $\dot{m}$', RMSE are 0.000455 and 0.0000427; and R^2 are 0.9998 and 0.9996. Those outcomes are quite tolerable, hence the above ANFIS model is validated.

It is seen in figure 5 that the predictions from ANFIS agreed very well with the CFD results for both the flow rate and the penetration depth, even for the cases without a reverse flow (D_p=0.0 m). It is noted that the data set in figure 5 was randomly picked from the CFD results for different heights and heat fluxes. Therefore, the good agreement between the two results in figure 5 also shows the ability to predict the mass flow rate and the penetration depth of the ANFIS model to overcome the major drawbacks of the CFD model, i.e. expensive computational cost.

4. Conclusion

This study attempted to introduce the precision of Adaptive Neuro-Fuzzy Inference System (ANFIS) model for evaluating the reverse flow in a solar chimney in order to maximize ventilation performance, which is the mass flow rate in this study. For saving resources in simulation and experiment efforts, the numerical data obtained from the CFD modeling for range of heat fluxes, channel gap, and height were used for training the ANFIS. By comparing penetration depth and mass flow rate between CFD and

ANFIS model for both training and testing, it can be confirmed that the ANFIS model is validated with acceptable RMSE and R^2. Consequently, reverse flow with different scenarios can be estimated using ANFIS technique simultaneously. Therefore, the optimal design of a solar chimney can be achieved to maximize the advantage for ventilation as well as energy saving.

References

[1] Bansal N K, Mathur R and Bhandari M S 1994 A study of solar chimney assisted wind tower system for natural ventilation in buildings *J. Building Environ.* **3** pp 373–377

[2] Zavala-Guillén I, Xamán J, Hernández-Pérez I, Hernández-Lopéz I, Jiménez-Xamán C, Moreno-Bernal P and Sauceda D 2018 Ventilation potential of an absorber-partitioned air channel solar chimney for diurnal use under Mexican climate conditions *J. Appl. Therm. Eng.* **149** pp 807-821

[3] Chungloo S and Limmeechokchai B 2009 Utilization of cool ceiling with roof solar chimney in Thailand: the experimental and numerical analysis *Renew. Energy* **34** pp 623–633

[4] Miyazaki T, Akisawa A and Kashiwagi T 2016 The effects of solar chimneys on thermal load mitigation of office buildings under the Japanese climate, *Renew. Energy* **31** (7) pp 987–1010

[5] Zamora B and Kaiser A S 2009 Optimum wall-to-wall spacing in solar chimney shaped channels in natural convection by numerical investigation *Applied Thermal Engineering* **29** pp 762–769

[6] Khanal R and Lei C 2012 Flow reversal effects on buoyancy induced air flow in a solar *J. Solar Energy* **86** pp 2783-2794

[7] Nguyen Y Q 2018 Studying convective flow in a vertical solar chimney at low rayleigh number by lattice boltzmann method: A simple method to suppress the reverse flow at outlet *Proceedings of the International Conference on Advances in Computational Mechanics 2017* pp 807-820

[8] Enayatollahi H, Fussey P and Nguyen B K 2020 Modelling evaporator in organic rankine cycle using hybrid GD-LSE ANFIS and PSO ANFIS techniques *Thermal Science and Engineering Progress* **19** 100570

[9] Pourtousi M, Sahu J N, Ganesan P, Shamshirband S and Redzwan G 2015 A combination of computational fluid dynamics (CFD) and adaptive neuro-fuzzy system (ANFIS) for prediction of the bubble column hydrodynamics *Powder Technology* **274** pp 466–481

[10] Beigzadeh R 2019 The CFD providing data for adaptive neuro-fuzzy to model the heat transfer in flat and discontinuous fins *Iranian J. Chem. Engineering* **16** (2) 57-69

[11] Nguyen Y Q and Wells J C 2020 Effects of wall proximity on the airflow in a vertical solar chimney for natural ventilation of dwellings *Journal of Building Physics* **44** (3) pp 225–250

[12] Nguyen Y Q and Wells J C 2020 A numerical study on induced flowrate and thermal efficiency of a solar chimney with horizontal absorber surface for ventilation of buildings *Journal of Building Engineering* **28** 101050

[13] Jang J-S R 1993 ANFIS: adaptive-network-based fuzzy inference system *IEEE Transactions on Systems, Man, and Cybernetics* **23** (3) pp 665-685

[14] Mamdani E H, and Assilian S 1975 An experiment in linguistic synthesis with a fuzzy logic controller *Int. J. Man-Mach. Stud.* **7** (1) pp 1-13

[15] Takagi T and Sugeno M 1985 Fuzzy identification of systems and its applications to modeling and control *IEEE Transactions on Systems, Man, and Cybernetics* **1** pp 116-132.

IOP Conf. Series: Materials Science and Engineering 1109 (2021) 012037 doi:10.1088/1757-899X/1109/1/012037

Velocity analysis of a six wheel modular mobile robot using MATLAB-Simulink

A Fernando[1], L GanLim[1]

[1] Mechanical Engineering Department, De La Salle University, Manila Philippines

arvin.fernando@dlsu.edu.ph

Abstract. This paper presents the modelling and simulation of a six wheeled three body modular robot in MATLAB Simulink. The modules are modelled as an independent differential drive wheeled mobile robot. Velocity and distance were measured using three different configurations: two wheel, four wheel and six wheel configuration. The simulation result was able to predict the response of the modular body in different configurations in terms of velocity and distance with contact forces acting upon it.

1. Introduction

Modular reconfigurable robots or MRR are made up of modules that can rearrange themselves to different configurations to perform certain tasks. Modular self-reconfigurable systems, size, robustness and performance have been continuously improving however there are challenges for these systems to realize their promise of adaptability, robustness and affordability. MRR faces hardware design and Software control challenges [1-3]. Modular Robots are consisting of several independent modules that have different capabilities that can connect or form into a different configuration. These hardware and mechanical structures such actuators, a sensing communication device, structural strength, connection mechanism, and others are required in order for it to function properly. To perform certain task a modular system, have complex software requirements that help them run independently, as a swarm, form an organism, task shape matching, reconfiguration, planning and control.

A Six-wheel independent drive is considered to have better performance, efficient power utilization & stability over other wheeled vehicles such as four wheel drive. Due to the advantages over conventional axle drive systems; All wheel independent drive vehicles are getting popularity in special purpose as well as commercial purpose vehicles. This configuration is good in obstacle climbing, off-road maneuvering, failure handling of few wheels, because of independent wheel motion control, used in exploration, search and rescue [4].

Several researches use a six or more wheel configuration design for climbing, transport and exploration [5-7]. The concept of [5] is to develop a detachable modular robot capable of cooperative climbing and multi agent exploration and will improve the task and exploration coverage but there are a variety of optimal control problems that might pose in this system. [7] Develop an end to end system for addressing tasks with modular robots and demonstrate that it can accomplish challenging multi part tasks. They use a SMORES-EP Modular robot that has 6 different configurations depending on the number of connected modules and the stair climber configuration has a total of 4 modules. [8] Uses an S-bot in a chain of three or more to transport the object. With these different studies they use 3 modules to accomplish certain tasks and a six wheel configuration is ideal in an exploration and rover application. There has been a lot of interest in optimizing, formulation and control of actuators for six-wheel independent drive vehicles to maximize its exploration capabilities [4].

A six wheeled three body modular mobile robot was modeled, and the dynamics were simulated. To understand and analyse its capability and see the effect of several interconnecting module bodies to the whole system such as velocity, distance, climbing capacity, pulling capacity and other parameters that will affect its performance on different tasks. To understand if the configuration would yield an advantage or disadvantage to the modular system.

2. Methodology

2.1. Design Model

A differential drive design model was used per module that consisted of two wheels and was then combined to form a six wheel. Equation 1 shows the differential drive modular body system. The system state x can be described by the robot's location xR and yR, forward velocity v, heading θ, and angular velocity ω. Mathematically, this is described as x = [xR; yR; v; θ; ω] T. Taking our inputs as ul and ur, the forces exerted by the left and right wheels respectively, system.

$$x = f(x) = \begin{bmatrix} v\cos\theta \\ v\sin\theta \\ \dfrac{u_r + u_l}{M} - bv \\ \omega \\ \dfrac{B(u_r - u_l)}{Iz} - b_r\omega \end{bmatrix} \quad (1)$$

where M is the vehicle mass, b is the linear drag constant, B is the length from the center of the vehicle to the wheel, Iz is the moment of inertia, and br is the turning drag constant. We assume the magnitude of each input u_{rl} is bounded by u_{max} and the velocity is bounded by v_{max}. The proponent will use a differential ground drive robot for each module and will be simulated in a six wheeled configuration.

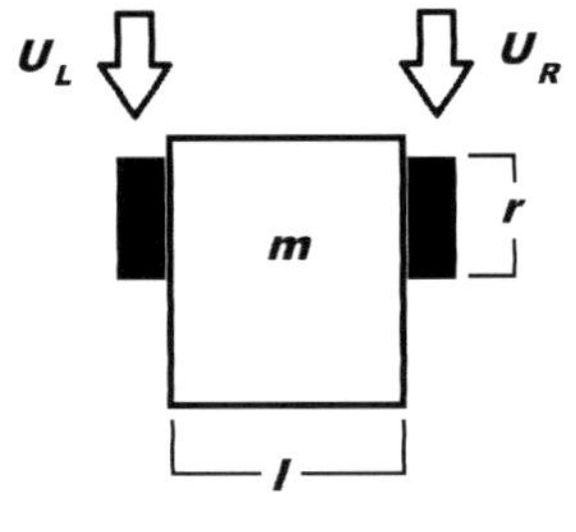

Figure 1 Two wheel differential drive module

A single module body will be modeled as a differential drive robot. L as a wheelbase body and r as the wheel radius. The robot will be controlled by moving the two wheels. The input on the wheel on the left and the wheel on the right. As the configuration changes from four wheel to six wheel they will be simulated and check the response of the configuration pertaining to the action of each module. If they will increase the module capacity of the whole system in terms of total velocity or distance. The configuration set-up which is the four wheel, or six wheels will check the effect of velocity and distance travel of different configurations. As mentioned in [5-7] they have certain advantages and disadvantages when they are in a collective manner or as an individual. They will be simulated by checking the response of the system when the whole module has inputs on their wheel and varying each module's wheel input to check the interaction of the modules in the simulated configuration.

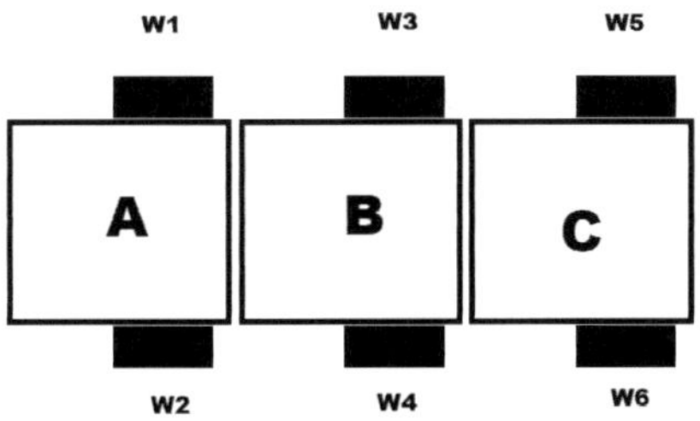

Figure 2 Propose six wheel 3 module setup

Contact modelling was used in the modular body to approximate the physical phenomena of the mass of the module as well as the rolling of the tires. This helps in the analysis of the mechanical system.
A sphere to plane contact force library was used to simulate the design model of the modular body and compare the result of the individual and whole configuration. This contact force library implements a contact force between two bodies which is the sphere (wheels) and the plane(floor). To check the movement of the wheel with respect to different forces such as gravity and friction forces. Static and Kinematic Friction are set up in this library as well as the density of the body to signify the effect of the gravitational phenomena.

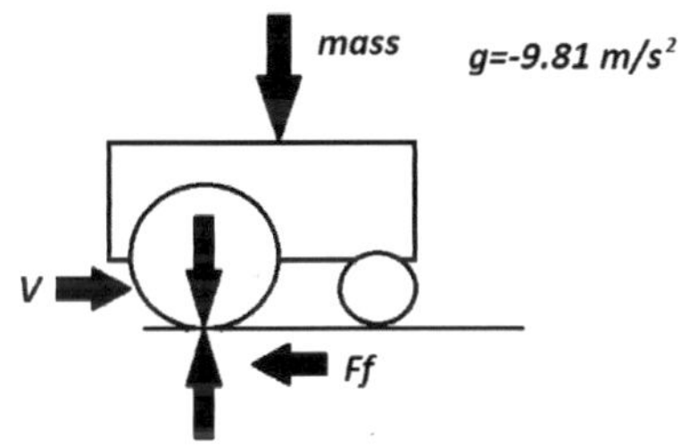

Figure 3 Contact Forces acting on a body

2.2. MATLAB Simulink
Contact force library was used in the simulation [9]. Figure 3 shows a Simulink model of a two wheel differential module. Each component of the system has a force acting on the body from the weight of the plate to the ground; Rotational motion of the wheel to the body; The surface contact and weight of the wheel to the ground. A ramp up signal is used each wheel to simulate its movement; the floor is used as a world frame reference for rigid transformation in order to set the gravitational force direction. The wheel is connected to the body via a revolute joint transformation. The wheel is in contact in the floor via prismatic joint transformation. Velocity and distance are measured via a Simulink transformation sensor relative to the ground and wheel in contact. In simulating different configurations for a four wheel and six wheels, the mass is changed and depends on the number of modules with each wheel having an independent motion. To evaluate the effect of the different configuration on velocity and distance.

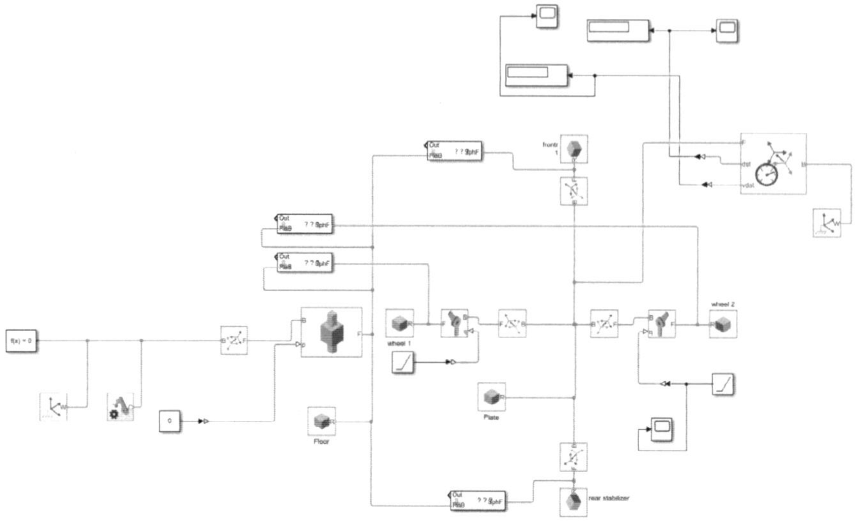

Figure 4 Simulink Model of a two wheel module

3. Results and Discussion

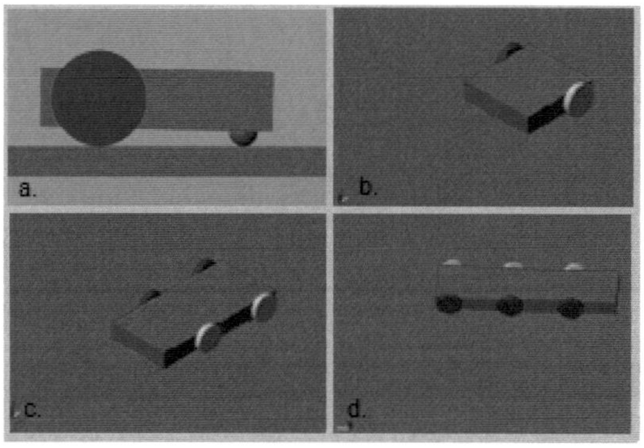

Figure 5 MATLAB Simulink Design Model

The six wheel three module body was modelled and simulated in MATLAB Simulink pertaining to the movement of wheels. A total of three configurations were tested in two wheel, four wheel and six wheel set-up and then tested on one module moving the body. Figure 5 shows the design model of the module in the MATLAB Simulink. It was modeled in the MATLAB Simulink with estimated parameters of one module will have L = 80mm and R = 16mm with thickness of 7mm as the module size is reference to a Micromouse platform [15] and other modular reconfiguring robots are at the micro size level. Contact forces are set to default such as gravity, kinetic and static friction. The density of the material is set to 1000 kg/m^3 to simulate the mass of the module. A ramp up signal was introduced to both wheels and it yielded a simulated velocity of 0.1596 m/s in a one module differential drive set-up. It was tested on a 10 second time trial and the measure the distance that it recorded on the Simulink scope is 1.589 m that can be seen in figure 6. Dimension such as body and wheel size will affect the output result due to weight. In this set up we investigate the effect of the module configuration to the capacity of the system which we define as velocity and distance. Different parameters and different set up will yield different results.

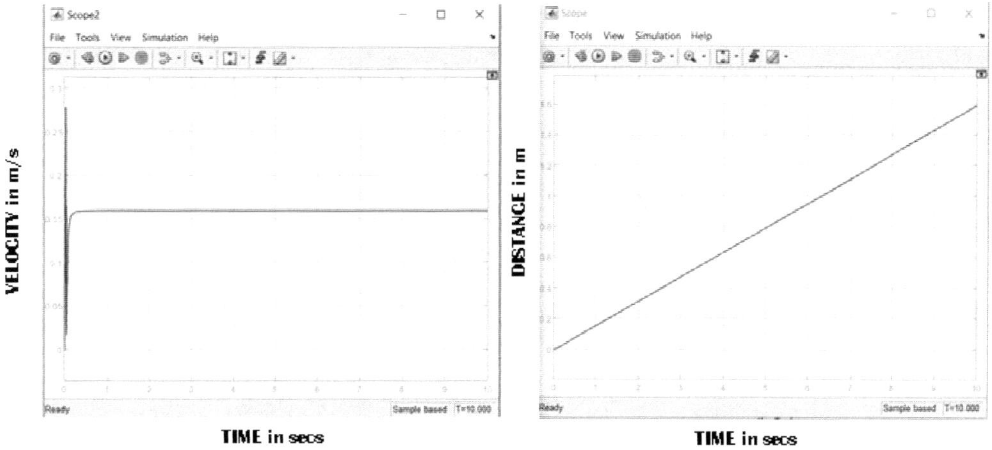

Figure 6 Single Module Velocity and Distance Result

Velocity and distance were tested for the two wheel one module body, four wheel two module body and a six wheel three module body. Every module is identical. The simulated output results of the four wheel and six wheels can be seen in figure 7 and 8. The four wheel two module set up resulted in an increase of velocity and distance travel. There is a slight increase from 0.1596 to 0.16 m/s as simulated while distance travel increases from 1.589 to 1.595m. The initial start-up as seen in the graph is faster compared to the single module configuration. However, results from the 3 module configuration resulted in a decrease in velocity and distance due to the total weight having a velocity of 0.1594 and distance travel of 1.57m in 10 seconds. The initial start-up of velocity was also slow in comparison to the two module set up even though it is in six wheel configuration This result give a in terms of velocity and distance travel however if we use the stability, terrain capabilities and other advantages of a six wheel such as climbing, pulling capacity and other things will be different.

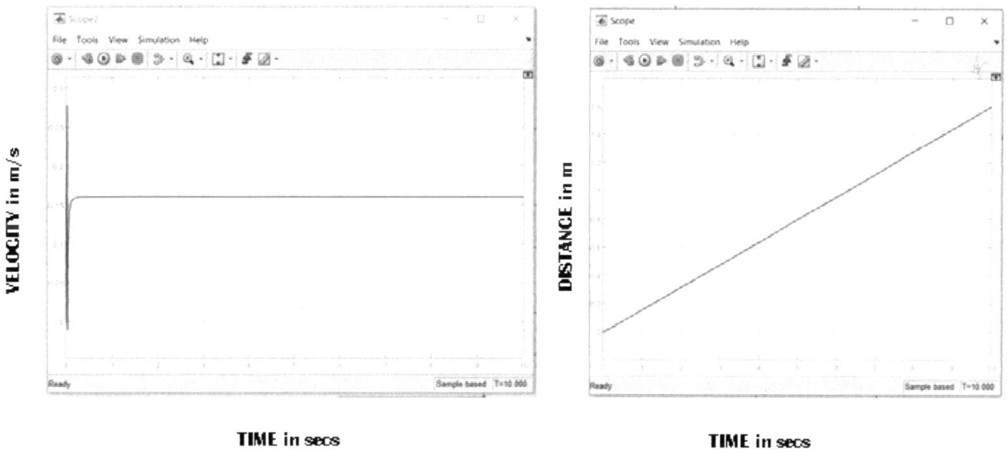

Figure 7 Two Module Velocity and Distance Result

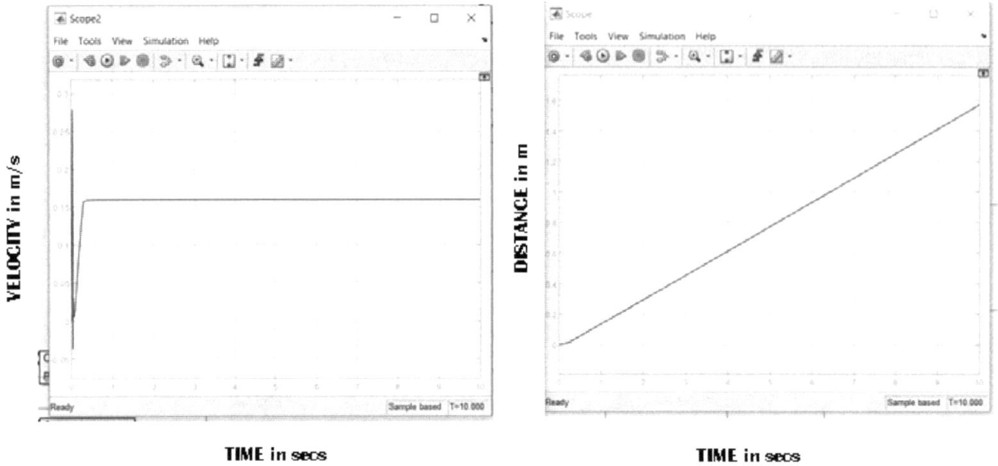

Figure 8 Three Module Velocity and Distance Result

4. Conclusion

The study was able to model and simulate the dynamics of a six wheel three body modular robot with interaction forces acting upon it such as gravity and contact surfaces. Measuring the different configuration velocity and distance and understanding its effect on the system. The simulation is used as a tool to check the proposed strategy in measuring the dynamics of a six wheeled modular mobile robot that will aid in measuring other capacities such as climbing, torque, push, pulling and other capabilities on different configurations. Further simulation and validation of the study will be continued and analysed by the researcher.

References

[1] Seo J, Paik J and Yim M 2019 Modular reconfigurable robotics *Annual Review of Control, Robotics, and Autonomous Systems* **2** pp 63-88

[2] Brunete A, Ranganath A, Segovia S, de Frutos J P and Hernando M 2017 Current trends in reconfigurable modular robots design International *Journal of Advanced Robotic Systems* **14** (3) 1729881417710457

[3] Yim M, Shen W M, Salemi B, Rus D and Moll M 2007. Modular self-reconfigurable robot systems [gran challenges of robotics]. *IEEE Robotics & Automation Magazine* **14**(1) pp 43-52

[4] Prasad R and Ma Y 2019 Hierarchical control coordination strategy of six wheeled independent drive (6WID) skid steering vehicle IFAC 2019

[5] Turlapati S H, Srivastava A, Krishna K M and Shah S V 2017 Detachable modular robot capable of cooperative climbing and multi agent exploration *2017 IEEE International Conf. on Robotics and Automation (ICRA)* 255-260

[6] Turlapati S H, Shah M, Teja S P, Siravuru A and Shah S V 2015 Stair climbing using a compliant modular robot *2015 IEEE/RSJ International Conf. on Intelligent Robots and Systems*

[7] Jing G, Tosun T, Yim M and Kress-Gazit H 2016 An end-to-end system for accomplishing tasks with modular robots, *Robotics: Science and Systems*

[8] Groß R, Tuci E, Dorigo M, Bonani M and Mondada F 2006 Object transport by modular robots that self-assemble Proceedings *2006 IEEE International Conf. on Robotics and Automation*

[9] Miller S 2020 Simscape Multibody Contact Forces Library (https://www.mathworks.com/matlabcentral/fileexchange/47417-simscape-multibody-contact-forces-library), MATLAB Central File Exchange

[10] Christensen A L 2008 Fault Detection in Autonomous Robot, Université Libre de Bruxelles

[11] Fourlas G K, Karkanis S, Karras G C and Kyriakopoulos K J 2014 Model based actuator fault diagnosis for a mobile robot *2014 IEEE International Conf. on Industrial Technology (ICIT)* 79-84

[12] Bandala A A, Dadios E P, Vicerra R R P and Lim L A G 2014 Swarming algorithm for unmanned aerial vehicle (uav) quadrotors–swarm behavior for aggregation, foraging, formation, and tracking *Journal of Advanced Computational Intelligence and Intelligent Informatics*

[13] Bandala A A and Dadios E P 2012 Development and design of mobile robot with IP-based vision system *TENCON 2012 IEEE Region 10 Conf.*

[14] Del Rosario J R B, Sanidad J G, Lim A M, Uy P S L, Bacar A J C and Cai M A D 2014 Modelling and characterization of a maze-solving mobile robot using wall follower algorithm *Applied Mechanics and Materials* **446** pp 1245-1249

[15] Gavin Nielsen 2017 Micro Mouse Modelling Simscape MultiBody

[16] Ramakrishnan R, Elayaraja D and Ramabalan S 2014 Adaptive control of nonholonomic wheeled mobile robot *ICAET.*

[17] Angel L, Hernández C and Díaz-Quintero C 2013 Modeling, simulation and control of a differential steering type mobile robot *32nd Chinese Control Conf.*

[18] Khnissi K 2019 3D Simulator for Navigation of a Mobile Robot Using Simscape-SIMULINK ICCAD

Visual effects creation to improve user performance in fully immersive virtual environment of automobile engine assembly

L L Win[1,2], A A Faieza[1, 7], A A Hairuddin[1], L N Abdullah[3], H J Yap[4], H Saito[5], and S Hisham[6]

[1] Department of Mechanical and Manufacturing Engineering, Faculty of Engineering, Universiti Putra Malaysia, 43400 UPM Serdang, Selangor, Malaysia
[2] Department of Mechanical Engineering, Government Technical High School (Loikaw), 09011 Loikaw, Kayah State, Myanmar
[3] Department of Multimedia, Faculty of Computer Science, and Information Technology, Universiti Putra Malaysia, 43400 UPM Serdang, Selangor, Malaysia
[4] Department of Mechanical Engineering, Faculty of Engineering, University of Malaya, 50603 Kuala Lumpur, Malaysia
[5] Department of Information and Computer Science, Keio University, 3-14-1 Hiyoshi Kohoku-Ku Yokohama, 223-8522, Japan
[6] Department of Engineering Technology, University Kuala Lumpur, Malaysia Italy Design Institute (UNiKL MIDI), 119, Jalan 7/91, Taman Shamelin Perkasa, 56100 Kuala Lumpur, Wilayah Persekutuan Kuala Lumpur

[7]faieza@upm.edu.my

Abstract. Virtual reality (VR) technologies and computer-generated technologies are very useful for many industries, such as medical, education, gaming, mining, engineering, and other sectors. In engineering sector specifically for assembly department, the major problem they are facing was that traditional components of automotive engine are complex and very difficult to assemble using current method which use paper manual. The goal of this project is to create an interactive VR application such that users can quickly integrate and perform the work easily. In this work, the authors designed and developed an immersive VR application with several stages, such as 3D model creation, visual effects creation and gaming scene building. The application was built using Unreal and HTC vive headset. As this is an on-going project work, further functionality will be added to render the application immensely for users and assist the sense of pedagogy in terms of the digital method of learning.

1. Introduction

Virtual reality (VR) is currently a potentially strong tool that can transform the simple act of conventional learning through active interactions into an unforgettable experience. Most teachers believe VR has had positive effects in the learning process. A blended teaching methods program contributes to significant changes in educational outcomes [1]. VR systems can be divided into three categories such as no immersion systems, semi-immersion systems, and fully immersion systems [2] by form of display such as desktop displays [3], projection-based screens [4], and helmet displays [5].

The VR system usually consists of three elements, such as VR input device, VR application developing software, and output device. VR systems typically need at least one input device for the input device section, such as VR gloves [6], bodysuit [7], Phantom Omni [8], vive controllers [9], keyboard and mouse [10], touchscreen keyboard [11], joysticks [12], and voice recognition [13], and several other [14-15]. Major components of VR system need only one VR development software, such as Unreal, Unity 3D, Delta3D, OpenSimulator platform and many others.

Typically, there are three types of VR displays for the VR output system segment, such as desktop displays, projection-based screen, and helmet displays. VR helmet displays have also been classified into two types which are mobile headsets [16] and mobile-less headsets [17]. For adaptive production engineering, Giorgio et al. [18] performed human-machine collaboration in the virtual reality. They assembled the cubes in their application into three levels: posing, recording, and playing with built virtual ABB IRB 120 robot manipulator attached to the HTC Vive headset and controllers. They mentioned that their assembly of cubes was not intended for real practical use. They suggested technologies of virtual reality to be used for real application to be more helpful for the user.

Therefore, in this study, the authors aim to assemble virtual automobile engine components for actual functional assembly in virtual environment. In this work, we used virtual two-hand controllers with both HTC Vive headset and two vive controllers. In this project, all 3D engine components were designed by CATIA V5 software, and the immersive virtual automobile engine components assembly application was developed by using Unreal Engine and HTC vive headset. Two vive controllers were used as input devices. The development of application consists of several features such as teleportation, physics and collision, hologram set up, materials and textures, creation of animation, and building gaming scene. In future, experiments will be carried out to determine the effectiveness of developed immersive VR application compare to the traditional method using learning motivational tool of instructional materials motivation survey.

2. Materials and methods

The entire process of designing and developing virtual environment with holographic guideline was detailed in figure 1. The virtual environment was built mainly using the Unreal game engine, with CATIA V5 and Blender software being used to develop automobile engine assembly required for this module. The main aim of technical advancement was to provide an effective way to help and enhance the learning encouragement session in the challenging age of the present time.

This immersive VR technology aims at finding an alternative method of learning for assembly of engine components in tertiary education. Relative to the complexity of the parts and the difficulties of memorizing the assembly process without clear guidance, this engine components assembly scheme was selected as a tool for developing learning ability. This method for assembling automobile engine consisted of forty-four components and fort-four stages.

This VR system consists of two main items, such as a high-end laptop (Laptop ALIENWARE 15 R3) and HTC vive. The high-end laptop has been used for modeling 3D models, and for building and animating 3D elements. Five key software programs were used to create this VR application, including CATIA V5, (CAD software) for engine components modeling, Blender (a device to translate from native CAD file format (. CATProduct or .wrl) to Unreal file format (. FBX script)), Unreal for VR application creation, Steam software and Vive software have been introduced to the HTC vive for immersion in the VR. All forty-four 3D engine components were modeled using Catia V5. The models made in Catia were file format for. CATProduct or .wrl and were imported into Blender to change from . CATProduct or .wrl file to .FBX file as Unreal was incompatible with obtaining the . CATProduct or .wrl file. After that, all were exported into Unreal.

3. Development of virtual environment

In this application, many virtual components and many features were added to feel like a real workshop. We built the workshop with around two hundred and thirty virtual 3D components including automobile

IOP Conf. Series: Materials Science and Engineering 1109 (2021) 012038 doi:10.1088/1757-899X/1109/1/012038

engine components, tables, ceiling lamps, table lamps, drilling machine, milling machine, CNC machine, cupboards, shelves, chairs, and other several workshop appliances.

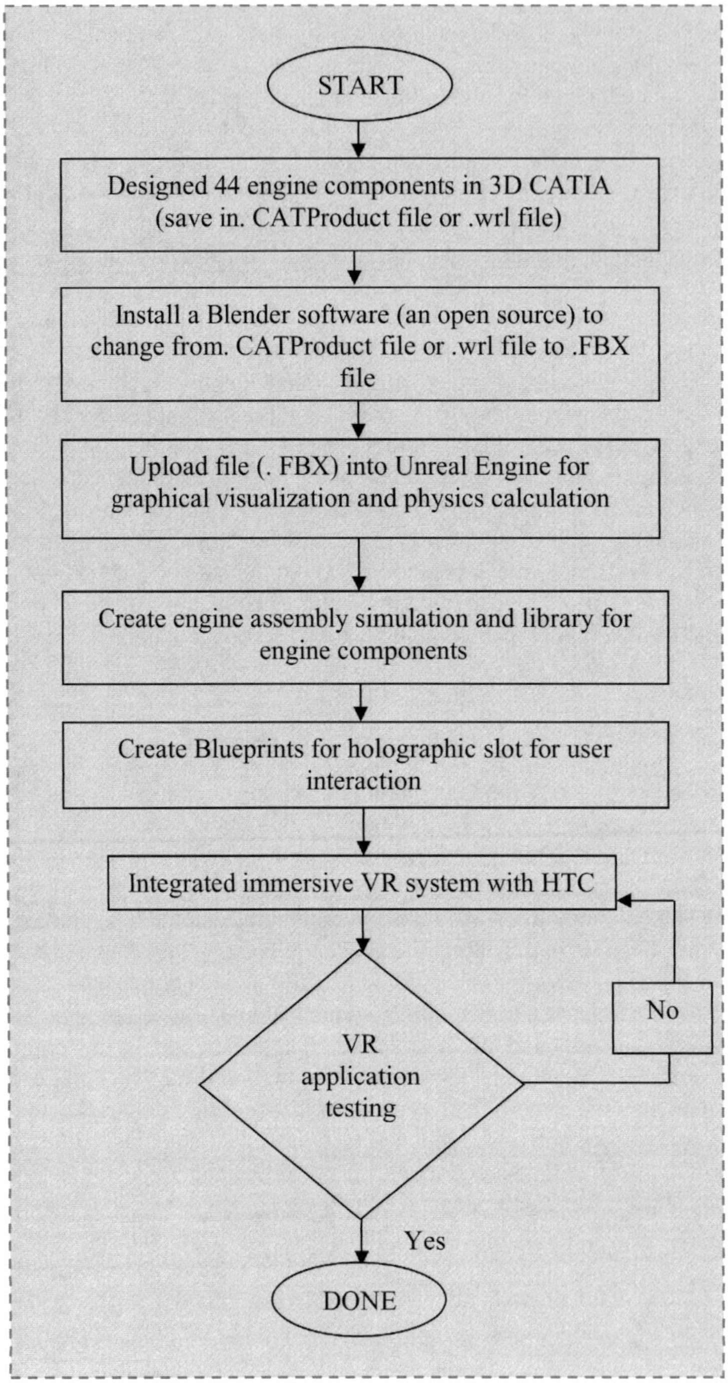

Figure 1. Process flowchart of immersive VR application development.

Much more features such as teleportation, blueprints visual scripting, hologram setup, detail setting of actors (actors are components used in this application) such as transform, static mesh, materials, physics, collision, and rendering as shown in figure 2 were also added and explained in the following sections so that the user can feel like real environment and improve user satisfaction in the virtual environment as shown in figure 3.

Figure 2. Fully application with several features.

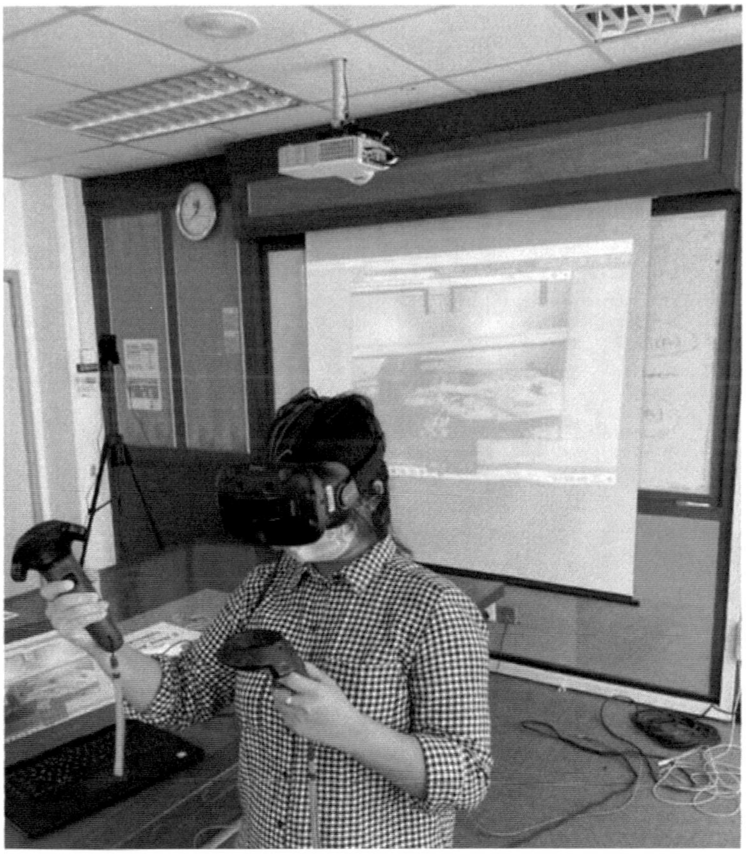

Figure 3. Developed virtual reality application demonstration.

3.1. Teleportation
In VR development of Unreal, teleportation was fantastic. It could control the actor from current place to a new location. That means Blue Laser could track components of the automobile engine components (see figure 4). For teleportation and grabbing components assignation, two vive controllers were used. Trackpad was assigned for teleportation and trigger was assigned for grabbing components.

Figure 4. Teleportation.

3.2. Blueprints visual scripting
In the Unreal 4.24, the framework was deployed with visual scripting of the blueprints. The visual scripting framework for the blueprints was a full scripting system based on the node-based interface model. This programming was incredibly versatile and efficient that provided designers with the ability to use the entire range of concepts. A blueprint scripting could also be used to change all the detail setting of actors.

3.3. Hologram setup
Hologram was a photographic technique that recorded the light reflected from an object in a way that appears three-dimensional, and then revealed. It helped the users to communicate more effectively with everything. An engine component was assigned by blue colour of hologram was seen in figure 5.

Figure 5. Holographic engine component.

3.4. Transform and Static Mesh
The part of Transform section included the position, rotation, and scale of each object. In this segment, the actor's transform could be modified. Static Mesh was a piece of geometry consisting of using the

graphics card to render. 3D models produced in external modelling applications could also be imported via the Content Explorer to the Static Mesh Editor. In our application, Transform and Static Mesh setting of all forty-four virtual engine components were adjusted even though they had their different native settings.

3.5. *Materials*
A material was an element which can be added to a mesh to monitor the scene's visual look good. It could be coated in a color and can then made it shiny too. It could also be inputted from different images, such as textures and math expressions, into the Content. It also had the different property settings including Base Color, Metallic, Specular, and Roughness. In our application, the workshop setting of all holographic surfaces were applied with blue color.

3.6. *Collision and Rendering*
Collision could manipulate the actors' set-up to block one another. The Collision setting of all virtual forty-four components were adjusted the same. Rendering section was modified to enable and inactivate the actor. The Rendering feature of all virtual components were also adjusted the same.

4. Conclusions
It has profoundly affected the practices and technologies of humanity in the accelerated development of modern science and technology. This is particularly beneficial for training which needs high costs, room for training, equipment, much time consuming, and so on. As a result, the latest growth of VR technology is very useful for students to be able to learn more comfortably and quickly memorize complex assembly of engine components. This interactive virtual reality (VR) technology was therefore planned and built for the learning inspiration of the students, with the application consisting of both around three hundred virtual 3D models and many features such as teleportation, hologram setup, details of actors such as transform, static mesh, materials and textures, physics collision, lighting, rendering. This ongoing study aims to perform an experiment in the future, and to measure the efficacy of the established application on respondent satisfaction and learning motivation variable for students compared to the current existing method using learning motivational tool of instructional materials motivation survey.

Acknowledgments
The authors sincerely appreciate the full financial support of JICA's AUN/SEED-Net at UPM (UPM CR 2001: Vote Number: 6282000), and the Hyper Version Research Laboratory, Keio University, Japan for its kind partial facility support.

References
[1] Innocenti E D, Geronazzo M, Vescovi D, Nordahl R, Serafin S, Ludovico L A and Avanzini F 2019 Mobile virtual reality for musical genre learning in primary education *Comput. Educ.* **139** 102-17
[2] Mujber T S, Szecsi T and Hashmi M S J 2004 Virtual reality applications in manufacturing process simulation *J. Mater. Process. Technol.* **155** 1834-38
[3] Peng G, Wang G, Liu W and Yu H 2010 A desktop virtual reality-based interactive modular fixture configuration design system *Comput. Des.* **42** 432-44
[4] Sharples S, Cobb S, Moody, A and Wilson J R 2008 Virtual reality induced symptoms and effects (VRISE): comparison of head mounted display (HMD), desktop and projection display systems *Displays* **29** 58-69
[5] Wu C M, Hsu C W, Lee T K and Smith S 2017 A virtual reality keyboard with realistic haptic feedback in a fully immersive virtual environment *Virtual Real.* **21** 19-29
[6] Tseng J L 2018 Development of a 3D intuitive interaction interface for head-mounted virtual reality system *Int. J. Adv. Stud. Comput. Sci. Eng.* **7** 19-24

[7] Walker C and Babak B 1999 System for tracking body movements in a virtual reality system *United States Pat* 1-16
[8] Rohidatun M W, Faieza A A, Rosnah M Y, Hayati S N and Rahinah I 2016 Development of virtual reality (VR) system with haptic controller and augmented reality (AR) system to enhance learning and training experience *Int. J. Appl. Eng. Res.* **11** 8806-09
[9] Buttussi F and Chittaro L 2020 Locomotion in place in virtual reality: a comparative evaluation of joystick, teleport, and leaning *IEEE Trans. Vis. Comput. Graph.* **2626** 1-12
[10] Robertson G, Card S K, Mackinlay J D and Parc X 1993 Three views of virtual reality: nonimmersive virtual reality *Computer (Long. Beach. Calif).* **26** 81-83
[11] Grubert J, Witzani L, Ofek E, Pahud M, Kranz M and Kristensson P O 2018 Text entry in immersive head-mounted display-based virtual reality using standard keyboards *25th IEEE Conf. Virtual Real. 3D User Interfaces* (Germany IEEE) p 159-66
[12] Bachman P and Milecki A 2009 MR haptic joystick in control of virtual servo drive *J. Phys. Conf. Ser.* **149** 1-4
[13] Carruth D W, Hudson C R, Bethel C L, Pleva M, Ondas S and Juhar J 2019 Using hmd for immersive training of voice-based operation of small unmanned ground vehicles *Int. Conf. on Human-Computer Interaction* (Denmark Springer) p 34-46
[14] Casey P, Baggili I and Yarramreddy A 2019 Immersive virtual reality attacks and the human joystick *IEEE Trans. Dependable Secur. Comput.* **5971** 1-13
[15] Barrera S, Takahashi H and Nakajima M 2004 Hands-free navigation methods for moving through a virtual landscape walking interface virtual reality input devices *Proc. Comput. Graph. Int. Conf. CGI* (Greece IEEE) p 388-94
[16] Wiederhold B B K, Miller I T, Wiederhold M D, and Care H 2018 Using virtual reality to mobilize health care *IEEE Consum. Electron. Mag.* **7** 106-09.
[17] Niehorster D C, Li L and Lappe M 2017 The accuracy and precision of position and orientation tracking in the HTC vive virtual reality system for scientific research *Iperception.* **8** 1-23
[18] Giorgio A D, Romero M, Onori M and Wang L 2017 Human-machine collaboration in virtual reality for adaptive production engineering *Procedia Manuf.* **11** 1279-87

IOP Conf. Series: Materials Science and Engineering 1109 (2021) 012039 doi:10.1088/1757-899X/1109/1/012039

100kW grid-connected renewable energy system design for an educational campus in Tarlac: A step towards green energy campus development

Adam Rombaoa

Mechanical Engineering Department, De La Salle University-Manila, Metro Manila, Philippines

adam_rombaoa@dlsu.edu.ph

Abstract. This paper discussed the design of a grid-connected renewable energy system for an education institution. One-year energy consumption and the hourly load of the institution were determined to design the energy system. HOMER Pro and What's*Best!* 16.0 for Microsoft Excel optimization tools were used in the design, consisting mainly of wind turbines and solar PV. Weibull probability function and the simplified irradiance model were used to predict the wind turbine and solar PV power output in the Microsoft Excel spreadsheet. Results showed that the mean velocity at 50 m reference height is 3.886 m/s. The annual solar PV system energy produced based on the HOMER Pro simulation and What's*Best!* was found to be 170,939 kWh/yr and 189,454 kWh/yr, respectively. HOMER Pro and What'sBest! iteration results showed that the optimum number of components are 303 and 0 for solar PV and wind turbine, with net present costs of Php 23,388,612 and Php 22,769,478, respectively. With this optimum result, emissions can be reduced by up to 51%, a financial saving of at least Php 1, 970, 145 per year can be achieved with payback period of 5 years.

1. Introduction

For the last couple of centuries, the progress of man - industrialization in the nineteenth and the large-scale urbanization in the twentieth - has given rise to global modernization. However, there is another price the Earth has paid for modernization rightly requires immediate attention. This is global warming. For this generation, with the devastating effect of global warming and climate change looming over us, there is a need for the energy industry to find energy sources free of carbon dioxide pollution. The Renewable Portfolio Standard (RPS), a regulation that requires the increased production of energy from renewable energy resources, places an obligation on electricity supply companies to produce a certain fraction of their electricity from renewable energy resources. Among these renewables, wind and solar energy are at the forefront of the drive to significantly reduce the greenhouse gasses to meet the carbon dioxide (CO_2) limit. This is largely because we know that if we can replace fossil fuel with wind and solar energy for generated electricity, we can significantly reduce CO_2 emissions [1]. Small-scale shifts to renewable energy utilization are a milestone. Different researchers have done studies to optimize energy production and reduce ozone-depleting substances to date. In the year 2011, a study of a hybrid alternative power system appropriate for a remote area in Orissa State, India, consists mainly of micro-hydro plant, wind turbines, diesel generator units, solar PV panels, and battery storage was conducted [2]. Furthermore, different optimization techniques and available applications are being used nowadays for ease of better design of renewable energy systems because oversizing causes higher system costs, and under-sizing causes insufficient power supply [3].

On the other hand, this study considered Tarlac State University (TSU) to design a renewable energy system considering net-metering for renewable energy at maximum power reliability and minimum cost. Its San Isidro campus is the potential location for solar PV and wind turbine installation because it has the largest available land area amongst the other TSU campuses.

2. Material and Methods

The method of research used is descriptive-evaluative and the plan of action of the study was divided into four sections.

2.1. Assessment of TSU-San Isidro electric load & energy usage for one year

The area under study's one-year energy usage was determined to estimate the load to serve for the optimum renewable energy system design. The statement of account (SOA), which contains the electric bill and energy consumption, was requested to the TSU administration and the connected loads were then determined based on the actual rating of devices, appliances, and equipment being used by all rooms.

2.2. Solar PV & wind turbine power computations and renewable energy system configuration

Before solving for the solar PV and wind turbine power output, solar and wind resource assessments were performed. HOMER Pro was used for solar resource assessment and Windographer for wind resource assessment. PV array power output P_{pv} was then calculated using equation (1), also called the simplified irradiance model [4], [5] expressed as

$$P_{pv} = n_{pv}A_{pv}G_T \tag{1}$$

where n_{pv} is the conversion efficiency and A_{pv} is the solar PV area. The efficiency and area of the solar PV used are 16.97% and 1.94432 m^2, respectively. G_T is the hourly total incident radiation on the solar PV (inclined surface). G_T is the sum of beam radiation (G_b), ground-reflected radiation (G_r), and diffuse radiation (G_d). The mathematical models used for these three variables were based on the review of the previous works [6]–[14]. Equation (2) was used for the beam radiation on inclined surface expressed as

$$G_b = R_b G_{bH} \tag{2}$$

where R_b is the ratio of hourly radiation received by an inclined surface to beam radiation on a horizontal surface. G_{bH} is the hourly direct radiation on a horizontal surface expressed as

$$G_{bH} = G_H - G_{dH} \tag{3}$$

G_H is the global horizontal solar radiation (also called total radiation), and diffuse radiation (G_{dH}) on horizontal surfaces. G_H and G_{dH} where solve in terms of the hourly clearness index which was gathered during solar resource assessment.

The ground-reflected radiation model used is given by equation (4)

$$G_r = G_H \rho_g \frac{1 - \cos \cos \beta}{2} \tag{4}$$

where ρ_g is the ground reflectance (also called ground albedo) and is typically equal to 0.2 [15]. Angle β is the tilt angle (in degree) of the solar PV. According to the study of [16], the optimum tilt angle of solar PV in the Philippines can be calculated by multiplying 0.812117 correction rate to the latitude, which is equal to 12.59°. While HOMER Pro uses HDKR (Klucher & Reindl, and Hay & Davies) model, the present study used the Perez model for G_d as it was found to be the most accurate model for Singapore [17], the country near the Philippines, and cited in other study [8]. Perez model for G_d is given by equation (5)

$$G_d = G_{dH}\left[\frac{1 + \cos \cos \beta}{2}(1 - F_1) + F_1\frac{c'}{d} + F_2 \sin \sin \beta\right] \tag{5}$$

The annual energy and total power that the turbine will deliver considering its $V_{cut\text{-}in}$ and $V_{cut\text{-}out}$ velocity, therefore, are given by equations (6)-(8) respectively [18]:

$$E_{WT} = Total\ P_{WT} \cdot 8760hr/yr \tag{6}$$

$$Total\ P_{WT} = \sum_{cut-in}^{cut-out} P_v\, f(v) \tag{7}$$

$$f(v) = \frac{k}{c}\left(\frac{v}{c}\right)^{k-1} \cdot e^{\left[-\left(\frac{v}{c}\right)^{k}\right]} \tag{8}$$

The Windographer software used in this study uses 50m as reference height, the value of $f(v)$ at the turbine hub height H was determined using resolve values of k and c using equations (9) and (10) given in [19] expressed as

$$k = \left(\frac{\sigma_v}{V_m}\right)^{-1.090} \tag{9}$$

$$c = \frac{V_m\, k^{2.6674}}{0.184 + 0.816\, k^{2.73855}} \tag{10}$$

where σ_v is the standard deviation of velocities. V_m is the mean velocity (m/s). The σ_v and V_m values at the turbine hub height were computed using equation (11)

$$\frac{V_m}{(V_m)_{H_0}} = \left(\frac{H}{H_0}\right)^{\alpha} = \frac{\sigma_v}{(\sigma_v)_{H_0}} \tag{11}$$

where α is the friction coefficient, sometimes called Hellman exponent or shear exponent, equal to 0.15 for tall grass on level ground [18]. The hub height H of the turbines considered is 12m while the reference height H is 50m. After solving for wind turbine and solar PV power outputs, the energy system configuration was done, using only the available components from HOMER Pro's library, as shown in Figure 1.

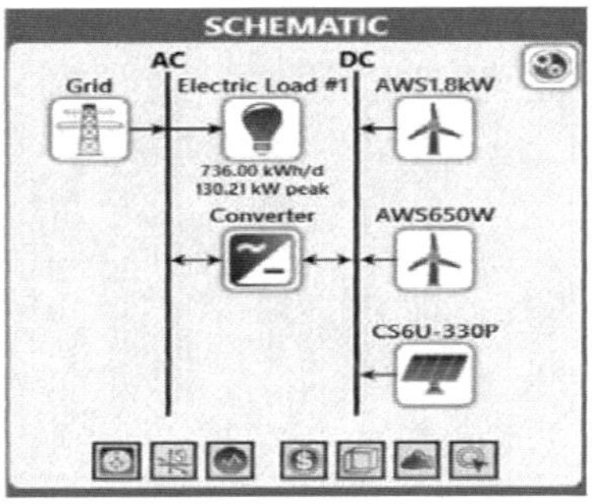

Figure 1. Renewable energy system components model in HOMER.

2.3. *Simulation using HOMER Pro and What'sBest!*

After configuring the system, simulations were performed using HOMER Pro student version and What's*Best!*. The HOMER Pro 'Setup Assistant' was used as a basis for creating and simulating the energy system under study. For What's*Best!* optimization process, the net present cost (NPC) formula was set up within the Microsoft Excel as it is what to minimize. The formula used to minimize NPC is given by equation (12)

$$(NPC) = \sum_{all\ components} \left[C_{0,i} + \sum_{n=1}^{t} \frac{C_{n,i}}{(1 + r)^n}\right] \tag{12}$$

where $C_{0,i}$ is the capital cost of the component i at year 0. The annual net cash flow $C_{n,i}$ includes the annual cost (O&M, replacement, and cost of buying electricity from the grid) of the components minus the revenue generated (e.g. salvage value and grid sale). The cost of buying electricity from the grid was ₱ 12.87 per kWh ($0.2475/kWh) which is based on the statement of account issued by the distribution unit to San Isidro campus from July 2018-June 2019 while the selling price of excess energy is based on Energy Regulation Commission (ERC) rules on net-metering [20], which was assumed half of the grid purchase, of Php 6.435/kWh. The components of the grid-connected renewable energy system of the present study considered are wind turbine, solar PV, grid, and converter. The discount rate r used in the present study is 10% [21].

3. Discussion of Results

The results of this study are divided into four sections, presented as follows:

3.1. TSU San Isidro Campus educational buildings energy consumption and electric load

The energy consumption of San Isidro campus is shown in figure 2 with the highest energy consumption at the month of September.

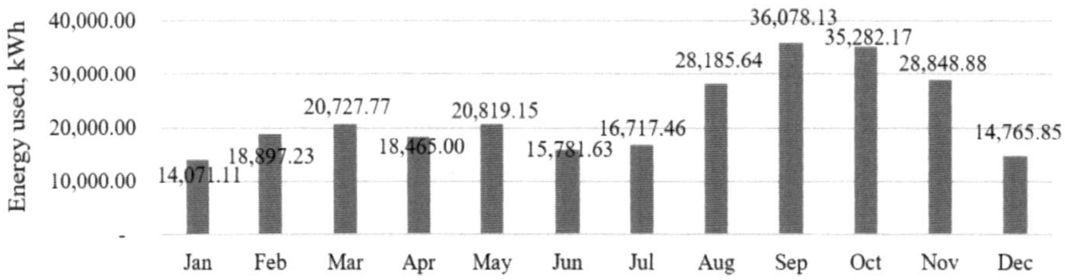

Figure 2. Monthly energy Consumption.

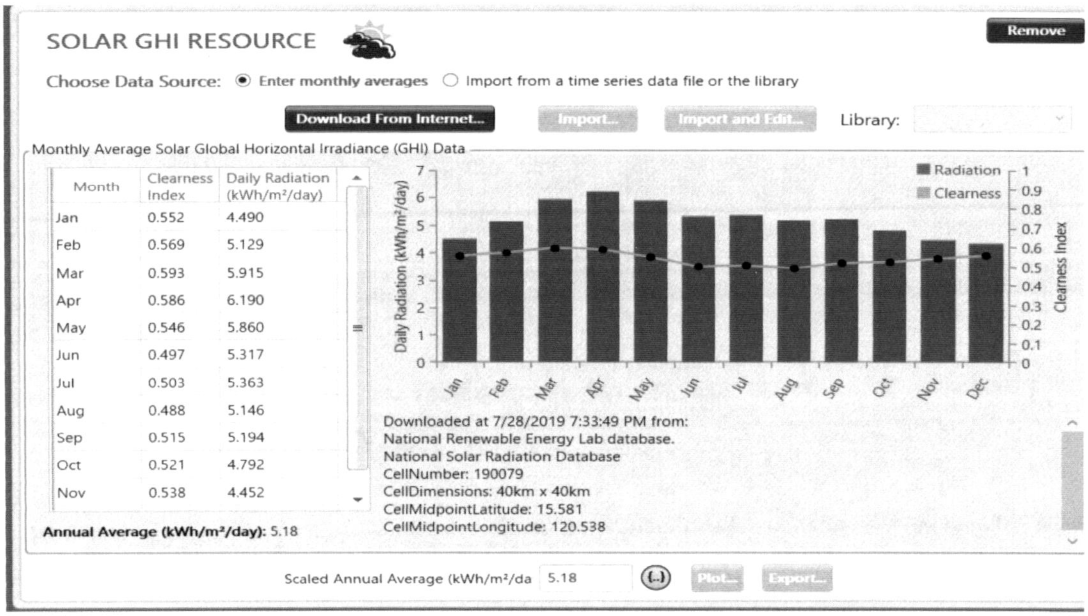

Figure 3. Solar resource assessment using HOMER.

3.2. TSU San Isidro campus solar and wind resource assessments result using HOMER and Windographer

The result of solar resource assessment of the study is shown in see figure 3. It can be observed that the highest daily radiation is in April and the lowest one is in December. Meanwhile, using Windographer, the monthly average wind speed and direction frequency at 50 m reference height H_0 are shown in figure 4. The value of the Weibull scale parameter c_{H_0} = 4.38 m/s and shape factor k_{H_0} = 1.94, the mean velocity of $(V_m)_{H_0}$ = 3.886 m/s, and a standard deviation of $(\sigma_v)_{H_0}$= 2.019 at 50 m reference height H_0

These results indicate that the site is only good for small-scale wind turbines. Thus, considering the small-scale wind turbines of the same brand inside HOMER Pro library, Aeolos and AWS wind turbines of 10m and 12m hub heights, respectively, were considered. The power curves of each turbine were redrawn as shown in figure 5. To estimate the wind turbine power output at 10m and 12m hub heights, the probability density functions were solved and is shown in figure 6.

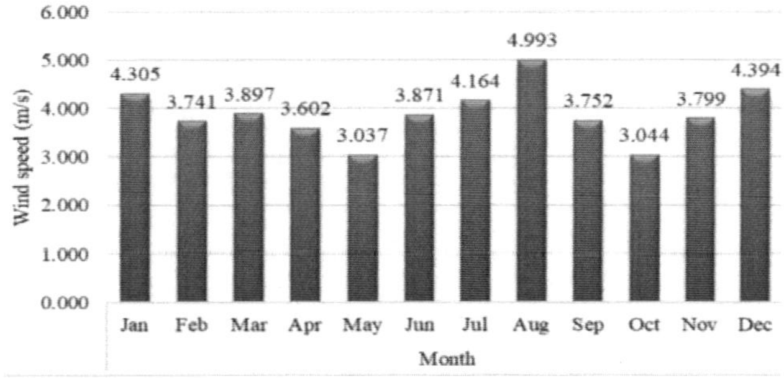

Figure 4. Monthly average wind speed gathered using Windographer.

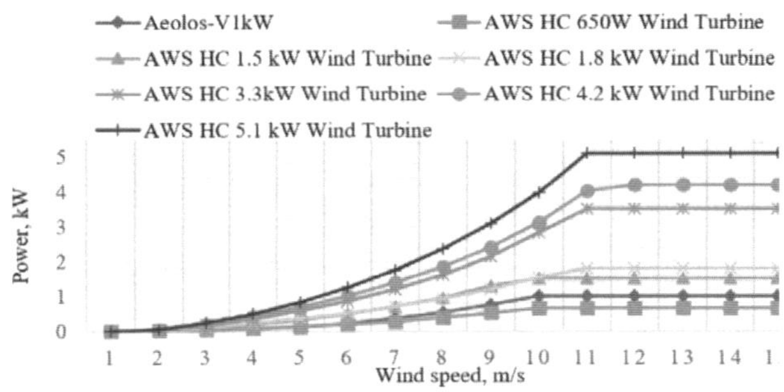

Figure 5. Wind turbines power curve.

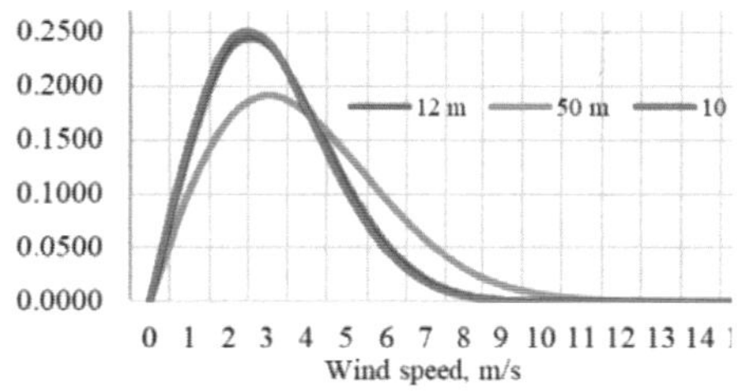

Figure 6. Probability density function at different height.

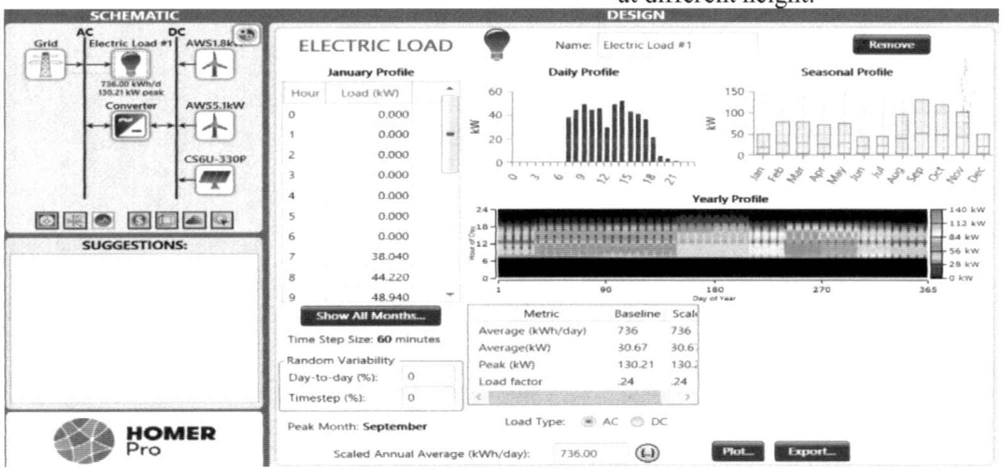

Figure 7. Renewable energy system components model and load profiles sample in HOMER Pro.

3.3. The design, simulation and optimization of the grid-connected renewable energy system using HOMER Pro and Microsoft Excel spreadsheet

The components considered and the electric load profile of San Isidro campus grid-connected renewable energy system design are illustrated in figure 7.

IOP Conf. Series: Materials Science and Engineering 1109 (2021) 012039 doi:10.1088/1757-899X/1109/1/012039

3.4. Comparison of HOMER simulation and spreadsheet iteration result

Results were compared based on the optimum number of components, NPC, and energy produced, considering the net-metering. Emission reduction is also presented in the following.

3.4.1. Optimization result for the system with net-metering. For systems with net-metering (limited to 100kW capacity), HOMER Pro simulation result showed 100 kW rated capacity solar PV system, equivalent to 303 units, with an inverter capacity of 89.6 kW and zero turbines. This component's combination has the least NPC of $449, 781 (Php 23, 388, 612). See figure 8. In terms of energy production, solar PV has 170, 939 kWh/yr while the grid purchase is 129, 386 kWh/yr.

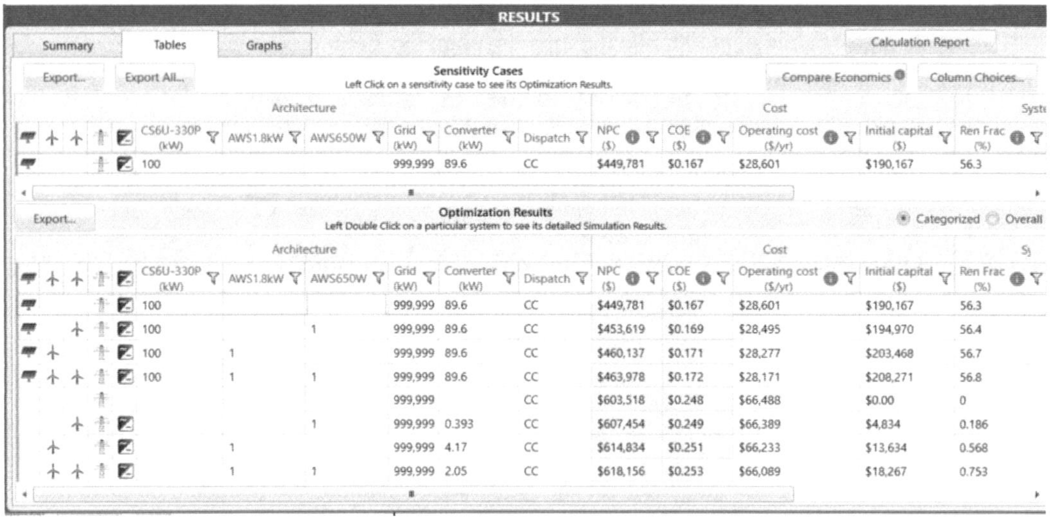
RESULTS

Summary | Tables | Graphs | Calculation Report

Export... | Export All... | Compare Economics | Column Choices...

Sensitivity Cases
Left Click on a sensitivity case to see its Optimization Results.

CS6U-330P (kW)	AWS1.8kW	AWS650W	Grid (kW)	Converter (kW)	Dispatch	NPC ($)	COE ($)	Operating cost ($/yr)	Initial capital ($)	Ren Frac (%)
100			999,999	89.6	CC	$449,781	$0.167	$28,601	$190,167	56.3

Export...

Optimization Results
Left Double Click on a particular system to see its detailed Simulation Results.
Categorized Overall

CS6U-330P (kW)	AWS1.8kW	AWS650W	Grid (kW)	Converter (kW)	Dispatch	NPC ($)	COE ($)	Operating cost ($/yr)	Initial capital ($)	Ren Frac (%)
100			999,999	89.6	CC	$449,781	$0.167	$28,601	$190,167	56.3
100		1	999,999	89.6	CC	$453,619	$0.169	$28,495	$194,970	56.4
100	1		999,999	89.6	CC	$460,137	$0.171	$28,277	$203,468	56.7
100	1	1	999,999	89.6	CC	$463,978	$0.172	$28,171	$208,271	56.8
			999,999		CC	$603,518	$0.248	$66,488	$0.00	0
		1	999,999	0.393	CC	$607,454	$0.249	$66,389	$4,834	0.186
	1		999,999	4.17	CC	$614,834	$0.251	$66,233	$13,634	0.568
	1	1	999,999	2.05	CC	$618,156	$0.253	$66,089	$18,267	0.753

Figure 8. HOMER Pro optimization result with net-metering.

MODEL INFORMATION:

CLASSIFICATION DATA	Current	Capacity Limits
Total Cells	289204	
Numerics	289199	
Adjustables	2	Unlimited
Continuous	2	
Free	0	
Integers/Binaries	0/0	Unlimited
Constants	201584	
Formulas	87613	
Strings	0	
Constraints	5	Unlimited
Nonlinears	17520	Unlimited
Coefficients	205722	

Minimum coefficient value: 0.0022180061406674 on NPC!H8563
Minimum coefficient in formula: NPC!I8563
Maximum coefficient value: 384228 on NPC!Y3
Maximum coefficient in formula: NPC!AA3

MODEL TYPE: Nonlinear (Nonlinear Program)

SOLUTION STATUS: LOCALLY OPTIMAL (see messages below)

OPTIMALITY CONDITION: SATISFIED

OBJECTIVE VALUE: -22,769,478.163407

Figure 9. HOMER Pro optimization result with net-metering.

On the other hand, What's*Best!* iteration result showed an NPC of $ 437, 875 (Php 22, 769, 478) as shown in WB!Status sheet of figure 9. Three hundred three (303) units of solar PV was found to be the optimum number of solar PV, equivalent to 100 kW rated power, the number of wind turbine is equal to zero, and the converter capacity is 84.53 kW. In terms of energy production, the solar PV system produced 189,454 kWh/yr while the grid energy purchase is 110,894 kWh/yr. Comparison of optimization results of both HOMER and What'sBest! for Microsoft excel spreadsheet, with net-metering, is summarized in table 1. The simple payback period was computed by dividing the initial investment by the annual savings.

Table 1. Summary of comparison between HOMER Pro and What's*Best!* for excel with net-metering

Parameter	HOMER Pro	What'sBest!	Unit
NPC	449,780.92 (23, 388, 612)	437, 875 (Php 22, 769, 478)	\$ (Php)
No. of solar PV	303	303	
No. of wind turbine	0	0	
PV system rated capacity	100	100	kW
Annual $\boldsymbol{E_{PV}}$	170, 939	189,454	kWh/yr
Excess Electricity	492	0	kWh/yr
Investment	190, 167 (9, 888, 684)	189, 744 (9, 866, 688)	\$ (Php)
Simple Payback Period	5.02	4.85	yr

3.4.2. Optimization result for the system without net-metering. For systems without net-metering (no grid sales), HOMER Pro simulation result gave 379 units of solar PV and zero turbines which has an NPC of \$ 603, 518 (Php 31, 382, 936). Meanwhile, What'sBest! iteration result without net-metering showed the least NPC of Php 24, 386, 651.83 with the number of solar PV of approximately 308 units with equivalent NPC of \$ 468, 975 (Php 24, 386, 691).

3.4.3. Emission and cost comparison based on the optimum result. The amount of emission can be found by simply multiplying the energy purchased from the grid by the kilograms of emission per kWh (CO_2 = 0.632, SO_2 = 0.00274, NO_2 = 0.00134). The amount of emissions can be reduced by up to 51%. See table 2 for the summary of emission comparison and annual costs in table 3.

Table 2. Emissions comparison of the optimum result.

Quantity	Grid Only	HOMER Pro	What'sBest!	Unit
Carbon Dioxide (CO_2)	169, 780	81, 772	70, 085	kg/yr
Sulfur Dioxide (SO_2)	736	355	304	kg/yr
Nitrogen Dioxide (NO_2)	360	173	149	kg/yr

Table 3. Annual cost comparison of the optimum result.

Grid Only	HOMER Pro	What'sBest!	Unit
3,457,396.86	1, 487, 252	1, 421, 475.54	Php/yr

4. Conclusion

The energy consumption, electric load, and hourly load of Tarlac State University San Isidro campus were determined and generate the hourly load profile. Meanwhile, solar and wind resource assessments in the San Isidro campus were successfully conducted using HOMER Pro and Windographer software, respectively. Wind resource assessment showed that an average of 3.886 m/s wind speed prevails on the campus at 50 m height. Using these values, HOMER Pro then was used to simulate and analyzed the renewable energy system design, which yields a purely solar PV system as the best optimization results in terms of NPC. Similarly, the What's*Best!* optimization tool was used to minimize the NPC, but results showed that there is a significant difference between that of the HOMER Pro. HOMER Pro gave a higher NPC of Php 23, 388, 612 compared to What's*Best!* in the spreadsheet of Php 22, 769, 478 but with the same 100 kW rated solar PV power and no wind turbines. Since NPC is lower when net-metering is considered, therefore, 303 units of solar PV and no wind turbine are the optimum numbers of components. With these numbers of components, a reduction of carbon dioxide, sulfur dioxide, and nitrogen oxides of up to 51% are feasible. Moreover, a financial saving of at least Php 1, 970, 145 per year can be achieved.

5. Recommendations

The researcher of the study would like to recommend the application of net-metering if the installation of solar PV in the university studied will push thru. The optimum size of solar PV and converter may not be available in the market, and it is therefore recommended to use the nearest (higher) standard sizes. To further improve the study: (1) hourly load profile from distribution companies might be better to use for a more accurate result; (2) the use of other different mathematical models/techniques for solving solar PV energy and other variables; and (3) use of other optimization tools is recommended to have a comparison of wind and solar energy estimation.

References

[1] Letcher T M 2017 *Wind Energy Engineering: A Handbook for Onshore and Offshore Wind Turbines* (London EC2Y 5AS: Academic Press) ed Lisa Reading pp 26-31
[2] Lal D K, Dashand B B and Akella A K 2011 *Int. J. Electr. Eng. Informatics* **3** 3 pp 307–25
[3] Sinha S and Chandel S S 2015 *Renew. Sustain. Energy Rev.* **50** 755–69
[4] Liu Y, Luand Z and Yang F 2018 The investigation of solar PV models *IEEE Power Energy Soc. Innov. Smart Grid Technol. Conf. ISGT* pp 1–5
[5] Diaf S, Notton G, Belhamel M, Haddadi M and Louche A 2008 *Appl. Energy* **85** 10 pp 968–987
[6] Iqbal M 1983 *An Introduction to Solar Radiation* **53** 9 (Ontario M3C 2A1: Academic Press, Inc.) pp 321-49
[7] Spencer J 1971 Fourier series representation of the position of the Sun *Search* **2** 5 p 172
[8] Maleki S A M, Hizam H and Gomes C 2017 *Energies* **10** 1 p 134
[9] Duffie J, Beckman W and McGowan J 2013 *Solar Engineering of Thermal Processes* (New Jersey : John WIley & Sons, Inc.) pp 43-91
[10] Kuo C W, Chang W C and Chang K C 2014 *Renew. Energy* **66** pp 56–61
[11] Gorjian S, Hashjin T and Ghobadian B 2012 *2nd Iran. Conf. Renew. Energy Distrib. Gener. ICREDG* pp 172–177
[12] Chapman R 1990 *IEEE Conf. Photovolt. Spec.* pp 965–70
[13] Liu B and Jordan R 1960 *Sol. Energy* **4** 3 pp 1–19
[14] Kalogirou S 2014 *Solar Energy Engineering Processes and Systems* (Oxford OX 5 1GB: Academic Press) pp 75-105
[15] Gueymard C 2009 *Sol. Energy* **83** 3 pp 432–44
[16] Malicdem E 2015 Optimal tilt of solar panels in the Philippines
[17] Khoo Y S, Nobre A, Malhotra R, Yang D, Ruther R, Reindland T and Aberle A G 2013 *IEEE J. Photovoltaics* **4** 2 pp 647–53
[18] Masters G 2013 *Renewable and Efficient Electric Power Systems* (New Jersey: John Wiley & Sons, Inc.) pp 430-63
[19] Mathew S 2006 *Wind Energy Fundamentals, Resource Analysis and Economics* (Berlin Heidelberg: Springer Science & Business Media) pp 75-76
[20] Energy Regulation Commission 2013 Rule Enabling the Net-Metering Program for Renewable Energy
[21] National Economic Development Authority 2016 Revisions on ICC Guidelines and Procedures

IOP Conf. Series: Materials Science and Engineering 1109 (2021) 012040 doi:10.1088/1757-899X/1109/1/012040

Comparative analysis of Ormat and Kalina Cycle for the geothermal resource in Mabini, Batangas using thermoeconomic analysis

G E E Dela Cruz III and M C E Manuel, PhD

School of Mechanical and Manufacturing Engineering
Mapua University, Muralla St., Intramuros, Manila, Philippines

geedelacruz@mapua.edu.ph and mcemanuel@mapua.edu.ph

Abstract. The Philippines is amongst the countries located in the ring of fire, thus makes it rich in geothermal energy. The Department of Energy is considering the low-temperature resource for power generation to address the continuous increase in electricity demand. The binary plants designed by Ormat technologies are currently in use for power production for the Philippine low-temperature heat source. Another technology used for the low-temperature heat source is the Kalina power cycle. In this paper, both cycles are evaluated using the parameters from the Mabini, Batangas geothermal resource, and principles of thermoeconomics are applied to determine the performance of the two cycles under Philippine conditions.

1. Introduction

The Philippines is blessed with an abundance of natural resources. One resource in specific is geothermal energy since it is situated in an area called the "Ring of Fire". It has made use of its geothermal resource mostly for power generation and used to be rank as the world's second-largest producer of geothermal energy but has been recently overtaken by Indonesia in the-global ranking in which the United States is leading [1].

The Department of Energy (DOE) estimated the geothermal resource potential of the country at around 4,000 MWe [2] while the installed capacity of the Philippines is only 1944 MWe [3]. There is still a large number of resources that can be harnessed and help in the increasing energy dependence of the country. The DOE added that the country has already been developed most of the high-enthalpy geothermal sources into commercial operations but there are marginalized low enthalpy resources that can be tapped for exploitation [4]. Optimization of the utilization of the geothermal resource through cascade use and development of the low enthalpy direct use is one of the goals of the DOE [5].

Significant support for the geothermal energy resource development of the country was realized through the passage of Republic Act No. 9513 also known as the Renewable Energy Law of the Philippines which took effect on January 30, 2009, and resulted in a total of 42 contracts for geothermal projects [6]. Geothermal companies also increased from 2 to 11 because of attractive government incentives [7].

The development of low enthalpy geothermal resources in the Philippines has been hindered because of the high cost and high risks involved in the exploration, production drilling, and construction phase. Intermediate to low-enthalpy resources in the Philippines are mostly used for nonelectrical and direct utilization like hot springs, geothermal salt making, and drying [8]. These low enthalpy resources can be used to generate power using a binary power cycle or ORC [9]. One prospect for development is located in Mabini, Batangas which is an intermediate-temperature with a reservoir temperature of at least 180°C with possible 20MWe [10]. The development of these resources is in line with the government's energy plan in which a total of 253MW will be available on 2030 based on the outlook of the DOE for a balanced energy mix and reducing dependence on fossil fuels.

At present, binary plants in the Philippines are used together with the traditional flash system to maximize the efficiency of the heat recovery from the geothermal wells that are still producing. The Binary plants operate on an Organic Rankine Cycle (ORC) which is a steam cycle system where refrigerants and hydrocarbons as a replacement to water to utilize the low-temperature heat resource for power production [11]. Ormat Technologies Inc. supplied the technology for the binary plants in the Philippines using heat from the waste brine [12].

Unfortunately, geothermal power plant efficiency is lower than the conventional power plants like coal-fired power plants. Increasing the current plant efficiency of geothermal plants becomes important to maximize the energy conversion from the current resource without an increase in cost due to drilling and exploration that sometimes yield unfavorable results when no productive steam is drilled. Another option is to use emerging technologies for low-temperature resource development. One of these technologies for low-temperature geothermal resources is the Kalina cycle. This technology was conceptualized by Dr. Alexander Kalina using ammonia-water in the cycle and modifying the Rankine cycle to achieve higher efficiency for low heat energy sources [13].

With the government pushing for cleaner energy, the development of the low to medium enthalpy geothermal areas will be a significant part of the future of geothermal energy in the Philippines. Using the appropriate technology for the development and efficient extraction of energy from these resources becomes imperative. The study of the thermodynamic efficiencies and economic viability of the available technologies namely the ORC together with the Kalina cycle system 34 will be the focus of this study. The study of both exergetic and economic analysis is known as thermoeconomics and consists of detailed exergy and economic analyses of the components of the cycle after which an evaluation of each component is conducted

1.1. Binary Cycle

With the advent of innovative technology, it has become possible to also produce power from low-to-medium enthalpy reservoirs with temperature below 150°C with the use of a binary power plant or an Organic Rankine cycle [14]. The Organic Rankine cycle uses the geothermal water to heat a secondary fluid with lower boiling temperature, also called as the working fluid, such as propane, isobutane, isopentane, pentaflouropropane, and ammonia. Heat exchangers are used for transferring energy from the hot brine to the secondary fluid which turns the turbines to produce electricity. The geothermal water is also re-injected in a closed loop. There is no direct contact of the geothermal fluid and the hydrocarbon in this type of plant unlike the direct and flash system.

One notable binary power plant is the 125 MW Upper Mahiao Power Plant in Leyte which utilized the Ormat air-cooled Geothermal Combined Cycle technology designed for high enthalpy fluid other than pure dry steam. The cycle has a back-pressure steam turbine for the high-pressure steam followed by the Ormat Energy Converter for preheating and recovered in a binary plant or Organic Rankine Cycle [15].

1.2. Kalina Cycle

An efficient novel cycle was developed during the mid-1980s by a Russian engineer named Dr. Alexander Kalina. The cycle named after Dr. Kalina was used on a combined-cycle system as the bottoming cycle that replaced the use of the Rankine cycle. The key feature of this novel cycle is the use of a combination of ammonia and water (70/30 mass percent mixture) as the working substance. An increase of about 20% system efficiency and a 50-52% increase in a conventional gas turbine cycle was achieved [16]. The cycle provides varying temperatures for heating and boiling which matches the cooling curve of the heat source reducing entropy generation and unique heat rejection system that permits the working fluid composition to be less volatile [17]. The flexibility of the Kalina cycle is attributed to the ammonia-water properties but the limitation of the cycle is also seen in the fluid chemistry of Ammonia which is unstable at higher temperatures greater than 300°C where hardening or nitride corrosion becomes a problem [18].

Mlcak [13] discussed the difference of the ammonia-water mixture used by Kalina as compared to pure water (H2O) or ammonia (NH3). The following are the main differences between the mixture and pure substances.

1. Both pure H2O and NH3 have constant boiling / condensing temperatures which is not the case for the ammonia-water solution [19].
2. The thermophysical properties of the ammonia-water solution can be improved by adjusting the concentration which is not true for pure water and pure ammonia as their properties are fixed.
3. The temperature of pure water and pure ammonia will not change without energy while the mixed fluid temperatures of ammonia-water can change without heat content.
4. Ammonia-water solution freezing temperature is very low compared to water freezes at 0°C and ammonia at -78°C.

There are several proprietary Kalina Cycle Systems that are modifications of the cycle designs that make use of the flexibility of the ammonia-water concentration. The first cycle used as gas turbine bottoming cycle mentioned earlier, this was later named the Kalina Cycle System 1 (KCS 1) and was used for the US DOE-ETEC at Canoga Park, California demonstration [19]. Other notable Kalina cycles are KCS 6 which has the highest efficiency applies to gas turbine and KCS 5 used for direct-fired plants [19].

The United States DOE funded a research and development grant to POWER Engineers, Inc to perform pilot testing of the Kalina cycle to determine if the viability of the cycle for heat recovery of a silica-rich geothermal brine with a temperature of 171°C in Roosevelt hot spring in Utah. It was found that installing a 13 MW Kalina cycle power plant is feasible in the said geothermal area. The plant is estimated to generate power at a high availability greater than 95%.

Low to medium temperature geothermal resources are by far the most commonly available type of geothermal resource in the world. The binary power plant is the technology that is used to exploit this type of source of energy. The Kalina cycle system is suitable for this application. KCS 11 and KCS 34 are the common system designs for geothermal applications. KCS 11 can be used for resources up to 204°C while KCS 34 is utilized for lower temperatures [19]. The Kalina cycle is composed of the separator, recuperators, evaporator, feed pump, condenser, and turbine generator. Geothermal brine heats the ammonia-water which produces a rich ammonia vapor that is extracted in the separator and utilized in the turbine to generate power. After expansion, the ammonia-rich liquid is mixed with the liquid from the separator to the recuperator before going to the condenser for further cooling and then returns to the evaporator to re-start the cycle [20].

According to Mlcak [21], The KCS 34 consists of a vapor turbine-generator, evaporator, separator, condenser, recuperator exchangers, and feed pump. KCS 34 was designed for generating power from low-to-medium-enthalpy geothermal resources which were used for district heating topping cycle and this is the cycle that is being utilized in the Orkuveita Husavikur Geothermal Power Plant system located in the town of Husavik, Iceland which has been in service since July of 2000. The plant performance ranges from 20% to 25% compared to the Organic Rankine Cycle that was considered for the plant. The KCS 34 cycle also proved to have an advantage over the ORC as the plant actual power output of 2,060 kW at 121°C brine inlet temperature during the operation was greater than that of the proposed ORC at 124°C with output ranging from 1,550 to 1,610 kW guaranteed [22].

Arslan [20] conducted a study of the KCS 34 using the Simav geothermal resource conditions in Turkey. The Simav field has temperatures up to 148°C with drilled depths at 169m to 725m. The waste heat is recommended to be used for heating. The electricity generated from the Simav geothermal field was determined to vary from 32.3 to 43.4MW depending on the ammonia concentration used. The net plant electricity output ranged from 243 to 346 GWh per year while the efficiency of the KCS 34 varied from 9.7 to 14.8%. The KCS 34 exergetic efficiencies were found to be at 36.2% for a temperature of 90°C and 34.6% for 100°C.

Mergner and Weimer [23] performed a comparative thermodynamic analysis on a geothermal power plant using KCS 34 and compared it to a Siemens' patent geothermal cycle Kalina SG-1 similar to the Kalina cycle. The main difference between the KCS 34 and SG-1 cycles being the internal heat recovery location. KCS 34 utilizes bypassed liquid while SG1 uses the expanded steam and the bypassed liquid for the recovery. Calculations suggest that the KC SG1 attains a high flow rate through the turbine compared to the KCS 34 which operates at high turbine enthalpy difference. However, the efficiency increase could not be validated due to challenges with the absorption and resorption processes. They recommended further study focusing on validating the simulation results of the cycles used.

Tiangco [24] recommended the study of the Kalina Cycles, such as KCS34, for industrial waste heat and low-temperature low enthalpy geothermal resources in the Philippines which can be beneficial for power generation.

Nasruddin et al. [25] modeled the KCS 34 using cycle tempo 5.0 and validated the model for the installation in Husavic, Iceland, which resulted in electrical output of 1,959 MW compared to the actual electrical power output of the plant at 1,950 MW. Energy and exergy analysis was performed, and it was concluded that the system power output and performance will increase if the ammonia mass fraction is held constant while decreasing the exit pressure from the turbine or by increasing the fraction with the turbine exhaust pressure is constant. They also studied the Kalina cycle application in Indonesia where the highest electrical power output is reached with a 78% mass fraction and 7.4 bar exit pressure from the turbine.

Salman [26] worked on a computational investigation of the Kalina Cycle System 34 for power generation using low-medium heat source from industrial waste heat. Engineering Equation Solver (EES) was used for the thermodynamic analysis of the KCS 34. The maximum cycle efficiency was found to be about 29.5% at mass-fraction of 0.76 and inlet pressure of 100 bars. 26% efficiency was the lowest recorded at a mass fraction of 0.77 and inlet pressure of 60 bars. The pressure ranges were selected to avoid critical conditions

2. Methodology

The following steps are undertaken for the analysis, the first step is to conduct a detailed thermodynamic analysis using mass balances, traditional energy balanced and exergy balances for the selected system or plant. This is followed by a detailed economic analysis in which the investment and operating costs for each component of the plant is determined. The next step is to calculate the component's exergy unit cost and then the cost allocated to exergy loss for the whole system.

2.1. Thermodynamic Analysis

The thermodynamic analyses of the two cycles used the parameters from the Ormat Bulalo power plant Single OEC Unit gathered from Ormat Technologies, Inc. [27]. The turbine and pump efficiencies were based on the recommendation of Karlsdottir, Palsson, and Palsson [28].

P_{high}	19.48 Bar	Pressure (High-side)
P_{low}	1.4 Bar	Pressure (low-side)
T_{high}	180°C	Geothermal Water inflow Temperature
T_{low}	140°C	Geothermal Water outflow Temperature
m_w	361 kg/s	Geothermal water flowrate
η_t	85%	Turbine isentropic efficiency
η_p	50%	Pump isentropic efficiency

The main assumptions for the simulation are as follow:
1. The cycle works under the steady-state
2. The potential energies and kinetic energies are negligible
3. The temperature losses and pressure losses of the geothermal fluid, heat exchangers, and pipes are considered as negligible.

The thermodynamic principles of mass balance, energy balance, and exergy were used to determine the thermodynamics equations of the components of the plant. The energy and exergy relations of the subsystems are listed below and were used for the analysis of the binary cycle.

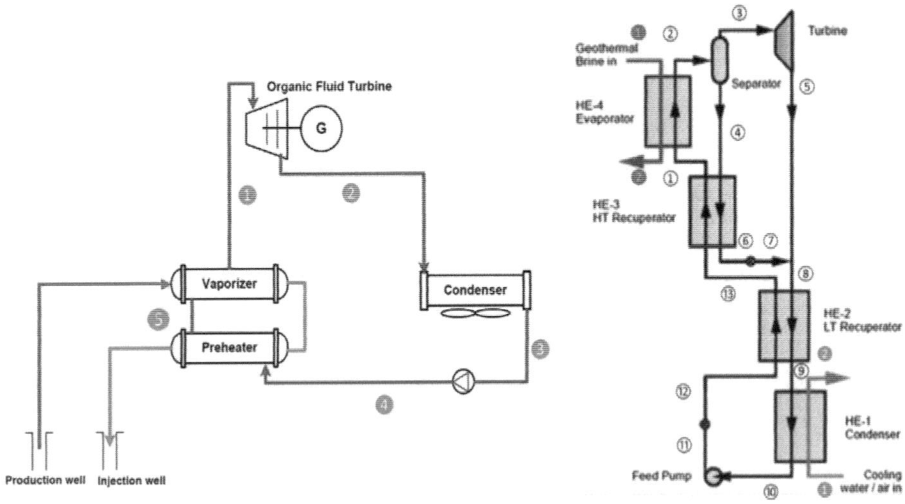

Figure 1. Ormat Binary Cycle and KCS 34 Configuration (source: [19])

The calculation of the properties, energy, and exergy streams was conducted using the EES program with the parameters from Ormat and n-pentane is the motive fluid used for the Upper Mahiao geothermal power plant situated in Leyte [15]. The equation below was used to calculate the exergy streams and considering the dead state conditions of 27°C and 101.325 kPa.

$$\dot{E}_k = \dot{m}[(h - h_o) - T_o(s - s_o)] \quad (1)$$

Table 1. Binary Cycle Energy Balance and Exergy Relation

Subsystem	Energy Relation	Exergy Relation
Evaporator	$\dot{m}_w(h_{1w} - h_{iw}) = \dot{m}_{wf}(h_1 - h_5)$	$\dot{E}_{in} - \dot{E}_{int} = \dot{E}_1 - \dot{E}_5 + \dot{E}_{evap,D}$
Pre-heater	$\dot{m}_w(h_{iw} - h_{2w}) = \dot{m}_{wf}(h_5 - h_4)$	$\dot{E}_{int} - \dot{E}_{out} = \dot{E}_5 - \dot{E}_4 + \dot{E}_{phtr,D}$
Turbine	$\dot{W}_t = \dot{m}_{wf}(h_1 - h_2); \eta_t = \frac{h_1-h_2}{h_1-h_{2s}}$	$\dot{E}_1 = \dot{W}_t + \dot{E}_2 + \dot{E}_{turb,D}$
Condenser	$\dot{m}_{wf}(h_2 - h_3) = \dot{m}_{cw}(h_{cw,out} - h_{cw,in})$	$\dot{E}_2 - \dot{E}_3 = \dot{E}_{cw,out} - \dot{E}_{cw,in} + \dot{E}_{cond,D}$
Pump	$\dot{w}_p = \dot{v}_3(P_4 - P_3); \eta_p = \frac{h_3-h_{4s}}{h_3-h_4}$	$\dot{W}_p = \dot{E}_4 + \dot{E}_3 + \dot{E}_{pump,D}$

The KCS-34 is then evaluated using the same parameters from the Binary cycle and applying the known thermodynamics principles. The cycle has additional components namely, the separator, Low (LT) and high (HT) temperatures recuperators, and the throttling valve. NH3H2O (ammonia-water) is used by the Kalina cycle instead of the n-pentane. For this paper, the concentration is assumed at 50%.

Table 2. Kalina Cycle Energy Balance and Exergy Relation

Subsystem	Energy Relation	Exergy Relation
Evaporator	$\dot{m}_w(h_{1w} - h_{iw}) = \dot{m}_{mix}(h_2 - h_1)$	$\dot{E}_{in} - \dot{E}_{int} = \dot{E}_2 - \dot{E}_1 + \dot{E}_{evap,D}$
Separator	$\dot{m}_{mix}h_2 = \dot{m}_{vap}h_3 + \dot{m}_{liq}h_4$	$\dot{E}_2 = \dot{E}_3 + \dot{E}_4 + \dot{E}_{sep,D}$
Turbine	$\dot{W}_t = \dot{m}_{vap}(h_3 - h_5); \eta_t = \frac{h_3-h_5}{h_3-h_{5s}}$	$\dot{E}_3 = \dot{W}_t + \dot{E}_5 + \dot{E}_{turb,D}$
Low-Temp Recuperator	$\dot{m}_{mix}(h_8 - h_9) = \dot{m}_{mix}(h_{12} - h_{11})$	$\dot{E}_8 - \dot{E}_9 = \dot{E}_{11} - \dot{E}_{12} + \dot{E}_{LTR,D}$
Condenser	$\dot{m}_{mix}(h_9 - h_{10}) = \dot{m}_{cw}c_{p,cw}(t_{w,out} - t_{w,in})$	$\dot{E}_9 - \dot{E}_{10} = \dot{E}_{cw,in} - \dot{E}_{cw,out} + \dot{E}_{Cond,D}$
Pump	$\dot{w}_p = \dot{v}_3(P_4 - P_3); \eta_p = \frac{h_3-h_{4s}}{h_3-h_4}$	$\dot{W}_p = \dot{E}_{11} - \dot{E}_{10} + \dot{E}_{pump,D}$
High-Temp Recuperator	$\dot{m}_{liq}(h_4 - h_6) = \dot{m}_{mix}(h_1 - h_{12})$	$\dot{E}_4 - \dot{E}_6 = \dot{E}_{11} - \dot{E}_{12} + \dot{E}_{HTR,D}$
Throttling Valve	$\dot{m}_{vap}h_6 = \dot{m}_{vap}h_7$	$\dot{E}_6 = \dot{E}_7 + \dot{E}_{TV,D}$

To better evaluate the performance of the process or the system, the exergetic efficiency ε, should be calculated. This is the relative exergy value between the product E_P and the fuel E_F [29]. The E_P is the combination of the exergy losses E_L and exergy destruction $E_D = T_0\dot{S}_{gen}$.

$$\varepsilon = \frac{E_P}{E_F} = 1 - \frac{E_D + E_L}{E_F} \tag{2}$$

2.2. Economic Analysis

The Purchase Equipment Cost (PEC) is only a part of the total costs needed for the estimation of the capital needed for the binary plant. The operation and maintenance costs, direct and indirect costs are all part of the investment costs that are needed for the completion of the plant cost estimation. The TCI is defined as the summation of the Fixed Capital Investment (FCI), which is the capital for the purchase of land, all the facilities needed, and installation of the different types of machinery and set of equipment for the system, and all other outlays. The outlays include the startup costs after the construction but before operation (SUC), working capital to sustain the operating expenses before payment (WC), the lump-sum cost for licensing, past research, and system or process development (LRD), and allowance or interest during the time period of construction (AFUDC) [30].

The following are the assumptions made for the estimation of the FCI based on the recommendation of Bejan, Tsatsaronis, and Moran [30]. The piping cost is assumed to be 35% of the PEC, the cost for the instrumentation is at 12%, the electrical cost is at 13%, the civil works are at 21% [31] while the land cost is at 10% of the PEC [30].

Another major cost of geothermal energy development is the drilling cost of the wells. In the Philippines, geothermal resources are mostly found in mountainous regions and the drilling of exploratory wells requires the opening of new roads and pad preparations which will include cutting of trees (requires permits from DENR). The cost also includes the drilling rig and third-party contractors specialized for the drilling activities. 3 exploratory wells will be assumed for the estimation at $3,200,000 per well [32].

The SUC includes expenditures to be used only during the startup time while the WC represents the needed funds for the operations of the plant. The AFUDC is calculated as 15% of the FCI minus the LRD. Dorj [33] provided a simple formula to determine the cost for the LRD which can be obtained 150 x turbine power.

With all of these, total net capital investment together with the total net capital investment, non-depreciable and depreciable capital investments can be estimated [30].

According to Tsatsaronis [29] the economic analysis that is performed in a thermoeconomic analysis is conducted on each component and determines the capital cost of the kth component Z_k in the cost balance.

2.3. *Thermoeconomic Evaluation*

Thermoeconomics or exergoeconomics analysis is based on this exergy costing of the component of the plant. This type of costing involves cost balances for the parts of the system and is determined using the exergy costs of both the inlet and outlet rates. The complete cost rate related to the total exergy flow rate is denoted by $\dot{C}_n^{TOT}$ [29].

$$\dot{C}_n^{TOT} = c_n^{TOT} \dot{E}_n^{TOT} \tag{3}$$

The cost balance for the plant at steady state is written as the total rate related to the product equal to the fuel and the total capital cost rate minus the loss. The cost rate $\dot{Z}$ is the combined value of the capital investment $\dot{Z}^{CI}$and the operating and maintenance cost $\dot{Z}^{OM}$. The exergy loss cost rate $\dot{C}_L$ embodies the system loss to the surrounding. So the cost balance of the plant becomes

$$\dot{C}_p = \dot{C}_F + \dot{Z} - \dot{C}_L \tag{4}$$

The rate of exergy destruction is an unseen cost that can only be exposed through the use of thermoeconomic analysis [30]. The cost of destruction can be proven by relating the cost formula of loss stream and exergy destruction and exergy loss:

$$\dot{C}_{D,k} = \dot{C}_{D\ F,k} = c_{F,k}\dot{E}_{D,k} \quad when\ \dot{E}_{P,k} = constant \tag{5}$$
$$\dot{C}_{D,k} = \dot{C}_{D\ P,k} = c_{P,k}\dot{E}_{D,k} \quad when\ \dot{E}_{F,k} = constant \tag{6}$$

It is between these two equations that the true cost of exergy destruction is calculated.

The exergoeconomic factor provides the relative significance of the non-exergy-related costs which consists of the investment costs (including O&M) and the exergy destruction and loss costs.

$$f_k = \frac{\dot{Z}_k}{\dot{Z}_k + c_{F,k}\left(\dot{E}_{D,k} + \dot{E}_{L,k}\right)} \tag{7}$$

A low-value factor indicates that improvement of the efficiency may be necessary to realize savings. Whereas a high-value represents a lower component investment but with exergetic efficiency affected.

3. Results and Discussion

The summary of the results of the simulation of the two cycles considered for the geothermal resource in the vicinity of Mabini, Batangas is shown below. The Kalina Cycle System 34 has slightly higher efficiency as compared to the Binary since it has higher power output. The total costs of the Kalina cycle are also higher due to the additional equipment purchase costs that affected the capital investment but with higher power output it has more income compared to binary and with that, the payback period is slightly faster.

Table 3. Binary and Kalina Cycle Performance

Performance Criteria	Binary Cycle	Kalina Cycle
Power Output, kW	9,502.00	10,257.70
1st Law Efficiency	15.13%	16.33%
2nd Law Efficiency	51.10%	63.55%
Mass flow, Kg/s	118.70	30.87
Total Surface Area, m^2	4,066.10	5,064.37
Total Capital Investment, US$	30,177,390.00	31,183,224.00
Income, US$	4,413,000.00	4,756,000.00
Payback Period, Years	7.0230	6.7380
Cost flow Rate, US$/s	0.0250	0.0368
Overall Destruction and Loss Ratio	20.8130%	30.0693%

Table 4. Binary Cycle Thermoeconomic Evaluation

Component	Exergetic Efficiency, ϵ_k, [%]	Investment Cost $\dot{Z}$, US$/s	Destruction Cost $\dot{C}_{D,k}$, US$/s	Cost Sum Sum, US$/s	Exergoeconomic Factor, f_k, [%]
Condenser	90.41	0.000914	0.02667	0.02758	60.84
Evaporator	84.65	0.001209	0.00007398	0.001283	99.91
Pre-heater	66.32	0.000717	0.000594	0.001311	89.02
Pump	41.63	0.008614	1.203	1.212	68.99
Turbine	22.58	0.013555	0.002259	0.01581	98.29

Table 5. Kalina Cycle Thermoeconomic Evaluation

Component	Exergetic Efficiency, ϵ_k, [%]	Investment Cost $\dot{Z}$, US$/s	Destruction Cost $\dot{C}_{D,k}$, US$/s	Cost Sum Sum, US$/s	Exergo-economic Factor, f_k, [%]
Condenser	97.88	0.0015227	0.00005206	0.00157476	38.26
evaporator	83.43	0.0015729	0.002364	0.0039369	95.6
High Temp Recuperator	9.082	0.0000823	0.004907	0.0049893	22.42
Low Temp Recuperator	27.01	0.00005583	0.003987	0.00404283	19.69
Pump	49.89	0.009177	0.0007151	0.0098921	99.47
Separator	99.4	0.000014147	0.00007529	8.94E-05	15.81
Turbine	90.83	0.014332	0.00337	0.017702	92.79
Throttle Valve	2.325	0.00000198	4.74E-06	6.72E-06	29.31

The sum of the cost rates indicates the component that requires design changes and for the Ormat cycle, the pump registered the highest sum with 1.212 US$/s while for the Kalina cycle the turbine has the highest sum amongst the components at 0.017702 US$/s. Both components must be considered for re-design to enhance the performance of the respective cycle they are involved with. Lastly, the exegoeconomic factor of the Ormat cycle evaporator and the KCS 34 turbine recorded the highest value at 99.91% and 99.47%, respectively. With both components also included in the redesign consideration, the exergoeconomic factor of these sets of equipment will be affected once the cost and energy components are re-evaluated in order to find the balance between the energy and cost performance.

4. Conclusion

The Kalina cycle system 34 1st efficiency is higher compared to the Ormat binary cycle but it is only about 1.2% higher. This is due to the higher power output that the Kalina Cycle can produce. This is also true for the 2nd law efficiency of the KCS 34 at 63.55% while the Ormat cycle has 51.10%.With the higher power generated of the Kalina cycle, The total capital investment of the plant at 31.18 M US$ is also higher than the Binary cycle at 30.18 M US$. It also has a higher income at 4.76 M US$ with the power output generated compared to the 4.41 M US$ of the binary cycle. The payback period is slightly faster by almost a quarter of a year than that of the Ormat cycle.

Based on the results of the thermoeconomic analysis, the Kalina cycle turbine and the Ormat Cycle pump required re-design as the two equipment registered high values of the sum of the cost rates. Improvements in the performance of the two cycles can be achieved by re-designing the said equipment and using simulation and optimization. The main difference of the Kalina and the Ormat cycle is the working fluid. Kalina Cycles are designed for ammonia-water mixture while the Ormat cycle uses n-pentane. The use of the ammonia-water mixture provides flexibility in terms of adjustable concentration.

References

[1] Richter A 2017 Indonesia to surpass the Philippines as world's 2nd largest geothermal producer
[2] Pastor M S, Fronda A D, Lazaro V S and Velasquez N B 2010 Resource assessment of Philippine geothermal areas *Proc. World Geotherm. Congr.* no April pp 25–29
[3] Department of Energy (DOE) 2019 2018 Power Statistics no. March p 2019
[4] Department of Energy 2015 Philippine Energy Plan 2016-2030 Update p 101
[5] Peñarroyo F S 2015 Renewable energy act of 2008: Hits and misses for the Philippine geothermal industry no April pp 19–25
[6] Fronda A D, Marasigan M C and Lazaro V S 2015 Geothermal development in the Philippines: The country update *Proc. World Geotherm. Congr.* no April pp 19–25
[7] Balangue-Tarriela M I R D and Mendoza J P 2015 Updates on the geothermal energy development in the Philippines *World Geotherm. Congr. 2015* no April p 4
[8] Ulgado A F and Gular M T M 2005 Status of direct use of geothermal energy in the Philippines no April pp 24–29
[9] Halcon R M et al 2015 Detailed resource assessment of selected low-enthalpy geothermal areas in the Philippines *World Geotherm. Congr. 2015* no April p 11
[10] Benito F A, Ogena M S and Stimac J A 2005 Geothermal energy development in the Philippines: Country update *Proceedings World Geothermal Congress 2005*
[11] Tchanche B F, Lambrinos G, Frangoudakis A, and Papadakis G 2011 Low-grade heat conversion into power using organic Rankine cycles: A review of various applications *Renew. Sustain. Energy Rev.* vol 15 no 8 pp 3963–3979
[12] Vasquez N C, Javellana S P, and Ferrer H P 1998 Present and future geothermal development in the Philippines *Geotermia* vol 14 no 3 pp 153–159
[13] Mlcak H A 1996 An introduction to the Kalina cycle *Proc. Int. Jt. Power Gener. Conf.* vol 30 pp 1–11
[14] Gupta S 2007 Harsh; Roy *Geothermal Energy An alternative Resource for the 21st Century*
[15] Forte N T 1996 The 125 MW upper Mahiao geothermal power plant *Conference: Annual meeting of the Geothermal Resources Council, Portland, OR (United States), 29 Sep – 2 Oct 1996; Other Information: PBD: 1996; Related Information: Is Part Of Geothermal*

development in the Pacific rim. Transactions, Volume 20; PB: 886 p., 1996
[16] Kalina A I 1984 Combined-cycle system with novel bottoming cycle *J. Eng. Gas Turbines Power* vol 106 no 4 pp 737–742
[17] Bliem C J 1988 Kalina cycle and similar cycles for geothermal power production
[18] Eliasson L and Valdimarsson P P 2003 Kalina technology: Where and why to utilize *Sochi*
[19] Zhang X, He M, and Zhang Y 2012 A review of research on the Kalina cycle *Renew. Sustain. Energy Rev.* vol 16 no 7 pp 5309–5318
[20] Arslan O 2011 Power generation from medium temperature geothermal resources: ANN-based optimization of Kalina cycle system-34 *Energy* vol 36 no 5 pp 2528–2534
[21] Mlcak H A 2001 Design and start-up of the 2 MW Kalina cycle ® Orkuveita Húsavíkur geothermal power plant in Iceland *Eur. Geotherm. Energy Counc.*
[22] Mlcak H A and Mirola M 2002 Notes from the North: A report on the debut year of the 2 MW Kalina Cycle ® geothermal power plant in Húsavík, Iceland pp 1–8
[23] Mergner H and Weimer T 2015 Performance of ammonia e water based cycles for power generation from low enthalpy heat sources *Energy* vol 88 pp 93–100
[24] Tiangco C C 2001 Design of a Kalina cycle system for geothermal power generation (Australia: The University of New South Wales)
[25] Nasruddin, Usvika R, Rifaldi M, and Noor A 2009 Energy and exergy analysis of Kalina cycle system (KCS) 34 with mass fraction ammonia-water mixture variation *J. Mech. Sci. Technol.* vol 23 no 7 pp 1871–1876
[26] Salman A M 2018 Low power organic Kalina cycle driven by low-grade waste heat vol 5 no 9 pp 154–159
[27] Bar-Sadeh A 2020 S-H&M Balance.pdf
[28] Karlsdottir M R, Palsson H, and Palsson O P 2010 Comparison of methods for utilization of geothermal brine for power production *Proc. World Geotherm. Congr.* no April pp 25–29
[29] Tsatsaronis G 1993 Thermoeconomic analysis and optimization of energy systems *Prog. Energy Combust. Sci.* vol 19 no 3 pp 227–257
[30] Bejan A, Tsatsaronis G, and Moran M 1996 *Thermal design and optimization* vol 21 no 5
[31] Chao L and District T 2014 Thermodynamic and thermoeconomic optimization of a geothermal power plant in Sichuan, China no. 18
[32] Dolor F M 2006 Phases of geothermal development in the Philippines *Work. Decis. Makers Geotherm. Proj. Cent. Am.* no December
[33] Dorj P 2005 Thermoeconomic analysis of a new geothermal utilization CHP plant

Design of small scale, axial flux, permanent magnet wind generator in Vietnam

Le Hung Tien

Faculty of Engineering, Van Lang University

Tien.lh@vlu.edu.vn

Abstract. This paper presents the design of light weight, small scale, disk type, axial flux, permanent magnet (AFPM) generator for wind applications. This article also shows the partial works done under state projects in renewable energy in Vietnam to find out the most suitable, reliable, self – made wind generator types for rural and remote areas. The collaboration project of Hochiminh City University (HCMUT) – Hanoi University of Science and Technology (HUST) – Berlin University of Technology (TU Berlin) resulted in some small scale, axial flux, slow speed, permanent magnet (PM) generators which best suited for Vietnam countryside areas. The researches in rare earth PM from HUST, the electrical machine design knowhow of TU Berlin and the renewable energy research experiences of HCMUT, especially in slow speed wind turbines have shown many promising applications. The small scale, permanent magnet wind generator for remote and rural areas as a cheap source of power supply is necessary in Vietnam. Though there are many large wind farms built in many provinces recently but the needs for telecommunication, family lighting in the heights, mountain, unreachable electric grid communes are still necessary. Further research should be done to match the generator for hydro power stations, diesel coupled internal combustion engine. These prospects are very promising.

1. Overview of Wind Power application in Vietnam

According to Wind energy potential in Vietnam [5], in Vietnam, several wind measurement studies are conducted. Vietnam is considered to have the best wind resources in Southeast Asia, especially in the nearshore/offshore and onshore coastal regions in the south of Vietnam. In these areas yearly average wind speeds of 9 to 10 meters per second are measured. Generally, wind speeds are declining further inland. Noteworthy is that Vietnam is vulnerable to extreme weather events like storms and typhoons, especially in July, August, September and October. The possibility of a typhoon is relatively high in these months, which can have severe consequences for the exploitation of wind farms.
Vietnam has paid attention to renewable energies a long time ago with many projects, especially in solar and wind energy [1].

1.1. Research gaps

The demands for power supply in remote, rural, unreachable power grid areas in Vietnam are high, however there are no wind generators matching the low wind speed profiles of Vietnam provinces. Furthermore, the imported, expensive, high wind starting speed wind generators mostly from China are not suitable for Vietnam applications. The research for the low cost, self - made, reliable wind generator type utilizing in fierce exposed weather conditions needs to be done.

1.2. *Small scale wind power in Vietnam*

In the year 1992-1995, Research Center for Thermal Equipment and Renewable Energy (RECTERE), Ho Chi Minh city University of Technology (BKU, HCMUT) cooperated with some US research agencies to facilitate the pilot measurement for wind power in Can Gio district, Hochiminh city. Thousands of ferrite magnet wind generators with hundred watts in power have been manufactured and installed in Thieng Lieng Islets, Ho Chi Minh City and islands in Nha Trang city, Khanh Hoa province [1].

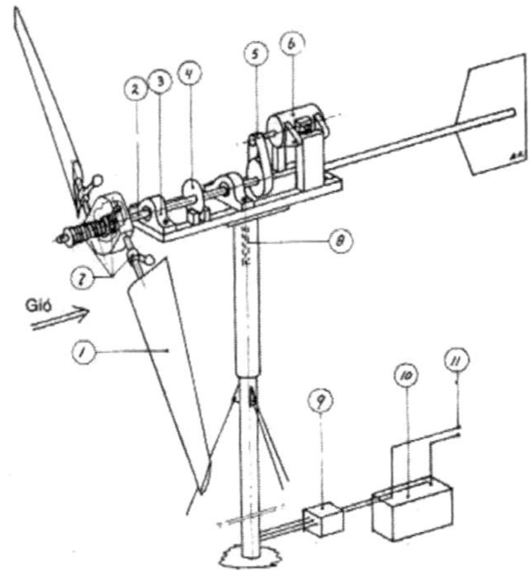

Figure 1. A typical small scale wind generator [1]

1.3. *Large wind farm*

Vietnam's first high-capacity wind farm, with 30 MW, was inaugurated by the Vietnam Renewable Energy Joint Stock Company (REVN), in the central province of Binh Thuan on April 18.[5] The next two projects—the Phú Quý island hybrid grid with 6 MW and the near-shore Bac Lieu phase 1 with 16 MW—both completed in 2013. No new capacity was added in 2014 or 2015.

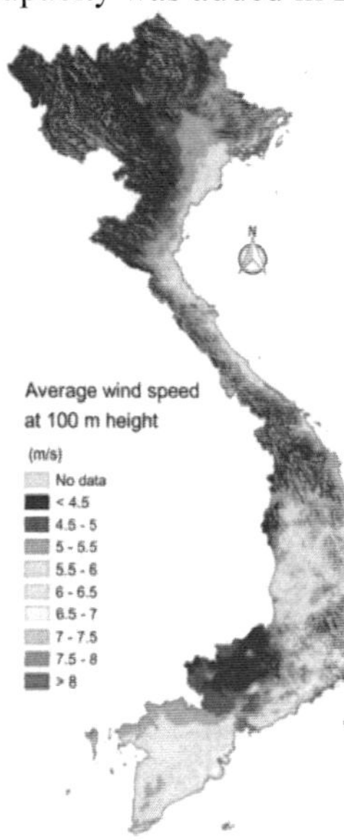

Figure 2. Wind speed at 100 m height in Vietnam [4]

The wind power industry is taking off in 2019 in Vietnam, although not as spectacularly as the solar power industry. By the end of May 31, 2019, 7 wind power plants were in operation, for a national installed capacity of 331 MW.

No.	Name of province	Capacity (MW)	
		2020	2030
1	Bình Thuận	700	1.570
2	Ninh Thuận	220	1.429
3	Sóc Trăng	200	1.470
4	Quảng Trị	110	447
5	Trà Vinh	270	1.608
6	Bến Tre	150	1.520
7	Bạc Liêu	401	2.507
8	Cà Mau	350	3.607
9	Thái Bình	40	70
10	Đăk Lăk	138	1.382
11	Bà Rịa Vũng Tàu	34	107
	Total	2.613	15.717

Figure 3. Provincial Wind Power Development Plans in Vietnam [6]

Up to now, wind power has been deployed enormously in many big wind farms as listed in [3].

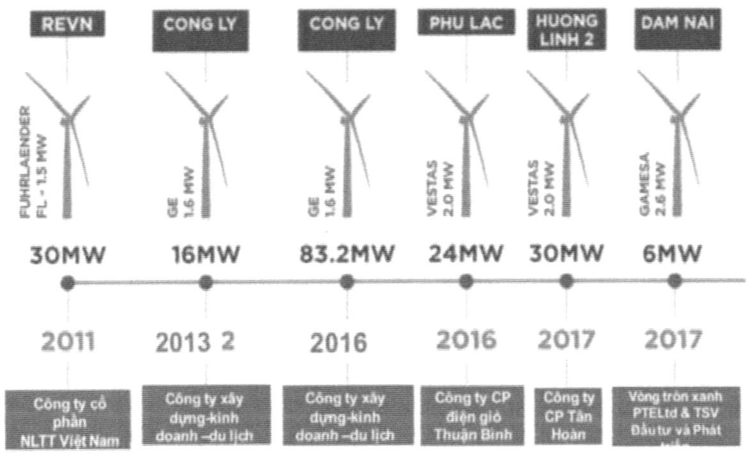

Figure 4. Operational wind farm in Vietnam [6]

2. Overview of Wind Power application

2.1. Advantages and disadvantages of permanent magnet generator (PMG)

The most important component of wind – electric conversion systems is a generator, the development of permanent magnet materials recently pose the potentials of compact, lightweight PMG applications. The advantages and disadvantages of PM excited generators [2,4,7] are listed in table 1.

Table 1. Advantages and disadvantages of PM excited generators

1. **Advantages**	2. **Disadvantages**
High efficiency	Few types of generators from the manufacturers.
No brush losses	Uncontrolled magnetic exciting current.
No external exciter	Controlled via output power
No magnetic exciter losses	Heavily damaged if having armature winding defects.

2.2. Torus PMG configuration

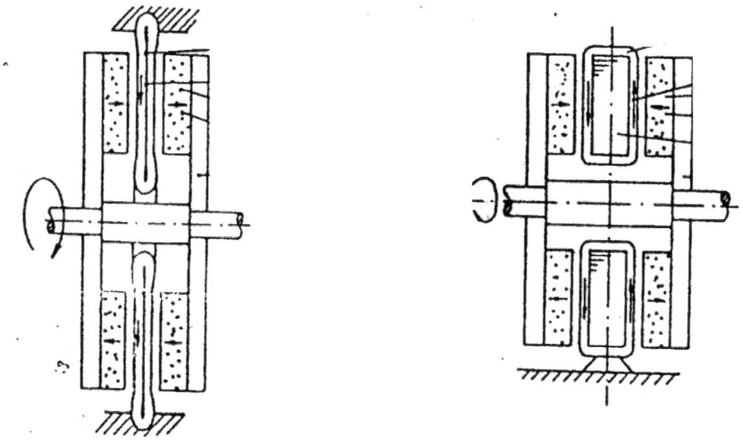

Figure 5. Coreless and Slotless Axial Flux PMG [1]

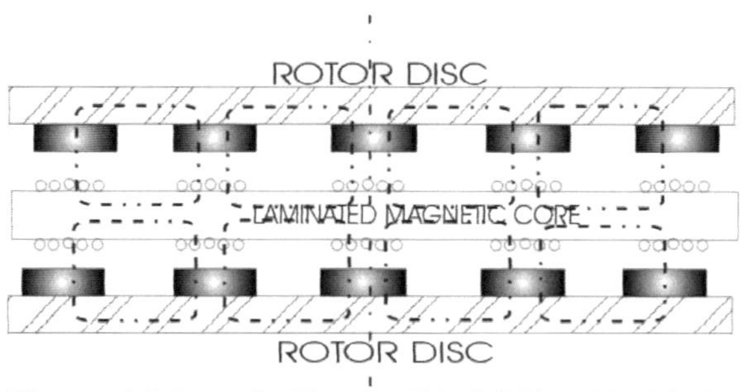

Figure 6. Magnetic Circuit of Axial Flux PMG [1]

2.3. Slottless and slotted toroidal generator

The slottless prototype is made with rectangular rare earth based magnet NdFeB (NeoDelta Magnet) from IBS Magnet, Germany with BxLxh = 20x30x6 (mm), remanence $B_r = 1.2$ T, coercivity $H_c = 930$ kA/m and $(BH)_{max} = 280$ kJ/m^3. The slotted prototype is made with circular anisotropic Barium ferrite magnet from IBS Magnet, Germany with diameter d = 15 mm, remanence $B_r = 0.4$T, coercivity $H_c =$ 175 kA/m and $(BH)_{max} = 25.5$ kJ/m^3. The generator's parameters are shown in Figure 7.

Figure 7. N35UH permanent magnet properties

Magnetic Properties	Characteristic	Units	min.	nominal	max.
	Br, Residual Induction	Gauss	11,700	12,100	12,500
		mT	1170	1210	1250
	H_{cB}, Coercivity	Oersteds	10,800	11,400	12,000
		kA/m	860	907	955
	H_{cJ}, Intrinsic Coercivity	Oersteds	25,000		
		kA/m	1,990		
	BHmax, Maximum Energy Product	MGOe	33	36	38
		kJ/m^3	263	283	302

Table 2. Slotless and slotted toroidal generators

Machine parameters	Slotless	Slotted
Stator core OD in mm	146	146
Stator core ID in mm	86	86
Core axial thickness in mm	20	20
Winding lavers	6	6
Winding coils	18	18
Phases	3	3
Turns/coil	11	11
Wire diameter in mm	0.7	0.7
Magnet thickness in mm	6	6
Air gap flux density in T	0.5	0.33
Air gap flux in mm	8	3
Length of machine in mm	50	50
Output power at rated speed 500 rpm in Watts	300	130

2.4. *Torus PMG calculation*

2.4.1. *Nomenclature:*

- T_e – the electromagnetic torque in [N.m].
- ω_e - the electrical angular speed in [rad/s].
- R_0 – outer radius.
- R_i - inner radius.
- 2p - number of pole pair.
- K_l - flux leakage factor.
- $\emptyset$ - magnetic flux.

2.4.2. *Electromagnetic torque*

For the axial flux, slottless, permanent magnet generator, the basic parameter can be calculated respectively in the following [2,4].

The output power P [W] can be calculated by a simple formula:

$$P = T_e . \omega_e \quad (1)$$

Or

$$P = T_{\Sigma} . \omega_m \quad (2)$$

One of most important factors, the rated angular speed ω_m of the Torus is determined considering the motive source coupled to the generator. In the case of wind energy application, the rated wind rotor should be chosen how to use the wind energy in the low wind speed.
The electromagnetic torque is determined by following equations:

Due to the double side magnet polarization arrangement, the total electromagnetic torque can be calculated as:

$$T_{\Sigma} = 2T_e$$

$$T_e = K_p \int_{Ri}^{R0} F_i r dr \tag{3}$$

With:
- K_p – the percentage of armature winding under pole flux
-$F_i = B_{av}.l_i.I_i$ – Lorentz tangential force on current carrying conductor element
- B_{av} - Average flux density in the air gap.
- l_i - Length of a current carrying conductor element in radio direction: $l_i = dr$.
- I_i – Current generated in the conductor element. In polar coordinates: $I_i = J_{av} r d\Theta$.

Then $T_e = K_p \int_0^{2\pi} \int_{Ri}^{R0} B_{av} J_{av} r^2 d\Theta dr$

Thus:

$$T = 2K_p \int_0^{2\pi} \int_{Ri}^{R0} B_{av} J_{av} r^2 d\Theta dr$$

But the current in other parts of the conductor is equal to the current in the conductor at inner radius Ri, therefore, Javr can be replaced by JiRi.

$$T = 2K_p \int_0^{2\pi} \int_{Ri}^{R0} B_{av} J_i R_i r d\Theta dr$$

$$T_{\Sigma} = 2\pi pmn B_{av} J_i R_0^3 (1 - a^2) \tag{4}$$

To maximize the output power, the outer R_0 and the inner R_i radius should satisfied the following equation:

$$\frac{R_0}{R_i} = \sqrt{3} \tag{5}$$

The number of pole pair 2p is determined as:

$$2p = \frac{60f}{n_1} \tag{6}$$

The electric current can be roughly calculated from the Eq.6 as:

$$I = \frac{P}{U}(A) \tag{7}$$

The current density (the electric loading) Ji can be chosen as follows:
- For the long term working mode: $J_i = (2 - 4)A/mm^2$
- For the short repeated term working mode: $J_i = (5 - 12)A/mm^2$
- For the short circuit working mode: $J_i = (2 - 4)A/mm^2$

The permanent magnets can be rare earth based such as NdFeB, Sm_2Co_5, Sm_2Co_{17} or ferrite ones. The segment shape of permanent magnet is assumed (see Fig.1) to generate the constant flux along radii of armature winding core.

The flux leakage factor K_l defined based on magnetic flux $\emptyset$ must be optimized. For the surface mounted, segment shape magnet, the flux leakage is determined in Eq.8:

$$K_l = 1 + \frac{2\mu_0 p\alpha H_{magnet} l_{magnet}^2 \left[ln\, ln\left(\frac{R_0}{R_i}\right) + 2\alpha\right]}{\pi(1-K_a)} x \frac{2\left(1+\frac{l_{airgap}}{l_{magnet}}\right)}{B_r\beta(R_0^2 - R_i^2)} \quad (8)$$

$$\frac{\emptyset_{leakage}}{\emptyset_{total}} = \frac{2\mu_0 p H_{magnet} l_{magnet}^2 \left[ln\, ln\left(\frac{R_0}{R_i}\right) + \frac{4\pi K_a}{p}\right] x2\left(1+\frac{l_{airgap}}{l_{magnet}}\right)}{\pi(1-K_a)\frac{2\pi K_a B_r}{p}(R_0^2 - R_i^2) + 2\mu_0 p H_{magnet} l_{magnet}^2 \left[ln\, ln\left(\frac{R_0}{R_i}\right) + \frac{4\pi K_a}{p}\right] x2\left(1+\frac{l_{airgap}}{l_{magnet}}\right)} \quad (9)$$

The optimum K_a is then obtained when the ratio in Eq.9 achieves minimum value. This happen when $K_a = 0.8$

For the optimum design the following relation between the magnet thickness and the armature winding thickness is of help:

$$P_{copper} = \frac{K\psi}{l_{magnet}^2(l_m - 2t_{core} - 2\Delta - 2l_{nc})} \quad (10)$$

The constant in Eq.9 are found as:

$$K = \frac{2\rho|(R_0 - R_i) + t_{core}|}{d} \text{ and } \psi = \frac{T_\Sigma^2(l_m - 2t_{core})^2}{\left[2\pi pmn B_r R_0^3 a(1-a^2)\right]^2}$$

With the given torque, a fixed machine length and core length, the copper loss is minimized if the term $K(x) = l_{magnet}^2(l_{machine} - 2t_{core} - 2\Delta - 2l_{magnet})$ is maximized. This happens if $dK(x)/dx = 0$ when $x = l_{nc} = 2t_{winding}$. Then the air gap flux density $B_{airgap} = 2B_r/3$, with given current density J, the torque is maximized when B_{av} maximized, that means $l_{magnet} = l_{airgap}$. The airgap flux density is then $B_r/2$.

The number of winding turn per phase is:

$$n = \frac{T_\Sigma}{2\pi pmn B_{av} J_i R_0^3(1-a^2)} \quad (11)$$

The number of coils per phase is:

$$n_2 = \frac{n}{n_3} \quad (12)$$

The number of winding layer per coil n_l and coil width b_c is given by:

$$b_c = \frac{nd}{n_1 n_2} \quad (13$$

The coil width and number of coil can be checked by:

$$3mb_c + K_q \leq 2\pi R_i \tag{14}$$

Where K_q is reserved factor for overall space between coils. K_q can be chosen 10% perimeter at inner radius.

$$b_c \leq \frac{2\pi R_i - K_q}{3m} \tag{15}$$

3. Conclusion

In this article, the prototype of slotted and slottless axial flux permanent magnet generators have been investigated due to its simple configurations and easy manufacturing technology. All calculations for the new proposed AFPM generator are performed, these calculations are necessary for Vietnam to design wind generators at very low cost. These generators promise the potential applications in remote and rural areas in Vietnam. The further studies will be carried out for small hydro power stations and diesel engine coupled generators.

4. Acknowledgement

This project was carried out under the financial support of Nuffic scholarship, The Netherlands and technical support from Institute for electrical machines, Berlin University of Technology, Germany. The prototypes of wind generators were manufactured for wind applications in Vietnam. The author would like to express special thanks go to International Training Institute for Materials Science, Hanoi University of Science and Technology, Vietnam.

References

[1] Tien L H 2003 Applied research of magnetic materials in wind generator manufacturing in Vietnam Ph.D Thesis (Hanoi)
[2] Nguyen C H 1996 Electrical engineering 1 (Hochiminh City National University)
[3] McCaig 1995 *Permanent Magnet in Theory and Practice*
[4] List of wind projects in Vietnam (http://inchamhanoi.vn/wp-content/uploads/2019/07/KEY-List-of-wind-projects-in-Vietnam.pdf)
[5] Campbell P, Rosenberg D J and Staton D P 1981 The computer design and optimization of axial-field permanent magnet motors *IEEE Transactions on Power Apparatus and Systems* **PAS**(4) pp 0-100
[6] Renewable Energy in Vietnam https://en.wikipedia.org/wiki/Renewable _energy_in_Vietnam More references
[7] Khanh H 1997 Wind energy potential in Vietnam (Electrical Machine Design, Science and Technology Publishing) (https://www.rvo.nl/sites/default/files/2019/02/ Wind-Energy-Potential-Vietnam.pdf)

IOP Conf. Series: Materials Science and Engineering 1109 (2021) 012042 doi:10.1088/1757-899X/1109/1/012042

Energy and exergy evaluation of an onshore solar dryer for seaweeds

A Culaba[1*], A H Atienza[1], A Ubando[1], A Mayol[1] and J Cuello[2]

[1] Mechanical Engineering Department, De La Salle University Manila, Philippines
[2] Department of Agricultural and Biosystems Engineering, University of Arizona

*alvin.culaba@dlsu.edu.ph

Abstract. The possible sources of the energy losses in the drying systems can occur in the drying chamber and solar collector. The energy from these parts were not properly utilized so it is evident to further search for improvements. Energy and exergy analyses were being used to identify the possible source of inefficiencies quantitatively and qualitatively in the drying system and aid in recommendation for its improvement. This study aimed to evaluate the performance of an onshore drying system for seaweeds by conducting energy and exergy analyses. It has shown that the dryer efficiency was 33.58% and its solar collector efficiency was 68%. The exergy efficiency was obtained as 41.38% relates to an exergy improvement potential of 256.30 W. It is recommended to perform partial to full exhaust air recirculation and provide thermal storage may boost the exergy efficiency hence minimizing energy losses in the drying system.

1. Introduction

A solar dryer is composed of two major parts namely the solar collector and the drying chamber. The solar collector gathers energy from the sun in the form of solar radiation. This energy is used in heating the moist air that enters the pipes of the solar collector. The air comes from a fan or air blower to force in air into the solar collector. Heated air exits the solar collector and eventually enters the drying chamber. However, energy losses occur due to convection and evaporation inside the drying chamber [1]. In the case of the solar collector, the incident solar radiation strikes the surface of the glass cover of the collector. Around 10% of the radiation is reflected, 5% absorbed the glass cover and 85% transmitted to the piping. Approximately 5% of the reflected radiation is still present in the pipings and 80% of the heat is collected by the absorber plate. Heat loss on the absorber plate is estimated to be 35%. The collector only utilizes 45% of the incident solar radiation [2]. Due to the losses from the drying chamber and solar collector it is necessary to improve the drying system. One of the important concepts that must be focused to lessen energy losses is analyzing exergy. Exergy is a property that measures the potential for use of energy. It is considered as the maximum theoretical work of the system in equilibrium with the environment and is destroyed in the process due certain irreversibilities [3]. Conducting an exergy analysis on the drying system can help pinpoint sources of inefficiencies and can provide possible improvement in the system [4]. Few literatures discuss exergy and energy analyses in seaweed drying. The study of [5] focuses mainly on energy and exergy analysis of a forced convection indirect type drying system for red seaweeds. The study has shown that the average solar, dryer and exergy efficiencies are 35%, 27% and 30%, respectively. A 247 W average exergy improvement potential was obtained. For this study, a forced convection indirect type dryer will also be employed, however the air distribution system has different configuration. This study also mainly focuses on the exergy and energy analyses of the onshore solar dryer for Eucheuma cottonii seaweeds. From the data obtained from the analyses, this also highlights the possible improvements that can be implemented to increase productivity and efficiency of the onshore solar dryer.

2. Materials and Methods

2.1. Onshore solar dryer

The onshore solar dryer as seen in Figure 1 consists of a solar collector and drying chamber. The drying chamber is 1.20m x 0.92m x 0.62m length, width, and height, respectively. Its body frame is composed of 1.5mm thick 1" x 1" square tube. The trays made up of stainless-steel expanded sheets with 1" x 1" x 1/8" angle bars serve as support. The materials for the walls of the drying chamber are galvanized steel gauge 20 for the inner and outer walls, and polyurethane insulation aluminized in between. The trays made up of stainless-steel expanded sheets with 1" x 1" x 1/8" angle bars serve as support. The air distribution system in the dryer is modified. It consists of a ¾" copper pipe works as a main pipe connected to ½" copper distribution pipes. The main pipe is connected to the solar collector. The solar collector has serpentine piping configuration made up of ¾" copper tube with aperture area of the solar is 0.43m^2. The absorber plate is made up of matte black aluminum and the insulation is composed of polyurethane foam aluminized. The cover of the solar collector is acrylic glass.

Figure 1. The onshore solar dryer

2.1.1. Operation of the Onshore Solar Dryer

The dryer can be classified as an indirect solar forced convection type where it consists of a solar collector that heats up incoming air going to the drying chamber. The air is brought in by the forced circulation of the blower. The air then circulates into the main pipe branching to the distribution pipes that delivers the flow to the tray locations until it goes out of the exhaust pipes.

2.2. Experimental Procedure

The seaweed drying experiment set-up involves the installation of instrumentations, fine tuning and testing proper. Load cell-based weighing scales, Arduino Uno Software, temperature sensors, pyranometer and digital anemometer were used for data gathering. For the first phase of the seaweed drying experiment, the initial moisture content of the seaweeds is to be determined. Placing 100 grams of seaweeds in aluminum foils and brought into the microwave oven with temperature setting at 105°C based. The aluminum foil is pulled every 10 minutes to observe and record the mass reduction of the

IOP Conf. Series: Materials Science and Engineering 1109 (2021) 012042 doi:10.1088/1757-899X/1109/1/012042

seaweeds. Three same successive mass readings can safely assume that the seaweeds are totally dried. The initial moisture content of the seaweeds is obtained using the formula below:

$$\text{Initial Moisture Content}=\frac{m_i-m_f}{m_i}\boldsymbol{x100\%} \qquad (1)$$

The experiment started from 8am-5pm but to provide accurate reading the 800 W blower was turned on 30 minutes prior to data gathering for forced air circulation into the drying chamber. One kilogram of *Eucheuma cottonii* seaweeds was placed per tray location. The moisture content reduction of seaweeds per tray location was monitored and the target final moisture content is 40% [6]. The temperatures for the drying system and solar irradiance were monitored.

2.3. Energy Analysis

The collector efficiency η_{SC} is described as the ratio of the useful heat transfer Q_u of the collector and the incident solar energy over a time. The incident solar energy is the product of solar irradiance S and solar collector area A_{SC}

$$\eta_{SC} = \frac{Q_u}{SA_{SC}} \qquad (1)$$

Since the useful heat transfer $Q_u = \dot{m}C_p(T_{outlet} - T_{inlet})$,the solar collector efficiency formula in (1) can be modified with equation in (2) resulting to

$$\eta_{SC} = \frac{\dot{m}C_p(T_{outlet}-T_{inlet})}{SA_{SC}} \qquad (2)$$

$\dot{m}$ is the air mass flow rate, C_p is the specific heat at constant pressure of air while T_{outlet} and T_{inlet} are the outlet and inlet solar collector temperatures, respectively. This study will use (2) based on the study of [5]

Below are the set of equations that will serve as performance parameters in the dryer energy analysis used by [5].

The dryer efficiency is described by the formula
Dryer efficiency

$$\eta_{dryer} = \frac{ML}{(A_cS+P_{blower})t} \qquad (3)$$

Mass of water evaporated, M in kg:

$$M = m\frac{M_o-M_f}{100-M_f} \qquad (4)$$

m is the mass of the seaweeds, Mo = initial moisture content of the seaweeds, Mf=final moisture content of the seaweeds, Lis the latent heat of vaporization of water at an exit air temperature in J/kg, P_{blower} is the power input of the blower in W and t = time in seconds

2.4 Exergy Analysis

The exergy analysis under steady state of the dryer from the study of [5] is composed of the exergy inflow, outflow, and losses from the drying chamber and described by the following formulas

For exergy inflow in the dryer,

$$Ex_{dryer,i} = \dot{m}C_p\left[T_{i,dryer} - T_{amb} - T_{amb}\ ln\ ln\ \frac{T_{i,dryer}}{T_{amb}}\right] \qquad (5)$$

For exergy outflow in the dryer,

$$Ex_{dryer,o} = \dot{m}C_p\left[T_{o,dryer} - T_{amb} - T_{amb}\ ln\ ln\ \frac{T_{o,dryer}}{T_{amb}}\right] \qquad (6)$$

$\dot{m}$ is the mass flow rate of air, C_p is the specific heat at constant pressure, $T_{i,dryer}$ and $T_{o,dryer}$ are the inlet temperature and outlet temperatures of the dryer, respectively, finally T_{amb} is the ambient temperature

For the exergy losses in the dryer, this would be the difference of exergy inflow and outflow

$$Ex_{losses} = Ex_{dryer,i} - Ex_{dryer,o} \qquad (7)$$

The exergy efficiency of the dryer is expressed as the ratio of the exergy outflow to inflow as expressed by [5, 7] below

$$\eta_{Ex} = \frac{Ex_{dryer,o}}{Ex_{dryer,i}} = 1 - \frac{Ex_{losses}}{Ex_{dryer,i}} \quad (8)$$

It is highly necessary to include in the exergy analysis of a system the exergy improvement potential. The exergy improvement potential, EIP, of a system pertains to the maximum improvement on the exergy of the system and can be achieved if the exergy losses were reduced [5, 8]. The formula for exergy improvement potential is provided by the equation

$$EIP = (1 - \eta_{Ex})Ex_{losses} \quad (9)$$

3. Results and Discussion

3.1 Seaweeds

The initial moisture content was determined to be 89.8% in 5 hours and 19 minutes to reach 40% moisture content. The velocity of air into the drying chamber is about 12.5 m/s. It is found out that top tray has reduced the maximum moisture content of the seaweeds in a drying rate of 1.384 g/min while the lowest rate of 1.126 g/min was observed in the bottom tray as seen in Figure 2. The increase in the moisture reduction in the top tray is caused by the hot air flow circulation which is maintained at the top due to the chimney effect [9].

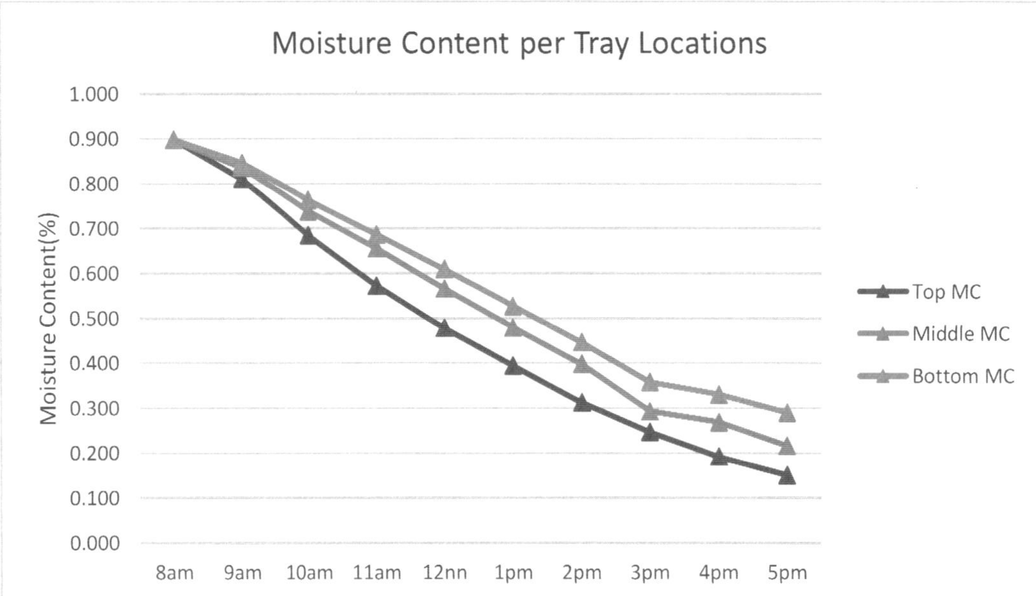

Figure 2. Moisture Content per Tray locations

3.2. Solar Collector

The trend in inlet and outlet temperatures have been monitored and shown in Figure 3. It can be observed that the peak solar inlet and outlet temperatures were 41.54°C and 51.30°C, respectively. These temperatures both occurred at 12nn with solar irradiance of 465W/m^2. The outer temperature is observed to be higher compared to the inlet due to heat transferred from the serpentine tubes into the air [10].This result agrees with study of [11].

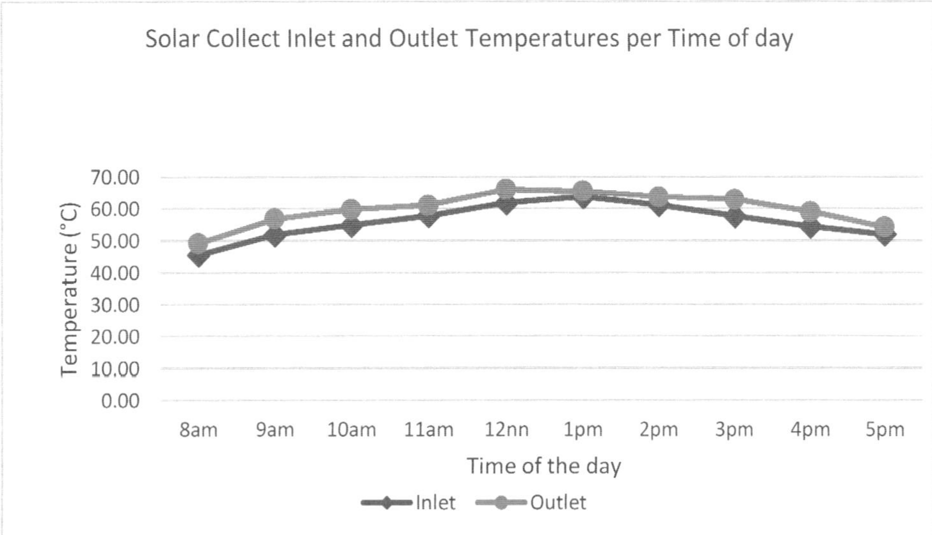

Figure 3. Solar Collector's Inlet and Outlet temperatures per time of day

In reference to the time of day, the highest average efficiency went as high as 68% at a solar irradiance of 427.72 W/m^2 as seen in Figure 4. The average solar collector efficiency is 43.40% at an average solar irradiance of 351.20 W/m^2. The value of solar collector efficiency is higher compared to the results of [5] and [12] due to the turbulent flow provided by the serpentine configuration.

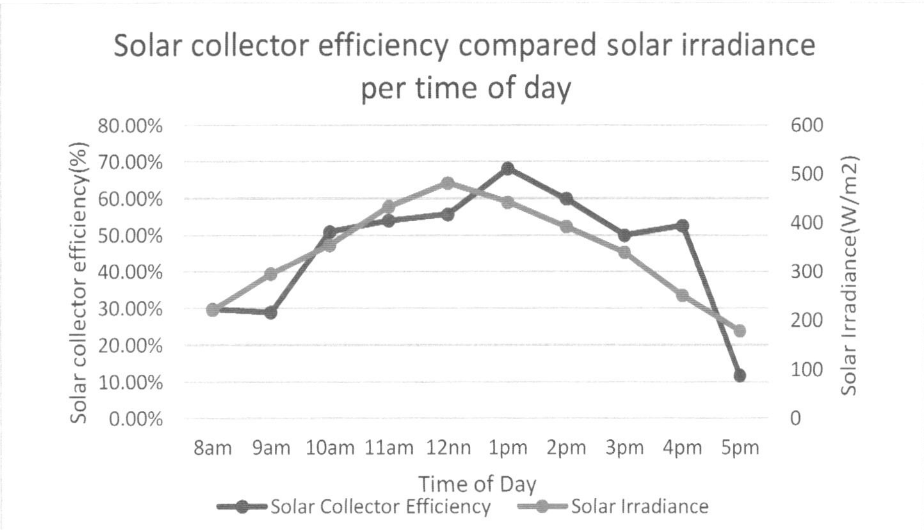

Figure 4. Solar Collector efficiency and solar irradiance per time of day

3.3. Dryer

The outlet and inlet temperatures were monitored per time of day. It can be observed that the highest inlet temperature was 46.8°C at 12nn with 471.9 W/m^2 solar irradiance while the outlet temperature peaked to 51.3°C at 1pm corresponds to a solar irradiance of 473.3 W/m^2 as seen in Figure 5.

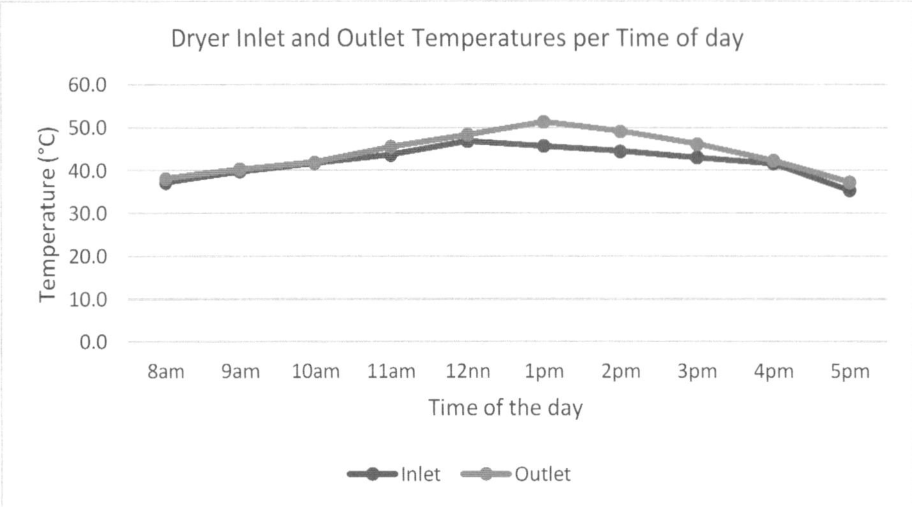

Figure 5. Dryer inlet and outlet temperatures per time of day

The highest dryer efficiency was reached at 33.58% at 2pm. The lowest drying efficiency was observed as 27.08% at 9am 1 hour after the drying process started. The average dryer efficiency is 32.79%.at an average solar irradiance of 351.20 W/m^2 shown in Figure 6. The value of the dryer efficiency is relatively high compared to the study of [5] and [12]. The sudden spike of the dryer efficiency from 9am to 10am is caused by less solar irradiance yielded since the experiment started thirty minutes before 8am, the heat is not that enough to increase its temperature.

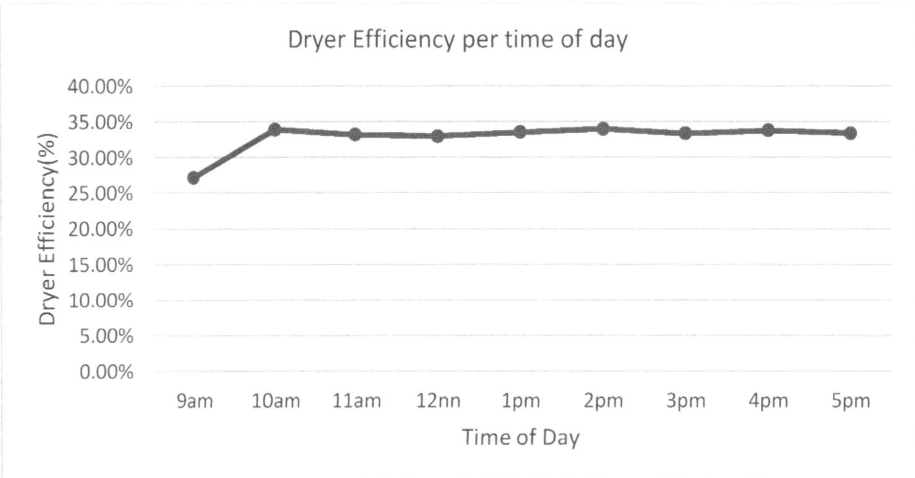

Figure 6. Dryer efficiency per time of day

3.4 Exergy Efficiency and Improvement Potential

Exergy analysis for the dryer has been carried out to determine the exergy inflow, out flow and losses. Figure 7 shows the exergy efficiency per time of day. The lowest observed value of 16.60% at 10am while the average exergy efficiency is 28.72%. The value of exergy efficiency is lower compared to 30% of Fudholi et al. (2014).

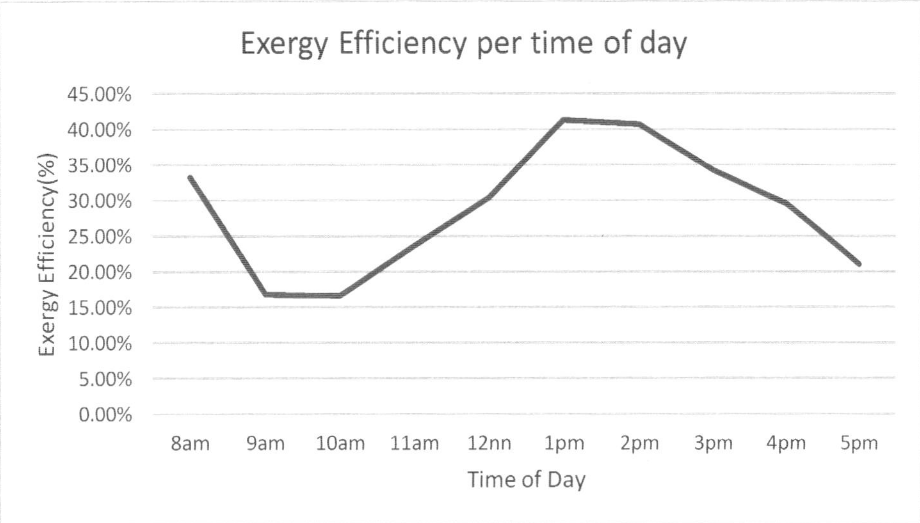

Figure 7. Exergy efficiency of the dryer per time of day

The improvement potential of the dryer as shown in Figure 8 has been obtained per time of day.The peak exergy improvement potential of 256.30W was observed at 10am.The lowest observed value of 225W was observed at 8am. The average exergy improvement potential is 350.03W. It is has higher exergy improvement potential compared to 247W of Fudholi et al. (2014). Introduction of partial to full recirculation of exhaust air can be implemented to increase the exergy efficiency and exergy potential of the solar dryer. This would increase the exergy efficiency by 33% to 58% as observed in the study of [13]. This is being supported by the study of [14] that states that energy savings can be increased up to 70% by recirculation of exhaust air. Also, providing thermal storage medium such as sodium sulfate decahydrate and sodium chloride will further increase the exergy efficiency by 29.3% [14,15].

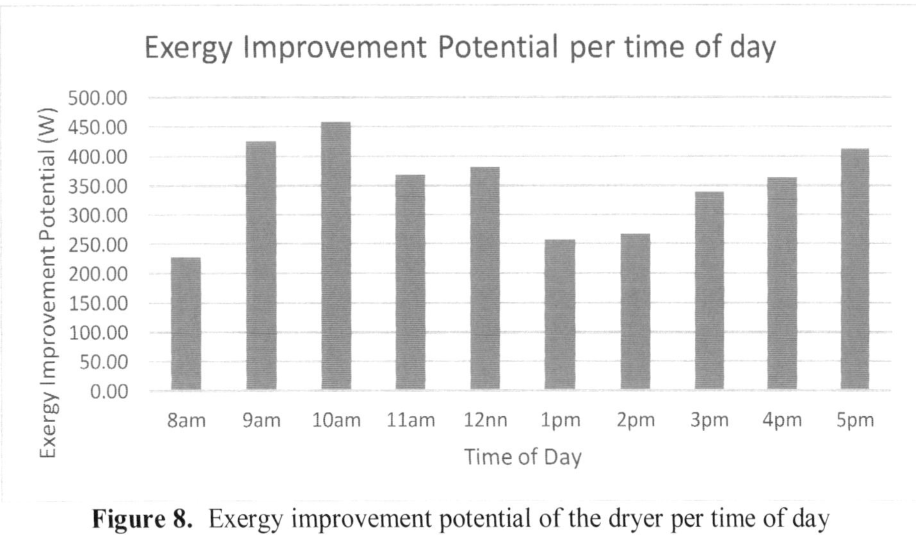

Figure 8. Exergy improvement potential of the dryer per time of day

4. Conclusion and Recommendation

The energy and exergy analyses were conducted in the onshore solar dryer. It is found out that the peak dryer efficiency was 33.58% and highest solar collector efficiency was 68%. The highest exergy efficiency was determined to be 41.38% that corresponds to an exergy improvement potential of 256.30 W. Future works would focus on further improvements that can be conducted in the onshore solar dryer such as partial to full exhaust air recirculation and provide thermal storage medium in the system.

Acknowledgement
The researchers would like to express their appreciation to DOST-CRADLE Project Grant and URCO Funding for the financial support for this research.

References

[1] Sharma A, Chen C R, & Vu Lan, N 2009 *Solar-energy drying systems: A review.* Renewable and Sustainable Energy Reviews, 13(6-7), 1185–1210. doi:10.1016/j.rser.2008.08.015
[2] Struckmann F 2008 *Analysis of a Flat-plate Solar Collector*, Retrieved from http://www.ht.energy.lth.se/fileadmin/ht/Kurser/MVK160/Project 08/Fabio.pdf;
[3] Moran M, Shapiro H, Boettner D & Bailey M 2014 *Fundamentals of Engineering Thermodynamics* 8th ed John Wiley & Sons, Inc.
[4] Aghbashlo M, Mobli H, Rafiee S & Madadlou A 2013 *A review on exergy analysis of drying processes and systems. Renewable and Sustainable Energy Reviews,* ***22,*** *1–22.* doi:10.1016/j.rser.2013.01.015
[5] Fudholi A, Sopian K, Othman M Y & Ruslan M H *2014 Energy and exergy analyses of solar drying system of red seaweed. Energy and Buildings*, **68**, 121–129. doi:10.1016/j.enbuild.2013.07.072
[6] Philippine National Standard 2007 Seaweed Specification. Retrieve form https://seaplant.net/bimpeaga/images/downloads/PNCS.pdf
[7] Akbulut A & Durmus A 2010 *Energy and exergy analyses of thin layer drying of mulberry in a forced solar dryer*, Energy **35** 1754-1763
[8] Van Gool W 1997 *Energy Policy: Fairy Tales and Factualities. Innovation and Technology Strategies and Policies*, 93–105. doi:10.1007/978-0-585-29606-7_6
[9] Al-Neama M A & Farkas I 2018 *Thermal efficiency of vertical and horizontal-finned solar collector integrated with forced air circulation dryer for Apple as a sample. Drying Technology*, 1–13. doi:10.1080/07373937.2018.1488260
[10] Tian Y & Zhao C Y *2013 A review of solar collectors and thermal energy storage in solar thermal applications*. Applied Energy, 104, 538–553. doi:10.1016/j.apenergy.2012.11.051
[11] Korres D N & Tzivanidis C 2019 *Thermal Analysis of a Serpentine Flat Plate Collector and Investigation of the flow and convection regime. Thermal Science* **23**(1), pp. 47-59
[12] Poblete R, Cortes E, Macchiavello J & Bakit J 2018 *Factors influencing solar drying performance of the red algae Gracilaria chilensis.* Renewable Energy, **126**, 978–986. doi:10.1016/j.renene.2018.04.042
[13] Rabha D K & Muthukumar P *2017 Performance studies on a forced convection solar dryer integrated with a paraffin wax–based latent heat storage system.* Solar Energy, **149**, 214–226. doi:10.1016/j.solener.2017.04.012
[14] Mathew A A & Venugopal T 2018 *Solar power drying system: a comprehensive assesment on types, trends, performance, and economic evaluation.* International Journal of Ambient Energy, 1–67. doi:10.1080/01430750.2018.1507933
[15] Ndukwu M C, Bennamoun L, Abam F I, Eke A B & Ukoha D 2017 *Energy and exergy analysis of a solar dryer integrated with sodium sulfate decahydrate and sodium chloride as thermal storage medium.* Renewable Energy, **113**, 1182–1192.

Energy equity analysis in Metro Manila using household expenditure data

Christian Roice Tayag[1,2], **Neil Stephen Lopez**[1]

[1]Mechanical Engineering Department, De La Salle University, 2401 Taft Ave, Malate, Manila, 0922 Metro Manila

[2]christian_roice_tayag@dlsu.edu.ph

Abstract. Energy poverty in the Philippines has been prevalent, but current energy equity metrics in the Philippines are done only on a national level. Lorenz curve and Gini coefficient are concepts traditionally applied in economics to measure wealth inequality. In this paper, the authors will adapt the methodology used by Jacobson to measure energy inequality in Metro Manila for 2015 using energy expenses of households retrieved from the Family Income and Expenditure Survey conducted and published by the Philippine Statistics Authority every three years containing energy expenses, income, and socio-economic identifiers of a household. The main energy resources listed in the survey are electricity, LPG, kerosene, fuelwood, charcoal, and biomass residue. To translate energy expenses to energy consumption, historical costs available from public records are used in conjunction with the fuels' calorific values. From there, Lorenz curves and Gini coefficients are calculated for districts and cities in Metro Manila. Further comparison reveals the relationship of income and energy consumption Gini coefficients in the region. Based on the results, the authors include recommendations to improve the use of Lorenz curve and Gini coefficient as a metric in quantifying energy equity.

1. Introduction

One of the major global problems being faced today is energy poverty. It can be defined as the inability to acquire modern energy services, which can be in the form of cooking, lighting, transport, and many others. This inability to access energy can lead to adverse consequences on people's health, social activities, education, and overall quality of life.

Various approaches have been done in recent years to further analyze energy poverty. One example is the integrated approach by Che et al, generating an indicator system and assigning weights by rough set and a large-scale survey [1]. Their study also presented a global assessment of the existence of energy poverty in 125 countries. Another study conducted by Scarpellini et al used spatial analysis of energy poverty in Spain in conjunction with a statistical analysis of related financial aids provided to households. The authors found that energy poverty is less acute in rural households in the region [2]. Meanwhile, a study on the effect of government energy efficiency programs to energy poverty in Alaska is conducted by Pride et al [3]. Petrova, on the other hand, examined the effect of energy poverty on the consumption of light and the overall quality of life of the people [4]. Finally, an analysis on the connection between urban form and city densification to energy equity was considered by Poruschi and Ambrey in Australia [5].

Measurements such as kilowatt-hour (kWh) per capita or kWh/GDP may be used to measure economic performance of a country, but there is no consensus as to what the most efficient way is in assessing the corresponding energy poverty. Over the past years, several attempts to quantify energy equity have been done by researchers. Templet in 1994 defined energy equity as the ratio of energy prices of the residential sector to the industrial sector [6]. Hall et al had a more comprehensive scope, defining energy equity as equality in terms of access to affordable, safe, and reliable energy, and the distribution of the risks and benefits of new technology [7]. Moreover, Adams and Bell limit their definition to three key issues: the distribution of impacts, responsibility, and costs and benefits [8]. From an interesting perspective, Kemmler and Spreng proposed the use of energy-based indicators to measure poverty and equity in a country [9]. While the existing literature on energy equity is largely from a social science standpoint, Jacobson et al introduced a quantitative perspective by using the Lorenz curve and Gini coefficient in the analysis of energy distribution through parallel case studies in Norway, USA, El Salvador, Thailand, and Kenya [10]. The Lorenz curve and Gini coefficient have been developed and traditionally used to analyze wealth distribution in countries.

The Philippine economy has been steadfastly growing in recent years. In fact, the country recorded a Gross Domestic Product (GDP) growth of 5.9% in 2015, next only to India at 7.3%, China at 6.9%, and Vietnam at 6.7%. That same year, the six-year moving average of real GDP growth was recorded at 6.2%, its highest since 1978 during that time [11,12]. Philippines' National Capital Region (NCR) or more known as Metro Manila, also recorded a high regional growth of 6.6%, while having the highest contribution to the national growth with 2.4 percentage points. NCR also had the largest share of the country's GDP with 36.5% in 2015 [12].

Despite the growing national economy, energy poverty in the Philippines remains. According to the 29th EPIRA Status Report published by the Department of Energy (DOE), the national electrification rate reached 89.61% in 2016, leaving approximately 2.4 million Filipinos with no access to electricity [13]. This is close to figures found in the 2016 Annual Poverty Indicators Survey (APIS) Report published by the Philippine Statistics Authority (PSA), indicating that 91.4% of families in the country had access to electricity [14]. The report also indicates that only 79.8% of the total households belonging to bottom 30% income earners had electricity connection, while 96.4% of the top 70% had electricity connection. These numbers suggest that the level of income affects electricity distribution among families in the country. However, a study looking at the socio-economic aspect of energy consumption in the Philippines, specifically in Metro Manila, using household data is yet to be conducted. Severe inequity in energy use in the country is evident, but there is currently no means to assess its degree and scientific evidence to influence policy. This paper aims to quantify and assess the energy poverty in Metro Manila by calculating the Gini coefficients and providing the Lorenz curves using the household energy expenditure data extracted from Family Income and Expenditure Survey (FIES) report published by PSA.

The study shall cover only Metro Manila and will be analyzed on city and district levels using household data. The data to be used will be derived from the FIES report for year 2015 [15]. FIES is conducted in the Philippines every three years and serves as the key source of data on family income and expenditure, indicating the levels of consumption by item of expenditure and sources of income. FIES divides households into categories, from the region, the district, the city, up to the barangay that it belongs to; however, for this study the energy consumption distribution will be analyzed on city level only. That is due to several barangays having only one participant in the survey; thus, analysis on barangay level cannot be done. PSA defines barangay as the smallest political unit in the country, usually having an enumerator assigned for identification [16].

2. Methodology

FIES provides energy expenses per household. These expenses can be classified under direct fuel purchases or indirect fuel expenses. Direct fuel expenses for LPG, kerosene, fuelwood, charcoal, and other fuels are indicated separately in the FIES. Expenses pertaining to fuel directly burned by the households may be converted to energy units by dividing with the historical fuel price per liter or kilogram and multiplying it to the calorific values of corresponding types of fuel, such as in [17]. Expenses pertaining to indirect fuel consumption such as electricity will be divided by the historical cost per kilowatt-hour (kWh). The calculated energy consumption per household for these expenses shall then be used to measure energy distribution by calculating the Gini coefficients and plotting the Lorenz curves. Results will then have to be analyzed and compared to the prevailing economic and energization conditions of the region. A simple block diagram showing the conceptual framework of this study is provided in Figure 1.

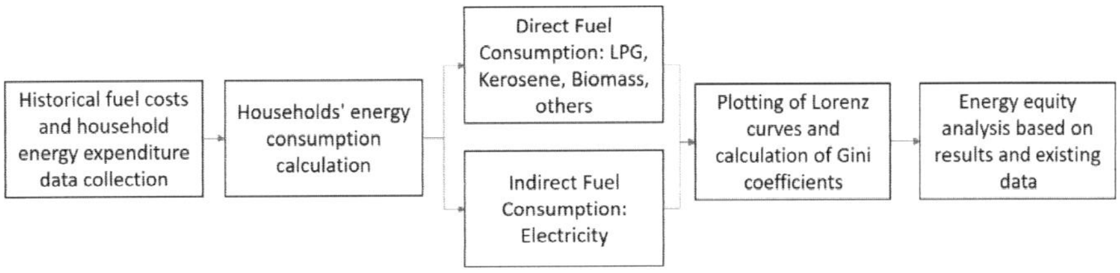

Figure 1. Conceptual framework for energy equity analysis.

2.1 Energy Consumption Calculation
Electricity and fuel historical costs are important in accurately quantifying the energy consumption in a region. Unfortunately, most of the data provided by the government publicly are already aggregated and might not account for the variability of the costs relative to time. Requests for unaggregated data are limited, if not totally inaccessible to the public due to laws surrounding data privacy.

2.1.1 Electricity In this study, it is assumed that the electric consumption of all households included in the FIES for Metro Manila are supplied by Manila Electric Railroad and Light Company (MERALCO) which is the sole utility distribution company in the region[1]. Due to varying subsidies given by the company to households with low consumption (0-100 kWh) mandated by the government, as well as the varying taxes imposed for specific consumption ranges, assuming one average annual price point for electricity will not give an accurate account of the energy used in a household. Fortunately, the company provides the unaggregated data used to calculate the monthly price point per kWh of a specific consumption range[2].

To simplify calculations, the authors assumed that each household utilizes the same amount of power per month. The first step in quantifying energy use is to find within which consumption range a household's electricity expense falls. These can be seen on Table 1. Once determined, the corresponding consumption can be obtained using equation (1).

$$EC = 12 \cdot \frac{E-B}{A+C} \; kWh/household \tag{1}$$

where: EC = Annual electricity consumption per household, kWh/household
E = Actual annual electricity expense per household, Php/household
A = Partial annual rate per kWh, Php/kWh/household
B = Partial annual rate per household, Php/household
C = Energy tax factor

[1] As confirmed by Distribution Utility Map provided by Philippine Energy Regulatory Commission online at https://www.erc.gov.ph/ContentPage/374
[2] MERALCO posts the typical household consumptions for each month, as well as the specific charges and taxes for the month which are used to calculate for the historical costs online at https://company.meralco.com.ph/news-and-advisories/rates-archives

Table 1 shows which values to use for equation (1) to calculate the energy consumption. E is the actual annual electricity expense for each of the households included in the 2015 FIES. Values for A, B, and C reflected on the table are calculated using the unaggregated data from MERALCO.

Table 1. Derived variables for energy consumption calculation.

Consumption per household (kWh/household)		Electricity Expense per household (Php/household)		Partial Annual Rate per kWh ((Php/kWh)/household)	Partial Annual Rate per household (Php/household)	Energy Tax Factor
min	max	min	max	A	B	C
0	20	67.50	128.13	3.03	67.50	
21	50	1,343.76	2,987.00	56.66	153.83	
51	70	3,911.20	5,293.81	72.77	199.98	
71	100	6,556.21	9,133.57	88.87	246.13	
101	200	11,598.20	22,665.16	111.79	307.67	
201	300	23,692.43	35,210.30	116.34	307.67	
301	400	36,620.33	48,563.70	120.64	307.67	
401	500	51,686.38	64,370.90	128.13	307.67	
501	650	64,499.02	83,589.87	128.13	307.67	
651	1000	83,719.19	128,854.13	128.13	-472.34	1.20
1001	1500	128,984.65	194,117.36	128.13	-1,672.34	2.40
1501	>1501	194,249.69	over 194,250	128.13	-4,372.34	4.20

In actual conditions, monthly power consumption of a household may vary due to several external factors. One of these factors is climate, wherein months in the summer season tend to increase energy use, while colder months tend to decrease the energy use. This may result in a household's annual electricity expense to not fall within any range initially calculated. Estimating the equivalent energy consumption for these households may be calculated by interpolation between consumption ranges in relation to electricity expense which is also provided in Table 1. Lastly, since it is indicated in the FIES if the household is connected to an electricity provider, data cleaning may be done as well. Households which are indicated to have no electricity connection but have expenses for electricity shall have an equivalent of zero (0) kWh consumption instead.

2.1.2 LPG, kerosene, fuelwood, charcoal, and bagasse The authors were able to request for the historical cost data on LPG, kerosene, and bagasse through the Freedom of Information (FOI) online facility[3]. For LPG and kerosene, the average price per month in the given year was provided. Based on this, the price point averages of respective fuels for 2015 are computed, as reflected on Table 2. Also included in the FIES is the expenditure for household fuels not specifically identified in the survey form. To quantify the energy consumption for this expense, the authors assumed the fuel to be bagasse, a form of biomass residue, in accordance with the Household Energy Consumption Survey (HECS) conducted in 2011 by the DOE, where it was reported that 22.3% of households in the Philippines used biomass residue as one of their fuels at home in that year [18]. DOE, through the FOI facility, was able to give only the price range of bagasse from monthly generation reports of existing biomass powerplants[4]. Despite this limitation, the authors decided to use the midpoint of this price range to account for the energy consumption of this fuel type for this study.

[3] Data not publicly available can be requested at Freedom of Information Facility. Access to the historical data for fuels utilized by the authors is listed.
For LPG:
http://www.foi.gov.ph/requests/aglzfmVmb2ktcGhyHQsSB0NvbnRlbnQiEERPRS0xMTE5ODA4MjMyNzIM
For kerosene at:
https://www.foi.gov.ph/requests/aglzfmVmb2ktcGhyHQsSB0NvbnRlbnQiEERPRS04NzgzMDQyMzEwMzEM
For bagasse at:
https://www.foi.gov.ph/requests/aglzfmVmb2ktcGhyHQsSB0NvbnRlbnQiEERPRS0yMTI1OTY2MTE5MjYM

The average retail prices per cubic meter (m^3) of fuelwood and per sack of charcoal for each region in the Philippines were provided in the 2015 Philippine Forestry Statistics published by the Forestry Management Bureau of the Department of Environment and Natural Resources (FMB-DENR) [19]. The methodology from Wood Fuel Surveys Report commissioned by the Food and Agriculture Organization of the United Nations (FAO UN) is used to convert these units to per unit mass [20]. The mass calculated is then further converted to energy units using conversion table provided by DOE[4]. Table 2 shows the price points per unit derived from the various extracted data, as well as the corresponding calorific values of the different fuel types based on DOE's energy conversion table which can be used for calculating the respective fuels' energy consumption in Equation (3). Fuel expenditure per household (e) is given in FIES.

$$EC = \frac{e_i}{P_i} \times CV_i \ kWh/household; \quad (2)$$

where: EC = energy consumption of a household for specific fuel type, kWh/household
e = expense of a household for specific fuel type i, Php/household
P = price per unit of specific fuel type, Php/unit
CV = calorific value of specific fuel type, kWh/unit

Table 2. Calorific values and historical costs of fuels in 2015.

Fuel type	**Unit**	**Price (Php)**	**Calorific Value (kWh)**
Kerosene	liter	50.34	9.29
LPG	liter	26.28	6.74
Charcoal	kilogram	8.45	6.98
Fuelwood	tonne	1,142.38	3,826.27
Bagasse	tonne	2,500.00	4,954.38

2.2 *Lorenz Curve and Gini Coefficient*

Lorenz curves and Gini coefficients are concepts in economics traditionally used to graphically represent the income inequality of a given population. Jacobson et al utilized the Lorenz curve and Gini coefficient to quantify energy poverty in five countries: Norway, USA, El Salvador, Thailand, and Kenya – with Norway having the least measured energy poverty [10]. Adapting Jacobson's methodology to measure energy equity, the Lorenz curve will be defined as a ranked distribution of the cumulative percentage of the population for NCR represented in the FIES 2015 on the x-axis, against the cumulative percentage of energy resource consumption distributed along the y-axis. Perfect equity is shown by a 45-degree diagonal line which extends from the origin to the point with coordinates of cumulative percentage population, x = 100%; and cumulative percentage energy resource y = 100%. The greater the distance of the actual curve to the perfect equity line, the greater the inequality in energy consumption. The distribution can be measured numerically by calculating the Gini coefficient of the corresponding Lorenz curve. The Gini coefficient measures the difference between a uniform distribution and the actual distribution of a resource. It is the ratio between the area enclosed by the diagonal line and the Lorenz curve and the total area under the diagonal line of uniform distribution. The adapted formula from Jacobson et al is shown in equation (3).

$$G_e = 1 - \sum_i \quad (Y_{i+1} + Y_i)(X_{i+1} + X_i); \quad (3)$$

[4] DOE provides basic energy conversion table at: https://www.doe.gov.ph/units-measurement-and-conversion-table

where: Ge = Gini coefficient
Xi = the number of energy users in population group i/total population
Yi = the quantity of energy used by population group i/total energy use

Before proceeding with the calculations, Yi must be ordered from lowest to highest consumption. Values for the Gini coefficient are from 0 to 1, with the value of 0 indicating perfect equity – as represented by the 45-degree diagonal line mentioned earlier. A value of 1 is indicative of complete inequity.

Lorenz curves are not able to give a direct measurement of different utility services, but they can measure quantitatively the differential amounts of energy consumption. That is, when given with the same amount of energy, a household may consume it differently in various forms of services such as lighting, heating, or powering appliances. The variations in amount of energy consumption can also be a result of the different efficiencies of the different technologies used for specific services. If the average overall efficiency among consumers can be assumed constant, as in the case of MERALCO providing electricity for Metro Manila, then quantities of energy can be considered a reasonable measure of utility. Same can be assumed when a unit of energy's marginal benefit is consistent from consumer to consumer. It should also be noted when these given conditions do not hold. For instance, using fuelwood or charcoal for cooking requires a greater amount of energy than when using LPG. In this situation, the lower level of energy consumption is preferred between the two. Energy consumption can be measured quantitatively to estimate energy equity in situations with which a single energy source is considered, or where there is constant average conversion efficiency per user.

3. Results Discussion

As mentioned in the previous section, respective energy consumption due to different energy sources cannot be added up quantitatively, unless efficiency is assumed constant as in the case of residential electricity consumption. It will be greatly inaccurate to simply just add up the respective energy consumption of a household calculated from the energy expenses in FIES. An illustration of this are households utilizing fuelwood for cooking versus households using electricity for the task. The smaller amount of energy needed to cook by using electricity might be misrepresented as an inequality. Thus, the authors decided to measure the Lorenz Curve and Gini coefficient for electricity consumption only. The method in determining the energy consumptions for fuels other than electricity is still presented, as data concerning the efficiency of an energy resource might emerge in the future.

Table 3. Income and energy Gini coefficients, annual income per capita, annual electricity consumption per capita.

Region/District/City/Municipality	Income Gini Coefficient	Energy (Electricity) Gini Coefficient	Total annual income per capita	Total annual electricity consumption per capita	% of population with income per capita lower than a	% of population with electricity consumption per capita lower than b
			Php per capita a	kWh per Capita b	%	%
NCR	**0.53**	**0.84**	**92,870.01**	**2,019.23**	**≈ 68**	**≈ 94**
District 1	**0.50**	**0.85**	**80,907.88**	**2,640.44**	**≈ 67**	**≈ 92**
City of Manila	0.50	0.85	80,907.88	2,640.44	≈ 67	≈ 92
District 2	**0.53**	**0.85**	**95,175.75**	**2,000.50**	**≈ 67**	**≈ 94**
City of Mandaluyong	0.45	0.83	81,640.13	1,789.53	≈ 63	≈ 96
City of Marikina	0.46	0.86	71,216.18	1,335.85	≈ 64	≈ 96
City of Pasig	0.51	0.85	81,732.98	1,722.92	≈ 69	≈ 95
City of San Juan	0.42	0.82	94,059.26	2,024.90	≈ 68	≈ 94
Quezon City	0.55	0.86	103,208.53	1,749.11	≈ 68	≈ 94
District 3	**0.46**	**0.81**	**77,871.96**	**1,242.10**	**≈ 65**	**≈ 96**
City of Caloocan	0.46	0.85	76,133.83	1,398.91	≈ 65	≈ 96
City of Malabon	0.46	0.83	71,541.14	1,476.53	≈ 66	≈ 97
City of Navotas	0.48	0.50	73,242.26	328.10	≈ 70	≈ 71
City of Valenzuela	0.45	0.79	88,374.94	1,156.38	≈ 63	≈ 94
District 4	**0.56**	**0.84**	**109,203.52**	**2,763.81**	**≈ 69**	**≈ 91**
City of Las Pinas	0.60	0.84	91,937.05	3,446.86	≈ 70	≈ 91
City of Makati	0.61	0.80	101,539.74	2,519.59	≈ 67	≈ 88
City of Muntinlupa	0.52	0.87	95,816.16	2,238.42	≈ 66	≈ 97
City of Paranaque	0.62	0.90	116,974.34	2,167.21	≈ 74	≈ 92
City of Taguig	0.48	0.85	82,025.13	2,627.10	≈ 63	≈ 92
Pasay City	0.65	0.84	103,926.51	4,650.74	≈ 65	≈ 85
Municipality of Pateros	0.60	0.86	128,679.00	3,387.63	≈ 74	≈ 90

Table 3 shows the Gini coefficients calculated for Metro Manila or National Capital Region (NCR) and the districts and cities belonging to the region. The table also includes the respective income and electricity per capita, as well as the percentages of the population having lower income and electricity than the computed total, respectively.

The income Gini coefficient computed in this paper is different from the income Gini coefficient of 0.3909 included in the FIES report for 2015 [11]. That is because the x-axis used for the Lorenz curve graph is the percent cumulative number of households and the y-axis is percent cumulative income for FIES 2015. In this study, it is assumed that electricity consumption in a household is distributed equally for each member, and the abscissa and ordinate for the Lorenz curve graphs and the computed Gini coefficients utilize computations per capita to further illustrate the existing inequality. The computed Gini coefficients for electricity are much closer to 1 than the coefficients for income. Further comparison of the results to the national electrification rate of 88% in 2015 [21] shows despite being connected to available energy infrastructures, many households are still not able to fully utilize electricity as their main source of energy, with more than 80% of the surveyed households still relying on other sources of energy to do household cooking, lighting, and other services. Another indicator of the uneven distribution of both income and energy is the large population percentage having lower income and energy income per capita than the average. Individual results for this can be seen on Table 3.

Figure 2 shows the Lorenz curve for the income inequality in NCR and its districts. Figure 3 shows the energy Lorenz curve for the electricity inequality in NCR and its districts. Notice in the figures that for a given income Gini coefficient, the corresponding energy (electricity) Gini coefficient is higher.

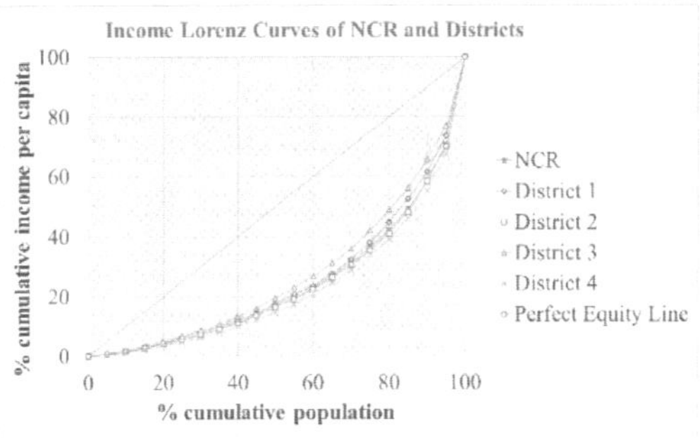

Figure 2. Income Lorenz curve of NCR and its districts.

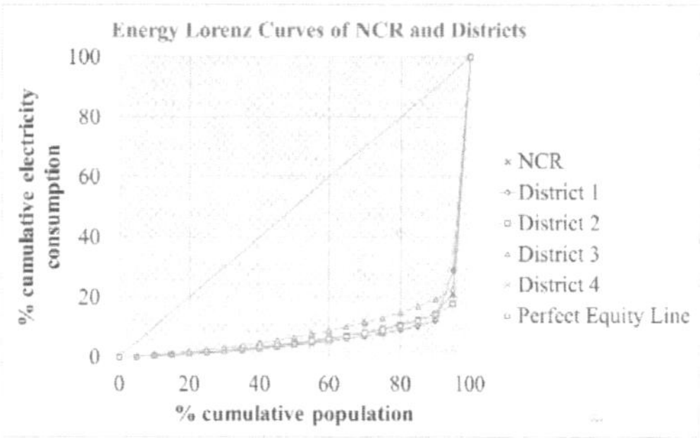

Figure 3. Energy (Electricity) Lorenz curve of NCR and its districts.

To confirm the relationship between income and energy distribution, a graph of cumulative income vs cumulative energy consumption per capita is shown on Figure 4. It can be seen from the figure that for a household to have a unit increase in electricity consumption, it should have an increase in income, indicating significant correlation between said variables.

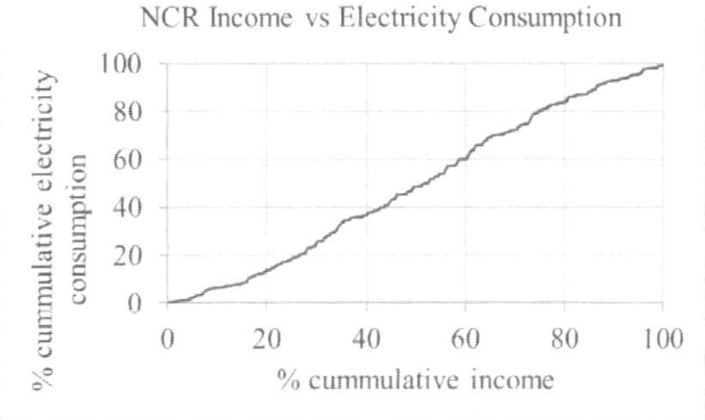

Figure 4. NCR cumulative income vs. electricity per capita.

4. Conclusion

In this study, the energy equity in the Philippines is measured using the Lorenz curve and Gini coefficients with household energy expenditure data source from Family Income and Expenditure Survey conducted by the Philippine Statistics Authority.

The authors were able to describe a methodology to convert household expenditure data into energy use. However, one serious limitation of the Lorenz curve and Gini coefficient is that you can only use this for energy resources with either constant efficiency or when efficiency of a technology employed for a specific energy resource is known. Because of this, energy consumption calculated from the methodology presented for other fuel types such as LPG, kerosene, and bagasse can only be used once the utilization level of energy services such as cooking, lighting, and others are known for each energy resource.

Despite the growing economy of the Philippines, specifically Metro Manila, resulting Gini coefficients and graphed Lorenz curves imply poor distribution of electricity among households in the region. The resulting Gini coefficients for the whole NCR and its districts do not vary much, ranging from 0.82 to 0.85 for energy and 0.46 to 0.56 for income. Looking at the city level, the calculated Gini Coefficients are more varied, ranging from 0.50 to 0.90 for energy and 0.42 to 0.65 for income. Even with relatively high electrification rate in the country, expenditure data from FIES imply that low income households still rely on other energy resources aside electricity for different services. It should also be noted that a low Gini coefficient does not immediately mean good energy equity. Results show that regardless of having a low energy coefficient of 0.50, the city of Navotas also has the lowest electricity consumption per capita. Thus, careful consideration should be done to properly assess the energy equity of an area in relation to the available data. Additionally, the graph of income vs electricity consumption distribution suggests that a household's electricity utilization can be dependent on its level of income.

Further studies should be done to investigate energy equity corresponding to certain time periods in order to show periodical comparisons. This should provide insights on the electrification progress of the country versus the periodical changes in distribution of energy. Currently, there are no studies setting the standard of acceptable values for energy Gini coefficients; thus, further analysis should be done to examine the relationship of Lorenz curve, Gini coefficients, and the quality of life of people. This should help decision makers in crafting policies for better energy distribution in the country. Lastly, additional studies should be done in order to integrate the different consumption of sources of energy so as to improve the Gini coefficients and Lorenz curves produced in this study.

References

[1] Che X, Zhu B and Wang P 2021 Assessing global energy poverty: an integrated approach *Energy Policy* **149** 112099

[2] Scarpellini S, Alexia Sanz Hernández M, Moneva J M, Portillo-Tarragona P and Rodríguez M E L 2019 Measurement of spatial socioeconomic impact of energy poverty *Energy Policy* **124** 320–31

[3] Pride D, Little J and Mueller-Stoffels M 2018 The value of residential energy efficiency in interior Alaska: a hedonic pricing analysis *Energy Policy* **123** 450–60

[4] Petrova S 2018 Illuminating austerity: Lighting poverty as an agent and signifier of the Greek crisis *European Urban and Regional Studies* **25** 360–72

[5] Poruschi L and Ambrey C L 2018 Densification, what does it mean for fuel poverty and energy justice? An empirical analysis *Energy Policy* **117** 208–17

[6] Templet P H 1995 Equity and sustainability: An empirical analysis. *Society and Natural Resources* **8** 509–523

[7] Hall S M, Hards S and Bulkeley H 2013 New approaches to energy: equity, justice and vulnerability *Local Environment* **18** 413–21

[8] Adams C A and Bell S 2015. Local energy generation projects: assessing equity and risks *Local Environment* **20** 1473–88

[9] Kemmler A and Spreng D 2007 Energy indicators for tracking sustainability in developing countries *Energy Policy* **35** 2466–80
[10] Jacobson A, Milman, A D and Kammen D M 2005. Letting the (energy) Gini out of the bottle: Lorenz curves of cumulative electricity consumption and Gini coefficients as metrics of energy distribution and equity *Energy Policy* **33** 1825–32
[11] National Economic and Development Authority 2016 *Socioeconomic Report 2015* (Pasig City)
[12] Philippine Statistics Authority 2016 *Gross Regional Domestic Product (GDRP) 2013-2015* (Quezon City)
[13] Department of Energy 2016 *29th Electric Power Industry Reform Act (EPIRA) Implementation Status Report* (Taguig City)
[14] Philippine Statistics Authority 2017 *2016 Annual Poverty Indicators Survey* (Quezon City)
[15] Philippine Statistics Authority 2017 *2015 Family Income and Expenditure Survey* (Quezon City)
[16] Philippine Statistics Authority 2017 *2015 Census of Population* (Quezon City)
[17] Lopez A K, Lopez N S, Chiu A and Biona, J B M 2020 Spatial decomposition of drivers of household energy use in Metro Manila using LMDI *Chemical Engineering Transactions* **81** 769–74
[18] Department of Energy 2013 *Household Energy Consumption Survey 2011* (Taguig City)
[19] Forestry Management Bureau - Department of Environment and Natural Resources 2017 *Philippine Forestry Statistics 2015* (Quezon City)
[20] FAO Forestry for Local Community Development Programme Sweden 1983 *Wood Fuel Surveys* (Rome: Food and Agriculture Organization of the United Nations)
[21] Department of Energy 2015 *27th Electric Power Industry Reform Act (EPIRA) Implementation Status Report* (Taguig City)

Enhancing induced flow rate through a solar chimney by a forced flow

Y Q Nguyen

Faculty of Engineering, Van Lang University, 45 Nguyen Khac Nhu St., Co Giang Ward, Dist. 1, Ho Chi Minh City, Viet Nam

Email: y.nq@vlu.edu.vn

Abstract. Solar chimneys absorb solar radiation for natural ventilation or cooling of buildings. In typical solar chimneys, the air flow is induced by the thermal effect which causes pressure gradient inside the chimney to deviate from that of the ambient air. The induced flow is employed to ventilate the connected building naturally. In this study, we aimed to boost the induced flow rate through a typical solar chimney by a forced flow at the inlet of the air channel which is divided into two ports for the forced and the induced flows. The air flow and heat transfer were modeled with a Computational Fluid Dynamics (CFD) model with the ANSYS Fluent CFD code. Changing factors included the forced velocity, the dimensions of the inlet port for the forced air flow and the length of the partition between the ports of the naturally induced and forced air flows. It is seen that the forced flow significantly reduces the pressure in the air channel compared to that of a typical chimney; hence increasing the induced flow rate. The flow rate was significantly enhanced as the forced velocity was above 0.2 m/s; the area of the inlet port for the forced flow was 12.4% of the total inlet area; and the partition height was less than 5.0% of the total height of the air channel. In this configuration, the forced flow can be the waste exhaust air flow from other mechanical ventilation systems of the building. Therefore, no additional energy is required for this solution.

Keywords: Solar chimney, natural ventilation, forced flow, induced flow rate, CFD.

1. Introduction

Thermal or chimney effect can be utilized for ventilation or cooling of a building with enclosed cavities on the building envelope, such as double – skin facades, which connect to the rooms in the building. As solar radiation is absorbed on the surfaces of the cavity, air inside the cavity is warmed, rises, and induces an air flow through the cavity and the connected room; hence natural ventilation. Such cavities are categorized as solar chimneys [1].

The naturally induced air flow through solar chimneys has been examined by a number of researchers [1-4]. These reports show that the induced air flow rate increases with the major dimensions of the air cavity, or air channel, including the height and the gap, and the heat flux on surfaces of the channel. With proper designs, solar chimneys can provide adequate ventilation rate and thermal comfort conditions for buildings [5].

Forced air flow can also assist the naturally induced one in solar chimneys. Elghamry and Hassan [6] considered both forced and natural ventilation and cooling of a room with the combination of a solar chimney, photovoltaic, and geothermal air tube. They reported that the daily ventilated air and heat released from the room are enhanced with the forced flow through the geothermal tube. Liu et al. [7] utilized an indoor exhaust fan to ventilate a façade. Their results show that the forced flow reduces up to 90% of the heat flux transferred through the façade. In addition, the optimal forced air velocity is in the range of 0.5 – 1.0 m/s.

In this study, we proposed a combined system of forced and natural flows in a solar chimney for natural ventilation of a building. The forced flow assists the natural flow which ventilates the connected room. Therefore, the forced flow does not go through the room as in [6]. In addition, the forced flow can be collected from the waste flow at the exhaust of other mechanical ventilation systems; hence no additional energy is required as in [7].

The proposed system was examined numerically. The induced flow rate and temperature rise were investigated under influences of different factors including the force air velocity, the lengths of the ports of the forced and natural flows, and the height of the partition between the ports.

2. Description of the problem and the numerical model

2.1. The problem

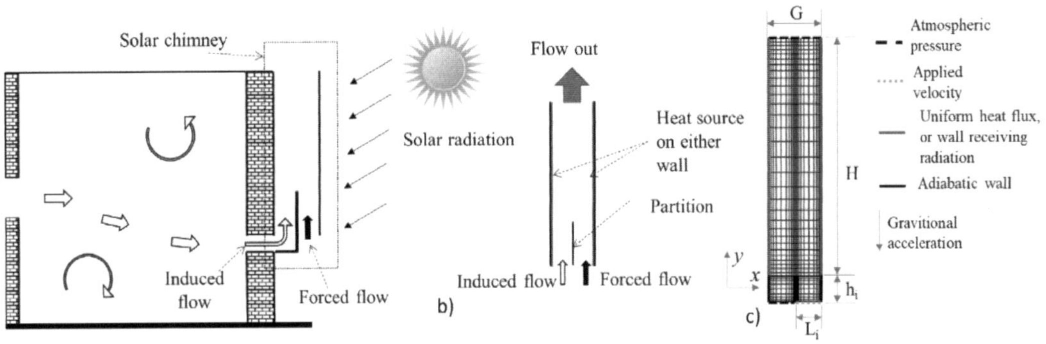

Figure 1. Schematic of: a) A solar chimney ventilating a building, b) A simplified model of the solar chimney, and c) The computational domain and mesh.

Figure 1a shows a solar chimney for natural ventilation of a building. The chimney is attached to a building wall. Either the building wall or the cover plate of the chimney can absorb solar radiation. The lower end of the air channel is divided into two portions by a partition: The left portion is connected to the room while the right part is supplied with a forced air flow. Without the forced flow, as the air in the channel between the cover plate and the building wall is warmed by the absorbed heat, it rises and induces an air flow through the connected room. The forced flow is aimed to assist the thermal effect and enhance the induced flow through the room.

The simplified model of the solar chimney in figure 1a is presented in figure 1b. It consists of a straight channel with two inlets at the lower end. The induced flow and the forced flow enter the channel from the left and the right inlets, respectively. The upper end of the chimney opens to the atmosphere. The absorbed solar radiation is described by a heat source on either the left or the right wall of the air channel. This simplified model is similar to those in the literature [3,4,7].

2.2. Governing equations

The computational model was based on the Computational Fluid Dynamics technique. The air flow and heat transfer in the air channel were approximated by the governing equations which state conservation laws of mass, momentum, and energy. For the chimney model in the vertical plane shown in figure 1b, a two – dimensional, steady, incompressible, and turbulent flow can be modeled with the Reynolds Averaged Navier – Stokes equations as follows.

$$\frac{\partial u_j}{\partial x_j} = 0 \quad (1)$$

$$\frac{\partial(u_i u_j)}{\partial x_j} = -\frac{1}{\rho}\frac{\partial p}{\partial x_i} - g_i\beta(T - T_0) + \frac{\partial}{\partial x_j}\left(\nu\frac{\partial u_i}{\partial x_j} - \underline{u_i' u_j'}\right) \tag{2}$$

$$\frac{\partial(T u_j)}{\partial x_j} = \frac{\partial}{\partial x_j}\left(\frac{\nu}{Pr}\frac{\partial T}{\partial x_j} - \underline{T' u_j'}\right) \tag{3}$$

where u and T stand for the time – averaged velocity and temperature; u' and T' are the fluctuating velocity and temperature; T_0 is the ambient temperature; p is the pressure; ν, ρ, β are the air kinematics viscosity, density, and thermal expansion coefficient; Pr and g are the Prandtl number and gravitational acceleration; $\underline{u_i' u_j'}$ and $\underline{T' u_j'}$ are the turbulence stress and turbulence heat flux which are solved with the RNG $k - \epsilon$ turbulence model. Other turbulence models, such as the standard $k - \epsilon$, standard $k - \omega$, standard $k - \omega$ with modifications for low Reynolds number effects were also tested. The RNG $k - \epsilon$ offered the best convergence rate. Therefore, it was selected.

2.3. *Numerical setup*

The above equations were discretized by the Finite Volume Method with the CFD (Computational Fluid Dynamics) code ANSYS Fluent (Academic version 2020R2). The computational domain and mesh are presented in figure 1c. The domain coincided with the air channel. The mesh was structured and rectangular. The mesh density was higher toward the solid walls. By changing the number of cells and the growth rate of the mesh cells from the solid surfaces, the induced flow rate through the chimney was compared. It was seen that the difference in the flow rate was less than 1.0% when the maximum non–dimensional distance of the first cell centre from the solid walls, or y^+, was less than 1.5. The according mesh resolution was more than 15000 cells in a solar chimney of H=0.9 m and G=0.2 m, and a heat flux of 500 W/m^2. For all tests in this study, the mesh was carefully checked to satisfy that required maximum y^+.

At the inlet for the induced flow, the pressure was assumed at the atmospheric value while at the inlet for the forced flow, a uniform air velocity was assigned. At the outlet of the air channel, the atmospheric pressure was also applied. Air entered the domain at both inlets at the ambient temperature which was fixed at 20°C in all tests. Inside the air channel, a uniform heat flux was applied on one wall while the opposite wall was able to receive radiative heat transfer from the other. Two cases of the heat source location were examined: heating either the left or the right wall. All solid walls applied non – slip conditions.

As seen in figure 1c, the height over which the heat source was applied is denoted as H. G indicates the gap of the air channel. L_i and h_i respectively are the length of the inlet for the forced velocity and the height of the partition. Radiative heat transfer was allowed between the two surfaces over the height H. Other walls were adiabatic.

For the numerical setups, radiation was computed with the S2S radiation model in ANSYS Fluent. The SIMPLE method for the coupling between the continuity equation (equation (1)) and the momentum equation (equation (2)), and the PRESTO! method for interpolation of the pressure on the staggered mesh were selected. Second order upwind scheme for equations (2) and (3), and first order upwind scheme for the equations for k and ϵ were utilized.

2.4. *Validation*

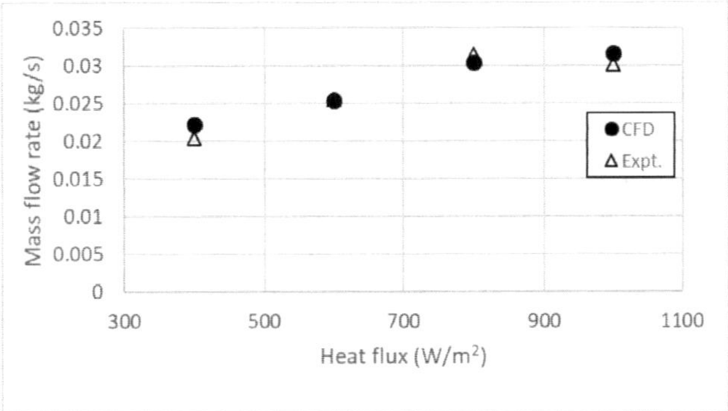

Figure 2. Computed vs. measured flow rate through the solar chimney in the experiment by Burek and Habeb [8].

In figure 2, the CFD model was validated against the experiment by Burek and Habeb [8]. The chimney had a length of 1.025 m, a width of 1.0 m, and a gap of 40 mm. The heat flux changed in the range of 400 – 1000 W/m^2. The CFD model slightly under – estimated the flow rate at 400 W/m^2 and 1000 W/m^2 and over – estimated at 800 W/m^2. However, the maximum difference between them was about 9.2% at 400 W/m^2. This discrepancy should be within the uncertainties of the measurement. Therefore, the prediction of the CFD model was acceptably accurate for the induced flow rate.

3. Results and Discussions

In the following tests, the chimney height H and the air gap G were fixed to 0.9 m and 0.2 m, respectively. Two heat fluxes of 250 W/m^2 and 500 W/m^2 were applied on either wall of the air channel. The induced air flow and temperature rise were examined as the forced air velocity V_i, length L_i of the inlet for the forced air flow, and the height h_i of the partition changed.

3.1. *Effects of the forced velocity*

In this test, L_i and h_i were fixed to 0.025 m and 0.1 m, respectively. The forced velocity, V_i, changed from 0.0 m/s to 1.0 m/s. When the forced air velocity was 0.0 m/s, the inflow along L_i was entirely induced by the thermal effect in the air channel. The heat source was on the left wall. The induced flow rate and the temperature rise through the chimney are presented in figures 3a and 3b, respectively.

In figure 3a, the flow rate was computed at the inlet of the induced flow rate, which is the left inlet in figures 1b and 1c. It is seen that the induced flow rate increased with the forced velocity when V_i was above 0.2 m/s. The induced flow rate was slightly affected for $V_i \leq 0.2\ m/s$ with changes of within 6.0% compared to the case of V_i=0.0 m/s. At V_i=2.0 m/s, the flow rate was up to about 3.1 times of that at V_i=0.0 m/s. The flow rate also increased with the heat flux, as expected.

The temperature rise, ΔT, through the chimney is plotted in figure 3b. In contrast to the flow rate, ΔT decreased with the forced velocity. However, similar to the flow rate, ΔT was almost unchanged when V_i was less than 0.2 m/s. ΔT at $V_i = 2.0\ m/s$ was about 70% of that at $V_i = 2.0\ m/s$.

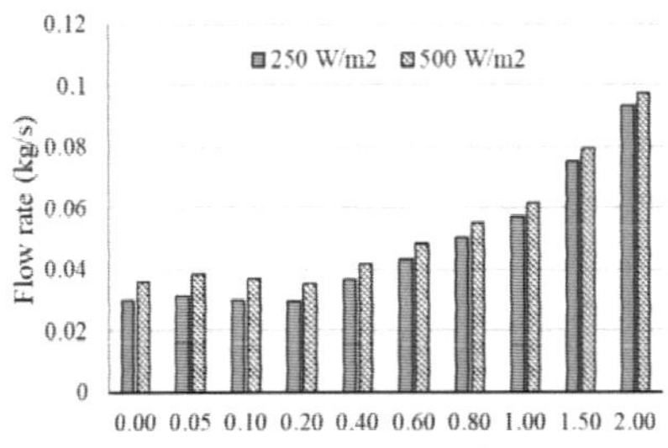

a)

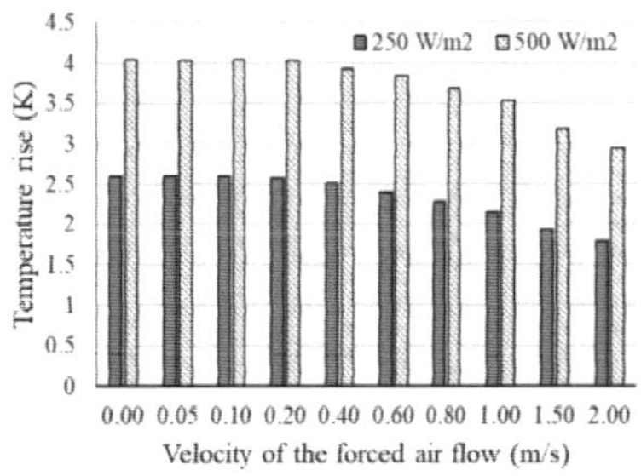

b)

Figure 3. Induced flow rate as the forced velocity changed.

To observe effects of the forced flow, the pressure, velocity, temperature distributions at $V_i = 0.0\ m/s$ and 1.0 m/s are presented in figure 4. Without the forced flow, the pressure distribution in the air channel followed the well – known thermal effect where the pressure increased along with the channel height. Near the inlet, the pressure was in the range of -0.0131 ÷ -0.0115 Pa. With the forced flow, the pressure on the lower half of the air channel was strongly reduced down to -0.0164 Pa. Lower pressure near the inlet achieved with the forced flow resulted in higher induced air velocity, as seen in figure 4c; hence higher flow rate, as seen in figure 3a.

It is seen in figure 4b for the temperature fields that in both cases of V_i, the thermal boundary layers along the heated (left) wall were identical. However, the thermal layer near the right wall, which received radiative heat from the left wall, was thinner with the forced flow. As a result, the temperature rise was lower for $V_i = 1.0\ m/s$, as seen in figure 3b.

3.2. *Effects of inlet length*

In this test, h_i and V_i were fixed to 0.1 m and 1.0 m/s, respectively. L_i changed from 0.0 m (no forced flow) to 0.15 m, which is 75% of the air gap. Figure 5 shows the flow rate obtained at two heat fluxes of 250 W/m^2 and 500 W/m^2. The heat source was either on the left (Left) or the right (Right) wall. It is seen that the flow rate varied strongly with L_i. The peak flow rate was at $L_i = 0.025\ m$, which is 12.5% of the air gap. The peak flow rate was about 200% of that at $L_i = 0.0\ m$. As L_i increased further, the flow rate decreased. At $L_i = 0.15\ m$, the flow rate was down to about 76% (at 250 W/m^2) to 64% (at 500 $W/m^2 - Right$) of that at $L_i = 0.0\ m$.

In figure 5, the flow rate is seen to increase with the heat flux. At the heat flux of 500 W/m^2, heating the right wall resulted in a lower flow rate which was about 6.0 to 10.0% less than that with heating the left wall. However, similar trends of variation of the data are seen for both cases of the heat flux and location of the heat source.

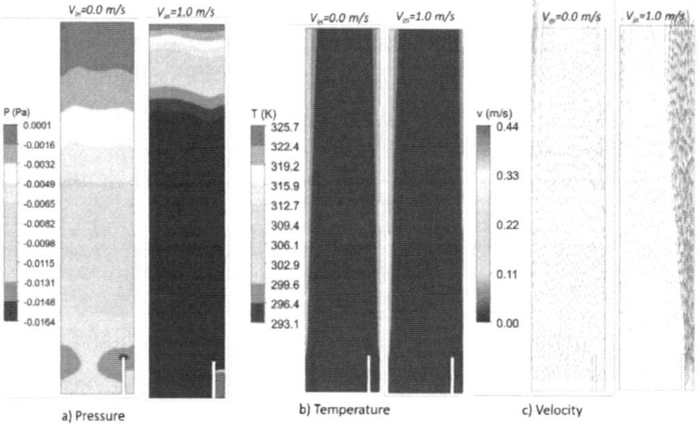

Figure 4. Pressure, velocity, temperature distributions at $V_i = 0.0\ m/s$ and 1.0 m/s.

3.3. *Effects of partition height*

In this test, L_i and V_i were kept to 0.025 m and 1.0 m/s, respectively. The heat source was on the left wall. The partition height changed from 0.025 m to 0.3 m at two heat fluxes of 250 W/m^2 and 500 W/m^2. The induced flow rate and the temperature rise are plotted in figure 6. It is seen that the flow rate and the temperature rise increased with the applied heat flux but they were almost unchanged as L_i increased from 0.025 m to 0.05 m. As L_i increased further, both the flow rate and the temperature rise decreased down to 83% and 78% of those at $L_i = 0.025\ m$. Examination of the pressure field reveals that the pressure near the inlet of the air channel increased with the partition height; hence less induced flow rate. In addition, as the partition as treated as an adiabatic plate, radiative heat transfer in the air channel was blocked more as h_i increased. This fact resulted in lower temperature near the right wall and less temperature rise through the air channel.

IOP Conf. Series: Materials Science and Engineering 1109 (2021) 012044 doi:10.1088/1757-899X/1109/1/012044

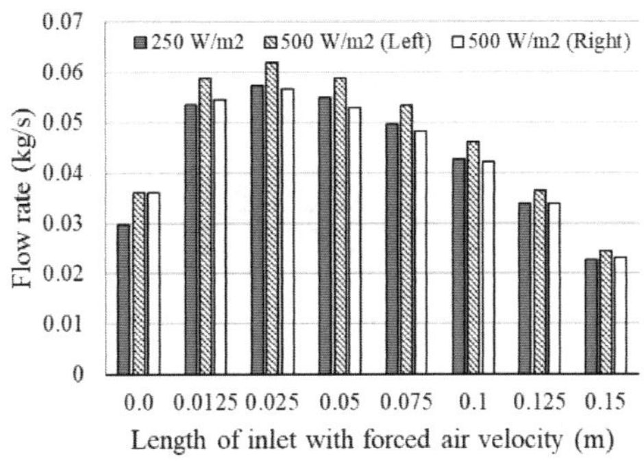

Figure 5. Induced flow rate obtained with different $L_i's$.

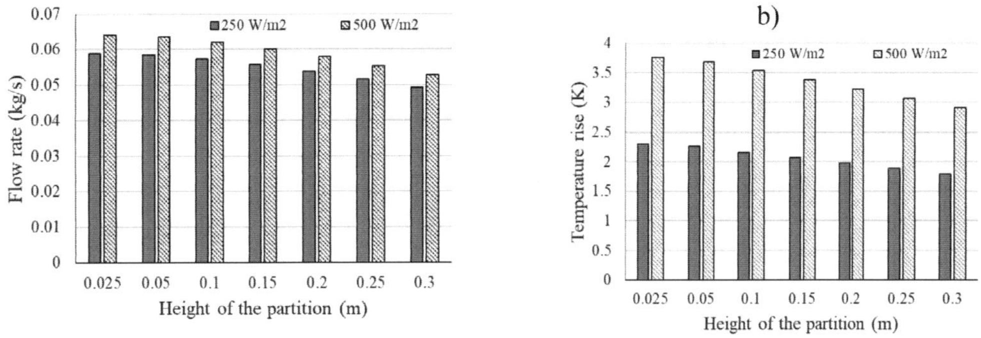

a)

Figure 6. Induced flow rate and temperature rise as the partition height changed.

4. Conclusions

From the results, the following findings are observed:

- The forced flow further decreased the pressure near the inlet of the air channel compared to that due to the thermal effect. The flow rate increased but the temperature rise decreased when the forced velocity was above 0.2 m/s. At $V_i = 2.0\ m/s$, the flow rate was up to about 3.1 times but ΔT was down to 70% of those at V_i=0.0 m/s. At $V_i \leq 0.2\ m/s$, the flow rate and temperature rise were almost unaffected.
- As the length of the inlet for the forced flow increased, at $L_i = 0.025\ m$, the flow rate increased to the peak value of about 200% of that at $L_i = 0.0\ m$. Further increase of L_i reduced the flow rate.
- Increasing the heat flux also resulted in an increase flow rate with similar trends at 250 W/m^2 and 500 W/m^2. At 500 W/m^2, heating the right wall reduced the flow rate up to 10% less than that of heating the left wall.
- Increasing the partition height did not affect the flow rate and temperature field of up to $h_i = 0.05\ m$. Afterward, the flow rate and ΔT decreased down to 83% and 78% of those at $L_i = 0.025\ m$.

In practical applications, exhaust air from the outlet of other mechanical ventilation systems can be collected for the forced air flow. Based on the findings in this study and depending on the required ventilation rate of the building, the dimensions of the solar chimney and the port of the forced flow, and the forced air speed can be selected. A damper for controlling the air speed at the outlet of the exhaust air may be required to achieve the optimal forced air speed according to the results in this study.

In future works, performance of the proposed system at different scales can be investigated both numerically and experimentally.

References

[1] Bansal N K, Mathur R and Bhandari M S 1993 Solar chimney for enhanced stack ventilation *Building and Environment* **28** (3) 373–377

[2] Chen Z D, Bandopadhayay P, Halldorsson J, Byrjalsen C, Heiselberg P and Li Y 2003 An experimental investigation of a solar chimney model with uniform wall heat flux *Building and Environment* **38** (7) 893–906

[3] Khanal R and Lei C 2012 Flow reversal effects on buoyancy induced air flow in a solar chimney *Solar Energy* **86** (9) 2783–2794

[4] Nguyen Y Q and Wells J C 2020 Effects of wall proximity on the airflow in a vertical solar chimney for natural ventilation of dwellings *Journal of Building Physics* **44** (3) 225–250

[5] Tan A Y K and Wong N H 2012 Natural ventilation performance of classroom with solar chimney system *Energy and Buildings* **53** 19–27

[6] Elghamry R and Hassan H 2020 An experimental work on the impact of new combinations of solar chimney, photovoltaic and geothermal air tube on building cooling and ventilation *Solar Energy* **205** 142- 153

[7] Liu L, Yu Z and Zhang H 2017 Simulation study of an innovative ventilated facade utilizing indoor exhaust air *Energy Procedia* **121** 126–133

[8] Burek S A M and Habeb A 2007 Air flow and thermal efficiency characteristics in solar chimneys and Trombe Walls *Energy and Buildings* **39** (2) 128–135

Experimental Study in Measuring Pressure Loss at Micro Channel

J Hendrarsakti[1*], R M Huda[1] and E Junianto[2]

[1]Faculty of Mechanical and Aerospace Engineering, Institut Teknologi Bandung, Bandung 40132, Indonesia
[2]Research Centre for Electrical Power and Mechatronics, Indonesian Institute of Sciences, Bandung 40135, Indonesia

*jooned@ftmd.itb.ac.id

Abstract. The objective of study was to conduct experimental study in measuring the pressure loss due to the major and minor loss in fluid flow at straight and bent micro channel. Test channel had circular cross-sections with hydraulic diameters of 118 μm to 1203 μm for straight and 348 to 1203 for bent micro channels. The results showed that losses for larger diameter of straight channel were well predicted. The bent channel showed that the loss coefficient values were higher if the Reynolds number got smaller.

Keywords: Micro channel, pressure drop, friction factor, bend loss coefficient

1. Introduction

Research on pressure drop in an internal flow has long been conducted. In the early nineteenth century, H Darcy [1] conducted experiments on the pressure drop of the flow in pipes with various materials and roughness. He introduced the concept of relative roughness especially for turbulent flow regime. Darcy found that the type of flow would depend on the roughness of the pipe, the diameter of the section, and also the slope of the pipe. Nearly 30 years later, J T Fanning [2] then proposed a correlation for pressure drop as a function of surface diameter and roughness. The quantification of the effect of surface roughness on pressure drop was first introduced by J Nikuradse [3], who proposed grain coarseness (ε) as relative roughness, equivalent to the diameter of sand particles used in his experiment as the main parameter affecting friction factors in laminar and turbulent flows. Variation in diameter and roughness was done for the first time 6 years after that. C F Colebrook [4] varied the pipe cross-sectional diameter (101.6 to 1524 mm) and surface roughness (0.043 to 0.254 mm) to see their effect on pressure drop. The pipes are tested with water flow in transition and turbulent regime. It was found that the friction factor decreases with increasing water velocity, but once the speed reaches a fully developed condition, the friction factor will remain constant for higher speeds. Then L F Moody [5] characterized the friction factor that had been studied by Darcy (f_{Darcy}) as a function of fluid velocity and channel cross-sectional diameter. Moody's diagram combined the results of Colebrook [4] and Nikuradse's [3] research to determine the pressure drop (Δp) and the relative roughness ε.

With the development of technology from macro to micro size, it requires accurate pressure loss calculations for smaller diameter channel. As the channel size becomes smaller, conventional theories for fluid and heat transfer at macro scale might need to be revisited to ensure these theories are applicable at smaller scale. Additional forces such as capillary and adhesion might contribute more at the micro channel. Other fundamental elements such as in modeling the modeling of fluid flow in small diameter channels may arise as a result of uncertainty about the application of empirical factors derived from experiments conducted on a larger scale, such as the loss coefficient due to angles in pipe bends or friction factors in pipes. The objective of current study was to conduct experimental study in measuring the pressure loss due to the major and minor loss in fluid flow at straight and bent micro channel with variety of pipe length and diameter at micro scale channel.

Peiyi and Little [6] conducted one of the preliminary investigations of fluid flow characteristics in microchannels. The study was conducted using variations of hydraulic diameter with variations of 45.46 to 83.08 μm. The fluid used for research is a gas and also varies, namely H_2, N_2 and Ar. The conclusion of this study was that the effect of surface roughness still affects the value of the friction factor even under laminar flow conditions. The rougher the surface, the faster the transition flow conditions were formed. For ducts with untreated glass material, the transition conditions were achieved at a Reynolds number of about 350. The friction factor for bends (angle 90° and 135°) were also found with respect to Reynolds number.

Mala & Li, [7] investigated the characteristics of water flow in micro channels with two different materials, fused silica and stainless steel. For fused silica lines the hydraulic diameter varies from 50 to 250 μm, while for stainless steel it is between 63.5 and 254 μm. Experiments were carried out with various Reynolds numbers up to 2300. Each channel with the same diameter had two different channel lengths, so that there were two pressure drop values for a long channel and a short channel for one type of diameter. The difference in pressure drop for each type of diameter was divided by the difference in length of the channel to produce a pressure gradient which was then plotted on the $\Delta P/\Delta l$ vs Re curve. The resulting pressure gradient value in the experiment was higher than the prediction using conventional theory.

Pfund [8] studied friction factors at laminar flow for water with high aspect ratio channels using depths varied from 128 to 521 μm and Reynolds numbers were between 60 and 3,450. By omitting the pressure loss at entrance and outlet, measurement was focused at the channel. Measurement uncertainty was calculated and propagated into the estimated friction constant. The study results showed the significant differences remained between the results and classical theory

Qu and Mudawar [9] conducted research of pressure drop on heat sinks with micro channels both experimentally and numerically. The channel cross-section is a rectangle with dimensions of 231 μm × 713 μm. The test was carried out with two variations of heat flux, namely 100 W/cm^2 with Reynolds numbers ranging from 139 to 1672 and 200 W/cm^2 with Reynolds numbers ranging from 385 to 1289. The test results show that the slope change in pressure drop to Reynolds's number can be related to the viscosity of water is temperature dependent. Based on this research, fluid flow with a hydraulic diameter greater than 350 μm and a Reynolds number below 1700 can still follow the conventional Navier-Stokes equation. Wu and Cheng [10] investigated the laminar pressure drop of water in 13 different trapezoidal silicon microchannels. Based on 168 experimental data points, dimensionless correlations for the apparent friction constant are obtained for the flow of water in trapezoidal microchannels having different geometric parameters, surface roughnesses and surface hydrophilic properties.

Hsieh et al. [11] carried study the pressure drop using deionized water inside a channel with hydraulic diameter of 146 μm. Using Reynolds numbers of 50,100, 470, and 900, mass flow rate and the pressure drop between inlet and outlet of the channel were measured. The results also showed the flow visualization of the micro flow at the mid channel in order to obtain the hydrodynamic entry length correlation for both laminar and turbulent flows.

Xiong and Chung [12] conducted an identical study as [7] did, namely examining the pressure gradient in a channel with the same hydraulic diameter. The test was carried out on three channels with a hydraulic diameter of 209μm, 395μm and 549μm. The transition flow conditions occured earlier than the macro channel theory, namely the Reynolds numbers ranging from 1500 to 1700. The bend loss coefficients have been calculated and it had three characteristics.

Qureshi et al. [13] conducted numerical simulation of blood flow and pressure drop in the pulmonary arterial and venous circulation. They studied the effects on flow and pressure associated with three classes of pulmonary hypertension expressed via stiffening of larger and smaller vessels, and vascular rarefaction. The results of simulating these pathological conditions were in agreement with clinical observations

Y Liu et al. [14] conducted a test regarding the effect of the relative roughness of the micro channel with air as a fluid. The effect of relative roughness on airflow has an impact on the difference in friction factor and is compared with conventional theory. The channel used is a micro channel with a hydraulic diameter of 400 μm with a relative roughness variation of 0.58% to 1.26% and tested at Reynolds numbers between 200 and 2100. The results of this study are that the resistance of flow increases with the proportion of roughness relatively. Dirker et al. [15] conducted an experimental investigation of pressure drop in the rectangle microchannels. It was determined that critical Reynolds number were between 1800 and 2000 for sudden contraction inlet type, and adiabatic friction factor were well predicted by laminar correlation formulated for rectangle macro channel.

2. Experimental Apparatus

The variations of the micro channel diameters in this study were 118 μm, 144 μm, 348 μm, and 1203 μm for straight micro channels. For bent micro channels, it only used the 348 μm, and 1203 μm. The sizes were picked randomly from available micro channel in the market. Angle variation for bent micro channel can be seen in Table 1. Measurement of the channel cross-sectional diameter was carried out using a Meiji Techno America microscope connected to a computer display.

Table 1. Angle variation for bent micro channel

Diameter (μm)	Angle (°)
1203	90, 120, 135
348	90, 120, 135

For channel measurements with very small cross-sections of 118 μm, 144 μm and 348 μm there were no obstacles when determining the length of the channel diameter, but when measuring the cross-sectional diameter of the channel above 1000 μm, the microscope had difficulty capturing the entire image with focus, so the resulting image was less than perfect. Especially for the channel with the largest cross-section, the image that can be captured is only about a quarter of the total image and it was decided to estimate the diameter by equation (1). The scheme for estimating the length of the diameter can be seen in Figure 1 with description of *W*, *H* and *R* as the measurement parameters.

$$R = \frac{W^2+H^2}{2H} \quad (1)$$

The variable that has been determined in this study is the volume of water flowing in the micro channel that is 60 ml for variations in diameters 1203 μm and 348 μm. Meanwhile, for the diameters of 144 μm and 118 μm, 10 ml of water were used. Determination of the volume of water injected into the micro channel triggered a difference in mass flow rate in data collection for each test. The selection of control variables was based on the length of time to collect data and the limitations of the syringe. Meanwhile, the dependent variable in this study was the value of pressure drop or the difference in pressure at the entry and exit of the test channel as measured on the micro channel.

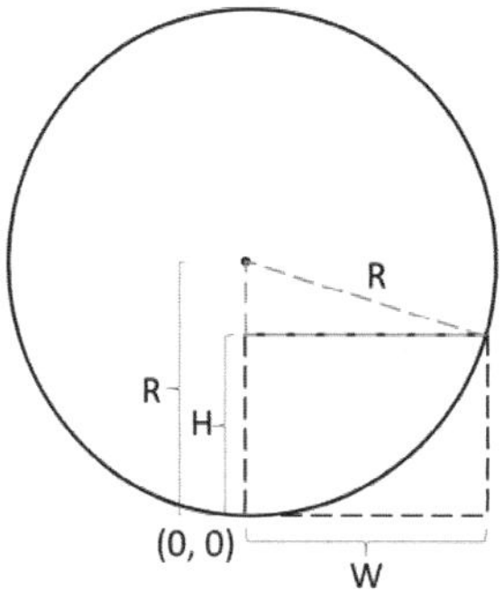

Figure. 1. Estimating the length of the largest channel cross-sectional diameter

Pressure measurement on the test line used two pressure transducers with different measuring ranges. In testing channels with a diameter of 348 µm to 1860 µm and a low Reynolds number, pressure transducer with a range of 0-5 psi was used for the inlet and a range of 0-15 psi pressure transducer for the outlet was used and vice versa for testing high Reynolds numbers. Meanwhile, for channels with a diameter of 118 µm and 144 µm, a pressure transducer with a range of 0-60 psi was used for high Reynolds numbers. This is because the smaller the diameter, even with a lower Reynolds number when compared to a larger diameter channel, will result in a greater pressure, thus requiring a sensor with a large range for measurement. The static pressures were measured at the diameter of micro channel respectively.

Pressure transducers calibration was carried out on using a pressure gauge with the same range. The pressure gauge firstly was calibrated, then pressure transducers were validated by using Pascal Law. The calibration was conducted by making a simple instrument that is able to suppress fluid in a closed channel with a pressure gauge and pressure transducer attached to the ends and sides. Validation was conducted manually by pushing the instrument and recording the results of the pressure transducer for every 1 psi increase. From the results collected, a calibration graph and a regression equation was generated.

Figure. 2. The test scheme uses a pressure transducer.

The method used was by measuring all channels in order to obtain the gross pressure drop value which was then reduced by the theoretical equation for pressure drop. Measurements were made using a pressure transducer that was placed on the connection between the tube from the injection and the micro channel. Figure 2 shows the experimental research scheme used to investigate the pressure drop of a water flow in a micro channel.

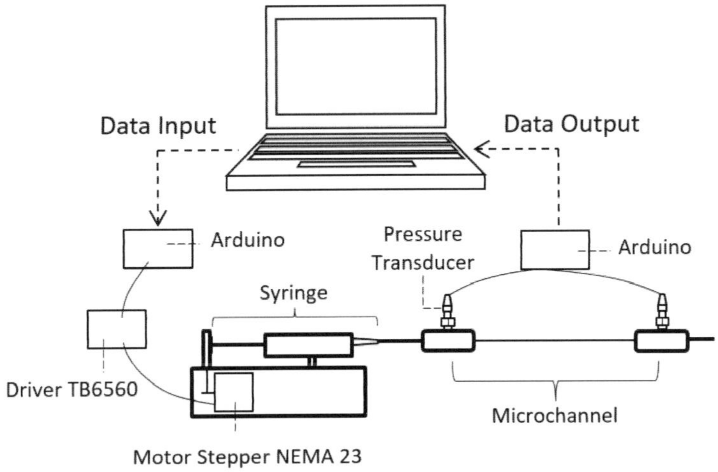

Figure 3. The experimental set up.

The overall research scheme can be seen in figure 3. This test used a custom syringe pump as a fluid booster. The stepper motor was connected to the syringe pump system and then controlled by the control program. The syringe was connected by a hose to the tee connector for input, one to the pressure transducer and the other to the test line. The test line connected to the tee output connector to the pressure transducer and the hose with atmospheric pressure (open system). The stepper motor used was a unipolar type with 6 pins, an input voltage of 12V, and a strong current of 0.8A. The stepper motor was connected to the TB6560 Driver before connecting to the control program. The voltage source to drive the motor used an adapter with an output voltage of 12V and a strong current of 1A. The speed of the stepper motor was controlled using the Arduino program.

Pressures were measured with pressure transducer before and after the test line, so the effect of the fluid at the entrance and exit cannot be neglected. The pressure transducer has a calibrated reading range of 0-5 psi, 0-15 psi and 0-60 psi. The reading and data acquisition of measured pressure values were also performed using the Arduino software.

3. Experimental Result and Discussion

Figure 5 shows the effect of Reynolds number over pressure drop per unit length (bar/cm) in microchannel, which has five diameter variations. All pressure drop tests were assumed in adiabatic conditions. The experimental results are compared to the theoretical calculation of Poiseuille equation as shown in equation 2.

$$\Delta p = \frac{128\mu LQ}{\pi D^4} \qquad (2)$$

Where P is pressure, D is diameter, L is length, Q is discharge, and μ is viscosity. As expected, the pressure drop per unit length increases as the Reynolds number increases. This is because the friction force increases with the increase in the average velocity of the fluid and the shear stress on the surface. As it can be seen, for channels with diameter 348 μm, 1203 μm and 1860 μm, pressure drop per length from the test results is almost similar to the Poiseuille theoretical calculation. This condition is exactly the same as the conventional diameter tube, at least in this study up to a channel with a diameter of 118 μm. Unlike three previous diameters, the two channels with the smallest diameter in this study, 144 μm and 118 μm, have a large gap between the test results and conventional theory. The experimental results of both are smaller than Poiseuille theoretical calculation; even so, the resulting trends are actually quite identical.

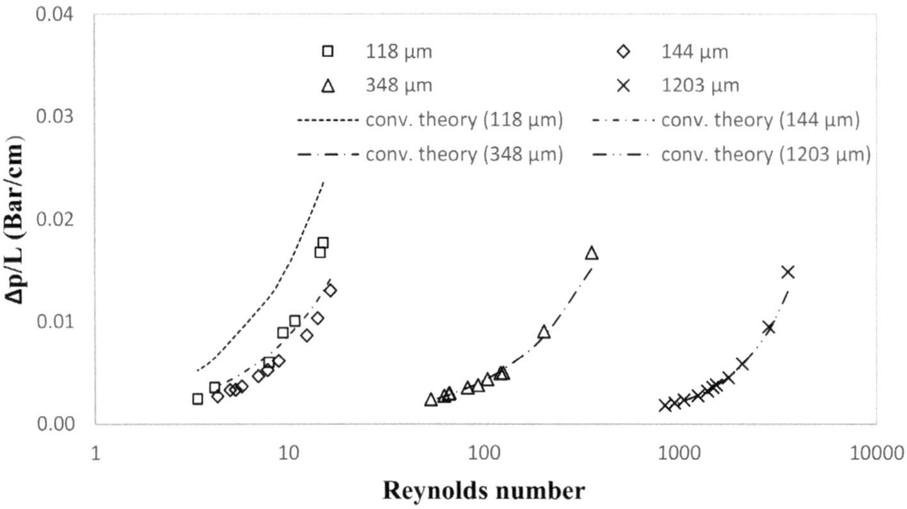

Figure 4. Comparison between theoretical and experimental ΔP with respect to Re

All the experimental data were also plotted in correlation of Reynolds number over friction factor in figure 5. The graph shows all measured data for all five channels. From the overall data it can be seen that the channels with larger diameters, i.e. 0.348 μm, and 1203 μm, the data agree well with laminar friction factor theoretical relation (f = 64/Reynolds number). The micro channel with a diameter of 144 μm has the same graph trend as conventional theory even though there is a sizable gap between them. The discrepancy between the experimental and theoretical results might due to the existing of other forces acting on the channel surface such interfacial tension and capillarity effect.

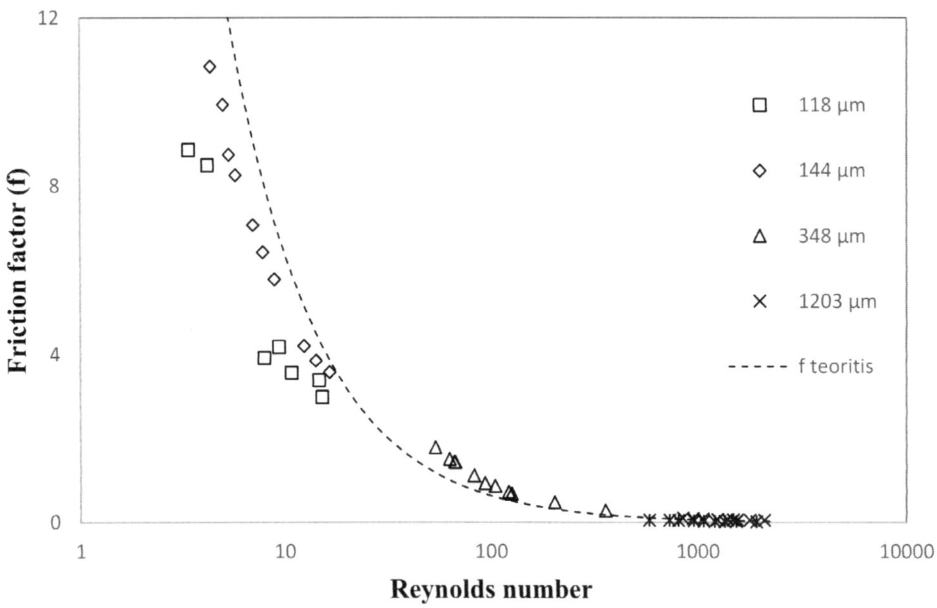

Figure 5. Comparison between theoretical and experimental *f* with respect to Re

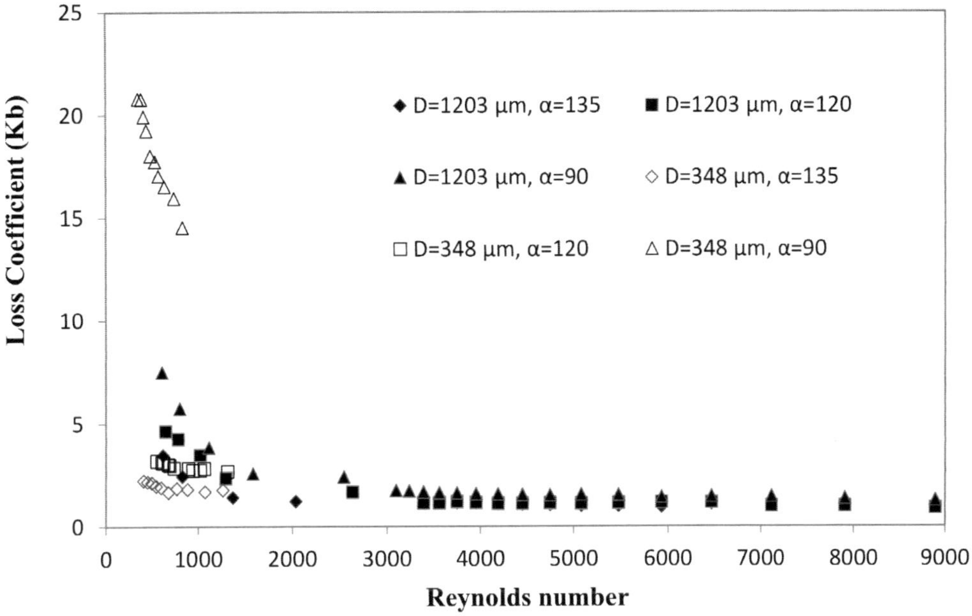

Figure 6. Relationship between the bent loss coefficient and the Reynolds number.

For the minor loss due to the bent channel, the results are presented in figure 6. The bent loss coefficients (k_b) for two diameter channel (348 and 1203) and three different angles (90^0, 120^0, 135^0) with Reynolds number from 500 to 9000 are presented. It is similar conclusion with Yamashita et al. [15] and Xiong and Chung [12] that the bent loss coefficient in laminar regime (according to macroscale theory) is larger than in turbulent regime. The results show that k_b is dependent of Reynolds number and decreases as Reynolds number increase, which is also different with turbulent flow. For macro channel turbulent flow at larger Reynolds number, K_b almost would not change with the increasing of Reynolds number. When Reynolds number is larger than some value in 1000, K_b almost keeps constant and changes in the range of ±10-15%.

4. Conclusion

In this paper, the experimental study of pressure drop on straight and bent microchannel were carried out. The microchannel had circular cross-sections with diameter 118 μm, 144 μm, 348 μm, and 1203 μm for straight and 348 μm, and 1203 μm for bent. The test has been carried out using a custom syringe pump and the fluid velocity is controlled using the Arduino software. Data acquisition is done by taking data for every second. From this study, following conclusion can be drawn.

- For straight microchannels, the ΔP/L experimental show agreement with conventional theory especially on larger diameter channels, however on channels of smaller diameter, there is a large enough gap. Friction factor for microchannels also shows agreement to the standard theory except for the channel with the smallest diameter, which actually have the same trend despite having a much larger gap.
- The bend loss coefficients for every test almost keeps constant and changes in the range of ±10-15% when Re > 1000, but increased significantly when Re goes to a smaller number.

Acknowledgement

We are grateful for receiving funding from the P3MI program at Institut Teknologi Bandung

References

[1] Darcy H 1857 Recherches Experimentales Relatives au Mouvement de L'Eau dans les Tuyaux *Mallet-Bachelier* Paris France

[2] Fanning J T 1886 A Practical Treatise on Hydraulic and Water Supply Engineering *Van Nostrand* New York

[3] Colebrook F C 1939 Turbulent flow in pipes, with particular reference to the transition region between the smooth and rough pipe laws J. Inst. Civ. Eng. Lond. **11** 133

[4] Nikuradse J 1933 Strmungsgesetze in Rauben Rohren, *Verein Deutsher Ingenieure, Forschungsheft*, p 361.

[5] Moody L F 1944 Friction Factors for Pipe Flow. Transactions of the American Society of Mechanical Engineers **66** 671-681

[6] Peiyi W and Little W A 1983 Measurement of friction factors for the flow of gases in very fine channels used for microminiature Joule-Thomson refrigerators *Cryogenics 23*(5) 273–277

[7] Mala M G and Li D 1999 Flow characteristics of water in microtubes *International Journal of Heat and Fluid Flow* **20(2)** 142–148

[8] D Pfund, D Retor, A Shekarriz, et al. 2000 Pressure drop measurements in a microchannel *AIChE J.* **46** 1496–1507

[9] Qu W and Mudawar I 2002 Experimental and numerical study of pressure drop and heat transfer in a single-phase micro-channel heat sink *International Journal of Heat and Mass Transfer* **45(12),** 2549–2565

[10] Wu H Y and Cheng P 2003 An experimental study of convective heat transfer in silicon microchannels with different surface conditions Int. J. Heat Mass Transf. **46** 2547–2556

[11] S.S. Hsieh, C.-Y. Lin, C.-F. Huang, H.-H. Tsai, Liquid flow in a micro-channel, *J. Micromech. Microeng.* **14** 436–445.

[12] Xiong R and Chung J N 2007 Flow characteristics of water in straight and serpentine micro-channels with miter bends *Experimental Thermal and Fluid Science* **31(7)** 805–812

[13] Qureshi M U, Vaughan G D A, Sainsbury C, Johnson M, Peskin C S, Olufsen M S and Hill NA 2014 Numerical simulation of blood flow and pressure drop in the pulmonary arterial and venous circulation *Biomech Model Mechanobiol.* **13(5)** 1137-1154

[14] Liu Y, Xu G, Sun J, and Li H 2015 Investigation of the roughness effect on flow behavior and heat transfer characteristics in microchannels *International Journal of Heat and Mass Transfer* **83** 11–20

[15] Dirker J, Meyer J P and Garach D V 2014 Inlet flow effects in micro-channels in the laminar and transitional regimes on single-phase heat transfer coefficients and friction factors Int. J. Heat Mass Transf. **77** 612–626

[16] H Yamashita, R Izumi, G Kushida and T Mizuno Fluid flow and heat transfer in a two-dimensional miter-bend (1st report, experiments and analysis) Bull. 1986 JSME **29**, 4164–4169

Investigation of solution differences on natural convection in a tall cavity simulated by using the Boussinesq approximation and the real air density equation

Uthai Prasopchingchana[1]

[1] Department of Mechanical Engineering, Burapha University, Chonburi 20131, Thailand

uthai@buu.ac.th

Abstract. Solution differences of natural convection in a tall cavity filled with air in which the density of the body force term are approximated by using the Boussinesq approximation (BA) and by using the real air density equation (RE) are numerically investigated. The main parameters for the investigation are Nusselt numbers and numbers of cells in the multi-cellular flow. The investigation is carried out for the Rayleigh number of 10^4, and the cavity aspect ratio equal to 60. The temperature differences of the vertical cavity walls are varied from 1 K to 200 K. These conditions are appropriate to form the multi-cellular flow in the tall cavity obviously, and close to the boundary of the laminar and turbulent flows. Consequently, the differences in both Nusselt numbers and numbers of the cells in the multi-cellular flow can be easily observed. The finite volume method with the Lagrange interpolating polynomial (LIP) scheme are selected to discretize the governing equations of the flow, and the Semi-implicit method for pressure linked equations (SIMPLE) algorithm is used to determine the pressure field of the flow. The maximum difference of the average Nusselt numbers of natural convection in the tall cavity simulated by using BA and RE is 0.436 % for the range of the investigation. This indicates that the difference of heat transfer solutions of natural convection in the tall cavity simulated by using BA and RE is insignificant for practical applications, but there is an obvious difference on the numbers of the cells in the multi-cellular flow, when the temperature difference of the vertical cavity walls is greater than or equal to 120 K.

1. Introduction

Natural convection in a tall cavity plays an important role for heat transfer mode in the air gaps of solar collectors and double-glazed windows. This phenomenon has been investigated by many researchers in both experimental implementations [1] and numerical simulations [2-13]. For the numerical simulations of the phenomenon, many researchers assumed that airflow in the cavity was incompressible, because the velocity of the flow was quite slow. But the air density of the body force term in the vertical momentum equation was not included in the assumption, because this term was the main factor to generate the buoyancy force for air circulation. For the approximation of the air density in the body force term, the numerous researchers [2-8, 10-13] adopted the Boussinesq approximation (BA), due to the approximation gives stability for computation at the small temperature difference and consumes low computing resources. BA defines the compressible air density as the function of the temperature. For the condition of the tall cavity with the temperature difference between the left and right walls, and the large pressure difference between the bottom and top areas of the cavity, using BA to simulate the phenomenon may obtain inaccurate solutions. To ensure that the solutions computed from the density by using BA at these conditions are consistent with the phenomenon, the solution accuracy must be investigated by comparing the solutions computed from the density by using BA with the solutions computed from the density by using the real air density equation (RE).

The objective of this paper is to numerically investigate the solution differences of natural convection in a tall cavity simulated by using BA and RE. The investigation emphasized heat transfer across the cavity and the pattern of airflow in the cavity.

The subsequent contents of this paper are organized as follows: Section 2. the mathematical formulation for simulation is depicted; Section 3. numerical implementation is described, and code validation is performed to ensure that the solutions computed from the code are correct; Section 4. results are displayed, and discussion is carried out; and in the final section, the conclusion of the results is presented.

2. Mathematical formulation

Two-dimensional simulation of natural convection in the tall cavity is defined by the continuity equation, the momentum equations and the energy equation. The equations are expressed as:

$$\frac{\partial u}{\partial x}+\frac{\partial v}{\partial y}=0 \tag{1}$$

$$\rho\left(\frac{\partial u}{\partial t}+u\frac{\partial u}{\partial x}+v\frac{\partial u}{\partial y}\right)=-\frac{\partial p}{\partial x}+\mu\left(\frac{\partial^2 u}{\partial x^2}+\frac{\partial^2 u}{\partial y^2}\right) \tag{2}$$

$$\rho\left(\frac{\partial v}{\partial t}+u\frac{\partial v}{\partial x}+v\frac{\partial v}{\partial y}\right)=-\frac{\partial p}{\partial y}+\mu\left(\frac{\partial^2 v}{\partial x^2}+\frac{\partial^2 v}{\partial y^2}\right)-\rho_{i,j}g \tag{3}$$

$$\rho c_p\left(\frac{\partial T}{\partial t}+u\frac{\partial T}{\partial x}+v\frac{\partial T}{\partial y}\right)=k\left(\frac{\partial^2 T}{\partial x^2}+\frac{\partial^2 T}{\partial y^2}\right) \tag{4}$$

where u, v, p, and T are the horizontal and vertical velocities, the pressure and the temperature, respectively. Then, ρ, μ, c_p, k, and g are the constant density, the viscosity, the specific heat at the constant pressure, the thermal conductivity and the gravitational acceleration, respectively. Also, $\rho_{i,j}$ is the compressible density. The compressible density is the main value resulting in airflow because of the buoyancy force. The compressible density can be defined from BA or RE.

For BA, the compressible density can be calculated from

$$\rho_{i,j}=\rho_0\left(1-\beta\left(T_{i,j}-T_0\right)\right) \tag{5}$$

where T_0, ρ_0 and β are the reference temperature, the density at the reference temperature, and the volumetric thermal expansion coefficient, respectively. The volumetric thermal expansion coefficient can be calculated from

$$\beta=\frac{1}{T_0} \tag{6}$$

For RE, the equation of the compressible density is the function of the temperature and pressure and can be written as

$$\rho_{i,j}=\sum_{mp=1}^{mpden}\left(\left(\sum_{mt=1}^{mtden}\left(a_{mt,mp}T_{i,j}^{mt-1}\right)\right)p_{i,j}^{mp-1}\right) \tag{7}$$

where $a_{mt,mp}$ is the coefficient of the equation obtained from two variable regression. The data for the regression is received from the National Institute of Standards and Technology. The values *mtden* and *mpden* in the equation are 6. Consequently, the determination coefficient (R^2) equals to 0.999 for the equation.

Figure 1. shows the geometry of the "natural convection in a tall cavity problem. The letters *H* and *W* are the height and width of the cavity, respectively. And the letters T_{hot} and T_{cold} are the hot and cold temperatures on the vertical cavity walls. The horizontal cavity walls are well insulated. The velocities on the cavity walls are treated as a no-slip condition. For the "natural convection in a tall cavity" problem, the reference temperature in BA can be computed from

$$T_0 = \frac{T_{hot} + T_{cold}}{2} \quad (8)$$

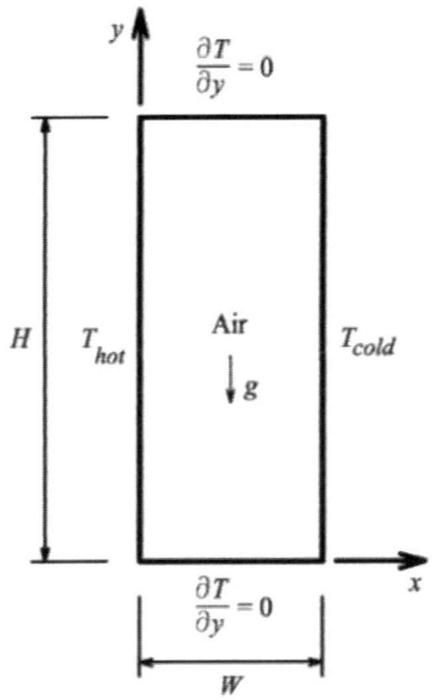

Figure 1. Geometry of the "natural convection in a tall cavity" problem.

3. Numerical implementation and code validation

To reach the purpose of this study, a new code was developed. The finite volume method with the Lagrange interpolating polynomial (LIP) scheme [14] and the Semi-implicit method for pressure linked equations (SIMPLE) algorithm were employed to generate the discretized equations from the governing equations of the airflow. The Gauss-Siedel iteration method was used to solve the discretized equations. Non-uniform grids were used for simulation based on fine grids closed to the cavity walls and coarse grids laid on the core area of the cavity. The proportion of the spaces between the grids for the coarse grids to the fine grids was 1.01. The dimensionless value of the time step was 0.001 for the code. The convergent criterion in each time step was that the relative difference of present loop solutions and previous loop solutions was less than or equal to 10^{-4} for ten consecutive loops. To ensure that the solutions computed from the code were correct, code validation must be performed. The comparison of the solutions computed from the code with the solutions obtained from the correlation equation of the experimental data and the published numerical solutions was selected for the code validation.

The main solution of the natural convection is the average Nusselt number. The average Nusselt number $\left(\overline{Nu}\right)$ is a dimensionless solution showing the ratio of convection heat transfer to conduction heat transfer. $\overline{Nu}$ is defined as

$$\overline{Nu} = \frac{\left(\left(\frac{1}{2H}\right)\int_0^H \left(q''\big|_{x=0} + q''\big|_{x=W}\right)dy\right)}{\left(\left(\frac{k}{W}\right)\left(T_{hot} - T_{cold}\right)\right)} \quad (9)$$

where q'' is the heat flux on the surfaces of the vertical cavity walls. q'' is calculated from

$$q'' = \frac{-k\,dT}{dx} \tag{10}$$

Table 1. shows the variation of $\overline{Nu}$ computed from the code due to the change of grid sizes. At the grid size 35 × 200, the variation of $\overline{Nu}$ is equal to zero. This means that the amount of $\overline{Nu}$ is independent of the grid sizes. The variation is computed from

$$\text{Variation}(\%) = \frac{\left(\left|\overline{Nu}_{\text{fine grid}} - \overline{Nu}_{\text{coarse grid}}\right|\right)\times 100}{\overline{Nu}_{\text{fine grid}}} \tag{11}$$

Table 2. displays the comparison of the solutions of the natural convection between $\overline{Nu}$ yielded by the code with $\overline{Nu}$ obtained from the correlation equation of the experimental data and $\overline{Nu}$ published in the literature at the Rayleigh number (Ra) equal to 1.1 × 10^4 and the cavity aspect ratio (AR) equal to 16. The maximum differences of $\overline{Nu}$ are 6.0912 % and 6.1037 % for comparing between the present study BA and RE with $\overline{Nu}$ reported in [1], respectively. Ra is yielded from

$$Ra = \frac{g\,\beta\,c_p\,\rho^2\left(T_{hot} - T_{cold}\right)W^3}{\mu k} \tag{12}$$

And, AR is calculated from

$$AR = \frac{H}{W} \tag{13}$$

Figure 2. shows the temperature and streamline contours at $Ra = 1.1\times10^4$ and $AR = 16$ for the natural convection. The contour patterns of the present study BA and RE are almost the same as the pattern shown in [13].

Table 1. Grid independence test of $\overline{Nu}$ of the natural convection computed from the code at $Ra = 1.1\times10^4$ and $AR = 16$.

Grid size	$\overline{Nu}$	Variation (%)
20 × 160	1.5235	-
20 × 180	1.5236	0.0066
20 × 200	1.5236	0.0000
25 × 200	1.5237	0.0066
30 × 200	1.5235	0.0131
35 × 200	1.5235	0.0000

Table 2. Comparison of the solution of the natural convection between $\overline{Nu}$ yielded by the code with $\overline{Nu}$ obtained from the correlation equation of the experimental data and $\overline{Nu}$ published in the literature at $Ra = 1.1 \times 10^4$ and $AR = 16$.

Study	$\overline{Nu}$	Implementation
Present study (BA)	1.5235	Numerical method
Present study (RE)	1.5237	Numerical method
[1]	1.4307	Experiment
[13]	1.5250	Numerical method
[2]	1.5351	Numerical method

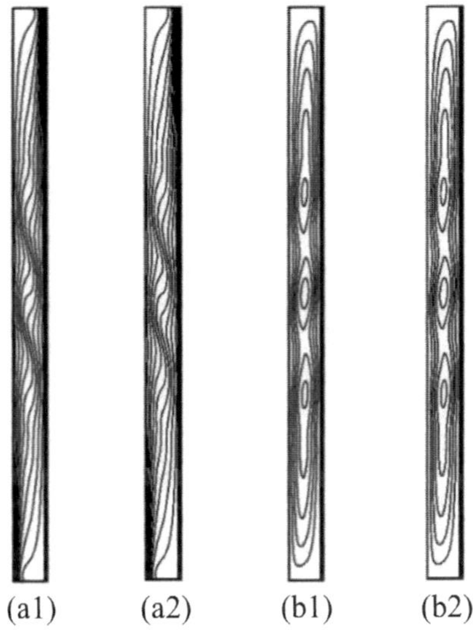

Figure 2. Contours of temperature and streamline at $Ra = 1.1 \times 10^4$ and $AR = 16$ for the natural convection
(a1) temperature contour BA,
(a2) temperature contour RE,
(b1) streamline contour BA, and
(b2) streamline contour RE.

4. Results and discussion

The investigation of the solution differences on the natural convection simulated by using the BA and RE is performed for $Ra = 10^4$. For $Ra = 10^4$ with the appropriate cavity aspect ratio, airflow in the cavity is a special flow called multi-cellular flow. Therefore, the solution differences on the natural convection simulated by using the BA and RE is easily observed by comparing the cell numbers and cell center locations of the flow. In addition, the main parameters that affect the differences of the solutions obtained from the simulation of the natural convection simulated by using BA and RE are the temperature difference between the left and right walls of the cavity and the pressure difference between the bottom and top areas of the cavity. Therefore, to obviously illustrate the differences of solutions obtained from the simulation of the natural convection simulated by using BA and RE, the simulation is performed until the highest value of the temperature difference of the vertical cavity walls and the aspect ratios of the cavity are at the boundary of the laminar and turbulent flows. Then the simulation of the natural convection cavity simulated by using BA and RE is carried out at 1 K to 200 K of the temperature difference between the left and right walls of the cavity (ΔT) and the aspect ratio of the cavity equal to 60. The grid independence test of the simulation is performed until the variation of the average Nusselt numbers is less than or equal to 0.1 %. Consequently, grid size 40 × 500 is selected for simulation.

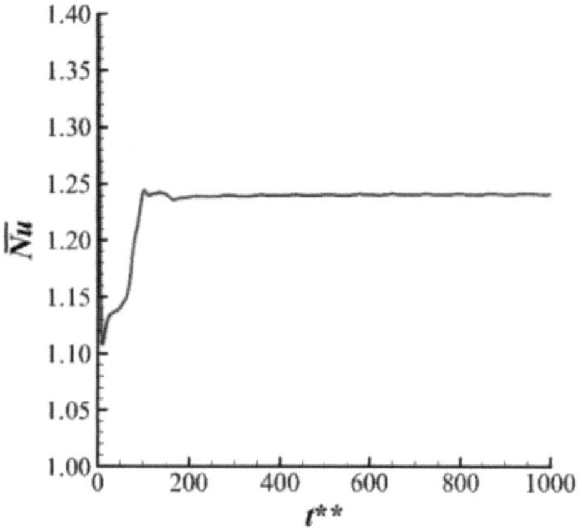

Figure 3. Variation of $\overline{Nu}$ with the dimensionless time which is simulated by using RE for $Ra = 10^4$, $\Delta T = 200$ K and $AR = 60$.

Due to the instability of the solutions obtained from the simulation of the natural convection simulated by using RE, the solutions at the dimensionless time (t^{**}) equal to 1,000 are assumed to be the solutions of the problem at the steady-state condition. Figure 3. shows that the solutions of the natural convection simulated by using RE at $Ra = 10^4$, $\Delta T = 200$ K and $AR = 60$ are almost constant at the dimensionless time greater than or equal to 200. Therefore, it ensures that the solutions at the dimensionless time equal to 1,000 are the steady-state solutions. t^{**} is obtained from

$$t^{**} = t\sqrt{\frac{g\,\beta\left(T_{hot} - T_{cold}\right)}{W}} \tag{14}$$

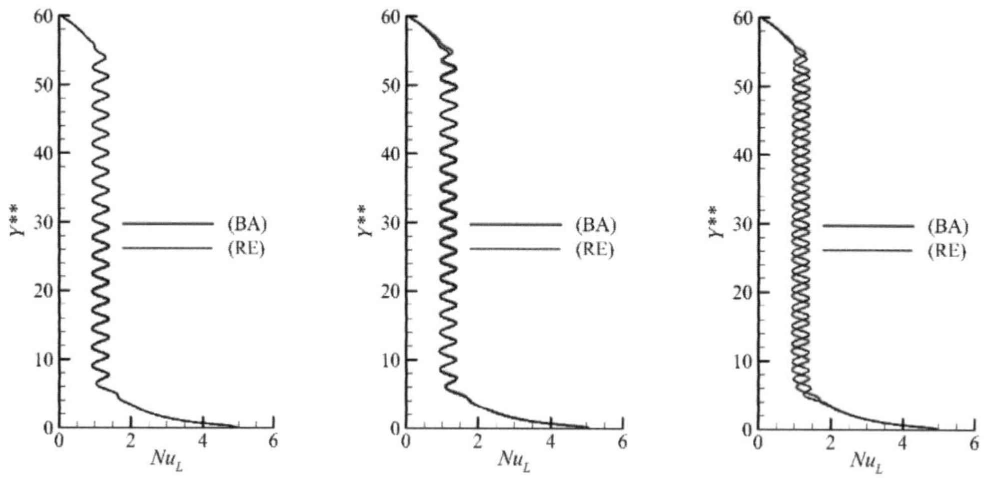

(a) (b) (c)

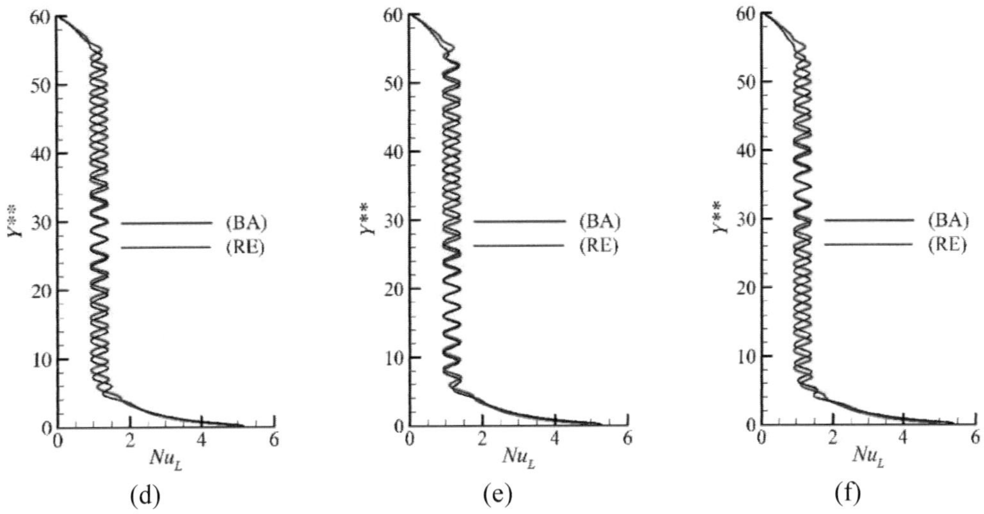

Figure 4. Nu_L along the left walls of the tall cavity for $Ra = 10^4$ and $AR = 60$
(a) $\Delta T = 1$ K, (b) $\Delta T = 40$ K, (c) $\Delta T = 80$ K,
(d) $\Delta T = 120$ K, (e) $\Delta T = 160$ K and (f) $\Delta T = 200$ K.

The local Nusselt numbers (Nu_L) calculated along the left wall of the cavity for $Ra = 10^4$ with $AR = 60$, and ΔT equal to 1 K and 200 K are shown in Figure 4., where Y^{**} is the dimensionless distance along the left wall. Graph lines of Nu_L computed from BA and RE are an oscillating pattern. There are some differences between the graph lines of Nu_L computed from BA and RE when ΔT is greater than or equal to 80 K. Y^{**} is given by

$$Y^{**} = \frac{y}{W} \tag{15}$$

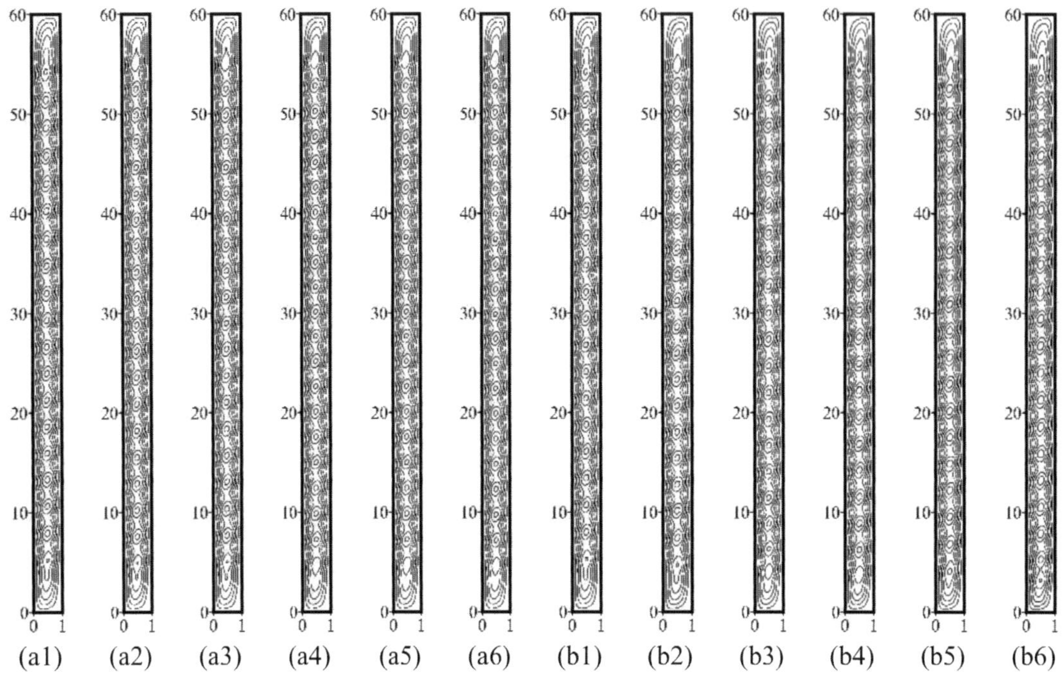

Figure 5. Streamlines of airflow in the cavity for $AR = 60$

(a1) BA, $\Delta T = 1$ K, (a2) BA, $\Delta T = 40$ K, (a3) BA, $\Delta T = 80$ K,

(a4) BA, $\Delta T = 120$ K, (a5) BA, $\Delta T = 160$ K, (a6) BA, $\Delta T = 200$ K,

(b1) RE, $\Delta T = 1$ K, (b2) RE, $\Delta T = 40$ K, (b3) RE, $\Delta T = 80$ K,

(b4) RE, $\Delta T = 120$ K, (b5) RE, $\Delta T = 160$ K and (b6) RE, $\Delta T = 200$ K.

Figure 5. exhibits the streamline contours of the airflow in the tall cavity which are computed from BA and RE for $AR = 60$ at the varied difference of the temperatures between the left and right walls of the cavity. The patterns of the airflow in the cavity show almost the same figure, but some locations of the cell centers are different.

Table 3. displays the comparison of $\overline{Nu}$ and the numbers of the cells in the multi-cellular flow of the natural convection, which are computed from BA and RE for $Ra = 10^4$ and $AR = 60$, where X^{**} and Y^{**} are the dimensionless Cartesian coordinate components in the horizontal and vertical directions, respectively. The maximum difference between $\overline{Nu}$ computed from BA and RE is 0.436 % occurring at $\Delta T = 120$ K. The difference between the numbers of the cells in the multi-cellular flow computed from BA and RE is one cell, when the temperature difference is greater than or equal to $\Delta T = 120$ K. X^{**} is computed from

$$X^{**} = \frac{x}{W} \tag{16}$$

Table 3. Solutions of the natural convection for $Ra = 10^4$ and $AR = 60$.

ΔT (K)	Air density in the body force term	$\overline{Nu}$	Number of cells in the multi-cellular flow	Cell center location of the lowest cell (X^{**}, Y^{**})	Cell center location of the highest cell (X^{**}, Y^{**})
1	BA	1.2325	19	(0.4856, 5.0001)	(0.4856, 54.3976)
	RE	1.2331	19	(0.4856, 5.3614)	(0.4856, 54.1566)
40	BA	1.2387	20	(0.4856, 5.0001)	(0.4856, 55.3614)
	RE	1.2381	20	(0.4856, 4.7590)	(0.4856, 55.3625)
80	BA	1.2390	20	(0.4856, 5.0001)	(0.4856, 55.3614)
	RE	1.2371	20	(0.4856, 4.2771)	(0.4856, 54.2772)
120	BA	1.2438	21	(0.4856, 4.8795)	(0.4856, 55.4819)
	RE	1.2384	20	(0.4856, 4.0361)	(0.4856, 54.3976)
160	BA	1.2438	21	(0.4856, 4.8795)	(0.4856, 55.4819)
	RE	1.2399	20	(0.4856, 4.0361)	(0.4856, 54.8795)
200	BA	1.2439	21	(0.4856, 4.8795)	(0.4856, 55.4819)
	RE	1.2412	20	(0.4856, 3.4337)	(0.4856, 53.5542)

5. Conclusion

The solution differences of the natural convection simulated by using BA and RE are investigated for $Ra = 10^4$ and $AR = 60$. The main parameters of the phenomenon, Nusselt number, and the flow pattern, multi-cellular, are compared between the solutions computed from the BA and RE. The difference of the average Nusselt numbers computed from BE and RE is quite small and insignificant in practical applications. The multi-cellular flow computed from BA and RE shows the same feature, and there are some bit differences of the cell center locations. But, there is an obvious difference in the numbers of the cells in the multi-cellular flow, when the difference of the temperatures between the left and right walls of the cavity is greater than or equal to 120 K.

Acknowledgment

The author appreciates Thai National e-Science Infrastructure Consortium for support on computing resources.

References

[1] ElSherbiny S M, Raithby G D and Hollands K G T 1982 Heat transfer by natural convection across vertical and inclined air layers *Journal of Heat Transfer* **104** 96-102

[2] Báez E and Nicolás A 2013 From cats eyes to multiple disjoint natural convection flow in tall cavities: A direct primitive variables approach *Physics Letters, Section A: General, Atomic Solid State Physics* **377** 2270-2274

[3] Christon M A, Gresho P M and Sutton S B 2002 Computational predictability of time-dependent natural convection flows in enclosures (including a benchmark solution) *Int. J. Numer. Methods Fluids* **40** 953-980

[4] Dunn T A and McCallen R C 2002 Parallel computations of natural convection flow in a tall cavity using an explicit finite element method *Int. J. Numer. Methods Fluids* **40** 1075-1081

[5] Kizildag D, Rodríguez I, Oliva A and Lehmkuhl O 2014 Limits of the Oberbeck-Boussinesq approximation in a tall differentially heated cavity filled with water *Int. J. Heat Mass Transf.* **68** 489-499

[6] Nicolás A, Báez E and Bermúdez B 2011 From cat's eyes to disjoint multicellular natural convection flow in tall tilted cavities *Physics Letters, Section A: General, Atomic and Solid State Physics* **375** 2683-2688
[7] Pons M 2008 Transition from single-to multi-cell natural convection of air in cavities with an aspect ratio of 20: A thermodynamic approach *Int. J. Thermodyn.* **11** 71-79
[8] Rathore S K and Das M K 2016 Investigation on the relative performance of various low-Reynolds number turbulence models for buoyancy-driven flow in a tall cavity *Heat and Mass Transfer/Waerme- und Stoffuebertragung* **52** 437-457
[9] Suslov S A and Paolucci S 1995 Stability of natural convection flow in a tall vertical enclosure under non-Boussinesq conditions *Int. J. Heat Mass Transf.* **38** 2143-2157
[10] Wang C, Gao G, Zhao L, Schuetz W and Xu Y 2017 Accurate numerical simulation of turbulent natural convection in an enclosed tall cavity *Proc. Int. Conf. on 10th International Symposium on Heating, Ventilation and Air Conditioning* **205** 1733-1740
[11] Xamán J P, Hinojosa J F, Flores J J and Cabanillas R E 2008 Effect of the surface thermal radiation on turbulent natural convection in tall cavities of façade elements *Heat and Mass Transfer/Waerme- und Stoffuebertragung* **45** 177-185
[12] Yang H and Zhu Z 2008 Numerical study of three-dimensional turbulent natural convection in a differentially heated air-filled tall cavity *Int. Commun. Heat Mass Transf.* **35** 606-612
[13] Zhu Z J and Yang H X 2003 Numerical investigation of transient laminar natural convection of air in a tall cavity *Heat and Mass Transfer/Waerme- und Stoffuebertragung* **39** 579-587
[14] Prasopchingchana U and Manewattana T 2015 A new scheme for the finite volume method verified with two dimensional laminar natural convection in a square cavity *Eng. J.* **19** 133-152

IOP Conf. Series: Materials Science and Engineering 1109 (2021) 012047 doi:10.1088/1757-899X/1109/1/012047

Multi-sector decomposition analysis of Philippine CO_2 emissions

E A S Ng[1], N S A Lopez[1]

[1]Mechanical Engineering Department, De La Salle University, Manila, Philippines

earle_anderson_ng@dlsu.edu.ph

Abstract. Global CO_2 emissions are rising due to the rising energy demand and the extent to which this is fed by fossil fuels. The LMDI method of IDA is a valuable tool for determining the contributing driving forces behind emissions in a particular time frame, region, or economy. This study applies the techniques of spatial LMDI across the industry sectors of the 2017 Philippine economy to understand the effects of value share, energy intensity, energy structure, and fuel types on carbon emissions per sector. A combination of IEA, PSA, and IPCC data are used in order to ascertain these effects based on the Kaya identity. Results and tabulated and visualized in Cartesian scatter plots. Trends identified include the negative relationship between value share and energy intensity effects and between oil use and natural gas use effects. Outliers to these trends and in the Figures were also highlighted and elucidated. The study recommended increased use of natural gas due to its lower emission factor vs oil and coal, government assistance to low-value sectors in procurement of more efficient equipment, and bespoke solutions per sector based on the given data herein.

1. Introduction

Energy is required by all economic and societal activities [1,2]. The global energy demand entails carbon dioxide (CO_2) and greenhouse gas (GHG) emissions to the extent that fossil fuels are used to fulfill them [2–4]. Furthermore, the US Energy Information Administration forecasts further increases to global energy consumption [4,5].

There is a need to understand the various drivers of carbon emissions in order to craft effective development policy for mitigating them [6]. The index decomposition analysis (IDA) techniques have enabled researchers to analyze the varying effect of a parameter to changes in carbon emissions over time [7]. For purposes of examining energy use and CO_2 emissions, IDA also requires less data than the other factor decomposition method: structural decomposition analysis [8]. Moreover, it has proven to be more frequently used among the 2 methods [7]. Among the various IDA techniques, the logarithmic mean Divisia index (LMDI) method has so far been the most accepted tool for the investigation of energy problems [9]. It is capable of expressing each contributing factor as a rate of change. Thus, it displays the effect of each factor to carbon emissions per year relative to a base year [10,11].

For an investigation across several places (e.g., countries or regions) rather than time periods, the LMDI method may still be used for the analysis of relative contributions of activity level, energy intensity, energy structure, and other factors contributory to carbon emissions [12–14]. In this method, the contribution of each effect to carbon emissions of a particular place are calculated relative to a base country/region. This basis is purely theoretical and represents the average value of all the places under consideration [12]. Furthermore, the advantage of the LMDI relative to other IDA methods (e.g., Laspeyres, Shapley-Sun, etc.) lies in its ability to handle zero values without leaving residuals in analysis, its simplicity in computation, and the ease at which equations are formed [15]. This method also incorporates the Kaya identity. It was originally posited in 1989 by Yoichi Kaya, a Japanese scholar, at a UN IPCC seminar. It summarizes the relative roles of population levels, energy, and economic scale towards carbon emissions generation in a single equation [16].

Being the largest emitter of carbon emissions in the world and the 2nd largest economy, China has been the subject of several Kaya-LMDI-based carbon emissions studies [11,13,17,18]. Several developed economies have also been studied through this method [19,20]. As the Philippine economy continues to grow alongside the ASEAN economy as a whole, there is a need to ascertain the degree to which certain industries are affected by energy structure, value added, among others. Such an investigation would aid decision makers in decoupling these economic activities from carbon emissions typically associated therewith [2,3,21].

As evidenced by the literature cited in preceding paragraphs, the Kaya-LMDI methods have been frequently applied on a temporal or spatial analysis. This novelty of this study is in the modification of these methods as applied to multiple sectors modeled in a similar fashion to places in the spatial LMDI analysis. This technique is applied here to the Philippine setting. The research gap is addressed through the new application of the LMDI methodology to this setting in particular. The general objective of this study is to determine the drivers of carbon emissions across the various Philippine sectors. Specifically, the study aims to use a multi-sector Kaya-LMDI method to determine the magnitudes of the various contributing factors affecting carbon emissions in the Philippines from 2017 data, to present an interpretation of these effects and trends across the industries, and to provide useful insights for decision makers regulating and leading these industries.

2. Methods and Data

2.1. *Accounting CO_2 emissions*

The CO_2 emissions from each sector were obtained from their energy sources the 2017 Final Consumption International Energy Agency (IEA) Sankey Diagram [22]. The electricity component of this Final Consumption energy was subsequently deconsolidated into the fuel mix of the Philippine electrical supply for that year as stated in the Philippine Statistical Yearbook (PSY) 2018 [23]. In doing so, final consumption of a particular fuel type could be obtained in terms of energy.

CO_2 emitted per sector was approximated directly from multiplying the energy consumed from the said source and multiplying this by the emission factor according to the Intergovernmental Panel on Climate Change (IPCC) Emission Factor Database (EFDB) based on the 2006 IPCC guidelines for national greenhouse gas inventories [24]. The following Equation (1) is used:

$$C_i^j = (E_i^j + S^j L_i)EF^j \tag{1}$$

where C refers to CO_2 emissions in kg; the subscript i refers to the sector and superscript j refers to the fuel type; E refers to energy of fuel j consumed by sector i in terajoules (TJ); S refers to the energy share of fuel j in the electricity supply in %TJ/TJ; L refers to the energy electricity used by sector i in TJ; EF refers to the emission factor of fuel type j in terms of kg of CO_2/TJ.

Table 1. Summary of emission factors used in the study.

Fuel Type	Emission Factor (kg CO_2/TJ)
Coal	94600[a]
Oil	73300[a]
Natural Gas	64200[a]
Biomass	N/A
Other Renewables (Hydro, solar, or wind)	N/A

[a] [24]

2.2. *Index decomposition analysis*

The IDA, in particular the LMDI method, is used in this paper to analyze the driving forces behind CO_2 emissions in each sector of the Philippines as delineated by the IEA. The sectoral decomposition analysis here is drawn from the radial-regional (R-R) spatial decomposition analysis model due to its simplicity over the bilateral-regional (B-R) model for large numbers of regions in consideration [12]. Whereas it was used for a multi-country analysis, it shall be applied here for a multi-sector analysis.

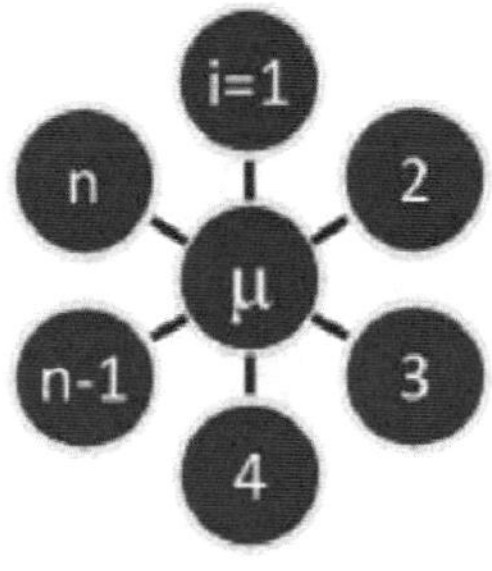

Figure 1. The R-R multi-sector decomposition model.

Figure 1 shows that the multi-sector decomposition model holds the various sectors against a base sector (μ) created from the average of all sector parameters. This theoretical base sector will now serve as the origin point in all the succeeding Cartesian plane scatter plots describing one effect in relation to another.

IDA identity used in this study is as shown in Equation (2):

$$C = \sum_{ij} C_{ij} = \sum_{ij} \left[(V) \left(\frac{V_i}{V} \right) \left(\frac{E_i}{V_i} \right) \left(\frac{E_{ij}}{E_i} \right) \left(\frac{C_{ij}}{E_{ij}} \right) \right] = \sum_{ij} \left[(V)(N_i)(I_i) \left(F_{ij} \right) \left(M_{ij} \right) \right] \quad (2)$$

where C refers to the CO_2 emissions in kg, V is the total gross value added among considered sectors in millions of Philippine pesos; V_i is the gross value added by sector i in millions of Philippine pesos; E_i is the total energy consumed by sector i in TJ; E_{ij} is the energy consumed by sector i originating from fuel type j also in TJ; C_{ij} refers to the emissions of sector i originating from consumption of fuel type j in kg of CO_2; N_i is the share of sector i in the gross value added by all sectors considered by the study in PHP/PHP; I_i is the energy intensity of sector i in terms of TJ/10^6 PHP; F_{ij} is the share of the fuel type j among the total fuel consumed by sector i in TJ/TJ; M_j is the emission factor of fuel type j used in sector i in kg/TJ.

In order to calculate the effect of each driving factor on the carbon emissions of a sector, the following equations are used:

$$\Delta C^S_{val} = \frac{C_i^S - C_i^0}{\ln ln\, C_i^S - lnC_i^0} \left(lnN_i^S - lnN_i^0 \right) \quad (3)$$

$$\Delta C^S_{int} = \frac{C_i^S - C_i^0}{\ln ln\, C_i^S - lnC_i^0} \left(lnI_i^S - lnI_i^0 \right) \quad (4)$$

$$\Delta C^S_{str} = \sum_j \left[\frac{C_{ij}^S - C_{ij}^0}{\ln ln\, C_{ij}^S - lnC_{ij}^0} \left(lnF_{ij}^S - lnF_{ij}^0 \right) \right] \quad (5)$$

The superscripts S and 0 in Equations (3), (4), and (5) refer to a particular sector and to the base sector, respectively. C refers to the CO_2 emissions in kg, and ΔC refers to the change in emission levels arising from the effect. The subscript i indicates sector, while the subscript j indicates the fuel type. Finally, the subscripts val, int, and str refer to value share effect, intensity effect, and energy structure effect, respectively.

2.3. Data sources
The data for this study has been taken from several reputable sources: the IEA, the IPCC, and PSY 2018 [22–24]. The gross value added across the sectors considered by this study along with their summation for 2017 have been taken from the PSY2018. In addition, the PSY2018 also provided the fuel mix for electric generation in the Philippines for the year 2017 [23]. Meanwhile, the annual consumption of energy, and fuel shares thereof have been obtained from the IEA Sankey Diagram of Final Consumption [22]. Lastly, the emission factors for calculating the emissions per sector were obtained from the IPCC Guidelines for Greenhouse Gas Inventories accessed via the IPCC EFDB [24].

3. Results and Discussion

Table 2. Driving forces of carbon emissions in billions of kg of CO_2 per industry sector.

	Sector	*Value* Share Effect	*Intensit y* Effect	*Energy Structur e Effect*	*Coal Use Effect*	*Oil Use Effect*	*Natural Gas Use Effect*	Sum of Effects
A	Iron and Steel	-5.2052	17.2598	0.1280	1.5778	-1.7003	0.2505	12.3106
B	Chemical and petro	1.0339	-5.2070	0.0147	0.0100	-0.0905	0.0952	-4.1438
C	Non-metallic minerals	-15.7227	-11.3686	2.5656	5.5653	-2.9298	-0.0700	-21.9602
D	Machinery	-4.2448	6.6127	-0.2385	0.6784	-1.2704	0.3536	1.8909
E	Mining and quarrying	-2.4921	-0.3829	0.0227	-0.3163	0.3074	0.0316	-2.8295
F	Food and tobacco	4.8745	-19.6265	-2.2371	0.2608	-2.5719	0.0740	-19.2261
G	Paper pulp and print	-3.3293	0.2160	-0.5186	1.0271	-1.6722	0.1265	-4.1505
H	Wood and wood products	-3.1651	-0.2891	-0.3578	0.3015	-0.7742	0.1149	-4.1700
J	Construction	2.1671	-6.5331	-0.0366	-0.5206	0.5039	-0.0198	-4.4392
K	Textile and leather	-1.6544	-2.1959	0.0830	0.2685	-0.1490	-0.0366	-3.6843
L	Non-spec industry	-3.3084	0.1722	-0.6133	0.3348	-1.2932	0.3451	-4.3628
M	Land transportation	-1.1785	-3.8531	2.5713	-1.5979	4.3291	-0.1599	0.1109
N	Domestic aviation	-3.2817	1.7482	0.8059	-0.0182	0.8683	-0.0442	0.0783

O	Domestic navigation	-7.9722	30.4771	0.5661	-0.5213	1.1330	-0.0456	23.6370
P	Agri forestry	2.4081	-6.5590	-0.2504	0.3828	-0.8167	0.1835	-4.6517
Q	Fishing	-1.0693	-2.4406	-0.0100	-0.4547	0.4504	-0.0058	-3.5300

Table 2 shows the driving forces for carbon emissions for each industry sector. The considerations here are the value contribution, energy intensity, energy structure, coal use, oil use, and natural gas use effects. The results displayed in Table 2 are visually represented in the succeeding Figures with their corresponding interpretations. The total effect shown shows the difference from the base sector in terms of billions of kg CO_2 emitted due to these driving forces. The domestic navigation shows the highest difference in emissions at 23.6370 billion kg CO_2 more than the base sector. On the other hand, the lowest reported emissions are from the non-metallic minerals industry at a value of -21.9602 billion kg CO_2 from the average of all sectors.

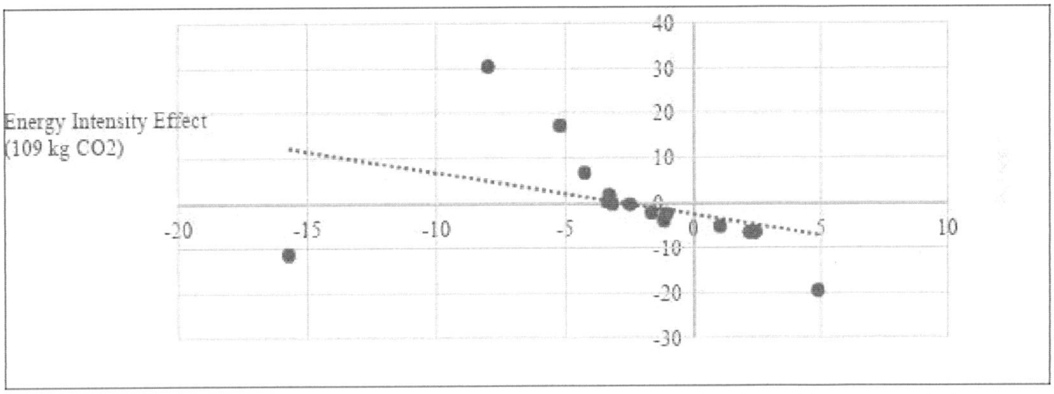

Figure 2. LMDI analysis of energy intensity and value contribution effects on emissions.

Figure 2 represents the CO_2 emissions performance of each sector in terms of energy intensity effects and effect from share of gross value added among considered sectors. It is clear that a general trend of higher energy intensity effect is present in sectors that experience lower value contribution effects to their CO_2 emissions. Perhaps this is due to the capability of higher-value sectors to procure more efficient and modern machinery/processes [25–27]. This also points out where the opportunities to innovate are – in traditional energy intensive sectors where alternative energy technologies are not yet available and in low-value sectors where technology adoption can be slow. Furthermore, it is notable that there are no sectors in the upper-right quadrant of the Cartesian plane. This is indicative that no sectors in this study have both parameters of value contribution and energy intensity that is above average. They are above average in at most 1 of these 2 parameters.

Notable outliers here are the non-metallic minerals, domestic navigation, and food & tobacco sectors (labelled as A, C, O, and F, respectively) for different reasons. The iron and steel industry experiences an increase in carbon emissions due to energy intensity effects. This may be due to the energy intensive processes involved in the processing of iron and steel. The non-metallic minerals industry has a notably lower value contribution effect along with the 2nd lowest energy intensity effect. Meanwhile, the domestic navigation sector has the highest energy intensity contribution towards its CO_2 emissions with the 2nd lowest value contribution effect on emissions. Lastly, we have the food & tobacco sector with the highest value contribution effect and lowest energy intensity effect on carbon dioxide emissions.

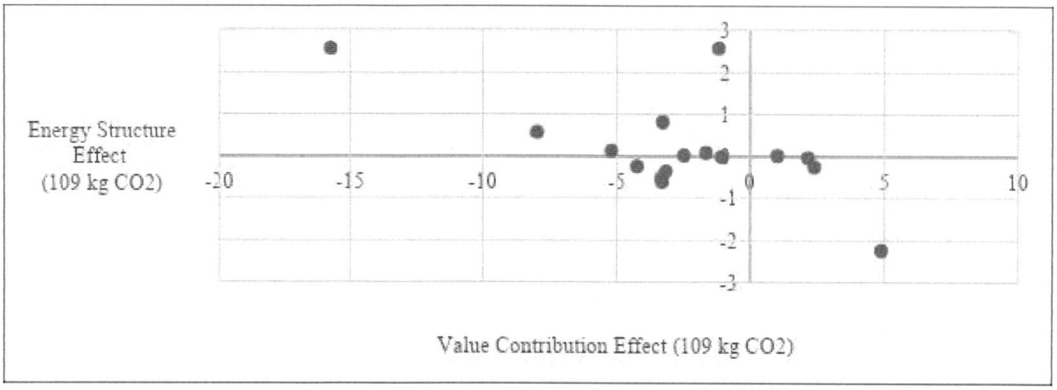

Figure 3. LMDI analysis of energy structure and value contribution effects on emissions.

As shown in Figure 3 and also on Table 2, the energy structure generally has an effect of <0.5kg of CO_2 for all but 3 industry sectors. The non-metallic minerals, land transportation, and food & tobacco sectors (labelled as C, M, and F, respectively). This means that deviations from the average energy structure have created relatively larger changes to their carbon emissions. For non-metallic minerals (due to coal dependence) and land transportation (due to oil dependence), the differences in energy structures with the rest of the sectors have yielded higher CO_2 emissions. Meanwhile, in the food & tobacco sector, the fuel structure difference caused an overall decrease in the sector's CO_2 emissions.

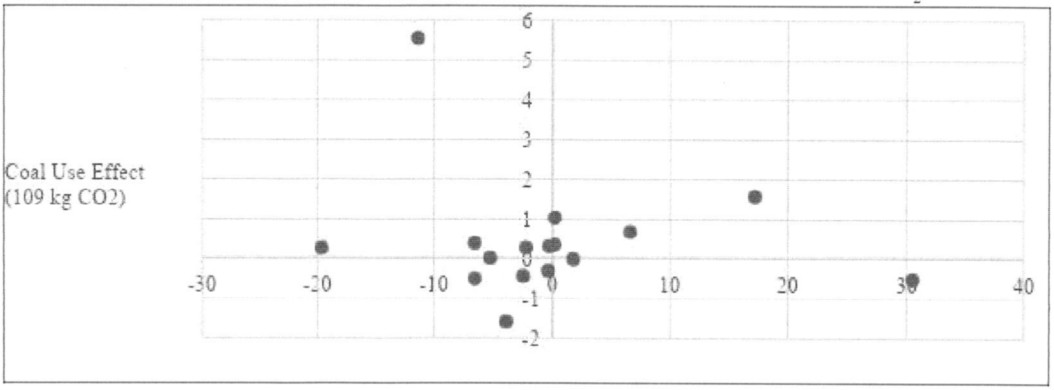

Figure 4. LMDI analysis of coal use and energy structure effects on emissions.

Figure 4 shows how the share of coal in a sector's fuel consumption mix drives its carbon emissions upward or downward. In the case of the non-metallic minerals industry, a coal use effect more than threefold of the next highest sector is observed. Perhaps this is due to the coal itself being a product of the non-metallic mineral industry. As such, coal is more accessible as an energy source for this industry and may be driving emissions upwards due to excessive use of their own product. It may also be observed that only 6 sectors experienced lower carbon emissions due to their relatively less use of coal.

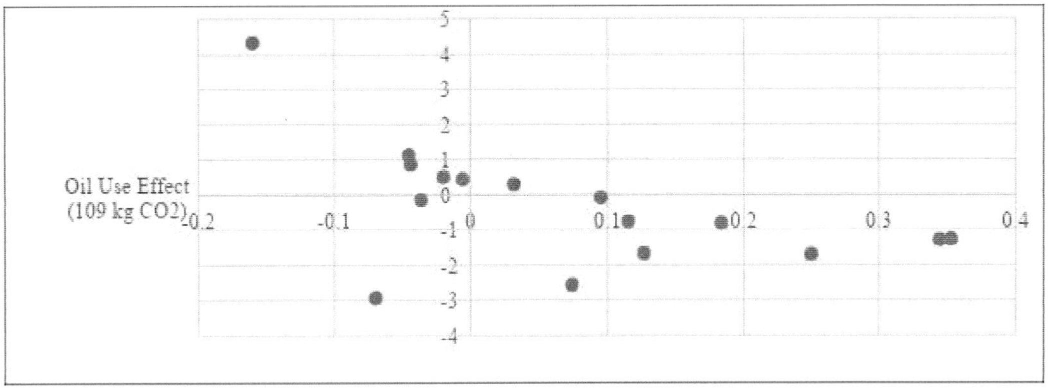

Figure 5. LMDI analysis of oil and natural gas use effects on emissions.

As shown in Figure 5, there appears to be an overall trend of greater natural gas use effects in sectors wherein oil use effects are less evident. This possibly suggests that natural gas is often used as a substitute to oil use, or vice versa. It must also be noted that the range of natural gas use effects are much lower than those of oil or coal use (in Figure 4). Some obvious outliers to this observed trend are the non-metallic minerals sector and the food & tobacco sector (labelled C and F, respectively). The reason for the negative oil use and natural gas use effects on the non-metallic minerals coincide with the one stated in the previous paragraph explaining its close relation to coal mining. The land transport sector (labelled M) shows a high oil use effect due to the prominence of the oil-based gasoline and diesel fuels as mobile fuel sources. Meanwhile, the machinery sector shows the highest natural gas use effects among all the fuel sources.

Among all the Figures shown above, there are insights available to each particular sector mentioned. These will form the basis for the recommendations in the subsequent section.

4. Conclusion and Recommendations

Therefore, it may be concluded that the main drivers of carbon emissions across Philippine sectors vary deeply on a sector-to-sector basis. Nonetheless, several trends have been found such as the negative relationship between oil and natural gas use effects, the relatively low energy structure effects for most sectors relative to their value contribution effects, the negative relationship of energy intensity and value contribution effects, and lastly, the low effect of natural gas use effects on CO_2 emissions are evidence of its low contribution to the overall national fuel supply.

Given the magnitude of these various contributing factors to carbon emissions in the country for 2017, several insights may be created accordingly. First, the relatively similar effects of energy structure on the sectors mentioned indicate potential for turning these into negative effects so as to reduce emissions across the entire economy. Second, as sectors with relatively higher value shares appear to have lower energy intensity, perhaps it would be worthwhile to invest in less energy intensive equipment (higher efficiency) across all industries. Low-value sectors may require certain government intervention to afford such equipment. Third, the dependence of the Philippines on oil and coal may be hampering growth in the natural gas sector. This is hinted at by the minimal impact of natural gas use effects on CO_2 and the comparison of oil and natural gas use effects on a Cartesian plane. This presents an opportunity for reduction of emissions through the use of greater amounts of natural gas throughout the national energy supply. The average 73,300kgCO_2/TJ and 94,600kgCO_2/TJ emission factors for coal and oil, respectively, are higher than the 64,200kgCO_2/TJ for natural gas by 14.17% and 47.35%. In consideration of these emission factors only, using more natural gas would produce the same energy with less CO_2 emissions. Finally, as each row in Table 2 represents a particular industry sector and each sector is represented among all the LMDI Figures, there are tailor-made solutions that must be derived from this study by each sector and their respective regulators towards the decreasing of emissions.

Nomenclature

GHG	greenhouse gas emissions
IDA	index decomposition analysis
R-R	radial-regional
LMDI	logarithmic mean Divisia index
IPCC	Intergovernmental Panel on Climate Change
PSY	Philippine Statistical Yearbook
IEA	International Energy Agency
EFDB	Emission Factor Database
val	value contribution effect
int	energy intensity effect
str	energy structure effect

References

[1] San Juan J L G, Aviso K B, Tan R R and Sy C L 2019 A multi-objective optimization model for the design of biomass co-firing networks integrating feedstock quality considerations *Energies* **12**
[2] Deutch J 2017 Decoupling Economic Growth and Carbon Emissions *Joule* **1** 3–5
[3] Obama B 2017 The irreversible momentum of clean energy *Science (80-.).* **355** 126–9
[4] Patrizio P, Leduc S, Kraxner F, Fuss S, Kindermann G, Mesfun S, Spokas K, Mendoza A, Mac
Dowell N, Wetterlund E, Lundgren J, Dotzauer E, Yowargana P and Obersteiner M 2018 Reducing US Coal Emissions Can Boost Employment *Joule* **2** 2633–48
[5] Singpai B and Wu D D 2021 An integrative approach for evaluating the environmental economic efficiency *Energy* **215**
[6] Wang Z, Jiang Q, Dong K, Mubarik M S and Dong X 2020 Decomposition of the US CO2 emissions and its mitigation potential: An aggregate and sectoral analysis *Energy Policy* **147**
[7] Sumabat A K, Lopez N S, Yu K D, Hao H, Li R, Geng Y and Chiu A S F 2016 Decomposition analysis of Philippine CO2 emissions from fuel combustion and electricity generation *Appl. Energy* **164** 795–804
[8] Wang H and Ang B W 2018 Assessing the role of international trade in global CO2 emissions: An index decomposition analysis approach *Appl. Energy* **218** 146–58
[9] Li A, Zhou D, Chen G, Liu Y and Long Y 2020 Multi-region comparisons of energy-related CO2 emissions and production water use during energy development in northwestern China *Renew. Energy* **153** 940–61
[10] Ang B W 2005 The LMDI approach to decomposition analysis: A practical guide *Energy Policy* **33** 867–71
[11] Ma X, Wang C, Dong B, Gu G, Chen R, Li Y, Zou H, Zhang W and Li Q 2019 Carbon emissions from energy consumption in China: Its measurement and driving factors *Sci. Total Environ.* **648** 1411–20
[12] Ang B W, Xu X Y and Su B 2015 Multi-country comparisons of energy performance: The index decomposition analysis approach *Energy Econ.* **47** 68–76
[13] Song F, Su F, Mi C and Sun D 2021 Analysis of driving forces on wetland ecosystem services value change: A case in Northeast China *Sci. Total Environ.* **751**
[14] Wu Y, Tam V W Y, Shuai C, Shen L, Zhang Y and Liao S 2019 Decoupling China's

economic
growth from carbon emissions: Empirical studies from 30 Chinese provinces (2001–2015) *Sci. Total Environ.* **656** 576–88
[15] Mousavi B, Lopez N S A, Biona J B M, Chiu A S F and Blesl M 2017 Driving forces of Iran's CO2 emissions from energy consumption: An LMDI decomposition approach *Appl. Energy* **206** 804–14
[16] Yang J, Cai W, Ma M, Li L, Liu C, Ma X, Li L and Chen X 2020 Driving forces of China's CO2 emissions from energy consumption based on Kaya-LMDI methods *Sci. Total Environ.* **711**
[17] Ding Y and Li F 2017 Examining the effects of urbanization and industrialization on carbon dioxide emission: Evidence from China's provincial regions *Energy* **125** 533–42
[18] Yang L, Yang Y, Zhang X and Tang K 2018 Whether China's industrial sectors make efforts to
reduce CO2 emissions from production? - A decomposed decoupling analysis *Energy* **160** 796–809
[19] Kim S 2017 LMDI decomposition analysis of energy consumption in the Korean manufacturing sector *Sustain.* **9** 1245–53
[20] Karmellos M, Kopidou D and Diakoulaki D 2016 A decomposition analysis of the driving factors of CO2 (Carbon dioxide) emissions from the power sector in the European Union countries *Energy* **94** 680–92
[21] Jenkins J D, Luke M and Thernstrom S 2018 Getting to Zero Carbon Emissions in the Electric Power Sector *Joule* **2** 2498–510
[22] International energy agency (IEA) IEA Sankey Diagram Philippines Final Consumption 2017
[23] Philippine statistics authority (PSA) 2018 *Philippine Statistical Yearbook*
[24] Intergovernmental panel on climate change (IPCC) 2006 IPCC guidelines for national greenhouse gas inventories
[25] Le Brun N, Simpson M, Acha S, Shah N and Markides C N 2020 Techno-economic potential of low-temperature, jacket-water heat recovery from stationary internal combustion engines with organic Rankine cycles: A cross-sector food-retail study *Appl. Energy* **274**
[26] Li Y, Li K, Liu X, Wang Y and Zhang L 2021 Lithium-ion battery capacity estimation — A pruned convolutional neural network approach assisted with transfer learning *Appl. Energy* **285**
[27] Dai Y, Zheng H, Jiang Z and Xing B 2020 Comparison of different crop residue-based technologies for their energy production and air pollutant emission *Sci. Total Environ.* **707**

Numerical modelling of freestream cavitating flow through ship propeller using OpenFOAM

Truong Van Ngoc[1], **Mai Ngoc Luan**[2], **Ngo Khanh Hieu**[1,3]

[1] Aerospace Engineering Department, Ho Chi Minh City University of Technology
[2] TechnoStar Co., Ltd. – Ho Chi Minh City Representative Office
[3] VNU-HCM Key Laboratory for Internal Combustion Engine

ngoctr9898@gmail.com, ngokhanhhieu@hcmut.edu.vn

Abstract. In marine industry, cavitation has been a massive problem when dealing with ship propeller. One of the main effects caused by cavitation is thrust breakdown, which decreases the efficiency of propeller. It is difficult to completely avoid this phenomenon, but there are ways to reduce its occurrence. It is important to study the phenomenon in the process of designing propeller. For that reason, cavitation prediction on certain geometry is compulsory to give out the optimal ship propeller design. In this paper, the cavitation analysis on propeller VP1304 was conducted numerically utilizing OpenFOAM code. The motion of the propeller was represented by Arbitrary Mesh Interface (AMI) methodology with the aid of RANS turbulence models. The authors examined the differences between k-ε model and k-ω SST model in this case with the involvement of Schnerr-Saur, Merkle and Kunz cavitation model. The results achieved shows that the k-ω SST model performed more accurately in term of propeller thrust, torque and efficiency prediction. Furthermore, among the three employed cavitation models, Schnerr-Saur agrees best with empirical data, and hence, can be considered most suited for this analysis.

Keywords. AMI, cavitation, OpenFOAM, simulation

1. Introduction

The cavitation phenomenon has frequently appeared as a great concern in the field of rotating fluid due to its cause and effects. The phenomenon is the formation of vapor cavities because of phase change when the pressure drops down rapidly in high-speed flow. These cavitation bubbles will then collapse when arriving to higher pressure zone. In the case of propeller, cavitation can bring up damage to equipment with vibration and erosion, reducing the efficiency and producing noise. This raises up the problem in propeller design and for that reason, the studying of cavitation analysis has always been considerable.

With numerical analysis, cavitation behavior in different regimes can be predicted. Adding to the problem, cavitating flow simulation is rather complicated since it involves phase change process and works on the complex geometry of propeller. In order to develop the numerical simulation for this case, many studies have been done with different models and methods.

In this paper, the simulation is carried out using Potsdam Propeller Test Case (PPTC) propeller and validation data provided by SVA Potsdam. The cavitation phenomenon is simulated with VP1304 propeller using OpenFoam software and its meshing utility, snappyHexMesh. OpenFOAM supported two methodologies in simulating the rotating motion of propeller that are Arbitrary Mesh Interface (AMI) and Multiple Reference Frame (MRF). MRF technique approximates the transient motion of propeller into steady state, while AMI technique is applied for full transient simulation. Hence, AMI technique is expected to give better results than MRF in simulating complicated transient case like cavitation. In this study, AMI methodology is preferred to perform cavitation analysis on propeller VP1304. The simulation is performed with OpenFOAM, an open-source software for CFD (Computational Fluid Dynamics) development. The work will contribute to the research of cavitation modeling with the application of sliding mesh technique. For turbulence modeling, the k-ε and k-ω SST model are considered. The OpenFOAM's solver supports different cavitation models for cavitation simulation. For comparison, cavitation models by Schnerr and Saur [1], Merkle et al. [2] and Kunz et al. [3] are employed.

2. Numerical Methods

2.1. Governing Equations

For cavitation case, the flow is treated as a mixture of liquid and vapor. The VOF method introduced the volume fraction α to distinguish the phases. With volume fraction of liquid αl and volume fraction of vapor αv, the mixed density and mixed viscosity are written as [4]:

$$\rho = \alpha_l \rho_l + \alpha_v \rho_v \tag{1}$$

$$\mu = \alpha_l \mu_l + \alpha_v \mu_v \tag{2}$$

The continuity equation, momentum equation and mass transfer equation of liquid volume fraction for unsteady multi-phase flow are written respectively as followed [4]:

$$\frac{\partial \rho}{\partial t} + \nabla(\rho U) = 0 \tag{3}$$

$$\frac{\partial(\rho U)}{\partial t} + \nabla(\rho U) \cdot U = -\nabla p + \nabla\big((\mu + \mu_t)\nabla U\big) + S_M \tag{4}$$

$$\frac{\partial \rho}{\partial t} + \nabla(\alpha_l U) = \frac{\dot{m}}{\rho_l} \tag{5}$$

In the above expressions, ρ is the mixed density, μ is the mixed viscosity, μt is the turbulence viscosity, SM is the momentum source effected to the quantity and m˙ is the mass transfer rate. Cavitation models can be employed to model the mass transfer rate.

2.2. Cavitation Modelling

To describe the expression of mass transfer between liquid and vapor, many cavitation models have been developed. In this segment, the three following models that are already employed in OpenFOAM for solving multi-phase problem will be demonstrated: Schnerr-Saur model, Merkle model and Kunz model. For Schnerr-Saur model, the terms of evaporation and condensation of mass transfer appeared in one equation. Meanwhile, Merkle model and Kunz model treated the evaporation and condensation terms individually.

For Schnerr-Saur cavitation model, the mass transfer of liquid and vapor is described by the following transport equation [5]:

$$\frac{\partial}{\partial t}(\rho_v \alpha) + \nabla(\rho_v \alpha U) = R_e - R_c \tag{6}$$

Where Re and Rc stand for evaporation and condensation rate terms. Introducing pv as saturation pressure and RB as the bubble radius, Re and Rc can be described as followed [6]:

$$R_e = \frac{\rho_v \rho_l}{\rho_m} \alpha(1-\alpha) \frac{3}{R_B} \sqrt{\frac{2}{3} \frac{(p_v - p)}{\rho_l}} \quad if\ p_v \geq p \tag{7}$$

$$R_c = \frac{\rho_v \rho_l}{\rho_m} \alpha(1-\alpha) \frac{3}{R_B} \sqrt{\frac{2}{3} \frac{(p - p_v)}{\rho_l}} \quad if\ p_v \leq p \tag{8}$$

The mass transfer rate in Merkle model is written as [6]:

$$\dot{m} = \dot{m}^+ + \dot{m}^- \tag{9}$$

Where $\dot{m}^+$ and $\dot{m}^-$ represent the evaporation and condensation terms respectively. The two terms are specified as followed [6]:

$$\dot{m}^+ = \frac{C_c \rho_l}{\frac{1}{2} U_\infty^2 t_\infty} \alpha_l min[0, p - p_v] \tag{10}$$

$$\dot{m}^- = \frac{C_c \rho_l \rho_v}{\frac{1}{2} U_\infty^2 t_\infty} \alpha_l max[0, p - p_v] \tag{11}$$

From the two equations above, C_c and C_v are two empirical constants, t∞ is the mean flow time scale (=L/U∞) and U∞ is the reference velocity.

Kunz model is a modification to Merkle model. The main difference of Kunz model is that the condensation term becomes independent to pressure. The evaporation and condensation terms here are written as [6]:

$$\dot{m}^+ = \frac{C_v \rho_v}{\frac{1}{2} U_\infty^2 t_\infty \rho_l} \alpha_l min[0, p - p_v] \tag{12}$$

$$\dot{m}^- = \frac{C_c \rho_v}{t_\infty} \alpha_l^2 \alpha_v \tag{13}$$

3. Computational Mesh

The basic mesh domain for propeller analysis contains a stationary region and a rotating region. The stationary region size should be considered so that it is able to cover the outward flow after the propeller. In this case, the AMI method is applied with the same requirement. The region sizes are firstly chosen as reference from related research [7]. Some adjustments are made to obtain a suitable domain for this case (see figure 1).

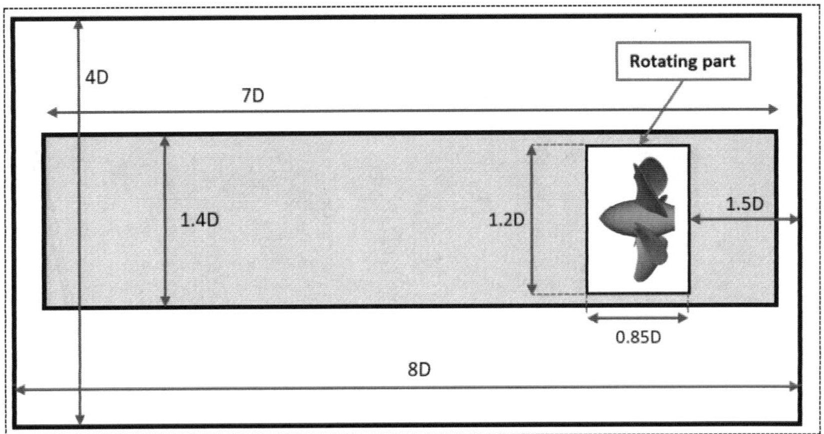

Figure 1. The dimensions of computational domain

For this case, the mesh was generated with approximately 1.58 million cells in total using OpenFOAM's snappyHexMesh tool. The mesh size is 0.0375 m in the domain, 0.0094 m in refinement region, 0.0047 m in rotor region and 0.0012 m on propeller surface. Between each region that has different mesh size, three buffer layers are added to avoid instabilities from sudden changes in size. At the propeller surface edge, the mesh size was refined up to 0.0006 m. The curve at the tip and edges of the propeller is complex and the geometry is very thin as well. For that reason, the cell size on propeller surface must be smaller at the edge to preserve the geometry. Most regions in the domain are made of hexahedral cells except the propeller surrounding. The snappyHexMesh treated the complex geometry of propeller by generating polyhedral cells in that region.

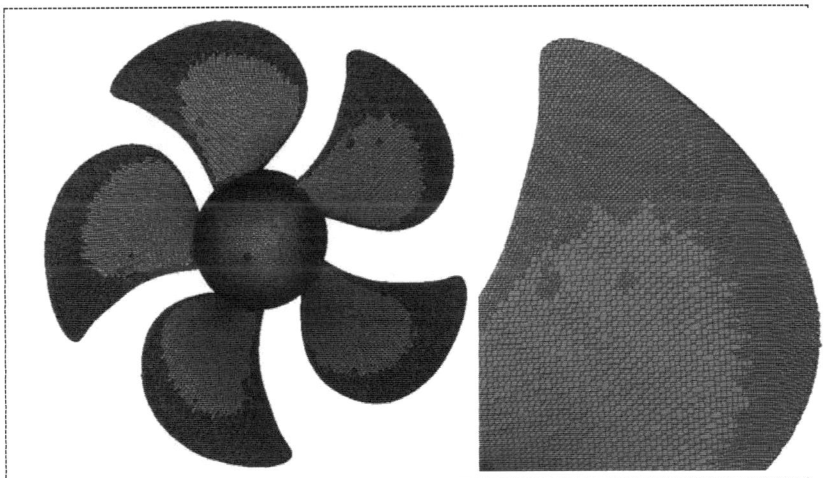

Figure 2. Mesh on propeller surface

In AMI method, the interface between fixed region and rotating region was created by two separated patches that belong to each region. These two patches are completely identical and overlapped each other. At this stage, createPatch function in OpenFOAM is used to create the two AMI patches from the rotor part. However, the meshes generated on two patches are not totally the same as in Figure 3. The edges were refined to 0.00235 m to maintain the conformity between two patches.

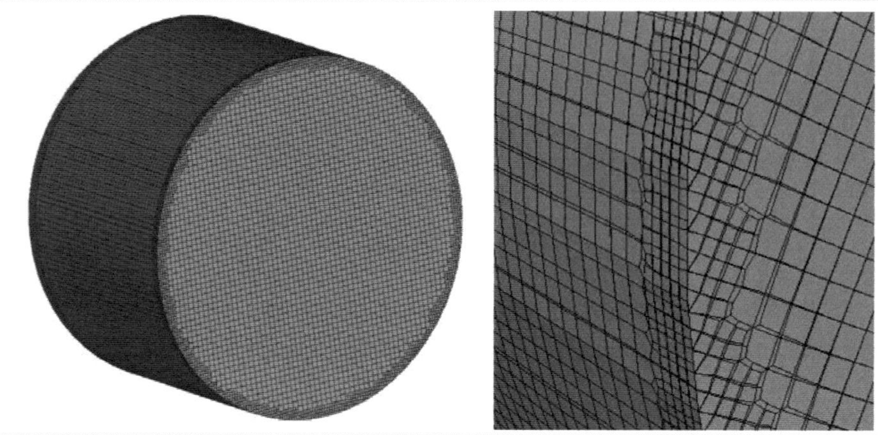

Figure 3. Rotating region mesh with overlapped interfaces

4. Results

The simulation was implemented for three test cases with all information summarized in Table 1. In all cases, the number of revolutions in 1500 RPM.

Table 1. Information of three test cases

Test case	Advanced coefficient, J	Cavitation number, σ_0	Vapor pressure, p_v (Pa)
1	1.019	2.024	2818
2	1.269	1.424	2818
3	1.408	2.000	2869

The inlet velocity was calculated from the advance coefficient and number of revolutions, and ambient pressure was attained from cavitation number and vapor pressure. As the input velocity and rotational speed are rather high, the solver failed in adjusting the timestep and Courant number exploded. In order to maintain the stability, the inlet velocity and rotational speed were set to increase linearly from zero value in the first 0.01 seconds. After that, the two values were kept unchanged. The case was set to run in 0.2 seconds to collect results when the solver was in stable state.

In this section, results of three test cases will be discussed. In each test case, two turbulence models were performed: k-ε model and k-ω SST model. For comparison between two turbulence models, Schnerr-Saur cavitation model was employed. The turbulence model with the best result was chosen to perform analysis with three cavitation models: Schnerr-Saur, Merkle and Kunz. All models were evaluated based on thrust coefficient results. Cavitation observation of each case will also be presented.

4.1. *Comparison of Turbulence Models*

For test case 1, both turbulence models achieved pretty good results. From thrust and torque coefficients, the propeller efficiency was obtained.

In test case 2 and test case 3, the advance coefficient is higher, so the results tend to have higher errors than test case 1. In high velocity condition, it is clearer to see the difference between k-ε model and k-ω SST model.

Table 2. Results of thrust and torque coefficients

Test case	Advanced coefficient, J	Thrust coefficient K_T			Torque coefficient K_Q		
		k-ε	k-ω SST	Exp.	k-ε	k-ω SST	Exp.
1	1.019	0.3690	0.3710	0.3735	0.0937	0.0937	0.0970
2	1.269	0.1867	0.1913	0.2064	0.0631	0.0575	0.0631
3	1.408	0.1223	0.1256	0.1362	0.0489	0.0447	0.0489

Table 3. Results of propeller efficiency

Test case	Advanced coefficient, J	Efficiency		
		k-ε	k-ω SST	Exp.
1	1.019	0.638	0.652	0.625
2	1.269	0.656	0.689	0.689
3	1.408	0.614	0.614	0.624

From Table 4, it is recognized that k-ε model has higher accuracy in predicting the efficiency. However, the simulation results of this model are less sensitive when performing at different advanced coefficient. The efficiency is higher than experimental value in test case 1, but in the other test cases, the efficiency appears to be lower. This tendency can be clearly observed from Figure 4. In contrast, for k-ω SST model, the results are higher than experiment in all cases, and the differences are around 4%. Despite having high errors, k-ω SST model seems to have a certain propensity where the numerical error rises as the advance coefficient increases.

Table 4. Results errors for comparison of turbulence models

Test case	Advanced coefficient, J	Thrust coefficient, K_T		Torque coefficient, K_Q		Efficiency	
		Error % (k-ε)	Error % (k-ω SST)	Error % (k-ε)	Error % (k-ω SST)	Error % (k-ε)	Error % (k-ω SST)
1	1.019	1.2	0.70	0.0937	0.0937	2.14	4.30
2	1.269	9.54	7.32	0.0631	0.0575	0.58	4.40
3	1.408	7.78	8.69	0.0489	0.0447	1.64	4.47

From Figure 5 to Figure 7, the cavitating regions on both sides of the propeller are shown for each case. Iso-surface of vapor fraction is created to illustrate the cavitating region. In this case, cavitation patterns with iso-surface of 50% vapor fraction is observed.

For test case 1, sheet cavitation with tip vortex cavities and a small area of root cavitation appeared on the suction side of the propeller. Comparing to the experiment, the sheet cavitation seems to be thicker in simulation. As for test case 2, it is easy to notice the large area of root cavitation in the blade root. In test case 3, the sheet cavitation on the leading edge of pressure side occurred in wider area and extended to the root. The phenomenon appears to be reasonable comparing to the one observed in experiment. Although, the simulation cannot predict the tip vortex cavitation which can be observed from experiment. The mesh surrounding tip vortex region should be fine enough in order to simulate the tip vortex cavitation.

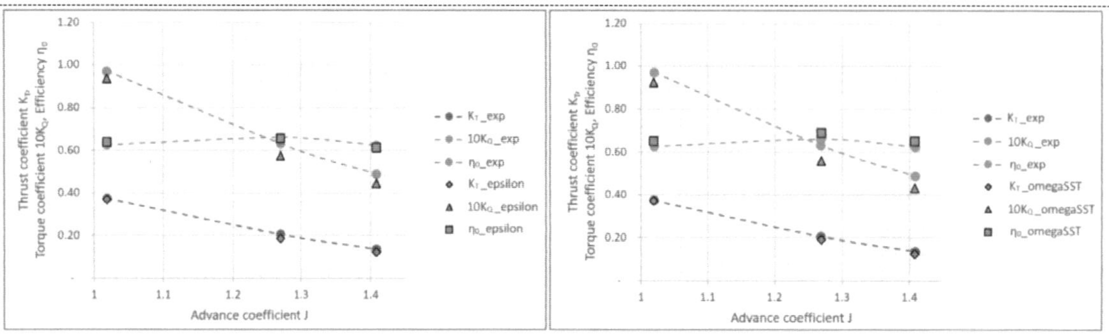

Figure 4. Comparison of simulation results to experiment
(Left: k-ε model results, right: k-ω SST model results)

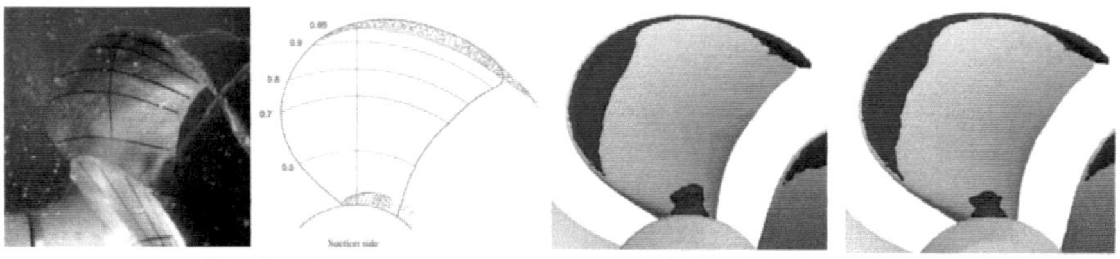

Experiment **k-ε** **k-ω SST**

Figure 5. Cavitation observed in test case 1

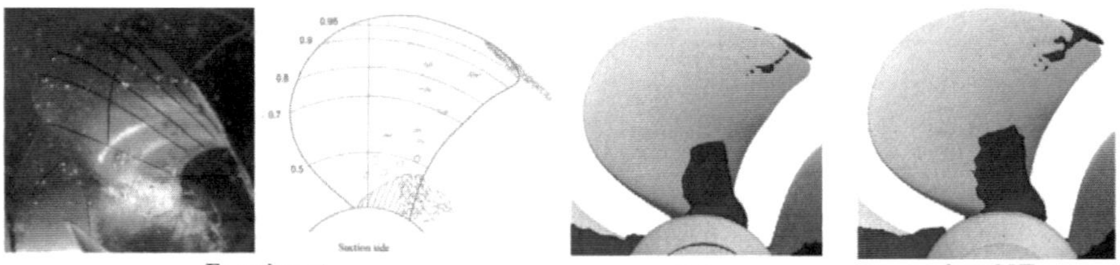

Experiment **k-ε** **k-ω SST**

Figure 6. Cavitation observed in test case 2

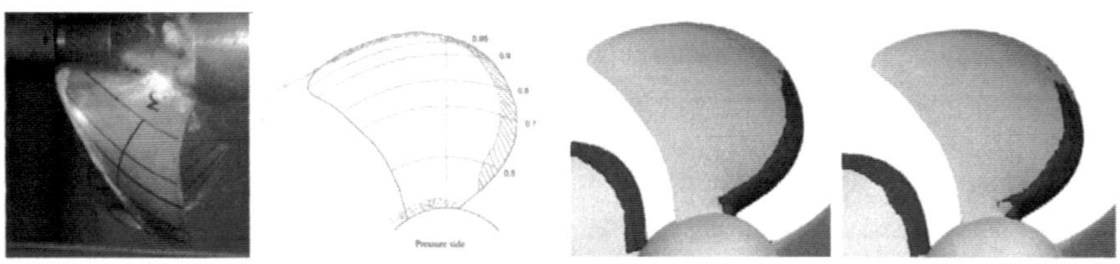

Experiment **k-ε** **k-ω SST**

Figure 7. Cavitation observed in test case 3

In general, it seems that two turbulence models gave quite similar results. However, the k-ε model is slightly incorrect when predicting the cavitating condition in test case 2. From Figure 6, cavity can be detected on the propeller hub in k-ε case, which is different from experiment. As k-ω SST model has proven its precision and high stability at different regimes of propeller, this model is certainly appropriate for cavitation analysis.

4.2. *Comparison of Cavitation Models*

At this stage, test case 1 (J = 1.019) and k-ω SST model was chosen to perform the simulation with Kunz and Merkle model.

Table 5. Comparison of different cavitation models

Cavitation models	Thrust coefficient K_T		Torque coefficient K_Q		Efficiency	
	Value	Error %	Value	Error %	Value	Error %
Schnerr-Saur	0.3710	0.67	0.0923	4.82	0.6519	4.30
Merkle	0.3710	0.67	0.0923	4.87	0.6520	4.32
Kunz	0.3735	0.88	0.0932	3.86	0.6550	4.80

Results of thrust coefficient, torque coefficient and efficiency of VP1304 propeller can be obtained from Table 5. Different cavitation models seem to have minor effects on the results. The efficiency obtained from the thrust and torque coefficients in three models are quite similar.

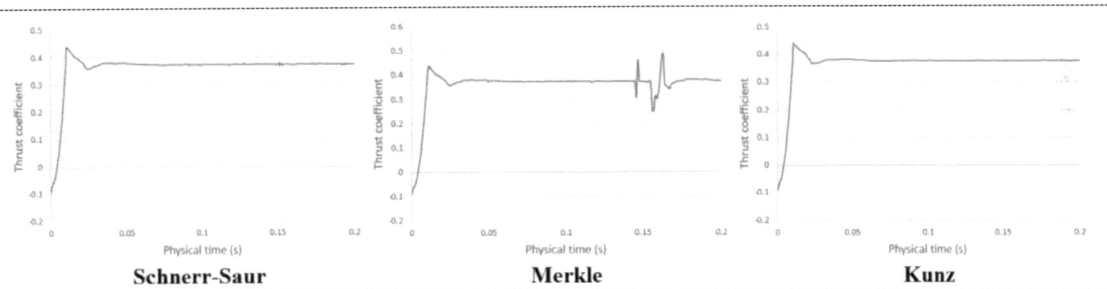

Figure 8. Thrust coefficient results over time

Figure 9. Cavitation observed for three cavitation models

The thrust coefficient results over time for three cavitation models are illustrated in Figure 8. Schnerr-Saur and Kunz models had achieved stability during the simulation. Difference from others, Merkle model seems to be stable at first; then it began to fluctuate in a short amount of time. For this case, it appeals that Merkle model is not reliable comparing to the others.

Schnerr-Saur model and Merkle model results in similar thrust coefficient; accordingly, the cavitation presences in both cases are also the same as seen in Figure 9. As for Kunz model, the sheet cavitation appears to be thinner with less cavities. Comparing to experiment, all models have some differences; even though, cavitating regions predicted by Schnerr-Saur and Merkle models are apparently more precise than Kunz model.

5. Conclusions

In this paper, the sliding mesh method has been successfully applied in the simulation of propeller cavitation. With the AMI application in OpenFOAM, the simulation has provided good results in all cases.

The assessment of k-ε model and k-ω SST model has proved that turbulence model has little effects on the observation of cavitating flow. Theoretically, k-ω SST model is more reliable with near-wall treatment. From analysis, it is true that this turbulence model has good performance and gives out high accurate result. In numerical analysis, it is also important to notice the consistency, that k-ω SST model has the benefit over k-ε model. Probably, k-ω SST model should be well-focused with more development for cavitation propeller simulation.

With k-ω SST model being chosen for further investigation, different cavitation models are applied and results in several manners. All cavitation models gave good results with low error percentages; however, there are some aspects that need to be considered. As a result, Schnerr-Saur model has the best agreement of accuracy and stability. The other models mostly depend on the empirical coefficients; especially Merkle model that is very sensitive with those parameters.

Acknowledgements
We sincerely thank the TechnoStar Co., Ltd. – Ho Chi Minh City Representative Office for great support on facility and tools to our research. This research is funded by VNU-HCM under grant number C2019-20-12. We acknowledge the support of time and facilities from Ho Chi Minh City University of Technology (HCMUT), VNU-HCM for this study.

References

[1] Schnerr G H and Sauer J 2001 Physical and numerical modeling of unsteady cavitation dynamics in *International Conference on Multiphase Flow* (New Orleans, USA)
[2] Merkle C L, Feng J Z and Buelow P E O 1998 Computational modeling of the dynamics of sheet cavitation *The 3rd International Symposium on Cavitation* (Grenoble, France)
[3] Kunz R F, Boger D A, Stinebring D R, Chyczewski T S, Lindau J W, Gibeling H J, Venkateswaran S and Govindan T R 2000 A preconditioned Navier-Stokes method for two-phase flows with application to cavitation prediction *Computers & Fluids* vol 29 no 8 pp 849-875
[4] Mehdipour R 2014 Simulating propeller and propeller-hull interaction in openFOAM (Stockholm, Sweden: Royal Institute of Technology)
[5] Paik K J, Park H G and Seo J 2013 URANS simulations of cavitation and hull pressure fluctuation for marine propeller with hull interaction *The Third International Symposium on Marine Propulsors - SMP'13* (Launceston, Australia)
[6] Long L V, Luan M N, Nghia N H, Thien D M and Hieu N K 2018 Propeller simulation in open-water condition with BKASM: A user interface based on SnappyHexMesh/OpenFOAM *Journal of Transportation Science and Technology* vol 30 pp 37-43
[7] Potsdam S 2011 PPTC smp'11 https://www.sva-potsdam.de/en/pptc-smp11-workshop/#workshop
[8] Zheng C, Shu L, Zhang Z and Liu D 2016 The numerical simulation of propeller cavitation in nonuniform wake based on OpenFOAM *The 12th International Conference on Hydrodynamics* (Delft, The Netherlands)

Design and performance analysis of 1KW ORC turboexpander with R245fa as working fluid

Asybel Bonar*, **Ari Darmawan Pasek, Harmen Burhannudin**

Faculty of Mechanical and Aerospace Engineering, Institut Teknologi Bandung, 40132, Bandung, Indonesia

Email: asybel.sidabutar@gmail.com

Abstract. Organic Rankine Cycle (ORC) is a reliable technology to utilize low and medium thermal energy and has been in development since the beginning of the 19th century and has rapid development in the last few decades. There are a couple of ways to optimize ORC. One of them is to develop an optimized turboexpander. This paper will explain in detail the preliminary and detailed design approach used to design turboexpander for 1kW ORC with R245fa as working fluid. The numerical simulation was performed using ANSYS CFX with turbulence model of Eddy Viscosity Reynold Shear Stress and Aungier Redlich-Kwong Equation of state to predict the material properties. A couple of adjustments had been made when deciding the rotor thickness and blade number, as the proposed method was not able to produce reasonable geometry. The CFD result of the preliminary design has 4.79% less efficiency and 22.7% less power produced than the 0D design. The off-design parameters of 16 Blade number, 20000 rpm, and mass flow of 0.32 kg/s able to produce power 1.38 kW with the efficiency of 76.44%

Keywords: Organic Rankine cycle, Radial inflow Turbine, CFD Analysis

1. Introduction

Organic Rankine Cycle (ORC) is an effective technology to utilize low and medium thermal energy sources such as geothermal, waste heat recovery, and solar power into electrical power since it has high reliability and simple configuration [1]. ORC has been in development since the early 19th century, and research has been growing rapidly over the last decade. A total of 2701 MW over 1754 ORC unit has been installed worldwide as of 31st December 2016, with 74,8% of the unit were installed for power generation from geothermal brines [2].

A microscale ORC has great potential to fulfill building energy needs or to improve engine efficiency by utilizing its heat waste using Combined Heat and Power system. A way to optimizing ORC technology is to develop the design of ORC turboexpander. Li et al. [3] studied the blade number and blade angle effect on radial inflow turbine performance. Most of the studies of Radial turbine design are for larger power generation (more than 50kWe). A micro-scale ORC with power generation of less than 10kWe has not shown its applicability at the commercial level. When building microscale ORCs, researchers mostly focus on utilizing scroll expanders, which converted from a compressor. However, the potential is limited in its efficiency (usually less than 60%) and several issues such as leakage, friction, and heat transfer losses. A radial turbine has larger efficiency compared to the scroll expander. However, designing a micro radial turbine for ORC applications is limited since most of the equations present made for larger size turbines. This study will use a previously developed 0D design method for radial ORC expanders made for 200kW ORC [5] and analyze the geometry made using ANSYS CFX tools.

2. Radial Turbine Design

2.1. Preliminary design

The basic geometry of the inflow radial turbine was shown in figure 1. The component is divided into several sections. The blade rotor of an inflow radial turbine is schematically shown in Figure 1, divided into several radial and axial sections. Station 1 indicates the inlet volute, station 2 indicates the inlet to the nozzle, station 3 indicates the nozzle's outlet, stations 4 and 5 indicate the inlet and outlet of the blades, and station 6 indicates the outlet of the exhaust diffuser.

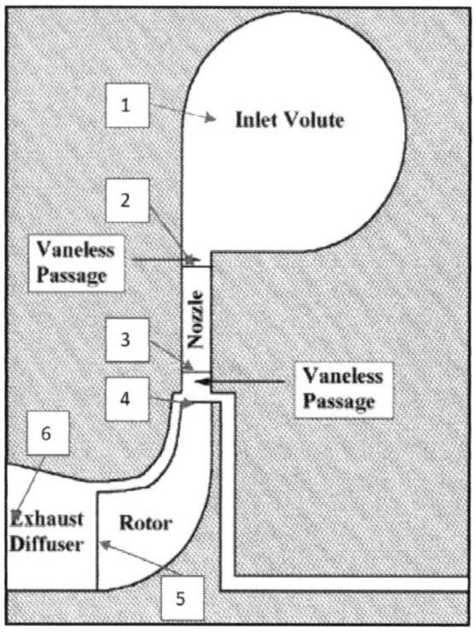

Figure. 1. Basic radial turbine configuration [6]

The design process was accomplished based on the predefined parameters, consist of inlet and outlet total temperature and pressure and turbine rotational speed. The relation of specific speed and rotational speed were shown in equation 1. The parameters used to design in this research are presented in table 1.

Table 1. Input Data of 1kW turboexpander

Parameter	Unit	Value
Fluid	-	R-245fa
Inlet pressure	KPa	730
Inlet temperature	K	368,87
Outlet pressure	KPa	560
Mass flow rate	kg/s	0.3
Power output target	kW	1
Rotational speed	rpm	20000
Efficiency	%	83.3
Expected Power	kW	1.292

ORC turbine required high-speed generator, especially in small geometry turbine. However, due to the limitation of available bearing system in the market, the turbine was designed to work around 20000 rpm. The results of preliminary design parameter are shown in table 2.

Table 2. 0D Design results

Parameter	Symbol	Value	Unit
Rotational Speed	N	20.000	rpm
Mass flow	$\dot{m}$	0,3	kg/s
Rotor discharge flow rate	Q_5	0,01	m^3/s
Rotor Outlet Pressure	ΔH_{id}	5,17	kJ/kg
Specific Speed	n_s	0,35	-
Velocity Ratio U_4/C_{0s}	v_s	0,599	-
Spouting velocity	C_{0s}	101,69	m/s
Inlct Tangential Speed	U_4	60,90	m/s
Rotor inlet radius	r_4	29,08	mm
Rotor inlet pressure	P_{t4}	722	kPa
Inlet tangential absolute velocity	$C_{\theta 4}$	70,70	m/s
Absolute inlet velocity angle	α_4	77,42	o
Inlet meridional absolute velocity	Cm_4	15,78	m/s
Relative inlet velocity angle	β_4		o
Inlet passage width	b_4	3,25	mm
Rotor axial length	ΔZ_R	10,47	mm
Rotor outlet radius	r_5	15,12	mm
Outlet passage width	b_5	6,98	mm
Outlet meridional absolute velocity	Cm_5	16,04	m/s
Relative outlet velocity angle	β_5	26,86	o

2.2. *Detailed Rotor Design*

The detailed rotor design was carried out using 2D and 3D model construction [6], [7]. The 2D model construction was carried out by drawing shroud contour using polynomial curve equation below [6], and hub contour by making a quarter circle with a radius of ΔZr with a center that located parallel to outlet station and adding straight line segment to the inlet. The quasi-normal line was constructed by splitting the curve of shroud and hub into the same size. These actions could be done easily with CAD software.

$$r = r_{s5} + (r_4 - r_{s5})\xi^n; 2 \leq n \leq 9 \tag{1}$$

$$\xi = \frac{(z - z_5)}{\Delta Z_R - b_4} \tag{2}$$

The first step to constructing a 3D model was by creating a meridional contour of the blade. The meridional of the rotor hub and shroud could be constructed from the following equations:

$$x_{s,i} = r_{s,i} \sin \theta_{s,i} \tag{3}$$

$$y_{s,i} = r_{s,i} \cos \theta_{s,i} \tag{4}$$

$$x_{h,i} = r_{h,i} \sin \theta_{h,i} \tag{5}$$

$$y_{h,i} = r_{h,i} \cos \theta_{h,i} \tag{6}$$

The thickness distribution along the blade was carried out using space vector, mentioned in the equation below, and the coordinates of blade surface were calculated using equation 39.

$$\begin{bmatrix}\vec{T}_x\\ \vec{T}_y\\ \vec{T}_z\end{bmatrix} = \vec{S}\times\vec{B} = \begin{vmatrix} i & j & k \\ \sin\theta_{j,i}\sin\phi_{j,i}\sin\beta_{j,i} + \cos\theta_{j,i}\cos\beta_{j,i} & \cos\theta_{j,i}\sin\phi_{j,i}\sin\beta_{j,i} - \sin\theta_{j,i}\cos\beta_{j,i} & \sin\phi_{j,i}\sin\beta_{j,i} \\ \frac{x_{s,i}-x_{h,i}}{L} & \frac{y_{s,i}-y_{h,i}}{L} & \frac{z_{s,i}-z_{h,i}}{L} \end{vmatrix} \quad (7)$$

$$x_{j,i}{}^{\pm} = x_{j,i} \pm \frac{1}{2} t_b T_{x,j,i} \quad (8)$$

$$y_{j,i}{}^{\pm} = y_{j,i} \pm \frac{1}{2} t_b T_{y,j,i} \quad (9)$$

$$z_{j,i}{}^{\pm} = z_{j,i} \pm \frac{1}{2} t_b T_{z,j,i} \quad (10)$$

In this research, the thickness of the blade was considered similar from leading until trailing edge. This action was taken because the thickness of the blade was calculated using equations proposed by Aungier [6], resulting in a very small dimension size. These happened because the equation was generally made for a larger sized turbine, with air as its working fluid [5]. The blade number was calculated using the equation proposed by Glassman [8]; however, a couple of research noted that the optimum blade number, especially in micro radial turbine usually different from equations as these equations also generally made for larger size turbine. Thus, the turbine off-design blade number was analyzed in CFD sections.

Table 3. Rotor design result

Parameter	Unit	Value
Rotor inlet radius	mm	29,08
Rotor blade thickness	mm	1
Rotor blade number	-	16
Rotor outlet passage width	mm	6,98
Rotor inlet passage width	mm	3,25

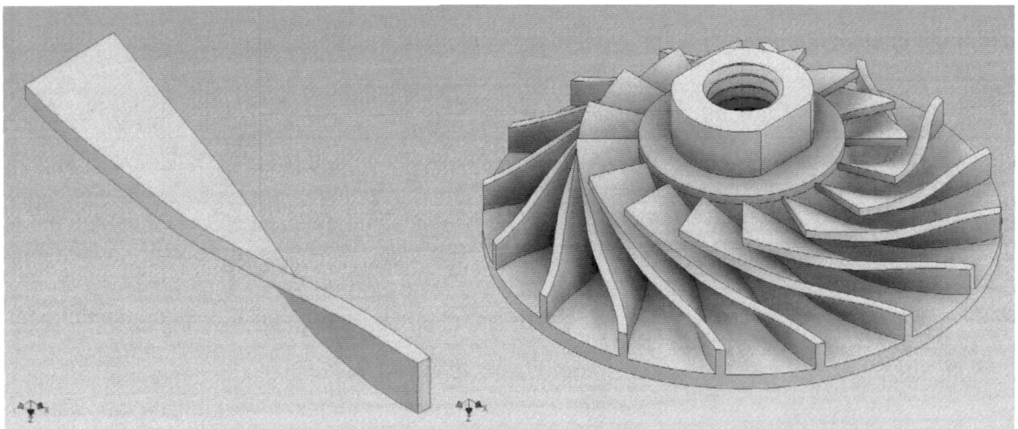

(a) (b)

Figure 2. 3D construction of rotor (a)Single Rotor blade (b) Final rotor geometry

2.3 *Nozzle Design*

The nozzle design input parameters were b_4, α_4, r_4, $C_{\theta 4}$, ρ_4, Nr, and $\dot{m}$. The design was started by building its meanline contour then constructing the top and bottom contour using thickness distribution along the meanline. The meanline contour and thickness were constructed using the equation below:

$$y_c = \frac{x_c(c - x_c)}{\left[\frac{(c-2a)^2}{4b^2} + \frac{c-2a}{b}x_c - \frac{c^2-4ac}{4b}\right]} \tag{11}$$

$$\frac{t_2}{c} = 0.03 \tag{12}$$

$$\frac{t_3}{c} = 0.015 \tag{13}$$

$$\frac{t_{max}}{c} = 0.06 \tag{14}$$

$$\frac{d}{c} = 0.4 \tag{15}$$

Then, the coordinate for top and bottom contour were constructed using:

$$x = x_c \pm 0.5t \sin \mathcal{K} \tag{16}$$

$$y = y_c \pm 0.5t \sin \mathcal{K} \tag{17}$$

Where

$$\mathcal{K} = \tan^{-1}\left(\frac{\partial y_c}{\partial x_c}\right) \tag{18}$$

Table 4. Nozzle design result

Parameter	Unit	Value
Nozzle inlet radius	mm	36,59
Nozzle outlet radius	mm	30,49
Nozzle maximum thickness	mm	1,56
Nozzle passage width	mm	3,25
Nozzle blade number	-	21
Nozzle chord length	mm	25,97

2.4. *Volute Design*

The volute type chosen in this research was the external volute type. The input parameters for volute design were c_{m2} and $c_{\theta 2}$. The distribution of volute cross section area was assumedd linear starting from largest at position angle of 360° and smallest at 0°. These considerations were taken to simplify the design and manufacture process.

Table 5. Volute design result

Position angle (°)	A_c (mm^2)	r_{max} (mm)	B (mm)	A (mm)
0	0,00	38,59	0,00	0,00
30	28,42	40,08	0,74	0,89
60	56,84	41,57	1,49	1,79
90	85,25	43,06	2,23	2,68
120	113,67	44,55	2,98	3,57
150	142,09	46,04	3,72	4,47
180	170,51	47,53	4,47	5,36
210	198,92	49,02	5,21	6,25
240	227,34	50,50	5,96	7,15
270	255,76	51,99	6,70	8,04
300	284,18	53,48	7,44	8,93
330	312,59	54,97	8,19	9,83
360	341,01	56,46	8,93	10,72

2.4. *Diffuser Design*

The diffuser were designed based on method proposed by Aungier [6], with a couple of design value were recommended by Merandy et.al [9]. to give optimal diffuser size. Those value were $2\theta_c = 11°$, $AR = 2.55$, $L/b_6 = 8$. The geometry of the diffuser could be calculated with equations 68 and 69.

$$2\theta_c = 2\tan^{-1}\left[\frac{(AR-1)b_6}{2L}\right] \tag{19}$$

$$AR = \frac{b_7}{b_6} \tag{20}$$

Table 6. Diffuser design result

Parameter	Unit	Value
Inlet diameter	mm	37,22
Outlet diameter	mm	94,91
Length	mm	297,76

3. Numerical Analysis

Three-dimensional modeling of rotor, nozzle, and volute are completed using Autodesk Inventor. The grid generated using ANSYS design modeler mesh. The numerical study was carried out from volute to rotor. ANSYS-CFX solver was chosen to solve the three-dimensional fluid problem.

The steady-state analysis was conducted using SST K-w turbulence model. The model boundary conditions were set for pressure and temperature inlet and outlet based on thermodynamic calculations with mixing plane interface set at rotor-nozzle interface. The material thermodynamic properties were calculated using Aungier Redlich Kwong EoS with input properties taken from NIST REFPROP 9.1 [10]. The fourth-order polynomial of specific heat capacity at zero pressure was shown in equation 70. The analysis was conducted at several blade numbers and rotational speed.

$$\frac{Cp}{R} = 5.327 + 1.317 \times 10^{-2}T + 1.223 \times 10^{-4}T^2 - 3.132 \times 10^{-7}T^3 + 2.585 \times 10^{-10}T^4 \quad (21)$$

Table 7. Parameter used to calculate R245fa Properties

Parameter	Nilai	Satuan
Real gas model	Aungier Redlich Kwong	-
Molar mass	134,05	g/mol
Critical temperature	427,16	K
Critical Pressure	3651	kPa
Critical volute	$0{,}260 \times 10^{-3}$	m^3/mol
Acentric factor	0,378	-
Boiling temperature	288,29	K

4. Simulation Results

The purpose of simulations was to evaluate the purposed method of geometry design method to generate a high efficiency turbine with producible geometry.

4.1. *Initial Design Results*

The initial design analysis was done in condition mentioned in table 1. The following results resulted from CFD analysis based on initial design parameters.

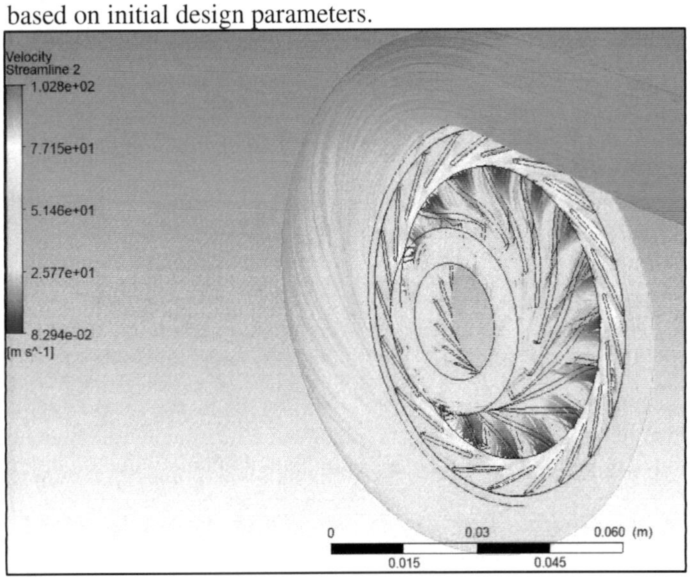

Figure 3. Initial design velocity streamline

1. The attempted parameter's power output was 0.998 kW, while the expected power produced was 1.292 kW, 22.7 % lower than the expected power. The working fluid mass flow usage for analysis results was lower than the preliminary design caused by a decreasing fluid passage area due to blade thickness modification, affecting the fluid's guidance by the blade.
2. Increasing velocity was observed in volute with small area passage, while an ideal volute design would not have this phenomenon and could affect the overall turbine efficiency.

The maximum Mach number from the design was observed at 0.69. Thus, the shockwave phenomenon did not occur in this analysis.

Table 8. Preliminary design results

Rotational Speed (rpm)	Massflow (kg/s)	Efficiency	Total Power (W)	Torque (N.m)	Maximum Mach Number
20000	0,2761	78,54%	998,54	0,4768	0,6896

4.2. Off Design Parameters

Due to unsatisfaction results of initial design parameters, some modifications of input parameters were considered. Rotor blade number plays a vital role in fluid guidance by the blade. This research use equation proposed by Glassman [8] to determine the rotor blade number. Several equations were also tested to calculate blade number, but they resulted in an excessive number of blades, resulting in unproducible rotor geometry. These results happened because equations to determine rotor geometry usually made for larger sized turbine [5]. Thus, the off-design blade number analysis was evaluated from 16 to 18 number of blades. Another essential aspect is the working fluid mass flow rate. Thus, the off-design mass flow rate parameter was also analyzed. The design was evaluated at the mass flow of working fluid at 0.3 to 0.33 kg/s at 20000 rpm.

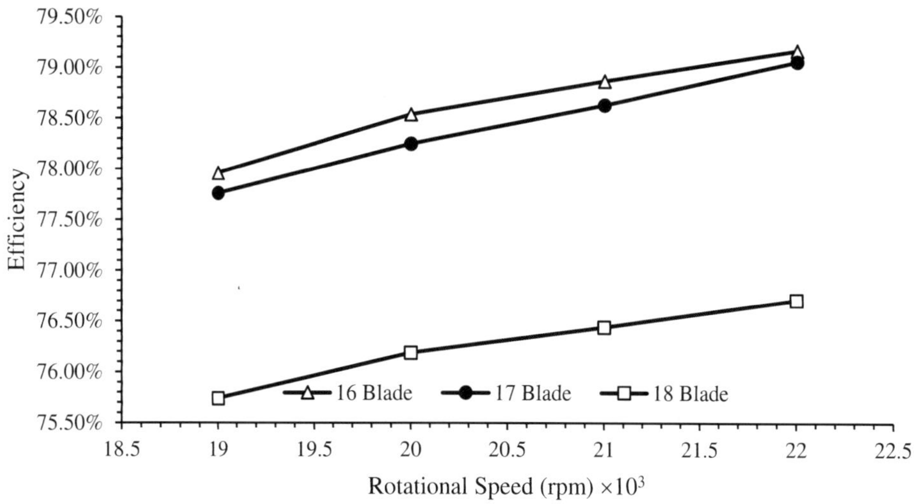

Figure 4. Rotational speed vs efficiency of different blade number

The highest efficiency was achieved at 16 blade configurations, while off design parameters at 17 and 18 had lower efficiency than the initial design. The turbine tends to have more vortex flow around the nozzle outlet at lower rpm, shown in figure 10. Based on figure 4, the turbine produced higher efficiency at higher rpm, similar to every blade number configuration. The increasing velocity across the volute small passage area is also noticed in every blade and rpm configuration. Maximum Mach number occurred at 16 blade passage, at Mach 0.698.

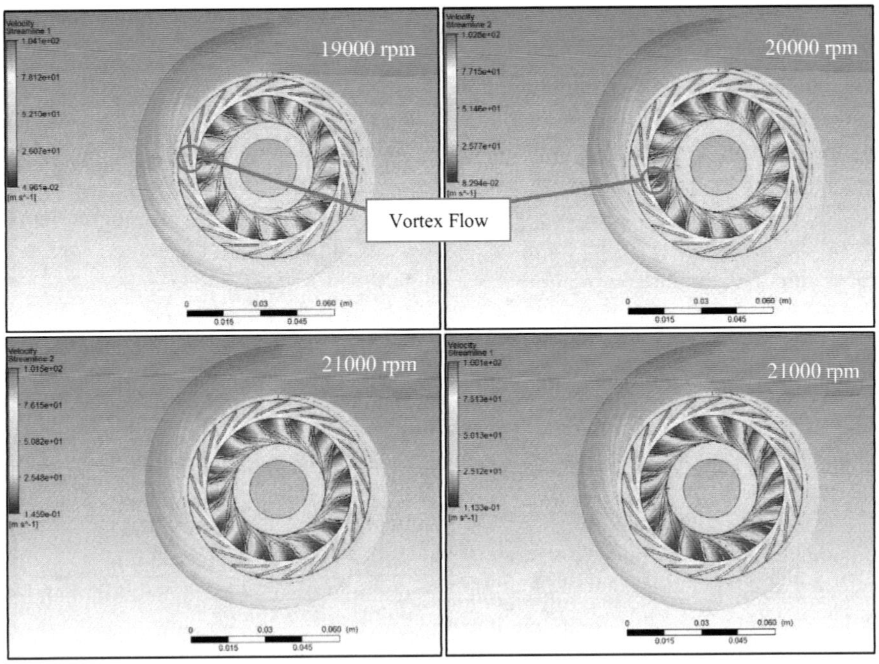

Figure 5. Velocity streamline with different rotational speed

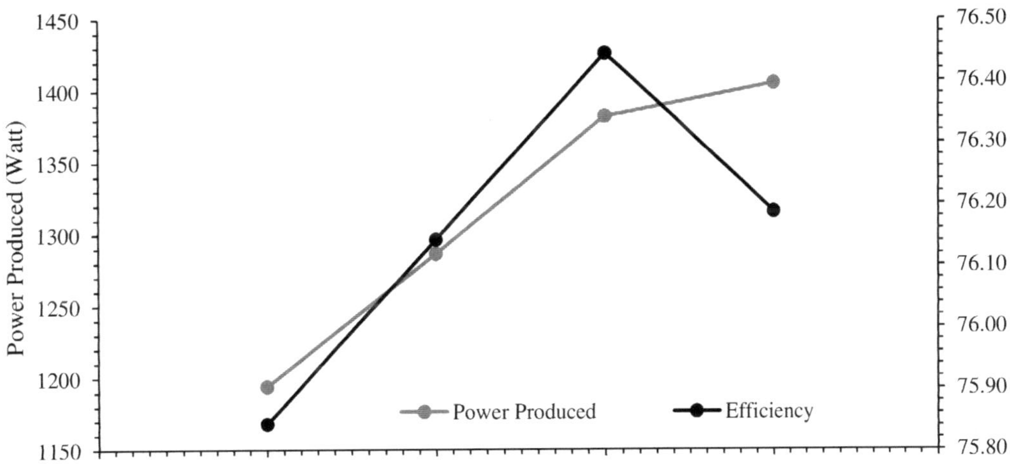

Figure 6. Power Produced vs Mass flow variation of working fluid

5. Conclusion

In this paper, the CFD analysis of the ORC radial inflow turbine for 1kW power production was carried out. The system was simulated using ANSYS CFX with fluid properties based on REFPROP 9.1 and calculated using Aungier Redlich Kwong EoS. From this study, the following conclusions were drawn.

- The preliminary design's efficiency drop was 4.79% from the CFD result to the initial design result, while the power produced is 22.7% lower. Thus, a couple of parameter adjustments are needed so that the turbine produces the expected power.

- Based on geometry results, the proposed design method could build a micro radial turbine with some adjustment when calculating rotor blade thickness and blade number, as the proposed method from the previous study was not capable of producing a reasonable micro turbine geometry.
- Based on the CFD simulation, the proposed method of geometry design could build a radial turbine for purposed power of 1.38 kW with an efficiency of 76,44% at off-design parameters. The parameters are 16 blade numbers, with rotational speed of 20000 rpm, and mass flow rate of 0.32 kg/s. The 20000 rpm parameters were taken even though having lower efficiency than higher rpm because of the limited available component such as bearing and generator in the market for desirable design. So, the turbine was designed as such to work on 20000 rpm.

References

[1] Bao J and Zhao L 2013 A review of working fl uid and expander selections for organic rankine cycle *Renew. Sustain. Energy Rev* **24** pp 325–342
[2] Tartière T and Astolfi M 2017 A world overview of the organic rankine cycle market *Energy Procedia*
[3] Li P, Han Z, Jia X, Mei Z, and Han X 2018 Analysis of the effects of blade installation angle and blade number on radial-inflow turbine stator flow performance *Energies*
[4] Han S, Seo J B, and Cho B S 2014 Development of a 200 kW ORC radial turbine for waste heat *recovery J. Mech. Sci. Technol.* **28** 12 pp 5231–5241
[5] Fiaschi D, Innocenti G, Manfrida G, and Maraschiello F 2015 Design of micro radial turboexpanders for ORC power cycles: From 0D to 3D *Appl. Therm. Eng.* **99** pp 402–410
[6] Aungier R H 2006 Turbine aerodynamics: axial-flow and radial-inflow turbine design and analysis (New York: ASME Press)
[7] Putri R P 2019 Studi perancangan turbin radial aliran masuk 250 kw pada siklus rankine organik dengan fluida kerja propana (Institut Teknologi Bandung)
[8] Glassman A J et al. 1994 *Turbine Design and Application* (Nasa Spec. Publ.)
[9] Pala'biran M 2020 *Pengembangan Lapangan Gas Raksasa East Natuna Dengan Metode Carbon Capture Storage (CCS) dan Memanfaatkan Rancangan Turboexpander Fluida Kerja Superkritis Untuk Meningkatkan Keekonomisan Lapangan* (Institut Teknologi Bandung)
[10] Lemmon E W, Huber M L, and McLinden M O 2019 *NIST Standard Reference Database 23: Reference Fluid Thermodynamic and Transport Properties-REFPROP, Version 9.1*

IOP Conf. Series: Materials Science and Engineering 1109 (2021) 012050 doi:10.1088/1757-899X/1109/1/012050

Power output and power coefficient calculations of a small HAWT with tubercles using Blade Element Momentum Theory

V Damirez[1*] and G Augusto[1]

[1]Department of Mechanical Engineering, De La Salle University, Manila, Philippines

*von_damirez@dlsu.edu.ph

Abstract. A 0.6 m diameter, three-bladed, untwisted, fixed-pitch, and small horizontal axis wind turbine with tubercles was analysed using the Blade Element Momentum (BEM) Theory. Calculation frameworks based from the Original BEM Theory, Wilson-Walker Method, Glauert's Empirical Formula, and Buhl's Theory were used in the prediction of the power output and power coefficient. Numerical results from the calculation frameworks were compared to the experimental data taken from the literature. The power output predicted by the Buhl's Theory was the closest to the actual results. In the case of the power coefficient, the Wilson-Walker Method produced an almost equal prediction at λ=4.06, showing only a percentage difference of 0.47%. Results also showed that the power coefficient predicted by the four calculation frameworks were more accurate at higher tip speed ratios.

1. Introduction

Wind turbine technology had been improving gradually and remarkable advances in the wind turbine design had been achieved by a handful of researchers. In contrast to large horizontal axis wind turbines (HAWT) which provided reliable, cost-effective, and competitive power, small HAWT were required to produce power without necessarily the best of wind conditions [1-2]. Despite its growth and popularity, small HAWT had not been studied as thoroughly as their large counterparts.

One of the most noteworthy of all the remarkable advances on wind turbine technology was the correlation of turbine blades with the flippers of a humpback whale. The humpback whale (*Megaptera novaeangliae*), as seen on Figure 1, was incomparable among the large baleen whales because of its ability to undertake aquatic manoeuvres to catch prey despite its large size. Humpback whales utilized extremely mobile, wing-like flippers with rounded scalloped bumps called tubercles. Turbine blades with tubercles performed better than a straight blade by providing constant power output under the stall conditions, an important factor for sites with harsh atmospheric conditions and unsteady wind [3]. Addition of tubercles could even increase power generated, prevent loss of lift, and decrease noise [4-6]. Past studies have proven the aerodynamic promise of the tubercles. But despite its potential, wind turbines blades integrated with tubercles is not yet commercially available.

The most common tool in evaluating the aerodynamic performance of a HAWT is the Blade Element Momentum (BEM) Theory. BEM Theory is one of the oldest and simplest yet most effective methods in analyzing the loads of a HAWT. Despite its simplicity, the BEM theory provides relatively accurate results. For the past few years, researchers and wind turbine designers have optimized and modified the BEM theory into various windmill brake state models like the Wilson-Walker method, Glauert's empirical formula, and Buhl's theory which incorporate various corrections to further increase the precision in the prediction.

Figure 1. Tubercles on the flipper of a humpback whale (*Megaptera novaeangliae*) and their novel application on wind turbines. Images adapted from [7], [8].

The related literature on BEM theory analysis of small wind turbines with tubercles is rather limited, if not inexistent. Wind turbine researchers had been studying the BEM theory since its conception and marine biologist had been fascinated by the aerodynamic promise of the tubercles of humpback whales, but the combination of those fields produced a novel concept that was not yet thoroughly explored in a lot of studies. This paper investigated a small horizontal axis wind turbine with tubercles by analyzing its aerodynamic performance using the calculation framework based from the original BEM Theory and its three corrected models (Wilson-Walker method, Glauert's empirical formula, and Buhl's theory).

2. Materials and Methods

This paper adapted the blade design and experimental results from the study of Alsultan [8]. The turbine has a 0.6 m diameter, three blades, no twist, and fixed pitch at 0°. The design originated first from a straight blade. The chords lengths were manipulated throughout the blade span to transform the smooth edge into sinusoidal protuberances, resembling the tubercles of a humpback whale. Each cross section of the blade was a NACA4412 airfoil, one of the most widely used airfoil profile for the wind turbine applications. The initial parameters and geometrical inputs were entered in the code in Microsoft Excel to calculate the power output and coefficient using the calculation frameworks based from the Original BEM Theory, Wilson-Walker method, Glauert's empirical formula, and Buhl's theory. These particular methods were also used by various authors in the aerodynamic analysis of different HAWT [9–11]. The results were then compared to the actual results from the wind tunnel experiment done on the same blade from the literature.

3. Blade Element Momentum Theory

BEM Theory was a relatively straightforward method for aerodynamic analysis of horizontal axis wind turbines. The theory of the BEM combines the method of the one dimensional momentum theory and the two dimensional blade element theory.

Taking the concept of conservation of angular momentum in an annular stream tube, the thrust and torque on the annular element gives

$$dF_{thrust} = 4a\pi\rho(1-a)V^2rdr \quad (1)$$

$$dT = 4a'(1-a)\rho V\Omega\pi r^3dr \quad (2)$$

Meanwhile, the thrust and torque evolved by the elemental blade length positioned at a radius of "r" in the blade element theory, while multiplying it with the number of blades B, are given by

$$dF_{thrust} = \frac{1}{2}\rho cBW^2(C_L cos\emptyset + C_D sin\emptyset)dr \quad (3)$$

$$dT = \frac{1}{2}\rho cBW^2(C_L sin\emptyset - C_D cos\emptyset)r\,dr \quad (4)$$

where ρ, c, W, C_L, C_D, and ϕ are the density, chord length, resultant velocity, lift coefficient, drag coefficient, and flow angle respectively. Power output can be solved by applying the trapezoidal rule of integration to determine the overall torque. The equation for predicted power output was,

$$P = T\Omega \quad (5)$$

where T was the total torque for the entire blade length and Ω was the angular speed of the rotor. If the unit for torque and angular speed are Newton-meter and radian per second, respectively, then the power output could be expressed as Watts. Another important parameter in analyzing the performance of a wind turbine, aside from the power output, was the power coefficient. Manwell's integral equation can be used to calculate the rotor's power coefficient by getting the summation of this expression,

$$C_p = \frac{8}{\lambda}\frac{R - R_h}{NR}\sum_{i=1}^{N} F_i \lambda_{r,i}^3 a_i'(1-a)\left[1 - \frac{C_{D,i}}{C_{L,i}}\cot\emptyset_i\right] \quad (6)$$

The subscript i denoted the element number, R was the rotor radius and R_h was the radius of the hub.

4. Original BEM Theory

Combining equations (1) and (3) to eliminate the elemental thrust force, the new axial induction factor gives

$$a = \frac{1}{\left[\frac{4\sin^2\emptyset}{\sigma_r C_n} + 1\right]} \quad (7)$$

$$C_n = C_L cos\emptyset + C_D sin\emptyset \quad (8)$$

$$\sigma_r = \frac{Bc}{2\pi r} \quad (9)$$

where C_n is the normal force coefficient and σ_r was the local solidity factor. Similarly, combining equations (2) and (4) to cancel out the elemental torque, the new tangential induction factor becomes

$$a' = \frac{1}{\left[\frac{4\sin\emptyset\cos\emptyset}{\sigma_r C_t} - 1\right]} \quad (10)$$

$$C_t = C_L sin\emptyset - C_D cos\emptyset \quad (11)$$

where C_t was the tangential force coefficient. This paper utilizes the iterative calculation feature in Microsoft Excel which eases the process of solving necessary variables for the analysis of the performance of the wind turbine. This study uses calculation frameworks to visualize the logical structure in solving the power output and power coefficient [12]. Figure 2 describes the logical structure needed in calculation.

IOP Conf. Series: Materials Science and Engineering 1109 (2021) 012050 doi:10.1088/1757-899X/1109/1/012050

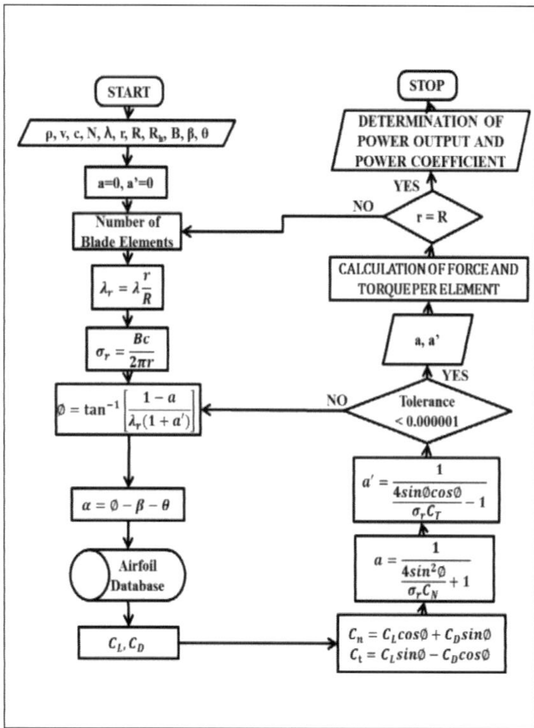

Figure 2. Calculation framework based from the Original BEM Theory

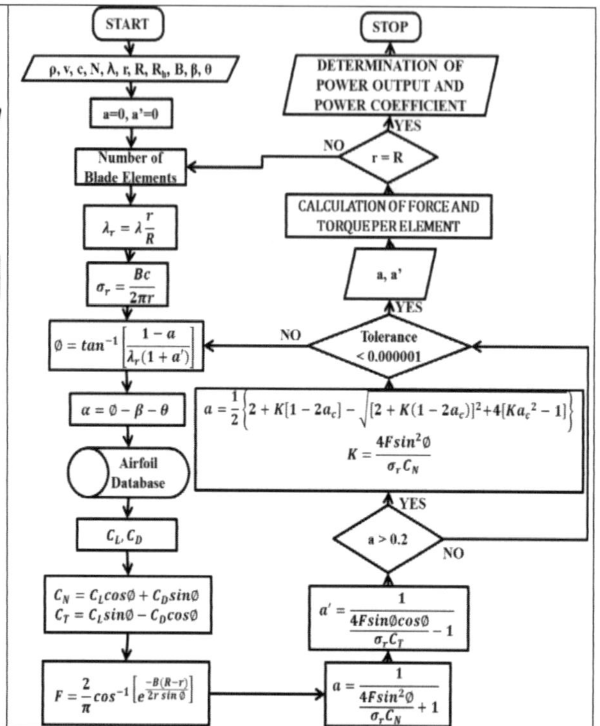

Figure 3. Calculation framework based from the Wilson-Walker Method

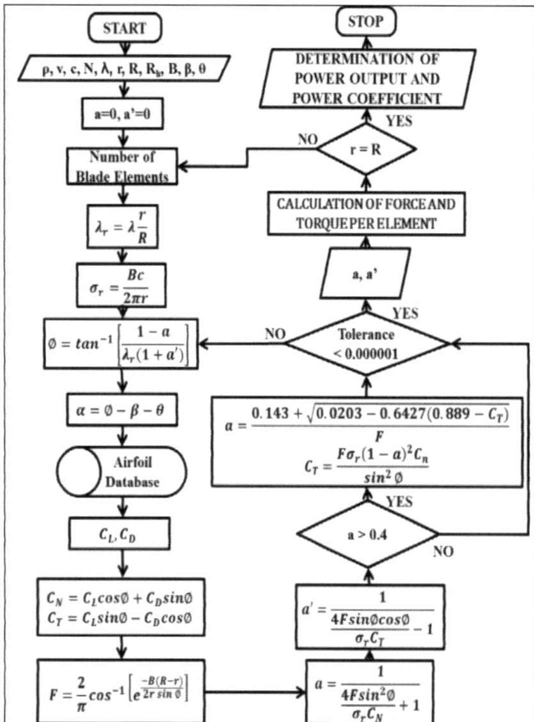

Figure 4. Calculation framework based from the Glauert's empirical formula

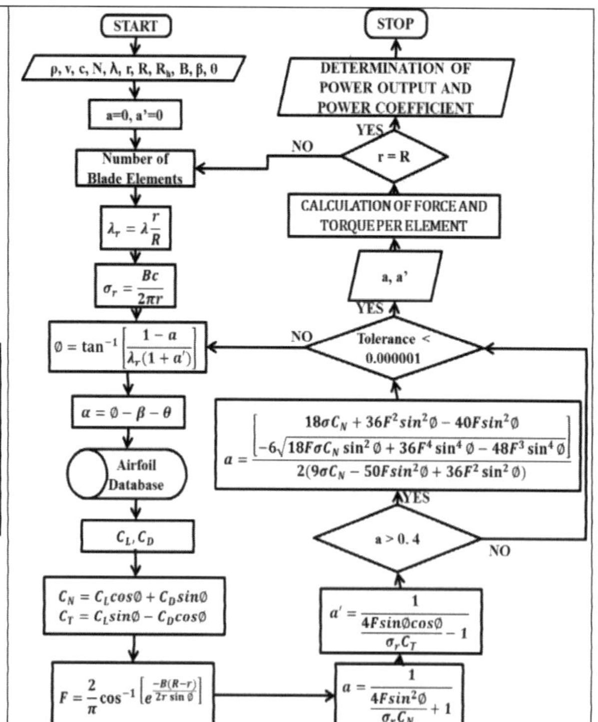

Figure 5. Calculation framework based from the Buhl's Theory

5. Brake state models

The original BEM Theory had no correction factors integrated in its formula. Many corrections had been proposed to increase the accuracy of the calculations. One of the factors that could be incorporated on the formulation was the Prandtl's tip loss factor (F).

$$F = \frac{2}{\pi}\cos^{-1}\left\{e^{\left[\frac{-B(R-r)}{2r\sin\emptyset}\right]}\right\} \tag{12}$$

With F defined, the axial and tangential induction factors became

$$a = \frac{1}{\left[\frac{4F\sin^2\emptyset}{\sigma_r C_n} + 1\right]} \tag{13}$$

$$a' = \frac{1}{\left[\frac{4F\sin\emptyset\cos\emptyset}{\sigma_r C_t} - 1\right]} \tag{14}$$

5.1. Wilson-Walker Method

In 1984, Wilson [13] reported the Wilson-Walker Method, which incorporated the Spera's correction for the thrust coefficient based on the axial induction factor. If a became greater than the critical axial induction factor value, ac, which was 0.2, the following equations could be used,

$$a = \frac{1}{2}\left\{2 + K[1-2a_c] - \sqrt{[2+K(1-2a_c)]^2 + 4[Ka_c{}^2 - 1]}\right\} \, and \; K = \frac{4F\sin^2\emptyset}{\sigma_r C_n} \tag{15}$$

The step by step procedure of the Wilson-Walker method for each element of the blade can be summarized by Figure 3.

5.2. Glauert's Empirical Formula

Glauert reported in 1926 that experimental results showing the thrust coefficients to be invalid if the axial induction factor exceeds approximately 0.4. Hence, Glauert came up with an empirical formula which addresses the validity of the numerical solution.

$$a = \frac{0.143 + \sqrt{0.0203 - 0.6427(0.889 - C_T)}}{F} \; and \; C_T = \frac{F\sigma_r(1-a)^2 C_n}{\sin^2\emptyset} \tag{16}$$

The procedure in Figure 4 describes the flow needed in calculation.

5.3. Buhl's Theory

Buhl[13] derived a new equation for the thrust coefficient that solved the numerical instability commonly experienced in solving the BEM Theory iteratively.

$$a = \frac{18\sigma C_n + 36F^2\sin^2\emptyset - 40Fsin^2\emptyset - 6\sqrt{18F\sigma_r C_n\sin^2\emptyset + 36F^4\sin^4\emptyset - 48F^3\sin^4\emptyset}}{2(9\sigma_r C_n - 50Fsin^2\emptyset + 36F^2\sin^2\emptyset)} \tag{17}$$

The procedure in applying the Buhl's Theory for each element of the blade was described in Figure 5.

6. Results and Discussion

In this study, a small HAWT with tubercles was analyzed by using the original BEM Theory and other three brake state models (Wilson-Walker Method, Glauert's Empirical Formula, and Buhl's Theory). The numerical results from the four calculation frameworks were compared with the experimental results from the literature.

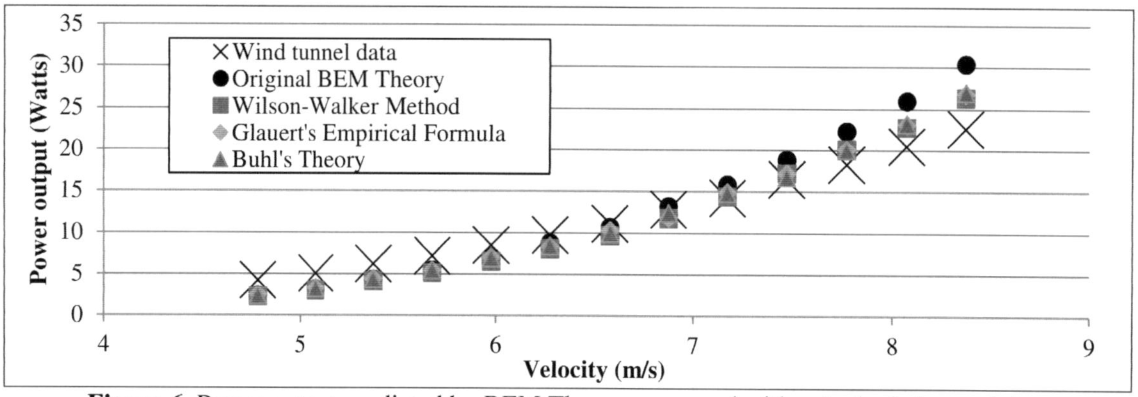

Figure 6. Power output predicted by BEM Theory compared with actual wind tunnel data

Figure 6 compared the numerical results of the original and corrected models of BEM theory to the experimental data taken from the wind tunnel test from the literature. The small HAWT with tubercles produced relatively small power outputs. Evidently, the four versions of the BEM Theory had underpredicted the power output from low to midrange wind velocities (4.7824m/s $\leq$ v $\leq$ 6.2769m/s). The predictions then became closer to the actual results as the wind velocity increased to 7.1736 m/s. But as the velocities were further increased, overprediction of the power output results started to become noticeable and the deviations of the predictions from the four models of BEM Theory to the actual results also started to increase. As can be seen from the graph above, the numerical results from the original BEM Theory were the most dissimilar as compared to the other three brake state models. Results from the original theory started to drift away with the increase in velocity, while the other three models stayed close with each other. This particular difference was expected since the original BEM Theory did not include factors and corrections, especially the Prandtl's tip loss factor, in the calculation of the axial and tangential induction factor. The original version of the BEM Theory includes assumptions of ideal flow, which could be the main reason of the power overprediction at larger wind velocities.

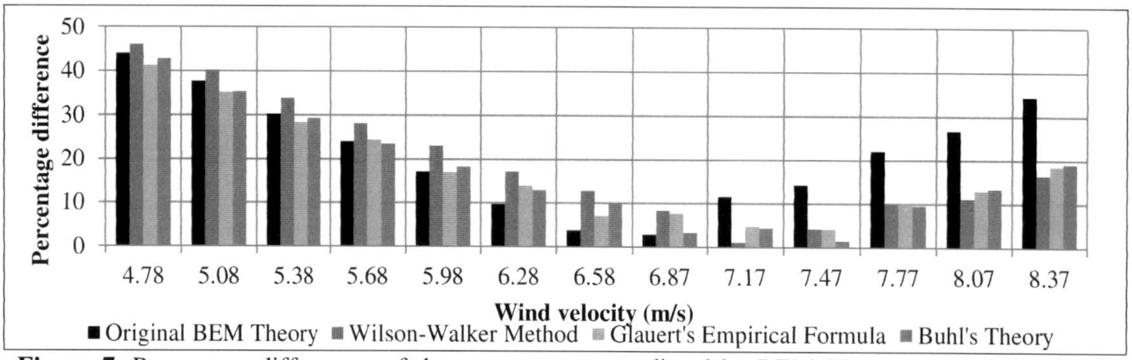

Figure 7. Percentage differences of the power output predicted by BEM Theory to the actual wind tunnel data

To properly visualize the accuracy of the presented methods, the percentage differences were also computed, as can be seen on Figure 7. Since the study was dealing with small HAWT and small power outputs, small differences in the predictions could reflect greatly on the computed percentage errors. As seen from the graph above, the largest values of percentage error for the four methods occurred at the starting speed. But those percentage errors diminished as the velocity increased. When the velocity reached a value of 7.47 m/s, the methods integrated with the Glauert's Empirical Formula and Buhl's Theory registered their lowest percentage difference of 4.21% and 1.56% respectively. A 1.17% error was the lowest value for the Wilson-Walker Method at a speed of 7.17 m/s. For the original BEM Theory, a percentage difference as low as 2.88% was calculated at 6.88 m/s. As those four methods reached their lowest values, the percentage differences started to increase hugely as the wind speed were also increased. This paper used the mean absolute percentage difference (MAPD) as a statistical tool in selecting the framework that provided the closest prediction of power output. Among the four calculation methods, the Buhl's Theory recorded the lowest MAPD of 17.21%, followed by the Glauert's Empirical Formula with an average of 17.34%. The original BEM Theory had the highest MAPD of 21.40%.

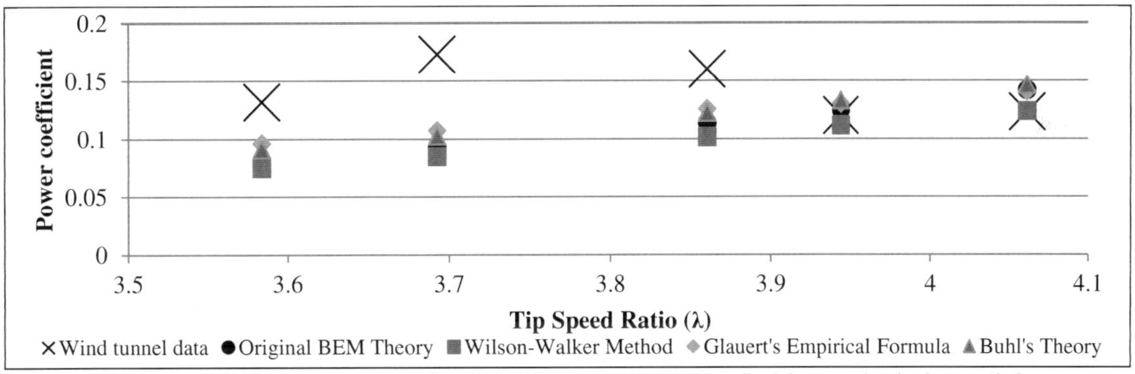

Figure 8. Power coefficient predicted by BEM Theory as compared with actual wind tunnel data

Figure 8 showed the comparison of the numerical results from the four calculation frameworks with the experimental results from the wind tunnel test from the literature. At λ=3.58, 3.69, and 3.86, the four methods had underpredicted the power coefficient and had a noticeable significant difference to the actual wind tunnel data. But with the remaining tip speed ratios (λ= 3.94, 4.06), the predictions of the four versions of the BEM Theory became closer to the actual data. The Wilson-Walker Method even predicted an almost equal value of the power coefficient (c_p=0.124) with the actual experimental data (c_p=0.123).

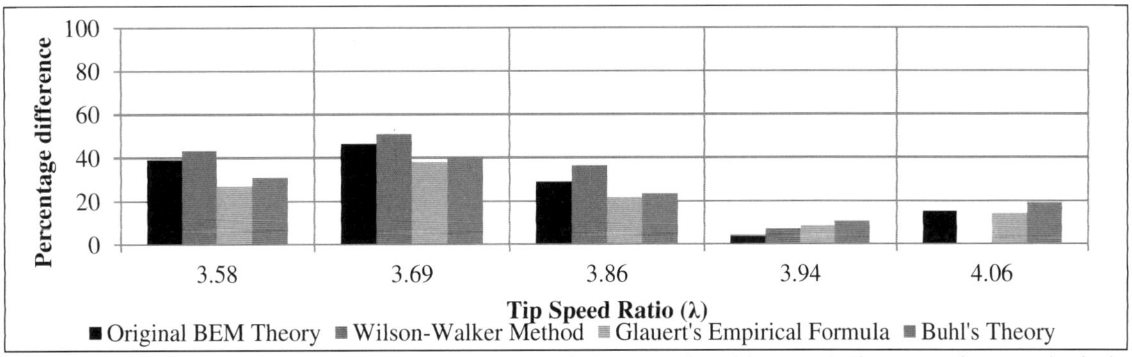

Figure 9. Percentage differences of the power coefficient predicted by BEM Theory to the actual wind tunnel data

The Wilson-Walker Method provided the closest prediction of the power coefficient at λ=4.06, showing only a percentage difference equivalent to 0.47%, but it also had the highest calculated percentage difference amounting to 51.06% (λ=3.69). The Original BEM Theory, Glauert's Empirical Formula, and Buhl's Theory had their lowest percentage differences at λ=3.94, with values of 3.79%, 8.56%, and 10.78% respectively, and their highest percentage differences at λ=3.69, with values amounting to 46.48%, 37.92%, and 40.66% respectively. Unlike the power output values which had thirteen actual values, there were only five power coefficient values from the wind tunnel experiment to be compared to the numerical results. This means that determining the MAPD were not as accurate as that of the power output.

7. Conclusion

The most common tool in evaluating the aerodynamic performance of a wind turbine is the Blade Element Momentum (BEM) Theory. This study focused on a small horizontal axis wind turbine with tubercles. Calculation frameworks based from the Original BEM Theory, Wilson-Walker Method, Glauert's Empirical Formula, and Buhl's Theory were used in the prediction of the power output and power coefficient. Numerical results from the calculation frameworks were compared to the actual wind tunnel experimental data taken from the literature. Since the study was dealing with small HAWT with tubercles, the power outputs calculated were relatively small. Because of that, slight differences in the predictions could reflect greatly on the computed percentage differences. The power output predicted by the Buhl's Theory was the closest to the actual results, with a mean absolute percentage difference (MAPD) of 17.21%, followed by the Glauert's Empirical Formula with a MAPD of 17.34%. In the case of the power coefficient, the Wilson-Walker Method produced an almost equal prediction at λ=4.06, showing only a percentage difference of 0.47%. Results also showed that the power coefficient predicted were more accurate at higher tip speed ratios. These findings proved that power output and power coefficient prediction were more accurate using the brake state models rather than the original BEM Theory, because these methods include formulas that addressed the instability in the calculation when the axial induction factor exceeded a certain value. Also, these brake state models integrated some corrections like the Prandtl's tip loss factor.

References

[1] M. Refan and H. Hangan, "Aerodynamic Performance of a Small Horizontal Axis Wind Turbine," **134**, no. May 2012, pp 1–7, 2013.

[2] R. K. Singh and M. R. Ahmed, "Blade design and performance testing of a small wind turbine rotor for low wind speed applications," Renew. Energy, **50**, pp 812–819, 2013.

[3] M. Ibrahim, A. Alsultan, S. Shen, and R. S. Amano, "Advances in Horizontal Axis Wind Turbine Blade Designs: Introduction of Slots and Tubercle," *J. Energy Resour. Technol.*, **137**, no. 5, pp 1–6, 2015.

[4] A. Dilimulati, T. Stathopoulos, and M. Paraschivoiu, "Wind turbine designs for urban applications: A case study of shrouded diffuser casing for turbines," *J. Wind Eng. Ind. Aerodyn.*, **175**, no. September 2017, pp 179–192, 2018.

[5] K. L. Hansen, R. M. Kelso, and B. B. Dally, "Performance Variations of Leading-Edge Tubercles for Distinct Airfoil Pro fi les," no. January, 2011.

[6] V. V Kumar and D. A. Shah, "Application of Tubercles in Wind Turbine Blades : A Review," **867**, no. 1, pp 254–260, 2017.

[7] I. H. Ibrahim and T. H. New, "Tubercle modifications in marine propeller blades," *10th Pacific Symp. Flow Vis. Image Process.*, no. June, pp 1–11, 2015.

[8] A. A. Alsultan, "Computational and Experimental Study on Innovative Horizontal-Axis Wind Turbine Blade Designs," no. May, pp 1–107, 2015.

[9] Q. Song and W. D. Lubitz, "BEM Simulation and Performance Analysis of a Small Wind Turbine Rotor," pp 381–399, 2013.

[10] Y. M. El Okda, "Design methods of horizontal axis wind turbine rotor blades," *Int. J. Ind. Electron. Drives*, vol. 2, no. 3, p. 135, 2015.
[11] P. Pratumnopharat and P. S. Leung, "Validation of various windmill brake state models used by blade element momentum calculation," *Renew. Energy*, vol. 36, no. 11, pp. 3222–3227, 2011.
[12] G. Augusto, A. Culaba, and W. Chen, "Power Output Prediction for LM Wind Turbine Blade using Blade Element Momentum Theory and GH Bladed Software," pp. 2–8, 2018.
[13] M. L. Buhl Jr, "A New Empirical Relationship between Thrust Coefficient and Induction Factor for the Turbulent Windmill State," *Tech. Rep. NREL/TP-500-36834*, no. August, 2005.

Seaweed drying characterization via serial statistical criteria analysis

A Culaba[1*], A H Atienza[1], A Ubando[1], A Mayol[1] and J Cuello[2]

[1] Mechanical Engineering Department, De La Salle University Manila, Philippines
[2] Department of Agricultural and Biosystems Engineering, University of Arizona

*alvin.culaba@dlsu.edu.ph

Abstract. In drying seaweeds, it is important to describe its drying behavior using semi-theoretical models. Predicting the best drying model for the seaweed was commonly done by testing the goodness of the curve for seaweed drying models by calculating the coefficient of determination R^2, the reduced chi square χ^2 , root means square error (RMSE) and mean bias error (MBE). However, it is essential to rank the drying models before selection by assessing its performance index Φ and mean relative deviation (MRD). It is also necessary to test the normality by evaluating the skewness and kurtosis statistics using the D' Agostino Pearson test which is the best test for goodness of fit. The simple sample run test statistics and confidence interval must also be satisfied These are the conditions that should be met under the serial statistical criteria analysis. This study utilized it to choose the best drying model for seaweeds. Employing this statistical analysis for testing commonly used drying models, Modified Page model emerged as the best model that depicts the drying behavior of the seaweeds with accuracy of 99.98% and highest performance index of 452.1967.

1. Introduction

Drying is composed of heat and mass transfer from the product. The drying process can be modelled mathematically in which derived equations can predict drying characteristics of air in various products [1]. The idea of thin layer drying applies to an assumed temperature distribution of the material, whole or part is exposed freely to drying air and there is variation in the material's thickness [2]. To determine drying characteristics of different types of porous materials, thin-layer equations are frequently used. The thin-layer equations can be categorized into theoretical, semi-theoretical and empirical models [3]. Theoretical models depend on Fick's second law which correlate diffusion coefficient to the local moisture content of the material assuming that the resistance to the moisture flow is uniformly distributed throughout the interior of the homogeneous isotropic material and the volume shrinkage is negligible. Simplifying the general series solution for Fick's second law, semi-theoretical models can be developed. These models can also be derived from simplified models with valid constraints on drying parameters such as temperature, relative humidity, air velocity and moisture content [4, 5]. Most of the common drying models used for seaweed drying were the Modified Page [6,7], Page [8,9], Wang and Singh [10], Newton [10,11] and Henderson and Pabis [11]. Inorder to pinpoint the best drying model, the behavior of the moisture ratio (MR) must be obtained, and its best fit curve must be estimated. Statistical analysis is conducted to test the goodness of the best fit curve of MR compared to existing models. The commonly used statistical tools in testing the goodness of the curve for seaweed drying models are coefficient of determination (R^2), the reduced chi square (χ^2), root means square error (RMSE) and mean bias error (MBE) [6, 8, 12]. However, to ensure goodness of fit, it is necessary to rank the drying models using other parameters such performance index Φ and mean relative deviation (MRD). The study will employ another statistical approach of testing for the best drying model using serial statistical criteria analysis. This analysis was used to determine the drying behavior of banana [13] and moisture sorption isotherms for chayote, almond powder, and yogurt [14]. This consists of four (4) criteria. The first criterion composed of the six statistical tools mentioned earlier. The D' Agostino - Pearson test of normality is the second criterion for testing the best fit curve and also considered as the best test for goodness of fit. [13, 14]. This involves solving for the skewness test statistic z_1 and kurtosis test statistic z_2. Using these the test statistic D' Agostino Pearson, χ^2_{DAP} can be calculated. The third criterion for serial statistical criteria is using single sample run test which uses z_r as its test statistics for randomness of moisture residual series. For the fourth and last criterion, the significance and precision of drying parameters via creation of confidence interval of individual parameters will be conducted.

There is lack of literature that employed serial statistical criteria analysis in thin dry modelling for seaweeds. This study will provide benchmark approach to analyze and test the goodness of fit of the curves and select the best drying model. The models should satisfy the four criteria and the top model will be chosen with the highest performance index in the rankings.

2. Materials and Methods

The seaweed species *Eucheuma cottonii* was used in the experiment. Inorder to obtain the initial moisture content, the seaweeds loss on drying procedure [15] was performed using microwave oven with temperature setting of 105°C. This is the same temperature used by [16]. 100g of seaweeds were placed in an aluminum foil and dried at constant temperature. For monitoring the mass reduction of the seaweeds, the aluminum foil is pulled every 10 minutes. Three same successive mass readings can safely assume that the seaweeds are totally dried. The initial moisture content of the seaweeds is obtained using the formula below:

$$\text{Initial Moisture Content}=\frac{m_i-m_f}{m_i}\,\boldsymbol{x100}\% \quad (1)$$

One kilogram of seaweeds was placed in the solar dryer in each tray location. Load cells were used to monitor and measure the mass of the seaweed with the use of Arduino Software for data gathering. The moisture ratio of the seaweeds has been calculated. The moisture ratio was plotted, and best fit curve was being approximated in describing each drying model using Minitab 18 and employing Microsoft Excel Real Function Tool for statistical calculations

2.1. Moisture ratio

The moisture ratio is defined as the ratio of the change of moisture at any given moisture M with the equilibrium moisture content M_e over the difference of the initial M_O to the equilibrium moisture content M_e and expressed in the equation:

$$MR = \frac{M-M_e}{M_o-M_e} \quad (2)$$

However due to relative humidity fluctuations in the dryer [17, 18] and the moisture ratio is simplified as

$$MR \approx \frac{M}{M_o} \quad (3)$$

2.2. Semi-theoretical drying models

The common widely used semi-theoretical models were given below and utilized in the study to approximate the best fit curve for MR. The Page Model is one of the commonly used thin drying model for seaweeds and expressed as

$$MR = e^{-kt^n} \quad (4)$$

where t is the time, M is the material moisture content, M_O is the initial moisture content and M_e= equilibrium moisture content with k and n as empirical coefficients

Modified Page model is one of the commonly used thin drying model for seaweeds. It is the revised version of the Page Model. Modified Page model are better fit to drying models for seaweeds compared to the Page Model [6]. The formula was stated as

$$MR = e^{-(kt)^n} \quad (5)$$

Henderson and Pabis model was the first term of the general series solution of Fick's Second Law and described by

$$MR = ae^{-kt} \quad (6)$$

where t is the time while a and k are drying parameters

The simplified version for the Henderson and Pabis model is the Lewis Model or sometimes called Newton model and described by the equation

$$MR = e^{-kt} \quad (7)$$

The Wang and Singh Model is described by the equation

$$MR = at^2 + bt + 1 \quad (8)$$

where t is the time a and b are drying constants

2.3. *Serial Statistical Criteria Analysis*

Table 1 shows the summary of Serial Statistical Criteria Formulas. For the first criterion MRexp,i is the ith experimentally observed moisture ratio, and MRpre,i is the ith predicted moisture ratio. N is the number of observations and n is the number of constants. M_i is the moisture content and $\underline{M_i}$ is the mean moisture content. Φ is ratio of R^2 to the product of RMSE and MRD.The second criterion is meant for the calculation of the D' Agostino test statistic χ^2_{DAP} . SE_m is the auxiliary variable for the computation of the normality test statistic where m=2, 3,4. The error about the residual mean is denoted by ε_i while $m_3$is the third moment of the mean for residual population and m_4 is fourth moment of the mean for residual population while s is the standard deviation of the individual mean. Equations (20) and (22) to (27) are applied in solving for skewness statistics z_1 in (33) while the kurtosis test statistics z_2 in (34) are obtained using equations (21) and (28) to (32). The resulting values in (33) and (34) are used to determine χ^2_{DAP}.To accept the drying model for curve fitting,χ^2_{DAP} in (35) should be compared with critical chi square χ^2_{crit} at α=.05 with df=2 or $\chi^2_{crit} = 5.99$. The drying model for best fit curve will be selected if $\chi^2_{DAP} < \chi^2_{crit}$. The simple sample run test is the third criterion in which the test statistics z_r is determined. This test statistics is depended to n1 which is the number of positive residuals, n2= the number of negative residuals and g = the number of times the sequence of residual changes sign. Using equations (36) and (37), z_r in (38) is calculated. The value of z_r will be compared with critical value of z, z_{crit}, at significant level α=.05 or $z_{crit} = 1.96$. If $z_r < z_{crit}$ hence accept the drying model for best fit curve otherwise discard it. The fourth and final criterion is based on the creation of the confidence interval of individual drying parameters. The drying model is accepted if the drying parameters are within the 95% confidence interval or else the model will be eliminated.

Table 1. Summary of Serial Statistical Criteria Formulas

Criteria	Parameters	Formulas	Interpretation
First Criterion	R^2, χ^2, RMSE, MBE, Φ, MRD	$R^2 = 1 - \frac{\sum_{i=1}^{N} (MR_{pre,i}-MR_i)^2}{\sum_{i=1}^{N} (MR_{exp,i}-MR_i)^2}$ (9) $\chi^2 = \frac{\sum_{i=1}^{n} (MR_{exp,i}-MR_{pre,i})^2}{N-n}$ (10) $RMSE = \sqrt{\frac{\sum_{i=1}^{N} (MR_{exp,i}-MR_{pre,i})^2}{N}}$ (11) $MBE = \frac{\sum_{i=1}^{N} (MR_{exp,i}-MR_{pre,i})}{N}$ (12) $MRD = \frac{1}{n}\sum_{i=1}^{n} \left\lvert\frac{E_i}{M_i}\right\rvert$ (13) $E_i = M_i - \underline{M_i}$ (14) $\underline{M_i} = \frac{1}{n}\sum_{i=1}^{n} M_i$ (15) $\Phi = \frac{R^2}{(RMSE)(MRD)}$ (16)	The higher value of R^2 and Lower values for χ^2, RMSE and MBE will imply better curve fit for the drying model [13, 14].
Second Criterion	χ^2_{DAP}	$SE_m = \sum_{i=1}^{n} \varepsilon_i^m$ (17) $\varepsilon_i = E_i - \underline{E_i}$ (18) $s = \sqrt{\frac{SE_2}{n-1}}$ (19) $m_3 = \frac{nSE_3}{(n-1)(n-2)}$ (20) $m_4 = \frac{n(n+1)SE_4}{(n-1)(n-2)(n-4)} - \frac{3{SE_2}^2}{(n-2)(n-3)}$ (21) $A_1 = \frac{m_3(n-2)}{s^3\sqrt{n(n-1)}}\sqrt{\frac{(n+1)(n+3)}{6(n-2)}}$ (22) $B_1 = \frac{3(n^2+27n-70)(n+1)(n+3)}{(n-2)(n+5)(n+7)(n+9)}$ (23) $C_1 = \sqrt{2(B_1-1)} - 1$ (24) $D_1 = \sqrt{C_1}$ (25) $E_1 = \frac{1}{\sqrt{ln(D_1)}}$ (26) $F_1 = \frac{A_1\sqrt{C_1-1}}{\sqrt{2}}$ (27) $G_2 = \frac{24(n-2)(n-3)}{(n+1)^2(n+3)(n+5)}$ (28) $H_2 = \frac{(n-2)(n-3)\vert m_4\vert}{s^4(n-1)(n+1)\sqrt{G_2}}$ (29) $J_2 = \frac{6(n^2-5n-2)}{(n+7)(n+9)}\sqrt{\frac{6(n+3)(n+5)}{n(n-2)(n-3)}}$ (30) $K_2 = 6 + \frac{8}{J_2}(\frac{2}{J_2} + \sqrt{1 + \frac{4}{J_2^2}})$ (31) $L_2 = 1 - \frac{2}{K_2}/(1 + H_2\sqrt{\frac{2}{K_2-4}})$ (32) $z_1 = E_1 ln\,(F_1 + \sqrt{F_1^2 + 1})$ (33)	The resulting values in (33) and (34) are used to determine χ^2_{DAP}. To accept the drying model for curve fitting, χ^2_{DAP} in (35) should be compared with critical chi square χ^2_{crit} at α=.05 with df=2 or $\chi^2_{crit} = 5.99$. The drying model for best fit curve will be selected if $\chi^2_{DAP} < \chi^2_{crit}$ [13, 14, 21].

		$z_2 = \left(\frac{2}{9K_2}\right)^{\frac{-1}{2}}\left(1-\frac{2}{9K_2}-\sqrt[3]{L_2}\right)$ (34) $\chi^2_{DAP} = z_1^2 + z_2^2$ (35)	
Third Criterion	z_r	$g_1 = \frac{2n_1n_2}{n_1+n_2}+1$ (36) $\sigma_r = \sqrt{\frac{2n_1n_2(2n_1n_2-n_1-n_2)}{(n_1+n_2)^2(n_1+n_2-1)}}$ (37) $z_r = \frac{\lvert g-g_1\rvert-0.5}{\sigma_r}$ (38)	The value of z_r will be compared with critical value of z, z_{crit}, at significant level α=.05 or $z_{crit} = 1.96$. If $z_r < z_{crit}$ hence accept the drying model for best fit curve otherwise discard it.
Fourth Criterion	CI	Confidence Interval	The drying model is accepted if the drying parameters are within the 95% confidence interval or else the model will be eliminated.

3. Results and Discussion

The mass of the seaweeds was reduced from 100g to 11.2g for 0.5920 g/min drying rate at 105°C with initial moisture content of 89.8%. The mass reduction of the seaweeds in the drying experiment was also determined. It took 311 minutes or approximately 5 hours and 19 minutes to reach the moisture content of 40%. The falling drying rate highly dominates at a rate of 1.42 g/min while the constant drying rate on the other hand is at 1.12 g/min. The moisture ratio was also calculated per drying time and plotted as shown in Figure 1.

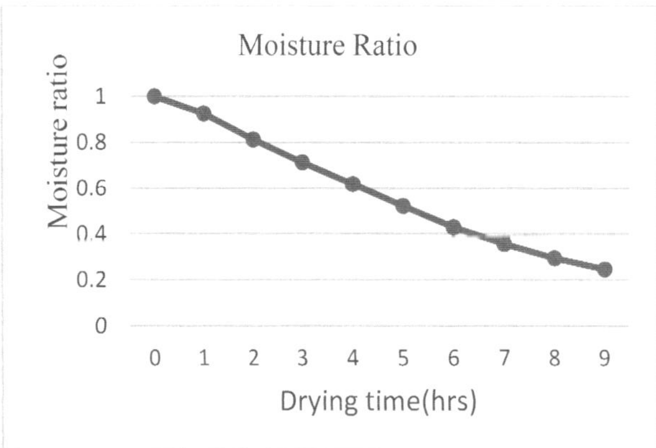

Figure 1. Actual Moisture Ratio per drying time

Table 2 Summary of Drying Model Curve Fitting via serial statistics criteria for the whole dryer

Thin drying models	Constants	Value	CI		R^2	RMSE	MBE	X^2	MRD	Φ	$\chi2_{dap}$	z_r
Modified Page*	k	0.1469	0.1458	0.148	0.9998	0.0034	-0.0002	0	0.6499	452.1967	0.7631	1.486
	n	1.2999	1.2779	1.322								
Page	k	0.0827	0.0794	0.0859	0.9998	0.0036	-0.0002	0.0000	0.6499	428.9897	2.1776	1.4288
	n	1.2999	1.2779	1.3220								
Henderson and Pabis	a	1.0519	1.0038	1.0999	0.9862	0.0298	-0.0022	0.0011	0.1183	279.6641	0	0.9932
	k	0.1486	0.1349	0.1624								
Wang and Singh	a	0.0023	0.0014	0.0032	0.9976	0.0125	0.0029	0.0002	0.6704	118.9297	0.0402	0.9774
	b	-0.1068	-0.1133	-0.1002								
Newton	k	0.1383	0.1269	0.1497	0.9782	0.0375	0.0072	0.0016	0.5714	45.6633	0.7631	0.9598

With this data, the curve fitting for each drying model has been conducted. Afterwards, serial statistical critera analysis was applied to obtain the best drying model. It can be observed in Table 2 that all drying models satisfied the four criteria. The most accurate were the Page and Modified Page model with R^2=0.9998. The Modified Page model has the lowest RMSE = 0.0034. Ranking Φ, it was evident that the Modified Page model has highest value of 452.1967 followed by the Page Model with Φ = 428.9897. The least accurate was the Newton model with R^2=0.9782 and Φ = 45.6633. The Modified Page model was the top choice for the best semi-theoretical seaweed drying model as denoted by the asterisk (*) symbol and agrees with the studies of [6, 7].

Using non-linear regression tool in Minitab 18, the best fit curve of MR for each drying model was plotted and shown in Figure 2.

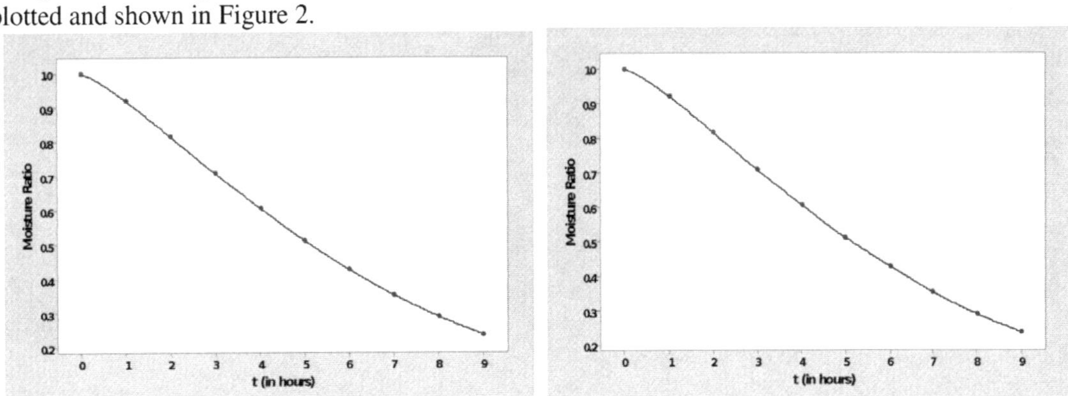

(a) (b)

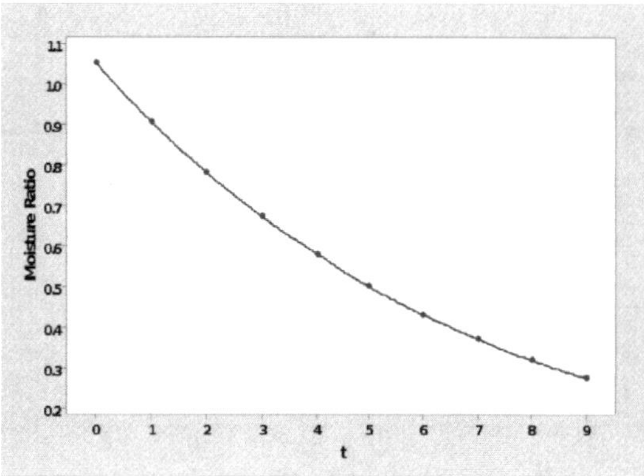

(c)

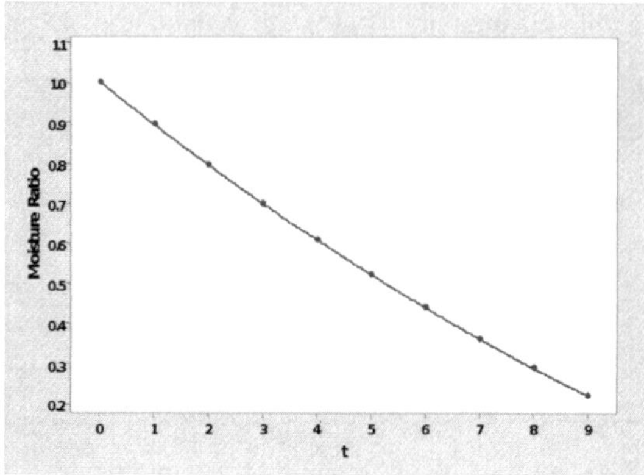

(d)

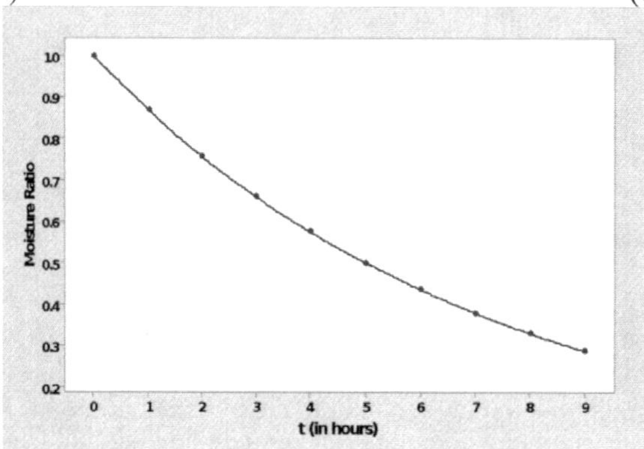

(e)

Figure 2. Best fit curve of MR for the drying model a) Modified Page b) Page c) Henderson and Pabis d) Wangh and Singh and e) Newton

The resulting equation for the Modified Page model with accuracy of 99.98% is expressed in the equation below

$$MR = e^{-(0.1469t)^{1.2999}} \quad (39)$$

From the above data, it can be noted that the Modified Page model is the top model for predicting the moisture content reduction in dried seaweeds under solar drying with an accuracy of 99.98%. This model was also used to observe drying behavior in mint leaves and basil leaves [2, 22]

4. Conclusion and Recommendation

It can be concluded in the study that the Serial Statistical Criteria Analysis is an effective way of selecting the potential drying model in seaweeds from which they were ranked in terms of performance indices. The Modified Page model was selected to be the best model for predicting the drying behaviour of the seaweeds having an accuracy of 99.98% with highest performance index of 452.1967. The researchers would like to recommend validating the findings of this study and compare them to the results of applying theoretical and empirical models.

Acknowledgement

The researchers want to give their utmost gratitude to DOST-CRADLE Project Grant and URCO Funding for financial support in this research.

References

[1] Yi X K, Wu W F, Zhang Y Q, Li J X & Luo H P 2012 Thin-Layer Drying Characteristics and Modeling of Chinese Jujubes. Mathematical Problems in Engineering*, 2012, 1–18.* doi:10.1155/2012/386214

[2] Onwude D I, Hashim N, Janius R B, Nawi N M & Abdan K 2016 Modeling the Thin-Layer Drying of Fruits and Vegetables: A Review. Comprehensive Reviews in Food Science and Food Safety, 15(3), 599–618. doi:10.1111/1541-4337.12196

[3] Kucuk H, Kilic A & Midilli A 2014 Common Applications of Thin Layer Drying Curve Equations and Their Evaluation Criteria. In: Dincer I., Midilli A., Kucuk H. (eds) Progress in Exergy, Energy, and the Environment. Springer, Cham. https://doi.org/10.1007/978-3-319-04681-5_63

[4] Aghbashlo M, Kianmehr M, Khani S & Ghasemi M 2009 Mathematical modelling of thin-layer drying of carrot. Int. Agrophys., 23(4), 313-317.

[5] Akbulut A & Durmuş A 2009 Thin layer solar drying and mathematical modeling of mulberry. *International Journal of Energy Research, 33(7), 687–695.* doi:10.1002/er.1504

[6] Fudholi A, Yogaswara R, Mardani D A, Ridho A, Hidayati S & Zen Z H 2020 Modified Page Model for Solar Drying of Seaweed. *International Journal of Advanced Science and Technology*, *29*(05), 7407-7413. Retrieved from http://sersc.org/journals/index.php/IJAST/article/view/18234

[7] Khan M, Wong J, Ruslan M H, Sulaiman J & Yasir S Md 2014 Effectiveness the Drying time and Kinetic of Seaweed Kappaphycus Alvarezii var. Tambalang in Green V-Roof Hybrid Solar Drier. *International Journal of Modern Mathematical Sciences*. 10. 125-136.

[8] Fudholi A, Ruslan M, Haw L, Mat S, Othman M, Zaharim A & Sopian K 2012 Mathematical modeling of brown seaweed drying curves, In: Proceedings of the WSEAS International Conference on Applied Mathematics in Electrical and Computer Engineering, USA, pp. 207–211.

[9] Djaeni M & Agustina Sari D 2015 Low Temperature Seaweed Drying Using Dehumidified Air. Procedia Environmental Sciences. 23. 10.1016/j.proenv.2015.01.002.

[10] Kadam S U, Tiwari B K & O'Donnell C P *2015* Effect of ultrasound pre-treatment on the drying kinetics of brown seaweed Ascophyllum nodosum. Ultrasonics Sonochemistry, 23, 302–307. doi:10.1016/j.ultsonch.2014.10.001

[11] Phang H K , Chu C-M, Kumaresan S, Rahman Md & Yasir S Md 2014 Preliminary Study of Seaweed Drying Under a Shade and in a Natural Draft Solar Dryer. *International Journal of Science and Engineering*. 8. 10-14. 10.12777/ijse.8.1.10-14.

[12] Fudholi A, Sopian K, Bakhtyar B, Gabbasa M, Othman M Y & Ruslan M H *2015 Review of solar drying systems with air based solar collectors in Malaysia.* Renewable and Sustainable Energy Reviews, 51, 1191–1204. doi:10.1016/j.rser.2015.07.026

[13] Mitrevski V, Lutovska M, Mijakovski V & Mijakovski N 2014 Statistical evaluation of thin layer drying of banana. *Journal of Hygienic Engineering and Design, 8*, 145-152.

[14] Ruiz-López I I & Herman-Lara E 2009 Statistical Indices for the Selection of Food Sorption Isotherm Models. Drying Technology, 27(6), 726–738. doi:10.1080/07373930902827932

[15] Philippine National Standard 2007 Seaweed Specification. Retrieve form https://seaplant.net/bimpeaga/images/downloads/PNCS.pdf

[16] Ali M K M, Fudholi A, Sulaiman J, Ruslan M H and Yasir S Md 2014 Sauna Technique, Drying Kinetic Modeling and Effectiveness on Solar Drying Compared with Direct Drying in Drying Process of Kappaphycus striatum in Selakan Island Malaysia. *Energy and Power Engineering*, **6**, 303-315

[17] El-Sebaii A A and Shalaby S M 2013 Experimental investigation of an indirect-mode forced convection solar dryer for drying thymus and mint. *Energy Conversion and Management,* **74**, 109–116.

[18] Akpinar E K 2010 Drying of Mint Leaves in a Solar Dryer and under Open Sun: Modelling, Performance Analyses. *Energy Conversion and Management*, 51, 2407-2418.http://dx.doi.org/10.1016/j.enconman.2010.05.005

[19] Fudholi A, Sopian K, Othman M Y, and Ruslan M H 2014 Energy and exergy analyses of solar drying system of red seaweed. *Energy and Buildings*, 68, 121–129. doi:10.1016/j.enbuild.2013.07.072

[20] Ali M K M, Rafiq S, Sakari M, Sulaiman J & Yasir S M 2013 Seaweed modeling for drying and the efficiency as heavy metal removal in 170 Malaysian Journal of Mathematical Sciences kappaphycus striatum variety sacol using solar dryer. International Journal of Environment and Bioenergy, 8:41–55.

[21] Sheskin D J 2004 *Handbook of parametric and nonparametric statistical procedures* (3rd ed.). Chapman & Hall/CRC.

[22] Akpinar E K 2006 Mathematical modelling of thin layer drying process under open sun of some aromatic plants. *Journal of Food Engineering, 77(4), 864–870.* doi:10.1016/j.jfoodeng.2005.08.014

IOP Conf. Series: Materials Science and Engineering 1109 (2021) 012052 doi:10.1088/1757-899X/1109/1/012052

Software development for sustainable design in Vietnam: Ecosystem – An approach

Le Hung Tien

Faculty of Engineering, Van Lang University, Vietnam

tien.lh@vlu.edu.vn

Abstract. The needs for accurate energy consumption estimation in different design alternatives are essential for all projects. The demands to achieve sustainable design drive not only the efforts of a large number of researchers and organizations all over the world but also for designers in Vietnam, especially for building design in Vietnam, whereas the green constructions are growing rapidly. The developed software in the first stage helps the designers to quickly calculate and select the alternatives for green building design. This paper is carried out under the collaboration project between Van Lang University (VLU, Vietnam) and Carnegie Mellon University (CMU, USA) to develop a software for sustainable design, which focuses on energy consumption optimization. This software is the first project towards the full software development life cycle for green design in Vietnam.

1. Background of the study

Traditionally, most of the building design in particular and construction projects in general just focus on the functions such as accommodation, utilization purposes. However, the efficient, effective and optimal design for energy consumption or environmentally friendly designs were not taken into account due to the limit of available technologies. For the last few decades, the world is facing fierce problems, especially climate change, which forces the world's governments, authorities and designers to take other approaches to achieve sustainable design. According to McKinsey studies recently, there are many reasons causing such issues:

- **Urbanization**: Over 50% of the world's population now live in cities and this trend will only continue. McKinsey estimates for the next 10 years, two third of the world population will live in cities or megacities. China and India alone will see an additional 500 million citizens move to their cities.
- **The availability of sacred natural resources** has disappeared. Energy and water in particular are major components in the cost of operations for corporations and municipal governments. So, investors, consumers, and communities all pay attention to obtain more efficient building constructions over their lifecycle.
- **Today's marketplace is intensely competitive**. The rapid changing of business environment requests the owners and AEC designers, consultants to achieve high efficiency to deliver projects faster with less environmental impact.

- **AEC projects are getting more complex.** With increasing requirements for sustainability or collaboration, projects are getting more and more complicated. You now need to coordinate and communicate design, materials, schedules, budgets, compliance, across a dispersed team of stakeholders.
- **There is a booming of new technology** – high speed, high computing capacity lead to fierce competitions between service providers.

For all these reasons, the demand for software development which can help the designers to trade - offs the alternative designs via energy consumption estimation using BIM technology is introduced and developed in the collaboration project between VLU and CMU in the software engineering field.

2. BIM background and prevailing trends

2.1. Worldwide

The rapid development of computer science recently and the applications of information technology on construction lead the trends that are driving change and opportunities in the Architecture – Engineering – Construction (AEC) industry. A component of the changes in technology we are witness to includes global trends in adoption of building information modeling (BIM) technologies.

According to Autodesk, the world No.1 in BIM technology, **BIM** originated as a term to describe a technology, but the term has evolved as industry has begun to understand the integrated nature of BIM. Today we describe BIM as a process, one that is model-centric and enabled by technology. BIM is not a product you can purchase or download, but is a methodology that is supported by a technology toolkit that helps business drive clarity.

Figure 1. Sustainable definition (Adapted from Autodesk documents)

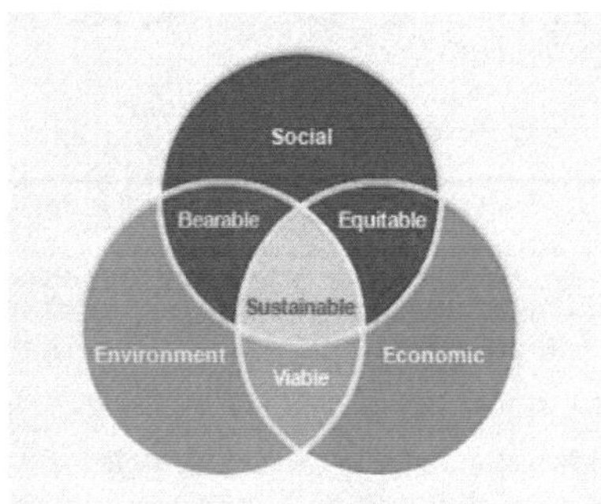

2.2. In Vietnam

The objective of achieving sustainable design, high performance construction facilities, low environmental impact are critical for all parties.

In Vietnam, decision No. 432/QD-TTg of the Prime Minister: To approve the Strategy for Sustainable Development Vietnam during the period 2011 to 2020. Vietnam BIM steering committee is responsible for BIM implementation in Vietnam though the results are still limited.

3. Ecosystem Software Development

3.1. Software development goals

The project goals are stated in the following:

- Develop the solution for Sustainable Design Calculation Intensive system.
- Develop and implement the software architecture which can adopt sustainable software development.

- Develop a framework that helps the other teams develop later to handle the scalability within the software development processes.
- Modify Heat Load & Thermal Equipment Design to fit with the architecture.
- Development Refrigeration Cycle & Automatic Drawing Generation.
- Integrate 4 modules above into the Cooling System to prove the software architecture that can handle many modules, much system to calculate the Energy.
- Develop the project within schedule and budget.
- Create the software development framework which is guided by SEI-CMU that ensures the software quality, also the sustainable software development.

3.2. Software requirements
The requirements for Ecosystem software development focus on the analysis and calculation in design phase:

- The analysis and calculation should be defined in the strong disciplines and principles.
- Allowing perform analysis of each thermal equipment individually, and then can be integrated these item calculation results to the whole system.
- The analysis and calculation should be optimal by the mathematics, algorithm and automatically, that reduce the effort for who performs the analysis.
- Reports should be generated automatically in the common standard form for many items and systems.
- Can have ability to integrate with the other vendors, such as Autodesk Revit, IES.
- Support analysis and calculation in design phase of construction projects, especially in Green Building Project.
- Meet the various rating systems, especially Vietnam standards.

3.3. Software development life cycle
For software development, the software engineering disciplines and principles must be strictly followed to achieve the quality product. In Ecosystem software development, the software development model is shown in figure 2 and 3.

IOP Conf. Series: Materials Science and Engineering 1109 (2021) 012052 doi:10.1088/1757-899X/1109/1/012052

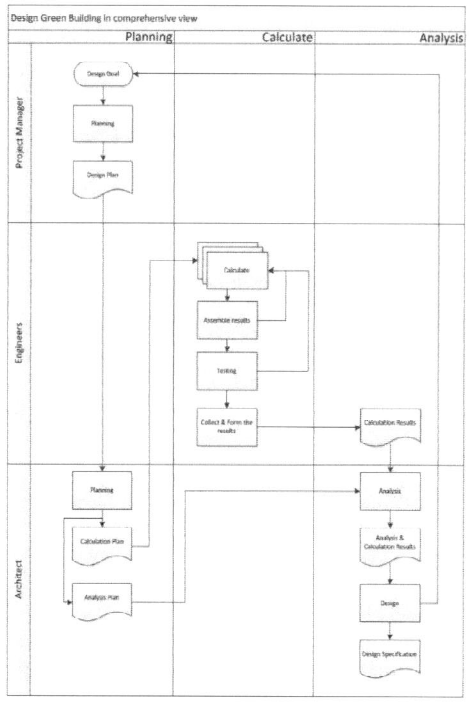

Figure 2. Ecosystem software in comprehensive view

There are also many software development models but for this project the V model is selected for the development.

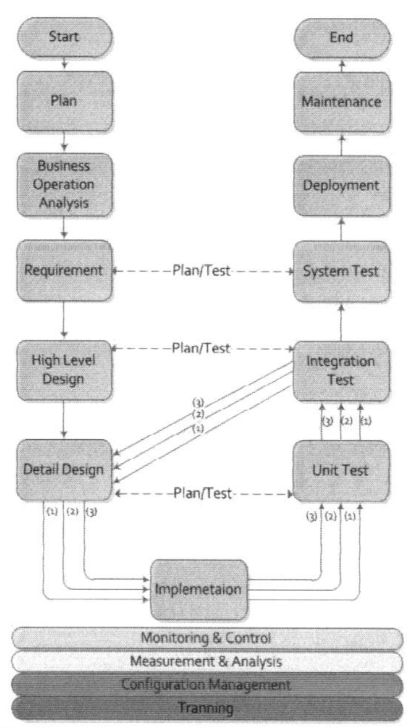

Figure 3. Ecosystem software – The V Model

3.4. *Software development processes*

Ecosystem software development life cycle consists of three main processes: Development Process – Management process and Teamwork process. All these processes are linked together in figure 4.

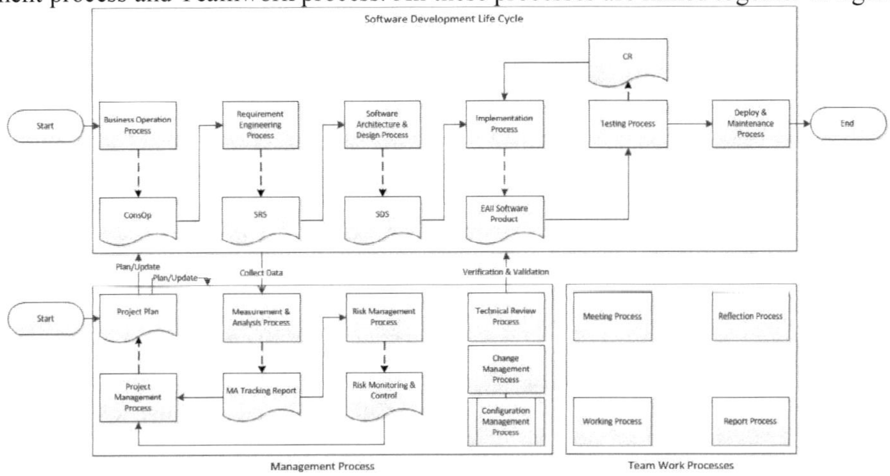

Figure 4. Ecosystem software – Development life cycle processes

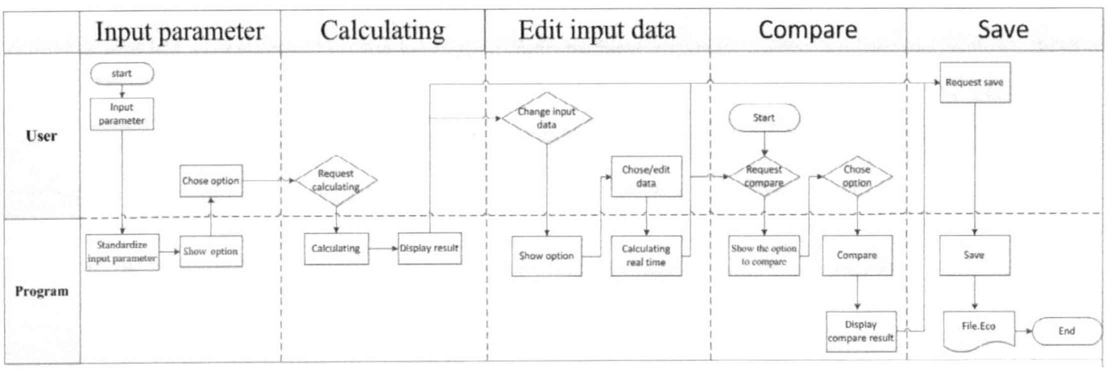

Figure 5. Ecosystem software – A typical calculation diagram

3.5. *Software development deliverables*

Table 1. Ecosystem software deliverables

Deliverable	Description
Process Description	Each process has been documented for description in detail.
Concept of Operation (ConsOp)	The document that describes ES Team's solution, includes the real problem of the customer, limited current situation and our team suggestion.
Software Requirement Specification (SRS)	The document that describes the software requirement for the ES development team.
Software Architecture Driver Specification (SAD)	The document that describes the software architecture driver which analyzed and filtered from SRS.
Software Architecture Specification (SAS)	The document that describes the architecture of ES.

Software Design Specification (SDS)	The document that describes the design of ES
Project Plan	The project plan of ES.
Project charter	The charter of ES.
Software Product	The stand-alone and installable application of ES.
User Guide & Installation Guide	The guide including with ES.
Presentation Slide	The presentation for capstone report & product promotion.
Measurement Plan (MP), Analysis Plan (AP), Measurement Report (MR).	The suite of documents on the Measurement & Analysis Program of ES.
Risk Management Plan (RMP), Risk Tracking and Report (RTR).	The suit of documents on Risk Management.
Testing Suit	Including Test Plan (STP), Test Cases (STCs) and Test Report (STR).

4. Ecosystem Software

The below pictures show few results from Refrigeration Cycle Calculation Module.

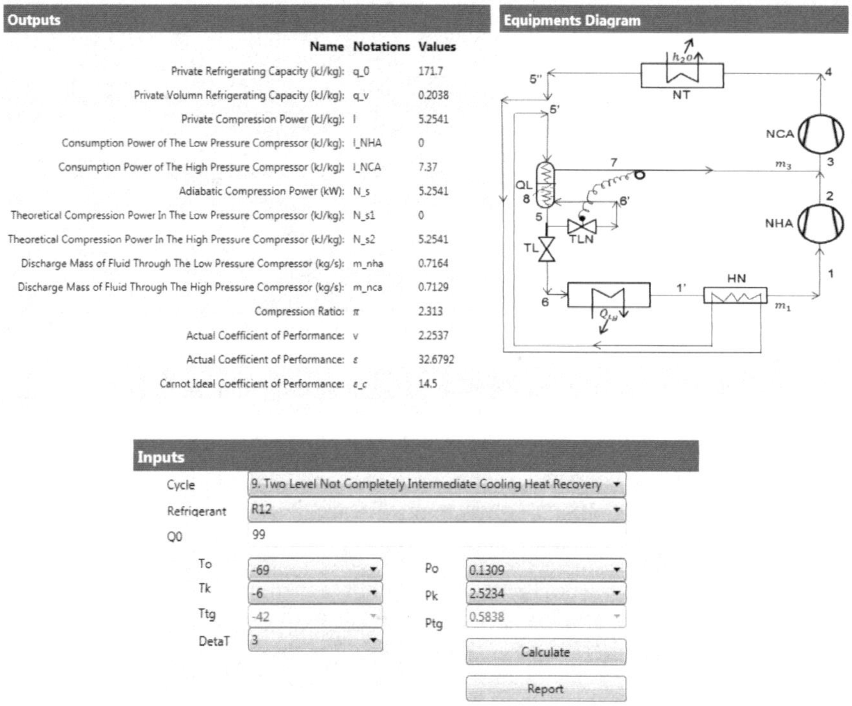

Figure 6. Ecosystem software – Refrigeration Cycle Calculation

5. Conclusion

This is the first step towards the full software development life cycle with full illustrated processes from the requirement phase to testing phase. Especially, this software development has both practical meaning in the academic environment and best practices for specific domain software development. The software helps the designers to perform the calculations rapidly in comparison with the traditional, time consuming methods.

The author will take the next investigation for integrated solution in sustainable solution optimization and automation while eliminating manual, time consuming calculation.

Acknowledgement

The author would like to express his deep gratitude and thanks to Van Lang University for financial support for this project, VLU's students and lecturers for the advices, positive contributions. Special thanks go to Institute for Software Research's professors and staff, Carnegie Mellon University, USA for their invaluable help.

References

[1] Hoang Dinh Tin 2012 Industrial Heat Engineering *National University Publisher*
[2] Nguyen Duc Loi 2011 Refrigeration system guide *Science and Technology Publisher*
[3] Nguyen Duc Loi and Pham Van Tuy 2011 Fundamentals for refrigeration technology *Science and Technology Publisher*
[4] Hoang Dinh Tin 2012 Fundamental of heat transfer *National University Publisher*
[5] Hoang Dinh Tin 2012 Heat transfer and heat exchanger calculation *National University Publisher*
[6] Autodesk developer network
[7] The Green Building XML (gbXML)
[8] Integrated Environmental Solution - VE-Pro
[9] Trace700 Guide
[10] Microsoft Developer Guide - Prism 4.1
[11] Microsoft Developer Guide - Application Architecture Guide v2
[12] Advanced MVVM Josh Smith
[13] Apress - Pro WPF and Silverlight MVVM
[14] Pro Expression Blend 4 – Apress
[15] MSDN Vietnam
[16] Apress - Practical WPF Charts and Graphics
[17] Apress - Foundations of WPF - Introduction to WPF
[18] Anthony Lattanze 2008 Architecting Software Intensive System – A Practitioner's Guide
[19] Professional Application Lifecycle Management with Visual Studio.
[20] Sustainable Software Development An Agile Perspective.
[21] Microsoft Press - Microsoft Project 2010 Inside Out.

The effect of the inlet temperature of the heat transfer fluid in a rectangular latent heat energy storage

Than Tun Naing[1], Mi Sandar Mon[2], and Sai Kham Twe[3]

[1,2,3]Department of Mechanical Engineering
Yangon Technological University, Yangon, Myanmar

konaing@ytu.edu.mm[1], misandarmon@ytu.edu.mm[2], saishanlayy@gmail.com[3]

Abstract. Solar energy is available during the day only. The excess heat energy collected during sunshine hours can be stored in the thermal energy storage and it can be reused during off sunshine hours. Latent heat energy storage systems by using phase change materials (PCMs) can be used effectively for storing thermal energy. It has high-energy storage density and the constant temperature storage process. In this paper, the performance of the latent heat storage system and phase change characteristics of PCM inside a rectangular latent heat energy storage unit will be presented. In this study, purified water is used heat transfer fluid (HTF) which is passed through the copper tubes inside the test unit and commercial-grade paraffin wax is used as PCM and indirect contact heat transfer method is used to heat exchange from the HTF to PCM. It was found that the PCM at the upper layer of the test section first melted during the melting process. Moreover, the effects of inlet temperature of HTF on the charging process was investigated. It can be concluded that increasing the inlet temperature of HTF effectively decreases the total charging time of the melting process. The amounts of heat storage for the inlet temperature of HTF 70 °C and 80 °C are 7400 kJ and 6555 kJ respectively during the melting process.

***Keywords* -** Latent Heat Energy Storage, Paraffin wax, Test section, Melting process, Effect of inlet temperature

1. Introduction

Thermal energy storage can be used effectively in solar energy storage, building cooling or heating systems, and industrial waste heat recovery [1]. The phase change process of PCM can be performed at the desired temperature level in the latent heat energy storage system [2]. The performance of the solar air heater collector with phase change material was investigated to improve thermal efficiency [3].

Un-used waste heat energy from factories is effectively stored in thermal energy storage, and this stored energy can be used again at the desired place and time [4-5]. Kousksou et al. investigated the geometry of the PCM container for the application in solar water heating [6]. The cylindrical latent heat storage system is used to determine the heat transfer characteristics of PCM [7]. The heat transfer characteristics of PCM in a rectangular storage container was experimentally developed [8-10].

Ambarita, H., et al. have investigated that paraffin wax is a good PCM because it can store more energy than stearic acid [11]. The range of melting temperatures of paraffin wax has from 23 to 67 °C [12]. The paraffin compounds C_nH_{2n+2} have good properties such as non-corrosive, non-toxic, repeatability, and no sub-cooling [13]. Paraffin is cheap and is easily available from many manufacturers [14-15]. Xinbin H. et al. experimentally investigated the different design parameters of TES system [16]. The effects of changing the inlet temperature and its mass flow rate were investigated on both the charging and discharging processes [17-19].

The objective of this study is to investigate the melting behaviour of PCM and the performance of the storage system by changing the inlet temperature of HTF. The 15 kg of Paraffin wax is used as a PCM with the melting temperature of 58~60 °C for the heat storage of the solar water heating system. Purified water is used as HTF.

2. Experimental setup and procedure

2.1. Experimental Setup

The experimental test rig was built in a Laboratory which is situated in Yangon Technological University. Figure 1 shows the schematic diagram of the experimental device. The experimental setup includes constant temperature control water bath, water circulation pump, test section, flowmeter, water tank, variable frequency drive, data logger, and thermocouples.

The constant temperature control water bath has the following specifications: 45 L, the maximum operating temperature 200 °C and the accuracy of temperature is ±1°C. The pump was responsible for the circulation of the hot water from the hot water bath to the thermal energy storage unit during the charging process and the cold water from the cold water tank to the TES unit during the discharging process. The pump is controlled by using a variable frequency drive to get the desired flow rate of heat transfer fluid. The acrylic EZ-View flowmeter with the model number (FL-9004) manufactured by OMEGA is used to measure the flowrate of HTF. The uncertainty of flow meter is within ± 5% of reading and it is installed on the exit pipeline of the test section. Figure 2 shows the actual setup of the experiment.

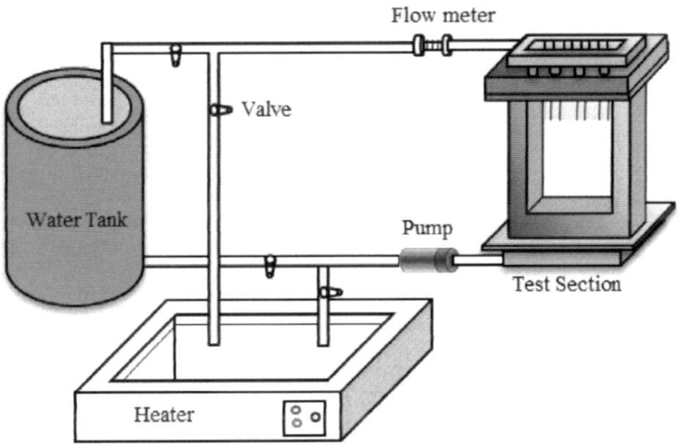

Figure 1. Schematic Diagram of Experimental Setup.

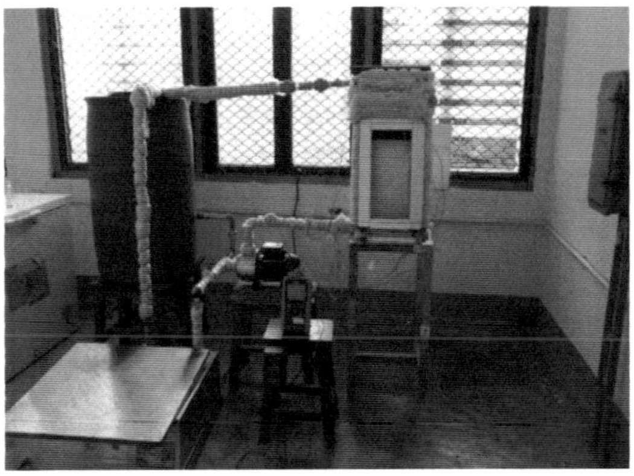

Figure 2. Photographic View of Experimental Setup.

The main component of this experiment is the test section. The test section includes HTF tubes connected with a small water reservoir that connected to the bottom part of the test section. It has one entry at the bottom and one exit for HTF flow at the upper part of the test section. Two acrylic sheets with 4 mm of thickness are attached on the front and back surfaces of the test section to capture the behaviour of PCM inside the storage unit by using a digital camera. The whole part of the test section is covered with glass wool (thickness of 40 mm) to prevent the heat loss from the test section to the surrounding. Purified water is used as HTF and passed through the pipes inside the test section in the upward flow direction.

The instantaneous temperature of the PCM is measured by using K-Type thermocouples and collected by using a data logger. Six thermocouples are inserted symmetrically at each side of the test section to measure the temperatures of PCM at six different places and 3 different levels of test section: T2 and T3 at the lower, T4 and T5 at the middle and T6 and T7 at the upper part (include this description to Figure 3). Locations of thermocouples inside the test section are shown in figure 3. To measure the inlet and outlet temperature of HTF, the two thermocouples were used. The commercial-grade paraffin wax with the melting temperature of 58~60 °C is used as PCM. The only 80% of PCM volume is filled in the latent heat storage unit (test section) for accommodating the volume expansion of PCM. The thermos-physical properties of paraffin wax are described in Table 1.

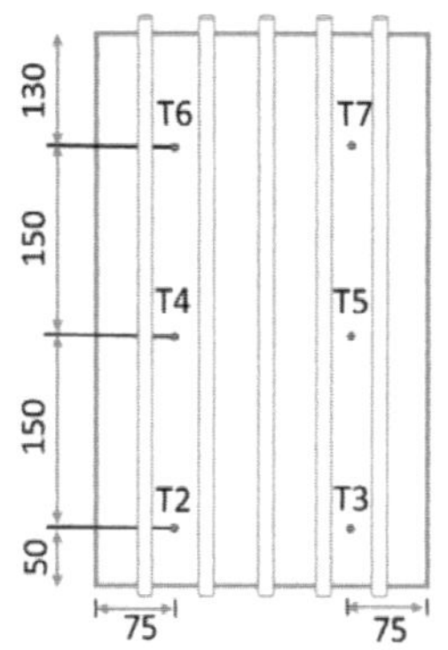

Figure 3. Locations of Thermocouples inside the Test Section.

Table 1. The Properties of Paraffin Wax [20].

Properties	Values
Melting Temperature	58~60 °C
Density (solid)	910 kg/m^3
Density (liquid)	795 kg/m^3
Latent heat of fusion	214 kJ/kg
Specific heat of PCM	1.85 kJ/kg K
Thermal conductivity	0.24 W/m K

2.2. Experimental Procedure

2.2.1. Preparation. Before running the experiment, the test section was filled with 15 kg of PCM. At the beginning of the experiment, the paraffin wax in the test section is in the solid-state. Before starting the melting (charging) process, the water from the temperature-controlled water bath is heated to reach desire inlet temperature. All of the pipelines and test section were covered with glass wool. Camera and stand are prepared to take photos and all thermocouples are connected to the data logger to record the temperature.

2.2.2. Experiment. The heated water flows from the water bath to the test section with a constant inlet temperature and constant flow rate of 9 L/min. The HTF was continuously circulated through the system during the whole experiment. The experiment was finished if the entire mass of solid PCM inside the test section was completely melted. When the charging process was completed, the discharging (solidification) process was immediately initiated by passing HTF from the cold water tank that has the room temperature to the system at a constant flow rate and a constant temperature. The water is continuously circulated through the system until the inlet and outlet temperature of HTF becomes the same. The temperatures of PCM that measured by the thermocouples are recorded with a data logger at the same time intervals of 30 seconds both in the charging and discharging periods.

2.2.3. Operating parameters. The constant flow rate of 9 L/ min purified water is used. The influences of inlet HTF temperature were investigated with 70 °C and 80 °C. The temperatures of PCM that measured by the thermocouples are recorded with a data logger at the same time intervals of 30 seconds. The experiment was finished if the entire mass of solid PCM inside the test section was completely melted.

3. Results and Discussions

3.1. Melting Process with 70 °C inlet temperature of HTF

For the first time, the inlet temperature of HTF was kept 70 °C and the initial temperature of PCM is at room temperature. Figure 4 shows the temperature profiles of PCM with inlet HTF temperature of 70 °C and experiment time in minutes at three different layers along the vertical direction of the test section. In this figure, the average values of two PCM temperatures at the same level were plotted.

The melting behaviour of PCM for the inlet HTF temperature of 70 °C is shown in figure 5. At the top of the test section, space must be provided to allow for volumetric expansion or contraction of PCM. Thus, the PCM is not fully filled inside the test section. Several pictures of phase changing characteristics of PCM are taken by the digital camera and they were shown in Figure 5.

It may be observed that PCM temperatures increase rapidly till its melting temperature is reached and remain constant during phase change and then liquid PCM temperature rises again gradually. The upper layer of the PCM was first to melt about 30 min and the phase change process of upper layer PCM can be seen in figure 4. From this figure, the PCM at the upper layer reached its melting point faster than that of the middle and lower layers. The melting process was finished at about 380 min. The amount of heat store can be calculated as equation (1).

$$Q_M = \int_{t_b}^{t_e} m_w C_{pw} (T_{in} - T_{out}) dt \quad (1)$$

Where,

Q_M = the amount of heat stored during the melting process(kJ)

t_b,t_e = the beginning and ending time of experiment (min)

m_w = mass of HTF (kg)

C_{pw} = specific heat of HTF (4.18 kJ/kg-K)

T_{in} = inlet temperature of HTF (°C)

T_{out} = outlet temperature of HTF (°C)

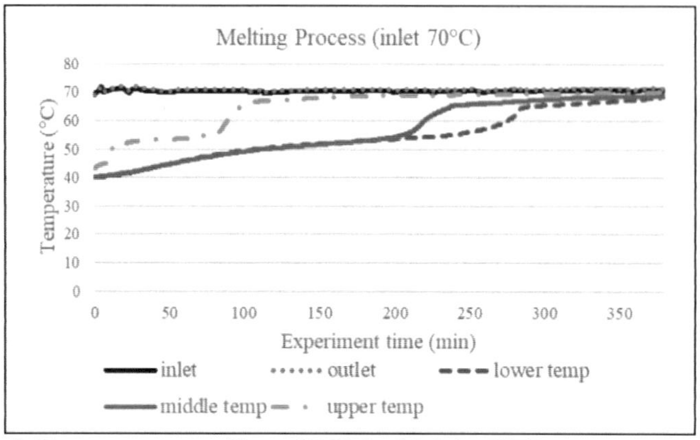

Figure 4. Temperature profiles of PCM with inlet HTF temperature of 70˚C

The height of the PCM was 380 mm in the solid phase before starting the melting process. The upper part of the paraffin wax has been melted about 30 min after starting the melting process. The volume of solid PCM decrease and the volume of liquid PCM increase by experiment time as shown in figure 5. Half of the solid PCM changed to liquid about 240 min. All of the PCM was melted about 320 min. At the end of the experiment, the height of the liquid PCM was 395 mm. Thus, the height of the PCM was change about 15 mm at the end of the experiment because of the volume expansion of the PCM. By using equation (1), the amount of heat stored for the inlet temperature of HTF 70 ˚C is 7400 kJ during the melting process.

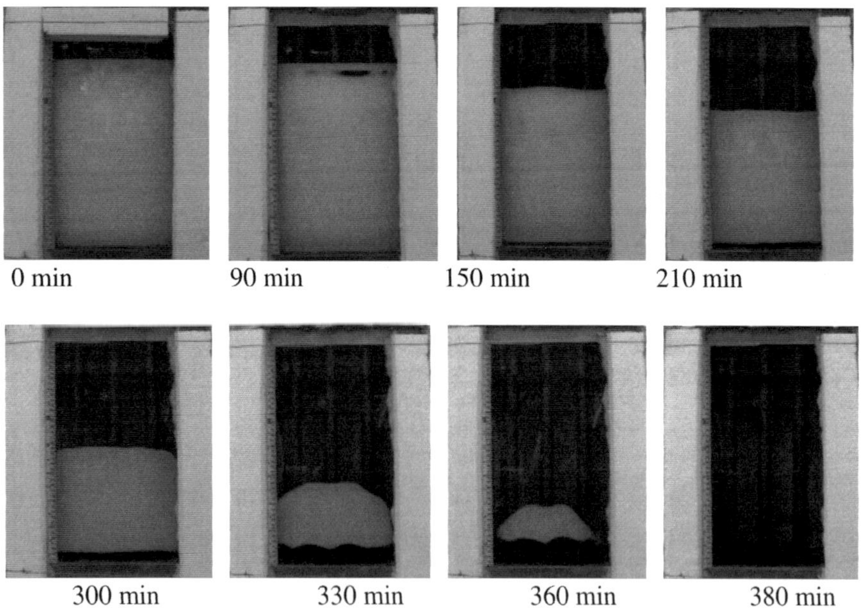

Figure 5. Melting behaviour of PCM for the inlet HTF temperature of 70˚C

3.2. *Melting Process with 80˚C inlet temperature of HTF*

In this time, the melting process with the inlet temperature of HTF 80˚C was carried out again. The constant flow rate of HTF was also used 9 L/min. Temperature profiles of PCM with inlet HTF temperature of 80˚C are shown in Figure 6 and it can be seen that the PCM at the upper layer reached its melting point faster than the PCM temperatures of middle and lower layers.

The melting behaviour of that is shown in figure 7. The solid PCM start melts from the upper layer and the volume of liquid PCM was increased by time. From figure 6 and 7, it was clear that the melting process was finished about 180 min. Thus, it can be concluded that the melting time is effectively decreased by changing the inlet temperature of HTF. The amount of heat stored for that condition is 6555 kJ. Although the amount of heat stored is less, the charging time is effectively reduced in that process. It can be concluded that the amount of heat stored 6555 kJ can be received at a short time of period than inlet temperature of 70˚C.

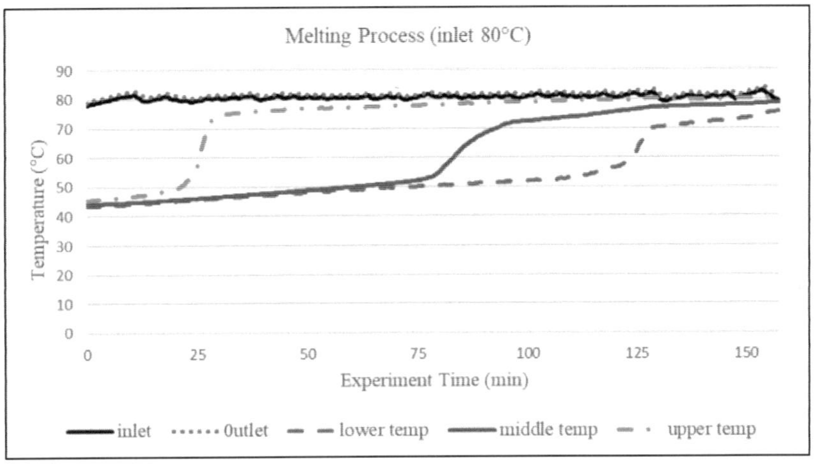

Figure 6. Temperature profiles of PCM with inlet HTF temperature of 80˚C

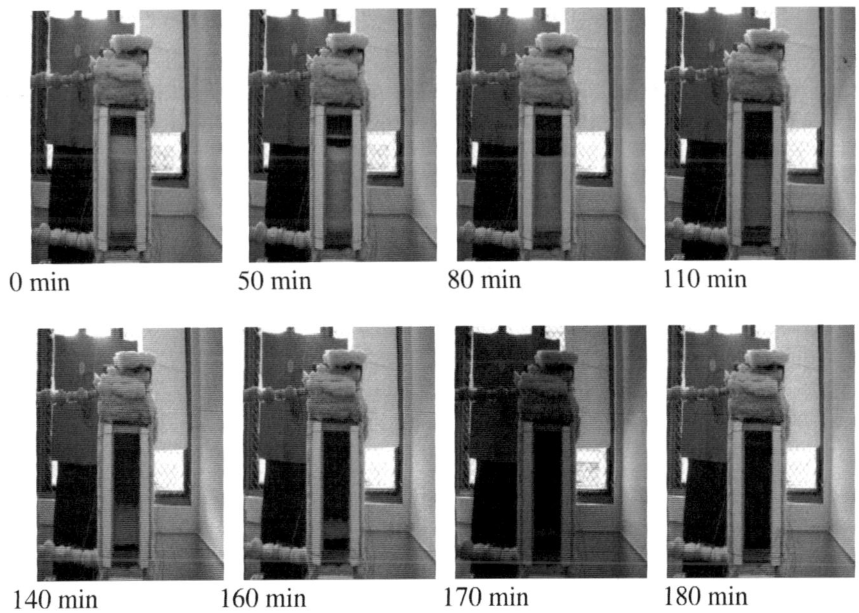

Figure 7. Melting behaviour of PCM for the inlet HTF temperature of 80˚C

4. Conclusions and Discussions

An experimental setup was built to study the performance of the latent heat storage system and the characteristic of PCM during the melting process for the two different inlet temperature of HTF in 70 °C and 80 °C. It was found that the PCM melting first occurred in the upper region of the PCM and then the melted PCM increased with time from top to bottom. Charging time for the melting process of PCM in the latent heat storage unit can be reduced by increasing the inlet temperature of HTF.

The charging time during the melting process with the inlet temperature of HTF 70 °C and 80 °C was 400 min and 180 min respectively. The increase of the HTF inlet temperature from 70 °C to 80 °C resulted in a reduction in charging time of about 52 %. The amount of heat stored for the inlet temperature of the HTF 80°C is 6555 kJ. Although the amount of heat stored is less in HTF inlet temperature of 80°C, the charging time is effectively reduced in that process than that 70°C. It can be concluded that the amount of heat stored 6555 kJ can be received at a short time of period than inlet temperature of 70°C.

Acknowledgements

The authors keep in mind with sincere thanks to JICA EEHE program and U Nyi Hla Nge Foundation for supporting the research fund to carry out this research. The authors would like to thanks to lab members who help the experiment.

References

[1] Zhang Z and Fang X 2006 Study on paraffin/expanded graphite composite phase change thermal energy storage material *Energy Conversion and Management* vol 47 pp 303-310
[2] Abhat A 1983 Low-temperature latent heat thermal energy storage: Heat storage materials *Journal of Solar Energy* vol 30 no 40 pp 313-332
[3] Arfaoui N, Bouadila S and Guizani A 2017 A highly efficient solution of off-sunshine solar air heating using two packed beds of latent storage energy *Solar Energy* **155** 1243-1253
[4] Murray R E and Groulx D 2015 Experimental study of the phase change and energy characteristics inside a cylindrical latent heat energy storage system: part 2, consecutive charging and discharging *Renewable Energy* **63** 724-734
[5] Bashar M and Siddiqui K 2017 Investigation of heat transfer during melting of a PCM by a U-shaped heat source *Int. J. Energy Res.* **41** 2091-2107
[6] Kousksou T, et al. 2014 Energy storage: Applications and challenges *Solar Energy Materials and Solar Cells* **120** 59-80
[7] Shen G, Wang X and Chan A 2019 Experimental investigation of heat transfer characteristics in a vertical multi-tube latent heat thermal energy storage system *Energy Procedia,* **160** pp 332–339
[8] Fadl M and Eames P C 2019 An experimental investigation of the heat transfer and energy storage characteristics of a compact latent heat thermal energy storage system for domestic hot water applications *Energy* vol 188 116083
[9] Naing T T, Horibe A, Haruki N and Yamada Y 2017 Melting and solidification heat transfer characteristics of a phase-change material in a latent heat storage vessel: Effects of a perforated partition plate and metal fiber *Journal of Power and Energy Engineering* **5** 13-29
[10] Naing T T, Horibe A, Haruki N and Yamada Y 2018 Reduction of the solidification height of phase-change material in direct-contact latent heat storage vessel *Thermal Science & Engineering* vol 26 no 1 19-27
[11] Ambarita H, et al. 2017 Experimental study on melting and solidification of phase change material thermal storage *IOP Conference Series: Material Science and Engineering* **180** 012-020

[12] Himran S, Suwono A, and Mansori G A 1994 Characterization of alkanes and paraffin waxes for application as phase change energy storage medium *Energy Source* **16** pp 117–128
[13] Or E, de Gracia A, Castell A, Farid M M and Cabeza L F 2012 Review on phase change materials (PCMs) for cold thermal energy storage applications *Applied Energy* 99 513533
[14] Sharma A, Sharma S D and Buddhi D 2002 Accelerated thermal cycle test of acetamide, stearic acid and paraffin wax for solar thermal latent heat storage applications *Energy Conversion and Management* **43** pp 1923-1930
[15] Tarish A L and Alwan N T 2017 Experimental study of paraffin wax-copper nanoparticles thermal storage material *International Journal of Modern Studies in Mechanical Engineering (IJMSME)* vol 3 issue 3 pp 11-17
[16] Cui H, Yuan X and Hou X 2003 Thermal performance analysis for a heat receiver using multiple phase change materials *Applied Thermal Engineering* **23** 2353–2361
[17] Khot S A and Sane N K 2011 Experimental investigation of phase change phenomena of paraffin wax inside a capsule *International Journal of Engineering Trends and Technology* vol 2 issue 2
[18] Khyad A and Samrani H 2016 State of the art review of thermal energy storage systems using PCM operating with small temperature differences: Focus on paraffin *J. Mater. Environ. Sci* **7** (4) 1184-1192
[19] Dukhan W A and Dhaidan N S 2020 Experimental investigation of the horizontal double pipe heat exchanger utilized phase change material *IOP Conf. Series: Materials Science and Engineering 3rd International Conference on Engineering Sciences* **671** 012148
[20] Jesumathy S P, Udayakumar M and Suresh S 2012 Heat transfer characteristics in a latent heat storage system using paraffin wax *Journal of Mechanical Science and Technology* vol 26 pp 959-965

IOP Conf. Series: Materials Science and Engineering 1109 (2021) 012054 doi:10.1088/1757-899X/1109/1/012054

The Efficiency Estimate of PV Cell under various wind velocities

Duy Tue Nguyen and Tri Nhut Do*

Van Lang University, 45 Nguyen Khac Nhu Street, Co Giang Ward, District No.#1, Ho Chi Minh City 700000, Vietnam

*trinhutdo@gmail.com

Abstract. It is a good chance to harness solar energy for Vietnam to geographic location is near to the equator with the high intense radiation. Photovoltaic (PV) cell is the device to convert directly solar energy into electrical energy. However, its efficiency is strictly impacted by the PV cell surface's temperature which depends on wind velocity, surrounding temperature, radiation and dust. For a long time, computational modelling has been a good choice for engineers to simulate their products. Currently, some softwares like Ansys Fluent, Comsol Multiphysics are chosen to simulate the thermal performance of product. In this paper, authors used Computational Fluid Dynamic (CFD) – Autodesk to estimates the PV cell temperature, then find out its corresponding efficiency. A 3 dimensions model of PV cell with some layers was created by Inventor software, and then simulated under various wind velocities like 0.5m/s, 1.5m/s, 2.5m/s, 3.5m/s, 4.5m/s, 5.5m/s and 6.5m/s to find out the PV cell corresponding temperatures. The solar radiation parameter in Ho Chi Minh city in April was established for simulation. With the same surrounding temperature at 30°C, the result shows that the highest PV cell surface's temperature of 84.8°C with wind velocity of 0.5m/s will drop to 49.81°C with the wind velocity of 3.5 m/s. When the wind velocity is continue increased, the drop of temperature is trivial. The PV cell efficiency at wind velocity of 6.5m/s is only 2.75% higher than that of wind velocity of 3.5m/s. Therefore, when choosing cooling PV cell by air, the air velocity should be chosen at 3.5 m/s to save energy.

Keywords: *Solar Energy, PV cell, CFD.*

1. Introduction

Viet Nam locates near the equator with the high intense radiation. Therefore, harnessing this free energy is the good way to save energy. Most of applications in solar energy is in the hot water supply. It can be exploited nearly every area of Viet Nam [1]. Besides, PV power is one of the viable energy to replace the traditional fossil fuels which strongly affect the environment. The PV cell is manufactured by following the Standard Test Condition (STC) which indicates that the solar radiation is 1000 W/m^2, panel temperature is 25°C and light spectrum with an air mass is 1.5. With this condition, the PV cell generates its maximum power named Watt-peak [2]. However, the local weather will affect these parameters and make its efficiency reduced. PV cell temperature is one of the major reasons which reduce PV cell's efficiency [3-4]. When the PV cell temperature rises, the rate of photon also rises. However, the reverse saturation current will be increased considerably and lead to the reduction of ban gap which makes the open circuit voltage (V_{oc}) affected [5]. With the PV cell temperature gain, its power output is gone down as a result of the negative temperature coefficient (-0.5%/°C rising in temperature). Therefore, PV cell has the high efficiency at the low temperature with the correct tilt angle. Radiation, surrounding temperature and wind velocity play an important role to the PV cell temperature. In term of wind velocity, some researchers study on this relationship between this one and PV cell temperature [6]. Latifa Sabri and Mohammad Benzirar made the experiment by changing the wind velocity through the PV cell, and they realized that when increasing wind velocity, PV cell temperature is also gone down, result in the enhanced electrical power. Khelifa [7] studied the PV cell performance by simulating the heat transfer between PV cells and cooling by using ANSYS Fluent software. The results shown that the lower PV cell temperature, the higher power will be gained [8].

In this paper, authors simulated the cooling the PV cell with 2 DC fans with air speed of 3.07 m/s, the surrounding temperature at 36.5°C and solar radiation of 1175 W/m^2, at 1.30pm by ANSYS software. The results shown that the highest PV cell can reaches 53.6°C in this condition. In [9], Jia Yang conducted the experiment about the impact of wind velocity to the PV cell temperature when he changed the wind velocity from 2 to 8 m/s. Consequently, he claimed that the increased velocity he made, the higher power out he had. S. Armstrong and W.G. Hurley in [10] analyzed the heat that transfer from the PV cell to the air with the velocity of 0.77m/s, 2.14 m/s and 5.76 m/s. The velocity of 5.6 m/s was the best value to have more heat transfer from PV cell to the air [11].

Authors simulated the PV cell temperature with different wind velocities of 0 m/s, 0.43 m/s, 2.5m/s and 6.95 m/s under the outdoor temperature at 35°C and the solar radiation of 1000W/m^2 by using ANSYS software; and obtained the record of maximum PV cell temperature at 60.885°C with 6.95 m/s compare to 91.651°C with 0m/s.

Moreover, authors estimated the PV cell temperature under Ho Chi Minh city solar radiation, with wind velocities of 0.5m/s, 1.5m/s, 2.5m/s, 3.5m/s, 4.5m/s, 5.5m/s and 6.5m/s by Computational Fluid Dynamic-Autodesk, then calculated the efficiency of PV cell in order to find out the most compatible wind velocity.

The rest of this paper is organized as follow: Model for simulation is in Section 2. Section 3 is Evaluation of Experimental Results. Section 4 concludes the paper.

2. Model for simulation.

The model in [12] was the PV cell which included 6x10 pieces of polycrystalline cells series (156mmx156mm per each piece) with parameters listed as in Table 1.

Table 1. PV cell parameters

Peak Power	240 Wp
Max. Power Voltage (Vmp)	30.18V
Max. Power Current (Imp)	7.96A
Open Circuit Voltage (Voc)	36.72V
Short Circuit Current (Isc)	8.99 A
Cell Efficiency	16.5%
Maximum System Voltage	DC 1000V
Temp. Coeff. Of Isc	+0.045%/K
Temp. Coeff. Of Voc	-0.34 %/K
Temp. Coeff. Of Pmax	240 Wp

For simulation, authors created the PV cell model (1560mm x 936mm) by Inventor software including 5 layers as: a/ Glass, b/ Encapsulated by Ethylene vinyl acetate (EVA), c/ PV cell, d/ EVA, e/ Back sheet by Tedlar. These material properties were used from [10] and listed as in Table 2.

Table 2. PV cell layer's properties

Layer	Thickness t,(mm)	Thermal conductivity, (W/m.K)	Density δ, (kg/m^3)	Specific heat Capacity c, (J/kg.^{o}C)
Glass	3	1.8	3000	500
PV cell	0.225	148	2330	677
EVA	0.5	0.35	960	2090
Tedlar	0.1	0.2	1200	1250

Subsequently, the model was imported to CFD-Autodesk 2018 in order to conduct the simulation. The following steps were carried out as following:

- ✓ Material assignment followed Table 2. Besides the PV cell layers, the air volume surround the PV cell was also made by CFD in order to simulate wind velocity.
- ✓ After finishing material assignment, the model was automatically meshed and some surfaces which were not meshed were modified manually.
- ✓ After meshing, the number of nodes were 577,176, the number of elements were 2,541,599, and the mesh was depicted as in Figure 1.
- ✓ Assign the boundary condition.

Figure 1. Meshing calculation on the model.

When the sunlight strike to the PV cell surface, it heats this surface and convert partly into electricity. Most of them become the heat that made the surface temperature increased. The heat loss is taken place by these types: radiation to surrounding air from the surface and the back of the PV cell, convection to surrounding air. The convection heat loss depends on the wind velocity with the Newton's law of cooling

$$Q=h.F.(t_s - t_f) \quad (1)$$
$$h=5.7 + 3.8v \quad (2)$$

where : Q is known as heat loss by convective (W), h is the coefficient of convective heat transfer ($W/m^2.K$), v represents the velocity of fluid (m/s) , t_s and t_f (°C) is temperature of the surface and temperature of the outside fluid, respectively.

The boundary conditions for determining in this simulation were: the variety of wind velocities were 0.5m/s, 1.5m/s, 2.5m/s, 3.5 m/s, 4.5m/s, 5.5m/s, 6.5m/s. Air temperature was at 30°C. Air volume surround the PV cell model with air emissive ε=0.3, the covering glass with the transmittance τ=0.91 and glass emissive ε=0.9; the PV cell with the absorbance α=0.9. The radiation was set at 01:00 pm on April 10th, 2020 with latitude of 10.769°, longitude of 106.662° in Ho Chi Minh City. Turbulence model: k-epsilon. The simulation was solved during in 3380s.

3. Evaluation of Experimental Results

With the air temperature at 30°C (the average temperature at Ho Chi Minh city in April yearly), solar radiation at 01:00 pm in April 10th, 2020 whose radiation is highest in a year was chosen, and the wind velocities was changed with values : 0.5m/s, 1.5m/s, 2.5m/s, 3.5m/s, 4.5m/s, 5.5m/s, 6.5m/s.

According to [13], with the PV cell area A_c (m^2), solar radiation G_T (W/m^2), and max power voltage (V_{mp}), max power current (I_{mp}), then the maximum power point efficiency (η_{mp}) of a module is given by:

$$\eta_{mp} = I_{mp}.V_{mp}/(A_c.G_T) \quad (3)$$

The dependence of the maximum power point efficiency (η_{mp}) to PV cell temperature (t_c) is expressed by this equation:

$$\eta_{mp} = \eta_{mp,STC}.[1+\alpha_p.(t_c-t_{c,STC})] \quad (4)$$

where :

$\eta_{mp,STC}$: the maximum power point efficiency under standard test condition.

α_p: the temperature co-efficiency of power, (%/°C)

$t_{c,STC}$: the cell temperature under standard test condition, (25°C)

3.1 PV panel layers average temperature distribution.

Solar radiation goes through the glass covering the PV cell, strikes on the PV cell surface and convert partly this energy into electricity. Most of them become heat and raise the PV cell temperature, and the temperature of PV cell is displayed in Figure 2.

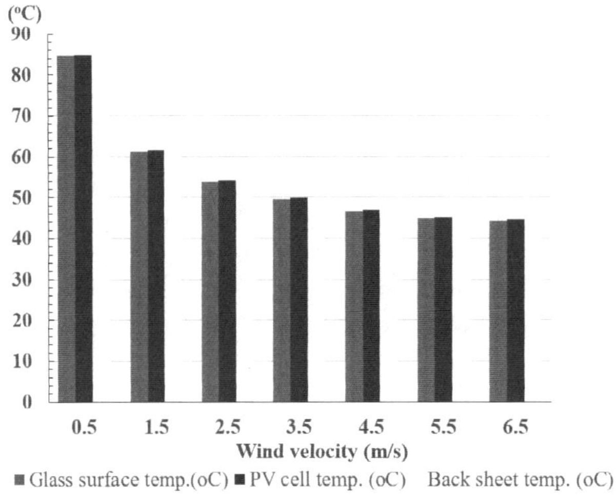

Figure 2. PV panel layers average temperature distribution.

With the lowest velocity of 0.5 m/s, PV panel layers average temperature are highest with the glass surface temperature of 84.7°C, PV cell average temperature of 84.83°C, the back of PV panel average temperature of 84.14°C, because the heat convection considerably depends on the velocity, so a few of heat is drawn. At the maximum velocity of 6.5m/s, the PV panel layers average temperature are lowest with the glass surface average temperature of 44.3°C, the PV cell average temperature of 44.67°C and the back sheet average temperature of 44.043°C. At velocity of 3.5m/s, the usual wind velocity of Ho Chi Minh city in April, the temperature is 49.56°C, 49.91°C, 49.28°C correspond to the glass surface average temperature, the PV cell average temperature, the back sheet average temperature, respectively.

When the wind velocity rises from 1.5 m/s to 3.5 m/s, the PV panel average temperature drop dramatically. However, the trend of this drop is slow with the velocity range from 3.5 m/s to 6.5 m/s. For instance, the rate of average temperature drop is about 6%, 4%, 1% with the range from 3.5m/s to 4.5m/s, 4.5m/s to 5.5 m/s, 5.5 m/s to 6.5 m/s, respectively. Among layers, the outside layers have the lowest temperature because it directly contacts with outdoor air. In contrast, the PV cell layers is higher than them because the heat must conduct through other layers to the outside air.

Figure 3 expresses the temperature distribution visually on the surface of the PV cell which is needed to estimate the efficiency of electrical power.

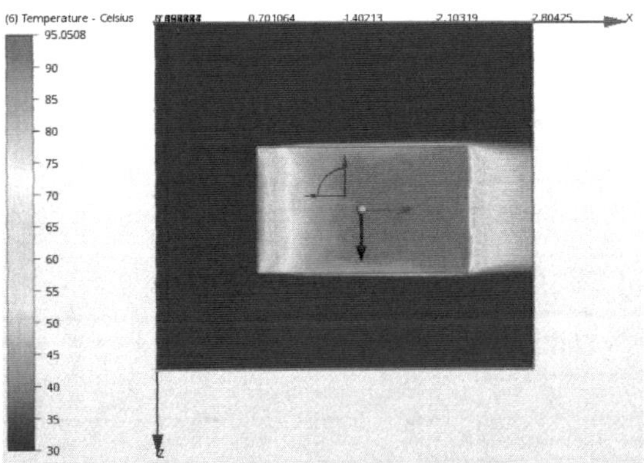

(a)

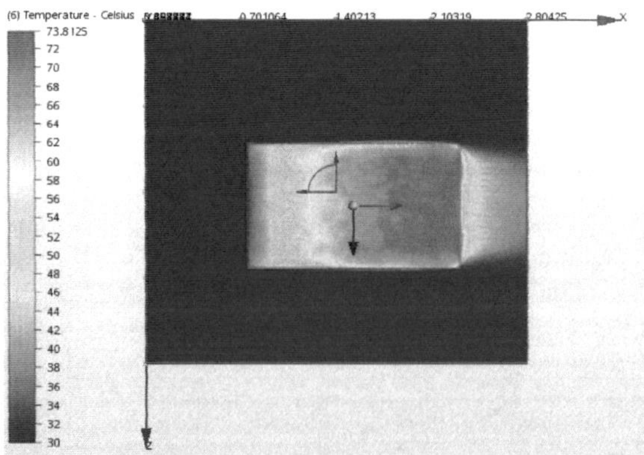

(b)

IOP Conf. Series: Materials Science and Engineering 1109 (2021) 012054 doi:10.1088/1757-899X/1109/1/012054

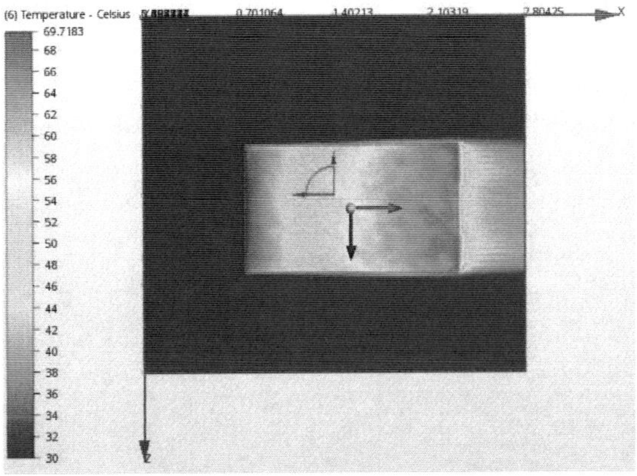

(c)

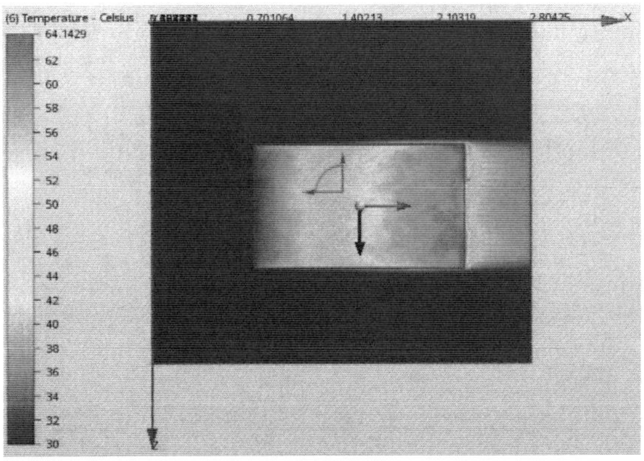

(d)

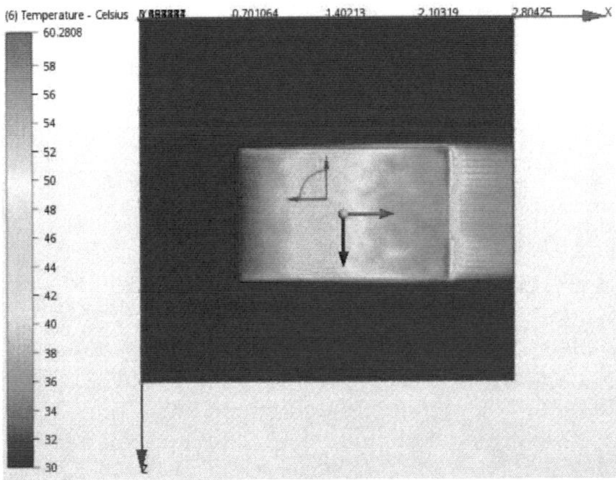

(e)

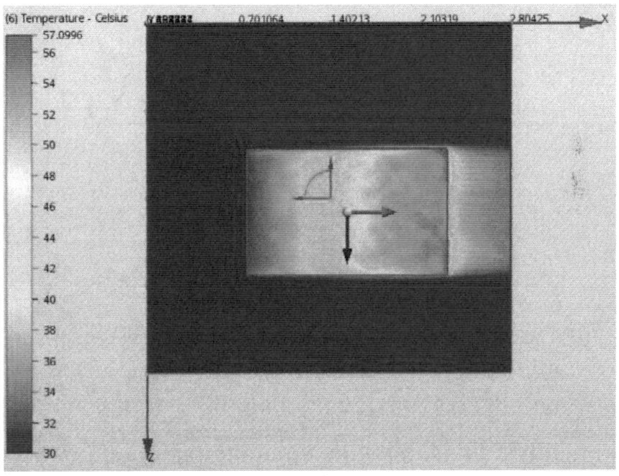

(f)

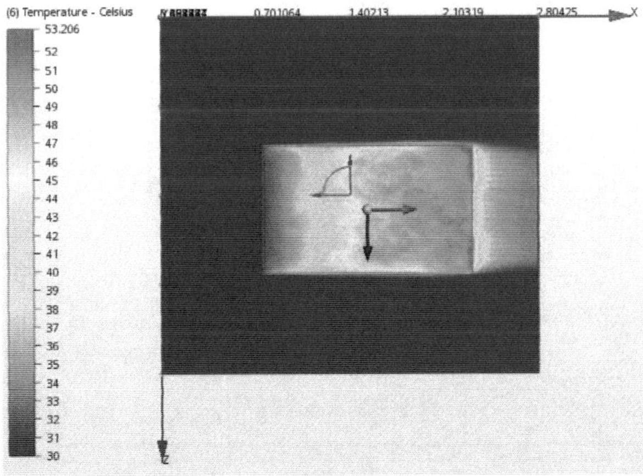

(g)

Figure 3. PV cell temperature distribution under wind velocities of 0.5 m/s (a), 1.5m/s (b), 2.5m/s (c), 3.5m/s(d), 4.5m/s (e), 5.5 m/s (f), 6.5 m/s (g).

In this simulation, temperature is lowest from the left and rises gradually to right due to the wind flow direction is from left to right. Therefore, the temperature distribution is not equal on the PV cell surface. Under the lowest wind velocity of 0.5 m/s in Figure 3(a), the red area (high temperature) is the biggest. When wind velocity is increased, the red area is decreased, become smaller and gradually to be replaced by the blue and the green areas which indicate the lower temperature.

3.2 Estimate the efficiency of PV cell power.

With parameters in Table 1, the average PV cell temperature in Figure 2 and the formula (4), authors estimate, and display values as shown in Figure 4.

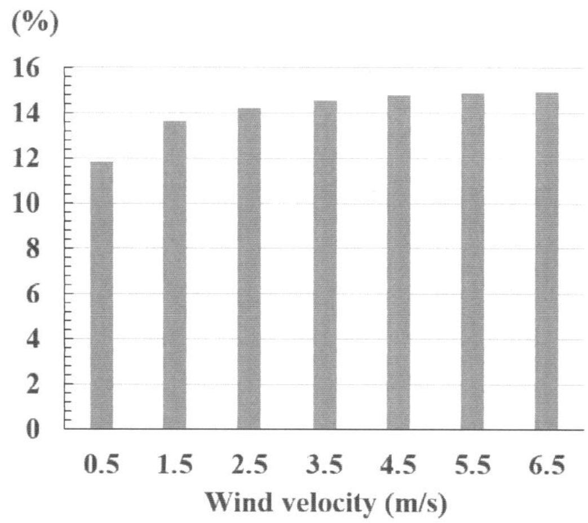

Figure 4. Efficiency of PV cell under various wind velocity.

Obviously, at the wind velocity of 0.5 m/s, the efficiency is the lowest value of 11.86%, and decrease by 28.12 % compare to the PV cell efficiency at Standard Test Condition. The PV cell efficiency increase accompany with the increase of wind velocity. When wind velocity reaches to 3.5m/s, the PV cell efficiency is 14.57%. However, when the wind velocity is continue raising, the increase in efficiency is inconsiderable. For instance, at the maximum wind velocity of 6.5m/s, the efficiency is only 2.75% higher than that of 3.5m/s. Therefore, if designing the cooling system for the PV cell by fan, wind velocity of 3.5m/s is the most suitable choice for saving energy and ensuring that the efficiency is not low.

4. Conclusion

This paper shows the results of PV cell efficiency under different wind velocities. The higher wind velocity, the lower PV cell temperature will it have. In fact, the PV cell efficiency also goes up to 14.57% at wind velocity of 3.5m/s compare with 11.86% at wind velocity of 0.5m/s. However, from the wind velocity of 3.5 m/s to 6.5m/s, the temperature reduction is very small (approximately 2.75%). Therefore, these efficiencies increase inconsiderably. For that reason, the wind velocity of 3.5 m/s is the best choice to choose for cooling system for the PV cell by fan in order to save energy.

Reference

[1] D. T. Nguyen and T.N. Do, 2020 "The efficiency of solar water heating system with heat pump software application designed for resorts in Vietnam," *Vietnam Journal of Science, Technology and Engineering* **62**(2), pp 56-64.

[2] A A N G Sapteka *et. al.* 2018 "Modelling of electric characteristics of 150-watt peak solar panel using Boltzmann sigmoid function under various temperature and irradiance" *J. Phys.: Conf.* Ser. 953 012048.

[3] A.A.E. Amin, M.A. Al-Maghrabi, 2018 "The Analysis of Temperature Effect for mc-Si Photovoltaic Cells Performance," *Silicon* **10**, pp 1551–1555.

[4] Vinod, R. Kumar and S.K. Singh, 2018 "Solar photovoltaic modeling and simulation: As a renewable energy solution," *Energy Reports* **4**, pp 701 – 712.

[5] D.R. Hiren, M. Subama, M. Ashish, 2014 "Computational fluid dynamics analysis and experimental validation of improvement in overall energy efficiency of a solar photovoltaic panel by thermal energy recovery," *Journal of Renewable and Sustainable Energy*, **6**, pp. 1-12.

[6] S. Latifa, B. Mohammed, 2014 "Effect of Ambient conditions on Thermal properties of Photovoltaic Cells: Crystalline and Amorphous Silicon," *International Journal of Innovative Research in Science, Engineering and Technology,* pp 17815-17821.

[7] A. Khelifa, K. Touafek, M.H. Ben, I . Tabet, C. E. H. H. Ben, H. Holoui, 2015 "Analysis of a Hybrid Solar Collector Photovoltaic Thermal (PVT)," *International Conference on Technologies and Materials for Renewable Energy, Environment and Sustainability*, **74**, pp 835-843.

[8] Z. Syafiqah, N. A. M. Amin, Y. M. Irwan, M. S. A. Majid and N. A. Aziz, 2017 "Simulation study of air- and water-cooled photovoltaic panel using ANSYS," *Journal of Physics*, Conference Series, Series **908,** pp. 012074.

[9] J. Yang, Y. Sun, X. Yang, 2013 "Modelling impact of environmental factors on photovoltaic array performance," *International Journal of Energy and Environment* **4**, pp 955-968.

[10] S. Armstrong, W.G. Hurley, (2010) "A thermal model for photovoltaic panels under varying atmospheric conditions," *Applied Thermal Engineering* **30**11-12, pp: 1488-1495.

[11] W.Z. Leow, Y.M. Irwan, M. Asri, M. Irwanto, A.R. Amelia, Z. Syafiqah, I. Safwati, 2016 "Investigation of Solar Panel Performance Based on Different Wind Velocity Using ANSYS," *Indonesian Journal of Electrical Engineering and Computer Science*. **1**(3), pp 456-463.

[12] BLD240-60P Solar Panel from Zhejiang BLD Solar Technology Catalogue.

[13] M. S. Buker, B. Mempouo, S. B. Riffat, 2015 "Experimental investigation of a building integrated photovoltaic/thermal roof collector combined with a liquid desiccant enhanced indirect evaporative cooling system," *Energy conversion and Management* **101**, pp 239-254.

Acknowledgments

This project is supported by Van Lang University (VLU), Electronics and Electrical Engineering Division, Ho Chi Minh City 700000, Vietnam.

Wind tunnel experiments on smoke diffusion from a chimney

X Inthavideth[1,3], S Nobumasa[1] and S Phommachanh[2]

[1] Department of Mechanical Engineering, Toyohashi University of Technology, 1-1 Hibarigaoka, Tempaku-cho, Toyohashi, Aichi, 441-8580, Japan
[2] Department of Mechanical Engineering, Faculty of Engineering, National University of Laos, Lao-Thai Friendship Rd, Sisattanak District, Vientiane Capital, P.O.BOX: 3166, Lao PDR

x199105@edu.tut.ac.jp

Abstract. The diffusion of heated and unheated jets from a chimney was investigated in cross flows with large-scale turbulence generated by a Makita-type active turbulence generator and with gird generated turbulence. The behavior of jet with smoke was observed by a high-speed camera. The smoke was dispersed more widely by meandering motion in the large-scale turbulence. The turbulent eddies corresponding to the integral scale should effectively contribute to the meandering smoke diffusion. The time-averaged smoke concentration distribution was obtained from the brightness values of instantaneous images. The stream-wise variations of the smoke diffusion width and the smoke flow center were evaluated from the distribution. Obtained data were compared with previous correlations.

1. Introduction

Air pollution and deposition of aerosol particles (PM2.5, PM10) caused by industrial plants, motor vehicle, ordinary houses where coal is used for heating apparatus, etc., seriously affect to ecosystems and human health. There are the several millions of people dying annually owing to the atmospheric pollution. Consequently, there is an increasingly high demand for a more precise assessment of particle diffusion, e.g., the diffusion of PM2.5, to mitigate air pollution problems. Due to increasingly strict standards, industrialists where impact of smoke released by chimneys on the surrounding area, are required to study the impact of chronic or accidental pollution caused by smoke released from the chimneys and assess the potential dangers to human health and the environment. The heat and mass transfer including smoke in a jet flow from a chimney is remarkably related to turbulent dispersion, one of the interests of turbulence researchers. In order to investigate the mechanism of the turbulent smoke diffusion, laboratory experiments using a wind tunnel is preferable. It is necessary for its wind tunnel testing to realize the smoke dispersion which is similar to that in an actual atmospheric boundary layer.

Many research [1-12] on smoke dispersion from a chimney have been conducted using a wind tunnel. The extensive results of these wind tunnel experiment have been presented and compared with theory, field measurement and numerical simulation results. Several theoretical research have been carried out on smoke dispersion. However, most of them lacked experimental data comparison [2-4]. Said et al. [9] used particle image velocimetry, PIV, to investigate the coherent structures in the near-wake region of a turbulent round-jet ejected perpendicularly from a chimney into a cross flow. Huang and Hsieh [10] classified the jet structure based on the momentum ratio between the cross wind and the jet velocity. Majeski et al. [11] proposed a phenomenological model for predicting the size of a low-momentum jet diffusion flame diluted with an inert gas in a cross flow. Huber and Snyder [12] reported concentration measurements for an elevated point source placed in the lee of a two-dimensional Gaussian model ridge, the height of which was about one-third of the boundary layer thickness. Their results demonstrated the occurrence of significant ground level concentration upwind of the stack source when it was placed at the ridge base within the lower part of the cavity region.

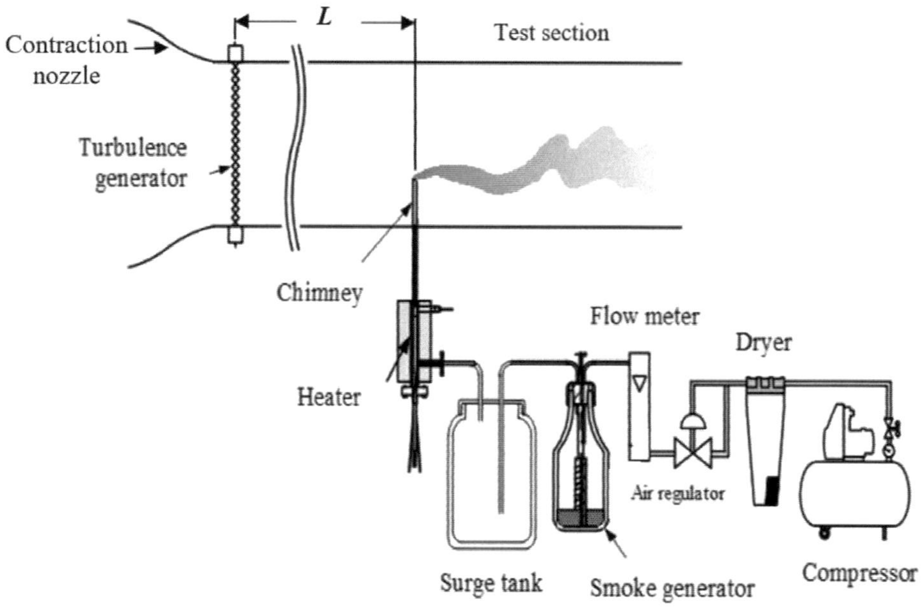

Figure 1. Wind tunnel and smoke generation system.

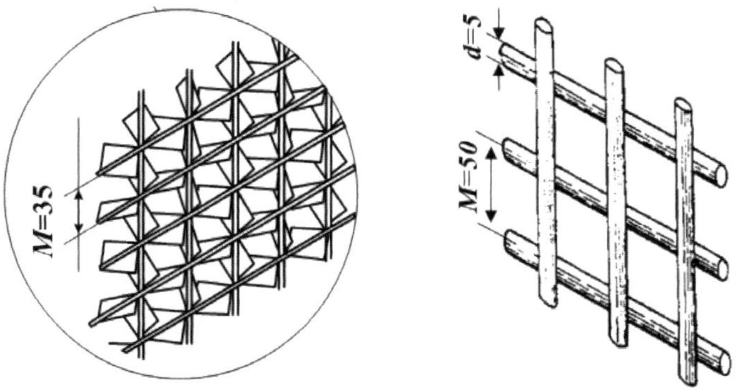

(a) active turbulence grid large-scale turbulence

b) rectangular grid grid turbulence

Figure 2. Turbulence generators.

However, it is difficult to reproduce an atmospheric turbulence phenomenon, such as smoke dispersion, in an ordinary-size wind tunnel using high-drag strakes, roughness blocks, a turbulence grid, etc. The wind tunnel experiments of a buoyancy jet with the smoke vertically ejected into the cross flow, which has large-scale turbulent eddies, are essential for the investigation and the precise estimation of smoke dispersion. Our final goal is to construct a turbulent diffusion model for the environmental assessment and prediction of atmospheric dispersion from a point source such as a chimney. In this study, the smoke of heated and unheated jets was observed in the cross flow with the large-scale and grid turbulence, the smoke diffusion width as well as the center of smoke flow was evaluated by processing the instantaneous images taken by the high-speed camera.

2. Apparatus and method

An atmospheric wind tunnel, shown in Fig.1, was employed in the present study and it had a test section of 0.7x0.7 m^2 in cross section and 6 m in length. A Makita-type active turbulence generator [13], shown in Fig. 2(a), was installed upstream of the test section. The turbulence generator could generate large-

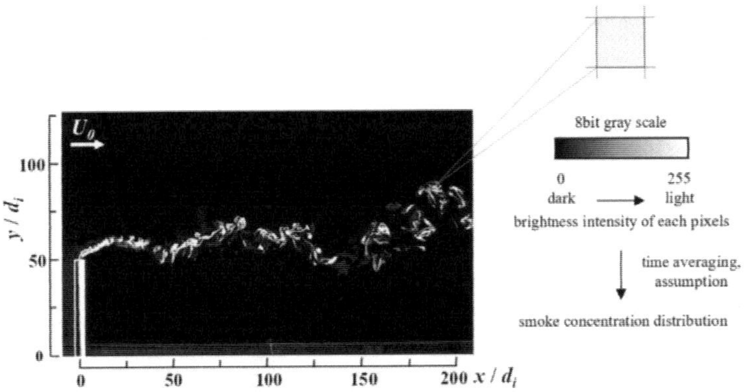

Figure 3. Method for evaluating smoke concentration distribution.

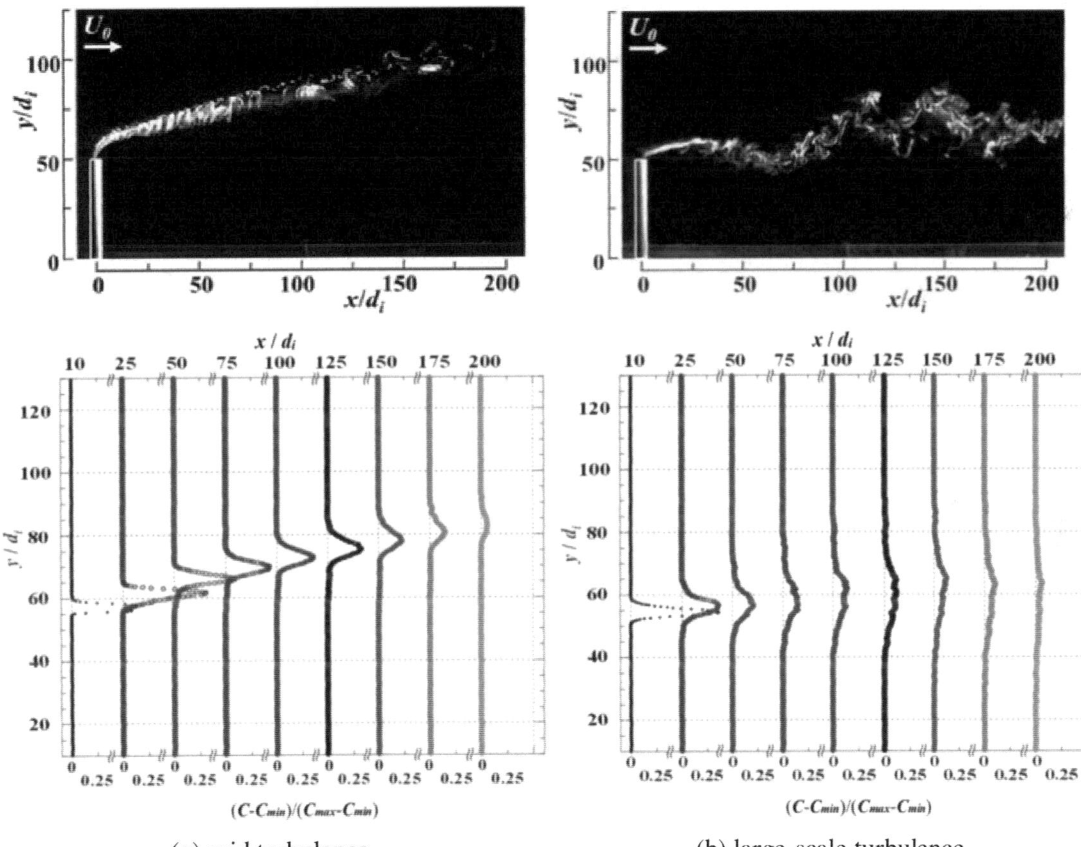

(a) grid turbulence
Mode II
U_0=0.3m/s, U_j=1.4m/s, $\Delta\theta$=100K

(b) large-scale turbulence
Mode V
U_0=0.6m/s, U_j=1.0m/s, $\Delta\theta$=200K

Figure 4. Dispersion structures and normalized smoke-concentration distributions.

scale turbulence with a high turbulence Reynolds number of R_λ ~ 390 at mean cross-wind velocity of U_0 = 5 m/s [13]. That is, a large integral scale, large turbulence fluctuations, and a wide inertial subrange in the energy spectra of velocity fluctuations were achieved. A conventional rectangular grid, shown in Fig. 2(b), was positioned in place of the active turbulence generator, to generate grid turbulence.

The chimney model of d_i = 4 mm in inside diameter, d_o = 19 mm in outside diameter and h = 200 mm in height was placed on the floor of the wind tunnel test section. The smoke generation system for ejecting heated or unheated air with the smoke from the chimney was composed of an air compressor, an air dryer, an air regulator, a flow meter, a smoke generator which was to smoke by using Ondina oil, a surge tank, a heater for generating the buoyancy jet and the chimney as shown in Fig.1. The origin of a coordinate system was the center of the chimney on the floor in the wind tunnel test section. The cross wind direction was $\boldsymbol{x}$, the vertical direction which is the direction of the jet velocity at the chimney exit was $\boldsymbol{y}$, and the span-wise direction was z. The cross-sectional average velocity of the jet at the chimney exit was controlled by the air regulator. The patterns of the smoke dispersion were visualized using a high-speed camera (Photron, FASTCAM SA3 model, 1000 frames/s) and a halogen light.

The present experiments were carried out in the experimental conditions of temperature differences between the cross wind and the jet from the chimney, $\boldsymbol{\Delta\theta}$, were 0 K, 100 K, 200 K, the jet velocity at the center of the chimney exit, $\boldsymbol{U_j}$, was from 0.5 m/s to 1.4 m/s, the mean velocity of the cross wind, $\boldsymbol{U_0}$, was from 0.3 m/s to 1.0 m/s. That is, the velocity ratio between the jet velocity and the cross wind velocity, $\boldsymbol{r}$, ranged from 0.5 to 4.7, the momentum ratio, $\boldsymbol{R}$, ranged from 0.145 to 21.8, and the density ratios between the heated jet and the cross wind, $\boldsymbol{S}$, were 0.59, 0.74, and 1.00. The mean velocity of the cross flow at the entrance of the test section was measured using a hot-wire anemometer. The vertical mean-velocity profile of the cross flow was uniform around the chimney exit in the large-scale turbulence case. Turbulence intensity, $\boldsymbol{u_{rms}/U_0}$, ranged from 9.5% to 11.0% and from 2.9% to 3.5% in the large-scale and grid turbulence cases, respectively.

The smoke concentration distribution was evaluated from smoke visualization images taken by the high-speed camera. By superposing 500 instantaneous images, shown in Fig.3, the time averaged brightness intensity distribution was obtained. Assuming that the averaged brightness intensity, $\boldsymbol{C}$, of each pixel, was proportional to the smoke concentration, the normalized distribution of smoke concentration, $\boldsymbol{(C-C_{min})/(C_{max}-C_{min})}$, was evaluated, where $\boldsymbol{C_{max}}$ and $\boldsymbol{C_{min}}$ were maximum and minimum pixel values of the average brightness intensity distribution, respectively.

3. Results and discussion

A detailed observation of the behaviors of the heated and unheated jets from the chimney in the cross flow was conducted. The representative images of smoke dispersion mode are shown in Fig.4. The smoke dispersion could be classified into six different patterns, Modes I-VI [14]. These smoke patterns in the downstream field were depended on the buoyancy force, turbulent motion, inertia force, etc.

The Mode V structure is affected of the meandering motion by the large integral scale in the large-scale turbulence. The smoke structure of the Mode V is dispersed more widely by the meandering motion which cannot be realized in a usual-size wind tunnel with roughness blocks, a conventional turbulence grid, etc. This fact can be noticed that the turbulent eddies corresponding to the integral scale effectively contribute to the meandering smoke diffusion. The Mode VI structure in the large-scale turbulence is a downwash with the hairpin vortices due to the wake of the chimney. Thus, the smoke does not rise above the chimney exit. This is the serious situation for an actual chimney.

In contrast, in the grid turbulence case, Mode I and II can be observed when the buoyancy effect is dominant. The structures of Modes I and II are composed of two longitudinal vortex tubes whose one end connected to the chimney exit. For Mode I, the two vortex tubes are apart from each other. For Mode II, these two vortex tubes are strongly interacted due to the strong buoyancy and its cross-sectional view looks like the inverse shape of a heart-type. Mode III structure comprise of connecting hairpin-type vortices generated by the Kelvin-Helmholtz instability near the jet exit where the velocity shear is high. Mode IV structure is composed of the developed coherent and turbulent vortices by the turbulent motion.

The smoke concentration distribution from the chimney in the large-scale turbulence and the grid turbulence was examined. The normalized distributions of the time-averaged smoke concentration are shown in Fig.4. Figure 4(a) shows the case where Mode II is observed under grid turbulence. Figure 4(b) shows the case where Mode V is observed under large-scale turbulence. The distributions were obtained based on the motion pictures of smoke taken by the high-speed camera. The peak of the normalized smoke concentration distribution in the large-scale turbulence case is lower than that in the

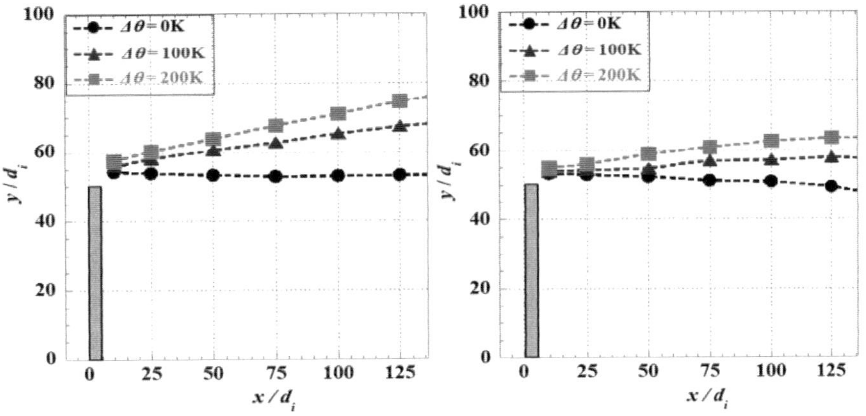

(a) grid turbulence
U_0=0.6 m/s, U_j=1.2 m/s

(b) large-scale turbulence
U_0=0.6m/s, U_j=1.2m/s

Figure 5. The trajectories of the smoke flow center.

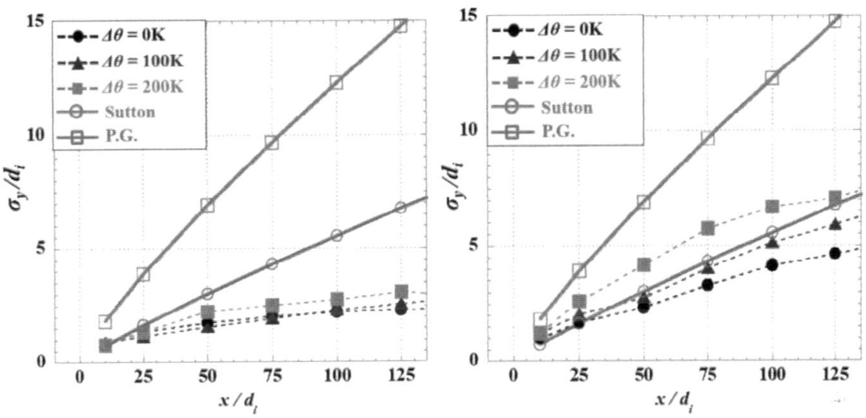

(a) grid turbulence
U_0=0.6 m/s, U_j=1.2 m/s

(b) large-scale turbulence
U_0=0.6m/s, U_j=1.2m/s

Figure 6. Variation of the diffusion width of smoke with an increase of the downstream distance.

grid turbulence case. The width of the peak in the case of large-scale turbulence is wider than that in the case of grid turbulence. In the case of the large-scale turbulence, the smoke is diffused more widely by the meandering motion of the cross flow, and the concentration peak disappear more rapidly than in the case of the grid turbulence. This can be explained by the significant fluctuations of the velocity and direction of the cross wind. That is, the large-scale turbulence is irregular, unsteady, and chaotic.

As can be seen from Fig.4, the vertical distribution of the time-averaged smoke concentration can be approximated by Gaussian distributions. By fitting a Gaussian distribution to the vertical distribution data of smoke concentration, the diffusion width was obtained as the standard deviation value of the Gaussian distribution, and the center of smoke flow was determined at the peak of the Gaussian distribution. The trajectories of the smoke flow center are shown in Fig.5 for various temperature differences. Fig.5(a) shows the case of grid turbulence and Fig.5(b) shows the case of large-scale turbulence. The jet velocity and cross wind velocity are the same in these charts. Although the smoke of the unheated jet flow down straight, the smoke of heated jet rise with an increase of the downstream distance in the grid turbulence case. In contrast of this, in the large-scale turbulence case, the smoke of the heated jet does not rise so much, and the smoke of unheated jet descend with an increase of downstream distance.

The variations of smoke diffusion width with an increase of downstream distance are shown in Fig.6 for various temperature differences. Fig.6(a) is the gird turbulence case and Fig.6(b) is the large-scale turbulence case. The conditions of cross wind velocity and jet velocity are the same in these charts. The diffusion width in the large-scale turbulence case is higher than that in the gird turbulence case. That is, the diffusion rate of smoke should be higher under the large-scale turbulence. The diffusion width is not affected so much by the temperature difference in the grid turbulence case, and the diffusion width increases gradually with an increase of downstream distance. On the other hand, the diffusion width increases more rapidly with an increase of downstream distance in the large-scale turbulence case. The diffusion width is large when the temperature difference is large.

The experimental data of dispersion width variations with downstream distance were compared with previous correlations of Sutton [15]

$$\sigma_y = \frac{C_y}{\sqrt{2}} x^{\left(1-\frac{n}{2}\right)} \quad (1)$$

and Passqill&Grifford [16]

$$\sigma_y = \gamma_y x^{a_y} \quad (2)$$

The results are shown in Fig.6 by bold solid lines. Constants, $\boldsymbol{C_y} = 0.07$, $\boldsymbol{n} = 0.25$, $\gamma_y = 0.1046$, and $\boldsymbol{a_y} = 0.826$, for neutral air stability were employed to evaluate the diffusion width from these correlations. The width variation estimated from the Sutton's correlation is close to the experimental results in the large-scale turbulence case, although the width estimated from the Passqill & Grifford's correlation is larger than the experimental results. This fact suggests that the smoke diffusion from the chimney in the cross flow under the large-scale turbulence is almost similar to the air pollution diffusion from a point source in the actual atmospheric boundary layer of neutral air stability case.

The present experiment was successful in the realization of the smoke dispersion from the chimney which can be observed in the actual atmospheric boundary layer using the atmospheric wind tunnel with the active turbulence grid. The present experimental results can serve as a reference for construction of turbulence diffusion models for the environmental assessment and prediction of atmospheric dispersion from the point source such as chimneys.

4. Conclusions

Flow visualization experiments were performed using a high-speed camera on the diffusing heated and unheated jets with smoke from a chimney in the cross flow with the large-scale turbulence and that with the grid generated turbulence. By analyzing the brightness value of instantaneous images, the time averaged smoke concentration was obtained. The smoke diffusion width and smoke flow center were evaluated from the concentration distribution. The following conclusions were deduced:

(1) The diffusion width of smoke concentration in the large-scale turbulence was higher than that in the grid turbulence. In the large-scale turbulence, the diffusion width increased rapidly with an increase of downstream distance, although the diffusion width increased gradually in the grid turbulence.

(2) In the large-scale turbulence case, the variations of the smoke diffusion width with an increase of downstream distance almost agreed with Sutton's correlation for neutral air stability case.

References

[1] Roaman R S and Ramachandran S 2010 Annual and seasonal variability of ambient aerosols over an urban region in western India *Atmospheric Environment* **44** 1200-1208

[2] Leelossy A, Molnar F Jr, Izsak F, Havasl A, Lagzl I and Meszaros R 2014 Dispersion modeling of air pollutants in the atmosphere: A review *Central European Journal of Geosciences* **6(3)** 257-278

[3] Taylor S G 1959 The present position in the theory of turbulent diffusion *Advances in Geophysics* **6** 101-112

[4] Gifford F 1959 Statistical properties of a fluctuating plume dispersion model *Advances in*

Geophysics **6** 117-137
[5] Brusca S, Famoso F, Lanzafameb R, Garrano A M C and Monforte P 2015 Experimental analysis of a plume dispersion around obstacles *Energy Procedia* **82** 695-701
[6] Bhouri I B, Bournot H, said N M, Mhiri H and Palec G L 2011 Experimental and numerical analysis of the jet dispersion from a bent chimney around an obstacle *Heat Mass Transfer* **47** 323-342
[7] New T H, Lim T T and Luo S C 2006 Effects of jet velocity profile on a round jet in cross flow *Experiments in Fluids* **40** 859-875
[8] Gupta A, Stathopoulos T and Saathoff P 2012 Wind tunnel investigation of the downwash effect of a rooftop structure on plume dispersion *Atmospheric Environment* **46** 496-507
[9] Said N M, Mhiri H, Palcc G L and Bournot P 2005 Experiment and numerical analysis of pollutant dispersion from a chimney *Atmospheric Environment* **39** 1727-1738
[10] Huang R F and Hsieh R H 2002 An experimental study of elevated round jets deflected in a crosswind *Experimental thermal fluid science* **27** 77-86
[11] Majeski A J, Wilson D J and Kostiuk L W 2010 Predicting the length of low-momentum jet diffusion flame in cross flow *Combustion Science and Technology* **176:12** 2001-2025
[12] Huber A H and Snyder W H 1976 Building wake effects on short stack effluents *Preprints, 3rd Symposium on Atmospheric Turbulence, Diffusion, and Air Quality.* (Boston, MA and Raleigh, NC: American Meteorological Society pp 235-242
[13] Makita H 1991 Realization of a large-scale turbulence field in a small wind tunnel *Fluid dynamic research* **8** 53-64
[14] Sekishita N, and Xangpheuak I and Sounthisack P 2019 Wind tunnel experiments on smoke structure dispersed from a chimney in a cross flow *10th AUN/SEED-Net Regional Conference on Mechanical and Manufacturing Engineering (RC MEManuE 2019)* pp 74-77
[15] Sutton O G 1932 A theory of eddy diffusion in the atmosphere *Proceeding of the Royal Society London, A* **135** 143-165
[16] Gifford F A 1961 Use of routine meteorological observations for estimating atmospheric dispersion *Nuclear Safety* **2** 47-51

IOP Conf. Series: Materials Science and Engineering 1109 (2021) 012056 doi:10.1088/1757-899X/1109/1/012056

Bamboo culm vertical lever flattening mechanism with sodium bicarbonate softening process

A H Atienza*, A Alba, D A Respicio and M A Baratang

Mapua Institute of Technology at Laguna, Malayan Colleges Laguna, Pulo Diezmo Road, Cabuyao City Laguna 4025 Philippines

*ahsatienza@mcl.edu.ph

Abstract. Some constructions require flattening bamboo culms to be used for constructing flat surfaces. With such need for flattened bamboo culms, the aim of this research is to develop a cost-effective and novel way of flattening bamboo culms for the species, *Schizostachyum lumampao* or buho, by integrating a softening process using sodium bicarbonate water solution and application of steam, then be pressed by a pressing mechanism using counter-bending technique to eliminate spring back effect of the bamboo culm. An open vessel was filled with sodium bicarbonate and water solution tested at concentrations between 15% and 26% by volume and then heated by 1.5 kW immersion heater. Pressure vessel partially filled with water and a 3-kW immersion heater were used to generate steam for the bamboo culm to maintain its flexibility as it is pressed. The bamboo culms reached a mean degree of flatness of 0.91. Thinner "Buho" (bamboo culm) submerged at around 20% of $NaHCO_3$ concentration was successfully flattened within an average flattening time of ten (10) minutes, similar with thicker bamboo culm but submerged at concentration greater than 20% of $NaHCO_3$. The machine had been incorporated with batch-type process and showed an efficiency of 90.75%.

1. Introduction

Bamboo, as a construction material, is a suitable alternative to wood products to scale down wood consumption [1]. Bamboos tend to have a faster growth rate and have greater mechanical strength compared to some species of trees [2]. Some processes utilizing bamboo requires them to be flattened for various usage, such as for flooring, walling, or any furniture that require a flat surface but most businesses focusing on bamboo products perform them by using a bolo knife as a cutting tool to cut them into flat strips. This method tends to sacrifice the material strength of the bamboo and is considered time and labor intensive. This paper aims to flatten bamboo culms by integrating a softening process using sodium bicarbonate water solution and then be flattened by pressing mechanism while applying steam generated from the pressure vessel. The target result is a flat bamboo culm retaining its flatness without cracks or any damage. This research focuses on the flattening of bamboo culms of the species, *Schizostachyum lumampao*, also known locally as buho, in the Philippines.

The foundation of this research requires understanding on the properties of bamboo, the effects of sodium bicarbonate solution on bamboos, and the approach or parameters to be considered on the flattening of bamboo culms. Testing several bamboos where different percent of humidity were applied, shows that shear strength at around 60%-80% humidity is high [3]. Though not directly, since shear stress test does not determine flexural strength, this can tell roughly at which humidity the bamboo has higher resistance to cracks from bending. Difficulty of flattening is mainly because of the bamboo's curvature. Previous studies tried reducing the circumference difference, while other researchers decreased the tangential stress through reducing of modulus of elasticity. One difficulty is the difference of circumference between the inner and outer layer of bamboo culm. Bamboo is a unidirectional fiber-reinforced bio-composite. Due to the weak longitudinal interface strength among bamboo cells, anti-cleaving, and anti-shearing strength along the grain of bamboo is relatively low [4]. Circumference difference, Cd, is a quick indicator of difficulty in flattening a bamboo culm. Cd is the circumference difference of outer and inner layer whereas the difference in radius is the culm's thickness. Thus, thicker culm means higher Cd, and higher Cd means the more difficult the flattening. A possible solution to counter the problem is by removing parts of inner zone in order to reduce the culm wall thickness. It was reported a better result of flattening a culm with thickness of 3mm compared to 4mm and 5mm [5]. Moreover, out of the three specimens, the 5mm wall thickness showed fractures during flattening. In the study of [6], pre-compressing method was introduced on the bamboo first to decrease the circumference of the outer zone. By pre-compressing, the outer circumference became smaller and the extension of compressed inner zone during flattening was compensated. Aside from removing the culm parts of outer zone, increasing the circumference of inner zone was also an option. This is possible by making non-penetrating cuts and indentations on inner surface. The study of [7] performed diagonal cuts on inner surface at angles 20° to 40° to bamboo longitudinal grain direction. Efficiency was low because at the time there were no special devices for cutting. This was improved in the work of [8] for a patented bamboo flattening with longitudinal grooves on inner surfaces. The grooves were marked by a gear with isosceles-triangle shaped racks, while the height of the racks was less than 2 mm and the distance between parallel racks was less than 3 mm. The grooves direction was parallel to longitudinal direction of bamboo culms.

Aside from the above mentioned, flattening bamboos can be done with softening agents or by heat treatment. A study included removal of outer node ridges and inner node diaphragms before softening with hot water [7]. Softened bamboo culms were pressed with hot press under three different processes; the softened culms were pressed directly, pressed gradually, and pressed gradually through a continuous press. However, many deep cracks were found, and efficiency was very low. Another study used hot water and high-pressure steam. Study presented by [6] developed a flattening process by pre-compressing bamboo culm before softening it using hot water. It was then pressed with a stainless-steel board, and then fixed with high pressure steam. The results indicated that the best steaming condition for circumferential compression of bamboo culm was 140°C for 30 min. The paper stated that the ideal compression ratio to flatten the bamboo cane was 14–19%, while the successful fixation treatment high-pressure was 180°C for 4 min. However, if pre-steamed bamboo is not softened well, bamboo culm might collapse during circumferential compression inside forming pipe. The variation of bamboo culm diameter was found to be a problem for this method.

Results show from the study of [9] that bamboo when simply submerging in water and heated at 120°C for 30 minutes, is compared to water added with $NaHCO_3$ submerged at the same setting for 30 minutes results to a 42% decrease in hardness of surface for outer culm and 54.6% decrease for inner culm. The chemical composition of bamboos is like that of wood which includes major components, such as cellulose, hemicellulose, and lignin. Sodium bicarbonate or $NaHCO_3$ solution with water reacts with the lignin, a rigid molecule of bamboos, and causes delignification or removal of lignin by a phenomenon called alkaline hydrolysis [10]. A good measure of flatness is the degree of flatness (DOF) that [11] took consideration, which is the ratio of difference between culm radius and culm height to the difference of culm radius to its thickness. Hence, a DOF of 1 pertains to a flat bamboo culm.

Briefly, the important methods of this research include an open vessel filled with sodium bicarbonate and water solution tested at concentrations between 15% and 26% by volume and then heated by 1.5 kW immersion heater. Then, to maintain the flexibility of bamboo as it is being pressed, a pressure vessel partially filled with water and a 3-kW immersion heater was used to generate steam and it was pressed with hot pressers called cartridge heaters. This study highlights the aid of both sodium bicarbonate and heat treatment process for softening process for batch production of flattened bamboo culms

2. Materials and Methods

The study patterned the experiment from [11] in which instead of linseed oil, water is used as one of the softening agents. To start off with the initial testing, bamboo from species of *Schizostachyum lumampao* are cut between two nodes and then were cut half along its length. Water with concentration of 10% by volume of sodium bicarbonate or baking soda had been prepared and filled at the open vessel. Water had been filled partially in the pressure vessel. Any safety precautions had been considered before operating the mechanism.

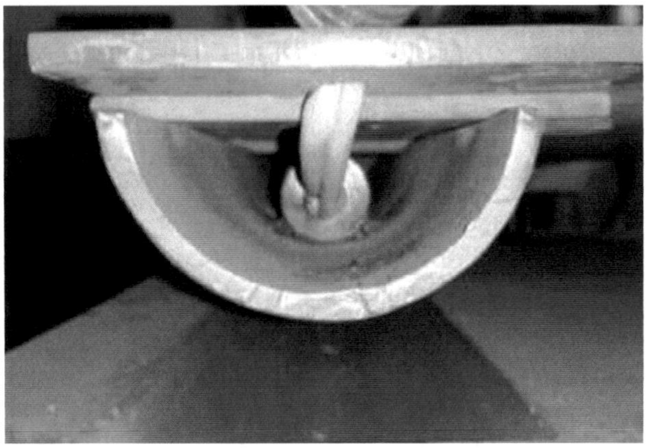

Figure 1. Bamboo culm in cartridge heater

The power cable was connected to the main breaker and the 3-kW heater at the pressure vessel was turned on with temperature set at 100°C as it takes a while to generate steam from the pressure vessel. The 1.5 kW heater at the open vessel was turned on as well. After it reaches the desired temperature of 100°C, the bamboo culms were then submerged into the solution. 1 kW cartridge heater set as shown in Figure 1 was turned on as well with temperature set at 150°C. At a certain amount of time, the bamboo culms were then transferred to the curved molders where they are to be pressed. Ball valve was then be opened to introduce steam to the bamboo culms. The bamboo culms were gradually pressed up until they are counter bended. The bamboo culms had been transferred to the flat molder where it was pressed further.

The flattening procedure had been repeated until no spring back effect was observed. The resulting bamboo culm had been measured again of its height. If the height is equal to the thickness of the culm, it is considered flat. The test is considered successful if it achieved flatness of bamboo culm without any visible cracks.

3. Results and Discussion

DOF required measurements of the bamboo culm, such as radius of halved-bamboo culm, its final height after pressing, and its thickness. The objective was to have a DOF equal to 1 which translates to having a flat bamboo culm. From several trials, the DOF attained was 0.9075 with a machine efficiency of 90.75%. Major factors affecting the flattening of bamboos had been its thickness, time pressed, and % concentration of softening agent. 20% concentration of solution had been the favorable concentration to flatten the thinnest group of bamboos having 4.5mm to 4.7mm.

The formula used to measure the degree of flatness of the bamboo was inspired by the equation by [6]:

$$Degree\ of\ Flatness\ (DOF) = \frac{(radius\ of\ bamboo - final\ height\ of\ bamboo)}{(radius\ of\ bamboo - thickness\ of\ bamboo)} \quad (1)$$

Table 1. Tabulation of Degree of Flatness with concentration and time

Trial	Radius (mm)	Thickness (mm)	Final Height (mm)	Concentration of Solution (%)	Time Elapsed (mins)	DOF
1	25.0	4.7	7.0	15.0	11.1	0.8867
2	26.0	4.5	5.4	22.0	6.9	0.9581
3	26.0	4.7	6.3	20.0	9.5	0.9249
4	25.0	5.0	7.5	15.0	10.5	0.8750
5	24.0	4.8	6.5	18.0	13.2	0.9115
6	25.0	4.7	6.5	18.0	13	0.9113
7	25.0	5.5	7.5	22.0	9.4	0.8974
8	24.0	5.0	7.0	20.0	11.5	0.8947
9	25.0	6.5	8.5	26.0	7.5	0.8919
10	26.0	4.5	6.5	18.0	10.6	0.9070
11	26.0	4.7	6.4	22.0	8.5	0.9202
12	26.0	4.5	6.4	20.0	8.3	0.9116
AVE	25.3	4.9	6.8	19.67	10	0.9075

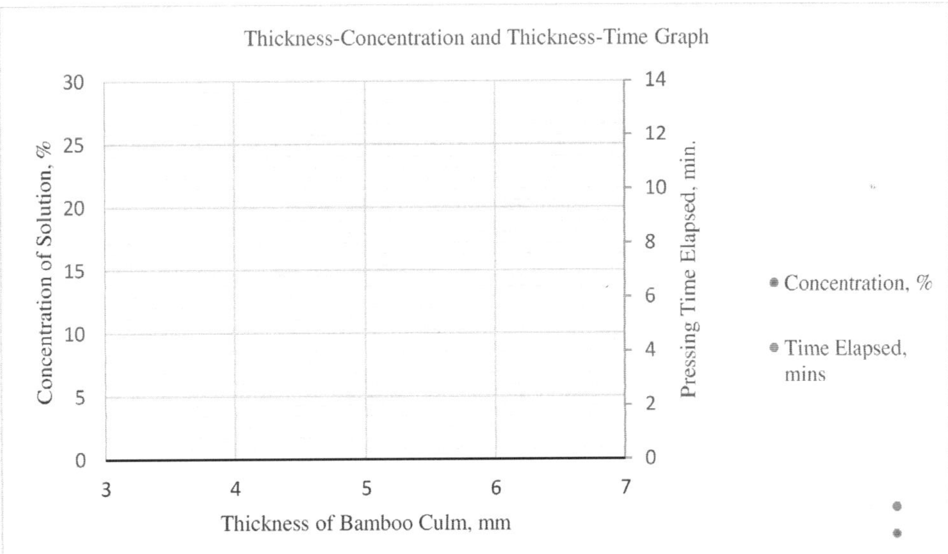

Figure 2. Relation between bamboo thickness and concentration of solution, and relation between bamboo thickness and pressing time elapsed

It was observed that thickness 4.925 mm as the mean, 5.495 mm as one standard deviation, SD, above, and 4.355 mm as one SD below as shown in Table 1. From the data gathered, the thickness had been divided in thin regime and thick regime. The thickness is lower than 4.925mm for the thin regime while those greater than mean thickness was categorized under the thick regime. Thicker bamboos above the mean tend to require longer time of pressing to flatten. Concentrations at less than 15% has negligible effect in the flattening process and caused several failed and split bamboo culms as seen in

Figure 2. Bamboo culms did not soften as much and most of them stiffened even more when flattened using heated pressers, and resulted to burnt, cracked, and split in half at less than 15% concentration as seen in Figure 3a. However, increased concentration of softening agent for more than 20% decreased the time needed to press the bamboo culm and still achieved about 0.9 DOF as shown in Figure 3b. The radius of the bamboo does not affect much on the flattening of bamboo as varying radii have varying results. Furthermore, the values of radius of bamboo on all the trials are close to one another.

From the trials done and results attained, thickness played a part on how easy the bamboo culms had been flattened. As shown in Figure 2a, the result of unsuccessful flattening results to cracked ones with radii. When the bamboo is properly flattened, this would result to better thickness as shown I Figure 2b with an average thickness from twelve (12) trials of 4.925 mm and one standard deviation, σ or SD of 0.57 mm, with 4.925 mm as the midpoint, 5.495 mm as one SD above, and 4.355 mm as one SD below. However, beyond one standard deviation, σ, too thin and it may crack but also may be easier to flatten if treated better. Too thick, such as Trial 9 with 6.5 mm thick sample, and it would be difficult to flatten. From the results, no trial has exceeded less than one SD of thickness, meaning no bamboo is too thin as of the experiment done as the least thickness among the trials are at 4.5 mm thick. On the other hand, one trial, Trial 9, has exceeded one SD of thickness, meaning the bamboo is too thick compared to the average

Notice that Trial 9 with the thickest culm tested has a larger DOF compared to Trial 1 and Trial 4. When the factor of concentration of solution comes in, Trial 1 and 4 have both 15% concentration of solution and have a time elapsed in pressing with 11.1 min. and 10.5 min, respectively, while Trial 9 has 26% concentration of solution and 7.5 min. time elapsed. This tells us that a slight increase in the percent concentration of the sodium bicarbonate solution cut off minutes of flattening of the semi-culm bamboo, either for bamboo that was thick or thin. 20% solution was the favorable concentration to flatten the thinnest group of bamboos having 4.5mm to 4.7mm, while higher concentration up to 26% was the favorable concentration for the thicker group of bamboos that were 4.8 mm and higher. This is supported by the fact that Trial 2 with 0.9581 DOF and 4.5 mm thickness, has the highest DOF.

(a) (b)

Figure 3. a) Cracked bamboo culms and b) flattened bamboo culms

Furthermore, 15% and higher up to 20% showed promising signs of flattening. The researchers observed that thinner bamboos flattened at range 15%-20%, while thicker bamboos flattened at range 20%-above. For an example, Trial 2 and Trial 9 had been compared. Trial 2 was 4.5 mm thin while Trial 9 was 6.5mm thick; however, both bamboos reached a degree of flatness close to 0.9 even though both had been pressed with the shortest time. A further explanation between thick and thin bamboos followed.

For the case of the group of thinnest bamboos with range 4.5 mm to 4.7 mm, the students observed that among the three trials with 4.5 mm thickness, an 18% concentration was enough to fully flatten the bamboo within the shortest time of ten (10) minutes. When the thickness increased to 4.7 mm, the bamboo flattened but took the longest time to flatten hence for thinner bamboos, a 15% $\leq$ percent concentration < 20% required more than ten (10) minutes to fully flatten, while a percent concentration of 20% flattened the bamboo at half the said time.

Concentrations at less than 15% has negligible effect in the flattening process and caused several failed and split bamboo trials. The researchers noticed that bamboos do not soften and most of them stiffened even more when flattened using heated pressers, and resulted to burnt, cracked, and split in half.

The radius of the bamboo does not affect much on the flattening of bamboo as varying radii have varying results. Furthermore, the values of radius of bamboo on all the trials are close to one another. The force applied to the bamboo is dependent on other parameters since other factors dictate how easy it is for the bamboo culms to flatten due to how much their temporary increase of flexural strength/modulus of rupture (in other words, withstanding bending force without cracking) is added from treatment of heat and softening agent.

4. Conclusion and Recommendation

Thinner "Buho" (bamboo culm) submerged at around 20% of $NaHCO_3$ concentration was successfully flattened within an average flattening time of ten (10) minutes, similar with thicker bamboo culm but submerged at concentration greater than 20% of $NaHCO_3$. No bamboo culms tested beyond 15% concentration had shown any signs of neither cracks nor scratch, and the bamboo culms reached a mean degree of flatness of 0.91. The flattening machine was fabricated using mostly materials used were mild steels, angle bars, GI sheets, and refrigerant tank, and installed with heaters for the solution, pressers, and to produce steam. The machine had been incorporated with batch-type process and showed an efficiency of 90.75%, determined through the relation of radius of the bamboo and its thickness.

The flattening process had involved submerging of bamboo in a sodium bicarbonate solution, and it was found that thickness of the bamboo affected both the percent concentration of sodium bicarbonate solution and the time to flatten the semi-circular bamboo culms. It was also found that the bamboo had been flattened at temperature of the solution at 60°C and temperature of the heated pressers at 150°C. Data had showed that an average of DOF of 0.91 or 91% was achieved with a 60°C heating of solution, a much lower needed temperature than the results of [11] where to achieve a DOF close to 1 or 100% required equal or more than 115°C linseed oil temperatures.

The researchers recommend for future studies to include density of bamboo culms as a parameter and study the effects of different densities in flattening bamboo culms. It will be also important to determine Modulus of Elasticity (MOE) and Modulus of Rupture (MOR) of the flattened samples. A study on the inclusion of heater design for the inner culm radius while being pressed and a way to maintain the humidity of the bamboo culms as it is being pressed, such as: steam application or submerging on water can be taken also consideration.

Acknowledgement
The researchers want to give thanks to the staffs at Jano Bahay Kubo Maker located in Calamba, Laguna for sharing them knowledge with regards in bamboo flattening and other related processes, Malayan Colleges Laguna A Mapua School for financing the presentation and paper publication, Mapua Institute Technology at Laguna for giving then unwavering support and motivation, and their families for providing them inspiration towards their work.

References
[1] Chaowana P 2013 Bamboo: An Alternative Raw Material for Wood and Wood-Based Composites. *Journal of Materials Science Research*. 2. 10.5539/jmsr.v2n2p90.
[2] Fang C H, Jiang Z H, Sun Z.J, Liu H.R, Zhang X B, Zhang R. & Fei B H 2018 An overview on bamboo culm flattening. *Construction and Building Materials,* ***171****, 65–74.* doi:10.1016/j.conbuildmat.2018.03.085
[3] Askarinejad S, Kotowski P, Shalchy F & Rahbar, N 2015 Effects of humidity on shear behavior of bamboo. *Theoretical and Applied Mechanics Letters*, *5*(6), 236-243. doi:10.1016/j.taml.2015.11.007
[4] Shao Z Fang C H & Tian G L 2009 Mode I interlaminar fracture property of moso bamboo (Phyllostachys pubescens). *Wood Science and Technology*. **43**. 527-536. 10.1007/s00226-009-0265-2.
[5] Parkkeeree T, Matan N, Matan N & Kyokong B 2018 Flattening of Half Tubular Bamboo Culms and Fixation of Bamboo Boards. Journal of Tropical Forest Science, **26**, 101-114. Retrieved from https://www.jstor.org/stable/23617019.10.060
[6] Sarula, Nakamura S, Yoshitani K & Tanahashi M 2012 Physical Properties of Bamboo Flattened by High-Pressure Steam Processing. *Mokuzai Gakkaishi*. 58. 201-208. 10.2488/jwrs.58.201.
[7] Zhang Q 1988 Studies of bamboo Plywood-I. the softening and flattening of bamboo, *J. Nanjing Fores. Univ*. 413-20
[8] Liu J, Zhang H, Chrusciel L, Na B & Lu X 2013 Study on a bamboo stressed flattening process. *European Journal of Wood and Wood Products, 71(3),* 291–296.
[9] Cheng R, Zhang Q & Sui S 2006. Improvement of softening treatment technology of bamboo. *Wood Sci Technol*, **40**, 327-335. Retrieved from DOI 10.1007/s00226-005-0034-9
[10] Li Z, Jiang Z, Fei B, Cai Z, & Pan X 2014 Comparison of bamboo green, timber and yellow in sulfite, sulfuric acid and sodium hydroxide pretreatments for enzymatic saccharification. *Bioresource Technology*, **151**, 91-99. doi:10.1016/j.biortech.2013
[11] Cherdchim B, Matan N & Kyokong B 2004 Effect of temperature on thermal softening of black sweet-bamboo culms (Dendrocalamus asper Backer) in linseed oil, 26(6). Retrieved from http://www.thaiscience.info/Journals/Article/SONG/10462484.pdf

CFD investigation of a combined biomass furnace and solar drying system for paver blocks

A H Atienza, L A Adorador, J A Hernandez and F J Vinagrera

Mapua Institute of Technology at Laguna, Malayan Colleges Laguna, Pulo Diezmo Road, Cabuyao City Laguna 4025 Philippines

ahsatienza@mcl.edu.ph

Abstract. Paver blocks have been widely used as concrete materials for exterior flooring. The process of producing paver blocks consumes a lot of time as the drying period takes at least eleven days. This paper had aimed to design a drying mechanism, a combined biomass furnace and solar collector drying system for paver blocks. and simulate the flow of the system through computational fluid dynamic analysis using SolidWorks Flow Simulation (SWFS) 2018. Desired results had been achieved as temperature, pressure, and velocity fields generated by simulation were accounted to observe fluid distribution in the drying system using contour plots. The temperature in the chamber had increased and uniformly distributed as the heat inputs from the two heat sources converged in the drying chamber. It was observed that the pressure varied from 101,331.72 Pa to 101,327.83 Pa with the highest temperature peaking at 600K and velocity was 1.5 m/s. The theoretical heat produced in the solar collector, biomass furnace and combine system were 191.70 W, 1290.66 W and 1482.36 W, respectively.

1. Introduction

Paver block manufacturing has been associated with construction-related industries extensively used in some applications. Due to modernization, materials used in making paver blocks had been modified in time. Having plastics as inseparable materials which are frequently used by the society, these objects have low density, long lifespan, and low cost of production which serve as key factors that support the rapid growth of plastic usage [1]. According to [2], these wastes and their industrial by-products should be considered as potentially valuable resources merely waiting for appropriate treatment and application. Given the situation, a number of researches were done to integrate plastic wastes as an aggregate material in block manufacturing and were introduced to compensate for the disadvantages in the old block manufacturing. Despite the improvement on the material science of paver blocks, its production is dependent on sun drying where large amounts of production time are consumed in the drying process. Sun drying is a slow process since it takes for about three (3) to four (4) drying days and is affected by weather conditions [3]. The importance of reducing the drying time of this is necessary and for it to happen, moisture should be removed. However, eliminating moisture in the blocks is critical as hasty moisture removal may result in the shrinkage on the surfaces of the paver blocks. The internal pressure experienced by the block material and immediate environment of the block within the system of the drying chamber may result in rupture. This has been proven by [4] as they have stated that severe stresses in bricks will result in cracking and rupturing. The drying phase is factored in by air temperature, relative humidity, and air velocity due to their contribution to moisture reduction [5,6]. Temperature rates are operated gradually to prevent brick cracking which is mainly due to the contraction rate difference between the core and the surface of the bricks. Safe ranges for air velocity need to be accounted for as well as high velocities aid low condensation but can lead to hasten drying. One of the efficient ways to evaluate the quality of airflow inside the dryer is by applying Computational Fluid Dynamics (CFD) to predict the flow pattern [7,8]. CFD simulations have been generally carried to process the most appropriate parameters to maximize efficiency and minimize production cost [9]. Moreover, the potential of CFD was strengthened by the fact that it can be adopted to minimize the costly experiment because users can generate visual results in similitude to real life situations [10].

In this study, CFD Analysis had been used to determine the drying conditions of the paver blocks by generating temperature, pressure, and velocity fields of the designed drying system– a combined biomass furnace and solar collector drying system under simulation, which appeared to predict significant heat in the system and quantify heat transfer observed at different drying configurations: use of the biomass furnace only, use of the solar collector only, and a combination of both drying systems. This study had determined the condition of the drying prototype before fabrication.

2. Materials and Methods

The project had been modified by the researchers to have forced convection while the absorber was to be coupled to a biomass furnace. Given this, the drying system subjected in simulation had been designed for handling paver blocks with plastic as its aggregates. The study focuses on atmospheric air as input and heater air as its output. The estimated temperature limits of the paver blocks ranges from 38°C to 204° C [11]. The biomass furnace was assumed to have constant heat while it varies in the case of the solar collector since it is dependent on time. The research had aimed to determine the condition of the drying system, to map heat patterns, and to account theoretical heat transfers from drying configurations of the system through CFD Analysis with the use of SolidWorks 2018 Flow Simulation. The flow analysis of the system had been done for monitoring pressure, temperature, and velocity fields. Moreover, as the system was simulated, heat mapping was done to determine the sections of the drying mechanism where significant heat values have been located. Considering drying configurations as biomass furnace only, solar collector only and both drying schemes combined, the system was simulated to determine each theoretical heat given. For the boundary conditions, there had been three (3) configurations of the system using the CFD analysis for the hybrid biomass for drying paver blocks. Physical parameters and air properties are seen below on Table 1.

Table 1. Physical Parameters and Properties for the Boundary Condition

Physical Parameters	Remarks
Fluid	Air
Velocity Magnitude (Outlet)	1.11 m/s
Gage Pressure inside the Chamber	101.325 kPa
Heat Produced by the Biomass Furnace	1290.66 W
Heat Produced by the Solar Collector	191.70 W
Mass Flow Rate in the Biomass Furnace	0.0022 kg/s
Mass Flow Rate in the Solar Collector	0.0119 kg/s
Specific Heat (Cp)	1 kJ/kg-K

3. Results and Discussion

The following had been the fields generated through simulation with heat produced at 1290.66 W by the furnace and a velocity of air at 1.11 m/s attached at the exhaust part of the system.

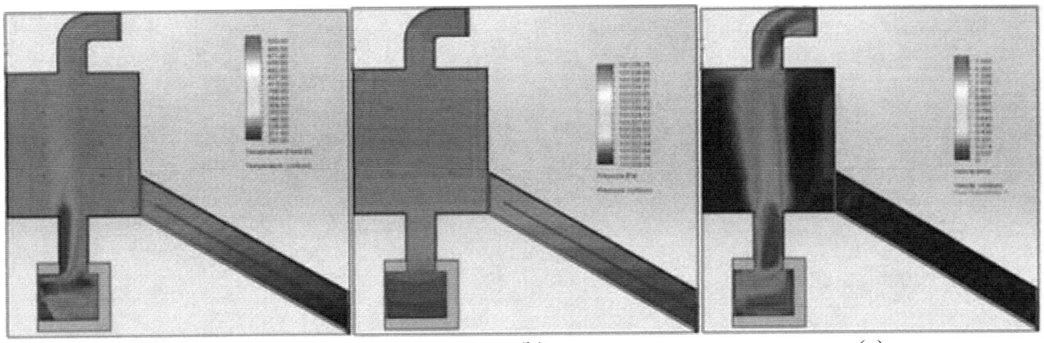

(a) (b) (c)

Figure 1. First Configuration: Biomass Furnace Only a. Temperature field b. Pressure field c. Velocity field of the biomass furnace as the only heat source of the drying system for paver blocks.

In Figure 1, the temperature, pressure, and velocity distributions inside the chamber without paver block were simulated. The contour plot in Figure 1a had shown that the heat was mostly concentrated in the middle of the drying chamber where the fire of the furnace was closest. The temperature of the other parts of the drying chamber were shown to be in uniform at 413 K. Figure 1b had shown that closing the solar collector had yielded to a lesser temperature in accordance with the reduced pressure in the drying chamber. Highest pressure had been seen in the furnace at around 101, 335.61 Pa and diminished to 101, 329 Pa at the top part of the drying chamber. Pressure difference was caused by the height and heat source placement in the system. With the solar collector outside in the field of computation, the velocity focused on the circulation of air near the exhaust and near the tube which had acted as a connection for the furnace and solar collector. A peak of 1.393 m/s can be seen at the said locations as shown Figure 1c. There was a peak velocity that had occurred in the connecting tube of the furnace and drying chamber that is caused by the change of area coming from the furnace to the tube itself. Since the heat produced by the solar collector was heavily dependent on time because of the effect of the temperature difference, the researchers had made five (5) trials from Figures 2 to 6a-c. Trials were time dependent to determine the heat produced at the given time. The outlet temperature of the solar collector was based on the study of [12] as shown in Table 2.

Table 2. Inlet and Outlet Temperature of the Solar Collector

Time	Temperature, °C	
	Inlet	**Outlet**
10:00 AM	36.11	100
11:00 AM	36.67	106
12:00 PM	37.77	90
1:00 PM	38.89	80
2:00 PM	38.89	55

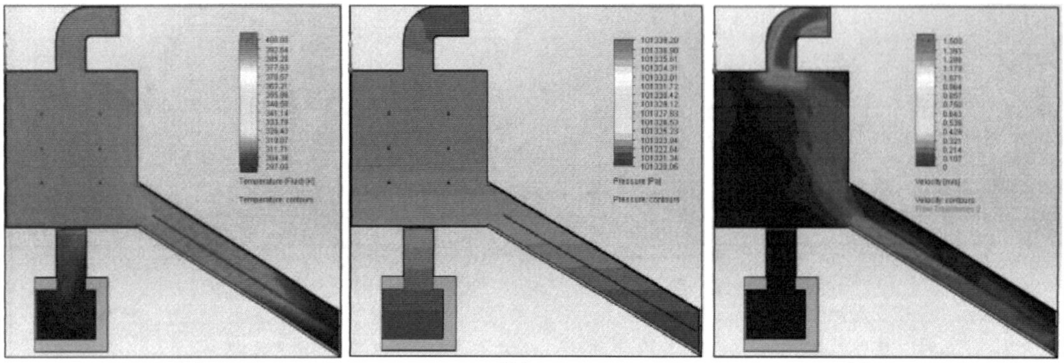

(a) (b) (c)

Figure 2. Second Configuration: Solar Collector Only at 10 AM a) Temperature field b) Pressure field and c) Velocity field of the solar collector as the only heat source of the drying system for paver blocks.

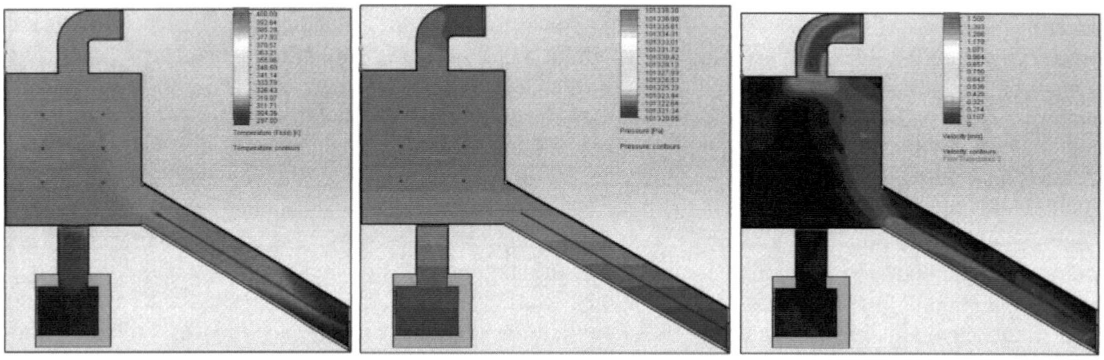

(a) (b) (c)

Figure 3. Second Configuration: Solar Collector Only at 11 AM a)Temperature field b) Pressure field and c) Velocity field of the solar collector as the only heat source of the drying system for paver blocks.

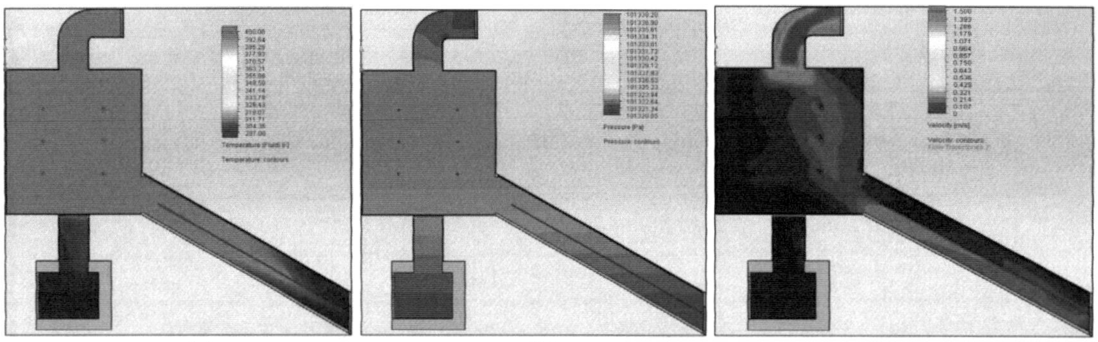

(a) (b) (c)

Figure 4. Second Configuration: Solar Collector Only at 12 NN a) Temperature field b) Pressure field and c) Velocity field of the solar collector as the only heat source of the drying system for paver blocks.

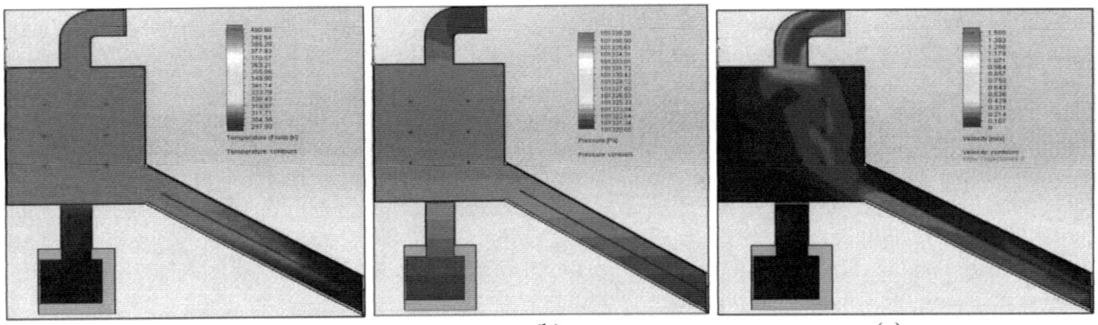

(a) (b) (c)

Figure 5. Second Configuration: Solar Collector Only at 1PM a)Temperature field b) Pressure field and c) Velocity field of the solar collector as the only heat source of the drying system for paver blocks.

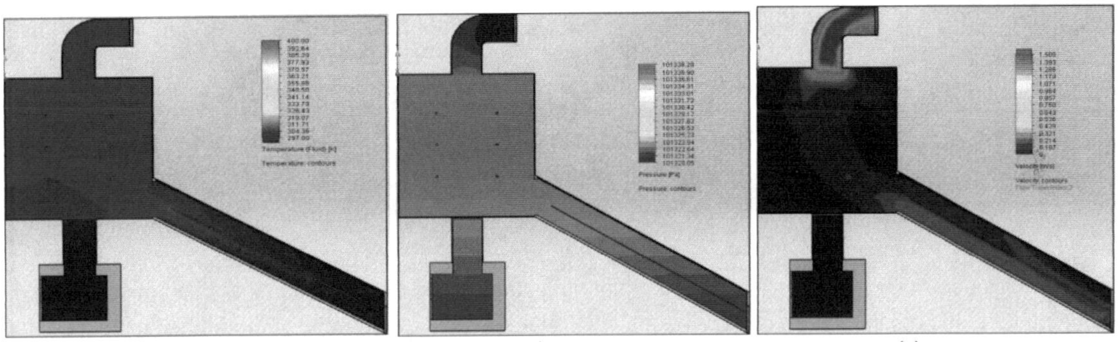

(a) (b) (c)

Figure 6. Second Configuration: Solar Collector Only at 2PM a)Temperature field b) Pressure field and c) Velocity field of the solar collector as the only heat source of the drying system for paver blocks.

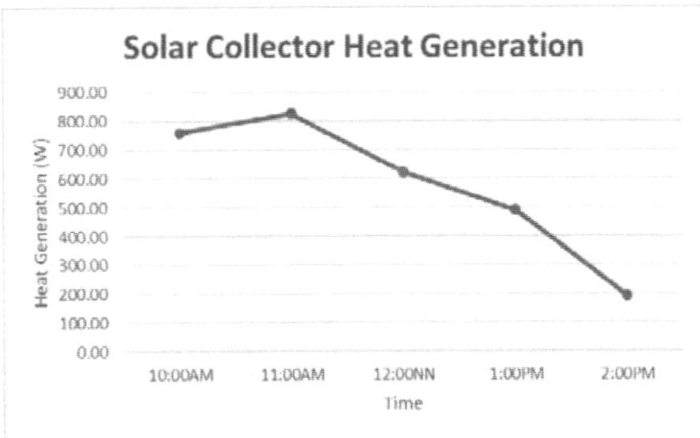

Figure 7. Solar collector heat generation.

Figure 7 above had shown that the heat generation of the solar collector with respect to time, being the time at the 2nd trial had yielded the largest generation of heat and the last or fifth trial was the lowest to yield heat. Hence, for the combined biomass furnace and solar collector drying configuration, the lowest heat yield had been considered to avoid overheating in the system as the heat input from the biomass is high. Both sources had produced heat into the system for the last configuration. Also, the temperature inside the drying chamber can be seen to be almost uniformly distributed. The pressure in the system heavily differs based on the placement of the brick in the drying chamber that might affect the drying time of the brick. The pressure inside the drying chamber ranges from 101,331.72 Pa to 101,327.83 Pa.

The exhaust part of the system was the area with significant result due to the fan attached. The flow rate inside the drying chamber had focused mainly on the middle part and had shown that there was forced convection in the drying chamber. Figure 8 had shown that both heat sources have significant temperature readings and as they have converged in the drying chamber, the temperature inside increased to a uniform value of 535 K. The highest temperature is 600 K which is located at the connecting tube as seen in Tables 3 and 4. It was also observed that the highest pressure occurred in the upper area of the chamber and the highest velocity attained in the dryer is about 1.50 m/s

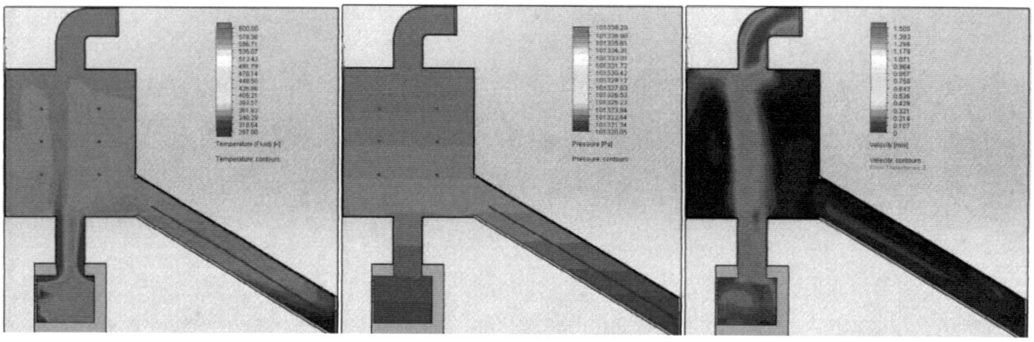

Figure 8. Third Configuration: Combined Biomass Furnace and Solar Collector for a) Temperature field b) Pressure field and c) Velocity field in the solar collector as the only heat source of the drying system for paver blocks.

Table 3. Parameter Readings Based on Simulation

Parameters	Temperature, K	Pressure, Pa	Velocity, m/s
Highest Value	600	101,331.72	1.50
Lowest Value	297	101,327.83	0

Table 4. Location of the Parameter Readings Based on Simulation

Location	Temperature, K	Pressure, Pa	Velocity, m/s
Highest Value	Connecting Tube	Upper area of the chamber to the exhaust	Exhaust
Lowest Value	Tip of the Solar Collector	Lowest area of both heat sources	Side areas of the chamber

The theoretical heat transfer based on the simulation was obtained for each configuration. The biomass furnace yielded 1290.66 W of heat while the solar collector provided 191.70 W. It was found out that the combine or hybrid system has the largest heat transfer produced at 1482.36W.

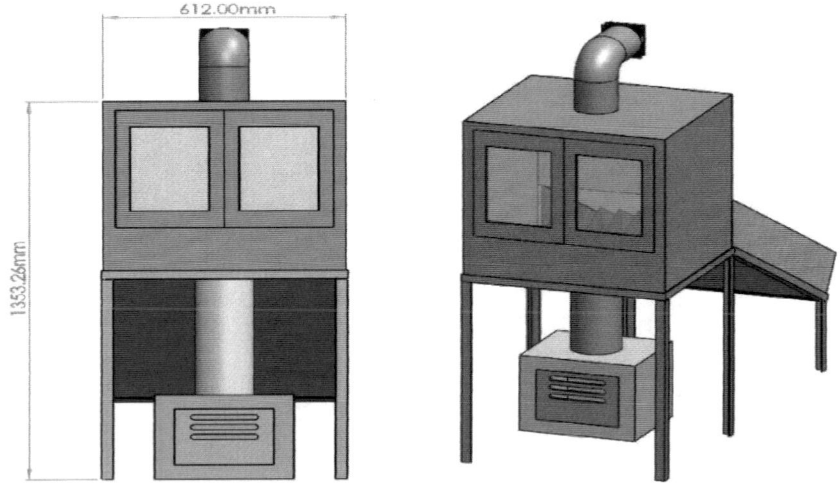

(a) (b)

Figure 9. Combined Furnace and Solar Drying System a) Front view b) Isometric View

After the simulation, the solar dryer was fabricated with material specifications and technical information. For the design of the system as shown in Figure 9, the solar collector had contained a solar heat absorber for air heating, a drying chamber for accommodation of wet products to dry, and a chimney for the air exhaust. The width of the dryer is 612 mm while its height is about 1.35m. The absorber had an area of 1.0 m^2 composed of two corrugated aluminum sheets painted with matt-black paint to increase its heat absorption. A 50 mm thick cork plate had been added which minimized heat loss at the bottom of the solar air collector. In addition, air heated had crossed the channels of the absorber to reach the drying chamber by natural convection. The drying chamber had been insulated from all sides with glass wool with a thickness of 5.0 mm. At the back of the drying chamber, a double-swing door had been installed for easy loading and unloading of products to dry. Four rectangular trays had been made of galvanized iron wire of dimensions 955 mm by 262 mm. The first tray had been fixed at 32.0 cm from the bottom of the drying chamber and the last at 13.5 cm from the top. The upper part of the drying chamber had been equipped with a vertical chimney made of aluminum of 70.0 cm height. The whole had been supported by a metal frame located at 86.0 cm above the ground.

4. Conclusions

The combined biomass furnace and solar collector drying system for paver blocks had been tested through computational fluid dynamic analysis with the use of software simulation. By means of SolidWorks Fluid Simulation, mapping was conducted to observe the behavior of specific parameters when the system is operated. Temperature, pressure, and velocity fields were accounted for in the simulation process to determine the condition that the paver blocks would be situated during operations. The highest temperature, pressure and velocity observed were 600K, 101,331.72Pa and 1.50 m/s, respectively. The simulation showed that the temperature inside the drying chamber increased with heat inputs from the biomass furnace and the solar collector as heat sources. The temperature differences in the system had been confirmed by the variation in contour plots. Moreover, the direct relationship of pressure and temperature in the system as shown in the simulation could affect the drying time of the paver blocks when the prototype would be fabricated. Pressure is non-uniform inside the drying chamber where it varies with elevation. For the velocity, forced convection had contributed significantly to improving the air circulation in the drying system. With the use of a fan in the design, forced convection can be initiated and enhanced. Mapping through simulation can greatly help a system under designing. The heat variations in the system are quantified which can improve or predict possible errors before having a prototype be fabricated. The researchers had observed the uniform distribution of heat in the system due to heat mapping. In connection, heat values from each configuration had been considered where the heat value of the hybrid system yielded the highest heat transfer among the three configurations. This could also help in the designing phase of a project.

Acknowledgement
The researchers want to give thanks to the staffs at Materials Recovery Facility (MRF) Ecopark, Timbao, Biñan City, Laguna for sharing their knowledge in paver block production and drying, Malayan Colleges Laguna A Mapua School for financing the presentation and paper publication, Mapua Institute Technology at Laguna for giving them unwavering support and motivation, and their families for providing them inspiration towards their work.

References
[1] Hossain M, Bhowmik P & Shaad, K 2016 Use of waste plastic aggregation in concrete as a constituent material. Progressive Agriculture, 27(3), 383–391. doi: 10.3329/pa.v27i3.30835
[2] Ismail Z Z and Al-Hashmi E A 2008. Use of waste plastic in concrete mixture as aggregate replacement. *Waste Management*, 28(11), 2041–2047. doi: 10.1016/j.wasman.2007.08.023
[3] Kumar M, Sansaniwal S K and Khatak P 2016 Progress in solar dryers for drying various commodities. Renewable and Sustainable Energy Reviews, 55, 346–360. doi: 10.1016/j.rser.2015.10.158
[4] Vasić M & Radojevic Z 2018 Comparison and evaluation of recently reported methods for optimization of industrial drying regimes. IOP Conference Series: Materials Science and Engineering. 400. 062030. 10.1088/1757-899X/400/6/062030.
[5] Mancuhan E 2009 Analysis and optimization of drying of green bricks in a tunnel dryer. *Drying Technology*. 27. 707-713. 10.1080/07373930902827692.
[6] Zhang M, Zhang S, Wang M, Lu Z and Jia W 2019 Effects of dehumidification drying parameters on physical and chemical properties of biomass brick. *Construction and Building Materials*, 215, 696–708. doi: 10.1016/j.conbuildmat.2019.04.216
[7] Tzempelikos D, Bardakas A, Alexandros V, Tsepenekas D, Christoloukas D, Filios A and Margaris, D 2013 Design, construction and performance evaluation of a new laboratory convective dryer. *5th International Conference on Experiments Process System Modeling Simulation and Optimization 5th IC-EpsMsO Athens*, 3-6 July, 2013
[8] Odewole M, Musliu Olushola S, Oyeniyi S and Adesoye O 2017 Computational fluid dynamics (CFD) simulation of hot air flow pattern in cabinet drying of osmo-pretreated green bell pepper. *Nigerian Journal of Technological Research.* 12. 36. 10.4314/njtr.v12i1.7.
[9] Jaskulski M, Atuonwu J, Tran T, Stapley A and Tsotsas E 2017 Predictive CFD modeling of whey protein denaturation in skim milk spray drying powder production. *Advanced Powder Technology*, *28*(12), 3140–3147. doi: 10.1016/j.apt.2017.09.026
[10] Mortensen LH, Woloszyn M, Hohota R and Rusaouen G *2005 CFD Modelling of Moisture Interactions between Air and Constructions. Building Simulation Ninth International IBPSA Conference Montréal, Canada* August 15-18, 2005.
[11] Sahu M K , Singh L & Choudhary S N 2016 Critical review on bricks. *International Journal of Engineering and Management Research* 6(5) pp 80-88
[12] Filli, F, Gebray P & Kebedom A 2018 Comparative study of antireflection coating materials for solar thermal collectors. *Momona Ethiopian Journal of Science*, *10*(1), 1. doi: 10.4314/mejs.v10i1.1

Classifying maturity of cherry tomatoes using Deep Transfer Learning techniques

Danh Phuoc Huynh[1], My Van Vo[1], Nghin Van Dang [1], and Tri Quoc Truong[1]

[1]Faculty of Engineering, Van Lang University, 45 Nguyen Khac Nhu street, Co Giang ward, District 1, Ho Chi Minh city, Vietnam

tri.truong@vlu.edu.vn

Abstract. This research studies a method to classify the tomatoes' maturity by using deep transfer learning techniques. We carry out sorting systems adopting three pre-trained convolutional neural networks of VGG16, VGG19, and ResNet101. The experimental results show that the VGG19 model obtains a high precision on both the train set and the test set.

1. Introduction

Tomatoes are one of the popular vegetable crops that are grown all over the world. They are healthy because of their good source of vitamin A and C. Along with the constant increase in demands for tomatoes, the consumer also pays more attention to their quality. Tomatoes maturity level plays an important role in product quality. However, the traditional classification of tomatoes is often based on the practical experience of farmers. It takes a lot of time and inaccuracy in harvesting the crop. There have been researches in the field of automatic classification of tomatoes over the last decade [1, 2, 3].

The rapid growth of recent techniques in image processing, computer vision and especially Machine Learning (ML) paved the way for high-quality agriculture. In [4], the authors implemented a fuzzy architecture based on RGB color to classify six categories of tomatoes. This approach obtained a good classification in RGB color space. The limitation of this paper was that the result is a suited prediction in the conditional environment with controlled lighting, fixed black background. In another study, by combining histogram feature extracting and Naïve Bayes Classifier, Kusuma et al. proposed a new tomato maturity classification [5]. The histogram feature acquired the pixel intensity values of the tomato's image, the information of a relative, brightness, and contrast. Hence, the Naïve Bayes Classifier learned the extracted values obtained from the histogram feature to make a more accurate prediction. According to the results, the proposed method produced an accuracy rate of 76%. In fact, the main concern of the classification method was accuracy and computational cost.

Recently, Deep Learning (DL) has attracted intensive research interest in the classification of tomatoes. By using a lot of labelled images of tomatoes as the input, the DL model is trained to learn the tomato's features. After a period of training, the "learned model" was then used to predict what type of tomatoes. However, the traditional DL techniques acquired big data to conduct a good classification. In recent years, many researchers have used transfer learning methods to overcome these problems [6, 7]. For instance, to detect the tomato crop disease, Ouhami et al. obtained deep learning models DensNet, 161 and 121 layers, and VGG16 [6]. The results show that transfer learning has a suitable architecture for the task of plant disease detection with less time-consuming and high accuracy.

In this study, we focused on the task of maturity classification of tomatoes. Based on the dataset of 1374 tomato images, we divided tomatoes into 3 categories: green, yellow, and red. By taking advantage of the transfer learning method, we employed three Deep Learning models of VGG16, VGG19, and ResNet101 to figure out the best model with great accuracy.

2. Dataset description

In this study, we acquired the dataset, named Fruits-360, which contained 90483 images of 131 popular fruits and vegetables in [8]. For simplicity, we chose 1374 images of tomatoes and divided into 3 classes: green, yellow, and red. The examples of 3 types of tomatoes were illustrated in Figure 1. All images in this dataset extracted the tomato from the background. Tomatoes were then scaled down to 100×100 pixels.

Figure 1. Three classes of tomato in scale of 100×100 pixels.

3. Deep Transfer Learning Model

3.1 Convolutional Neural Networks

[1] Convolutional Neural Networks (CNNs) are some types of Deep Learning (DL) models. The fundamental structure of DL model is illustrated in Figure 2. Here, the input the input $\boldsymbol{x} = [\boldsymbol{x_1}, \boldsymbol{x_2}, .., \boldsymbol{x_n}]^{\mathrm{T}}$, remarked by blue nodes, is fed into the network through the Input layer. In image processing $\boldsymbol{x}$, expresses the pixel matrix of the image. In image processing, $\boldsymbol{x}$ expresses the pixel matrix of the image. Then, $\boldsymbol{x}$ is combined with weight matrix $\boldsymbol{W}$ to generate the input $\boldsymbol{W}$ of the Hidden layer which is remarked by yellow nodes. We note that a deep neural network consists of numerous hidden layers and numerous nodes in each layer. Hence, the output of the previous layer becomes the input of the next layer. At the final output layer remarked by red nodes yields:

$$\hat{y} = f(W; x)$$

where, f is a function which describes the relationship between the input and the output. $\hat{y}$ is the predicted result. The algorithm will modify the weight matrix W such that $\hat{y}$ is close to the true output y. In the problem of classification of tomatoes' maturity, y is considered as red, yellow or green fruit.

Similarly, CNNs focus on solving the tasks relating digital image processing, computer vision. Such roles of CNNs are remarked as object classification, object detection and recognition or image segmentation [9]. In case of object classification tasks, the typical CNNs can be divided into two blocks: block of image's feature learning (called Block 1) and block of classification (called Block 2) as shown in Figure 3.

Block 1 consists of convolutional (abbr. *conv*) layer, max pooling layer and Rectified Linear Unit (ReLU) layer which is followed by a convolutional layer. The objective of the convolutional layer is to extract the feature of the image, such as edges, color or gradient orientation. Subsequently, the max pooling layer learns the dominant feature. Furthermore, the pooling layer is responsible for decreasing the size of convolved features expressed as a matrix. It makes the computational cost reduced, which is acquired to process the dataset.

Block 2 contains Flatten layer, Fully connected layer and Soft-max layer. Here, the image represented as a matrix is flattened into a column vector at Flatten layer. Then, the flattened output is fed into the Fully connected layer for classification purposes. At the rest of CNNs, by using the Softmax technique, the model is able to compute the highest score relating to the input image. We note that in our study, at the output we obtain three scores (between 0 and 1) that is predicted to be green, yellow, and red.

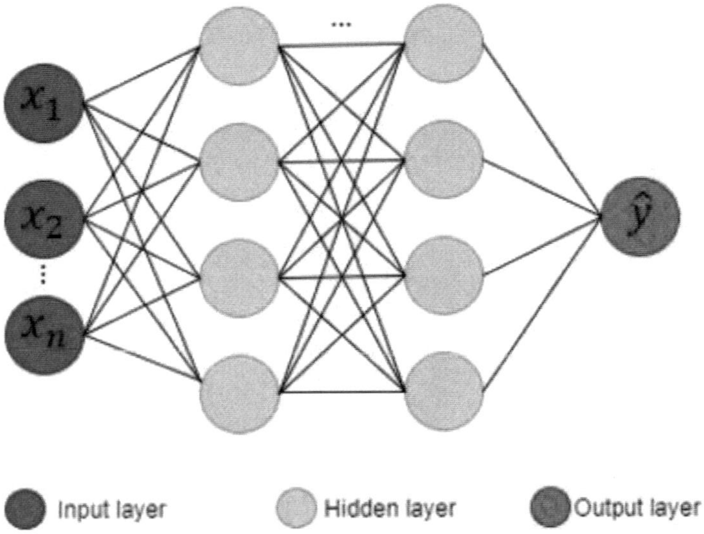

Figure 2. The structure of a typical Deep Learning model.

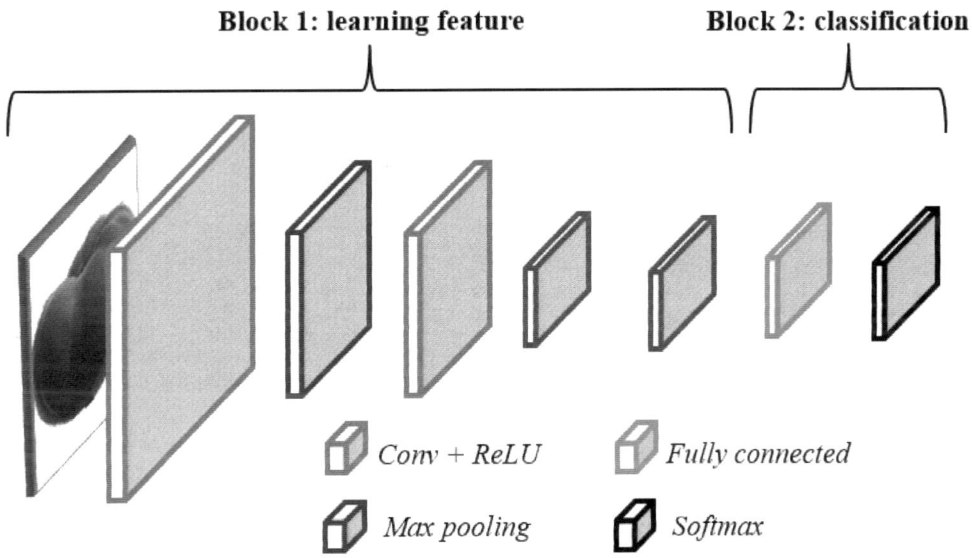

Figure 3. The structure of Convolutional Neural Networks.

3.2 Pre-trained model CNNs

In order to reduce the demand on large dataset and computational cost of DL model, we obtained a transfer learning method which takes advantage of pre-trained model. This study takes pre-trained weights from the VGG16, VGG19, and ResNet101 [10, 11]. These models trained on big labelled images from scratch for similar tasks of fruit classification. Hence, using a pre-trained model gains much faster and more accurate compared with models trained on a large dataset.

VGG16 is a famous deep CNNs which was proposed by Simonyan K and Zisserman A in [10]. Number 16 refers to the fact that this CNN has 16 layers consisting of 13 *conv layers* followed by the maxpooling and 3 *fully connected layer*. The architecture of the VGG16 model is shown in Fig 4. Meanwhile, VGG19 is a version of the VGG family model. As illustrated in Figure. 5, similar to that of VGG16 model, it contains 16 layers of *conv layers* and 3 *fully connected layers*.

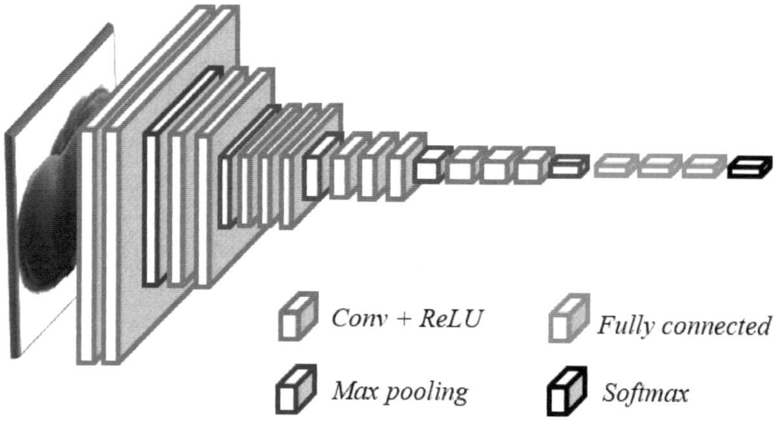

Figure 4. Architecture of VGG16 model which consists of 16 layers.

4. Results

The dataset is divided into two parts of 80% for training, named train test and 20% for testing, named test set. Figure 5 shows the accuracy rate (right) and the loss evaluation as training the model in case of VGG16 model. Consequently, Figure 6 and 7 respectively show the similar manner of VGG19 and ResNet101 models. We executed the training of all three models in 30 epochs. As shown in Table 1, we can observe that the VGG19 obtains excellent accuracy on both train set and test set.

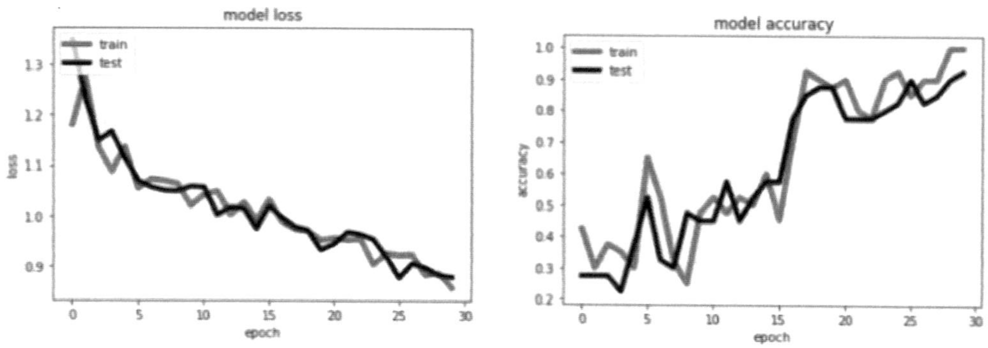

Figure 5. Accuracy and loss evolution of VGG16 model over 30 training epoches.

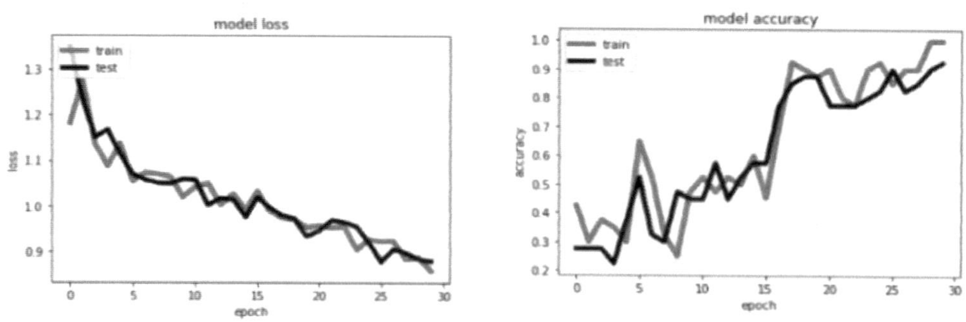

Figure 6. Accuracy and loss evolution of VGG19 model over 30 training epoches.

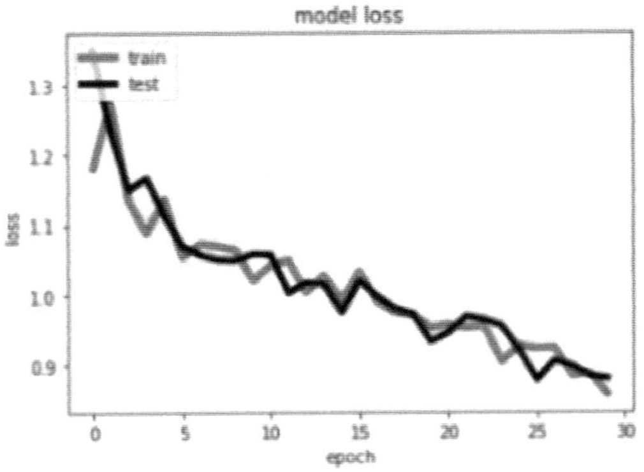

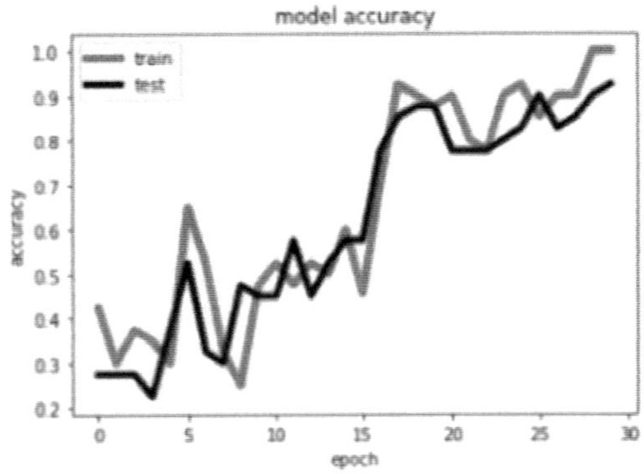

Figure 7. Accuracy and loss evolution of ResNet101 model over 30 training epoches.

Table 1. Classification results of three types of tomato using 3 pre-trained models.

Method	Training error	Testing error
VGG16	0.95	0.93
VGG19	0.97	0.9414
ResNet101	0.92	0.9

5. Conclusion

This research was conducted to study tomatoes classification by using deep transfer learning techniques. We employed 3 pre-trained CNNs models of VGG16, VGG19, and ResNet101. The experimental results show that the VGG19 model obtained a high precision of 94.14% accuracy in evaluating the level of how the cherry tomatoes ripen. However, in this classification task, we only considered 3 classes of tomatoes, so that it is easy to obtain high accuracy. In future work, we will consider enlarging classes of data and train the above CNNs model from scratch for real tomatoes classification.

References

[1] Fuentes A, Yoon S, Kim S C and Park D S 2017 A robust deep-learning-based detector for real-time tomato plant diseases and pests recognition *Sensors* **17**(9)

[2] Rangarajan A K, Purushothaman R and Ramesh A 2018 Tomato crop disease classification using pre-trained deep learning algorithm *Procedia computer science* **133** 1040.

[3] Zhang L, Jia J, Gui G, Hao X, Gao W and Wang M 2018 Deep learning based improved classification system for designing tomato harvesting robot *IEEE Access* **6** 67940.

[4] Villaseñor-Aguilar M J, Botello-Álvarez J E, Pérez-Pinal F J, Cano-Lara M, León-Galván M F, Bravo-Sánchez M G and Barranco-Gutierrez 2019 Fuzzy classification of the maturity of the tomato using a vision system *Journal of Sensors*.

[5] Kusuma A, and Putra M D M 2018 Tomato maturity classification using naive bayes algorithm and histogram feature extraction *Journal of Applied Intelligent System* **3**(1) 39.

[6] Ouhami M, Es-Saady Y, El Hajji M, Hafiane A, Canals R, and El Yassa M 2020 Deep transfer learning models for tomato disease detection *In International Conference on Image and Signal Processing* 65.

[7] Pattnaik G, Shrivastava V.K, and Parvathi K 2020 Transfer learning-based framework for classification of pest in tomato plants *Applied Artificial Intelligence* **34**(13) 981.

[8] Oltean M and Muresan H Fruits 360 dataset on github URL: https://github.com/Horea94/Fruit-Images-Dataset

[9] Girshick R, Donahue J, Darrell T, and Malik J 2015 Region-based convolutional networks for accurate object detection and segmentation *IEEE transactions on pattern analysis and machine intelligence* **38**(1) 142.

[10] Simonyan K, and Zisserman A 2014 Very deep convolutional networks for large-scale image recognition. arXiv preprint arXiv:1409.1556.VGG19.

[11] He K, Zhang X, Ren S, and Sun J 2016 Deep residual learning for image recognition *In Proceedings of the IEEE conference on computer vision and pattern recognition* 770.

IOP Conf. Series: Materials Science and Engineering 1109 (2021) 012059 doi:10.1088/1757-899X/1109/1/012059

Decision tree analysis of commuter mode choice in Baguio City, Philippines

I H V Gue[1,2], J Soliman[1], M De Guzman[3], R Cabredo[4], A Fillone[5], N S Lopez[1*] and J B M Biona[1,6]

[1]Mechanical Engineering Department, De La Salle University, Manila, Philippines
[2]Mechanical Engineering Department, Far Eastern University – Institute of Technology, Manila, Philippines
[3]Civil Engineering Department, Saint Louis University, Baguio City, Philippines
[4]Software Technology Department, De La Salle University, Manila, Philippines
[5]Civil Engineering Department, De La Salle University, Manila, Philippines
[6]Enrique K. Razon Logistics Institute, De La Salle University, Manila, Philippines

*neil.lopez@dlsu.edu.ph

Abstract. Transportation is a multidisciplinary system. Solving its issues would require the knowledge of social, economic, engineering, environmental, and technological disciplines. Emerging techniques used in problem-solving involve the use of machine learning techniques. In this study, a machine learning technique, decision tree, is used for mode choice analysis in Baguio City, Philippines. Using data from a household survey, the developed model uncovers the most significant factors affecting mode choice of residents in the city. The results highlight the role of income, which is related to the individuals' career level and stage in life. Interestingly, a mid-level income group seems to be highly inclined towards private vehicle use. To conclude, the authors note that the primary advantage of a decision tree is its simplicity and straightforward results interpretation, which is paramount in policymaking. For future work, the authors recommend exploring larger decision tree models for mode choice and conducting a validation interview of the insights obtained from the study.

1. Introduction

Recent research works are gaining interest in using machine learning (ML) techniques for modeling and pattern recognition. ML has applications in sustainability, including urban transportation [1]. ML algorithms such as Artificial Neural Network (ANN), Support Vector Machines (SVM), Cluster Analysis (CA), and Decision Trees (DT) are among the considered alternatives for choice modeling. Among the enumerated algorithms, DT shows significant potential as existing studies have indicated its superior performance. Paredes et al. [2] compared the different ML algorithms and discrete choice models. Their findings indicated better performance of DT after proper data treatment. Meanwhile, the review of Pineda-Jaramillo [3] identified that the random forest (RF) algorithm, a subset of DT, is better than the other algorithms and the multinomial logit model (MNL).

The application of DT for choice modeling focuses on the locality of transportation mode selection. Chapelau et al. [4] applied RF for commuter choice in Montreal, Canada. Their work attained accurate estimations on commuter preference for private cars, public vehicles, school buses, and walking.

Additionally, Pawar et al. [5] assessed transportation mode selection under restrictions caused by the COVID-19 pandemic. They used DT in modeling commuter preference in India. DT models developed from prior studies are limited to constraints in the locality due to differing socio-cultural characteristics. The models are not universally applicable; therefore, utilization of DTs requires data reflecting local socio-cultural characteristics. Ranosa et al. [6] studied the willingness of Baguio City residents to walk and their understanding of proper public transport stops. On the other hand, Estrella et al. [7] assessed the acceptability of a new transport mode, the Aerial Ropeway Transport, in Baguio City. It also looked into the current transport characteristics of the city to determine a suitable route for the new mode. From a more general perspective, Rith et al. [8] analyzed the relationship between household vehicle ownership and energy use, while on [9], the authors looked into the effects of individual type choice on energy consumption in Metro Manila.

Contributing to the present literature on transport mode choice modeling and machine learning, this study develops a DT model revealing commuter preference between public utility vehicles (PUVs) and private vehicles. The authors considered DT because of its high accuracy prediction and its ability to generate linguistic rules. The linguistic rules provide insights into the local commuter preference. Findings and insights generated by this work may guide the development of action plans for commuter switching behavior. The results also provide comparative preferences among different socio-cultural systems.

2. Methodology

2.1. *Study Area*

A household survey was developed to build a transport database for Baguio City. Participants ($N = 3216$) were randomly selected, while the number of participants per barangay was proportional to the ratio of barangay population relative to city population. The survey consisted of questions on socio-economic, origin-destination, and mode share parameters. However, the present study focuses on the socio-economic (income, number of household cars, etc.) and mode share (private or public) parameters only. It is important to take note that taxi services are also considered private modes in this study. Participant profile was compared to the 2015 Family Income and Expenditure Survey (FIES) of Baguio City [10]. A notable difference was observed in the percentage of lower age intervals, mainly because the present survey design only considered participants who are at least seven years old. Otherwise, participant profiles of other parameters such as monthly household income, number of cars in the household, and mode share were similar for both surveys. Hence, the data used in the present study was deemed sufficient to represent the population of Baguio City. According to Kotsiantis [11], rule-based algorithms perform better under discretized data; therefore, the data have been discretized as shown in Table 1.

2.2. *Decision Tree*

Morgan & Sonquist [12] was the first to present DT's initial variant, the Automatic Interaction Detection (AID) algorithm. The family of DT then grew and included algorithms such as Classification and Regression Trees (CART) [13] and C4.5 [14].

DT's general concept is partitioning a dataset to form a top-down structure of if-then rules describing recognized patterns. Pruning is then applied to the tree to avoid overfitting. Kotsiantis [15] provides a detailed pseudo code on the iterative partitioning of DT's growth phase. The initial step is in determining whether the current dataset has observations of the same category. If so, then a node is assigned according to the category. If not, the dataset is partitioned based on an identified feature. The feature is chosen based on its contribution to further partitioning. The algorithm creates a branch for each of the feature's values. From the branch, the process recursively iterates from the start of forming subtrees.

This study utilized the Statistics and Machine Learning Toolbox of MATLAB in the development of the DT model. The command *fitctree* of MATLAB uses the CART algorithm in growing the tree. We determined the required number of observations per node with the least classification loss from k-fold cross-validation for better generalization. The minimum number of observations per node was 75 with 0.215 classification loss. A 70:30 ratio was then applied to allocate the training and testing dataset.

Table 1. Participant Profile.

Demographic		No. of Participants	% of Participants	% Baguio*	Source
Age	< 25	293	9	53	PSA [10]
	25 to 29	675	22	9	
	30 to 34	530	17	7	
	35 to 39	495	16	6	
	40 to 44	282	9	5	
	45 to 49	279	9	5	
	> 49	572	18	14	
Monthly Household income	< 10,000	100	3	4	PSA [10]
	10,000 to 20,000	529	17	25	
	20,000 to 30,000	727	23	25	
	30,000 to 40,000	591	19	17	
	40,000 to 50,0000	351	11	9	
	> 50,000	828	26	20	
Number of cars in the household	0	2145	69	83	PSA [10]
	1	730	23	13	
	> 2	251	8	4	
Main mode of transportation	Private	941	30	27	PSA [10]
	Public	2185	70	73	

*Baguio City data is from 2015 FIES (N = 362).

3. Results and Discussion

The DT model attained accuracies of 78.53% during training and 75.53% during testing; therefore, the model achieved generalization of the datasets. Overall, the model has an accuracy of 77.63%. Figure 1 shows the confusion matrices of the DT model.

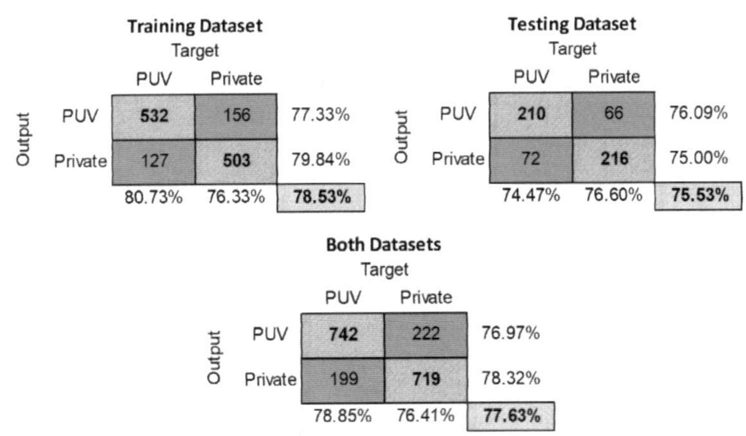

Figure 1. Confusion matrices

Figure 2 depicts the DT for the transportation mode selection. The values indicated below each node represent its classification counts, e.g., Node 1 correctly classified 444 observations as 'private' from the training dataset. The accuracy in the classification of each node differs from one another. Nodes 1 and 2 have the highest accuracy rating as Node 1 correctly classified 82.75% of its observations from both datasets, while node 2 attained 79.43% accuracy. Notably, node 4 attained an accuracy of 61.46%

during training and 53.90% for both datasets. The result indicates that datasets within the node are nearly homogenous.

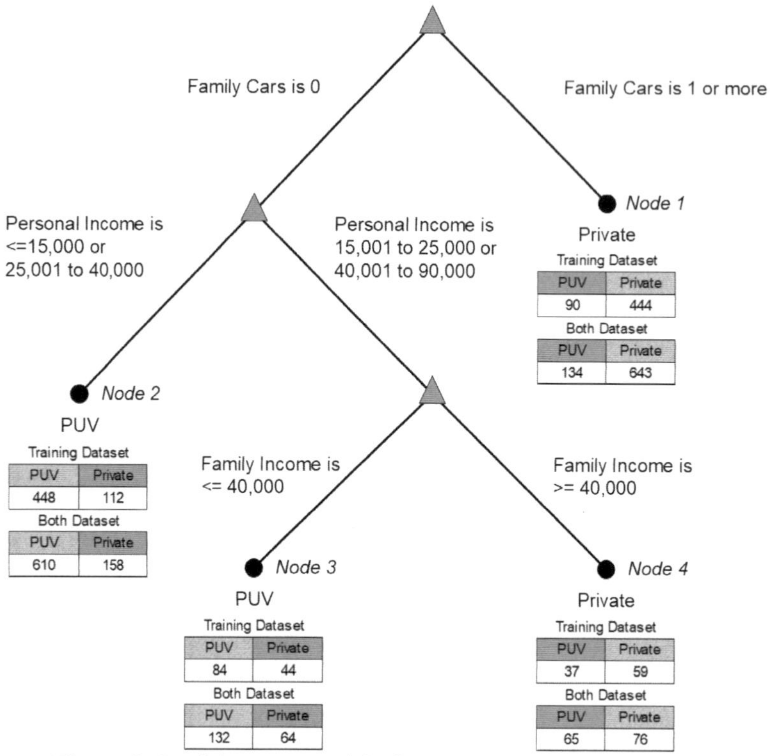

Figure 2. Decision tree model of transportation mode selection

To further understand the decision tree model, the age distribution of each classification of participants with no family car is illustrated in Fig. 3. It is assumed that the personal income level may explain an individual's decision to either use public or private transport. Participants with a personal income of less than Php 15,000 consist of individuals who are in the early stage of their career. This group tends to use public transportation due to their limited personal income. Likewise, participants with a personal income of Php 25,001 to Php 40,000 consist of individuals who are in the middle stage of their career. Although this group is more financially capable to use private transport, it is also susceptible to increasing financial responsibilities since individuals tend to start a family at this stage.

An interesting result in Fig. 2 is the inclusion of participants with a personal income of Php 15,001 to Php 25,001 together with those with a personal income of Php 40,001 to Php 90,000. Based on Fig. 3, this group is mostly composed of individuals who are between the early-stage and middle stages of their careers. This group tends to accumulate wealth as they gain more personal income without any significant financial burden. It is also observed that the age distribution of participants with a personal income of Php 15,001 to Php 25,001 is similar regardless of family income level. This is a possible explanation for the homogeneity of datasets in Node 4. Lastly, participants with a personal income of Php 40,001 to Php 90,000 consist of individuals who are in the late stage of their career. This group tends to accumulate more wealth due to higher personal and family incomes. Hence, the group may tend to use private transport more than public transport.

4. Conclusion

In this brief paper, the authors had demonstrated the potential application of DT models in mode choice modeling. While traditional mode choice models based on regression techniques can offer more insights, the advantage of DT is its simplicity and straightforward interpretation of results, which is paramount in policymaking. The results of this study provide interesting insights into the role of income to private or public mode choice selection, which seems to be related to an individual's career level and life stage. As previous studies have shown, having children and starting a family can affect a person's mobility choices.

For future work, larger DT models can be developed to consider other potential factors affecting this decision. Moreover, personal interviews may also be carried out to validate the insights obtained from the results.

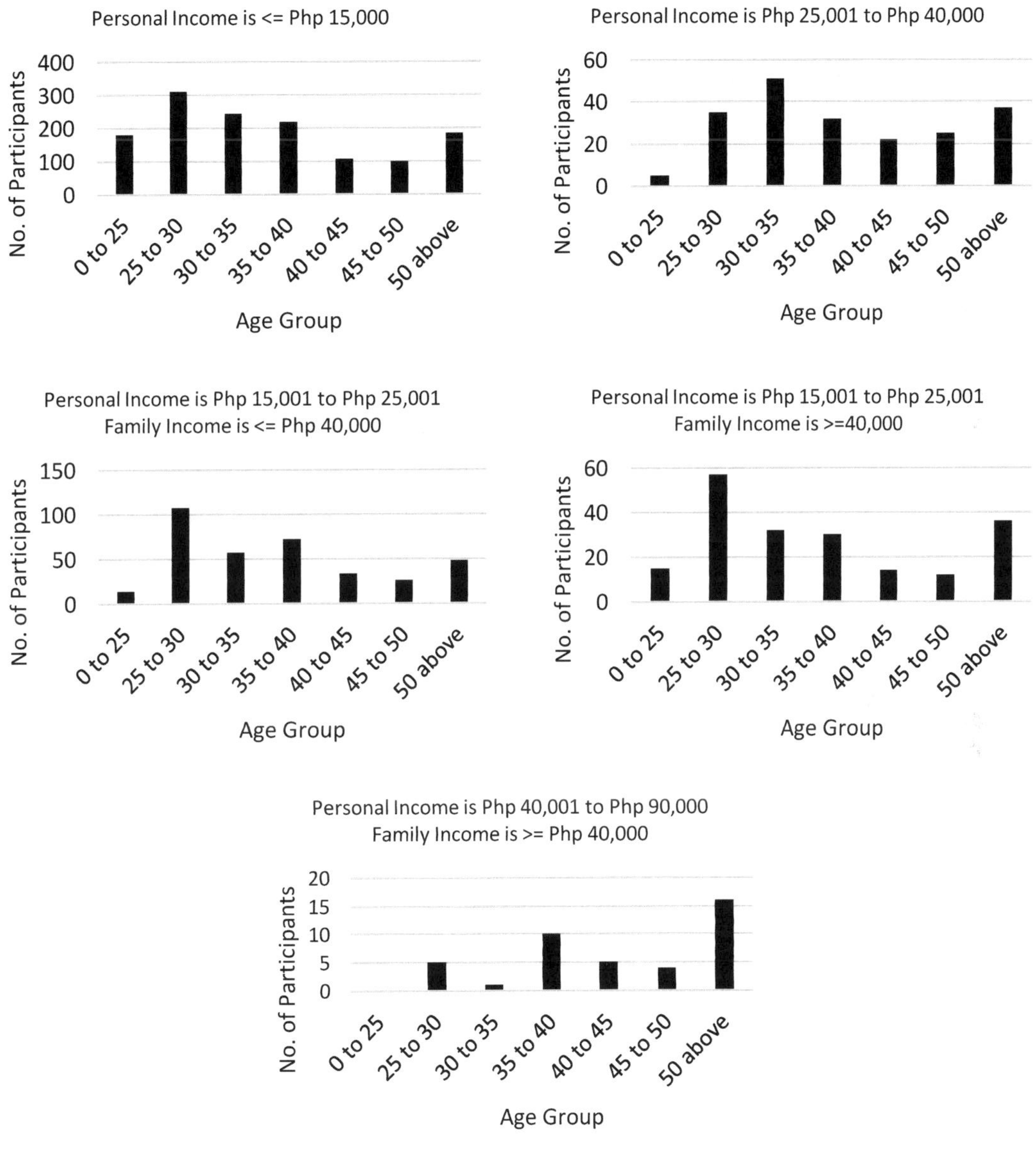

Figure 3. Age distribution of participants with no family car

Acknowledgments

The authors would like to express their gratitude to the Commission on Higher Education, Philippines for their generous funding in the conduct of this work, through the Grants-in-Aid program. The authors also received support from the University Research Coordination Office of De La Salle University, Philippines, through its Interdisciplinary Research Grants.

References

[1] Gue, I H V, Ubando A T, Tseng M-L and Tan R 2020 Artificial neural networks for sustainable development: a critical review. *Clean Technologies and Environmental Policy*, (0123456789). https://doi.org/10.1007/s10098-020-01883-2

[2] Paredes M, Hemberg E, O'Reilly U-M and Zegras C 2017 Machine learning or discrete choice models for car ownership demand estimation and prediction? *2017 5th IEEE Int. Conf. on Models and Technologies for Intelligent Transportation Systems* (MT-ITS) pp 780–785 IEEE. https://doi.org/10.1109/MTITS.2017.8005618

[3] Pineda-Jaramillo J D 2019 A review of Machine Learning (ML) algorithms used for modeling travel mode choice *DYNA* **86**(211) 32–41 https://doi.org/10.15446/dyna.v86n211.79743

[4] Chapleau R, Gaudette P and Spurr T 2019 Application of machine learning to two large-sample household travel surveys: A characterization of travel modes. *Transportation Research Record: Journal of the Transportation Research Board*, 2673(4), 173–183. https://doi.org/10.1177/0361198119839339

[5] Pawar D S, Yadav A K, Akolekar N and Velaga N R 2020 Impact of physical distancing due to novel coronavirus (SARS-CoV-2) on daily travel for work during transition to lockdown *Transportation Research Interdisciplinary Perspectives* **7** 100203 https://doi.org/10.1016/j.trip.2020.100203

[6] Ranosa L, Villaraza, C M, Fillone A, Guillen M D and Biona J B M 2017 Providers' and users' assessment of public transport and facilities: Baguio City, Philippines *2017 Proceedings of the Eastern Asia Society for Transportation Studies* **11**

[7] Estrella A F, Mendoza A K, Mendoza M A, Sanchez K, Fillone A, Paringit M C and Roxas N 2017 Impact analysis of aerial ropeway transport system as a form of mass transportation in Baguio City *Proceedings of the 24th Annual Conference of the Transportation Science Society of the Philippines*

[8] Rith M, Roquel K I D Z, Lopez N S A, Fillone A M and Biona J B M M 2020a Towards more sustainable transport in Metro Manila: A case study of household vehicle ownership and energy consumption *Transportation Research Interdisciplinary Perspectives* **6** 100163.

[9] Rith M, Lopez N S, Fillone A M and Biona J B M 2020b Determinants of individual vehicle type choice and energy consumption in a heavy traffic metropolis of Southeast Asia featuring the case of Metro Manila *Engineering and Applied Science Research* **47**(1) 56-65.

[10] Philippine Statistics Authority (PSA) (2017). 2015 Family Income and Expenditure Survey (FIES). https://psa.gov.ph/sites/default/files/FIES%202015%20Final%20Report.pdf

[11] Kotsiantis S B 2007 Supervised machine learning: A review of classification techniques *Informatica* **31** pp 249–268.

[12] Morgan J N and Sonquist J A 1963 Problems in the analysis of survey data, and a proposal *Journal of the American Statistical Association* **58**(302) p 415 https://doi.org/10.2307/2283276

[13] Breiman L, Friedman J, Stone C J and Olshen R A 1984 Classification and regression trees *CRC press*

[14] Quinlan J R 1992 C4.5 Programs for Machine Learning. San Mateo, CA: Morgan Kaufmann.

[15] Kotsiantis S B 2013 Decision trees: a recent overview *Artificial Intelligence Review* **39**(4) pp 261–283. https://doi.org/10.1007/s10462-011-9272-4

Decomposition analysis of Japan's CO_2 emissions

J A Dela Vega[1], J A Gonzaga[1], L A Gan Lim[1], N S Lopez[1]

[1] Mechanical Engineering Department, De La Salle University, Manila, Philippines

Email: john_christopher_delavega@dlsu.edu.ph

Abstract. Studies have been conducted to uncover the underlying factors of global CO_2 emissions. One method to quantify the driving forces of carbon dioxide emissions is decomposition analysis. In this paper, the logarithmic mean Divisia index (LMDI) method was used to investigate the driving forces of carbon dioxide emissions in Japan for the period of 1990-2017. As one of the world's leading economies, it also contributes one of the highest CO_2 emissions globally. Results highlighted the flaw of energy structure as one major factor affecting CO_2 emission intensity. Another notable finding is how the state of the economy played a pivotal role in the shift of carbon emission levels. Potential policy recommendations were also introduced to establish a better power structure for electricity generation in the country.

1. Introduction

As one of the world's leading economies, Japan had a nominal gross domestic product (GDP) of 5.15 trillion US dollars in 2019. It currently ranks third in terms of global economy and was projected to have a 0.7% growth in 2020 [1]. However, this economic development is countered by the country's contribution to the global CO_2 emissions. In 2019, Japan is the fifth largest contributor of CO_2 emissions worldwide [2]. As part of its initiative to mitigate its carbon emissions, Japan was one of the countries who ratified the Paris Agreement in 2016. In line with this, Japan vowed a 26% reduction of its greenhouse gas (GHG) emissions by 2030, as compared to its 2013 levels [3].

Several studies have already been conducted to mitigate carbon dioxide emission by investigating its driving factors. Most of these researches have employed the use of decomposition analysis to unravel the factors affecting CO_2 emissions. For instance, a study was conducted using structural decomposition analysis (SDA) to investigate the source of changes of CO_2 emissions in the United States [4]. A similar method was used to analyze how changes in demographics, industry structure, and urbanization drive regional CO_2 emissions in China [5]. Moreover, a paper also delved into the driving factors of emissions from all European Union countries due to electricity generation [6], while an identical approach was employed to analyze the efficiency of policies relating to carbon emissions adopted in Italy [7]. Meanwhile, a study using logarithmic mean Divisia index (LMDI) was presented to examine the carbon emission trends in the iron and steel industry in Mexico [8], whereas the same method was used to study the change in energy-related CO_2 emissions in China's industrial sector [9].

Other papers also used decomposition analysis to analyze CO_2 emission intensity in Japan. For example, a study used structural decomposition analysis to uncover the factors affecting the carbon emissions from the Japanese industry in 1985-1995 [10]. Another work evaluated the elements driving Japan's CO_2 emissions using input-output SDA for the periods 1995-2000 and 2000-2005 [11]. Also, a paper merged financial factors in formulating its identity function for a decomposition analysis of the CO_2 emissions from the Japanese manufacturing sectors [12].

In this study, the LMDI decomposition analysis method is used to investigate the driving forces of carbon dioxide emissions from Japan for the period of 1990-2017. While other approaches have been used in decomposition analysis, the LMDI method is preferred for its general advantages. This approach is easy to use in terms of calculation and data requirements, and its results are easy to interpret for policymakers [13]. Furthermore, it has the capability to deal with indicators having a value of "zero" and perhaps, its most desirable characteristic is its perfect decomposition – no residuals or unexplained quantities.

The rest of the paper is organized as follows: the LMDI methodology is discussed in the next section, together with the data sources; the results and significant findings are discussed in Section 3; lastly, the conclusion and possible policy recommendations are given in Section 4.

2. Methods and data

2.1. LMDI decomposition analysis

Index decomposition analysis (IDA) was first employed to investigate electricity consumption trends and has since been widely used in energy-related research. An in-depth review of this methodology as an approach in energy analysis and policy making is provided in the paper by Ang and Zhang [14]. Moreover, it is reported that the logarithmic mean Divisia index (LMDI) decomposition analysis is regarded as the most popular IDA method; being used in the majority of index decomposition analysis papers published in 2010-2014 [15].

The LMDI approach uses the Divisia index and logarithmic mean weighted value to decompose aggregate indicators such as energy consumption. It can either be classified as additive or multiplicative decomposition. In additive decomposition analysis, the arithmetic change in the quantity between two given points in time is decomposed [16]. On the hand, multiplicative decomposition analysis decomposes the ratio change of an aggregate with respect to the baseline year [17].

2.2. Identity function

The decomposition analysis identity used in this paper is based on the function (Eq (1)) provided by Kaya, which established the connection among human population, GDP per capita, energy intensity, and carbon intensity [18]. Furthermore, the identity function used was adapted from Sumabat et al. [19], as given in Eq (2). The annual contributions of each individual effect were solved using Eqs (3) to (6), where C refers to the CO_2 emissions, P signifies the human population, Q is the gross domestic product (GDP), E is for the energy usage, and i is the fuel type used to generate energy. In addition, G refers to GDP per capita, I is the energy intensity, S_i is shared in energy mix for fuel type i, and F_i is emission factor for fuel type i.

$$CO_2 = P\ x\ \frac{GDP}{P}\ x\ \frac{E}{GDP}\ x\ \frac{CO_2}{E} \tag{1}$$

$$C = \sum_i C_i = \sum_i P\ x\ \frac{Q}{P}\ x\ \frac{E}{Q}\ x\ \frac{E_i}{E}\ x \frac{C_i}{E_i} = \sum_i P\ x\ G\ x\ I\ x\ S_i\ x\ F_i \tag{2}$$

$$\Delta C_{pop}^{T} = \sum_i \frac{C_i^T - C_i^0}{\ln C_i^T - \ln C_i^0} \ln\left(\frac{P^T}{P^0}\right) \tag{3}$$

$$\Delta C_{act}^{T} = \sum_i \frac{C_i^T - C_i^0}{\ln C_i^T - \ln C_i^0} \ln\left(\frac{G^T}{G^0}\right) \tag{4}$$

$$\Delta C_{int}^{T} = \sum_i \frac{C_i^T - C_i^0}{\ln C_i^T - \ln C_i^0} \ln\left(\frac{I^T}{I^0}\right) \tag{5}$$

$$\Delta C_{str}^{T} = \sum_i \frac{C_i^T - C_i^0}{\ln C_i^T - \ln C_i^0} \ln\left(\frac{S^T}{S^0}\right) \tag{6}$$

The notation ΔC refers to the change in the emission levels from the baseline year, represented by the superscript 0, and the final year, represented by the superscript T, in that time interval. The subscripts *i, pop, act, int,* and *str*, refer to the fuel type, population effect, activity effect, intensity effect, and structure effect, respectively.

2.3. *Data sources*

The data used in this paper were taken from various sources. Japan's data for its population, along with its gross domestic product (GDP) in constant 2010 USD, were sourced from the database of the World Bank [20]. On the other hand, the figures for Japan's energy generation by source, electricity consumption by sector, and CO_2 emissions by sector, were obtained from the records of the International Energy Agency (IEA) [21].

3. Results and discussion

3.1. *Activity effect (ΔC_{act})*

Shown in Figures 1 and 2 are the summary of the changes in the CO_2 emission levels of Japan. The aggregated effects depicted in Figure 1 shows that the carbon emissions of Japan have been increasing at an average rate of approximately 19,650 ktons of CO_2 per year interval from 1990-2007. However, a significant decrease can be noted starting from 2008 up to the year of 2009, where a noticeable decline in the activity effect (ΔC_{act}) can be observed. Furthermore, looking at Figure 2, where the annual effects of CO_2 emissions is illustrated, it is shown that the activity effect has mostly been on the positive part of the graph, for the 1990-2007 period, indicating the contribution of the effect of economic activity in the emission levels. Nevertheless, a sharp turn of the activity effect into the negative region of graph is shown in the 2007-2009 period. The behavior of ΔC_{act} during the period 2007 to 2009 can be attributed as an effect of the Great Recession – an economic downturn after the global financial crisis.

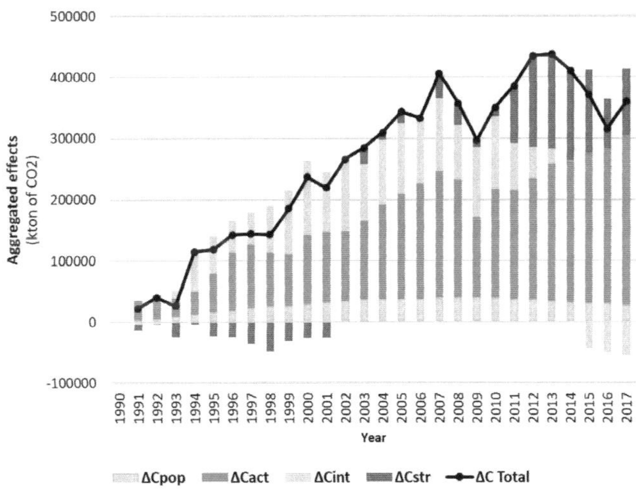

Figure 1. The aggregated CO_2 emissions level from various driving factors (1990-2017)

Elaborating, the global financial crisis refers to the extreme stress in the global economy in the period of mid-2007 to 2009 [22]. During this timeframe, major economies in the world experienced the worst recessions since the 1930 Great Depression, and many banking institutions depended on government support to prevent bankruptcy. In Japan, it caused a major collapse in the movement of exports and industrial productions [23], leading to its record high steepest fall in GDP at 15.2% since 1955 [24], thereby affecting the economic activity, and by extension, the share of activity effect in its CO_2 emissions.

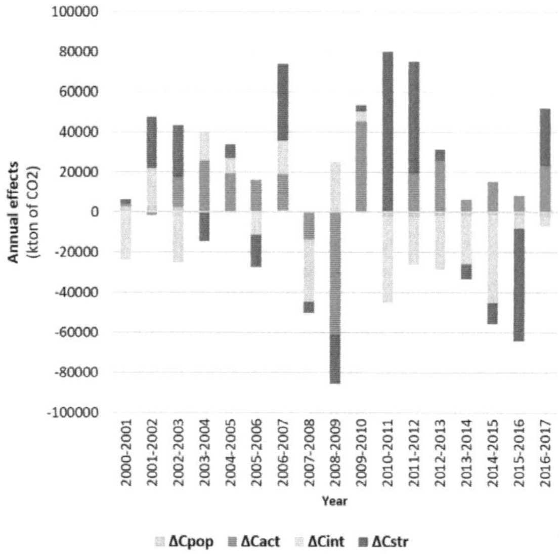

Figure 2. Japan's annual CO2 emission levels change due to quantified effects (2000-2017)

3.2. *Structure effect (ΔC_{str})*

On the other hand, Japan has a very diverse energy mix as shown in Figure 3. While it taps other energy generating facilities, the country is heavily dependent on zero emission nuclear power and emission intensive fossil fuels to address its energy demands. However, this will cause further serious implications as the country experienced one of the most disastrous calamities of the 21st century.

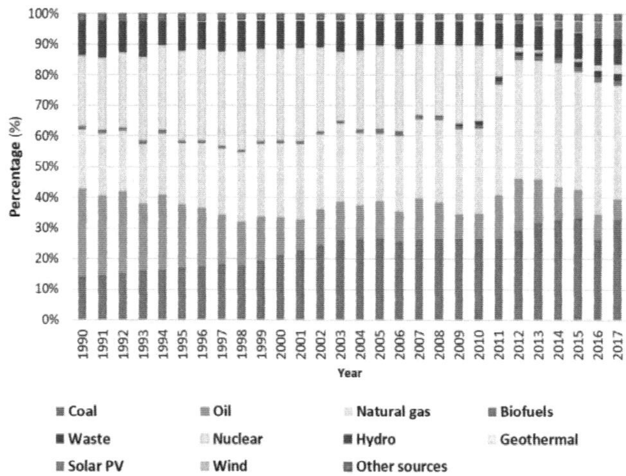

Figure 3. Japan's energy mix from 1990 to 2017

The 2011 Tōhoku earthquake and tsunami devastated the Pacific Coast of Japan causing immense casualties and damages, including the subsequent meltdown of the Fukushima Daichi Nuclear Facility [25]. It was followed by a massive shutdown of nuclear facilities all over Japan amidst the fear of further nuclear disaster [26]. The 2010-2012 trend in the structure effect (ΔC_{str}), as shown in Figure 2, is a consequence of the nuclear shutdown. As Japan has been left with no nuclear-derived electricity for the first time in 42 years, the country shifted towards cheaper, yet emission intensive fossil fuel-based facilities.

Moreover, in the 2010-2012 timeframe, the ΔC_{str} amounted to approximately 134,883 ktons of CO_2, the largest contribution of all the 4 quantified effects in the 1990-2017 study period. This is after the fact that Japan imported an estimated 30,000 barrels a day more of fossil fuels in 2011, and a further

80,000 barrels per day more than its usual consumption in 2012, to compensate for the loss of its nuclear-sourced energy generation [27].

3.3. *Intensity effect (ΔC_{int})*

The intensity effect (ΔC_{int}) in Japan's CO_2 emissions took a drastic constant change from 2010 onwards as shown in Figure 2. From an aggregated share of approximately 5,380 kton of CO_2 emission in the 2009-2010 period, it plummeted into the negative region of the graph in 2010-2011 period and has since been at an average yearly decline of 63,200 ktons of CO_2. Three factors can then be considered as plausible driving factors for this behavior of a quantified effect. First, the 2011 Tōhoku earthquake and tsunami prompted an energy crisis in Japan. Even with the production adjustment of manufacturing industries to save electricity, the Japanese government also mandated a reduction in electricity consumption in 2011 and 2012, aside from the rolling blackouts for several hours that were experienced in the first month after the disaster [28]. Another factor to be considered is that Japanese industries continually expand their overseas manufacturing operations. A survey report conducted by the Japan Bank for International Cooperation (JBIC) for the fiscal year of 2017 indicated that the overseas production ratio of Japanese manufacturing companies was at 35% in 2016 and is projected to increase up to 38.5% by 2020 [29]. Lastly, the constant increase of contribution of the services sector in Japan's gross domestic product (GDP), as shown in Figure 4, is also a factor worth considering. Additionally, it is reported that the country's capital expenditure is gearing towards the services sector and has risen to 9.2% mid-2018 [30].

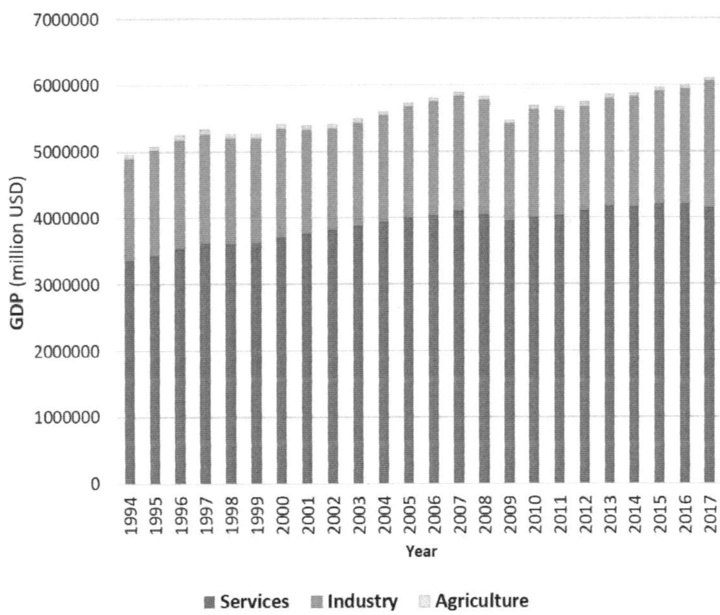

Figure 4. Gross domestic product (GDP) contribution per sector (1994-2017)

3.4. *Population effect (ΔC_{pop})*

The contribution of population growth to CO_2 emission levels remained positive (ΔC_{pop}) all throughout the 1990-2008 period, as shown in Figure 2. However, this behavior has changed from 2009 onwards, as Japan is currently experiencing a population crisis, due to a rapid decline in its population brought about by low fertility rates and an aging population. As of 2019, more than 20% of the Japanese populace is over 65 years old and its trend in fertility rates is heavily affected by various factors such as changing lifestyles, marriage dilemma, and economic insecurity of the present generation [31].

4. Conclusion and recommendation

4.1. Policy recommendation

Based on the presented data and analysis of the results, stated below are the recommendation options that the policy makers can explore:

1. Japan needs to ease its dependence on nuclear energy as its main energy source to prevent further energy crises that followed the 2011 nuclear disaster. More secure alternatives need to be explored.
2. In accordance with the previous recommendation, Japan can increase investment on renewable energy as a potential alternative to nuclear power.

The Great East Japan earthquake of 2011 exploited a weakness in Japan's energy security. Its heavy dependence on nuclear energy, amounting up to 31.3% before the disaster, proved to have serious implications. Moreover, the public perception on nuclear power has also been affected and protests against further use of nuclear energy were initiated as a yearning for a "zero nuclear power operation". Thus, the country needs to establish a better power structure in electricity generation. The barely tapped potential of renewable energy sources in the country is one probable solution to this problem. According to Japan's Ministry of the Environment in 2011, around 887 gigawatts (GW) of energy can be economically exploited in the country. It is almost 4 times the combined installed capacity of nuclear facilities and conventional thermal power plants in Japan in that period. In conclusion, since the energy structure has been the highlight of the study, planning of the best energy mix is key for a sustainable and energy-secure Japan.

4.2. Conclusion

This paper used the logarithmic mean Divisia index (LMDI) decomposition analysis to uncover the driving factors of carbon emission intensity in Japan. Particularly, this study focused on identifying specific strategies adopted by Japan to mitigate its CO_2 emissions, considering its unique economic and resource limitations. Results indicated that economic activity had a significant impact in the shift of CO_2 emissions level. Moreover, the energy mix of Japan for the 1990-2017 period was placed under scrutiny. Based on the analysis of data, Japan's energy structure needs to be diversified with more secure alternatives to alleviate its dependence on nuclear power. Potential policy recommendations were also given and discussed to ensure the country's energy security.

References

[1] World GDP Ranking 2020 [Internet]. Statisticstimes.com. [cited 25 April 2020]. Available from: http://statisticstimes.com/economy/projected-world-gdp-ranking.php

[2] CO2 emissions by country [Internet]. Statista. [cited 25 April 2020]. Available from: https://www.statista.com/statistics/271748/the-largest-emitters-of-co2-in-the-world/

[3] The Carbon Brief Profile: Japan [Internet]. Carbon Brief. [cited 25 April 2020]. Available from: www.carbonbrief.org/ carbon-brief-profile-japan

[4] Casler S and Rose A 1998 Carbon dioxide emissions in the US economy: a structural decomposition analysis *Environmental and Resource Economics* **11**(34) 349-63.

[5] Feng K, Siu Y, Guan D and Hubacek K 2012 Analyzing drivers of regional carbon dioxide emissions for China *Journal of Industrial Ecology* **16**(4) 600-11.

[6] Karmellos M, Kopidou D and Diakoulaki D 2016 A decomposition analysis of the driving factors of CO2 (Carbon dioxide) emissions from the power sector in the European Union countries *Energy* **94** 680-92.

[7] Andreoni V and Galmarini S 2012 Decoupling economic growth from carbon dioxide emissions: a decomposition analysis of Italian energy consumption *Energy* **44**(1) 682-91.

[8] Sheinbaum C, Ozawa L and Castillo D 2010 Using logarithmic mean divisia index to analyze changes in energy use and carbon dioxide emissions in Mexico's iron and steel industry *Energy Economics* **32**(6) 1337-44.

[9] Ouyang X and Lin B 2015 An analysis of the driving forces of energy-related carbon dioxide emissions in China's industrial sector *Renewable and Sustainable Energy Reviews* **45** 839-49.
[10] Yabe N 2004 An analysis of CO2 emissions of Japanese industries during the period between 1985 and 1995 *Energy Policy* **32**(5) 595-610.
[11] Akpan U S, Green O A, Bhattacharyya S and Isihak S 2014 Effect of technology change on CO2 emissions in Japan's industrial sectors in the period 1995–2005: An input–output structural decomposition analysis *Environmental and Resource Economics* **61**(2) 165–89.
[12] Yagi M and Managi S 2018 Decomposition analysis of corporate carbon dioxide and greenhouse gas emissions in Japan: Integrating corporate environmental and financial performances *Business Strategy and the Environment* **27**(8) 1476–92.
[13] Ang B 2004 Decomposition analysis for policymaking in energy: Which is the preferred method? *Energy Policy* **32**(9) 1131-9.
[14] Ang, B W and Zhang, F Q 2000 A survey of index decomposition analysis in energy and environmental studies *Energy* **25**(12) 1149–76.
[15] Ang B W 2015 LMDI decomposition approach: A guide for implementation *Energy Policy* **86** 233–8.
[16] Ang B W, Zhang F L and Choi K 1998 Factorizing changes in energy and environmental indicators through decomposition *Energy Policy* **23** 489–95.
[17] Ang B W and Liu F L 2001 A new energy decomposition method: Perfect in decomposition and consistent in aggregation *Energy Policy* **26** 537–48.
[18] Kaya Y and Yokobori K 1997 *Environment, energy, and economy* (Tokyo: United Nations University)
[19] Sumabat A, Lopez N, Yu K, Hao H, Li R, Geng Y and Chiu A 2016 Decomposition analysis of Philippine CO2 emissions from fuel combustion and electricity generation *Applied Energy* **164** 795-804.
[20] Japan Data [Internet]. Data.worldbank.org. [cited 22 February 2020]. Available from: https://data.worldbank.org/country/japan
[21] Japan - Countries & Regions - IEA [Internet]. IEA. [cited 22 February 2020]. Available from: https://www.iea.org/countries/japan
[22] The Global Financial Crisis [Internet]. Reserve Bank of Australia. [cited 2 March 2020]. Available from: https://www.rba.gov.au/education/resources/explainers/the-global-financial-crisis.html
[23] Kawai M and Takagi S 2009 *Why was Japan hit so hard by the global financial crisis?* (Tokyo: Asian Development Bank Institute)
[24] McCurry J. Japan suffers record fall in GDP. The Guardian [Internet]. 2009 [cited 2 March 2020];. Available from:https://www.theguardian.com/business/2009/may/20/japan-economy-gdp-fall
[25] Tohoku Earthquake and Tsunami [Internet]. National Geographic Society. [cited 2 March 2020]. Available from: www.nationalgeographic.org/thisday/mar11/tohoku-earthquake-and-tsunami/
[26] Batty D. Japan shuts down last working nuclear reactor. The Guardian [Internet]. 2012 [cited 2 March 2020];. Available from: https://www.theguardian.com/world/2012/may/05/japan-shuts-down-last-nuclear-reactor
[27] Lippert S. Changes in Japan's energy mix since the 2011 nuclear plant disaster [Internet]. Geopolitical Intelligence Services. 2018 [cited 2 March 2020]. Available from: https://www.gisreportsonline.com/gis-dossier-japans-search-for-an-energy-strategy-after-fukushima,energy,2750.html
[28] Umezawa S 2014 The great east Japan earthquake: Its damages, impacts on the labor-economy and restoration measures of the government *E-journal of International and Comparative Labour Studies* **3**(3)

[29] Survey Report on Overseas Business Operations by Japanese Manufacturing Companies [Internet]. Japan Bank for International Cooperation; 2018. Available from: https://www.jbic.go.jp/en/information/reference/reference-2017/contents/201803014_spot.pdf
[30] White S. Japan invests in service industry, reshaping its legendary hospitality. Reuters [Internet]. 2018 [cited 2 March 2020];. Available from: https://www.reuters.com/article/us-japan-economy-capex-services/japan-invests-in-service-industry-reshaping-its-legendary-hospitality-idUSKCN1N90RH
[31] Walia S. The economic challenge of Japan's aging crisis. The Japan Times [Internet]. 2019 [cited 2 March 2020];. Available from: https://www.japantimes.co.jp/opinion/2019/11/19/commentary/japan-commentary/economic-challenge-japans-aging-crisis/

IOP Conf. Series: Materials Science and Engineering 1109 (2021) 012061 doi:10.1088/1757-899X/1109/1/012061

Design, fabrication and testing of a portable gravity generator as a back-up power source.

R. Agustin[1], C Alonzo[1], K Ami[1], J M Brual[1], D N Viray[1], A Maglaya[1], A Fernando[1]

[1] Mechanical Engineering Department, De La Salle University, Manila Philippines

arvin.fernando@dlsu.edu.ph

Abstract. In this study a prototype small scale gravity driven generator was designed, tested, and evaluated. The prototype was designed using readily available components. It uses a chain sprocket system and a 24V DC generator. This study might help the needs for a new sustainable energy source, a gravity driven device that harness energy from a falling mass might provide an alternative solution for an increasing energy demand and supply electricity to remote areas.

1. Introduction

The need for a backup power source during calamities such as typhoons, earthquakes, tsunamis, and other natural phenomena is an issue that continues to grow. There is a need in finding an alternative backup power source during these situations. Gravity-driven generators are suitable for powering off-the-grid places; however, unlike conventional generators, gravity-driven generators only make use of mechanical parts and extract energy from falling objects with the use of potential energy however improvements in gearing trains, generator and gearing matching can be done to improve the generator's efficiency.

Hydroelectric power plant uses the gravitational potential energy of the water to harness store energy and has been long established. There have been recent developments in the field of using gravity energy as energy storage using a solid falling mass with several concepts and design [1] [2] [3] [4] [5]. There are only a handful of resources concerning the use of a solid mass gravity-driven system and even fewer data regarding their performance parameters particularly power output and system efficiency [1].

The study will develop a prototype and evaluate the performance of the Gravity-powered generator. The fabricated prototype will not be used for large scale power generation purposes, but only for the performance testing of the portable generator. Furthermore, the delimitation of the fabricated prototype should be able to withstand a practical/certain weight for the mechanism to function and to reduce stress on the system during testing.

2. Methodology

2.1. Design

Table 1 shows the base design parameter of a gravity generator. The target design power output is 5 watts however for a standard USB connection wattage, a 2.5 Watts is already sufficient. A fall time of 5 min is desired to decrease the user interaction when resetting and the unit is designed to be portable and can be lifted and hung when the need to use the device arises. The proponents used a chain and sprocket system over other gear design considerations. The chain and sprocket are easily industry available direct from manufacturers and suppliers while spur gears are usually designed and made to fabricate. They are relatively cheaper on the shelf and for the specific speed of application, chain and sprocket are better suited for low speed than spur gear. Chain and sprocket design have been chosen over belt drive due to the absence of slippage in sprocket designs and belts possess this disadvantage while sprockets allow synchronization of input and output gears. For the specific speed of application, chain and sprocket are better suited for low speed than belt drive systems. For the magnitude of load application chain and sprocket is better suited for sprocket gears then belts.

Table 1. Prototype Design Parameters

Specification	**Value**
Power output	**5 Watts**
Lifted Weight	**0.5 to 5 kg**
Time	**At least 2 minutes**
Device Weight	**Easily lifted and hung**

A 24V DC generator was used in this study and was procured online. It has a rating of 24V at 120 rpm. The expected transmitted revolutions per unit time and the respective voltage delivered was tested using a rotating device with a variable low and high-speed setting. It was attached to the motor and at each speed setting; a tachometer was used to observe the actual revolutions per minute that was delivered on the motor. A multi meter was then attached to the motor to record the voltage measured per setting of the variable speed drill. The result shows that a DC generator can produce higher voltage output at a higher rpm output but may result in a motor burn out. Through these measured data, several calculations are derived to compute for the required RPM for the target required output voltage.

Figure 1. CAD Model of the Prototype

Figure 2. Actual Prototype

The prototype was designed to have a feature of modularity in terms of the assembly of the general parts see figure 2 and 3. The main pillar posts that held each individual shafts and sprockets were designed to be modular and adjustable in order to accommodate the testing of the system with different gear pair's combination for an ease of attachment and replacement. The pillar post containing the motor is also adjustable in terms of positioning or alignment. A standard angled bracket was used per pillar post, as well as the same auxiliary parts such as nuts and bolts to make sure all components are easily assembled and disassembled if required.

A specific torque range will be determined and selected for the DC motor. After selection of the torque range, a gear train is designed such that the rotational speed at the suspended weight side falls slow enough to last 5 minutes and the rotational speed at the generator side is fast enough to generate sufficient power.

During the fabrication of the main shaft two options were made regarding the setup of the rotation of the shaft. The shaft could be either two main static non-rotating shafts that are independent from the sprockets attached or many individual rotating shafts that revolve with the actual rotation of the attached sprockets. The initial setup presented challenges in the welding of the outer bearing surface to the sprockets and resulted with a difficult fabricating process after a trial assembly. Due to this challenge, the fabrication moved to the second setup where the shaft is welded with the sprockets and welding to the bearing is not conducted anymore. The second shaft setup required each pair of the sprocket train system to have an independent shaft because these will differ in their rotating speed and torque transmitted.

2.2. *Testing*

2.2.1 Weight Weight Range: 6.85 - 7.9 kg Range Reason: From initial computation, a 5 kg falling weight would result to about 4.9 N-m torque. This weight was then set as the standard minimum weight to be used for testing. During the actual data collection, it was found that a 5 kg falling weight could not deliver enough torque for the drop to start. This can be seen as the effect of the mechanical losses in the system as well as losses due to the fabrication tolerance limitations and lack of machine accuracy in the assembly. Due to this event, the 5 kg weight was increased until it can be observed that the system will start the descent. The minimum weight that the drop starts was found to be 6.85 kg, and the computed torque from this result was found to be about 6.8 N-m. The upper limit of the weight range was then found to be 7.9 kg because a weight of 8 kg causes the DC motor to burn out, causing a burnt smell and unwinding the internal coil assembly of the motor.

2.2.2 Drop Height Constant Height, Height: 1.78 m. Constraint Reason: The highest altitude that was tested in the data-collecting phase was at 1.78 meters because this was the highest possible altitude that the assembly can be mounted in the testing facility. Any altitude higher would present challenges in the mounting of the brackets as well as the winding and rewinding procedure of the spool above ground level. It was also found that any height lower than 1.78 meters would result in a short--and in effect, a short generation period of the project. See figure 4 for the mounted set-up.

Figure 3. Mounted Set-up

2.2.3 Transmission Set-up Constant Gear Pair Amount Pairs Used: three pairs the amount of gear pairs was not varied for data collection because of the above constraints mentioned. A gear train with two pairs would result in a lessened drop time, therefore decreasing the generation period and increasing the frequency of winding and rewinding of the spool unit. A gear train with four pairs delivers a desirable gear ratio but the required torque for the assembly to start dropping would be high, increasing the falling weight required. Such increase would result in a large overall weight setup and the risk of burning out the motor is presented. Three pairs of the gear train are the optimal amount in terms of weight used and generating period.

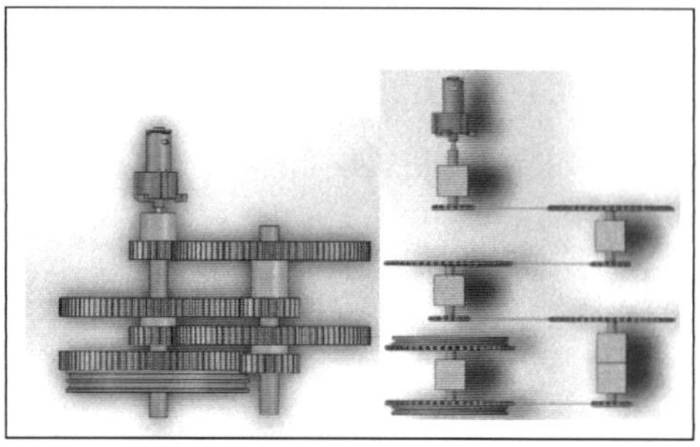

Figure 4. Shaft Rotation Set-up

The varying number of sprocket pairs in the series train setup seen in figure 6 was considered during the fabrication and assembly phase. During the initial computation, the gear train setup with four pairs resulted in a desirable gear ratio for the transmission of revolution of the falling weight to the motor. However, during the data-testing phase it was found that four pairs drastically required an increase in torque needed to rotate the spool. An increase in torque to be supplied would mean a higher weight magnitude for the falling weight. It was then realized that the setup involving three sprocket pairs is the optimal choice due to the required torque and falling weight magnitude being well contained in the set range to be used for the project.

Following the decrease of four pairs to three pairs used in the assembly, it was also tested how two pairs would affect the result. From the computation, the resulting gear ratio is not too effective. It was also found during testing that the 2-gear pair setup also fell short in the falling time duration because the weight traversed its drop course too quickly which would also mean an abrupt generating period.

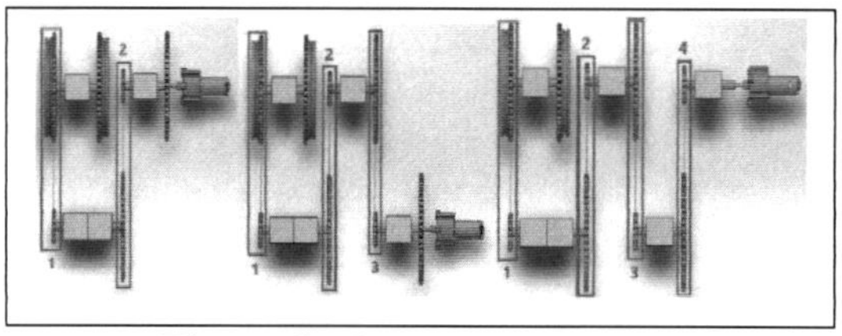

Figure 5. Gear pair Set-up

2.3. *Evaluation*

This study will analyze and evaluate the performance of the prototype design. The system consists of a weight drop mechanism that will generate a rotational mechanical energy to the chain and sprocket system that will drive the electric generator. The mechanism will be tested to a fixed height and varying suspended weights see Figure 7 for the Schematic Diagram of the Set- up.

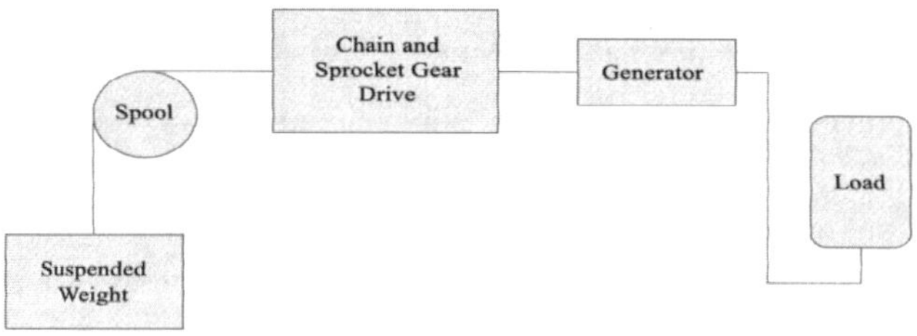

Figure 6. Schematic Diagram

The power input will be computed from the falling weight and the drop height using the potential energy equation. The power output will be measured at the electric generator. The efficiency of the system is given equation 1. Transmission Losses were also considered from the spool, bearing and chain sprocket.

$$Efficiency = \frac{Power\ Output}{Power\ Input} \tag{1}$$

3. Results and Discussion

Figure 8 shows the performance curve in the inlet side of the system comparing the spool rpm and input power to the torque. The red dots represent the power at specific torque while the blue dots represent spool rpm at specific weight and torque produced. The graph shows a directly proportional relationship between power input and spool rpm and the dots are almost overlapping as it is in the early phase of the system.

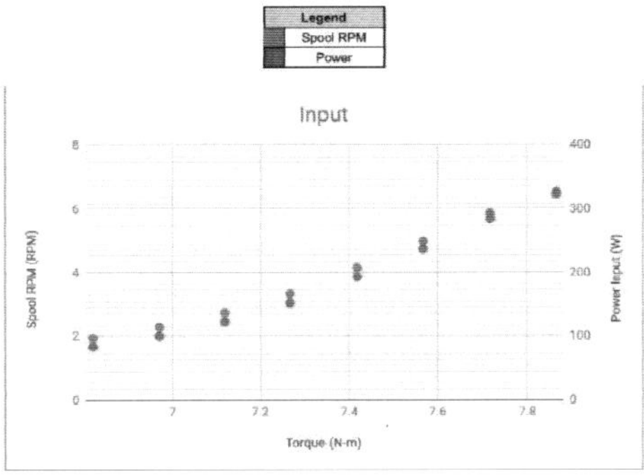

Figure 7. Performance Curve at the input

Figure 9 shows the relationship between generator rpm and power output. Although they both display a directly proportional relationship, the dots are further apart and it can be seen that the red dot is getting farther from the blue dot the higher the torque applied is.

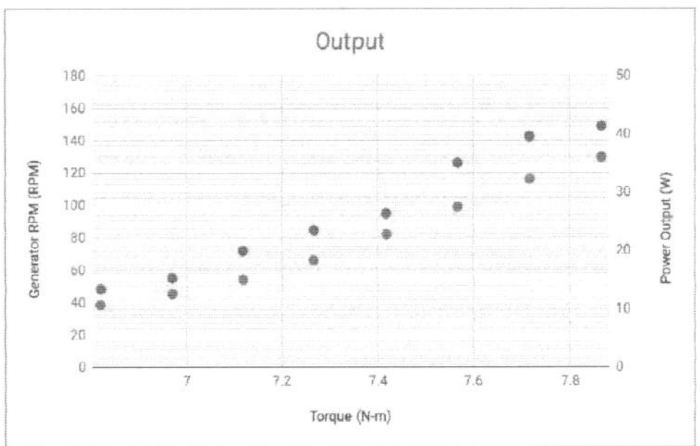

Figure 8. Performance Curve at the output

The overall efficiency of the system peaked at 16.31% and produced the lowest efficiency of 12.82% as seen in figure 10. The graph also shows that there is a specific torque or weight load that will produce the highest efficiency. It can also be noted that once the maximum value is reached, the efficiency drops on the following tests following a trend except for one outlier value. It can be concluded that a proper pairing of weight and generator is essential in producing the highest efficiency possible rather than just increasing the weight load.

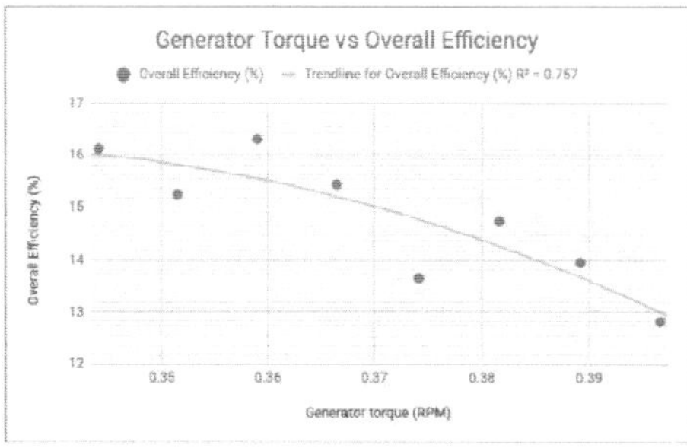

Figure 9. Generator Torque and Overall Efficiency

The system makes use of chains and sprockets and losses are naturally incurred from the friction between the chains and sprockets. Chain drives are highly efficient in terms of transmission and it can have efficiencies of about 98% [9]. In the system, three pairs of chains and sprockets are used. Assuming that each pair will have an efficiency of 98%, the total efficiency from the three pairs will be 94.12%. Within the generator itself, losses are incurred. For the system, a combined term for both the frictional losses and the core generator losses is considered. It will represent the generator's efficiency to convert the mechanical power exerted at the shaft to electrical power. For simplicity, 50% generator efficiency is considered at all torque values.

A linear and directly proportional relationship between the rotational speed and the torque/weight load can be extracted. It can also be observed that there is an increase in mechanical power since greater downward force is produced due to heavier load. The range of the rotational speed in the input side is between 1.95 RPM to 6.54 RPM while the range on the output side is between 38.74 RPM to 129.64 RPM. The results manifest a low speed rotation, which is a perfect use for a sprocket and chain drive compared to other gears. The low speed rotational speed of the system is also affected by its low gear train ratio of 34.328125.

With the increase of rotational motion, increase in voltage and decrease in current occurs based on the experiment. This occurrence is explained by their relationship, the higher the rotational motion, higher voltage and temperature in the generator is produced. The increase of temperature is directly proportional to the increase of internal resistance, which resulted in a lower recorded current. The experiment recorded a peak voltage of 38 V and lowest voltage of 9 V while having the highest current of 1.5 A and lowest recorded current of 1.09 A.

4. Conclusion

The prototype design was found to have an efficiency rating between 12% to 16% efficiency. From the model and parameters set, an expected peak theoretical efficiency of 50.52% was calculated compared to 16.31% efficiency peak from the actual tests. The overall loss in the system might be due to the friction losses from the chain and sprockets and the losses from the generator itself. The generator attached to the gear system demands higher torque for the falling process to begin which requires higher weight loads. During the experimental phase, it can be observed that a higher efficiency can be found with the use of higher ranges of applied torque. In addition, the resulting efficiency graph from the experiments exhibits a trend that describes a fall off after reaching the peak efficiency.

References

[1] Tagulao T C, Manabat R A, Atienza J P, Lucero M L, Maglaya A, Fernando A, Gan Lim L and Vicerra R R 2018 Performance evaluation of a small-scale gravity powered DC generator *IEEE HNICEM 2018*

[2] Gravity Light Foundation. (n.d.). GL02 Technical Specification

[3] Natu P, Nadkar S and Badgujar A 2015 Generation of electricity using gravity *International Journal of Mechanical Engineering and Technology*, 6(7)

[4] Schoonmaker G 2014 DIY gravity powered generator demo

[5] Rogge B 2013 Mini gravity generator running

[6] Labre V, Bouffier J, Cancelier M, Gazay E and Grandin T 2016 Energy production from buoyancy-gravity forces effect on an object in a fluid
[7] Helmholz A C and Moyer B J 1973 MECHANICS (Berkeley physics course, v. 1). United States of America: McGraw-Hill, Inc
[8] Uicker J J, Pennock G R and Shigley J E 2003 *Theory of Machines and Mechanisms* Oxford University Press, New York. Retrieved
[9] Khurmi R S and Gupta J K 2005 Chain drives. In A textbook of machine design
[10] Laibag 2017 Emergency mobile charger using DC motor.
[11] Lindsay L, Gonder W and Montgomery G 2016 Gravity charger *Mechanical Engineering*
[12] MPowerUKHydroelectric power. Battery and energy technologies.
[13] Natu P, Nadkar S and Badgujar A 2015 Generation of electricity using gravity *International Journal of Mechanical Engineering and Technology*
[14] Pedraza J M 2014 ResearchGate Post Forums. Generate electricity using gravitational force.

Eutectic potassium chloride (KCl) solution as phase change material (PCM) for fish cold storage

R K Sukoco[1 *], Y S Indartono[1,2], D Mujahidin[3]

[1] Faculty of Mechanical and Aerospace Engineering, Bandung Institute of Technology, Jl. Ganesha no. 10, Bandung 40132, Indonesia
[2] Research Centre on New and Renewable Energy, Bandung Institute of Technology, Jl. Ganesha no. 10, Bandung, 40132, Indonesia
[3] Faculty of Mathematics and Natural Science, Bandung Institute of Technology, Jl. Ganesha no. 10, Bandung 40132, Indonesia

*kristianto.sukoco@gmail.com

Abstract. Traditional fishermen need to buy ice blocks every time they go fishing, thus impacting their expenses. To potentially save money, an alternative solution using phase change material (PCM) is proposed. PCM contained properly is reusable, thus reducing expenses for fishermen. Eutectic potassium Chloride (KCl) solution is utilized as PCM, which has melting temperature of −10.7 °C and heat of fusion of 253 kJ/kg. The PCM is used to substitute ice as the currently primary source to chill fish in storage. The solution is contained in an aluminium-foil zip-lock bags, each containing 670 grams of solution. The experiment uses two Polystyrene boxes, one containing 10 pouches of eutectic KCl solutions as PCM and the other containing ice cubes with the same mass. Two kg of fish is stored for 24 hours in each box. Thermal analysis shows that the box temperature containing KCl solution is steady for the 24 hours of experiment. Starting to be steady at −2.8 °C from room temperature, the box temperature increases slowly, reaching only 2.1 °C after 24 hours. Organoleptic test shows that the fish stored inside KCl box loses 1.39 quality points on average, compared to 1.89 in ice box after 24 hours. Therefore, the fish in the PCM box is slightly fresher than the one stored using ice cubes.

1. Introduction

Indonesia has a big potential in the fisheries sector. In fact, fisheries product is one of the primary export commodities of Indonesia. Indonesia produce around 11.5 million tons of fisheries commodities in 2016, 6.5 million tons of which are harvested fresh fish. Moreover, in 2017, Indonesia is the 9th largest country in terms of fisheries export value, amounting to $4.2 million of total exports [1].

Indonesia has 2.5 to 3.3 million people employed as fishers, with around additional 1 million work in fisheries processing and marketing. However, around 90-95% of fishers are small-scale fishers, who use fishing vessels with less than 10 GT (gross ton) power [2]. Motorized vessels are only 32% of all fishing vessels, with the rest are outboard-motor boats at 41% and still 27% with no motors.

These vessels used by traditional fishermen therefore have no refrigeration or freezing machines. Commonly, ice blocks, which are usually shattered to small pieces, are used to chill fish until they arrive onshore. Due to ice blocks need to be refilled every time they go fishing, their expenses in ice to chill fish becomes significant. Moreover, for fishermen in remote areas such as Karimunjawa island, oftentimes they need to go outside of the island to buy ice blocks, making their expenses much higher.

Phase Change Materials (PCM) are substances that are utilized to store and release heat using their latent heat transfer at their phase-change stage. Due to PCM's high heat of fusion, it is used widely as heat storage medium. Moreover, its phase-change stage often happens in steady temperature, thus making it able to stabilize temperatures.

The melting point of PCM is crucial point of considerations in choosing PCM and the configuration of PCM inside the cool box. Eutectic potassium chloride solution was used as phase change material for fish storage. Eutectic salt-water solutions are known for their high heat of fusion, subzero melting temperature, the presence of subcooling phenomena, and its problem with phase segregation. However, the segregation could be minimized significantly if the mixture is eutectic [3]. This is rather important to reduce the drop of cold storage performance and the reusability of the PCM as cold storage medium. Potassium Chloride (KCl) solution with 19.5% (wt.%) mixture is a kind of eutectics inorganic PCM. KCl eutectic solution has a melting point of −10.7 °C and high heat of fusion of 253 kJ/kg [4]. Its low melting temperature could be utilized to store perishable foods such as fish. PCMs usually need to be encapsulated or contained properly to prevent contamination to products or the environment. If contained properly, KCl eutectic solution then has the potential to be used as PCM to store fish. Therefore, with its ability to be reused or recharged, KCl eutectic solution could be a solution for fish storage.

2. Related Works

Gin and Farid [5] explored the use of PCM for ice cream and meat storage. They used a vertical freezer with 153-liter volume, controlled at −16 °C, and was equipped with PCM panels on its inner walls. Eutectic ammonium chloride-water solution PCM was utilized to hold the temperature inside the freezer, while the power was turned off for 2 to 3 hours. The freezer equipped with PCM panels had the temperature significantly slower. After 2 hours of power-off, the temperature is still below −10 °C, compared to −2 °C in the original freezer.

Ahmed et. al. [6] explored the area of transportation as the target for PCM utilization. They built two simulators of refrigerated truck trailers using polyurethane, sandwiched to aluminum panels and glass boards. One was equipped with copper wires which was filled with PCM, and the other was not filled with PCM. The temperature was controlled to 4 °C, and the heat flux to the inner walls was measured. They found that the simulator with PCM has 16.3% lower incoming heat flow compared to the other, on 24-hour average. This could be a method to save energy and reduce pollution from refrigerated trucks utilizing fossil fuels.

Johnston et. al. [7] developed an advanced technique to nano-encapsulate PCM for food storage. They use a method and technique to build a nano-structured calcium silicate (NCS) as a dry powder and an organic PCM, with 8 °C melting point. They use this PCM to store asparagus inside a box by cooling the box to 1-2 °C then put it on ambient temperature. The inner temperature of box containing reached 10 °C almost 6 hours longer than without using PCM.

Temperature is one of the most important parameters in keeping fish freshness and safety. Agustini et. al. [8] analyzed the effect of storage temperature for yellowfin tuna *(Thunnus albacares)* to the K-Value of the fish. K-Value is a well-known biochemical method in measuring fish freshness. K value is the ratio of inosine (Ino) and hypoxanthine (Hx) to the quantity of ATP-related compounds such as ATP (adenosine triphosphate), ADP (adenosine diphosphate), AMP (adenosine monophosphate), and IMP (inosine monophosphate) [9]. K-Value of 20% is considered to be *sashimi grade* and 50% is the limit of shelf-life expiration [9-10]. They found that storage temperature correlates strongly to the reaction rate to increase K-value. K-value of 30% was attained after 3.85 days, 7,2 days, and 15.5 days for fish stored at 5 °C, 0 °C, and −3 °C, respectively. Therefore, small difference in temperature can affect the fish quality and shelf-life significantly.

Guizani et. al. [9] conducted similar experiment to Agustini et. al. to measure the effect of temperature to freshness of yellowfin tuna. They also measure histamine content in the fish, a substance that is associated with seafood poisoning. They concluded that storage in 20 °C will result in rapid loss of quality and safety of fish, while storage in 8 °C results in fish shelf-life to be around 6 to 8 days.

Chang et. al. [10] analyzed the effect of temperature on the quality of sea bass, including its K-Value. The temperature ranges from −3 to 10 °C. K-Value of 50% was reached after 2, 3, 8, and 30 days, respectively for 10, 5, 0, and −3 °C. It was concluded that reaching 0 °C or below was crucial for keeping fish fresh and safe. Silva [11] measured the histamine formation in big eye tuna and skipjack at different storage temperature. The tuna reached toxic level of histamine (> 50mg/100g) rapidly, after only 1 day, if stored at room temperature of 22 °C. At refrigerated temperature, it reached toxic level after 3 and 6 days, for 10 and 4 °C, respectively. It also was found that skipjack has higher levels of histamine than big eye tuna at the same storage temperature.

Some institutions give a recommendation for fish storage temperature. FAO [12] gives 0 °C as the ideal storage temperature for chilled fish, although fish can also be stored in frozen temperature below −18 °C with different treatments. Traditional way of storing fish with ice, as FAO puts it, is favorable for both fish freshness and for practical considerations. FDA [13] gives 0 °C as the ideal storage temperature. The fish storage temperature should also not be higher than 4.4 °C (40 °F). FDA warns that higher storage temperature can result in food poisoning due to histamine formation.

3. Materials and Methods

3.1. Fish Samples

Pacific Chub Mackerel *(Scomber japonicus)* fish is used as the specimen for cold storage experiment using PCM. The fish were obtained from local traditional market in Ciroyom, Bandung, West Java in the morning. Fish were randomly separated and put into the box with equal mass.

3.2. Phase Change Materials

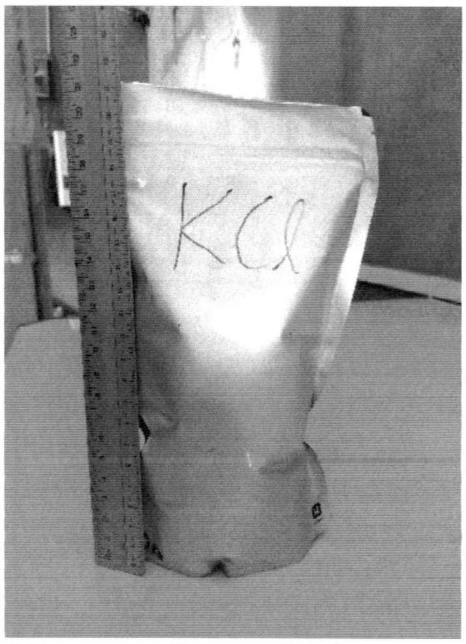

Figure 1. PCM packaging using aluminum foil zip bags

Eutectic potassium chloride solution has melting temperature of −10.7 °C and heat of fusion of 253 kJ/kg [4]. For the use of fish cold storage, the eutectic potassium chloride solution is suitable due to its subzero melting temperature, which compensates the thermal resistance from heat convection from the PCM to the surrounding air. Its high heat of fusion helps with the duration of the passive cold storage. Potassium chloride technical grade (99%) was used, mixed with distilled water with 19.5% (wt.%) mixture, which is its eutectic mixture [4]. The solutions were then contained inside an aluminum-foil zip-lock bags, as shown on Figure 1, where 670 grams of solution was contained in each bag. This packaging has the purpose to prevent contact between fish and the PCM. Moreover, the chosen container was proven to be tough enough to prevent any liquid leakages. Ice cubes, in the form of small cylinders, were obtained from local sellers.

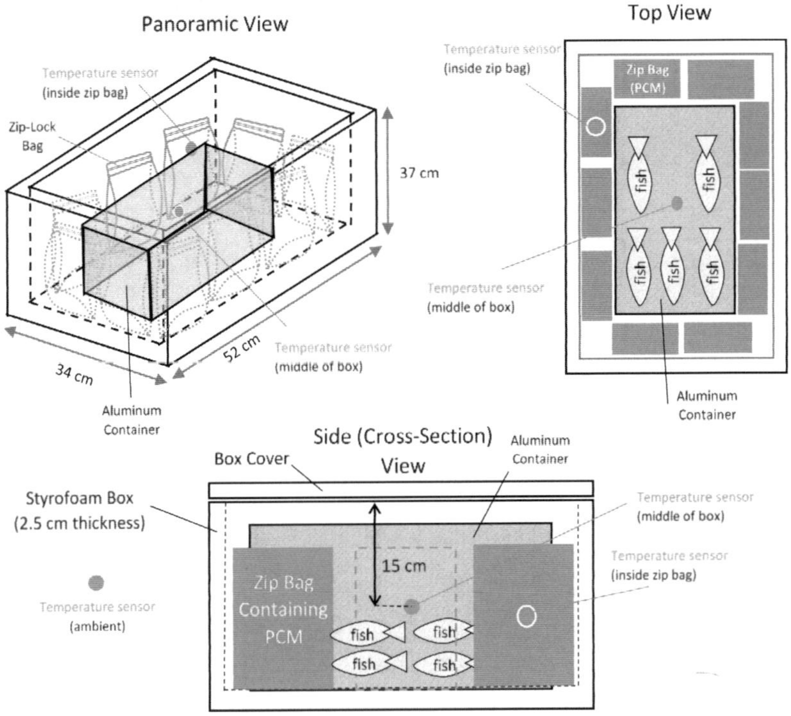

Figure 2. Sketch of fish storage box containing eutectic Potassium Chloride solution inside zip-lock bags.

3.3. *Cold Storage Experimental Procedure*

The objective of this cold storage experiment is to compare the thermal performance and the fish quality between the use of eutectic potassium chloride solution and ice cubes to store fish, which is commonly used by traditional fishermen to keep the fish fresh as it is delivered. To simulate real-life condition, two Polystyrene boxes, which have equal volume of 53 liters and equal dimensions, were used to store fish. One box contained eutectic potassium chloride solutions inside ten (10) pouches, thereby totaling 6.7 kg of PCM inside. The pouches were put around the inner walls of the Polystyrene box. The sketch is shown in Figure 2. The other box contained 6.7 kg of ice cubes, which is equal to the total mass of potassium chloride solutions inside the first box. The ice cubes were configured to be three layers, as shown in Figure 3, each has equal mass.

Each box contained two kilograms of fish inside. In the box with PCM, a thin aluminum container was placed to separate the pouches containing PCM and the fish stored inside. This was done to prevent any contamination to the fish products. The fish in this box were all put into this container forming around 3 to 4 layers. In the box containing ice cubes, the fish were configured into two layers, with each layer of fish was between two layers of ice cubes. The configuration of the Polystyrene box containing ice cubes is shown in Figure 3. Each layer of fish in this box has equal mass of one kilogram.

The potassium chloride solution and ice cubes were compared side-by-side at the same time. They were also taken out of the freezer into their respective boxes at the same time. After the PCM pouches were stored inside the freezer for around three days, then the pouches were put inside the Polystyrene box around its walls, as explained previously. Immediately, the fish were put into the aluminum box. At the same time, after ice cubes were taken out of the freezer, the ice cubes and fish in the other box were configured as explained. The experiment lasted for 24 hours. The duration was chosen in regards from the total duration from the fish harvesting to the fish delivery to market. This includes the fishing process by local fishermen, the duration from their onshore arrival up to delivering fish to the fish collectors, the duration of fish kept by fish collectors, and the duration to deliver fish to the market.

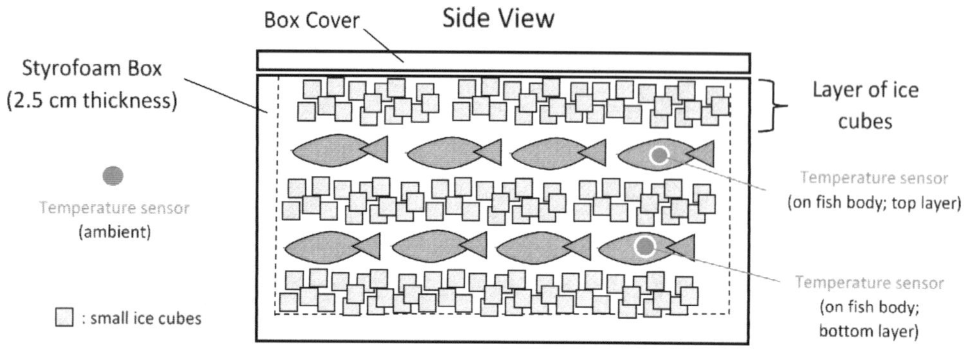

Figure 3. Sketch of fish storage box using ice cubes.

3.4. Temperature Measurement

Temperature inside the box and the PCM temperature itself are the most crucial parameters in evaluating cool box performance using PCM. Therefore, these two temperature points were measured for the box containing eutectic potassium chloride solution. The box temperature was measured at the center (vertically and horizontally) of the box. For PCM temperature, the temperature sensor was placed inside one of the pouches which contains PCM.

Inside the box containing small ice cubes, two temperature points were measured. These two temperature points were fish temperature which contacts directly to the ice cubes. The first and second temperature sensors each measured one fish temperature at the lower and upper fish layers. The temperature sensors were placed on the fish body. Additionally, ambient temperature from the environment nearby the cool box was measured. All these five temperatures were recorded every 5 seconds.

3.5. Organoleptic Test

Organoleptic test is one of the requirements to determine fish freshness, according to Indonesian Standard [10]. The test uses human sensory to measure fish quality. There are in total six parameters of the fish to measure in this test: appearances, where eye, gill, and slime qualities are measured, flesh, odor, and texture. The score for each parameter ranges from 1 to 9, with 1 being the worst and 9 the best. However, in each parameter, there is a guidance or description for the respective parameter. Hopefully, any subjectivity in fish quality measurement could be minimized.

Organoleptic test was done to compare fish quality before and after the experiment. One fish from each box was chosen randomly right before the experiment. Then, three people, with the authors not included, acted as evaluators or panels to measure fish quality for the organoleptic test. The evaluators will then fill an assessment sheet similar to the one in Indonesian standard for fish freshness, which is detailed in SNI 2729:2013 *Ikan Segar* [14] to measure fish quality based on six parameters. When the experiment is done, the fish quality was measured again by the evaluators by filling the same table. Additionally, to keep objectivity in the quality measurement, blind test method was employed in measuring fish freshness. All of the evaluators have no information as to which fish was stored inside the box containing eutectic KCl solutions, and which fish stored with ice cubes. The loss of quality, hence the difference of quality before and after the experiment is the main criteria to measure cold storage performance for fish.

4. Results and discussions

4.1. *Temperature Data*

The temperature data of fish storage experiment is shown in Figure 4. Temperature of the eutectic KCl solution inside the pouch (marked as "KCl Pouch") starts at around −30 °C (not shown) and increasing rapidly due to the sensible heat absorbed by the PCM. After the first 4 hours, then the temperature became steady at around −11.5 °C, which is close to the melting point of the eutectic KCl eutectic solution, which is −10.7 °C [3]. At this point, the eutectic KCl solution begins to utilize its latent heat capability. It slowly melts at its phase-change temperature. The eutectic KCl solution has high latent heat, thus the temperature is able to be steady for around 14 hours. After that, the temperature increases, marking the end of phase-changing stage and the beginning of sensible heat absorption while in the liquid phase.

The box temperature that contains the eutectic KCl solution (marked as "KCl Box" in Figure 4) starts at room temperature and rapidly declines to −2.8 °C due to the presence of PCM, which had low temperature. Then, the temperature starts increasing but rather gradually. In fact, after 24 hours of cold storage experiment, the box temperature only reaches 2.1 °C, which is quite impressive, increasing only around 5 °C during the 24-hour interval. This happens due to eutectic KCl solution's high latent heat and low melting point.

As discussed previously, the PCM temperature is steady after the 4th hour of storage. Therefore, the box temperature becomes even more steady after the 4th hour of storage, due to the utilization of PCM's latent heat storage, hence holding the box temperature as well. The box temperature is also still aligned with some recommendations on chilled fish storage, which is around ice's melting point (0 °C) and below 5 °C [7, 8].

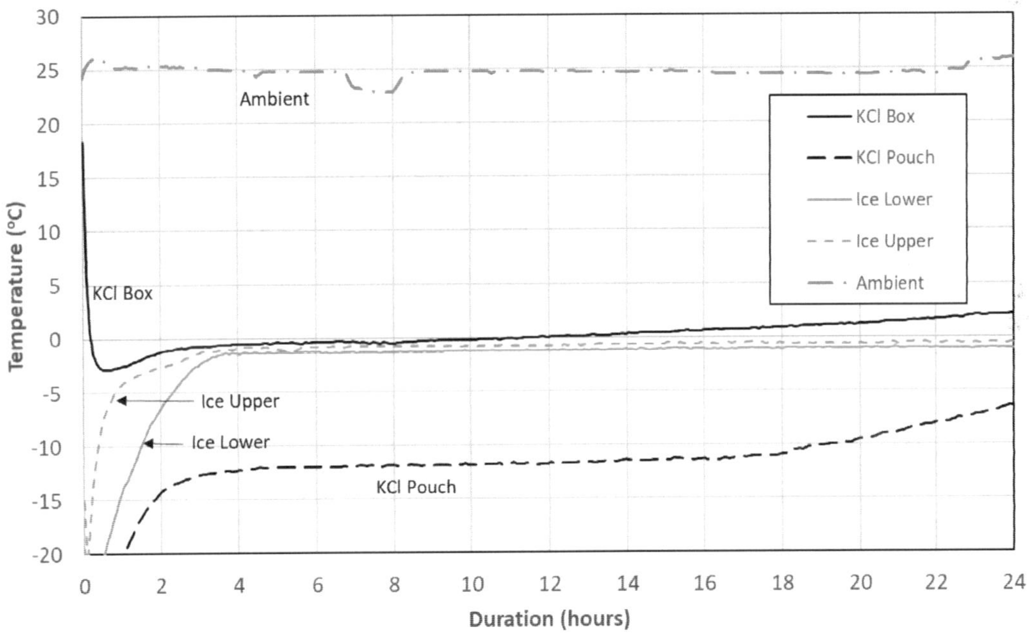

Figure 4. Temperature data from 24 hours of fish storage.

The box temperature is even comparable to fish temperature stored in the ice box. As stated previously, there are two temperatures measured: fish temperature on lower and upper layers. These two temperatures are quite identical. They begin at around − 30 °C (not shown) and increased rapidly, then it starts to be steady after the first 4 hours. As ice melts at 0 °C, the temperatures are steady at around −1 to 0 °C until the end of experiment. Therefore, the box temperature containing eutectic KCl solution, which is −2.8 to 2.1 °C is comparable to storage using ice cubes.

The main advantage of using PCM encapsulated in a container is that PCM as cold storage medium can be recharged, that is, it can be frozen again to be used for the next cooling occasions. On the other side, ice cubes will be melted and turn into water (liquid) after its use. Although the water could be yet frozen again, it doesn't use any packaging, therefore difficult to be reused. The difficulty in obtaining ice cubes or ice blocks and the need to buy them every time fishers go fishing are two of the problems facing fishermen, hence increasing their operational cost. Alternatively, if the ice is contained in the same container as the one used for eutectic KCl solution, then storage temperature will not be as low as if t eutectic KCl solution is used, which could affect fish quality. Therefore, thermally, the use of eutectic KCl solution with zip-locks of aluminum foil material could be an alternative solution to store fish for fishermen and fish cold-chain until it's delivered to the customer.

4.2. Organoleptic Test Result

The result of organoleptic test on fish stored in both boxes is shown in Table 1. The data shown are each averaged from the scores submitted by the 3 evaluators. As shown, on average, there is minuscule difference on the initial condition of fish stored in PCM box (8.72) and the one stored inside ice box (8.89). The parameters mostly have score of 9, and the lowest score is 8.33. The initial condition of both fish is therefore very fresh.

After 24 hours of storage, the fish stored in PCM box loses on average 1.39 points of quality, meanwhile fish stored in ice box lose 1.83. There is a little difference on the loss of average quality, indicating that the fish stored inside the box containing eutectic solution is slightly fresher, but not much, by only 0.44 points.

Table 1. Fish organoleptic test result.

Parameter	Fish in PCM box			Fish in ice box		
	Before	After	Diff.	Before	After	Diff
Eye	8.33	6.33	**2.00**	8.67	6.67	**2.00**
Gill	8.67	7.00	**1.67**	9.00	7.67	**1.33**
Slime	9.00	8.00	**1.00**	9.00	8.00	**1.00**
Flesh	8.67	7.00	**1.67**	8.67	6.00	**2.67**
Odor	8.67	7.67	**1.00**	9.00	8.00	**1.00**
Texture	9.00	8.00	**1.00**	9.00	6.00	**3.00**
Average	**8.72**	**7.33**	**1.39**	**8.89**	**7.06**	**1.83**

From Table 1, on the eye parameter, both fish lose 2 quality points, which is quite significant. However, there is no noticeable difference of the loss between the two fish. Both scored slightly below 7, which could be the eye qualities start to be stale. On the gill parameter, fish stored in PCM box loses 1.67 points, compared to the one stored in ice box with 1.33. The difference is minuscule, as both gills still show freshness. The gill parameter on both fish still score equal to or higher than 7.

Both fish contained in PCM box and ice box have good score in body slime, both still scoring 8 point after 24 hours of storage. The same score is also seen on the fish odor. Both fish have no noticeable difference in fish odor after the storage. From these parameters, the fish are still considered fresh.

Flesh quality score from fish stored in the box containing eutectic solution decreases 1.67 points into 7.00. Therefore, there is a noticeable difference after the storage. The flesh becomes slightly darker and its tissue is less strong. However, a bigger difference in flesh quality is from fish stored in ice box. The flesh quality score decreases 2.67 points into 6.00, even though they start at the same flesh quality.

In the texture parameter, fish stored in the box containing ice cubes undergoes a significant drop, from 9.00 to 6.00, which can be categorized as not fresh anymore. The fish texture becomes a bit tender, although if pushed by finger, the mark remains only for a while. Its initial condition is 9.00, which is dense, compact, and elastic. This is quite different from the fish stored inside PCM box. It only undergoes 1.00 point of loss in texture. The texture is still dense and compact, although not as elastic as the initial texture. This 3.00 points loss on fish inside ice box could happen due to its exposure to water, after some of the ice melts. Initially, when it is tested for organoleptic before experiment, the fish is sliced a bit on its body. Therefore, exposure with water could affect its texture and flesh quality.

In the end, on average from the six quality parameters, the fish stored inside the box containing eutectic solution with the container is a little bit fresher than the one stored inside the box containing ice cubes. Although the difference is not significant, there are some parameters in which the one contained in the first box are significantly better than the one in ice cubes. Looking at the organoleptic result, eutectic potassium chloride solution with proper container is able to keep fish still fresh with the loss of quality is slightly lower than the one stored in ice cubes. The average of quality loses slightly less, the quality in some parameters lose significantly less, and the main advantage of PCM with containers, that is its reusability for the next fish harvests, remains.

4.3. Practicality Aspect

This research gives conclusion that eutectic potassium chloride solution, contained in a zip-lock bags with aluminum foil material, can be an alternative solution for traditional fishermen. They could save more money by reducing their operational cost from buying ice repeatedly and keep their fish quality better. The thermal performance shows that the PCM box could be comparable to using ice to store fish. Moreover, the organoleptic test shows that on average, the fish stored in the PCM box loses less quality compared to the one stored in ice cubes. Moreover, potassium chloride is regarded Generally Recognized as Safe (GRAS) by FDA, and that "...the ingredient is used in food with no limitation other than current good manufacturing process." [15]. Therefore, any leakages will not make fish toxic.

However, the practice of traditional fishing should also be taken into account when considering alternative solutions like this. Shattered ice blocks and ice cubes are commonly used by fishermen to store their fish to keep fish fresh. For smaller-scale fishers, ice cubes are even commonly used. They are used to handle fish and ice in a certain method, which has been done for generations. Handling PCM inside zip-lock bags, however, is quite different. After using ice, the water is then thrown away and they will buy more ice. As opposed to that, if PCM inside pouches to be used, then it has to be stored inside a freezer for several days (depends on the size of freezers) to make it available again. Therefore, the availability of freezers along the cold chain should also be considered, especially for fishermen in remote areas. Moreover, the fishermen handle ice (e.g., for moving ice from one box to another) quite harshly, and there will be an impact load to the ice. Due to its hardness, the ice usually doesn't break, and even if it does, then it still can be used. This kind of method cannot be used to handle PCM pouches. Even if the container used is stronger (e.g., thicker aluminum or steel containers), there could be a leakage on the container and contamination to fish will occur. The PCMs will then be replaced, hence more cost is spent.

Environmental aspect is one concern when utilizing any chemicals for public use or common practices, including fishing sector. The PCM, which is a 19.5% KCl salt-solution, needs to be disposed and replaced with new KCl solution after certain cycles of usage. Although the potassium chloride itself is considered safe to use according to FDA [15], the disposal of the solution in huge amount could be dangerous to the environment. Due to its target use for traditional fishermen and fish collectors, the disposal of KCl solution (after its implementation) will likely occur as part of domestic/household waste. Therefore, a chemical assessment to the KCl solution should be done prior to its implementation to meet the quality standard of domestic wastewater in Indonesia. The quality standard is regulated in a decree by the Ministry of Environment and Forestry of Indonesia [16]. The pH, BOD, COD, TSS, fat/oil content, Ammonia content, the total of coliform, and wastewater debit are regulated and have their own limits according to the decree. Due to its nature as salt solution, the pH will be the parameter that is altered the most from domestic tap water. Nevertheless, a full assessment of the KCl 19.5% solution should be conducted to minimize any risks to the environment.

5. Conclusion

Passive cold storage using eutectic potassium chloride solution as phase change material is therefore a viable alternative solution for short-term fish storage. The thermal performance shows that the PCM in the pouch is able to maintain box temperature below 5 °C, which is the recommended chilled storage temperature for fish, for longer than 24 hours. The box temperature is also steady, starting at −2.8 °C (once the phase-change stage starts) and increasing slowly to 2.1 °C after 24 hours of storage. The organoleptic test also shows that the fish stored using PCM loses quality lower than the one stored in ice cubes. Although the difference on average is small, there are some parameters in which the difference in quality loss between the two is significant. The PCM ability to keep fish freshness is therefore evident.

However, to implement this solution in real-life commercial case has some challenges, especially in traditional fisheries. Traditional fishermen often have their own methods to handle ice cubes as cold storage medium. Therefore, to implement this solution to traditional fisheries, some tests in commercial settings need to be done, and trainings should be conducted to those who will utilize it. The fishermen should be taught how to handle PCM pouches, or any other PCM containers, and the availability of suitable freezers need to be increased along the fish cold chain.

Acknowledgement

We gratefully acknowledge the funding from *Riset Unggulan 2020* from Research Centre on New and Renewable Energy, Bandung Institute of Technology, Indonesia.

References

[1] Indonesia Eximbank Institute and UNIED 2019 *Proyeksi Ekspor Berdasarkan Industri: Komoditas Unggulan* (Jakarta: Indonesia Eximbank) (in Bahasa Indonesia)
[2] California Environmental Associates 2018 *Trends in Marine Resources and Fisheries Management in Indonesia: A 2018 Review* (San Francisco: California Environmental Associates)
[3] Mehling H and Cabeza L F 2008 *Heat and cold storage with PCM: an up to date into basics and applications* ed D Mewes and F Mayinger (Berlin, Heidelberg: Springer-Verlag)
[4] Li G, Hwang Y, Radermacher R, and Chun H 2013 *Energy* **51** 1-17
[5] Gin B and Farid M M 2010 *J. Food Eng.* **100** 372-6
[6] Ahmed M, Meade O, and Media M A 2010 *Energy Conv. Mgmt.* **51** 383-92
[7] Johnston J H, Grindrod J E, Dodds M, and Schimitschek K 2008 *Cur. Appl. Phys.* **8** 508-11
[8] Agustini T W, Suzuki T, Hagiwara T, Ishizaki S, Tanaka M, and Takai R 2001 *Fisheries Sci.* **67** 306-13
[9] Guizani N, Al-Busaidy M A, Al-Belushi I M, Mothershaw A, and Rahman M S 2005 *Food Res. Int.* **38** 215-22
[10] Chang K L B, Chang J, Shiau C, and Pan B S 1998 *J. Agric. Food. Chem.* **46** 682-6
[11] Silva C C G, Ponte D J B D, and Dapkevicius M L N E 1998 *J Food. Sci.* **63** 644-7
[12] Food and Agriculture Organization of The United States (FAO) 2009 *Code of Practice for Fish and Fishery Products* (Rome: Food and Agriculture Organization)
[13] Food and Drug Administration 2019 *Fish and Fishery Products Hazards and Controls Guidance Fourth Edition* (White Oak: Food and Drug Administration)
[14] Badan Standardisasi Nasional 2013 *Ikan Segar: SNI 2729:2013* (Jakarta: Badan Standardisasi Nasional) (in Bahasa Indonesia)
[15] 12 C.F.R (Code of Federal Regulations) part 184.1622
[16] Peraturan Menteri Lingkungan Hidup dan Kehutanan Republik Indonesia Nomor 68 Tahun 2016 *Baku Mutu Air Limbah Domestik* (in Bahasa Indonesia)

Evaluation of filtering facepiece respirators using chemical decontamination methods

Baderin Osman[1], Haalah Mahmud[2], Hussin Mamat[3*]

[1,2]National Institute of Occupational Safety and Health Malaysia, 43650 Bandar Baru Bangi, Selangor, Malaysia
[3]School of Aerospace Engineering, Engineering Campus, Universiti Sains Malaysia, 14300 Nibong Tebal, Penang, Malaysia

hussinm@usm.my

Abstract. Disposable filtering facepiece respirators (FFRs) are often recommended for use by healthcare workers and the general public for infection control. During pandemic, shortage in supply of FFRs is reported due to excessive demand. Hence, suitable decontamination methods of used FFRs are required to be identified. Various chemical decontamination methods (isopropyl alcohol (IPA), vaporized hydrogen peroxide (VHP) and detergent) have been explored in the present study. The performance of the FFRs after exposed to three types of chemical decontamination methods was evaluated based on average aerosol penetration and average filter resistance values. Face mask type N95 and KN95 with four different models were used in the study. Based on the normalized data for filter aerosol penetration of the three decontamination methods, VHP is recommended to be used for decontamination methods for FFRs. Model 3 FFRs showed good performance where less deviation in the normalized filter aerosol penetration and normalized filter air flow resistance for the three decontamination methods was observed.

Keywords: Decontamination, filtering facepiece respirator, N95 respirator, KN95 respirator, filter

1. Introduction

In Malaysia, all employees are covered under the Occupational Safety and Health Act (OSHA 1994) [1] and the Factories and Machinery Act (FMA 1967) [2]. The Department of Occupational Safety and Health (DOSH) in the Ministry of Human Resources (MOHR) implements these acts to ensure the safety and health of the workers. Occupational Safety and Health Act 1994 is a legislative framework which enacted to secure the safety, health, and welfare of persons at workplace, to protect others from safety and health risk in activities pertaining to their workplace and to promote occupational safety and health culture in workplace. Part IV Section 15 in OSHA 1994 stated the responsibility of every employer to ensure safety, health and welfare of the employees at work which include provision of information, instruction, training and supervision. Occupational Safety and Health Act 1994 is not only acted as a legislation; it also acts as a reference for employers to aware and promote the company's safety and health culture.

Occupational safety and health measures wearing suitable FFRs, it is important during epidemics or pandemics. Inadequate supply of FFRs become an issue especially in countries that are not the producer of FFRs. The pandemic of COVID-19 generates an unparalleled demand for N95 respirators in widespread usage [3]. The shortage of medical N95 respirators or other certified masks creates an urgent demand for suitable replacement [4]. The Centre for Disease Control and Prevention has issued guidance on maximizing stock supply by restricting use, reusing patients and providers, and making recommendations on alternative protective equipment [5]. Various decontamination techniques have been investigated to improve the security of FFRs reuse without compromising the capability and structural integrity of protective filtration. Testing of many other variants of N95 masks included in the strategic national inventory showed that they could resist sterilization process by using ultraviolet germicidal irradiation (UVGI), ethylene oxide, or vaporized hydrogen peroxide (VHP) while retaining the adequate protective role [6].

Studies on decontamination techniques of FFRs have gaining attention from previous researchers. Viscusi et al. [7] assessed the filtration performance of N95 FFRs after isopropyl alcohol (IPA) decontamination at 2 different times of submersion (1 second and 1 minute). Performance of FFRs was also evaluated using soap and water at 2 different time intervals, using a poly-dispersed sodium chloride aerosol test method. Changes in physical appearance, odour and laboratory performance of six N95 FFR models after 3 cycles of decontamination using VHP have been studied by Bergman et al. [8]. Carlos et al. [9] reported that UVGI and VHP appear to be the most positive decontamination methods for N95 FFRs, based on their biocidal efficacy, filtration performance, fitting characteristics, and residual chemical toxicity, as well as other practical aspects such as the equipment required for their implementation and the maximum number of decontamination cycles. In a recent study by Rafael et al. [10], they concluded that VHP and ultraviolet irradiation emerged to be the most favourable methods for N95 respirator decontamination.

Many researchers have extensively studied various methods of FFRs decontaminant in recent years. The pandemic of COVID-19 generates a great demand for the FFRs, thus suitable decontamination method is important to be identified. There is little attention has been paid to compare the performance of chemical decontamination method using FFRs produced by different manufacturers. This study is carried out to evaluate the effectiveness of three chemical decontamination methods; IPA, VHP and detergent. The performance of the FFRs after decontamination was measured by aerosol penetration and average filter resistance. Face mask type N95 and KN95 with four different models were used in the study.

2. Material and Methods

2.1. *Respirator Selection*

Four respirator models were used in this study, of which we referred as Model 1, Model 2, Model 3 and Model 4. Model 1 and Model 2 FFRs are approved by the USA-NIOSH while Model 3 and Model 4 FFRs are approved by the China-NIOSH. Figure 1 shows the four FFRs models used in the analysis. FFRs Models 1 and 2 are N95, and Models 3 and 4 are KN95. Table 1 shows description of the Models 1, to 4 used in the study. All FFRs consist of 3 layers with thicknesses ranging from 1.11 to 2.90 mm.

Table 1. Description of the FFRs Models 1- 4

Model	Manufacturer	Thickness (mm)	Layers	Filter	Shape
1	Singapore	2.90	3	Melt blown	Fold valve less
2	China	2.47	3	Melt blown	Fold valve less
3	China	1.11	3	Melt blown	Fold valve less
4	China	1.83	3	Melt blown	Fold valve less

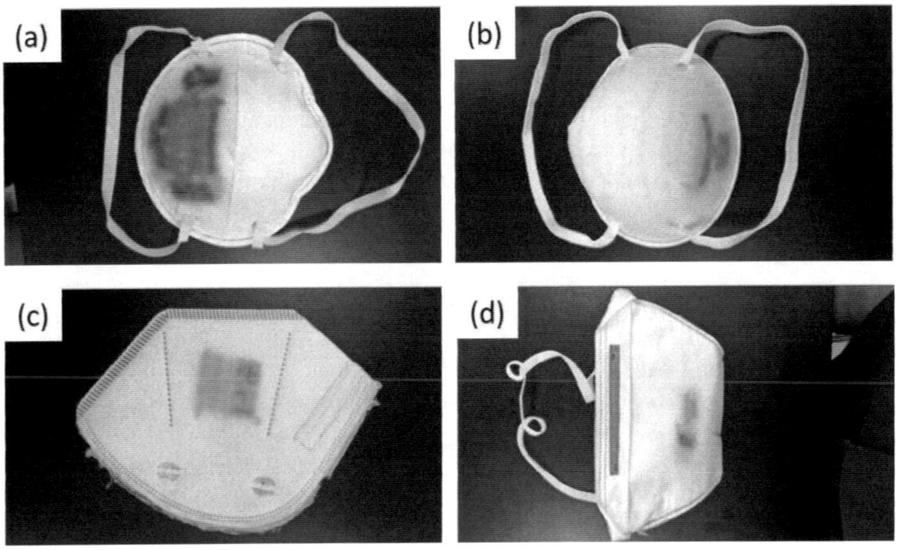

Figure 1. Four FFRs models considered in the experimental; (a) Model 1, (b) Model 2, (c) Model 3 and (d) Model 4

2.2. *Decontamination Method Selection*

Three chemical decontamination methods were used in the study; Isopropyl Alcohol (IPA), vaporized hydrogen peroxide (VHP) and detergent. Table 2 summarizes the experimental conditions and parameters for the three decontamination methods. All laboratory experiments were conducted under standard laboratory conditions (22 ± 2 °C and relative humidity of 50 ± 10%) on triplicate sets of FFRs. The test conditions of the three decontamination methods are selected based on previous study reported by Viscussi et al. [7]. The conditions were generally recommended by the literature and it was reported as effective conditions for the three decontamination techniques.

Table 2. Conditions of the three decontamination methods used in the study

Decontamination method	Recommended concentration /property	Recommended effective contact time	1st test condition Less aggressive	2nd test condition More aggressive
IPA	70%	10 to 30 min.	Dunk 1 second	Dunk 2 min.
VHP	58%	28 min.	1 cycle/36 min.	5 cycle/180 min.
Detergent	1 g/liter	2 min.	2 min.	20 min.

Control or 'as-received' sample refers to the untreated out-of-the-box FFRs for the four FFRs models used in the analysis. Liquid chemical treatment solutions were made by appropriate dilutions with tap water and reagents: IPA, 70%; Breeze detergent, 1g/L and diluted in tap water. FFRs samples were placed (submerged) into a pail containing 1 liters of treatment solution. FFRs were removed, hung on and air-dried for 72 hours, prior to filter penetration testing. For VHP, all samples were put in to the Plasmapp Sterlink Low Temperature Plasma Sterilizer for 2 stage cycle (1 cycle and 5 cycles). The respirators were air-dried 72 hours before filtering.

2.3. *Experimental Design*

A series of experiments were conducted on two N95 and two KN 95 FFRs model using the three decontamination methods described in Table 2. Two different conditions were considered in the study, which are chemical concentrations and time lengths. Control sample refers to the untreated out-of-the-box FFRs or no decontamination. At least three samples were used in airflow resistance and filter aerosol penetration tests, and the average was recorded.

The TSI 8130 instrument available at the National Institute of Occupational Safety and Health (NIOSH) Malaysia was used to identify the performance of the FFRs after decontamination process. The 72-hour air drying time was performed after decontamination to make sure the sample totally dry before penetration test. A visual inspection was done after decontamination to detect any noticeable changes in the respirator (e.g., straps, nose clips, exhalation valves, etc.), in addition to the filtering results.

2.4. Characterization

TSI Model 8130 Automated Filter Tester (AFT) (TSI, Inc., St. Paul, MN) was used to measure filter aerosol penetration and airflow resistance. The TSI 8130 delivers a solid polydisperse sodium chloride (NaCl) aerosol which meets the particle size and size distribution criteria set forth in 42 CFR 84 Subpart K, Section 84.181.(g) for NIOSH certification (CFR, 1995). The TSI Model 8130 AFT produces a particle concentration of 12 - 20 mg/m^3 and generates an initial instantaneous filter penetration result. The NaCl aerosol must have a count median diameter (CMD) of 0.075 ± 0.020 micrometer and a geometric standard deviation (GSD) not exceeding 1.86.

3. Results and Discussion

Table 3 shows the average aerosol penetration results. For comparison purpose, normalized data for all the decontamination methods for Model 1 to Model 4 are presented in Figure 2 (a-c). The normalized data was calculated based on the average filter aerosol penetration performance of each of the FFRs divided with that of control FFRs sample. Comparison on the control samples showed that Model 1 exhibited the lowest average filter aerosol penetration value than those other FFRs. Low average filter aerosol penetration value is required based on 95% filtration efficiency for N95 and KN95 FFRs. It is interesting to note that the thickness of the FFRs does not influence the filter aerosol penetration performance. This is proven by the inconsistency data shown by the FFRs models. Based on the data for IPA decontamination method, FFRs Model 4 showed good properties with the lowest normalized aerosol penetration values while Models 1, 2 and 3 showed higher values. This indicated FFRs Model 4 exhibited less deviation from the control FFRs. As expected, increase in soaking time results in increasing average filter aerosol penetration due to the filter structure degradation of the FFRs.

For VHP decontamination method, Model 2 and Model 3 showed the lowest normalized aerosol penetration values after sterilized for 1 cycle in 58% concentration. This indicated that less degradation of the filter structures was occurred in these two models compared to those Models 1 and 4. FFRs Model 3 which was decontaminated with detergent showed the lowest normalized aerosol penetration for test condition of 2 minutes, and the value increased for the 20 minutes test condition.

The increased of the normalized data for filter aerosol penetration trends shown in Figure 2 (a-c) was possibly due to the changes in the density and/or spatial distribution of the electret charges on the surface of the polymer fibres by the liquid phase application [11]. Based on the normalized data for the three decontamination methods, VHP is the most promising method since the normalized values observed were less than 5.2 compared to those 2 methods.

Table 3. Average filter aerosol penetration results of four types of face masks using IPA, VHP and detergent methods (note: in the bracket is the normalize value).

Decontamination method	Test condition	Type of Face Masks			
		Model 1	Model 2	Model 3	Model 4
IPA	Control	0.181 (1)	0.936 (1)	0.752 (1)	0.553 (1)
	Dunk 1 Sec	7.92 (43.7)	20.0 (21.4)	4.37 (5.8)	2.10 (3.8)
	Dunk 1 min.	8.91 (49.2)	22.1 (23.6)	12.4 (16.5)	4.24 (7.7)
VHP	Control	0.181 (1)	0.936 (1)	0.752 (1)	0.553 (1)
	1 Cycle	0.437 (2.4)	1.08 (1.2)	0.874 (1.2)	2.90 (5.2)
	5 Cycle	0.401(2.2)	1.92 (2.1)	1.56 (2.1)	1.63 (2.9)
Detergent	Control	0.181 (1)	0.936 (1)	0.752 (1)	0.553 (1)
	2 min.	7.35 (40.6)	35.6 (38)	19.6 (26.1)	24.3 (43.9)
	20 min.	41.6 (229.8)	41.4 (44.2)	29.8 (39.6)	26.5 (47.9)

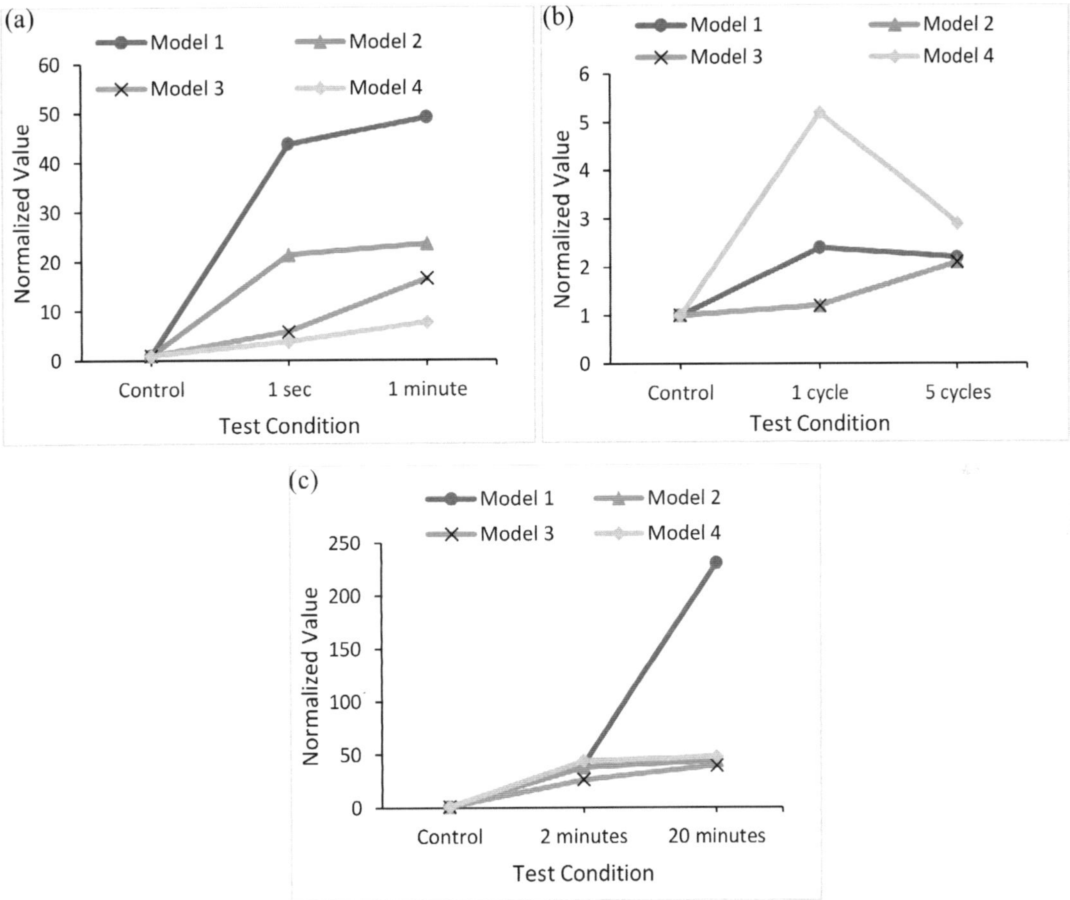

Figure 2. Normalized data of average filter aerosol penetration using decontamination (a) IPA, (b) VHP and (c) Detergent

Table 4 shows the average filter air flow resistance results and Figure 3 (a-c) shows normalized data for all the decontamination methods for Models 1 to 4. Results shown by the control samples indicated that Model 1 exhibited the lowest average filter air flow resistance value than those other FFRs. Low average filter air flow resistance value is desirable and indicates better performance of the FFRs. Based on the data in Table 4, it is found that insignificant changes were observed for the airflow resistance values using the three decontamination methods. Data shown by the normalized values in Figure 3 (a-c) indicated the small changes in the filter resistance performance.

Table 4. Average filter airflow resistance results of four types of face masks using IPA, VHP and detergent methods (note: in the bracket is the normalize value)

Decontamination method	Test condition	Type of Face Masks			
		Model 1 (mmH_2O)	Model 2 (mmH_2O)	Model 3 (mmH_2O)	Model 4 (mmH_2O)
IPA	Control	8.7 (1)	13.3 (1)	12.6 (1)	10.0 (1)
	Dunk 1 Sec	9.9 (1.14)	13.4 (1.01)	12.2 (0.97)	10.1 (1.01)
	Dunk 1 min.	9.0 (1.03)	12.1 (0.91)	12.5 (0.99)	10.3 (1.03)
VHP	Control	8.7 (1)	13.3 (1)	12.6 (1)	10.0 (1)
	1 Cycle	9.5 (1.09)	14.7 (1.11)	12.9 (1.02)	8.3 (0.83)
	5 Cycle	8.5 (0.98)	11.7 (0.88)	11.8 (0.94)	11.0 (1.1)
Detergent	Control	8.7 (1)	13.3 (1)	12.6 (1)	10.0 (1)
	2 min.	9.7 (1.11)	12.7 (0.95)	7.6 (0.60)	10.3 (1.03)
	20 min.	8.4 (0.97)	14.3 (1.08)	8.6 (0.68)	9.0 (0.9)

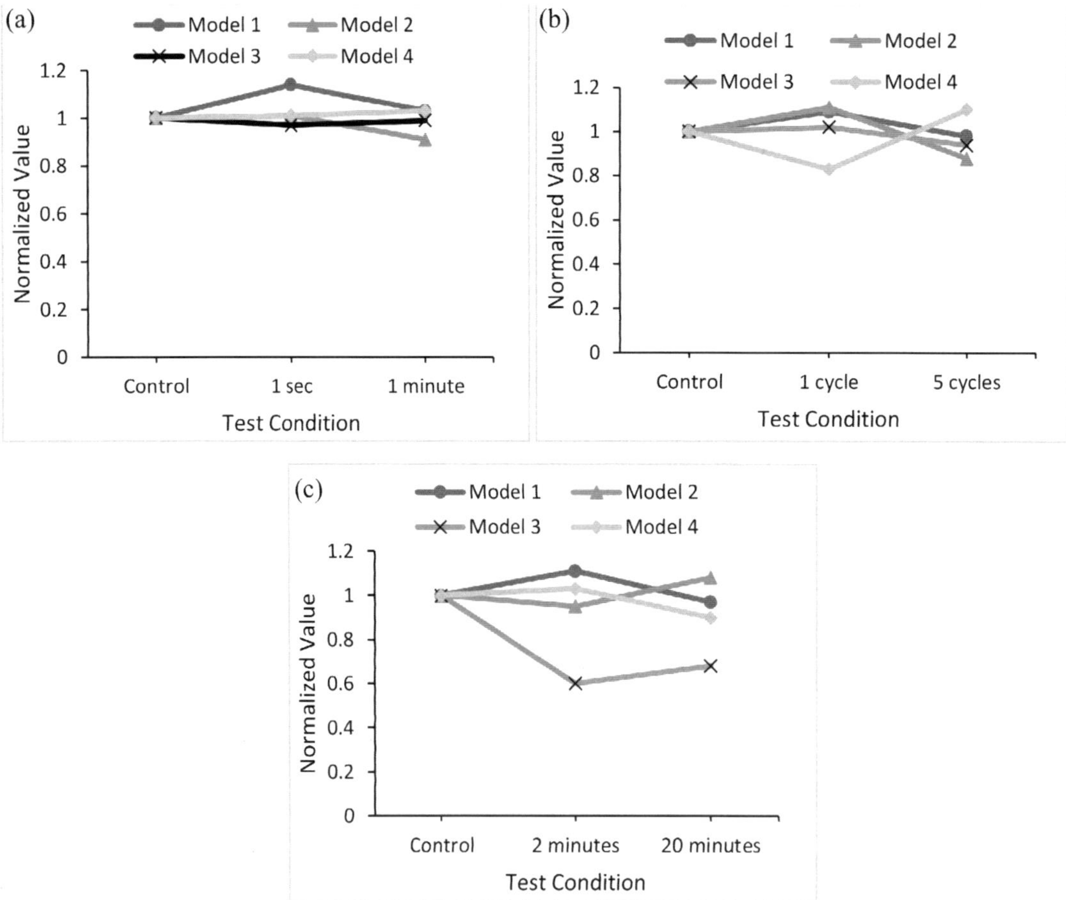

Figure 3. Normalized data of average airflow resistance using decontamination (a) IPA, (b) VHP and (c) Detergent

4. Conclusions

Based on the results of chemical decontamination methods used for four types of FFRs, the following conclusions were made;

(a) Comparison on the control samples, Model 1 exhibited the lowest average filter aerosol penetration and average filter air flow resistance than those other FFRs.
(b) Based on the normalized data for filter aerosol penetration of the three decontamination methods, VHP is recommended to be used for decontamination methods for FFRs.
(c) Model 3 FFRs showed less deviation in the normalized filter aerosol penetration and normalized filter air flow resistance for the three decontamination methods.

Acknowledgement

The authors would like to acknowledge contribution received from NIOSH Malaysia and Universiti Sains Malaysia (USM). The authors acknowledge the financial support from the USM through the Short Term Grant (Grant no. 6315193).

References

[1] 1994 Occupational safety and health https://www.dosh.gov.my/index.php/legislation/acts legislation/23-02-occupational-safety-and-health-act-1994-act-514/file. Accessed August 31, 2020
[2] Factories and machinery act 1967 https://www.ilo.org/dyn/natlex/docs/ELECTRONIC/14984/99465/F1817223295/MYS149 84%202016.pdf. Accessed August 31, 2020
[3] Nogee D and Tomassoni AJ 2020 Covid-19 and the N95 respirator shortage: Closing the gap *Infection Control & Hospital Epidemiology* https://doi.org/ 10.1017/ice.2020.124
[4] Katherine S, et al. 2020 An accessible method for screening aerosol filtration identifies poor performing commercial masks and respirators *Journal of Exposure Science & Environmental Epidemiology* https://doi.org/10.1038/s41370-020-0258-7
[5] Centers for Disease and Control Prevention Strategies for optimizing the supply of N95 respirators https://www.cdc.gov/coronavirus/2019-ncov/hcp/respirators-strategy/index.html. Published 2020. Accessed August 26, 2020
[6] Viscusi D J, Bergman M S, Eimer B C and Shaffer R E 2009 Evaluation of five decontamination methods for filtering facepiece respirators. *Ann Occup Hyg* **53** 815–27
[7] Viscusi D J, King W P and Shaffer R E 2007 Effect of decontamination on the filtration efficiency of two filtering facepiece respirator models *Journal of the International Society for Respiratory Protection* **24** 93-107
[8] Bergman M S, Viscusi D J, Heimbuch B K, Wander J D and Sambol A R 2010 Evaluation of multiple (3-Cycle) decontamination processing for filtering facepiece respirators *J Eng Fiber Fabr.* **5** 33–41.
[9] Carlos E R M, Monica P S B and Jorge A C 2020 Decontamination and reuse of N95 filtering facemask respirators: A systematic review of the literature *American Journal of Infection Control* **48** 1520-32
[10] Rafael S O et al. 2020 Decontamination of N95 respirators against SARS-CoV-2: A scoping review. Journal of Dentistry **104** 1-24
[11] Jasper W, Hinestroza J, Mohan A, Thompson D and Barker R 2005 Effect of phase of toluence on filtration performance of electret filter media against di-octyl-phthalate aerosols *Journal of the International Society for Respiratory Protection* **22** 97-105

Fuzzy-based automated nutrient solution control for a hydroponic tower system

J A Dela Vega[1], **J A Gonzaga**[1], **L A Gan Lim**[1]

[1] Mechanical Engineering Department, De La Salle University, Manila, Philippines

Email: john_christopher_delavega@dlsu.edu.ph

Abstract. Hydroponic farming is gaining significant interests in the field of agriculture due to its many advantages. Its design is modular thereby allowing it to be installed in limited spaces and urban applications. Also, the trend of controlled environment agriculture (CEA) is now prevalent in the study of hydroponics. This ensures that hydroponic setups are invulnerable to plant diseases, infestations, and variable weather systems. One of the challenges in its operation, however, is in the efficient and accurate maintenance of the nutrient solution. Sudden changes in the growth parameters such as ambient temperature, relative humidity, electrical conductivity (EC), and pH level negatively impact the growth of hydroponic plants. Several studies have proposed systems which can control the distribution of the nutrient solution and maintain the optimal levels of growth parameters. This paper discusses a proposed control framework for nutrient solution management in a hydroponic system. Moreover, this system is based on fuzzy logic and designed for the adjustment of the pH level and electrical conductivity (EC) of the mixture. Additionally, the complexities in the process and maintenance of the system were made simple to facilitate easy adaptation and familiarization that even non-educated growers can easily acclimate to the proposed system. Also, dedicated sensor modules were used for monitoring the growth environment. Four treatment solutions were administered using dosing pumps to correct parameter values that went beyond the desired setpoint ranges. This system is applied on a nutrient film technique (NFT) hydroponic tower farm to assess its performance on an actual physical setting.

1. Introduction

Hydroponics pertains to the technique developed by William Frederick Gericke in which plants are grown using nutrient solution instead of soil [1]. Given that its design can be scaled into vertical farms, it can produce greater yield per area as compared with horizontal farms [2]. However, the main challenge in its implementation is in the control of the nutrient solution. Regular monitoring of the parameters is necessary since sudden increase or drop in the pH value and electrical conductivity (EC) have adverse effects on plant health [3].

Several studies proposed methods on how to regulate the nutrient solution. Eridani et al. designed a prototype of an automated nutrient feeding system [4]. The system was able to detect changes in the parameter values and can correct the nutrient concentration back to normal. A similar approach was conducted by Sihombing et al. wherein an automated irrigation for a hydroponic system was developed [5]. Artificial intelligence, particularly fuzzy logic, was also involved in some works which attempted to control the nutrient solution. For example, Yolanda et al. proposed a nutrient adjustment system using fuzzy inference in which the system architecture is based on wireless sensor network [6]. Also,

Fuangthong and Pramokchon involved fuzzy logic in the decision-making process in nutrient solution adjustment [7]. The study also used HSV histogram to generate setpoint values for the sensors. However, some of these studies used experimental setups that were not entirely automated or intelligent. Moreover, in order to determine whether the control systems can manage the nutrient distribution system, there is a need for it to be applied in an actual physical farm. Lastly, the use of multiple stock solutions and treatment agents introduces further complexities in the control process.

This study, on the other hand, proposes an automated nutrient solution control based on fuzzy inference systems. This system is designed to manage the deployment of solution for a nutrient film technique (NFT) hydroponic tower growing cos lettuce. In this study, the complex operation and maintenance of the system were simplified so that even non-educated farmers can easily adapt to the proposed system which is user friendly. Moreover, the rest of the paper proceeds as follows. The related works are discussed in the next section. This is followed by the theoretical framework, proposed experimental setup and lastly, the conclusion.

2. Related Works

A hydroponic tower was developed by Tagle et al. for urban farming application [8]. It was installed with various sensor modules to monitor the growth parameters such as water temperature and level, ambient temperature, relative humidity, total dissolved solids, and light intensity. A subsequent study has seen the inclusion of an automated data acquisition system (DAQ) [9]. The DAQ program was written using Arduino software and was designed to gather and store data from 6 hydroponic parameters.

Another study used wireless sensor networks to detect the parameters in a greenhouse type hydroponic farm [10]. Moreover, the data from the sensor network was sent to cloud storage. Remote farm maintenance was also possible through the use of smartphones. In addition, a system was designed to cultivate plants in a hydroponic environment with minimal human intervention [11]. It also used the concept of Internet of Things to allow remote monitoring and control.

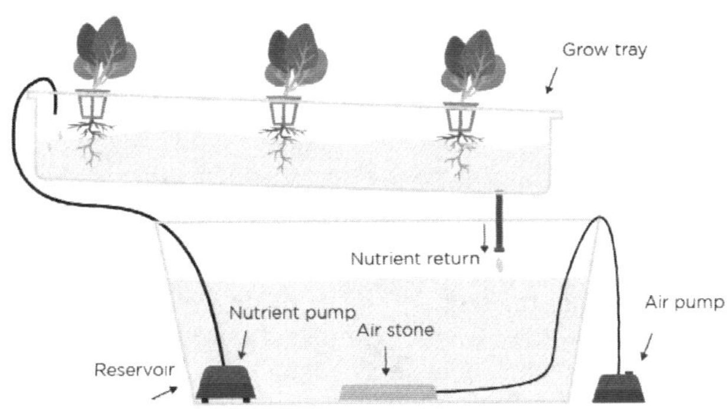

Figure 1. The basic setup of an NFT hydroponic system [12].

3. Theory

3.1. *The irrigation system*

Irrigation systems, often called as hydroponic setup, refer to the way the nutrient solution is administered to the crops. Currently, there are 6 general setups in hydroponic farming, namely: ebb and flow, deep water culture, wick system, drip system, aeroponics, and nutrient film technique.

Ebb and flow system and deep water culture are popular among home growers for its efficient use of resources and low start-up cost. The main difference between the two, however, is in its operation. While ebb and flow setup involve the 'flooding' and 'draining' of the plant roots, the deep water culture allows the roots to be fully submerged in the nutrient solution at all times. Wick system, on the other hand, is a passive system for it doesn't require the use of pumps to administer the nutrient solution. This setup uses the principle of capillary action to soak up the solution, which in turn dampens the growing medium where the plant roots are located. Moreover, the drip system administers the nutrients into the growing tray in slow, regular trickles and is ideal for plants with large roots. Meanwhile, aeroponics is a unique type of hydroponic setup for it doesn't require growing media. Plants are thereby cultivated in a mist-like environment where nozzles spray droplets of nutrient solution towards the exposed roots.

The nutrient film technique (NFT) is one of the frequently employed irrigation setups for growing leaf plants in hydroponics. This method is best suited for small, quick-growing plants. Since the hydroponic plant used in this study is lettuce – a fast-growing, vegetative plant, the NFT system is the most ideal setup.

Furthermore, its operation involves the flowing of a shallow stream of nutrient solution in growing channels where the roots of plants are exposed. As depicted in Figure 1, the nutrient solution is typically stored in a mixing tank just below the channels. A water pump drives the mixture from the tank to the top side of the NFT channel. Moreover, the flowing solution creates a shallow film or stream of fluid within reach of the plant roots exposed in the planting media. The plant roots absorb the nutrition from this continuous flow of nutrient-enriched water. On the other hand, the leftover solution is drained back to the tank to be recirculated again by the pump.

Table 1. The optimum range of pH and electrical conductivity (EC) for some hydroponic crops [13].

Plants	**Electrical Conductivity** [in micro Siemens per cm (mS/cm)]	**pH**
Cabbage	2.5 – 3.0	6.5 – 7.0
Celery	1.8 – 2.4	6.5
Cucumber	1.7 – 2.0	5.0 – 5.5
Eggplant	2.5 – 3.5	6
Lettuce	1.2 – 1.8	6.0 – 7.0

3.2. Control framework

The challenge in the maintenance of the nutrient solution lies in the accurate control of the electrical conductivity (EC) and the pH level, particularly that pH level frequently changes due to the ion absorption of plant roots [14] and EC is easily affected by dissolved solids in the water. This usually creates complexity and fuzziness in the control process. Conventional control can be employed yet it will encounter issues in dealing with fuzziness, while advanced controls such as PIDs involve a more complex control system [15].

Fuzzy logic is one ideal method in solving this problem. While traditional control can only represent true and false value, fuzzy logic handles the concept of partially acceptable truths. This is vital in designing the control parameters for the EC and pH values.

In addition, different plants grow best at different pH and EC values that are suitable for their development. As shown in Table 1, there are various recommended ranges of electrical conductivity and pH for hydroponic plants. These ranges are used as setpoint values for the control system.

The proposed approach can be visualized as a closed-loop system, as depicted in Figure 2. The setpoint values were the input or reference for the fuzzy-based control system and were the standard from which the actual or real values read by the sensors were compared. This comparison from actual to setpoint will generate an error value. The error value is directly proportional to the runtime of treatment pumps for the nutrient control system. Hence, the higher the error value, the longer the pumping time and vice versa. The actual values were read again by the sensors installed in the NFT hydroponic tower and fed back to the fuzzy inference system. This loop continues until the error value is within the desired range.

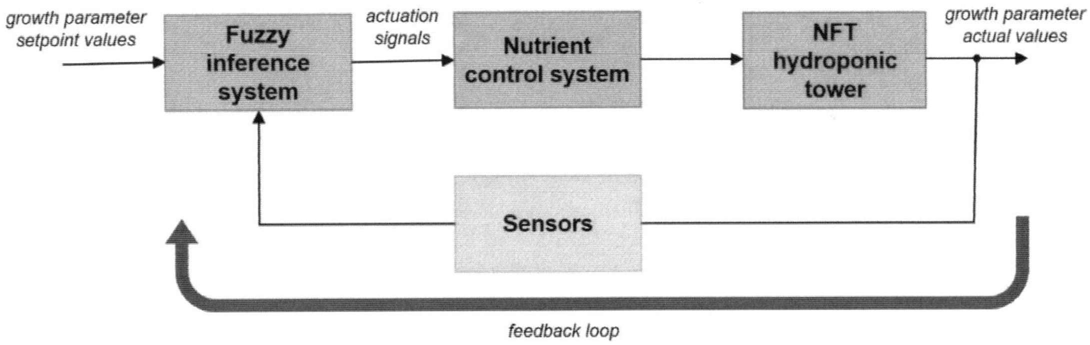

Figure 2. The fuzzy-based framework as a closed loop system.

4. Experimental setup

Shown in Figure 3 is the diagram of the proposed nutrient solution control system. The microcontroller processes the runtime of the dosing pumps for the treatment solutions as well as the main pump to recirculate the nutrient solution throughout the growing channels of the NFT hydroponic tower. The sensors were used to read and feed data into the control system. The EC and pH sensor were installed in the mixing tank. A temperature sensor, light sensor, and relative humidity (RH) sensor also monitored the conditions in the NFT tower. The data gathered by these sensors were sent into data storage.

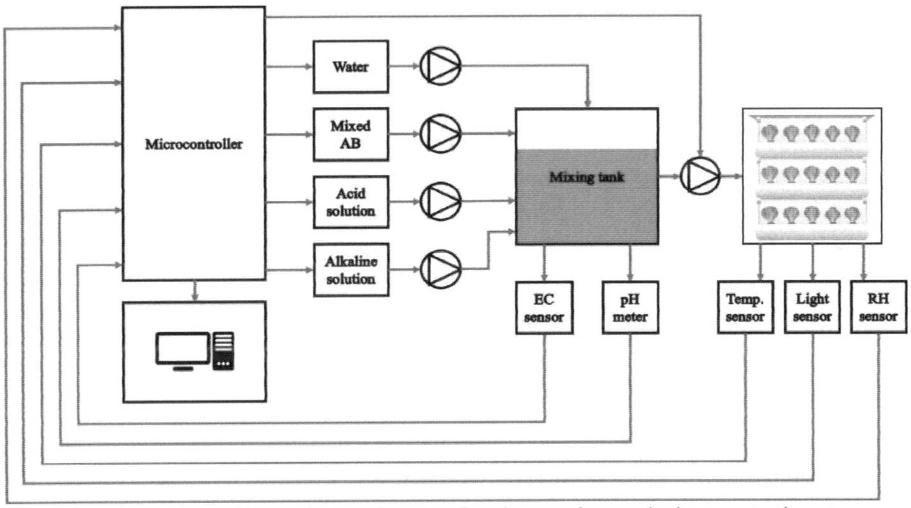

Figure 3. Proposed experimental setup for the nutrient solution control system.

4.1. *Treatment solutions*

In order to regulate the nutrient solution parameters within the setpoint ranges, four (4) treatment solutions were used, namely: water, AB mix solution, acid solution, and alkaline solution. Depending upon which parameter is in need of correction, these solutions were administered via dosing pumps into the mixing tank.

4.2. *Sensors*

For sensing the pH level, pH meter Pro Analog Sensor was used. It has a measuring range of 0 – 14 pH with an accuracy of ± 0.1 pH at standard temperature and pressure. In addition, it can measure the pH value of fluids within 0 – 60 °C temperature range in under 1-minute response time. On the other hand, the electrical conductivity (EC) was measured using DFRobot Gravity Analog Electrical Conductivity Meter. It can detect EC within the range of 0 – 20 mS/cm at 0 – 60 °C temperature range with ± 5% F.S. measurement accuracy.

A DHT11 sensor was used to detect the temperature and humidity in the NFT hydroponic tower. The module can read relative humidity from 20 – 80% range at 5% accuracy. Meanwhile, the support temperature range is from 0 – 50 °C with ± 2 °C accuracy. Other supplementary sensors for light intensity, water level, and water temperature were interfaced using Analog and Digital Output Light Sensor, HC-SR04 Ultrasonic Sensor, and DS18B20 Water Temperature Sensor, respectively.

4.3. Development board

The development board used in this study is Arduino Mega 2560. Its microcontroller is ATmega2560. It has 4 serial ports and 16 pins designed for analog input. Also, it holds 54 digital I/O pins; 15 of which provide pulse width modulation (PWM) function.

The program for the fuzzy inference system was written using Arduino Integrated Development Environment (IDE). The latest version of which can be downloaded in its dedicated website [16].

5. Conclusion

A proposed automated control system for nutrient solution of a hydroponic farm was discussed in this paper. Its framework is grounded on fuzzy logic to manage the adjustment of parameters of nutrient solution in a hydroponic tower system. Dedicated sensor modules were used to detect changes in the pH level and electrical conductivity of the solution. Moreover, 4 treatment solutions were used to correct the solution parameters back to ideal should its values go out of the desired ranges. This system was implemented in an actual NFT hydroponic tower to assess its performance.

Acknowledgement

The authors would like to acknowledge the support of the Department of Science and Technology (DOST) through the Engineering Research and Development for Technology (ERDT) program.

References

[1] Gericke W 1937 Hydroponics--crop production in liquid culture media *Science* **85** (2198) 177-8.

[2] Touliatos D, Dodd I and McAinsh M 2016 Vertical farming increases lettuce yield per unit area compared to conventional horizontal hydroponics *Food and Energy Security* **5** (3) 184-91.

[3] Wortman S 2015 Crop physiological response to nutrient solution electrical conductivity and pH in an ebb-and-flow hydroponic system *Scientia Horticulturae* **194** 34-42.

[4] Eridani D, Wardhani O and Widianto E 2017 Designing and implementing the arduino-based nutrition feeding automation system of a prototype scaled nutrient film technique (NFT) hydroponics using total dissolved solids (TDS) sensor *2017 4th Int. Conf. on Information Technology, Computer, and Electrical Engineering (ICITACEE).*

[5] Sihombing P, Karina N, Tarigan J and Syarif M 2018 Automated hydroponics nutrition plants systems using arduino uno microcontroller based on android *J. Phys.: Conf. Ser.* **978** 012014.

[6] Yolanda D, Hindersah H, Hadiatna F and Triawan M 2016 Implementation of real-time fuzzy logic control for nft-based hydroponic system on internet of things environment *2016 6th Int. Conf. on System Engineering and Technology (ICSET)*.

[7] Fuangthong M and Pramokchon P 2018 Automatic control of electrical conductivity and PH using fuzzy logic for hydroponics system *2018 Int. Conf. on Digital Arts, Media and Technology (ICDAMT).*

[8] Tagle S, Benoza H, Pena R and Oblea A 2018 Development of an indoor hydroponic tower for urban farming *Proc. of the 6th DLSU Int. Conf. on Innovation and Technology Fair.*

[9] Tagle S, Benoza H, Pena R, Oblea A, Ledesma N, Gonzaga J and Gan Lim L 2018 Development of an automated data acquisition system for hydroponic farming *2018 IEEE 10th Int. Conf. on Humanoid, Nanotechnology, Information Technology,Communication and Control, Environment and Management (HNICEM)*.
[10] Boonnam N, Pitakphongmetha J, Kajornkasirat S, Horanont T, Somkiadcharoen D and Prapakornpilai J 2017 Optimal plant growth in smart farm hydroponics system using the integration of wireless sensor networks into internet of things *Advances in Science, Technology and Engineering Systems Journal* **2** (3) 1006-12.
[11] Palande V, Zaheer A and George K 2018 Fully automated hydroponic system for indoor plant growth *Procedia Computer Science* **129** 482-8.
[12] What is the Nutrient Film Technique - NFT? How does it work? [Internet]. Trees.com. 2018 [cited 2 September 2020]. Available from: https://www.trees.com/nutrient-film-technique
[13] Singh H, Dunn B. Electrical Conductivity and pH guide for hydroponics [Internet]. Oklahoma State University Extension. 2017 [cited 2 September 2020]. Available from: https://extension.okstate.edu/fact-sheets/electrical-conductivity-and-ph-guide-for-hydroponics.html.
[14] Suhardiyanto H, Seminar K, Chadirin Y and Setiawan B 2001 Development of a ph control system for nutrient solution in ebb and flow hydroponic culture based on fuzzy logic *IFAC Proceedings Volumes* **34**(11) 87-90.
[15] Fakhrurroja H, Mardhotillah S, Mahendra O, Munandar A, Rizqyawan M and Pratama R 2019 Automatic pH and humidity control system for hydroponics using fuzzy logic *2019 Int. Conf. On Computer, Control, Informatics And Its Applications (IC3INA)*.
[16] Arduino - Software [Internet]. Arduino.cc. [cited 16 December 2020]. Available from: https://www.arduino.cc/en/Main/software

Numerical analysis on the influence of process parameters on the deep drawn product by using Taguchi method

Nay Chi Than Win[1,3], Phyo Wai Myint[1,2], Wutyee Nandar[1]

[1].Department of Mechanical Engineering, Yangon Technological University, Insein Township, The Republic of the Union of Myanmar.

[2]phyowaimyint@ytu.edu.mm, [3]ntwin14mech@ytu.edu.mm

Abstract. Deep drawing is a metal forming process that involves complex metal flow and widely used in various manufacturing. The process parameters such as punch corner radius and die corner radius affect the product quality. The punch velocity is also important to obtain the desired product shape. However, it is rare to observe the combination of process parameters effects such as punch and di corner radii and punch speed. In this study, the influences of punch corner radius, die corner radius, and punch speed on deep drawn formability and defect formation are analysed. Taguchi Method is applied to build the design of simulations, and the axisymmetric model is used to simulate with Finite Element Analysis-based ABAQUS software. The results showed that the punch corner radius and die corner radius mainly affect stress and strain distribution and the punch velocity affect the formability of the product and the production time.

1. Introduction

As the metal forming process is widely used in the mass production of automobile parts, aerospace industry, electrical components, and beverage applications, the quality is vital importance in industrialization. Deep drawing process can be defined as the process in which the depth of the blank is greater than half of the diameter of the blank. During the process, a blank sheet is experienced to plastic deformation in order to get the desired shape. Deep drawing involves complex material form and the presence of various nonlinearity such as geometric nonlinearity, material nonlinearity, and kinematic nonlinearity. Moreover, several defects like thinning, tearing, wrinkling, earing, and surface scratches are likely to develop during the process. Therefore, it is necessary to optimize the process parameters and geometric parameters to avoid defects in the final products.

In recent decades, a number of appropriate methods and analysis had been developed in sheet metal forming and still optimization of quality control is mandatory. Savas et al. [1] were interested in die and blank holder angle concerning the drawability. In this study, when the die angle α has increased, the deep drawing ratio has increased. Blank holder force is decreased by increasing α angle for a constant value of drawing ratio. Not only the production rate can be improved by optimization in punch speed and also affects draw ability. Yagami et al. [2] showed that the combination of variable punch speed (SPD) and variable blank holder force (BHF) technique can improve the deep drawability and product quality. It is established that the friction coefficient and the plastic anisotropy mainly effect on thickness and formability of the deep drawing process by Santhakumar, J et al. [3] Hadi et al. [4] observed that defects in workpieces vary in the different blank thickness by using FEM modeling. In this study, it was found that the blank thickness is greater than the threshold value then cracking may occur. If the blank thickness is less than the threshold value, wrinkling may occur. Kanttikar et al. [5] conducted ANSYS software, the finite element analysis of sheet metal forming was carried out, and the results showed that increasing in punch corner radius reduced the punch load along with the reduction of stresses. Reddy et al. [6] applied the Taguchi method to establish relationships between punch nose radius, die profile radius, and blank holder force in influencing sheet thickness. And it is identified that the influence of BHF, punch corner radius, die corner radius is 56.98%,30.12%, and 12.9% respectively in thickness variation.

Candra et al. [7] studied varying blank holder force in order to prevent the defects like cracking and improve formability by using FEM simulation. The material's microstructures and mechanical properties can be changed by heat treatment [8]. Lin et al. [9] focused on punch design with micro-ridges. Taguchi method was applied to identify the influence of ridge height, ridge distance, ridge nose radius, and ridge to punch nose distance on thinning rate. The drawing depth can be increased by using the ridged drawing punch more than 60% compared with the non-ridged drawing punch.

Mahmoodi et al. [10] determined the influence of punch corner radius, die corner radius, BHF, clearance between punch and die and the layer arrangement on the occurrence of wrinkling and thinning. And it was found that BHF has the strongest influence on wrinkling and punch corner radius has the greatest effect in thinning. Madi et al. [11] performed the variation of drawing depth with respect to the forming speed. And also presented determining the ideal forming speed can optimize the forming time. Dewang et al. studied the influence of punch velocity on the deformation behavior in deep drawing of aluminum alloy and it was found that the wrinkling was controlled by punch velocity [12]. Ju et al. [13] incorporated experimental and finite element analysis to analyze the drawing speed and the results described that the deeper the draw depth can be obtained by the higher punch speed. When the higher forming speed is used, significant reduction in punch load is found. Bohaâluca et al. [14] optimized the coefficient of friction between blank and die, blank and holder and blank and punch by using the combination of FE model ABAQUS and Taguchi method, so that to avoid the thinning. Numerous researchers performed the fundamental characteristics of drawing process. Optimization of the process parameters such as punch corner radius, die corner radius, punch speed, friction etc., can give the better-quality products which can meet the customer satisfactions.

Though the combination with the punch speed and process parameters are less studied in recent decades. Therefore, the influence of punch speed on the production time and formability, the predominant factor of defects and reduction of stress by punch corner radius and die corner radius are observed in this investigation. Energy consumption of deep drawing is mainly based on the punch velocity. The aim of this investigation is to determine the influence of process parameters such as die corner radius, punch corner radius and punch speed on stress and strain distribution by combining a statistical approach based on Taguchi Technique and Finite Element Method. ABAQUS software is adopted to perform the finite element analysis of deep drawing process.

2. Material and Methodology

The Taguchi method was used in studies to design experiments and determine the influence of process parameters on the properties of the formed parts [3]. The advantages of Taguchi (DOE) can reduce the number of tests and save manpower, material resources and financial resources. Taguchi Method treats optimization problems in two categories: Static Problem and Dynamic Problem. In static problem, there are three signals to noise ratios for common interests for optimization. 1. Smaller the better 2. Larger the better 3. Nominal is the best. In this investigation, the thickness distribution is interested and as it should be uniform in deformation, therefore, it involves in nominal the best category. The process parameters, punch corner radius, die corner radius and punch speed were studied and three levels were used for each parameter. Taguchi orthogonal L9 array was used to investigate the effect of three parameters. The high order interactions among the above three factors are assumed negligible and the information on the main effects can be obtained by running 3^3= 27 simulations. However, the appropriate Taguchi orthogonal array for the above three parameters with three levels is L9 to conduct nine simulations. The selection of the level values was performed by considering the shape and height of the drawn product and the conditions of convergence in simulation process. The level selection of the process parameters is shown in **Table 1** and L9 orthogonal array is presented in **Table 2**.

Table 1. Level Selection of the process parameters

Process parameters	Level		
	1	**2**	**3**
Punch Corner Radius(mm)	0.4	0.5	0.6
Die Corner Radius(mm)	1.2	1.5	1.8
Punch Speed(mm/s)	10	20	30

Table 2. Taguchi Orthogonal L9 Array for parameters level selection

Simulation No	Punch Corner Radius(mm)	Die Corner Radius(mm)	Punch Velocity (ms^{-1})
1	0.4	1.2	10
2	0.4	1.5	20
3	0.4	1.8	30
4	0.6	1.2	30
5	0.6	1.5	10
6	0.6	1.8	20
7	0.8	1.2	20
8	0.8	1.5	30
9	0.8	1.8	10

The geometrical dimensions in **Figure 2** are shown in **Table3**. In the drawing model, the punch, die and holder performed as rigid body and the blank, stainless steel SS304 was considered as the elastic plastic deformable body. The main material properties of the blank are presented in **Table 4**.

Table 4. Main material parameters of the blank

Parameters	**Value**
Density of material($kg/m^{3)}$	7800
Young's Modulus (GPa)	200
Yield Stress (MPa)	340
Poisson' ratio	0.29

Table 3. Geometric parameters in drawing process

Parameters	**Value**
Diameter of the blank(mm)	11
Thickness of the blank (mm)	0.5
Diameter of die(mm)	3.2
Diameter of punch(mm)	2.6
Die corner radius (mm)	1.2~1.8
Punch corner radius(mm)	0.4~0.8

The tensile test of SS 304 was conducted with a sheet thickness 2mm and punch speed of 5mm/min. The plastic strain and flow stress play the important role in plastic deformation of the material. The deep drawing process undergoes irreversible or permanent deformation and hence the elastic and plastic properties of the material are predominant. The flow stress and plastic strain of SS304 is presented in

Figure 1. The schematic diagram **Figure.2** shows the geometrical model of before and after the drawing process and geometrical dimensions. The coefficient of friction is set as 0.18 for all interaction properties between the punch, the holder, the die and the sheet.

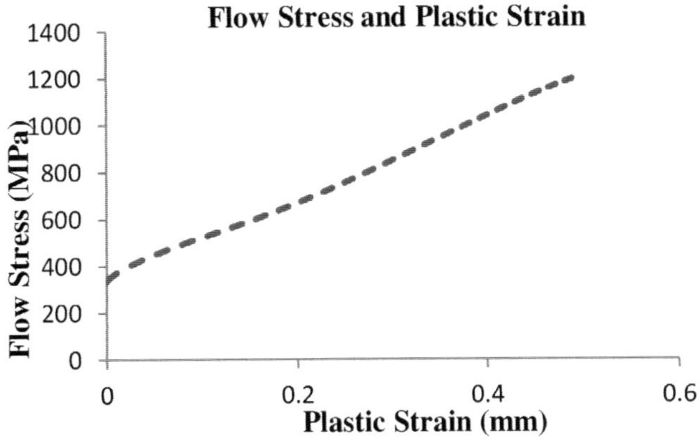

Figure 1. Flow stress vs plastic strain of SS304 with punch speed 5mm/min (0.083mm/s)

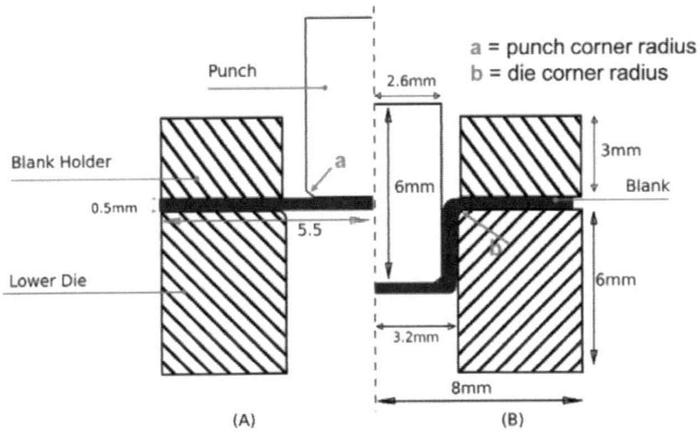

Figure 2. Schematic drawing of deep drawing process A: before drawing B: after drawing

3. Finite element simulations

The problems of nonlinear materials and geometrically nonlinear problems with plastic deformations can be solved by using ABAQUS software. ABAQUS software was employed to simulate the deep drawing process with dynamic explicit method. The boundary conditions are applied in forming process. The blank holder, punch, and die were considered as the rigid bodies. Surface to surface contact model and a penalty contact method was selected. To accelerate the computing speed, the quarter of model (axisymmetric geometry) was built. Sheet metal is a deformable solid C3DQR and meshed with the 8 nodes linear brick and C3D4 the 4 nodes linear tetrahedron presented in **Figure 3**. The die, punch and holder are R3D4 4 node and 3D bilinear rigid quadrilateral. The sheet metal is meshed with 1824 elements. The model is with total 4848 elements and 6018 nodes. The inertia forces are ignored and the model is quasi static, the kinetic energy of deforming part does not exceed the small fraction of internal energy[15]. When dynamic explicit method is employed to the model of quasi-static simulations, millions of time increments would be literally required and increasing the speed of process in the simulation is necessary to obtain an economical solution.

The nine simulations were conducted and the equivalent plastic strain and Von Mises stress were measured at nine locations, as show in **Figure 4**. While conducting nine simulations, Von Mises stress and equivalent plastic strain are observed according to nine measured positions of the deformed part in **Figure 8 and 9**.

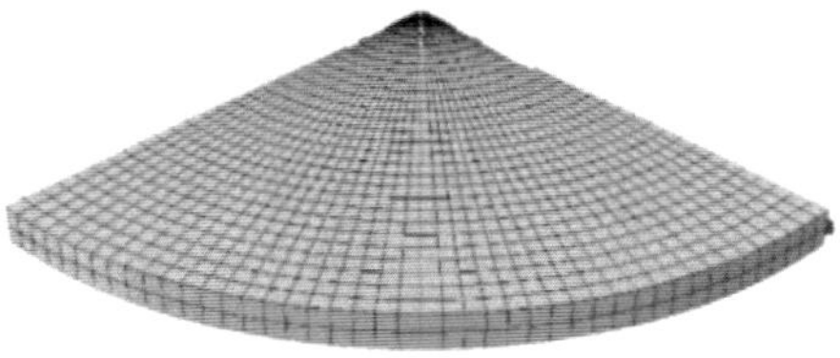

Figure 3. Blank sheet with 0.225 meshing and 1824 elements

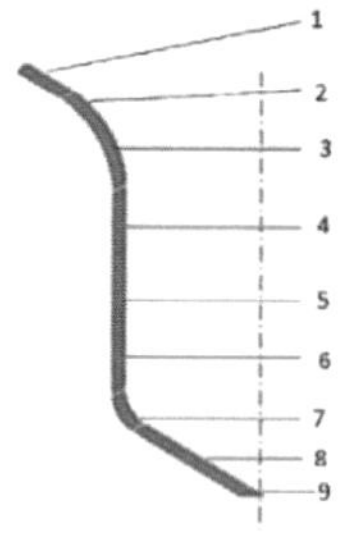

Figure 4. Measurement position of the deformed part from the flange to the bottom centre

4. Results and discussion

The results obtained from numerical analysis are described in below. The distribution of Von Mises stress and equivalent plastic strain for the simulation number 1 with the punch corner radius of 0.4 mm, die corner radius of 1.2 mm and punch velocity of 10 mm/s are shown in the **Figure 5** (a) and (b). The localized thinning at the right corner of the bottom is significant and failure can be occurred in that region due to stress localization. The punch velocity is 10 mm/s and according to time increment, the running time took much longer than others. The significant localized thinning effect was also occurred for the other simulation numbers 4, 5 and 9.

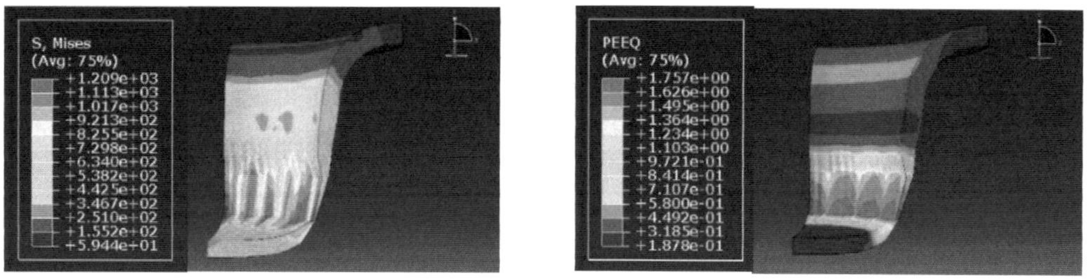

(a) (b)

Figure 5. (a)Von Mises stress and (b) equivalent plastic strain of simulation number 1

It was found that the formability is better in increasing the punch velocity from 10 mm/s to 20 mm/s and 30 mm/s. The results obtained from numerical analysis for the punch velocity of 20 mm/s, punch corner radius of 0.8 mm, die corner radius of 1.2 mm are shown in the **Figure 6** (simulation number 7). It was found that the places occurred the maximum stress and maximum strain are changed under the same die corner radius and the different punch corner radius and punch speed.

The results obtained from the deep drawn simulation with the punch speed of 30 mm/s, punch corner radius of 0.8 mm and die corner radius of 1.5 mm are shown in the **Figure 7** (simulation number 8).

The maximum stresses are occurred on the wall for both cases and the maximum strain was occurred at the bottom of punch nose for the punch velocity of 20 mm/s and at the flange area in the case of the punch velocity of 30 mm/s. The similar results with simulation number 8 are obtained for the simulation numbers 2, 3 and 6. The larger height of the deep drawn product was obtained from the simulation number 7. The results described that it is important to choose the correct parameters to get the desired shape of the deep drawn product.

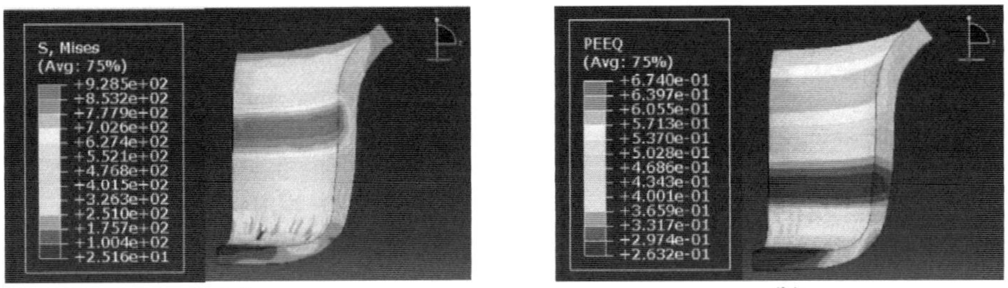

(a) (b)

Figure 6. (a) Von Mises stress (b) equivalent strain of simulation number 7

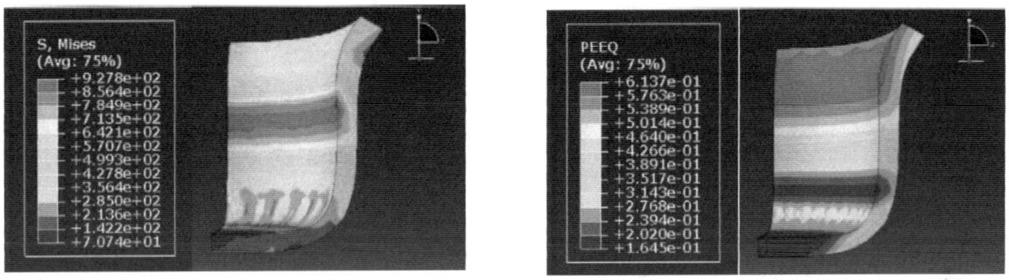

(a) (b)

Figure 7. (a) Von Mises stress (b) equivalent plastic strain of simulation number 8

The stress distribution on the nine positions of deep drawn part from flange area to the bottom centre is shown in the **Figure 8**. The results showed that the maximum stress localization was found at two places or three places for some simulation and mostly at the flange area and punch nose. The distribution of equivalent plastic strain is shown in the **Figure 9**. The maximum strain distribution was found at the places of wall, punch nose and bottom area. The simulation results described that the quality of the desired product is controlled by the punch corner radius and die corner radius and punch velocity.

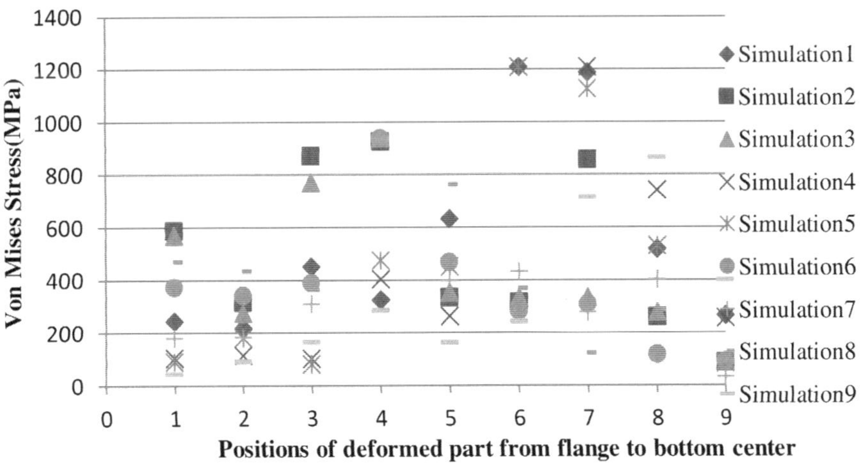

Figure 8. Stress distribution on the deformed part from flange area to bottom centre

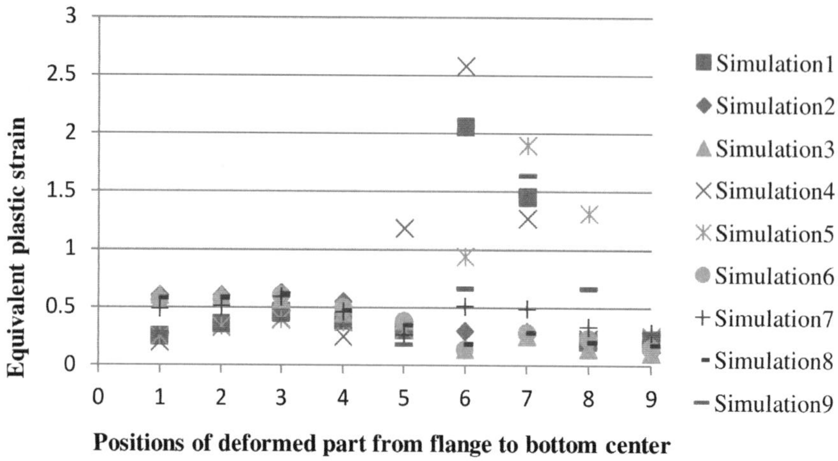

Figure 9. Strain distribution on the deformed part from flange area to bottom centre

The smaller the best on S/N ratio of Taguchi technique was applied by using MINITAB software, which shown in **Table 5** and **Table 6** to analyse the influence of factors on the stress and strain distribution. The results in the **Table 5** show that the punch corner radius has the strongest effect on the stress distribution and the die corner radius effect is more significant than the punch velocity.

Table 5. Response for signal to noise ratios of stress distribution

Level	Punch Corner Radius	Die Corner Radius	Punch velocity
1	-55.31	-55.11	-54.41
2	-54.97	-55.06	-54.71
3	-53.03	-53.13	-54.18
Delta	2.29	1.98	0.53
Rank	1	2	3

Table 6. Response for signal to noise ratios of strain distribution

Level	Punch Corner Radius	Die Corner Radius	Punch velocity
1	5.298	2.500	5.436
2	2.780	5.299	3.335
3	5.998	6.277	5.305
Delta	3.217	3.778	2.102
Rank	2	1	3

The results in the **Table 6** show that the die corner radius has the strongest effect on the strain distribution. The results obtained from simulation show that the increase in the punch speed reduces the processing time and the decrease in production time. And it was found that the part can be formed with good formability by applying the higher punch speed.

5. Conclusion

The cylindrical cup section deep drawing process was performed with numerical analysis to prevent localized thinning and stress and strain localization by appropriate selection of the influence parameters such punch corner radius, die corner radius and punch speed. Taguchi method was used to design the finite element simulations and analyse the effects of the influence parameters on the deep drawing process by using MINITAB software.

The punch corner radius and die corner radius are mainly significant on the stress and the strain distribution. The punch velocity can improve production time and effects on formability of the final part. The obtained results described that the quality of deep drawn product was affected by the punch corner radius, the die corner radius and the punch velocity. It can be concluded that the optimized parameters to obtain the desired deep drawn product with minimum stress and strain distribution and localized thinning are punch corner radius of 0.8 mm, die corner radius of 1.2 mm and punch velocity of 20 mm/s.

References

[1] Savas V 2007 *Materials & Design* doi:10.1016/j.matdes.2006.01.004.

[2] Control B H 2007 *FE Analysis on Deformation Mechanism of Strain-Rate-Sensitive Materials in Cylindrical Deep-Drawing with Combination Punch Speed* vol 1 no 12 pp 1385–1396 doi: 10.1299/jmmp.1.1385.

[3] Gopal A 2017 *Optimization of Deep Drawing Process Parameters using Design of Experiments*

[4] Hadi S, Yu H, Tieu K and Lu C 2015 *Simulation of Defects in Micro-Deep Drawing of an Aluminium Alloy Foil* pp 9–16 doi: 10.1063/1.4806838.

[5] Kanttikar R, Kodli B S and Chikmeti R B 2014 Analysis on deformation process in deep drawing of sheet metal part by FEM *IOSR J. Mech. Civ. Eng.* vol 11 no 4 pp 06–16 doi: 10.9790/1684-11440616.

[6] Reddy A C S, Rajesham S, Reddy P R, Kumar T P and Goverdhan J 2016 An experimental study on effect of process parameters in deep drawing using Taguchi technique *Int. J. Eng. Sci. Technol.* vol 7 no 1 p 21 doi:10.4314/ijest.v7i1.3.

[7] Candra S, Batan I M L, Berata W and Pramono A S 2015 Analytical study and FEM simulation of the maximum varying blank holder force to prevent cracking on cylindrical cup deep drawing *Procedia CIRP* vol 26 pp 548–553 doi: 10.1016/j.procir.2014.08.018.

[8] Liang L, Zhengyi J, Dongbin W, Ken-ichi M and Hideki S 2015 Experimental and numerical study of micro deep drawing doi:10.1051/matecconf/20152109003.

[9] Lin B and Yang C 2017 Applying the Taguchi method to determine the influences of a microridge punch design on the deep drawing *Int. J. Adv. Manuf. Technol.* pp 2109–2119 doi:10.1007/s00170-016-8911-y.

[10] Mahmoodi M and Sohrabi H 2017 Using the Taguchi method for experimental and numerical investigations on the square-cup deep-drawing process for aluminum/ steel laminated sheets vol 4 pp 169–177 doi:10.22075/MACS.2017.11051.1108.

[11] Madi M, Júnior M V, Antônio R, Filho C, Victor P and Marcondes P 2018 An analysis of the forming speed variation with relation to deep drawing depth of steel DP 600 sheets pp 2417–2424

[12] Dewang Y, Sharma V and Batham Y 2020 Influence of punch velocity on deformation behavior in deep drawing of aluminum alloy *J Fail. Anal. and Preven.* doi.org/10.1007/s11668-020-01084-5.

[13] Ju L, Mao T, Patil S and Altan T Investigation of forming speed and friction on drawability of Al 5182-O using a servo press with CNC cushion *Center for Precision Forming* (Columbus, USA: The Ohio State University, Promotional Literature)

[14] Bohaâluca K, Ghanameh M F, Faqir M, Mada M and Essadiqi E 2019 Evaluation of the effect of contact and friction on deep drawing formability analysis for lightweight aluminum

lithium alloy using cylindrical cup *Procedia Manuf.* vol 46 pp 623–629 doi: 10.1016/j.promfg.2020.03.089.
[15] 2005 Quasi-Static Analyses *ABAQUS/Explicit: Advanced Topics* ABAQUS, Inc.

Plant growth chamber with integrated air conditioning system, periodic irrigation, and adjustable photosynthetic light wavelength

A H Atienza, R T Cervantes, J A Dionisio and T A Tan

Mapua Institute of Technology at Laguna, Malayan Colleges Laguna, Pulo Diezmo Road, Cabuyao City Laguna 4025 Philippines

ahsatienza@mcl.edu.ph

Abstract. The increase in the average temperature due to climate change affects the crop yield production and it can lead to food insecurity. Rapid breeding helps increase crop yield production and improve the overall crop quality. The researchers designed a plant growth chamber with integrated air conditioning system, periodic irrigation, and adjustable photosynthetic light wavelength. The growth rate of the representative plant, green span lettuce, grown in the plant growth chamber is compared to conventional planting. Factors such as growth period rate, dry weight, and overall produce quality are considered. It was found that implementing a 24-hour photoperiod, drip irrigating 50 mL per plant every 2 days, adding vermicast to the cocopeat at a ratio of 1:1 and maintaining a temperature range of 20 to 25 degrees Celsius produced a higher relative growth rate. In consideration of additive nutrients, the addition of vermicast added a substantial increase in the relative growth rate. Quality was also improved due to the correlation of the relative growth rate and plant quality.

1. Introduction

Lettuce is considered a cool weather vegetable that should be planted during late summer in the Philippines [1]. The study of [2] also found that fancy lettuce varieties had an increased demand due to fast-food chains that influenced the eating habits of the Filipino people, but these fancy lettuce varieties require modern greenhouse technologies for production which requires a large capital that cannot be supported by all farmers. To induce speed breeding, a Plant Growth Chamber is required which is a fully enclosed environment capable of controlling the environmental parameters such as temperature, humidity, CO2 composition, light intensity, and light wavelength [3]. Numerous studies found that better quality crops may be grown faster by providing purple light as photosynthetic light. However, the light wavelength may vary as different crops have different responses to different wavelengths. The combination of red and blue light or purple stands as an effective light source for a plant's growth because it provides increased chlorophyll content, increased biomass and phytochemical content, and overall enhanced nutritional value [4][5][6]. Purple light using red and blue LED light arrays allows lettuce production in conditions where sunlight is unavailable and that purple LED lighting improved lettuce crop breeding speed by increasing the height and root ratio of lettuce faster than regular sunlight exposure [7]. Aside from the lighting, a periodic irrigation system is necessary to provide daily water requirements for the plants and soil moisture. It is observed to have yield increase, easier harvesting, reduced water consumption and better-quality lettuce when lettuce is drip irrigated [8].

This study primarily focused on the design and fabrication of a plant growth chamber for the speed breeding with romaine lettuce (Lactuca sativa L. var. longifolia) as the representative plant. Several parameters were considered to enhance the growth and quality of the plant. The proposed design of the plant growth chamber controlled and maintained the environmental parameters such as indoor temperature, humidity level and light wavelength. In addition to this, a periodic irrigation system is installed to maintain the amount of water used and less human intervention. Thus, this research designed and fabricated a plant growth chamber with integrated air conditioning system, periodic irrigation, and adjustable photosynthetic light wavelength. Also, the seed germination rate and quality of the representative plant, romaine lettuce must be increased and enhanced. The study highlights design and testing of a plant growth chamber with integration of air conditioning system and adjustable photosynthetic light wavelength. It also focuses on the drip irrigation system.

2. Materials and Methods

To implement the plant growth chamber concept as shown in Figure 1, a closed space was designed with vapor compression air-conditioning system with 2 HP unit, drip irrigation system, smart LED lighting system and a built-in temperature and humidity meter. To carry out the testing procedures, a seedling tray was used and planted with lettuce seeds and allowed to sprout in the plant growth chamber environment with a purple led light exposure at 100% intensity and a 24-hour cycle. To further explain the environmental parameters, the ideal relative humidity inside a controlled-environment chamber is in the range of 60 to 70% [9]. The growth rate of lettuce is stable at a temperature range of 20-25°C [10] and purple light is the best option as lighting for plant growth chambers [4]. In addition, an inlet and exhaust fan were installed that ran for 5 minutes every 3 hours to maintain the natural CO_2 level of the environment. These parameters were repeated for other trials, but the lighting system and irrigation system were varied throughout the testing cycles.

Figure 1. Plant Growth Chamber

The researchers conducted tests to make sure that the temperature and relative humidity inside the plant growth chamber is within the required parameters for the green span lettuce. Temperature and relative humidity sensors were placed inside the enclosure. The digital meter for temperature and relative humidity were placed outside for ease in monitoring. A thermostat controls the temperature inside the plant growth chamber's enclosure. On the other hand, the water flow for the periodic irrigation was monitored using a flow meter. The green span lettuce sprout planted in a Styrofoam cup was placed inside the plant growth chamber and the researchers observed its growth until harvesting stage. The researchers documented its growth through pictures and compared it with the germination rate of green span lettuce at Zacky's farm. The place has an ambient temperature ranging from 21- 26°C with relative humidity of around 84%.

The researchers conducted two simultaneous testing for the prototype testing of the plant growth chamber. Simultaneous testing means that green span lettuce grown with and without fertilizer or nutrients are placed inside the plant growth chamber at the same time. The ratio of vermicast to cocopeat is 1:1 while the ratio of nutrients from the vermicast shows 12:1 carbon to nitrogen ratio. A second trial was done with the same arrangement where half of the lettuce sprouts are grown with fertilizer or nutrients. The fertilizer used is vermicast which is an organic fertilizer made from castings of worms. The plant quality and relative growth rate for both tests were compared and analyzed. The researchers used the formula by [11] for determining the relative growth rate of plants between two-time intervals. Relative Growth Rate (RGR) is a well-known indicator of plant strategy with respect to crop yield as identified with environmental stress and disturbance regimes. In specific to testing of the growth of green span lettuce, two trials of simultaneous testing were done. The objective of the simultaneous testing was to investigate if there are any advantages or disadvantages of growing plants in the plant growth chamber with and without nutrient or fertilizer supplement. At different time intervals, the researchers weighed the lettuce plant including its roots to determine its dry weight.

To calculate the Relative Growth Rate of lettuce between two- time intervals., the researchers used the equation of Relative Growth Rate by [11] described by equation (1)

$$RGR = \frac{(ln ln\, W_2 - ln ln\, W_1)}{t_2 - t_1} RGR = \frac{(ln ln\, W_2 - ln ln\, W_1)}{t_2 - t_1} \quad (1)$$

Where, RGR = relative growth rate

W_1 = Initial dry weight of plant at time t_1, grams

W_2 = Final dry weight of plant at time t_2, grams

t_1 and t_2 = time, days

3. Results and Discussion

For the initial testing of the plant growth chamber, the parameters such as temperature, relative humidity, light intensity, and light duration were observed. The initial tests were done to guarantee that the plant growth chamber will work within the required parameters for the green span lettuce. The temperature inside the enclosure is within 20°C to 25°C and the humidity is kept at a range of 60% to 70%. With continuous observation, it is determined that the plant growth chamber can work within these parameters. The researchers tried different light intensity levels and light duration to observe its effect to the growth of green span lettuce. The seedlings showed considerably slow growth for the first 3 trials (seedlings without nutrients used). The seedlings grown in conventional farming will start to show its one true leaf within 7 days but the seedlings in the plant growth chamber took 9 days.

There were two trials done for green span lettuce seedlings with nutrients. The temperature and humidity range are kept at the same range with the first 3 trials done on seedlings without nutrients. A 24-hour photoperiod increases the biomass of the lettuce [12]. The ratio of the soil and nutrients were also varied for these trials. The researchers first tried the recommended 1:1 ratio by mass of soil and nutrients, and for trial 2 a ratio of 3:1 was tested. The green span lettuce seedlings showed a favorable result with the settings of the trial with 100% light intensity (24-hour photoperiod) and a ratio of 1:1 for the soil and vermicast. Therefore, it was decided to operate the plant growth chamber at these set parameters. The drip irrigation system is not utilized for the initial testing because the seedlings need to be watered using a mist applicator. In addition, the seedlings are sensitive to extra soil moisture that slows the seedling's growth rate.

A simultaneous testing was done for green span lettuce with and without nutrients. There were six test data taken at different time intervals as shown in Table 1. For this testing, the plant growth chamber is operating at a temperature within 20°C to 25°C, a relative humidity range of 60% to 70%, and a 100% light intensity with 24-hour photoperiod. The periodic irrigation system consists of an automatic watering pump and drip irrigation tubing. The drip irrigation system supplies each lettuce sprout 50 mL every 2 days so that it gives moisture content to the soil [13]. The interval of the days for weighing the lettuce are in parallel with the test data taken from the beneficiary which is Zacky's farm. As shown on Table 2, Relative Growth Rate (RGR) for the first trial, the RGR for the lettuce grown with nutrients is the highest with an average relative growth rate of 0.0657. It was followed by the lettuce grown without nutrients and the lowest average relative growth rate value is for the test data taken from the farm.

Table 1 *Results of Green Span Lettuce for Trial 1*

	Lettuce w/ Nutrients		**Lettuce w/out Nutrients**		**Farm**	
	Dry weight (g)	Growth Period	Dry weight (g)	Growth Period	Dry weight (g)	Growth Period
1	9	7	9	7	13	7
2	30	14	28	14	36	16
3	97	26	95	26	98	22
4	129	39	124	39	131	43
5	154	45	143	45	154	55
6	166	55	163	55	168	62

Table 2 *Relative Growth Rate (RGR) for Trial 1*

	Lettuce w/ Nutrients	**Lettuce w/out Nutrients**	**Farm**
1 & 2	0.1720	0.1621	0.1132
2 & 3	0.0978	0.1018	0.1669
3 & 4	0.0219	0.0205	0.0138
4 & 5	0.0295	0.0238	0.0135
5 & 6	0.0075	0.0131	0.0124
Average RGR	**0.0657**	**0.0643**	**0.0640**

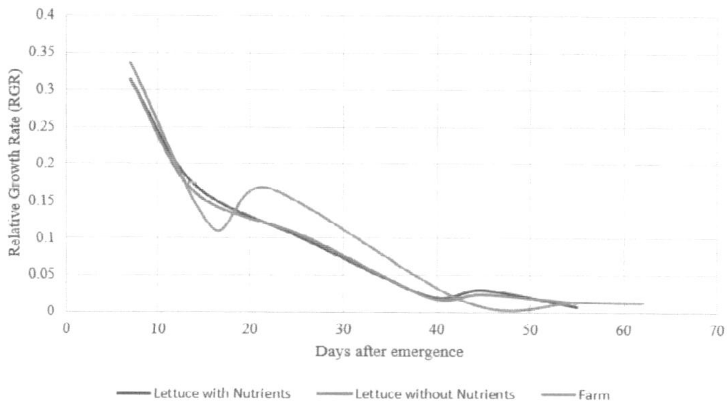

Figure 2. Relative Growth Rate for Trial 1

From Figure 2, it is noticeable that the relative growth rate values are close to each other. The decline in the three curves shown is also expected throughout the growth of the green span lettuce. As stated from the study of [14], a plant's growth is exponential, but it becomes linear once the critical mass is reached. A high relative growth rate value is also associated with high net assimilation rate (NAR) and high specific leaf area (SLA) as these two are general growth components of RGR [15]. The same procedure is done for the second trial of the plant growth chamber. It showed a similar result where the lettuce with nutrients has the highest average relative growth rate while the lettuce grown in the farm is the lowest as shown in Table 3, Table 4, and Figure 3. The actual photos of the lettuce grown in these conditions can be seen in Figure 4.

Table 3. Results of Green Span Lettuce for Trial 2

	Lettuce w/ Nutrients		**Lettuce w/out Nutrients**		**Farm**	
	Dry weight (g)	Growth Period	Dry weight (g)	Growth Period	Dry weight (g)	Growth Period
1	9	7	9	7	13	7
2	32	14	30	14	37	16
3	98	26	94	26	98	22
4	132	39	127	39	133	43
5	157	45	145	45	155	55
6	168	55	164	55	170	62

Table 4. Relative Growth Rate (RGR) for Trial 2

	Lettuce w/ Nutrients	Lettuce w/out Nutrients	Farm
1 & 2	0.1812	0.1720	0.1162
2 & 3	0.0933	0.0952	0.1623
3 & 4	0.0229	0.0231	0.0145
4 & 5	0.0289	0.0221	0.0128
5 & 6	0.0068	0.0123	0.0132
Average RGR	0.0666	0.0649	0.0638

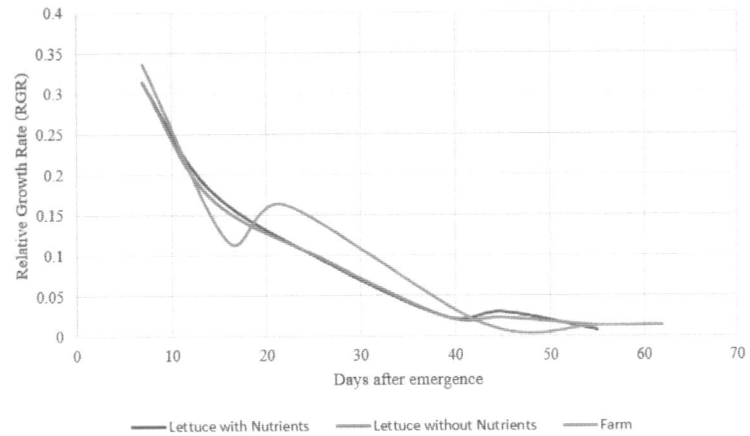

Figure 3. Relative Growth Rate for Trial 2

(a) (b) (c)

Figure 4. Actual photo of lettuce after 55 days a) in the plant chamber without nutrients b) in the plant chamber with nutrients and c) in the farm

4. Conclusion

The plant growth chamber was designed and fabricated using materials such as aluminum sheets, aluminum frames, square aluminum tubes, and polyethylene foam insulation to compensate for the internal heat gain and indoor-outdoor temperature differences. In testing of the plant growth chamber, considerable green span lettuce growth environments were exhibited, the chamber has shown that it can grow at a temperature range of 20 to 25 degrees Celsius, at a 24-hour photoperiod and through drip irrigation system.

In measuring the growth rate, the use of the relative growth rate (RGR) analysis has shown efficacy in determining the growth rate of the lettuce plant from sprout to harvest. In the trials done, the relative growth rate was augmented through implementing drip irrigating 50 mL per plant every 2 days and addition of vermicast to the cocopeat at a ratio of 1:1 The relative growth rate was successfully augmented with these conditions. In the use of additive nutrients, the addition of vermicast added a valuable increase to the relative growth rate in comparison to planting. In terms of quality, quality was improved due to the correlation of the relative growth rate to overall plant quality.

The use of a plant growth chamber is not limited to a specific type of plant. The researchers used lettuce as the representative plant since it is sensitive to temperature and other environmental conditions. Future researchers may use other plants that require a different operational range. It is preferable to test plants that are also sensitive to changes in temperature and can only be planted on specific areas. The use of the plant growth chamber is advantageous for trying out other crops since the temperature inside the enclosure can be varied. The light intensity and light duration as well as the drip irrigation system can also be changed depending on the plant.

Acknowledgements

The researchers want to give thanks to the staffs at Zachy's farm located in Amadeo, Cavite for sharing their knowledge in conventional soil farming and hydroponics, Malayan Colleges Laguna A Mapua School for financing the presentation and paper publication, Mapua Institute Technology at Laguna for giving them unwavering support and motivation, and their families for providing them inspiration towards their work.

References

[1] Jose D n d Lettuce production guide. *Department of Agriculture* Bureau of Plant Industry Retrieved from http://bpi.da.gov.ph/bpi/images/Production_guide/pdf/Lettuce.pdf

[2] Fang-Asan M L D, Dagupen M K T and Digal L N 2011 Lettuce supply chains and marketing margins in benguet, philippines. *Banwa*, *6*(2). doi: 10.3860/banwa.v6i2.1962

[3] Kapiel T 2018 Speed breeding: A powerful innovative tool in agriculture. *Innovative Techniques in Agriculture*, 2(3), 413-415.

[4] Naznin M, Lefsrud M, Gravel V and Azad M 2019 Blue light added with red LEDs enhance growth characteristics, pigments content, and antioxidant capacity in lettuce, spinach, kale, basil, and sweet pepper in a controlled environment. *Plants*, *8*(4), 93. doi: 10.3390/plants8040093

[5] Yeh S C, Lee H P, Kao S T and Lu J L 2016 Effects of supplementary lighting by natural light for growth of brassica chinensis. *Journal of Physics: Conference Series*, *710*, 012021. doi: 10.1088/1742-6596/710/1/012021
[6] Li Q and Kubota C 2009 Effects of supplemental light quality on growth and phytochemicals of baby leaf lettuce. *Environmental and Experimental Botany*, *67*(1), 59–64. doi: 10.1016/j.envexpbot.2009.06.011
[7] Marin G H, Ixtaina P and Ruscitti M 2017 Quick growth of lettuce plants using purple LED illumination devices. *Journal of Agronomy, 16: 87-93.*
[8] Bozkurt S and Mansuruglu G S 2011 Lettuce yield responses to different drip irrigation levels under open field condition. *Journal of Cell & Plant Sciences*, *2*(2), 12-18.
[9] Ghosh S, Watson A, Gonzalez-Navarro O E, Ramirez-Gonzalez R H, Yanes L, Mendoza-Suárez M and Hickey L T 2018 Speed breeding in growth chambers and glasshouses for crop breeding and model plant research. *Nature Protocols*, *13*(12), 2944–2963. doi: 10.1038/s41596-018-0072-z
[10] Hong P H, Kwon O H, Lee D I J R P, Ha J M, Jeong D U and Kim B 2011 Effects of LED light and temperature on lettuce growth. *Agribusiness and Information Management*, *3*(1), 67–74.
[11] Fisher R A 1921 Some Remarks On The Methods Formulated in A Recent Article On "The quantitative analysis of plant growth.". *Annals of Applied Biology*, *7*(4), 367–372. doi: 10.1111/j.1744-7348.1921.tb05524.x
[12] Koontz H & Prince R (1986). Effect of 16 and 24 hours daily radiation (light) on lettuce growth. HortScience, 21(1), 123-124.
[13] Sykes, D J 1964 The availability of soil moisture to plants. *Retrospective Theses and Dissertations*. 3011.Retrieved from https://lib.dr.iastate.edu/rtd/3011
[14] Turnbull L A, Paul-Victor C, Schmid B & Purves D W 2008 Growth Rates, Seed Size, And Physiology: Do Small-Seeded Species Really Grow Faster. Ecology, 89(5), 1352–1363. doi: 10.1890/07-1531.1
[15] Shipley B 2006 Net assimilation rate, specific leaf area and leaf mass ratio: which is most closely correlated with relative growth rate? A meta-analysis. Functional Ecology, 20(4), 565–574. doi: 10.1111/j.1365-2435.2006.01135.x

Prediction of flow field in a solar chimney using ANFIS technique

Minh-Thu T Huynh*, Tri Q Truong, Thinh N Doan, Trieu N Huynh, Tung V Nguyen, Viet T Nguyen, Y Q Nguyen

Faculty of Engineering, Van Lang University,
45 Nguyen Khac Nhu St., Co Giang Ward, Dist. 1, Ho Chi Minh City, Viet Nam

*thu.htm@vlu.edu.vn

Abstract. Solar chimneys have been intensively studied as an effective method for natural ventilation of buildings. Though numerical methods, such as Computational Fluid Dynamics (CFD), have been widely utilized in such studies, they usually require extensive computational resources. Moreover, experimental study is quite complicated and costly. In recent years, machine learning has started to be used as a tool in the thermal-fluid field. In this study, in order to save time and cost, Adaptive Neuro-Fuzzy Inference System (ANFIS) technique, a class of adaptive networks that incorporate both neural networks and fuzzy logic principles, is combined with CFD. A simulation model was first validated by experiment from another study in the field. The result was documented as a dataset using CFD code ANSYS Fluent (Academic version 2020 R2). Then, they are used to train and validate the ANFIS model. In particular, the study is to predict the fluid flow field in a 2-dimensional typical solar chimney when heat flux changes in the range of 400 to 1000 W/m^2. Inputs of the ANFIS model are position and heat flux, while outputs are temperature and velocity at that location. As a result, the 2 ANFIS models could achieve R^2 values of 0.997, 0.97 (training set) and 0.994, 0.9715 (testing set); RMSE are 1.009, 0.00224 (training set) and 1.074, 0.0204 (testing set) for outputs of temperature and velocity, respectively. Those results are acceptable. By using the ANFIS model, large amounts of flow fields with different scenarios can be estimated simultaneously. Therefore, it is expected that engineers and architects can have a quick tool in the process of design.

Keywords: solar chimney, CFD, machine learning, ANFIS

1. Introduction

Solar chimney is one of the most interesting passive methods in ventilation for buildings. Chimney effect due to higher temperature air heated by solar energy makes air flow out of the channel; hence, induces natural ventilation. Since the first study on solar chimneys carried out in 1993 by Bansal et al. [1], ventilation potential is proven to increase in a well-designed solar chimney. The study of Zavala-Guillen et al. [2] indicated that when the absorber-partitioned air channel solar chimney (SC-AP) is attached to a building, it is able to generate the air changes per hour (ACH) complying with the requirements recommended by ASHRAE. Studying glazed solar chimney walls combined with other solar chimney wall and roof configurations, Chantawong et al. [3] proved that high air ventilation rates can be induced, allow the substitution of stagnant room air with fresh outside air for a healthy and comfortable interior environment, and maintain indoor temperature at comfortable level. Comparing between a room in

Northern China during summer, with and without Trombe wall, a kind of solar chimney, the temperature in the room with a Trombe wall decreased by 0.5–3 °C and wind velocity increased to approximately 0.2–0.5 m/s [4]. Charvat et al. [5] also showed that with the increase of thermal mass, air velocity increases during the night; while during day the air velocity is increased by 25%.
In addition, energy-saving is another advantage of solar chimney [6-7]. Up to 50% of the annual ventilation energy required can be saved for a Tokyo's office building [8]. Because of the above benefits, a variety of numerical and experimental studies in solar chimney have been done so far [9-12]. However, those studies are for limited scenarios due to either limited resources in computation or expensive experimentation.

Recently, Machine Learning (ML) techniques have attracted intensive research interest in Fluid Dynamics problems. By using limited data obtained from the CFD or experiment as the input, the ML model is trained to learn the dynamical system's characteristics. Different computational intelligence approaches, i.e. artificial neural network (ANN), genetic programming (GP), and adaptive neuro-fuzzy inference system (ANFIS) have been evaluated and compared for predicting the energetic performance of a building integrated photovoltaic thermal system [13]. Likewise, ANN and ANFIS model are used to predict performance of solar chimney power plants [14]. In the study of Tian et al. [15], the permeability is predicted based on pore structure parameters of porous media using ANN with genetic algorithm as an optimization technique. Pourtousi [16] and his team have combined CFD and ANFIS for the bubble column hydrodynamics study. Likely, Beigzadeh [17] used CFD data for Adaptive Neuro-Fuzzy to study the heat transfer in flat and discontinuous fins.

In this research, we employed an ANFIS model to predict the flow and temperature fields in a solar chimney, which have not been studied in the literature. A CFD model was built, validated, and run to provide data for training the ANFIS model, which was then used to predict the flow and temperature fields in other cases. Thus, engineer or architect can use as a quick tool to optimize their design without complicated computation or experiment. Then, comparisons between the ANFIS and CFD results were then conducted and validated the ANFIS model.

2. Modeling setup

2.1. *Modeling in ANSYS*

Solar chimney model in this study is sketched in Figure 1. The major dimensions of the solar chimney are the gap (*G*) and the height (*H*), which respectively are 0.1 m and 1.025 m. The left wall is heated with uniform heat flux *I*. The right wall receives radiative heat transfer from the left one. Atmospheric pressure is set for air flow at inlet (lower end of the air channel) and outlet (the upper end).

RANS (Reynolds Averaged Navier – Stokes equations) method in CFD is used to estimate the governing equations of air flow and heat transfer. It consists of continuity, momentum, and energy equations. The computational domain and mesh are coincided the air channel as in figure 1. The mesh was structured and finer as it is closer the walls. For all the cases, the Mesh – refinement tests lead the numerical results changed less than 1.0% when the non–dimensional distance y^+ of the first grid points near the solid surface was less than 1.5. Finite Volume Method are also deployed by the CFD code ANSYS Fluent (Academic version 2020 R2), with following assumptions and setups:

- steady and two-dimensional flow and heat transfer; incompressible air flow in the channel.
- radiative heat transfer from the heated wall to the opposite wall in the air channel described by the S2S model;
- using RNG $k - \epsilon$ turbulence model;
- using SIMPLE method for the coupling between the continuity equation and the momentum equation;
- using PRESTO! Method for interpolation of the pressure on the staggered mesh, second order upwind scheme for momentum equation and energy conservation equation, and first order upwind scheme for the equation for k and ϵ.

Since mass flow rate is the most important parameter to evaluate performance of solar chimney, it is used to validate the solar chimney simulation model. Figure 2 shows that mass flow rate's discrepancy for simulation and experiment in the research of Burek and Habeb [9] is less than 10%, which is acceptable due to other minor factors during experimentation, such as heat loss through walls.

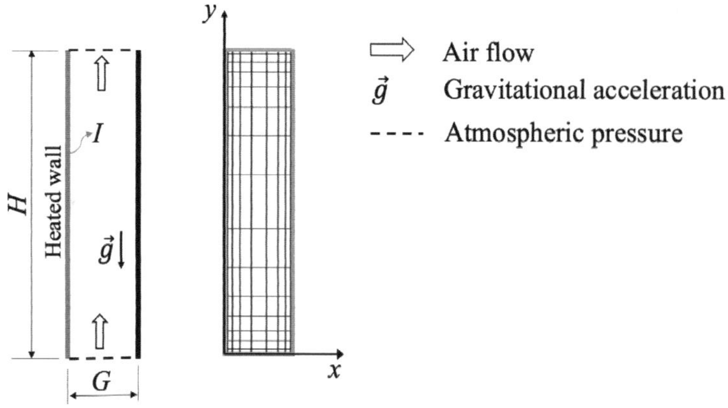

Figure 1. Modelling and meshing in ANSYS.

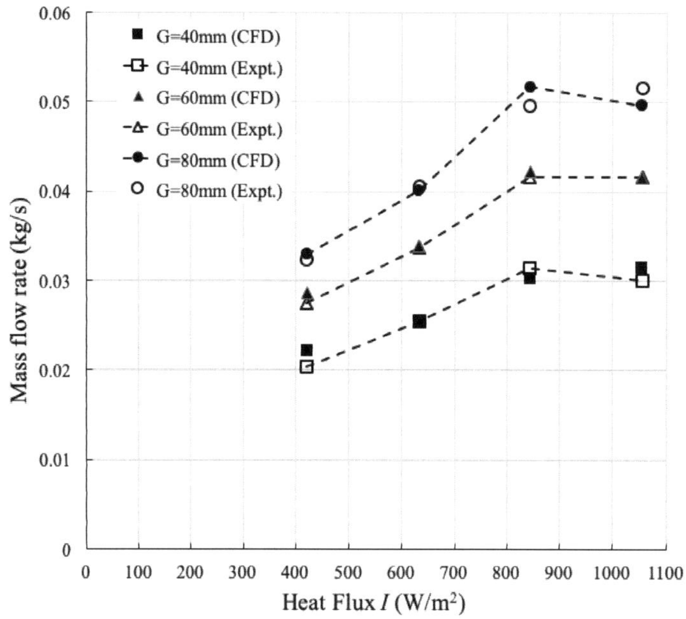

Figure 2. Computed (CFD) and measured (Expt.) flow rate in the experiment by Burek and Habeb [9] at different gaps and heat fluxes.

2.2. *ANFIS technique*

The general ANFIS structure [19], assuming to comprise two inputs x, y and 5 layers, is illustrated in Figure 3. Note that, the output is marked as the fifth layer. Based on the Sugeno fuzzy model [20], the model is defined by 2 rules as follows:

- Rule 1: If x is equal A_1 and y is equal B_1 then $f_1 = p_1x + q_1y + r_1$
- Rule 2: If x is equal A_2 and y is equal B_2 then $f_2 = p_2x + q_2y + r_2$

where p, q and r are ANFIS prediction parameters of each rules.

At the Layer 1, we define 4 membership functions that can be transformed the two inputs into linguistic labels. The membership function in the current study is selected as Bell-shaped functions:

$$\mu_i = \frac{1}{(1 + \left|\frac{x - c_i}{a_i}\right|^{2b_i})}$$

where a_i, b_i and c_i with $i = (1, 2)$ are the fuzzy set parameters.

The Layer 2 performs the combination of two membership functions expressed by two inputs. Subsequently, the Layer 3 is used to calculate the ratio of the i-th rule's firing strength to the sum of all rules' firing strengths. The ratio is labeled as $\overline{w_i}$ with $i = (1, 2)$. The Layer 4 gives the weighted output combining the previous layer and first-order based on Sugeno fuzzy model as $\overline{w_i}f_i$ with $i = (1, 2)$. The Layer 5 contains one node presenting the output of the network as the sum of all weighted output in Layer 4.

To evaluate the accuracy of produced outputs, two statistical indices of R^2 and RMSE are used.

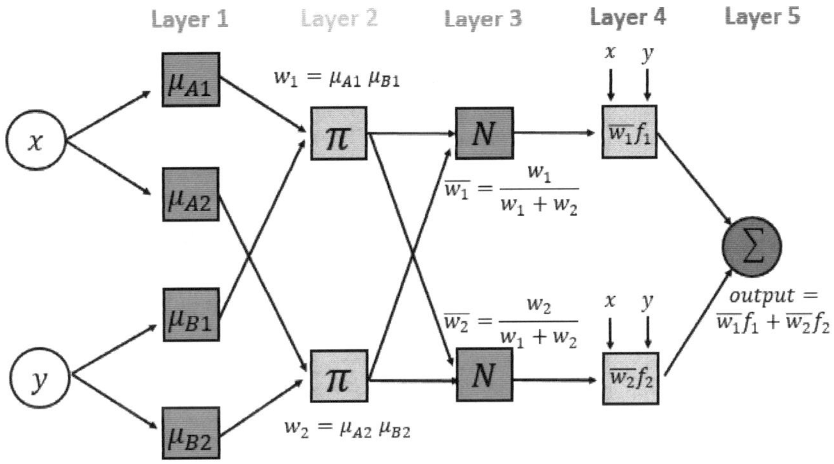

Figure 3. The general ANFIS structure.

3. Results and discussion

3.1. ANFIS model training

Dataset is generated using ANSYS with the above validated CFD model to include: location, *x* (m), *y* (m); heat flux, *I* (W/m^2); temperature *T* (K) and velocity *V* (m/s). The velocity is assumed as velocity on *y*-direction since the velocity on *x*-direction is very small; hence negligible.

As in all machine learning methods, training a successful network requires a lot of training data. More than 130 values of *I* in the range of 400 – 1000 W/m^2 have been simulated. In each case, the total number of mesh nodes is 20,300 (100 x 200 mesh elements).

The inputs of the ANFIS are *x*, *y* and *I*, while 2 outputs are *T* and *V*. The 4 G-bell functions are used for membership functions of all 3 inputs. From dataset of more than 2.8 million points created from ANSYS, we randomly chose a part of them to train the ANFIS model. By increasing the number of data points used for training, we found that a training set of 100.000 data points offered the best R^2 and RMSE. The fuzzy set parameters of each membership function are in Table 1.

The training results, respectively for temperature *T* and velocity *V*, give RMSE of 1.009 and 0.00224; and R^2 of 0.99666 and 0.97. Those results are quite reasonable for the outputs in this study.

Table 1. ANFIS model inputs

Input	MF	Type of MF	a	b	c
x (m)	in1mf1	Gbellmf	0.007	1.200	-0.015
	in1mf2	Gbellmf	0.029	2.000	0.027
	in1mf3	Gbellmf	0.028	2.000	0.072
	in1mf4	Gbellmf	0.010	2.000	0.115
y (m)	inf2mf1	Gbellmf	0.171	2.000	0.000
	inf2mf2	Gbellmf	0.171	2.000	0.341
	inf2mf3	Gbellmf	0.171	2.000	0.683
	inf2mf4	Gbellmf	0.171	2.000	1.025
I (W/m^2)	inf3mf1	Gbellmf	61.530	2.000	421.90
	inf3mf2	Gbellmf	61.532	2.000	545.003
	inf3mf3	Gbellmf	61.532	2.000	668.067
	inf3mf4	Gbellmf	61.532	2.000	791.130

3.2. ANFIS model validation

In order to validate the above ANFIS model, some values of heat fluxes in the remaining dataset after training are used for test. In this section, the results for three heat fluxes of 429, 532 and 615 W/m^2, which were selected randomly in the tested range, are presented. Table 2 shows RMSE and R^2 of T and V for each heat flux. It can be seen that the result for all 3 cases are acceptable.

Table 2. RMSE and R^2 for testing data.

I (W/m^2)	Outputs	RMSE	R^2
429	*T* (K)	1.00697	0.99454
	V (m/s)	0.01703	0.97730
532	*T* (K)	1.11894	0.99411
	V (m/s)	0.02014	0.97225
615	*T* (K)	1.09676	0.99334
	V (m/s)	0.02411	0.96372

Figure 4 shows comparisons of the temperature and velocity distributions in the channel obtained from CFD and ANFIS models for the cases presented in table 2. It is seen that the ANFIS model was able to capture well the main flow and temperature fields in the channel, particularly near the wall, at all heat fluxes. However, there are slight deviations among the two results near the channel center. The ANFIS model slightly over – predicted the temperature fields and under – predicted the velocity distributions. Since the channel center had the lowest temperature and velocity, the differences between the two results are seen in accord with the RMSE values in table 2.

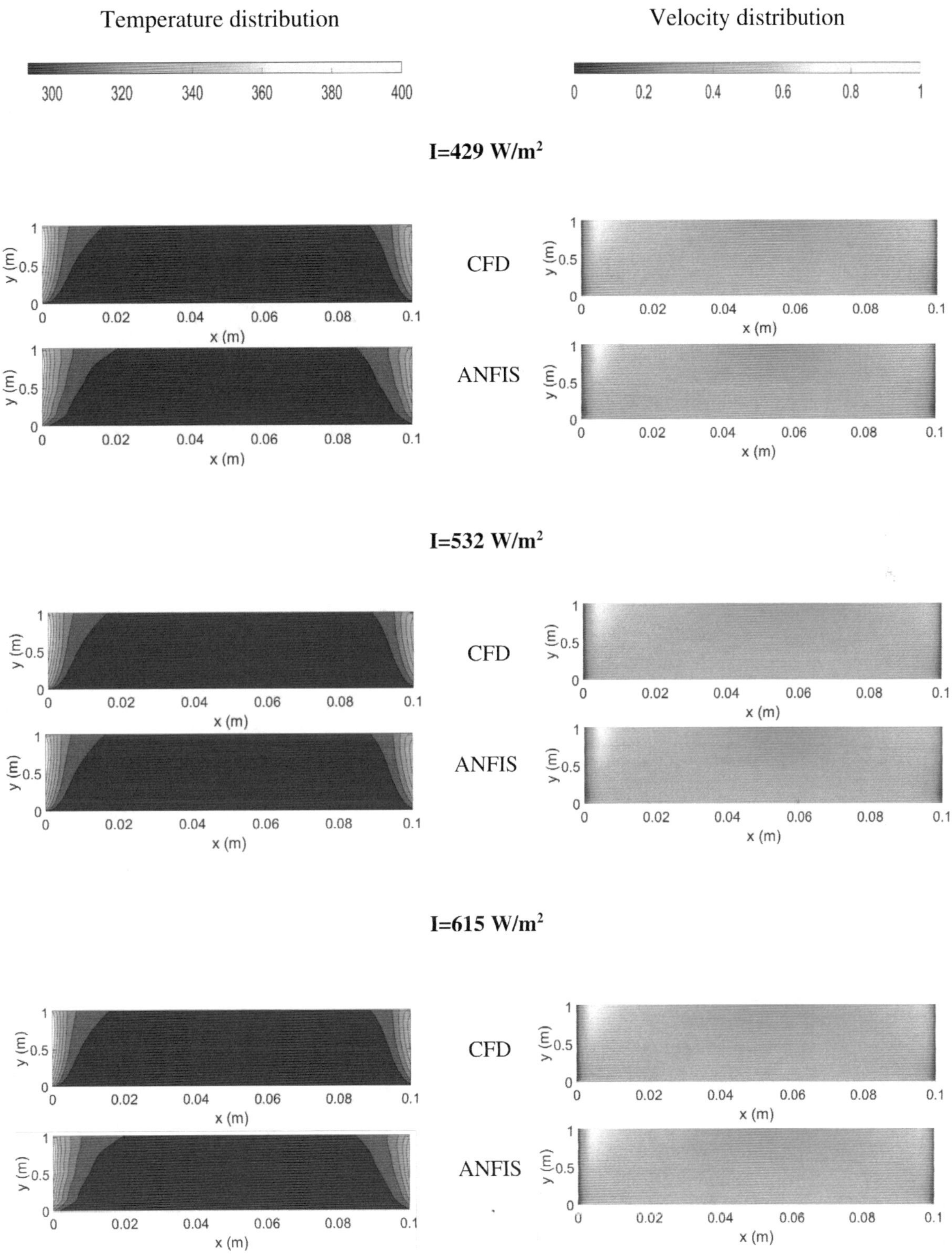

Figure 4. CFD vs. ANFIS model for temperature and velocity distributions with different heat fluxes (vertical and horizontal directions are not at the same scale).

4. Conclusion

This study attempted to introduce a test on combination of Adaptive Neuro-Fuzzy Inference System (ANFIS) model for evaluating the flow field in a solar chimney. By comparing flow fields with several testing heat fluxes between CFD and ANFIS model, it can be confirmed that the ANFIS model is validated with acceptable RMSE and R^2. This ANFIS model was able to predict flow field at any point in the tested solar chimney channel for any heat flux in the range of 400 – 1000 W/m^2. Then, the result can provide a fast tool for engineers or architects in their design process.

In future works, the present model can be extended to either different dimensions of the solar chimney or other types of solar chimney, such as inclined or combine vertical – inclined solar chimneys. In addition, computational time should also be compared between CFD and ANFIS models.

References

[1] Bansal N K, Mathur R and Bhandari M S 1994 A study of solar chimney assisted wind tower system for natural ventilation in buildings *J. Building Environ.* **3** 373–377
[2] Zavala-Guillén I, Xamán J, Hernández-Pérez I, Hernández-Lopéz I, Jiménez-Xamán C, Moreno-Bernal P and Sauceda D 2018 Ventilation potential of an absorber-partitioned air channel solar chimney for diurnal use under Mexican climate conditions *J. Appl. Therm. Eng.* **149** 807-821
[3] Chantawong P, Hirunlabh J, Zeghmati B, Khedari J, Teekasap S and Win M M 2006 Investigation on thermal performance of glazed solar chimney walls *Solar Energy* **89** (3) 288-297
[4] Zhou Y, Wang Z, Yang C, Xu L and Chen W 2019 Influence of Trombe wall on indoor thermal environment of a two-story building in rural Northern China during summer *Science and Technology for the Built Environment* **25** (4) 438–449
[5] Charvat P, Jicah M and Stetina J 2004 Solar chimneys for ventilation and passive cooling *Denver (USA): World Renewable Energy Congress.*
[6] Al Touma A and Ouahrani D 2018 Performance assessment of evaporatively-cooled window driven by solar chimney in hot and humid climates *Solar Energy* **169** 187–195
[7] Briga Sá A C, Martins A, Boaventura-Cunha J, Carlos Lanzinha J and Paiva A 2018 An analytical approach to assess the influence of the massive wall material, thickness and ventilation system on the Trombe wall thermal performance *Journal of Building Physics* **41** (5) 445–468
[8] Miyazaki T, Akisawa A and Kashiwagi T 2016 The effects of solar chimneys on thermal load mitigation of office buildings under the Japanese climate, *Renew. Energy* **31** (7) 987–1010
[9] Burek S A M and Habeb A 2007 Air flow and thermal efficiency characteristics in solar chimneys and Trombe Walls *J. Energy Build.* **39** 128–135
[10] Chen Z D, Bandopadhayay P, Halldorsson J, Byrjalsen C, Heiselberg P and Li Y 2003 An experimental investigation of a solar chimney model with uniform wall heat flux *J. Building Environ.* **38** (7) 893–906
[11] Khanal R and Lei C 2015 A numerical investigation of buoyancy induced turbulent air flow in an inclined passive wall solar chimney for natural ventilation *J. Energy Build.* **93** 217–226
[12] Shi L, Ziem A, Zhang G, Li J and Setunge S 2020 Solar chimney for a real building considering both energy-saving and fire safety – a case study *J. Energy Build.* **221** 110016
[13] Gao W, Moayedi H and Shahsavar A 2019 The feasibility of genetic programming and ANFIS in prediction energetic performance of a building integrated photovoltaic thermal (BIPVT) system *Solar Energy* **183** 293-305
[14] Amirkhani S, Nasirivatan S, Kasaeian A B and Hajinezhad A 2015 ANN and ANFIS models to predict the performance of solar chimney power plants *Renew. Energy* **83** 597-607
[15] Tian J, Qi C, Sun Y, Yaseen Z M and Pham B T 2020 Permeability prediction of porous media using a combination of computational fluid dynamics and hybrid machine learning methods *Engineering with Computers*
[16] Pourtousi M, Sahu J N, Ganesan P, Shamshirband S and Redzwan G 2015 A combination of

computational fluid dynamics (CFD) and adaptive neuro-fuzzy system (ANFIS) for prediction of the bubble column hydrodynamics *Powder Technology* **274** 466–481

[17] Beigzadeh R 2019 The CFD Providing Data for Adaptive Neuro-Fuzzy to Model the Heat Transfer in Flat and Discontinuous Fins *Iranian J. Chem. Engineering* **16** (2) 57-69

[18] Enayatollahi H, Fussey P and Nguyen B K 2020 Modelling Evaporator in Organic Rankine Cycle using Hybrid GD-LSE ANFIS and PSO ANFIS Techniques *Thermal Science and Engineering Progress* **19** 100570

[19] Jang J-S R 1993 ANFIS: adaptive-network-based fuzzy inference system *IEEE Transactions on Systems, Man, and Cybernetics* **23** (3) 665-685

[20] Takagi T and Sugeno M 1985 Fuzzy Identification of Systems and Its Applications to Modeling and Control *IEEE Transactions on Systems, Man, and Cybernetics* **SMC-15** (1) 116-132.

Strength and fire resistance characteristics of geopolymers synthesized from volcanic ash, red clay and waste pen shells

Martin Ernesto Kalaw[1*], Jose Mari Adiarte[1], Richard dela Cruz[1], Kylie Mae Martinez[1], Christian Adrian Vega[1], and Michael Angelo Promentilla[2, 3]

[1]Mechanical Engineering Department, De La Salle University, 1004 Manila, Philippines
[2]Chemical Engineering Department, De La Salle University, 1004 Manila, Philippines
[3]Center for Engineering and Sustainable Development Research, De La Salle University, Manila, 1004 Manila, Philippines

[*]martin.kalaw@dlsu.edu.ph

Abstract. This study utilized volcanic ash and red clay, as well as calcined waste pen shell (Baluko) in the production of geopolymer-based materials. The geopolymers were formed by activating the mixture of these raw materials (as the alumina-silica rich materials) with activating solution of 12M $NaOH/Na_2SiO_3$ (w/w: 2.5:1). Two sample types, a cube type and a slab type, were used in the study in order to conform to test standards for compressive strength and fire resistance test. The cube type molds were for the compressive strength tests while the slab type was used for the fire resistance tests. Material testing such as Fourier Transform Infrared (FTIR) spectroscopy was used to analyze the chemical characteristics of both the raw materials and the geopolymer specimens. The mixture containing 45% volcanic ash- 45% red clay-10% calcined waste pen shell powder (by weight) was observed to have the highest compressive strength out of all the samples tested. The fire resistance of the geopolymers formed from a ternary mixture of 16% volcanic ash-66.67% red clay- 16% calcined waste pen shell powder (by weight) was also observed to be comparable to that of ordinary Portland Cement (OPC). Furthermore, the FTIR results of both raw materials and geopolymer showed evidence that geopolymerization occurred in the samples, indicating that the selected precursors are viable for use in the formation of geopolymers.

Keywords: Geopolymer, volcanic ash, red clay, waste pen shell.

1. Introduction

Geopolymers, also referred to as alkali-activated cement, geocement, hydroceramic, etc., are inorganic polymers synthesized by the addition of an alkali solution to an aluminosilicate-rich precursor material. It is dubbed as a next-generation cement but with a lower carbon footprint and embodied energy as compared to ordinary Portland cement (OPC) [1-3]. Some of its properties are high compressive strength, good chemical resistance and high temperature resistance attributed to its three-dimensional structure of aluminate and silicate tetrahedra joined by oxygen corners [4, 5]. Some recent studies show that geopolymers could also be utilized for wastewater treatment, as a heavy metal adsorbent, and as a heterogeneous catalyst [6-9].

Geopolymer technology allows for waste valorization of industrial and agro-industrial by-products and waste materials that have high concentration of alumina and silica. These include coal ash, rice hull ash, red mud, blast furnace slags, among others [10-12].

Volcanic ashes, natural pozzolans rich in alumina and silica, have been used in geopolymer studies [13-14]. In this study, red clay, which is also naturally rich in minerals, is combined with volcanic ash as the aluminosilicate raw material mixture in the geopolymer synthesis. The addition of calcined Baluko shells, a type of pen seashells rich in calcium, is done in consideration of the calcium-silicate-hydrate, CaO-SiO_2-H_2O or C-S-H, reaction path in the presence of water [15, 16]. C-S-H is the primary product of hydration of silica and calcium and is the major factor in the strength of OPC-based concrete.

Thus, this study explores the Philippine-locally available materials of volcanic ash, waste pen shells, and red clay to form geopolymer-based materials. These three materials were sourced from the Bicol Region where Mayon Volcano is located, red clay is abundant and Baluko (Pinnidae) pen shell meat is a local delicacy. Seashells, such as Baluko shells, are good sources of calcium and the calcined shell has been utilized to make a composite geopolymer with fly ash [17]. The use of such indigenous resources and waste (as are the Baluko pen shells) as raw materials for geopolymer precursor mitigates not only the waste's environmental footprint but also contributes towards producing eco-friendly materials.

2. Materials and Method

2.1 Preparation of raw materials

The three materials, volcanic ash, red clay and waste pen shells (Baluko), sourced from Bicol region in the Philippines, are shown in Figure 1. The volcanic ash and red clay, being initially damp, rocky, and clumped together, underwent drying and grinding to achieve a particle size of not greater than 150 μm. The waste pen shells had to be washed first with sodium hypochlorite to remove contaminants that cling onto the shell before oven drying at 110°C. The dried shells were then calcined at 700°C in a muffle furnace.

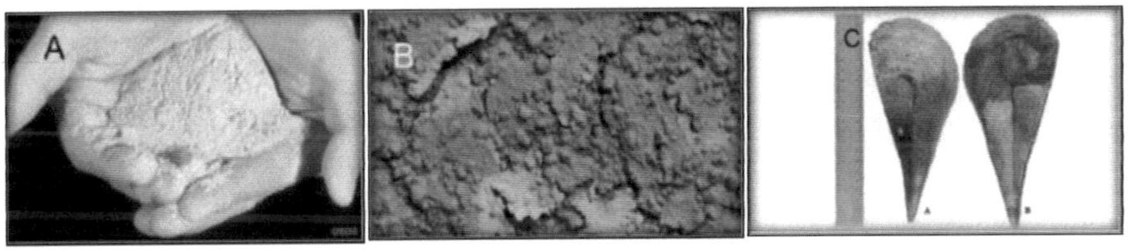

Figure 1. Raw materials: A volcanic ash, B red clay and C Baluko pen shells

X-ray Fluoroscopy (XRF) analysis show the composition of the component materials in Table 1. The volcanic ash and the red clay contain high alumina (63.91% and 65.55%, respectively) and silica (17.06% and 26.48%, respectively) and the calcined Baluko shell is 84.37% calcium oxide (CaO).

Table 1. Composition of raw materials (percentage by mass).

Component	Volcanic Ash	Red Clay	Baluko pen shell
SiO_2	63.91	65.55	8.01
Al_2O_3	17.06	26.48	4.12
Fe_2O_3	8.61	7.77	0.71
Na_2O	2.46	0.06	2.12
MgO	3.15	0.02	0.67
CaO	4.81	0.12	84.37
Total	100.00	100.00	100.00

2.2 Synthesis and testing of geopolymer samples

The activating solution was prepared by mixing 12M sodium hydroxide (NaOH) solution with water glass solution (Na_2SiO_3) in a ratio of 2.5:1. The solid-to-liquid ratio of 0.25 by mass was maintained in all mixtures following the mix design described in Table 2. The fresh mix was then placed in the appropriate molds to form 50 mm x 50 mm x 50 mm cubical samples and 200 mm x 200 mm x 50 mm slab type samples. The cube type samples were subjected to compressive strength tests while the slab type samples were used for the fire resistance tests. One set of samples was pre-cured at 80°C for one day and then fully cured for a total of 28 days. The other set was cured at ambient conditions for the full 28 days. Compressive strength was measured using a Universal Testing Machine (UTM) whereas the fire resistance was evaluated following ASTM E119: Standard Test Methods for Fire Tests of Building Construction and Materials (Standard Fire Test).

Table 2. Mix proportions of the geopolymer samples (percentage by mass).

Mixture Code	Volcanic Ash (VA)	Red Clay (RC)	Baluko shells (BS)
UCL008	75	25	0
UCL009	50	50	0
UCL010	45	45	10
UCL011	66.67	16.67	16.67
UCL012	16.67	66.67	16.67
UCL013	75	0	25
UCL014	25	75	0
UCL015	0	75	25

3. Results and Discussion

3.1 Fourier Transform Infrared (FTIR) spectroscopy analysis

Figure 2 shows the infrared spectrum of the red clay. The values indicated in the waveforms are the peaks in which there are bonds and minerals essential in geopolymerization. The peaks from 3624 -3693 cm^{-1} indicate that kaolinite is present. Kaolinite, $Al_2Si_2O_5(OH)_4$, is a major constituent of red clay which gives sharp absorption bands in the 3600-3700 cm^{-1} region [18]. The band at 1040 cm^{-1} is due to asymmetric stretching vibrations of silicate tetrahedron [18]. The peak at 793 cm^{-1} indicates the presence of quartz which is explained by the Si-O stretching vibration. On the other hand, the peak at 677 cm^{-1} indicates a Si-O symmetric bending vibration due to low level of Al for Si substitution. The band at 538 cm^{-1} is due to the presence of hematite. It overlaps into one broad adsorption band centered at 535 cm^{-1} assignable to Fe-O present in kaolinite [18]. The region between 2850 - 3000 cm^{-1} indicates that there is a presence of organic matter. More specifically, the peaks at 2860 and 2927 cm^{-1} correspond to the C-H stretching vibrations of some organic contribution [19].

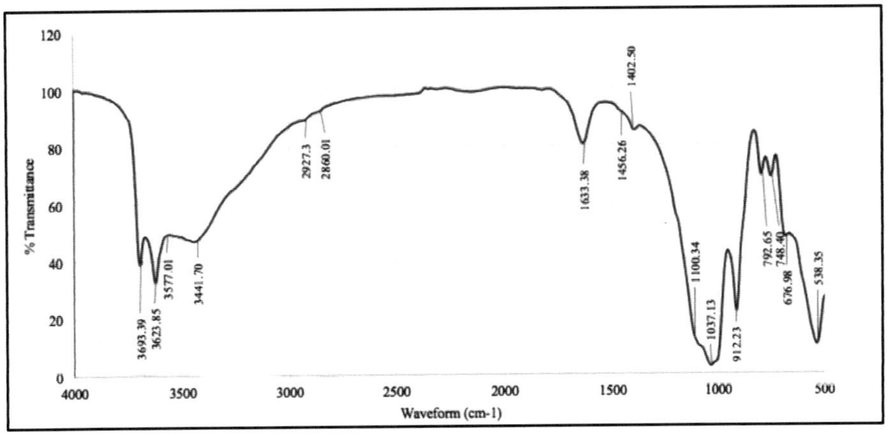

Figure 2. FTIR spectrum of red clay

Figure 3 shows the infrared spectrum of the uncalcined Baluko shell and the calcined Baluko shell. For the uncalcined Baluko shell, a broad absorption around 3421.55 cm^{-1} is caused by the O-H stretching vibration. The peak at 3642.18 cm^{-1} indicates a sharp O-H stretching band upon calcination. The peak at 1453.6 cm^{-1} indicates a bending vibration of the O-Ca-O functional group. The Baluko shells, being primarily composed of calcium carbonates, only act as aggregates and do not fully contribute to the geopolymerization process [15]. On the other hand, FTIR pattern from the calcination process indicates a new peak which appears at 3,620 cm^{-1}. This indicates the formation of basic OH groups which attached to the calcium atoms, which could make the calcined clay more reactive to form cementitious-like structure [20].

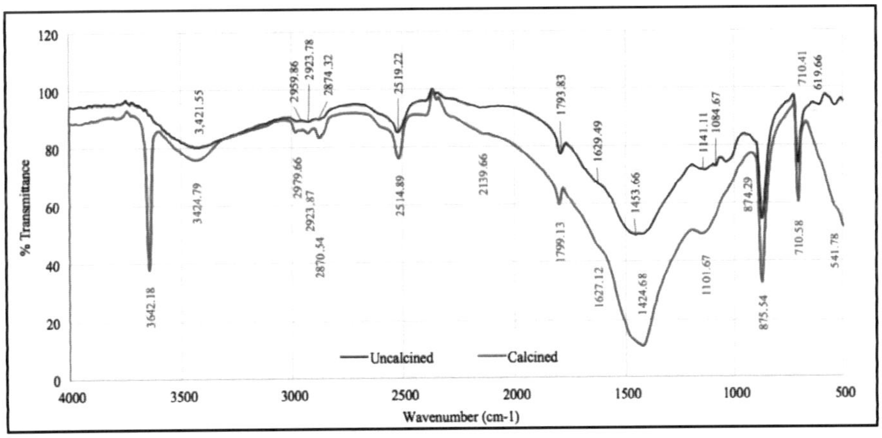

Figure 3. FTIR spectra of uncalcined and calcined Baluko pen shells.

Figure 4 shows the infrared spectrum of the volcanic ash. The bands at 726 cm^{-1} and 592 cm^{-1} indicate ring vibrations of Si-O bonds of the silicate network. The main band related to the vibration of SiO/Al–O bond of aluminosilicate framework is located at around 1014 cm^{-1}. The peaks at 2926 - 2959 cm^{-1} and 2860 cm^{-1}may relate to C-H vibrations bond of methane. The bands observed at 3420 and 1631 cm^{-1} are characteristic of stretching and bending vibrations, respectively, of O–H bonds from silanol group and water molecules [21].

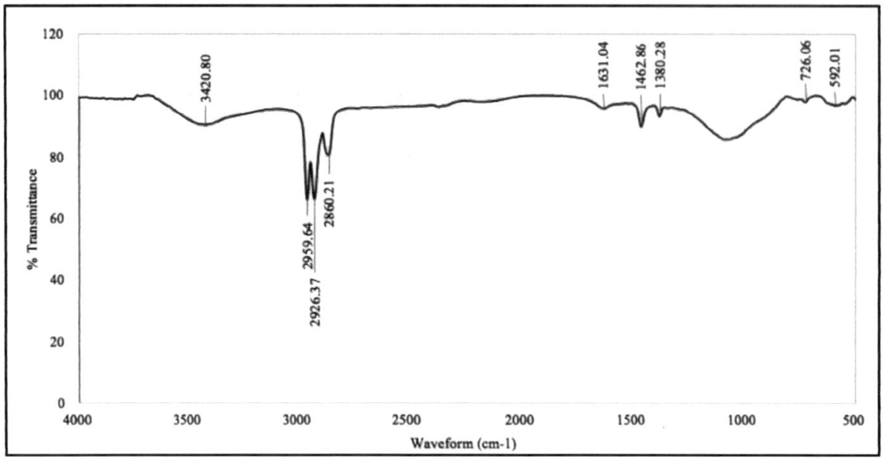

Figure 4. FTIR spectrum of volcanic ash.

Figure 5 shows spectra of some of the geopolymer specimens produced from this experiment. The peak at around 1600-1630 cm^{-1} indicate a H-OH bending vibration and is typical for polymeric structures including aluminosilicate [22]. At around 1400 cm^{-1}, O-C-O stretching of carbonates occur. The peaks ranging from 950-1200 cm^{-1} indicate that there is a T-O-Si asymmetric stretching, in which T can either be Al or Si. The key feature of the spectra are the bands around 1000 cm^{-1} to 900 cm^{-1} which indicates the presence of geopolymeric structure. As far as evidence of C-S-H reaction, the presence of the absorption hump at 970–1100 cm^{-1} due to polymeric silica and at 800-970 cm^{-1} due to the dissolution of calcium silicate, is correlated with the development of water bending vibration bands (1500–1700 cm^{-1}) due to the formation of calcium silicate hydrate, C-S-H [23, 24].

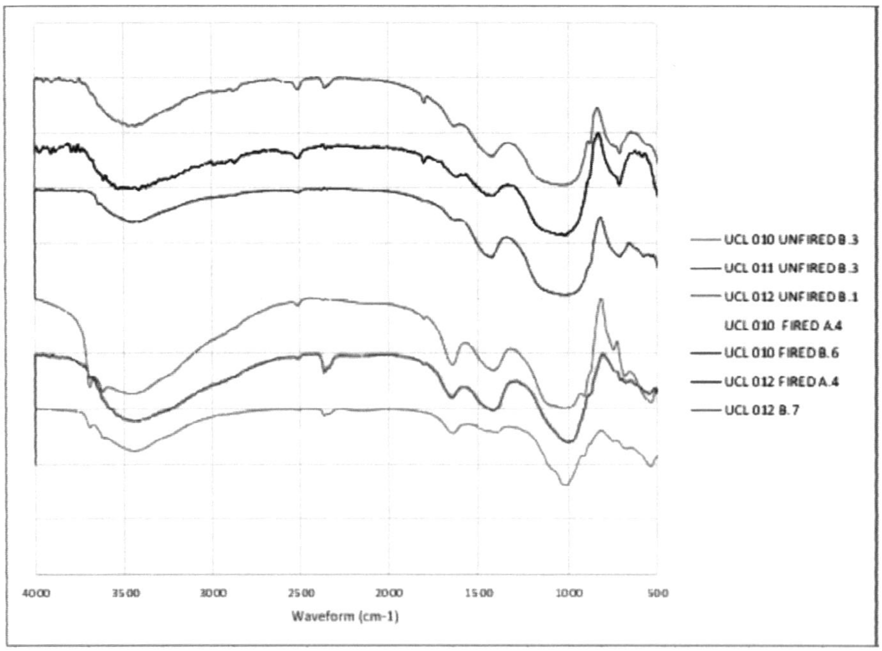

Figure 5. FTIR spectra of geopolymer samples formed

3.2 Thermogravimetric analysis (TGA) of Baluko seashells

As shown in Figure 6, the Baluko shells experienced a minimal loss of mass from 0°C-420°C of the test. However, at within the 500°C range, the weight of the sample began to decrease. At about 690°C, a dramatic reduction of weight can be observed. The decrease in mass was due to the thermal decomposition reaction that occurred when $CaCO_3$ was exposed to extreme temperatures, which resulted in the production of the compounds calcium oxide (CaO) and carbon dioxide (CO_2). At about 700°C, a chemical reaction occurred converting $CaCO_3$ into CaO.

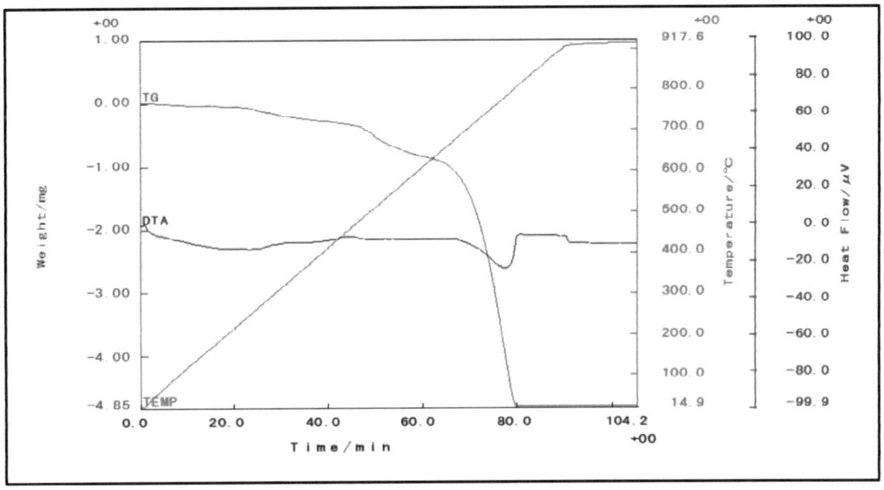

Figure 6. Thermograph of Baluko seashell.

3.3 Unconfined compressive strength

The mean compressive strengths of the geopolymer samples produced varied over a range of 0.06MPa to 3.61MPa for the samples cured at ambient conditions, whereas the mean compressive strengths of the samples pre-cured at 80°C varied over a range of 0.25MPa to 7.19MPa. These results indicate that the pre-curing of the samples at an elevated temperature has resulted in an overall gain in the mean compressive strength of the geopolymers [22]. The results of compressive strength test are summarized in Figure 7.

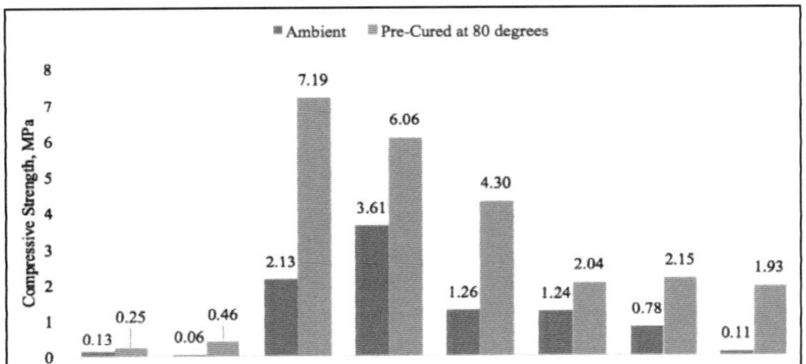

Figure 7. Unconfined compressive strength of geopolymer samples.

Moreover, the samples formed from ternary mixtures of volcanic ash, red clay, and calcined Baluko seashell were observed to have higher mean compressive strengths when compared to the samples formed from binary mixtures. It can be observed that the mean compressive strength of the ternary geopolymer binder mixtures ranges from 4.3MPa to 7.19MPa, attaining the highest recorded value when equal parts of volcanic ash content (45%) and red clay (45%) content are incorporated into the mixture, with an amount of calcined Baluko seashell not exceeding 10% of the whole mixture. This implies that the addition of calcined Baluko shells contribute to increasing the strength.

3.4 Fire resistance tests

The fire resistance tests were conducted following the Fire Test Standard ASTM E119 which exposes one side of the samples to a prescribed temperature vs. time profile as shown in Figure 8. According to the ASTM E119, the material is considered to have failed if the temperature of the unexposed surface has risen 140°C above its initial temperature. Material failure is also considered if visual cracking takes place. The time at which material failure has occurred is taken as the material fire resistance rating (FRR) [25, 26].

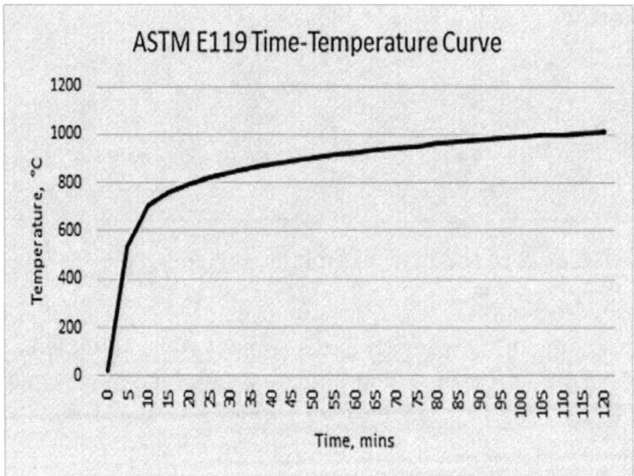

Figure 8. Time-temperature curve for the fire resistance test according to ASTM E119 [23].

In the fire resistance tests, the sample UCL 010 (45-45-10 VA-RC-BS) showed the fastest increase in temperature reaching its failure temperature within 17 minutes of direct fire exposure. The samples UCL 011 (66.67-16.67-16.67 VA-RC-BS) and UCL 012 (16.67-66.67-16.67 VA-RC-BS) have approximately 30 minutes fire resistance rating as shown in Table 3. It is seen here that the samples with higher calcined seashell content produces geopolymers with comparable fire resistance rating to OPC concrete and also with higher percentage of residual strength as seen in Table 4. Figure 9 seems to indicate that higher volcanic ash content has a higher fire resistance than red clay based on the actual temperature vs. time profile of the unexposed side of the samples and on the physical appearance of the unexposed surfaces before and after fire testing, as shown in Figure 10, which shows more visible cracks for the sample with higher red clay content.

Table 3. Fire resistance ratings of geopolymer samples vs standard concrete.

Sample (50 mm thickness)	Mix Proportion (VA -RC-BS)	Mean Fire resistance rating (FRR)
UCL 010	45-45-10	17 minutes
UCL 011	66.67-16.67-16.67	28.5 minutes
UCL 012	16.67-66.67-16.67	32.75 minutes
OPC Concrete	with Type A aggregate	31.3 minutes
NZS 3101-1	(standard for 50 mm thick concrete)	30 minutes

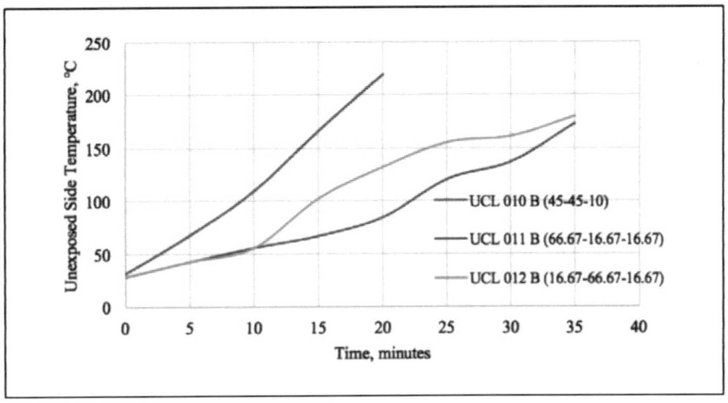

Figure 9. Temperature vs time of unexposed side of selected geopolymer specimens during fire resistance tests.

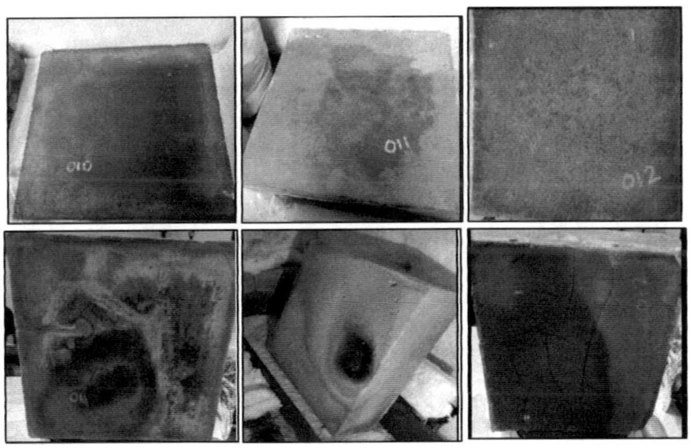

Figure 10. Unexposed surfaces of geopolymer pre-firing (top), post-firing (bottom).

Table 4. Residual strength after fire resistance test

Sample	Mix Proportion (VA - RC - BS)	Mean compressive strength before fire test	Mean compressive strength after fire test
UCL 010	45-45-10	5.96 MPa	5.30 MPa
UCL 011	66.67-16.67-16.67	2.10 MPa	2.62 MPa
UCL 012	16.67-66.67-16.67	8.14 MPa	7.84 MPa

4. Conclusions

Volcanic ash, red clay and waste Baluko shell were used to produce geopolymer-based materials. A mix proportioning which contains 45% volcanic ash- 45% red clay-10% calcined waste shell was observed to have the highest compressive strength out of all the samples. The fire resistance of the geopolymers formed from a ternary mixture of 16% volcanic ash-66.67% red clay- 16% calcined waste shell powder was observed to be comparable to that of OPC. Furthermore, the FTIR results showed evidence that geopolymerization occurred from the activation of these raw materials indicating that the selected precursors are viable for use in the formation of geopolymers. Also, better strength properties were observed for the samples that contain higher calcined waste seashells which may be attributed to the C-S-H reaction path.

With these results, it is envisioned that more studies will look into naturally occurring mineral-rich materials such as volcanic ash and red clay as additional alternative or replacement materials in the production of eco-friendly binders for the building sector. Thus, moreover, in this specific formulation, future work will investigate further the optimization of the properties of the geopolymer product.

Acknowledgement

This study was supported by the grant from Philippine Council for Industry, Energy and Emerging Technology Research and Development (PCIEERD) and Advanced Device and Materials Testing Laboratory (ADMATEL) of the Department of Science and Technology (DOST). The researchers also acknowledge the support provided by URCO-IR of De La Salle University (DLSU).

References

[1] Singh B, Ishwarya G, Gupta M and Bhattacharyya S 2015 Geopolymer concrete: a review of some recent developments *Constr. Build. Mater.* **85** *78–90*

[2] Duxson P, Fernandez-Jimenez A, Provis J, Lukey G, Palomo A and van Deventer J 2007 Geopolymer technology: the current state of the art *J. Mater. Sci.* **42** 2917–2933

[3] Provis J 2018 Alkali-activated materials *Cem. Concr. Res.* **114** 40–48

[4] Davidovits J 1991 Geopolymers *J. Therm. Analysis* **37** 1633–1656

[5] Sazama P, Bortnovsky O, Dědeček J, Tvarůžková Z and Sobalík Z 2011 Geopolymer based catalysts - New group of catalytic materials *Catal. Today* **164** 92–99

[6] Novais R, Seabra M and Labrincha J 2017 Porous geopolymer spheres as novel pH buffering materials *J. Clean. Prod.* **143** 1114-1122

[7] Ge Y, Cui X, Kong Y, Li Z, He Y and Zhou Q 2015 Porous geopolymeric spheres for removal of Cu (II) from aqueous solution: Synthesis and evaluation *J.Hazard. Mater.* **283** 244-251

[8] Ge Y, Cui X, Liao C and Li Z 2017 Facile fabrication of green geopolymer/alginate hybrid spheres for efficient removal of Cu (II) in water: Batch and column studies *Chem. Eng. J.* **311** 126-134

[9] Tang Q, Ge Y, Wang K, He Y and Cui X 2015 Preparation and characterization of porous metakaolin-based inorganic polymer spheres as an adsorbent *Mater. Des.* **88** 1244–1249

[10] Hardjito D, Wallah S, Sumajouw D and Rangan B 2004 On the development of fly ash-based geopolymer concrete *ACI Mater. J.* **101** 467-472

[11] Kalaw M, Culaba A, Hinode H, Kurniawan W, Gallardo S and Promentilla M 2016 Optimizing and characterizing geopolymers from ternary blend of Philippine coal fly ash, coal bottom ash and rice hull ash *Materials* **9** 580-609

[12] Promentilla M, Thang N, Kien P, Hinode H, Bacani F and Gallardo S 2016 Optimizing ternary-blended geopolymers with multi-response surface analysis *Waste Biomass Valori.* **7** 929-939

[13] Puput R, Adjib K, Ekaputri J, Poppy P and Febriano K 2016 Mechanical properties of volcanic ash based geopolymer concrete *Mater. Sci. Forum* **857** 377-381

[14] Tchakoute H, Elimbi A, Yanne E and Djangang C 2013 Utilization of volcanic ashes for the production of geopolymers cured at ambient temperature *Cem. Concr. Compos.* **38** 75–81

[15] Moon J, Bae S, Celik K, Yoon S, Kim K, Kim K S and Monteiro P 2014 Characterization of natural pozzolan-based geopolymeric binders *Cem. Concr. Compos.* **53** 97–104
[16] Canfield G, Eichler J, Griffith K and Hearn J 2014 The role of calcium in blended fly ash geopolymers *J. Mater. Sci.* **49** 5922–5933
[17] Dela Cerna K, Janairo J and Promentilla M 2018 Development of nanosilver-coated geopolymer beads (AgGP) from fly ash and baluko shells for antimicrobial applications *MATEC Web of Conferences* **268** 05003
[18] Manoharan C, Sutharsan P, Dhanapandian S and Venkatachalapathy R 2012 Spectroscopic and thermal analysis of red clay for industrial applications from Tamilnadu, India *J. Mol. Struct.* **1027** 99–103
[19] Socrates G 2004 *Infrared and Raman characteristic group frequencies* (New York: Wiley)
[20] Khachani M, Hamidi A, Halim M and Arsalane S 2014 Non-isothermal kinetic and thermodynamic studies of the dehydroxylation process of synthetic calcium hydroxide $Ca(OH)_2$ *J. Mater.Environ. Sci.* **5(2)** 615-624
[21] Djobo J, Elimbi A, Tchakoute H and Kumar S 2016 Reactivity of volcanic ash in alkaline medium, microstructural and strength characteristics of resulting geopolymers under different synthesis conditions *J. Mater. Sci.* **51** 22
[22] Bakharev T 2005 Geopolymeric materials prepared using Class F fly ash and elevated temperature curing *Cem. Concr. Res.* **35** 1224– 1232
[23] Ylmen R, Jaglid U, Steenari B and Panas I 2009 Early hydration and setting of Portland cement monitored by IR, SEM and Vicat techniques *Cem. Concr. Res.* **39** 433–439
[24] Xiaolu G, Huisheng S and Xue W 2017 Pore properties, inner chemical environment, and microstructure of nano-modified CFA-WBP (class C fly ash-waste brick powder) based geopolymers *Cem. Concr. Compos.* **79** 53-61
[25] American Society for Testing Materials, *Standard test methods for fire tests of building construction and materials (ASTM E119)*, ASTM International, West Conshohocken, Pennsylvania, USA, Current edition approved July 10, 2000. Published October 2000
[26] Sarker P, Kelly S and Yao Z 2014 Effect of fire exposure on cracking, spalling and residual strength of fly ash geopolymer concrete *Mater. Des.* **63** 584–592

IOP Conf. Series: Materials Science and Engineering 1109 (2021) 012069 doi:10.1088/1757-899X/1109/1/012069

Transport Emissions Modeling using Google Maps: An alternative approach for vehicle flow analysis

J E Rito[1,2], **N S Lopez**[1]

[1]Mechanical Engineering Department, De La Salle University, 2401 Taft Ave, Malate, Manila, 0922 Metro Manila

[2]joshua_rito@dlsu.edu.ph

Abstract. Continuous measurements of transport emissions are considered key issues for air pollution management in the transportation sector. In some instances, researchers may come across difficulties in doing transport emissions modeling such as overcomplexities and laborious methodologies. In this regard, the authors are introducing a novel method of doing a transport emissions modeling by utilizing Google Maps data. By getting the average travel time of a road segment and the corresponding length, the average speed can be obtained. This speed will be used to identify the flow of vehicles in terms of Passenger Car Unit (PCU) through a speed-flow curve on the basis of a Roadside Frictions Index (RSFI). PCU percentages are derived from the actual counting of vehicles using the street view features of Google Maps. Once the PCU count and the PCU percentages are established, the actual number of vehicle flow per type can now be determined. Consequently, emissions loads are calculated by multiplying the vehicle flow and road length to emissions factors derived from reliable sources.

1. Introduction

According to IEA[1], the global energy-related CO_2 emissions have hit a record high of 33.1 Gt CO_2 in 2018 or an increase of 1.7%. This is effectuated by the rising global energy consumption demand which increased by 2.3% in 2018 or almost twice the average growth rate since 2010. Such occurrence is quite alarming despite the implementation of the Paris Agreement which intends to limit the global warming below 2°C in the pursuing efforts of Greenhouse Gas (GHG) emissions reduction. The GHG emissions and the effects of global warming and climate change are some of the critical issues the global community is facing today [2]. The Paris Agreement is a treaty which requires all the 196 participating countries to develop and submit long-term plans of lowering GHG emissions. These plans are called nationally determined contributions, and each country will have to communicate each action they perform in reducing GHG emissions in order to achieve the goals of the agreement. The conference of the parties will be held every five years and all countries should find a way to evaluate their progress "based on the best available science" [3].

The transportation sector being one of the most fossil product-consuming sectors in the world accounts for around 21% share of the total global carbon emissions [4]. It should be noted that the transport sector alone covers approximately 24% share in the global CO_2 emissions by sector in 2018 [5]. In the report of Hall and Lutsey [6], road transport dominates the share of GHG emissions in the overall sectoral emissions by mode in 2018. In addition, all road vehicles when combined was estimated to be 77% of the distribution of GHG emissions by the transportation sector worldwide. However, the relevancy of these data is topical and there is a need to regularly quantify, monitor and evaluate whether emission reduction goals are properly monitored and met. From year to year, studies pertaining to quantification of emissions are done for the purpose of evaluating vehicular emissions and air quality assessment. As regards transportation, assessment of mobile emissions is important in air quality improvement programs [7].

In the interest of exploring the dynamics, scale and magnitude of emissions from the transport sector, transport emissions modeling provides useful emission data which may be essential for the theoretical and conceptual development of a particular area [8]. Moreover, continuous estimation and measurement of emissions by road traffic are considered significant issues for air pollution management in the transportation sector. However, it is not easy to do a road transportation emissions modeling and there are difficulties faced in this regard. Doing one may sometimes depend on the assessment of the researcher as well as the methodology that will better fit the situation. For instance, vehicle flow analysis is one part of transportation emissions modeling. The more disaggregated the data gathering is, the more time consuming the study will be [9]. Additionally, the process becomes labor-intensive.

In this study, the authors are introducing a novel method for transportation emissions modeling. The analysis is performed through available data generated from Google Maps with respect to Epifanio de los Santos Avenue (EDSA), a major road located in Metro Manila, Philippines. EDSA will be divided into a number of segments where each segment will undergo emissions load calculation. The objective of this study is to quantify GHG emission shares specifically the emission shares of air pollutants in the transport sector coming from the various types of vehicles plying along each defined road segments in EDSA. By acquiring the average travel time of vehicles from Google Maps in a certain segment and the corresponding road length, the average speed of vehicle flow is obtained. Furthermore, the traffic flow as passenger car unit (PCU) count per time unit is determined through speed-flow curves in reference to a roadside friction graph. In a general concept, emissions are calculated by multiplying emission factors of a pollutant and a rate of an activity [10]. With regards to transportation systems, the estimation of emission load is calculated by multiplying activity data on the basis of PCU count and emissions factors [11].

2. Methodology

EDSA serves as a major thoroughfare connecting the northern and southern parts of Metro Manila while interlinking five large cities [12]. EDSA as the object of this study was divided into 12 segments. For simplification purposes, each segment was bounded by two points in reference to the successive Metro Rail Transit Line 3 (MRT-3) stations. Starting from the northernmost station of MRT-3 in EDSA, the first 5 stations which are located in Quezon City are the following: North Avenue Station, Quezon Avenue Station, GMA Kamuning Stations, Araneta Cubao Station and Santolan-Annapolis Station. Ortigas Avenue station is the sixth station situated in Pasig City. Mandaluyong City has Shaw Boulevard Station as the seventh station and Boni Avenue Station as the eighth station. Makati City has the ninth, tenth and eleventh stations which are Guadalupe Station, Buendia Station and Ayala Avenue Station, respectively. Lastly, Taft Avenue Station as the only station located in Pasay City is the twelfth station [13]. Both the southbound and northbound directions of the road are considered in the emission load calculation.

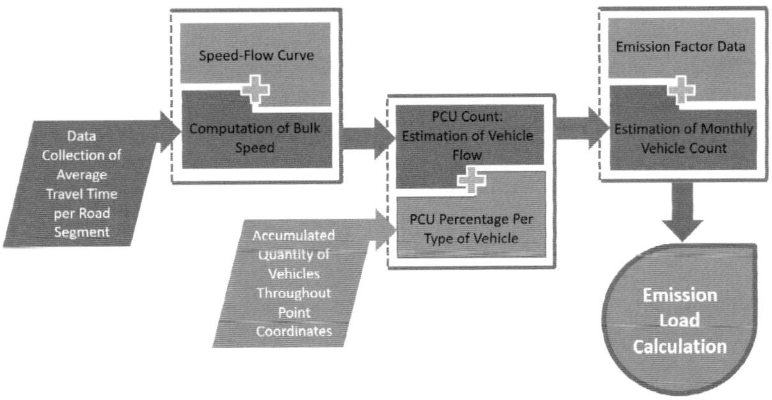

Figure 1. The conceptual framework in the estimation of emission load.

To estimate the emission load for each segment, the conceptual framework for this study is presented in Figure 1. Initially, the average travel time data and the road length per road segment were collected from Google Maps. These factors were converted into the average speed of all the vehicles passing through a segment for each hour in a 24-hour time. Using a Speed-Flow Curve, the estimated hourly vehicle flow in terms of Passenger Car Unit (PCU) is determined by plotting the computed bulk speed into the curve. The hourly PCU counts were consolidated in one day followed by converting the rate of periodicity from daily to monthly PCU count on the basis of a classified PCU count. The PCU mix is derived from actual classified counting of vehicles using the street view of Google Maps. Finally, the emission load is estimated by multiplying the monthly vehicle count and the corresponding emission factor per vehicle type. Each concept will be further discussed in the succeeding subsections.

2.1 Average Travel Time per Road Segment

The average travel time data was collected from Google Maps. One of the considerations of the authors was to perform the study in the month of June when typically, most schools in the country begin their school years. The month of June is also considered the onset of the rainy season in the Philippines[14]; however, the impacts of weather conditions on vehicle flow characteristics were not considered in this study. As regards to the collection of hourly average travel time data, a 24-hour time frame was used by recording all the full hours travel time data starting from 0:00 to 23:00. For each day, the average travel time was calculated every hour in a 24-hour time frame by getting the mean value of the minimum and maximum travel time data generated in Google Maps. In the interest of saving time during data collection, all the days in June 2019 were grouped into five classifications depending on the traffic situation per day. The classifications were Monday, Weekday, Friday, Saturday and Sunday. The traffic situation during Tuesday, Wednesday and Thursday were assumed to be identical; therefore, these days comprise the Weekday classification. June 1 (Saturday), June 2 (Sunday), June 3 (Monday), June 4 (Weekday) and June 7 (Friday) were the specific dates of the data collection. One difficulty encountered during the actual data collection was that the travel time collected in Google Maps had slight variations from one collection day to another notwithstanding the fact that the reference date is the same. To deal with this constraint, the authors conducted five sets of data collection on five different days attributed to the specific dates of collection and calculated the average of these data as a standard data value for each day classification.

2.2 Bulk Speed

Bulk speed is the average speed of all vehicles in a segment. To calculate for the segment average speed, the segment length or the distance between two coordinate points of a segment was divided by the corresponding average travel time [15]. The formula for bulk speed is expressed in Eq 1. Vehicle bulk speed V_B is in terms of kilometers per hour (km/hr), L_S is in terms of kilometer (km), and T_{Ave} is in terms of minutes (min).

$$V_B = \frac{L_S}{T_{Ave}} x\, 60 \tag{1}$$

2.3 PCU Count

The calculated bulk speed for each segment was used in determining the PCU Count. PCU is the conversion of a traffic volume having a characteristic of vehicle-type heterogeneity by the use of a relative weightage factor to an equivalent homogeneous quantity [16]. The calculated average speed was plotted horizontally from the y-axis of a Speed-Flow Curve which in return gave the volume of vehicle flow. By plotting a line vertically from the intersection of the horizontal line and the Speed-Flow Curve to x-axis, the vehicle flow was determined in PCU/hr.

Speed-Flow Curves are the variations of speed profiles of the operating vehicle volume. Various friction elements are used to develop these Speed-Flow curves. In actual scenarios, these friction elements known as side friction factors are activities that occur on the side of the roads or on the roads themselves which affect the normal traffic flow of vehicles [17]. These side friction factors are based on the static and dynamic characteristics and positions on a road. On the basis of quantifying side friction factors, an index called Roadside Friction Index (RSFI) is introduced. For this study, the authors adopted the graph of the Speed-Flow Curve in the study of Pal and Roy [18] which is presented in Figure 2. This study utilized the Speed-Flow Curve with an RSFI value of less than 40.

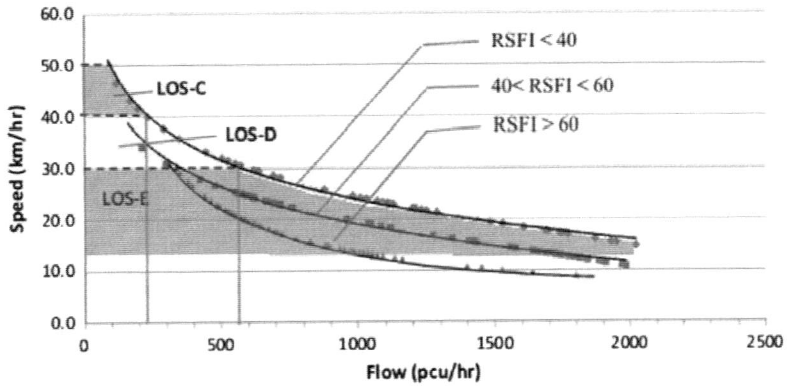

Figure 2. Speed-Flow Curve [18]

To accurately calculate the hourly vehicle flow, multiple points were fitted on a second-degree curve based on the referred Speed-Flow Curve with a less than 40 RSFI value as seen in Figure 2. Thus, a quadratic equation as a function of speed was developed.

The equation for computing the hourly vehicle flow is shown on Eq 2 where the variable PCU_{Count} is the hourly vehicle flow in PCU/hr. The number of lanes is assumed to be 5 for both northbound and southbound carriageways, and this is multiplied in the PCU_{Count} value in order to get the total vehicle volume.

$$PCU_{Count} = 2843 - 180.09\, V_B + 3.26\, V_B^2 \tag{2}$$

However, the authors hypothesize that the derived parabolic equation only applies from 5:00 to 22:00 during the 24-hour time frame. This is due to the assumption that there is a significant drop in vehicle flow along EDSA starting from 23:00 to 5:00. Correction factors were introduced in the computation of vehicle flow as constants of proportionality in reference to an actual counting of vehicles for all the hours of the day. These correction factors were multiplied to the summation of the PCU count values from 5:00 to 22:00 hours on the same reference day. The hours of the day with significant drops in vehicle volume and the respective constant of proportionality are presented in Table 1.

Table 1. The correction factor of each hour with significant drop in vehicle flow

Hours with Significant Vehicle Volume Drop	Correction Factor
0:00	0.02
1:00	0.01
2:00	0.01
3:00	0.01
4:00	0.02
23:00	0.03

2.4 Monthly Vehicle Count

To determine the share of each vehicle type in a segment, Google Maps is capable of showing street view on a selected point coordinate. Actual counting of vehicles was done to break down the vehicle volume by vehicle type. This was accomplished by selecting point coordinates within a segment. The number of point coordinates per 1 km of EDSA is 4. Additionally, the points were selected with approximately equidistant spacing one point after another. After counting all the vehicles per type for all the selected point coordinates, the PCU percentage per vehicle type was established by integrating the array of the total number of vehicles per type and recommended passenger car equivalence factors (PCEF). Table 2 shows the summary of different PCEF for each vehicle type [19]. According to the reference, tricycle, car, jeepney, bus, light-duty truck and heavy-duty truck have the following respective PCEF: 1.0, 1.0, 1.5, 2.0, 2.0 and 2.2. The authors of this paper assumed that a motorcycle has a 0.05 PCEF while a taxi and a sport utility vehicle (SUV) both have the same PCEF as a car.

Table 2. Summary of the Recommended PCEF for each vehicle type

Type of Vehicle	PCEF
Tricycle	1
Motorcycle	0.5
Jeepney	1.5
Taxi	1
Car	1
SUV	1
Light duty	2
Bus	2
Heavy duty	2.2

The PCU percentage for each vehicle type was used to calculate the monthly vehicle count. In sub-section 2.3, the calculation of the hourly PCU count was explained. The hourly PCU counts were consolidated into total daily PCU count by integrating all the calculated hourly PCU counts in reference to the 24 full hours starting from 0:00 and ending at 23:00. The rate of periodicity is then converted from daily to monthly per type of vehicle by getting the product sum of the daily PCU count to a factor pertaining to the specific number of days according to the day classification.

2.5 Emission Factors

Various factors were considered by the way the emission factors had been derived. First, identification of different variants for cars and SUVs were considered in the study. Mini-compact sedan, subcompact sedan, compact sedan and full-size sedan were identified as the variants of car as a vehicle type. The supposition of the authors with regards to SUVs was that vans and pickups were included in the variant share of SUVs. As a result, the specific elements included in the SUV variant were multi-purpose vehicle (MPV), sports pick-up, crossover/compact SUV, mid-size SUV, full-size SUV and van. Alternatively, motorcycle and tricycle variations were distinguished according to the engine type, in particular, two-stroke and four-stroke engines. It is worth noting that other types of vehicle such as taxis, jeepneys, buses, light-duty trucks and heavy-duty trucks remain without variations. The authors had also assumed values for emissions standards share such as the share of Pre-Euro, Euro 1, Euro 2 and Euro 4 as well as the fuel type share in each vehicle type and corresponding variants. In order to arrive from the final emission factor values for vehicle type per GHG and other air pollutant emission as referred in Table 3, interplay between factors such as variant share, emission standard share, fuel share, fuel economy in g/km and specific fuel consumption in g/gfuel transpire. For the data utilized in the variant share, emission standard share and fuel share, own estimates of the authors were used although alternative data concerning the transport sector in a country could be available from the websites of government institutions or agencies. For example, the Land Transportation Office (LTO) is the one responsible for all land transportation matters in the Philippines. On the other hand, the data utilized in fuel economy and fuel consumptions were taken in databases of Argonne National Laboratory which are the Greenhouse gases, Regulated Emissions, and Energy use in Technologies (GREET) Model [20] and the Alternative Fuel Life-Cycle Environmental and Economic Transportation (AFLEET) Tool [21].

3. Results and Discussion

As discussed previously in this paper, the emission loads are calculated through emission factors and the activity data of vehicles. As response to the research objectives of this study, emissions of air pollutants were quantified in accordance with the different types of road transport vehicles. The processes in computing or quantifying transport emission were applied for both southbound and northbound roads in EDSA. The analysis of this study involved disintegrating EDSA to a number of segments in reference to the MRT-3 stations. For each segment, emission loads were calculated, and the results are presented in succeeding the sub-chapters. The relative amount for each transport air pollutants emission particularly with regards to carbon monoxide (CO), non-methane volatile organic compounds (NMVOC), nitrogen oxide (NOx), Sulfur oxide (Sox) and particulate matter (PM) are presented in charts.

Table 3. Summary of the computed air emissions per vehicle type

Types of Emission	Emission Factors g/km								
	Tricycle	Motorcycle	Jeepney	Taxi	Car	SUV	Light duty	Bus	Heavy duty
NMVOC	4.090558	2.302180	0.235721	0.300046	0.740802	0.353766	0.364807	1.287302	1.023751
CO	25.664082	13.167092	3.120396	3.726704	9.062073	3.110799	4.127887	4.419186	3.884681
NOx	0.257945	0.252788	0.243801	0.245913	0.699392	0.961852	2.696264	15.797963	20.101508
PM10	0.056193	0.033614	0.846569	0.001137	0.022093	0.143038	0.751884	0.753899	0.673059
PM2.5	0.056193	0.033614	0.846569	0.001137	0.022093	0.143038	0.751884	0.753899	0.673059
SOx	0.003073	0.002390	0.021766	0.003425	0.009180	0.009430	0.027374	0.045460	0.054021
CH4	4.090558	2.302180	0.235721	0.300046	0.740802	0.353766	0.364807	1.287302	1.023751
N2O	0.002080	0.001499	0.031586	0.003893	0.009883	0.006310	0.022603	0.022228	0.024668
CO2	66.974662	60.098277	668.741525	41.920420	109.895772	92.403869	842.085172	1406.230130	1672.436324

3.1 Transport emissions of air pollutants per road segment

Table 4 summarizes the results that emerged from computing the amount of air pollutant emissions by road segment for both southbound and northbound EDSA. The figures for the southbound EDSA emissions show that CO has the biggest share in the transportation emissions. The 39.38 tons of CO corresponds to approximately 64% of the total emissions. Furthermore, 14.76%, which is considered the largest proportion, of the CO share came from GMA Kamuning-Araneta Cubao Station segment. Although GMA Kamuning-Araneta Cubao segment does not have the largest road stretch in all the southbound segments, the large share in CO is justifiable since this particular road segment has a lot of commercial establishments nearby. It is also worth mentioning that this road segment contains several bus terminals which some of these terminals operate 24 hours a day and 7 days a week.

In relation to the emissions on northbound EDSA, CO still has the largest share in the computed emissions. A 68.7% share is attributed solely to CO. The most notable road segment of all segments in northbound EDSA is the Santolan Avenue-Ortigas Avenue segment. Referring to Table 4, it has the largest Shares in CO, NMVOC and SOx, second largest in NOx and third largest in PM. Aside from having the highest road length among all with 2.3 km stretch, there are a lot of occurrences taking place in this area. One of the largest shopping malls in the Philippines namely, SM Mega Mall, is located at the start of this road segment, and several interchanges exist particularly along the EDSA-Ortigas avenue intersection. All these features affect the side friction of the road and therefore affect the flow of vehicles through vehicle deceleration. If the flow of the vehicles or the average speed is affected due to deceleration, then the average speed is reduced and the PCU count is increased based on Figure 2.

3.2 Transport emission share per type of vehicle

The total shares in emissions per vehicle type in EDSA are summarized in Figure 3 and Figure 4 for southbound and northbound EDSA, respectively. For southbound, motorcycles which use petrol as fuel provided the largest share in CO and NMVOC emissions, respectively, 35.67% and 45.1% of the total. Buses, on the other hand, which operate through diesel fuel contributed the largest share of NOx, SOx and PM emission, respectively, at 75.23%, 44.11% and 51.85%. The same is the case for northbound wherein shares in CO and NMVOC emissions were highest in motorcycles relative to 38.87% and 43.55%, while shares in NOx, SOx and PM emissions were highest in buses relative to 68.17%, 34.96% and 43.58%. Second to motorcycles in CO and NMVOC emissions shares were the cars whereas SUVs are second largest NOx, SOx and PM emissions contributor. In reference to the charts in Figure 3 and Figure 4, it may be regarded as certain that the type of emissions is dependent on the fuel type used by the vehicle. Vehicles utilizing petrol are expected to release more CO and NMVOC emissions in contrast to vehicles utilizing diesel which are expected to release more NOx, SOx and PM after fuel combustion.

IOP Conf. Series: Materials Science and Engineering 1109 (2021) 012069 doi:10.1088/1757-899X/1109/1/012069

Table. 4 Transport emissions of air pollutants in tons per road segment

Road Segment	Southbound Emissions ton					Northbound Emissions ton				
	CO	NMVOC	NOx	SOx	PM	CO	NMVOC	NOx	SOx	PM
North A - Quezon A	2.471001	0.337895	1.054745	0.004945	0.066745	4.073651	0.517722	1.079694	0.006723	0.092190
Quezon A - G Kamuning	2.017639	0.274553	0.214561	0.002228	0.026995	2.519460	0.311280	0.583175	0.003485	0.038591
G Kamuning - A C Cubao	5.813515	0.921262	1.950281	0.008155	0.113317	5.194919	0.789036	1.618692	0.007167	0.095377
A C Cubao - Santolan A	4.481821	0.611804	2.674337	0.011404	0.163559	5.489553	0.740080	2.454482	0.011346	0.153220
Santolan A - Ortigas A	5.459905	0.719022	1.978995	0.010398	0.137982	7.282521	0.890840	2.034889	0.012499	0.155190
Ortigas A - Shaw B	2.475599	0.299337	0.656823	0.004446	0.057265	3.985917	0.517703	1.397273	0.007089	0.087752
Shaw B - Boni	2.475599	0.299337	0.656823	0.004446	0.057265	3.985917	0.517703	1.397273	0.007089	0.087752
Boni- Guadalupe	1.576624	0.218005	0.558684	0.002887	0.038647	3.588510	0.439801	0.883447	0.005377	0.060093
Guadalupe - Buendia	4.037532	0.500287	1.741180	0.008922	0.116692	6.530134	0.852339	1.691332	0.010013	0.121766
Buendia - Ayala	1.971459	0.276435	0.564844	0.002959	0.042790	1.829786	0.233469	0.654341	0.003440	0.045606
Ayala - Magallanes	2.893416	0.497639	2.484726	0.008514	0.129292	3.116394	0.418325	0.761915	0.004940	0.067525
Magallanes - Taft A	3.706428	0.490241	0.983436	0.006869	0.123897	4.677068	0.547275	1.214540	0.009330	0.172108
Total	39.380538	5.445817	15.519434	0.076173	1.074445	52.273832	6.775574	15.771053	0.088500	1.177169

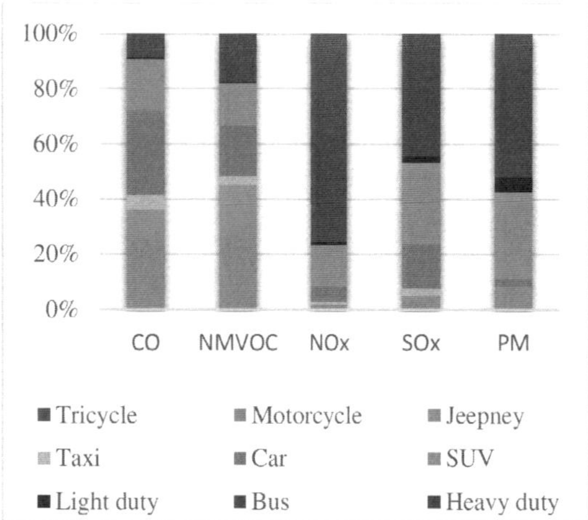

Figure 3. Shares in transport emissions of air pollutants per vehicle type in southbound EDSA

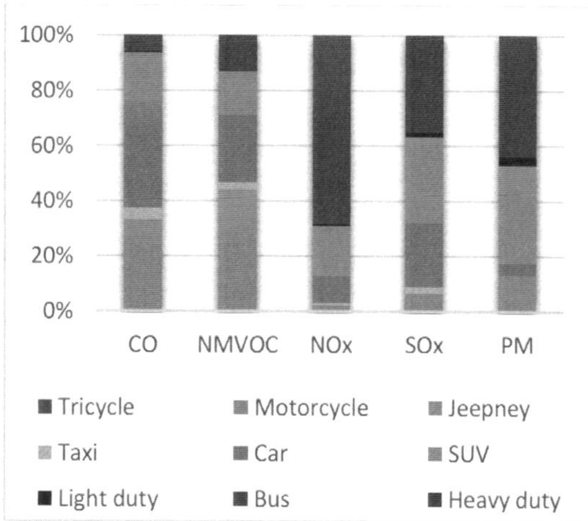

Figure 4. Shares in transport emissions of air pollutants per vehicle type in northbound EDSA

4. Conclusions and Recommendations

A novel method for quantifying transport emissions is introduced in this study. Through the use of Google Maps data, an alternative method for transport modeling was created. EDSA, a major thoroughfare connecting several roads in Metro Manila, Philippines, was chosen to be the interest of the study. The study was done by dividing EDSA into 12 successive segments bounded by the sequential MRT-3 stations which are North Avenue Station, Quezon Avenue Station, GMA Kamuning Stations, Araneta Center-Cubao Station, Santolan-Annapolis Station, Ortigas Avenue, Shaw Boulevard Station, Boni Station Avenue, Guadalupe Station, Buendia Station, Ayala Station and Taft Avenue Station. The end goal for this study was to calculate the monthly transport emission load for both southbound and northbound EDSA in reference to the whole month of June 2019. This month was specifically chosen by the authors in order to analyze transport emission load during a normal school month.

In the results with regards to emissions per road segment, approximately 64% of the total transport emissions were carbon monoxide for southbound EDSA. Furthermore, GMA Kamuning-Araneta Cubao Station was the segment with the highest share of CO having 14.76% of the total. Concurrently, the highest share in the total emissions for northbound EDSA is CO with 68.7% share. Based on Table 4, the Santolan Avenue-Ortigas Avenue segment had the largest Shares in CO, NMVOC and SOx among all other road segments, second largest in NOx and third largest in PM. In relation to the transport emission share per vehicle type, motorcycle which uses a petrol had the highest share in CO (35.67%) and NMVOC (45.1%) emissions for southbound EDSA, while buses which utilize diesel fuel had the highest share in NOx (75.23%), SOx (44.11%) and PM (51.85%) emissions. In the same case, motorcycles had the highest emission share in CO (38.87%) and NMVOC (43.55%) for northbound, while buses had the highest emission share in NOx (68.17%), SOx (34.96%) and PM (43.58%). This led to the conclusion that vehicles with petrol engines are expected to release more CO and NMVOC emissions; meanwhile, vehicles with diesel engines are expected to release more NOx, SOx and PM emissions.

In order to further establish the verified findings of this paper, the next step for this research would be the validation of the modeling results utilizing actual on-road data. Another recommendation from the authors is to take into account the effects of weather conditions on vehicle flow characteristics such as the bulk speed and the PCU count for the future progress of this research. In the Philippines, the month of June is considered the onset of the rainy season.

References

[1] International Energy Agency 2019 Global Energy & CO2 Status Report (Paris)
[2] Stern N 2008 The economics of climate change *The American Economic Review* **98** pp 1-29
[3] United Nations 2016 *Paris Agreement* (Paris)
[4] Lopez N S, Chiu A S F and Biona J B M 2018 Decomposing drivers of transportation energy consumption and carbon dioxide emissions for the Philippines: the case of developing countries *Frontiers in Energy* **12** pp 389-99
[5] International Energy Agency 2020 *Statistics report CO_2 Emissions from Fuel Combustion* (Paris)
[6] Hall D and Lutsey N 2019 *Estimating the Infrastructure Needs and Costs for the Launch of Zero-emission Trucks* (Washington, D C: International Council on Clean Transportation) pp 1-31
[7] National Research Council 2000 *Modeling Mobile-Source Emissions* (Washington, D C: National Academies Press)
[8] Linton C, Grant-Muller S and Gale W F 2015 Approaches and techniques for modelling CO_2 emissions from road transport *Transport Reviews* **35** pp 533-53
[9] Tsanakas N, Ekström J and Olstam J 2020 Estimating emissions from static traffic models: problems and solutions *J. Adv. Transp.* **2020** pp 1-17
[10] Olaguer E P 2017 Emission inventories *Atmospheric Impacts of the Oil and Gas Industry* (Cambridge: Academic Press/Elsevier) Chapter 7 pp 67-77
[11] Klein J, Hulskotte J, Ligterink N, Dellaert S, Molnár H and Geilenkirchen G 2018 *Methods for Calculating the Emissions of Transport in the Netherlands* (The Hauge) p 62
[12] Boquet Y 2013 Battling congestion in Manila: the EDSA problem *Transport and Communications Bulletin for Asia and the Pacific* **82** pp 45-59
[13] Ganiron Jr T U 2015 Exploring the emerging impact of metro rail transit (MRT-3) in Metro Manila *Int. J. Adv. Sci. Technol.* **74** pp 11-24
[14] Cruz F T, Narisma G T, Villafuerte II M Q, Cheng Chua K U and Olaguera L M 2013 A climatological analysis of the southwest monsoon rainfall in the Philippines *Atmospheric Research* **122** pp 609-16
[15] Jiménez-Meza A, Arámburo-Lizárraga J and de la Fuente E 2013 Framework for estimating travel time, distance, speed, and street segment level of service (LOS), based on GPS data *Procedia Technol.* **7** pp 61-70
[16] Sharma M and Biswas S 2020 Estimation of passenger car unit on urban roads: a literature review *Int. J. Transp. Sci. Technol.*
[17] Salini S, George S and Ashalatha R 2016 Effect of side frictions on traffic characteristics of urban arterials *Transp. Res. Procedia* **17** pp 636-43
[18] Pal S and Roy S K 2016 Impact of roadside friction on travel speed and LOS of rural

highways in India *Transp. Dev. Econ.* **9** p 12
[19] Japan International Cooperation Agency 1987 Feasibility Study of the Road Improvement Project on the Pan-Philippine Highway (Philippine-Japan Friendship Highway) (Tokyo)
[20] Argonne National Laboratory 2020 *The Greenhouse gases, Regulated Emissions, and Energy use in Technologies (GREET) Model* (Lemont)
[21] Argonne National Laboratory 2018 *Alternative Fuel Life-Cycle Environmental and Economic Transportation (AFLEET) Tool* (Lemont)

IOP Conf. Series: Materials Science and Engineering 1109 (2021) 012070 doi:10.1088/1757-899X/1109/1/012070

Using Logarithmic Mean Divisia Index method (LMDI) to estimate drivers to final energy consumption and emissions in ASEAN

E B Lisaba[1], N S Lopez[1]

[1]Mechanical Engineering Department, De La Salle University, Manila, Philippines

edwin_lisaba@dlsu.edu.ph

Abstract. The ASEAN is currently trying to mitigate the amount of CO_2 emissions per country. This is not only to contribute to the worldwide goal of lessening greenhouse gases, but to also ensure the safety of its member states by combatting the effects of climate change. The effects as stated by the IPCC, such as rising water levels, rising ambient temperature, and an increase in the frequency of tropical storms, can greatly endanger members of the ASEAN. It must be noted that a majority of its members are archipelagic, have areas of extreme poverty, and rely on agriculture for production. Therefore, the temporal logarithmic mean Divisia index (LMDI) has been conducted in order to determine the possible drivers that cause the changes in CO_2 emissions per year for all ASEAN member countries. The drivers of CO_2 were divided into the population, economic activity, energy intensity, and energy structure effects. CO_2 emissions per fuel type were calculated using their respective emission factors. Knowing which drivers constitute the most CO_2 emissions in a country would help pinpoint which areas could be worked on. It was found that Indonesia, Brunei Darussalam, and Lao PDR had the most impressive decrease in CO_2 emissions due to energy intensity and structure effects. A majority of the other countries have either stayed constant or even increased CO_2 emission output per year from each effect. Therefore, general policies based on the success of these countries that could be applied for each member country are recommended in this study.

1. Introduction

The Association of Southeast Asian Nations (ASEAN) is a collection of 10 countries that was established back in the 8th of August, 1967. One of the main purposes of the ASEAN is to accelerate and promote economic growth and cultural development within the region. It is also meant to promote active support and collaboration between its member countries in the areas of economics, science, culture, and education [1]. The members are Indonesia, Malaysia, Philippines, Singapore, Thailand, Brunei Darussalam, Vietnam, Lao PDR, Myanmar, and Cambodia. Outlined in the Treaty of Amity and Cooperation in Southeast Asia (TAC), the ASEAN was built upon the promise of continued support and unity between its nations, while maintaining mutual respect for their individual independence, territory, and national identity.

Climate change is a growing problem that can greatly affect countries and regions all around the world. As addressed in the Fourth Assessment Report of the Intergovernmental Panel on Climate Change (IPCC), the increase in average air and ocean temperatures, along with the melting of the polar ice caps, indicate the global warming effect [2]. This effect is then accelerated by the emission of greenhouse gases such as carbon dioxide, methane, and nitrous oxide. It was only in 2014 when the Fifth Assessment Report of the IPCC determined that the main driver for climate change was carbon dioxide [3]. The IPCC has projected that without preventative measures, the global average temperature will rise from 1.8 to 4 degrees Celsius, and that there will be a rise in sea levels by about 18 to 59 centimetres. The occurrence of natural disasters such as typhoons, heat waves, floods, and droughts will become more frequent as well.

Given that the ASEAN members are generally located to the west of the Pacific Ocean, and that a number of its members are archipelagic states, the effects of climate change will become more apparent and devastating if preventative measures are not implemented. Additionally, the region has areas of extreme poverty, and that the people are heavily reliant on agriculture. A study from Nunti et al. has proven that climate change has an overall negative effect on agricultural areas using the copula-based stochastic frontier approach [4]. This constitutes not only an escalated vulnerability to natural disasters, but also an increased risk for numerous communities in terms of livelihood and survival. For example, Typhoon Vamco has hit the Philippines, an archipelagic country, in November 2020. This caused massive flooding and displaced numerous communities, resulting in around 102 deaths and 402 million dollars in damages [5]. Another example would be the Central Vietnam Floods that was caused by a number of seasonal monsoons and tropical depressions. The country was also affected by Typhoons Goni and Vamco, resulting in 233 fatalities and about 1.52 billion dollars in damages.

Throughout the years the ASEAN nations have collaborated on many programs that addressed climate change. This is done not only to mitigate the loss of life and property, but also to contribute to the overall goal of decreasing CO_2 emissions in the world. For example, Indonesia has volunteered to reduce emissions by 26% to 41% by the end of 2020. Given the same deadline, Malaysia has also pledged to reduce emissions by 40%, Philippines by 20%, and Singapore by 16% [6]. Given the evident dangers of climate change, especially its devastating effects on the ASEAN region, as well as the ever-increasing carbon dioxide emissions, this study aims to estimate the drivers for CO_2 emissions in the ASEAN region. Additionally, historical data such as past policies and laws will be discussed in order to determine their efficacy in mitigating carbon dioxide emissions. This study aims to not only highlight the most effective policies that were enacted, but also infer and derive other possible policies based on the collated data.

Decomposition analysis will be used to study the acquired data, as this method is most commonly used to determine drivers that cause significant changes in terms of environmental impact and energy over time [7]. This method can then be divided into structural decomposition analysis (SDA) and index decomposition analysis (IDA). SDA is most commonly used for studying the impacts brought on by the consumption of a particular indicator, while IDA is meant for studying the impacts of producing a particular indicator. It must be noted that decomposition analyses have already been conducted by previous studies. For example, a study by Sumabat et al. conducted a decomposition analysis using the logarithmic mean Divisia index method (LMDI) to determine CO_2 emissions from the consumption of fuel and generation of electricity in the Philippines [8]. Another study also used the LMDI method based on energy allocation to determine drivers of CO_2 emissions in Malaysia [9].

The paper will utilize IDA, particularly the logarithmic mean Divisia index method to estimate drivers that caused the biggest change in CO_2 emissions from 2010 to 2017 due to the final consumption of each ASEAN country. The CO_2 emissions will be calculated using the fuel type consumed by the industries of a particular country in a particular year. This study will be organized as follows: the equations, discussion, and data will be provided in Section 2; the results will be presented in Section 3; and the conclusions and proposed policies will be discussed in Section 4.

2. Methods and Data

2.1. Fuel Consumption per Country

As stated earlier, the amount of fuel consumed by all 10 countries was taken from the International Energy Agency (IEA). The final industrial energy consumption (excluding power generation) by a country has been further divided into 8 types; namely coal, crude oil, oil products, natural gas, nuclear, hydro, biofuels and waste, and wind and solar. The consumption of each fuel type per year from 2010 to 2017 was gathered in order to estimate the amount of carbon dioxide released after use. Figure 1 below shows the final industrial consumption for each type in the whole ASEAN.

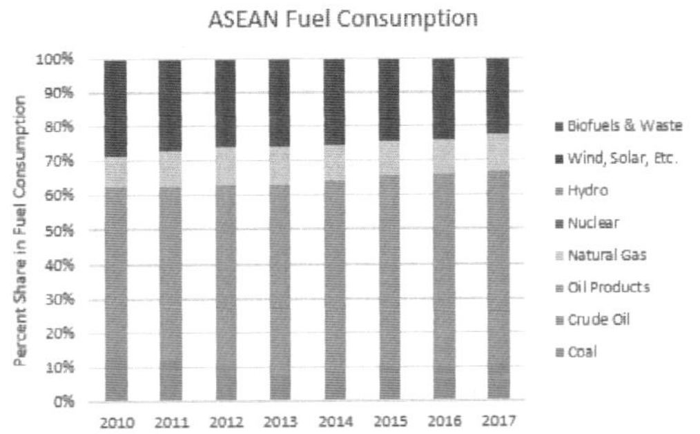

Figure 1. ASEAN percent share in industrial fuel consumption from 2010 to 2017

2.2. CO_2 Emissions per Country

The annual carbon dioxide emissions per fuel type was calculated using Eq. 1 below, where C is the CO_2 emissions in kilotons of CO_2 ($ktCO_2$); the subscript i stands for the fuel type; the superscript j refers to the year; A is the amount of fuel i consumed in kilotons of oil equivalent (ktoe) in the year j; and EF is the emission factor for a particular fuel type i. The EF was converted to $ktCO_2$ per ktoe using the units in Table 1 below.

$$C_i^j = A_i^j EF_i \tag{1}$$

The emission factors used for the corresponding fuel types are summarized in Table 1. It must be noted that there are no emission factors for nuclear, hydro, wind, solar, and biofuels. Biofuels were treated to be carbon-neutral in this study. Additionally, a majority of the countries have not used particular fuel types at all. In order for the LMDI to work however, small values of about $1 \times 10^{-6}\ ktoe$ were substituted instead.

Table 1. Emission Factors

Fuel Type	Fuel Combustion	Units
Coal	965[a]	tCO_2/GWh
Crude Oil	74.54[b]	$kgCO_2/mmBtu$
Oil Products	161.3[c]	$lbCO_2/mmBtu$
Natural Gas	117[c]	$lbCO_2/mmBtu$
Nuclear	N/A	-
Hydro	N/A	-
Wind, Solar, Etc.	N/A	-
Biofuels and Waste	N/A	-

[a] Sources: [10]
[b] Sources: [11]
[c] Sources: [12]

2.3. Index Decomposition Analysis

As stated earlier, the logarithmic mean Divisia index Method was used in order to determine the drivers of CO_2 emissions in the ASEAN region. A paper by Ang provided a practical tutorial to setting up the needed formulas and steps in conducting the method [13]. To begin with, the aggregate must first be determined, which in this case was the total CO_2 emissions denoted as C. Next, the IDA identity function must be outlined, shown in Eq. 2 below. In doing so, it is important that the function must equal the aggregate to be studied, namely C. The variables used in the identity must pertain to the economic activity of a country with respect to its energy usage. Therefore, P refers to the population, G is gross domestic product (GDP), and E is the amount of fuel used where i is the fuel type. Once simplified, M

is the GDP per capita, I is the energy intensity in ktoe per GDP, and S_i is the share in energy mix for a particular fuel type i.

$$C = \sum_i C_i = \sum_i P \times \frac{G}{P} \times \frac{E}{G} \times \frac{E_i}{E} \times \frac{C_i}{E_i} = \sum_i P \times M \times I \times S_i \times EF_i \quad (2)$$

Given that the LMDI used in this study was based on the change of certain drivers over the years, it must be noted that certain effects can take on negative values. A negative effect simply means that the CO_2 emissions from a particular fuel type in that year was lower than the last, and vice versa. The sum of a particular effect from the baseline year to the final year dictates the overall influence of that effect to the CO_2 emissions.

The effects used in this study were namely; the population effect (ΔC_{pop}), the economic activity effect (ΔC_{act}), the energy intensity effect (ΔC_{int}), and the structure effect (ΔC_{str}). The population effect referred to the contribution of the increase or decrease in the country's population, and it was calculated using P. The economic activity effect referred to the change in the economic activity, measured in million USD per capita or M in the IDA identity. The energy intensity effect referred to the net increase or decrease in the needed energy to produce the same amount of product. It was measured in ktoe per million USD, or I. Any decrease in the energy intensity means that the production process has either gotten more efficient or there has been a change in production. Finally, the structure effect referred to the fuel changes adopted by the country after a year has passed. This was calculated using S_i in the IDA identity. Eqs. 3 to 6 show the formulas for each effect.

2.4 Data Gathering

The amount of fuel consumed by each country was taken from the International Energy Agency [14]. For consistency, the amount of fuel used was expressed in kilotons of oil equivalent. Other pertinent data such as the population and Gross Domestic Product (GDP) were taken from The World Bank [15]. It must be noted that the GDP was in terms of millions of USD, and that it was kept at a constant 2010 USD value for all countries in order to take inflation into account.

$$\Delta C_{pop}^T = \sum_i \frac{C_i^T - C_i^0}{lnC_i^T - lnCI_i^0} \ln\left(\frac{P^T}{P^0}\right) \quad (3)$$

$$\Delta C_{act}^T = \sum_i \frac{C_i^T - C_i^0}{lnC_i^T - lnCI_i^0} \ln\left(\frac{M^T}{M^0}\right) \quad (4)$$

$$\Delta C_{int}^T = \sum_i \frac{C_i^T - C_i^0}{lnC_i^T - lnCI_i^0} \ln\left(\frac{I^T}{I^0}\right) \quad (5)$$

$$\Delta C_{str}^T = \sum_i \frac{C_i^T - C_i^0}{lnC_i^T - lnCI_i^0} \ln\left(\frac{S^T}{S^0}\right) \quad (6)$$

3. Results and Discussion

3.1 Total CO_2 Emissions per Country

After acquiring the amount of fuel consumed per country, the total CO_2 emissions were solved using Eq. 1. The emission factors were simply converted to $ktCO_2$ per *ktoe*. Figures 2a to 2j show the total CO_2 emissions of each fuel type per country.

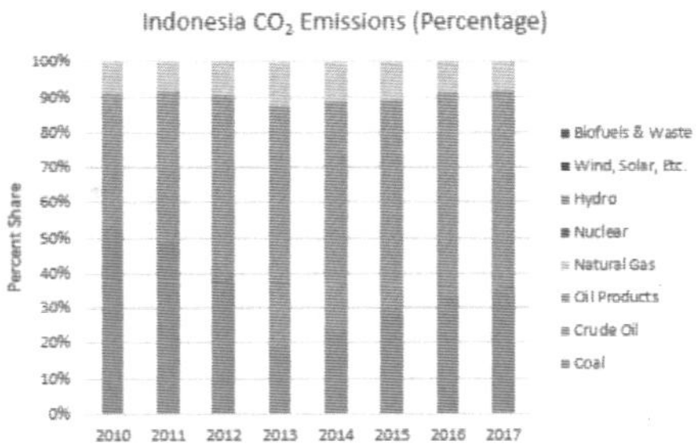

Figure 2a. Indonesia CO_2 Emissions

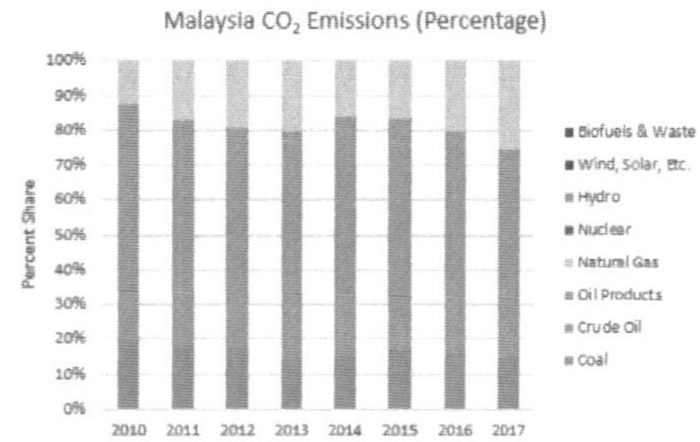

Figure 2b. Malaysia CO_2 Emissions

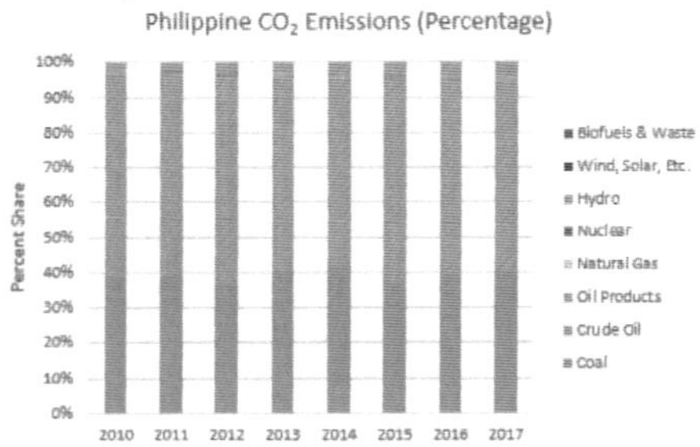

Figure 2c. Philippines CO_2 Emissions

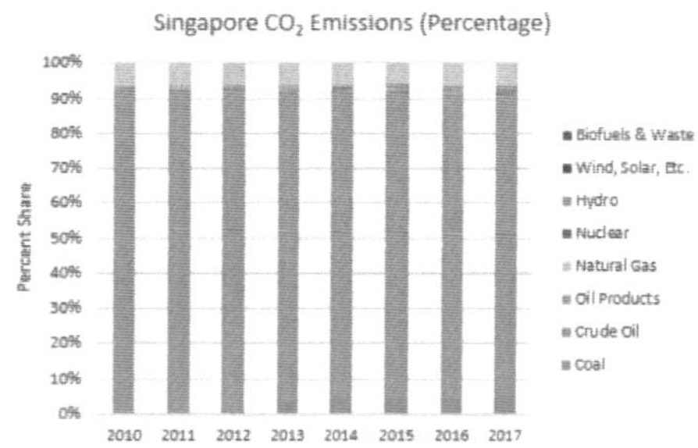

Figure 2d. Singapore CO_2 Emissions

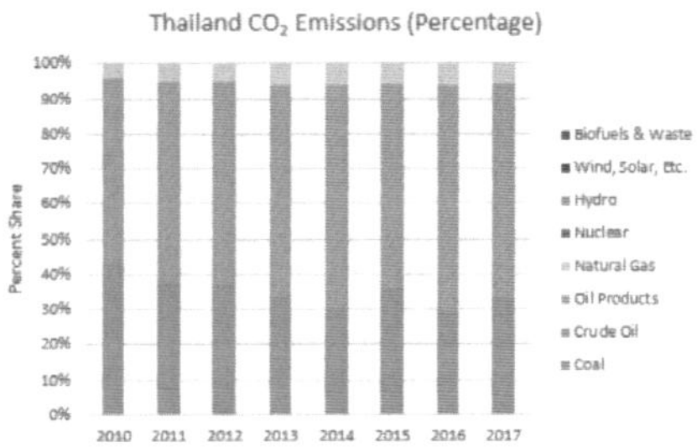

Figure 2e. Thailand CO_2 Emissions

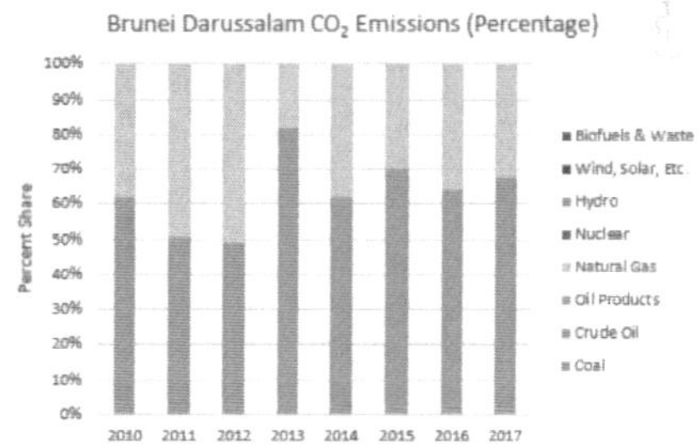

Figure 2f. Brunei Darussalam CO_2 Emissions

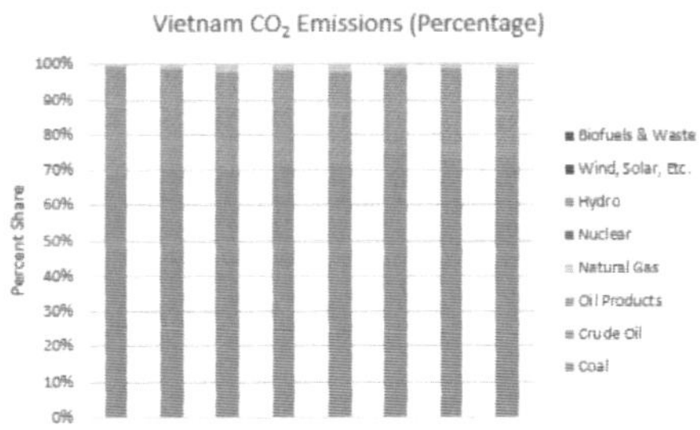

Figure 2g. Vietnam CO_2 Emissions

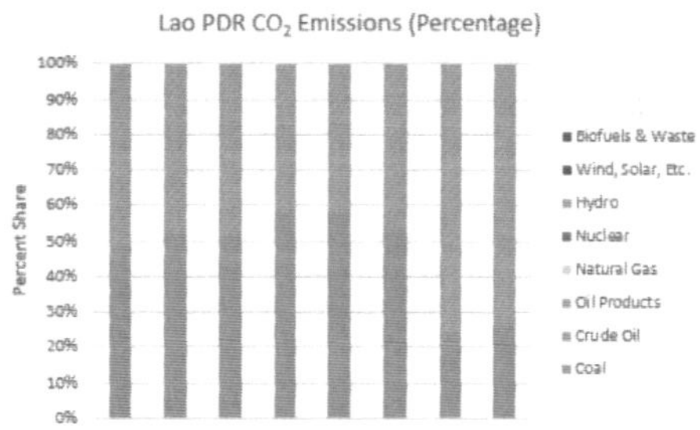

Figure 2h. Lao PDR CO_2 Emissions

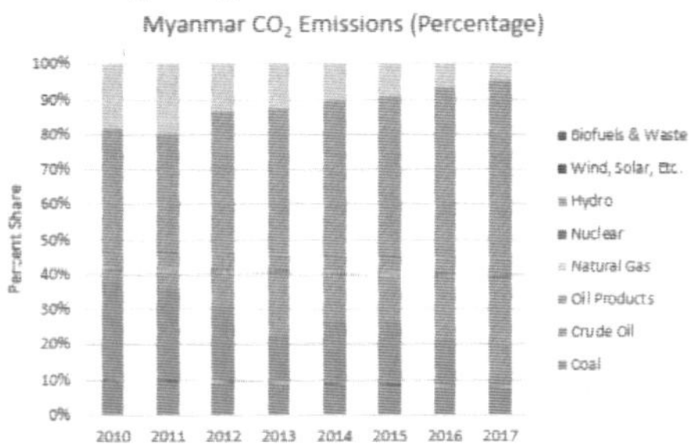

Figure 2i. Myanmar CO_2 Emissions

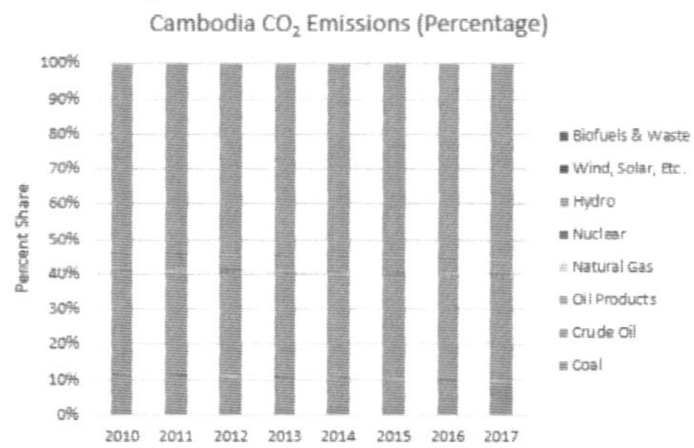

Figure 2j. Cambodia CO_2 Emissions

As can be seen from the graphs above, the countries of Philippines, Vietnam, and Cambodia seem to stay consistent with the types of fuel being used. A majority of the carbon dioxide emissions from the industrial sectors of these countries come from the consumption of oil products and coal, with relatively small amounts of other fuel types. Indonesia, Malaysia, Thailand, Lao PDR, and Myanmar have been able to reduce their consumption of coal products as they transition to the use of oil products instead. Contrarily, Singapore can be seen transitioning to a small amount of coal from oil products in more recent years. Lastly, Brunei Darussalam has been recorded to have no dependence on coal.

In studying the drivers, Figures 3a to 3j are presented below. It can be seen that population and economic activity effects will always be positive contributors to CO_2 emissions. However, all countries can be seen to have negative energy intensity and structure at certain years. A negative value is good, signifying that whatever policy or shift in energy use that was implemented in that year allowed the country to emit less CO_2 compared to the previous year. It can be seen that Indonesia, Malaysia, Singapore, Thailand, Brunei Darussalam, and Lao PDR have achieved an impressive decrease in CO_2 emissions in certain years. In contrast, the Philippines has produced more CO_2 from energy intensity and structure effects. Additionally, the energy structure effects in both Vietnam and Cambodia seem to play a big role in their CO_2 emissions despite having a negative and consistent energy intensity effect. Finally, a majority of Myanmar's CO_2 emissions stem from an excessively high energy structure effect.

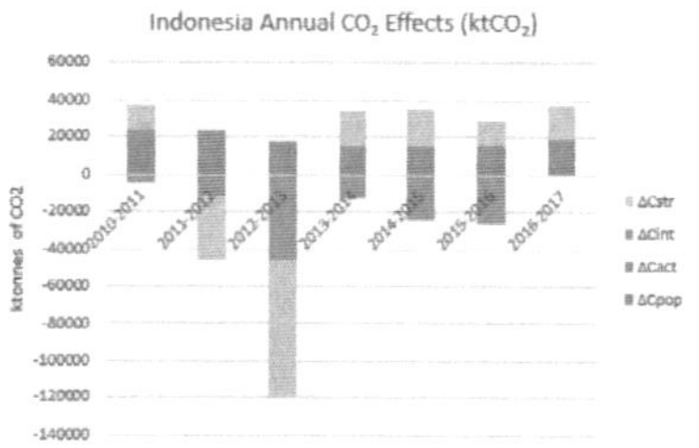

Figure 3a. Indonesia CO_2 Effects

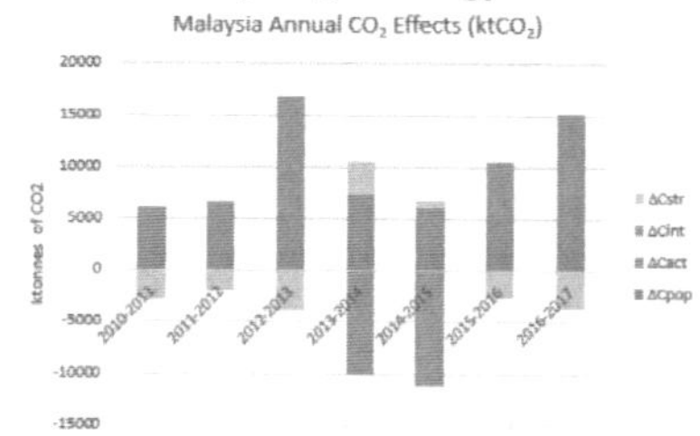

Figure 3b. Malaysia CO_2 Effects

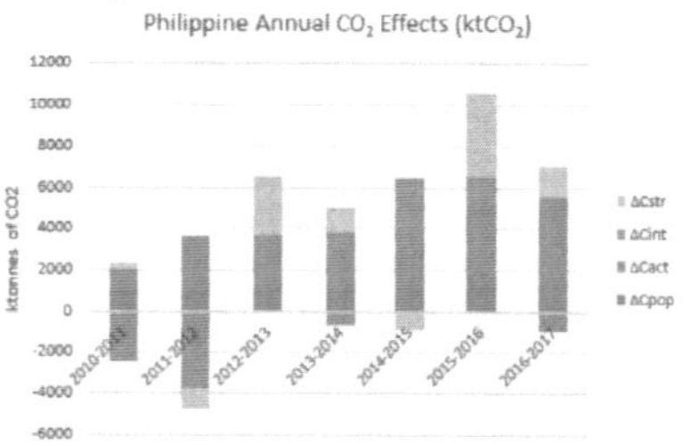

Figure 3c. Philippines CO_2 Effects

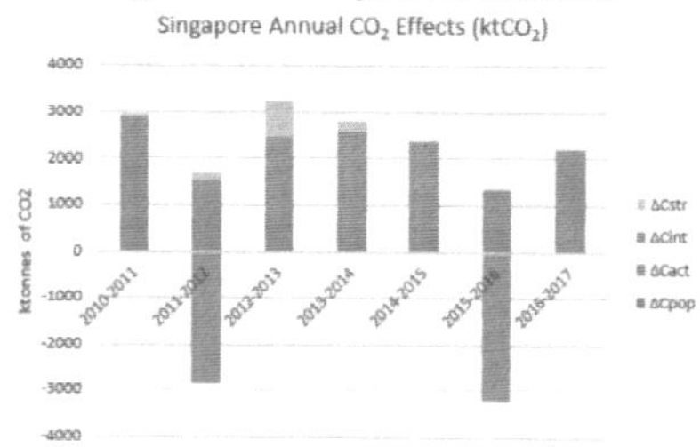

Figure 3d. Singapore CO_2 Effects

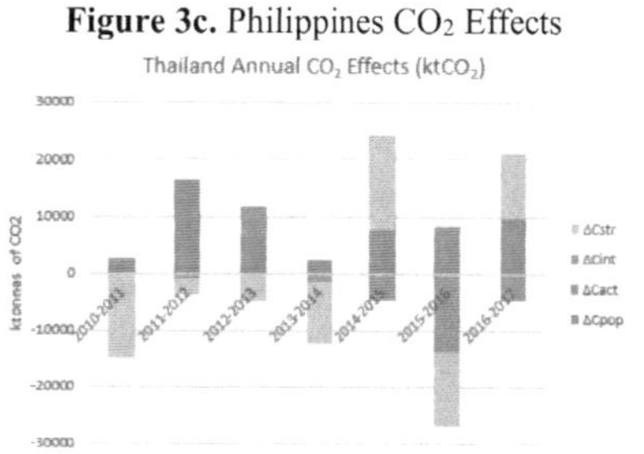

Figure 3e. Thailand CO_2 Emissions

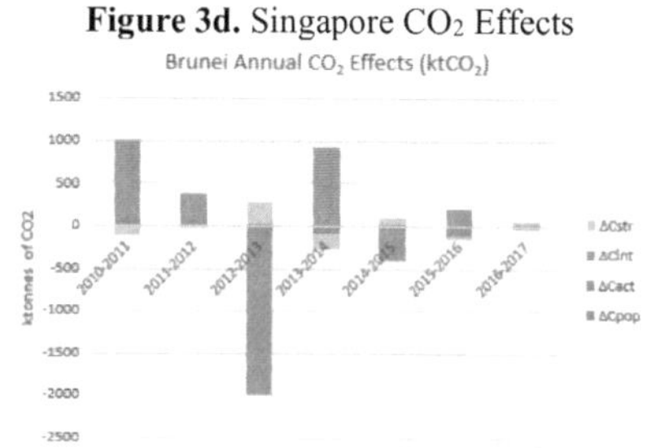

Figure 3f. Brunei Darussalam CO_2 Emissions

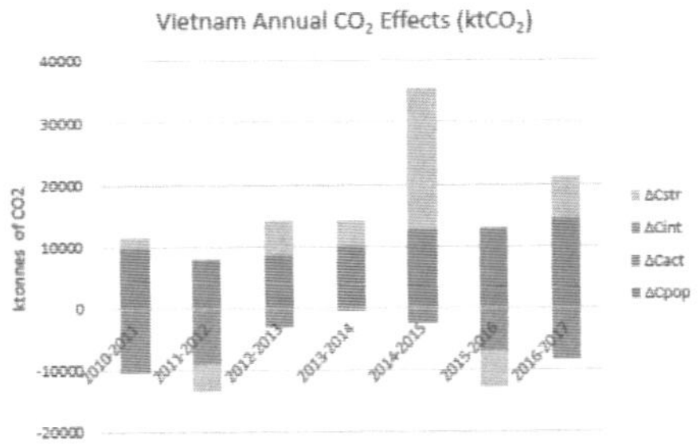

Figure 3g. Vietnam CO_2 Emissions

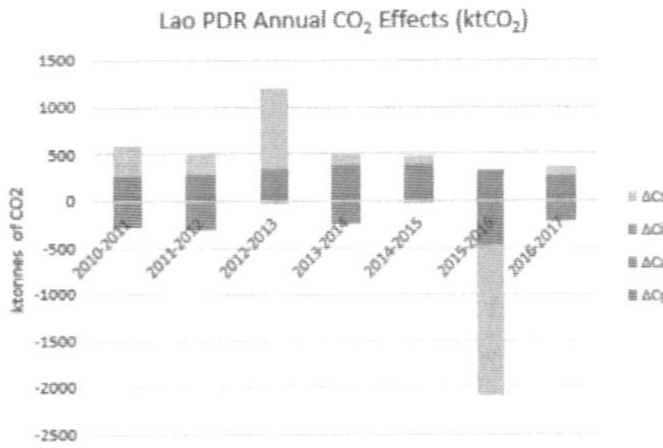

Figure 3h. Lao PDR CO_2 Emissions

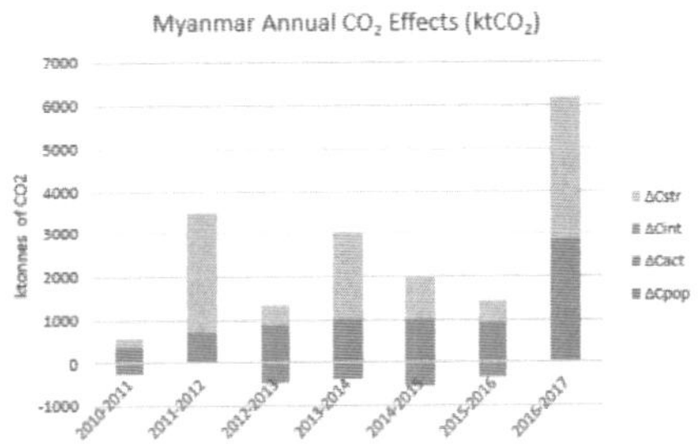

Figure 3i. Myanmar CO_2 Emissions

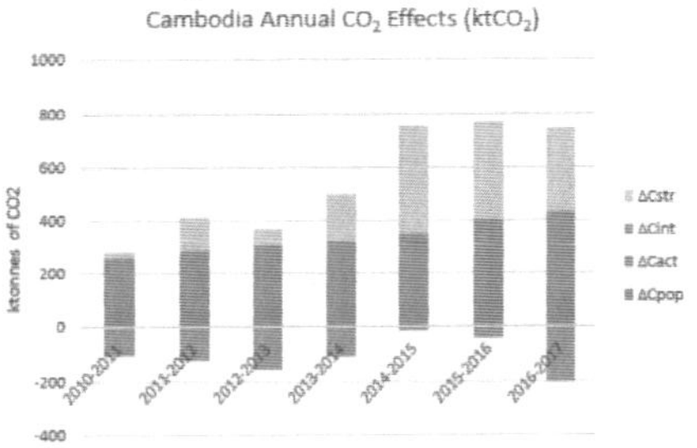

Figure 3j. Cambodia CO_2 Emissions

4. Conclusion and Policy Recommendations

From using index decomposition analysis, it can be seen that Indonesia, Brunei Darussalam, and Lao PDR have yielded the most impressive results. These countries were able to maintain the energy intensity at a constant negative effect in more recent years, and that CO_2 emissions were lessened by either maintaining or improving their energy structure.

To begin with, Indonesia has enacted a presidential regulation for a National Action Plan for Reducing Greenhouse Gas Emissions (RAN-GRK) that outlines the needed activities and policies for various institutions within the country to follow in order to reduce the amount of greenhouse gas (GHG) emissions in the year 2009 [16]. This is to follow-through with its promise of reducing greenhouse gases by 26% by the end of 2020. For this regulation, various factors were taken into account, such as national development priorities, feasibility, and sources of funding. Additionally, Brunei Darussalam has followed the Energy White Paper (EWP) in 2014. This set of policies published from the United Kingdom has set various regulations that use energy efficiency and conservation (EEC) initiatives to lessen CO_2 emissions in the energy sector [17]. Finally, Lao PDR has followed a series of policies that outlined its GHG emissions from 2011 to 2015 [18]. Some of these are the National Climate Change Strategy of 2010, the Renewable Energy Development Strategy of 2011, and the Climate Change Action Plan of Lao PDR in 2013, to name a few.

Taking from their example, it must therefore be noted that in creating the policies needed for all countries to be able to follow, certain actions that require the mitigation of GHG emissions need to be incentivized and fully supported by the government. Such methods can be done through:

1. Implementing tax and penalties to serve as a warning for industries and other sectors to maintain allowable GHG emissions.
2. Developing unique peer-reviewed standards and regulations that take into consideration the strengths of the country in terms of production, funding, and feasibility. These standards must be met by the different sectors of that particular country.
3. Actively campaigning the mitigation of GHG emissions. This can be done through informing the public and incentivizing citizens and businesses.
4. Promote research and development sectors to conduct studies that not only monitor the GHG emissions in their country, but to also develop more efficient and environmentally-friendly technology.

References

[1] Overview, Association of Southeast Asian Nations. <https://asean.org/asean/about-asean/overview/> [Retrieved December 26, 2020].

[2] Intergovernmental panel on climate change (IPCC). Climate Change 2007: Synthesis Report. Contribution of Working Groups I, II, and III to the Fourth Assessment Report of the Intergovernmental Panel on Climate Change

[3] Intergovernmental panel on climate change (IPCC). Climate Change 2014: Synthesis Report. Contribution of Working Groups I, II, and III to the Fifth Assessment Report of the Intergovernmental Panel on Climate Change

[4] Nunti C, Somboon K and Intapan C 2020 The impact of climate change on agriculture sector in ASEAN in *Journal of Physics: Conference Series, November 25, 2020, Guilin, China.* Available: Scopus. http://scopus.com. [Accessed: January 21, 2020].

[5] National Disaster Risk Reduction and Management Council 2020 Sitrep No. 26 re Preparedness Measures and Effects for Typhoon "ULYSSES" (I.N. VAMCO). <http://ndrrmc.gov.ph/attachments/article/4138/SitRep_no_26_re_TY_ULYSSES_as_of_06 DEC2020.pdf> [Retrieved December 26, 2020].

[6] ASEAN Cooperation on Climate Change, Association of Southeast Asian Nations. <https://environment.asean.org/asean-working-group-on-climate-change/#:~:text= Several%20ASEAN%20Member%20States%20have,by%202020%20compared%20 to%202005> [Retrieved December 26, 2020].

[7] de Boer P and Rodrigues J 2020 Decomposition analysis: When to use which method? *Economic Systems Research* **32**(1) pp 1-28 [Online]. Available: Taylor & Francis Online, http://www.tandfonline.com. [Accessed December 26, 2020].

[8] Sumabat A K, Lopez N S, Yu K D, Hao H, Li R, Geng Y and Chiu A S F 2016 Decomposition analysis of Philippine CO_2 emissions from fuel combustion and electricity generation *Applied Energy* **164** pp 795-804 [Online]. Available: ScienceDirect, http://www.sciencedirect.com. [Accessed December 26, 2020].

[9] Chong C H, Tan W X, Ting Z J, Liu P, Ma L, Li Z and Ni W 2019 The driving factors of energy-related CO_2 emission growth in Malaysia: The LMDI decomposition method based on energy allocation analysis *Renewable and Sustainable Energy Reviews,* vol. 115, 2019. [Online]. Available: ScienceDirect, http://sciencedirect.com. [Accessed December 26, 2020].

[10] Intergovernmental panel on climate change (IPCC). 1996 IPCC guidelines for national greenhouse gas inventories; 1996.

[11] Environmental protection agency. Emission Factors for Greenhouse Gas Inventories; 2014.

[12] U.S. Energy Information Administration (EIA). How much carbon dioxide is produced when different fuels are burned? < https://www.eia.gov/tools/faqs/faq.php?id=73&t=11> [Retrieved December 27, 2020].

[13] Ang B W 2005 The LMDI approach to decomposition analysis: A practical guide *Energy Policy* **35** pp 867-871, 2005. [Online]. Available: Elsevier, https://www.elsevier.com. [Accessed December 26, 2021].

[14] Data and Statistics, International Energy Agency. < https://www.iea.org/data-and-statistics/data-tables?country=WORLD> [Retrieved December 26, 2020].

[15] World Bank Open Data, The World Bank. < https://data.worldbank.org/> [Retrieved December 26, 2020].

[16] Thamrin S 2011 Indonesia's national mitigation actions: Paving the way towards NAMAs *CCXG/Global Forum on Environment Seminar on MRV and Carbon Markets, March 28-29, 2011, Paris, France*. Available: OECD. http://oecd.org. [Accessed: December 26, 2020].

[17] "Brunei Darussalam's second national communication: Under the United Nations framework convention on climate change," Energy and Industry Department, Bandar Seri Begawan, Brunei Darussalam, 2017.

[18] "Intended nationally determined contribution," Lao People's Democratic Republic, 2015.

AUTHOR INDEX